THE COMING PLAGUE

THE

COMING

PLAGUE

NEWLY EMERGING DISEASES

IN A WORLD OUT OF BALANCE

LAURIE

GARRETT

Farrar, Straus and Giroux / New York

Copyright © 1994 by Laurie Garrett
All rights reserved
First printing, 1994
Printed in the United States of America
Published simultaneously in Canada by HarperCollinsCanadaLtd

Library of Congress Cataloging-in-Publication Data
Garrett, Laurie.
The coming plague: newly emerging diseases in a world out of balance / Laurie Garrett.
p.   cm.
Includes index.
1. Epidemiology—Popular works.   2. Communicable diseases—Popular
works.   I. Title.
RA651.G37   1994   614.4—dc20   94-26285   CIP

Grateful acknowledgment is made to the following for permission to reprint
material from: The Power Unseen by Bernard Dixon. Copyright © 1994
by Bernard Dixon. Used with permission of W. H. Freeman and Company.
Earth in the Balance by Al Gore. Copyright © 1992 by Senator Al Gore.
Reprinted with permission of Houghton Mifflin Co. The Plague by Albert
Camus. Copyright © 1948 by Alfred A. Knopf, Inc. Rats, Lice, and
History by Hans Zinsser. Copyright 1935 by Little, Brown & Co. Excerpt
from "Sexually Transmitted Diseases" by H. H. Handsfield reprinted from
Volume 17, issue 1 of Hospital Practice. Excerpts from Arrowsmith by
Sinclair Lewis, copyright © 1925 by Harcourt Brace & Company and
renewed 1953 by Michael Lewis, reprinted by permission of the publisher.
Excerpt from "Lives of a Cell," copyright © 1971 by the Massachusetts
Medical Society, from The Lives of a Cell by Lewis Thomas, used by
permission of Viking Penguin, a division of Penguin Books USA Inc.

To the people of Bukoba, Lasaka, Dar es Salaam, and dozens of other African locales who so generously over the years have shared their lives and wisdom with an inquiring white Western woman. Consider this a down payment on an enormous debt.

Africa: Asante sana, Mwalimu.

Contents

Preface   *xi*

Introduction   *3*

1 / Machupo
BOLIVIAN HEMORRHAGIC FEVER   *13*

2 / Health Transition
THE AGE OF OPTIMISM—SETTING OUT
TO ERADICATE DISEASE   *30*

3 / Monkey Kidneys and the Ebbing Tides
MARBURG VIRUS, YELLOW FEVER, AND THE
BRAZILIAN MENINGITIS EPIDEMIC   *53*

4 / Into the Woods
LASSA FEVER   *71*

5 / Yambuku
EBOLA   *100*

6 / The American Bicentennial
SWINE FLU AND LEGIONNAIRES' DISEASE   *153*

7 / N'zara
LASSA, EBOLA, AND THE DEVELOPING
WORLD'S ECONOMIC AND SOCIAL POLICIES   *192*

8 / Revolution
GENETIC ENGINEERING AND THE
DISCOVERY OF ONCOGENES   *222*

9 / *Microbe Magnets*
URBAN CENTERS OF DISEASE *234*

10 / *Distant Thunder*
SEXUALLY TRANSMITTED DISEASES
AND INJECTING DRUG USERS *260*

11 / *Hatari: Vinidogodogo (Danger: A Very Little Thing)*
THE ORIGINS OF AIDS *281*

12 / *Feminine Hygiene (As Debated, Mostly, by Men)*
TOXIC SHOCK SYNDROME *390*

13 / *The Revenge of the Germs, or Just Keep Inventing
New Drugs*
DRUG-RESISTANT BACTERIA, VIRUSES,
AND PARASITES *411*

14 / *Thirdworldization*
THE INTERACTIONS OF POVERTY, POOR HOUSING,
AND SOCIAL DESPAIR WITH DISEASE *457*

15 / *All in Good Haste*
HANTAVIRUSES IN AMERICA *528*

16 / *Nature and* Homo sapiens
SEAL PLAGUE, CHOLERA, GLOBAL WARMING,
BIODIVERSITY, AND THE MICROBIAL SOUP *550*

17 / *Searching for Solutions*
PREPAREDNESS, SURVEILLANCE,
AND THE NEW UNDERSTANDING *592*

Afterword 621

Notes 623

Acknowledgments 729

Index 731

Maps

Amazonia   15

West Africa   74

Sub-Saharan Africa   207

Global Per Capita Earnings   249

*Key Gay Communities and Centers
of Injecting Drug Use, 1980s   275*

Central East Africa   388

Preface

We always want to believe that history happened only to "them," "in the past," and that somehow we are outside history, rather than enmeshed within it. Many aspects of history are unanticipated and unforeseen, predictable only in retrospect: the fall of the Berlin Wall is a single recent example. Yet in one vital area, the emergence and spread of new infectious diseases, we can already predict the future—and it is threatening and dangerous to us all.

The history of our time will be marked by recurrent eruptions of newly discovered diseases (most recently, hantavirus in the American West); epidemics of diseases migrating to new areas (for example, cholera in Latin America); diseases which become important through human technologies (as certain menstrual tampons favored toxic shock syndrome and water cooling towers provided an opportunity for Legionnaires' Disease); and diseases which spring from insects and animals to humans, through man-made disruptions in local habitats.

To some extent, each of these processes has been occurring throughout history. What is new, however, is the increased potential that at least some of these diseases will generate large-scale, even worldwide epidemics. The global epidemic of human immunodeficiency virus is the most powerful and recent example. Yet AIDS does not stand alone; it may well be just the first of the modern, large-scale epidemics of infectious disease.

The world has rapidly become much more vulnerable to the eruption and, most critically, to the widespread and even global spread of both new and old infectious diseases. This new and heightened vulnerability is not mysterious. The dramatic increases in worldwide movement of people, goods, and ideas is the driving force behind the globalization of disease. For not only do people travel increasingly, but they travel much more rapidly, and go to many more places than ever before. A person harboring a life-threatening microbe can easily board a jet plane and be on another continent when the symptoms of illness strike. The jet plane itself, and its cargo, can carry insects bringing infectious agents into new ecologic settings. Few habitats on the globe remain truly isolated or untouched, as tourists and other travelers penetrate into the most remote and previously inaccessible areas in their search for new vistas, business, or recreation.

This new global vulnerability is dramatically illustrated by the history of HIV/AIDS. While its geographical origins remain uncertain, it is clear that the global spread of HIV was underway by the mid-1970s. By 1980, about 100,000 people worldwide were infected with HIV. Yet the discovery of AIDS, in California in 1981, and the subsequent identification of the causative virus, HIV, in 1983, resulted from a series of very fortunate circumstances. Put another way, AIDS could have easily remained unrecognized for at least another five to ten years, with devastating global health consequences. Delay in discovering AIDS could have resulted from any or all of the following:

- if HIV took longer to cause detectable, clinical illness (AIDS);
- if the immunodeficiency of AIDS resulted in an increase of more typical infections rather than the easily recognized, unusual opportunistic infections (*Pneumocystis carinii* pneumonia) or cancers (Kaposi's sarcoma);
- if AIDS had not clustered among active, self-identified gay men, but rather had been spread more broadly within society;
- if AIDS had not occurred in a country (U.S.A.) with a highly developed disease surveillance system, capable of linking reports of cases from many different geographical areas;
- and if the science of human retrovirology had not been recently developed, including techniques for detection.

With AIDS, a combination of chance and circumstance relatively quickly led scientists to consider that a new health threat had arisen.

AIDS is trying to teach us a lesson. The lesson is that a health problem in any part of the world can rapidly become a health threat to many or all. A worldwide "early-warning system" is needed to detect quickly the eruption of new diseases or the unusual spread of old diseases. Without such a system, operating at a truly global level, we are essentially defenseless, relying on good luck to protect us.

Laurie Garrett has written a pioneering book. She provides us with a history, full of real people, sweat and grit, of the discoveries which have led us to realize that infectious diseases have not been vanquished—quite the contrary. It was in these places, in Bolivia, Sudan, Sierra Leone, and Zaire, that a group of highly trained, dedicated, and courageous people met the enemy on its own ground. Facing the unknown, at the frontiers of science, they struggled and wrested from nature an insight which Laurie Garrett shares with us—that diseases will remain a threat, that disease and human activity are inextricable, and that nature has many hidden places and surprises still in store.

The voyage that Ms. Garrett proposes is full of heart. I have been privileged to know many of the people in this book. They are heroes of a special kind: bonding science, curiosity, and humanitarian concern, combined with a very practical, "let's get it done" attitude. Not everyone could go, as Joe McCormick has done, into the field armed only with his will, intelligence, and confidence that a way forward would be found.

They have pioneered on our behalf. We owe them our thanks. Laurie Garrett has done us the great service of introducing them and their work to a large audience. And to those who sleep peacefully, unaware of the emerging global threat of infectious disease, and to those who through this book will be introduced to the new global realities, it is important to meet these men and women who confront disease along its frontier with society.

This book sounds an alarm. The world needs—now—a global early-warning system capable of detecting and responding to new emerging infectious disease threats to health. There is no clearer warning than AIDS. Laurie Garrett has spelled it out clearly for us. Now we ignore it at our peril.

JONATHAN M. MANN, M.D., M.P.H.
François-Xavier Bagnoud Professor of Health and Human Rights
Professor of Epidemiology and International Health
Harvard School of Public Health
Director, International AIDS Center
Harvard AIDS Institute
Cambridge, Massachusetts

THE COMING PLAGUE

Introduction

■ By the time my Uncle Bernard started his medical studies at the University of Chicago in 1932 he had already witnessed the great influenza pandemic of 1918–19. He was seven years old when he counted the funeral hearses that made their way down the streets of Baltimore. Three years earlier Bernard's father had nearly died of typhoid fever, acquired in downtown Baltimore. And shortly after, his grandfather died of tuberculosis.

In his twelfth year Bernard got what was called "summer sickness," spending the long, hot Maryland days lying about the house, "acting lazy," as his mother put it. It wasn't until 1938, when he volunteered as an X-ray guinea pig during his internship at the University of California's medical school in San Francisco, that Uncle Bernard discovered that the "summer sickness" was actually tuberculosis. He had no doubt acquired consumption from his grandfather, survived the disease, but for the rest of his life had telltale scars in his lungs that were revealed by chest X rays.

It seemed that everybody had TB in those days. When young Bernard Silber was struggling his way through medical studies in Chicago, incoming nursing students were routinely tested for antibodies against TB. The women who came from rural areas always tested negative for TB when they started their studies. With equal certainty, they all tested TB-positive after a year on the urban hospital wards. Any ailment in those days could light up a latent TB infection, and tuberculosis sanitariums were overflowing. Treatment was pretty much limited to bed rest and a variety of hotly debated diets, exercise regimens, fresh air, and extraordinary pneumothorax surgical procedures.

In 1939 Uncle Bernard started a two-year residency in medicine at Los Angeles County Hospital, where he met my Aunt Bernice, a medical social worker. Bernice limped and was deaf in one ear, the results of a childhood bacterial infection. When she was nine, the bacteria grew in her ear, eventually infecting the mastoid bone. A complication of that was osteomyelitis, which left her right leg about an inch shorter than her left, forcing Bernice to walk knock-kneed to keep her balance. Shortly after they met, Bernard got a nasty pneumococcal infection and, because he was a physician, received state-of-the-art treatment: tender loving care and oxygen.

For a month he languished as a patient in Los Angeles County Hospital hoping he would be among the 60 percent of Americans who, in the days before antibiotics, survived bacterial pneumonia.

Bacterial infections were both common and very serious before 1944, when the first antibiotic drugs became available. My Uncle Bernard could diagnose scarlet fever, pneumococcal pneumonia, rheumatic fever, whooping cough, diphtheria, or tuberculosis in a matter of minutes with little or no laboratory support. Doctors had to know how to work quickly because these infections could escalate rapidly. Besides, there wasn't much the lab could tell a physician in 1940 that a well-trained, observant doctor couldn't determine independently.

Viruses were a huge black box in those days, and though Bernard had no trouble differentiating between German measles, influenza, St. Louis encephalitis, and other viral diseases, he had neither treatments nor much of an understanding of what these tiniest of microbes did to the human body.

Uncle Bernard was introduced to tropical medicine during World War II, when he served in the Army Medical Corps at Guadalcanal and other battlefields of the Pacific. That's when he learned firsthand about diseases of which he'd heard very little in medical school: malaria, dengue (breakbone fever), and a variety of parasitic diseases. Quinine did a good job of curing malaria, but there was little he could do for GIs afflicted with the other tropical organisms that were rife in the Pacific theater.

Two years into the war the Army issued its first meager supplies of penicillin, instructing physicians to use the precious drug sparingly, in doses of about 5,000 units (less than a third of what would be considered a minimal penicillin dose for minor infections in 1993). In those early days before bacteria became resistant to antibiotics, such doses were capable of performing miracles, and the Army doctors were so impressed with the powers of penicillin that they collected the urine of patients who were on the drug and crystallized excreted penicillin for reuse on other GIs.

Years later, when I was studying immunology in graduate school at UC Berkeley, Uncle Bernard would regale me with tales of what sounded like medicine in the Dark Ages. I was preoccupied with such things as fluorescence-activated laser cell sorters that could separate different types of living cells of the immune system, the new technology of genetic engineering, monoclonal antibodies, and deciphering the human genetic code.

"I always liken the production of antibiotics to the Internal Revenue Service," Uncle Bernard would say when I seemed less than interested in the pre-antibiotic plights of American physicians. "People are always looking for loopholes, but as soon as they find them, the IRS plugs them up. It's the same way with antibiotics—no sooner have you got one than the bacteria have become resistant."

During the summer of 1976 I had reason to reconsider much of my Uncle Bernard's wisdom. As I tried to make sense of my graduate research project at Stanford University Medical Center, the news seemed overfull of infectious disease stories. The U.S. government was predicting a massive influenza epidemic that some said would surpass that of 1918—a global horror that claimed over 20 million lives. An American Legion group met in a hotel in Philadelphia on the Fourth of July, and something made 182 of them very sick, killing 29. Something else especially strange was going on in Africa, where, according to garbled press accounts of the day, people were dying from a terrifying new virus: in Zaire and the Sudan, something called Green Monkey Virus, or Marburg, or Ebola, or a mix of all three monikers was occupying the urgent attention of disease experts from all over the world.

In 1981 Dr. Richard Krause of the U.S. National Institutes of Health published a provocative book entitled *The Restless Tide: The Persistent Challenge of the Microbial World*,[1] which argued that diseases long thought to have been defeated could return to endanger the American people. In hearings a year later before the U.S. Congress, Krause was asked, "Why do we have so many new infectious diseases?"

"Nothing new has happened," Krause replied. "Plagues are as certain as death and taxes."[2]

But the shock of the AIDS epidemic prompted many more virus experts in the 1980s to ponder the possibility that something new was, indeed, happening. As the epidemic spread from one part of the world to another, scientists asked, "Where did this come from? Are there other agents out there? Will something worse emerge—something that can be spread from person to person in the air?"

The questioning grew louder as the 1980s dragged on. At a Rockefeller University cocktail party, a young virologist named Stephen Morse approached the institution's famed president, Nobel laureate Joshua Lederberg, and asked him what he thought of the mounting concern about emerging microbes. Lederberg characteristically responded in absolute terms: "The problem is serious, and it's getting worse." With a sense of shared mission, Morse and Lederberg set out to poll their colleagues on the matter, gather evidence, and build a case.

By 1988 an impressive group of American scientists, primarily virologists and tropical medicine specialists, had reached the conclusion that it was time to sound an alarm. Led by Morse and Lederberg of Rockefeller University, Tom Monath of the U.S. Army's Medical Research Institute of Infectious Diseases, and Robert Shope of the Yale University Arbovirus Research Unit, the scientists searched for a way to make tangible their shared concern. Their greatest worry was that they would be perceived as crybabies, merely out to protest shrinking research dollars. Or that they would be accused of crying wolf.

On May 1, 1989, the scientists gathered in the Hotel Washington, located across the street from the White House, and began three days of discussions aimed at providing evidence that the disease-causing microbes of the planet, far from having been defeated, were posing ever-greater threats to humanity. Their gathering was co-sponsored by the National Institutes of Allergy and Infectious Diseases, the Fogarty International Center, and Rockefeller University.

"Nature isn't benign," Lederberg said at the meeting's opening. "The bottom lines: the units of natural selection—DNA, sometimes RNA elements—are by no means neatly packaged in discrete organisms. They all share the entire biosphere. The survival of the human species is *not* a preordained evolutionary program. Abundant sources of genetic variation exist for viruses to learn new tricks, not necessarily confined to what happens routinely, or even frequently."

University of Chicago historian William McNeill outlined the reasons *Homo sapiens* had been vulnerable to microbial assaults over the millennia. He saw each catastrophic epidemic event in human history as the ironic result of humanity's steps forward. As humans improve their lots, McNeill warned, they actually *increase* their vulnerability to disease.

"It is, I think, worthwhile being conscious of the limits upon our powers," McNeill said. "It is worth keeping in mind that the more we win, the more we drive infections to the margins of human experience, the more we clear a path for possible catastrophic infection. We'll never escape the limits of the ecosystem. We are caught in the food chain, whether we like it or not, eating and being eaten."

For three days scientists presented evidence that validated McNeill's words of foreboding: viruses were mutating at rapid rates; seals were dying in great plagues as the researchers convened; more than 90 percent of the rabbits of Australia died in a single year following the introduction of a new virus to the land; great influenza pandemics were sweeping through the animal world; the Andromeda strain nearly surfaced in Africa in the form of Ebola virus; megacities were arising in the developing world, creating niches from which "virtually anything might arise"; rain forests were being destroyed, forcing disease-carrying animals and insects into areas of human habitation and raising the very real possibility that lethal, mysterious microbes would, for the first time, infect humanity on a large scale and imperil the survival of the human race.

As a member of a younger generation trained in an era of confident, curative medicine and minimal concern for infectious diseases, I experienced such discussion as the stuff of Michael Crichton novels rather than empiric scientific discourse. Yet I and thousands of young scientists also reared in the post-antibiotic, genetic engineering era had to concede that there was an impressive list of recently emergent viruses: the human immunodeficiency virus that caused AIDS, HTLV Types I and II which were linked to blood cancers, several types of recently discovered hepatitis-causing viruses, numerous hemorrhage-producing viruses discovered in Africa and Asia.

In February 1991 the Institute of Medicine (IOM), which is part of the U.S. National Academy of Sciences, convened a special panel with the task of exploring further the questions raised by the 1989 scientific gathering and advising the federal government on two points: the severity of the microbial threat to U.S. citizens and steps that could be taken to improve American disease surveillance and monitoring capabilities. In the fall of 1992 the IOM panel released its report, *Emerging Infections: Microbial Threats to Health in the United States*,[3] which concluded that the danger of the emergence of infectious diseases in the United States was genuine, and authorities were ill equipped to anticipate or manage new epidemics.

"Our message is that the problem is serious, it's getting worse, and we need to increase our efforts to overcome it," Lederberg said on the day of the report's release.

After the release of the report, the U.S. Centers for Disease Control and Prevention in Atlanta began a soul-searching process that would, by the spring of 1994, result in a plan for heightened vigilance and rapid response to disease outbreaks. The slow response to the emergence of HIV in 1981 had allowed the epidemic to expand by 1993 to embrace 1.5 million Americans and cost the federal government more than $12 billion annually in research, drug development, education, and treatment efforts.

The CDC was determined that such a mistake would not be repeated.

But there were dissident voices in 1993 who protested both the American scientific community's often narrow emphasis on viruses and its focus on threats posed solely to U.S. citizens. Disease fighters like Joe McCormick, Peter Piot, David Heymann, Jonathan Mann, and Daniel Tarantola argued forcefully that microbes had no respect for humanity's national borders. Furthermore, they said, in much of the world the most dangerous emerging diseases were not viral, but bacterial or parasitic. A far larger view was needed, they argued.

Other critics stressed that a historical perspective on mankind's bumbling, misguided attempts to control the microbes would reveal that much of the fault lay with the very scientific community that was now calling for vigilance. What seemed to make sense as microbe control action, viewed from the academic and government offices of the world's richest country, argued the likes of Uwe Brinkmann, Andrew Spielman, and Isao Arita, could prove disastrous when executed in the planet's poorer nations.

The critics charged that Americans, by virtue of their narrow focus on the appearance of disease within the United States, were missing the real picture. It was a picture captured in the sight of a little Ndbele girl wrapped in a green *kanga*. She lay on the hardened clay floor of a health care clinic outside Bulawayo, Zimbabwe. Her mother sat beside her, casting beseeching looks at every stranger who entered the two-room clinic. The four-year-old girl cried weakly.

"That is measles," said the clinic director, pointing a stern finger at the child. The director led an observer outside to show the local innovations in toilet hygiene and efforts to increase the protein content of village children's diets.

When he returned an hour later to the wattle-clay clinic, the mother was rocking back and forth on the balls of her feet, tears silently streaming down her face; the child's soft crying had ceased. A few hours later the mother and her husband placed across bicycle handlebars a rolled straw mat containing their little girl's body and, staring blankly at the horizon, forlornly walked the bike down the red clay road.

At a time when mothers of the world's wealthiest nations arranged to have their children "immunized" by deliberately exposing the youngsters to measles, mumps, even chicken pox, these diseases were forcing parents in some of the world's poorest nations to find ways to cope with the expected deaths of more than half their children before the age of ten.

The long list of vaccines and prescription drugs that American physicians urged their patients to take before traveling south of Tijuana was ample testimony to the health impact of the world's wide gulf in wealth and development. In the 1970s Americans and Europeans who were distressed by the poverty of the Southern Hemisphere poured money into the poorest countries for projects intended to bring their populations into "the modern age." The logic of the day was that the health status of a population would improve as the society's overall structure and economy grew to more closely resemble those of the United States, Canada, and Western Europe.

But by 1990 the world's major donor institutions would be forced to conclude that modernization efforts seemed only to have worsened the plight of the average individual in the Third World, while enhancing the power, wealth, and corruption of national elites and foreign-owned institutions. Bucolic agricultural societies were transformed in the space of a single generation into countries focused around one or more vast urban centers that grew like ghastly canker sores over the landscape, devouring the traditional lifestyles and environments of the people and thrusting young job seekers into sprawling semi-urban slums that lacked even a modicum of proper human waste disposal or public health intervention.

In the industrialized free market world of the 1970s, people at all societal strata became increasingly conscious of the link between environmental pollution and personal health. As the dangers of pesticide misuse, lead paint, asbestos fibers, air pollution, and adulterated foods became apparent, residents of the world's wealthiest countries clamored for regulations to curb contamination of the environment and the food supply.

With the discovery of Earth's ozone holes, the world's scientists initiated a debate about global responsibility for preventing further pollution destruction of the planet's protective gaseous layer. Similarly, marine biologists argued with increasing vehemence that all the nations of the world shared responsibility for the sorry state of Earth's oceans and the near-extinction or endangerment of its fish, coral, and mammal populations. Conservationists turned their attention to global wildlife protection. And biologists like Harvard's E. O. Wilson and the Smithsonian's Frank Lovejoy warned of a global mass flora and fauna extinction event that would rival that of the great Cretaceous period dinosaur die-off.

Citing the fossil evidence for five great extinction events in Earth's ancient history, Wilson asked how much more environmental destruction at man's hand the world could tolerate: "These figures should give pause to anyone who believes that what *Homo sapiens* destroys, Nature will redeem. Maybe so, but not within any length of time that has meaning for contemporary humanity."[4]

As humanity approached the last decade of the twentieth century, the concept of a Global Village—first elucidated in the 1960s by Canadian philosopher Marshall McLuhan as a description of the sense of worldwide interconnectedness created by mass media technology—had clearly entered mass consciousness in the context of Earth's ecology. Environmentalists were thinking on the macro level, plotting ways to change the whaling policies of places as disparate as Japan, Alaska, Russia, and Norway. The World Bank decided to include ecological concerns in its parameters for issuing loans to developing countries. The Chernobyl nuclear accident proved, in the eyes of many scientists, that it was folly to consider toxic risk control a problem whose solutions were always constrained by issues of national sovereignty.

And in 1992 the United States elected a Vice President who espoused an ambitious global Marshall Plan to protect the environment. Albert Gore warned that nothing short of a massive worldwide shift in human perspective, coupled with elaborate systems of international regulation and economic incentives, would be adequate to ensure the survival of the planet's ecology. And he adopted the rhetoric of critical environmentalists, saying, "Those who have a vested interest in the status quo will probably continue to stifle any meaningful change until enough citizens who are concerned about the ecological system are willing to speak out and urge their leaders to bring the earth back into balance."[5]

At the macro level, then, a sense of global interconnectedness was developing over such issues as economic justice and development, environmental preservation, and, in a few instances, regulation. Though there were differences in perspective and semantics, the globalization of views on some issues was already emerging across ideological lines well before the fall of the Berlin Wall. Since then it has sped up, although there is now considerable anxiety expressed outside the United States about American domination of the ideas, cultural views, technologies, and economics of globalization of such areas as environmentalism, communication, and development.

It wasn't until the emergence of the human immunodeficiency virus, however, that the limits of, and imperatives for, globalization of health became obvious in a context larger than mass vaccination and diarrhea control programs. From the moment it was discovered in 1981 among gay men in New York and California, AIDS became a prism through which the positive lights by which societies hoped to be viewed were fractured into thousands of disparate and glaring pieces. Through the AIDS prism it was possible for the world's public health experts to witness what they considered to be the hypocrisies, cruelties, failings, and inadequacies of humanity's sacred institutions, including its medical establishment, science, organized religion, systems of justice, the United Nations, and individual government systems of all political stripes.

If HIV was our model, leading scientists concluded, humanity was in very big trouble. *Homo sapiens* greeted the emergence of the new disease first with utter nonchalance, then with disdain for those infected by the virus, followed by an almost pathologic sense of mass denial that drew upon mechanisms for rationalizing the epidemic that ranged from claiming that the virus was completely harmless to insisting that certain individuals or races of people were uniquely blessed with the ability to survive HIV infection. History, they claimed, would judge the 1980s performance of the world's political and religious leaders: would they be seen as equivalent to the seventeenth-century clerics and aristocracy of London who fled the city, leaving the poor to suffer the bubonic plague; or would history be more compassionate, merely finding them incapable of seeing the storm until it leveled their homes?

Over the last five years, scientists—particularly in the United States and France—have voiced concern that HIV, far from representing a public health aberration, may be a sign of things to come. They warn that humanity has learned little about preparedness and response to new microbes, despite the blatant tragedy of AIDS. And they call for recognition of the ways in which changes at the micro level of the environment of any nation can affect life at the global, macro level.[6]

Humanity's ancient enemies are, after all, microbes. They didn't go away just because science invented drugs, antibiotics, and vaccines (with the notable exception of smallpox). They didn't disappear from the planet when Americans and Europeans cleaned up their towns and cities in the postindustrial era. And they certainly won't become extinct simply because human beings choose to ignore their existence.

In this book I explore the recent history of disease emergence, examining in roughly chronological order examples that highlight reasons for microbial epidemics and the ways humans respond, as cultures, scientists, physicians, bureaucrats, politicians, and religious leaders.

The book also examines the biology of evolution at the microbial level, looking closely at ways in which disease agents and their vectors are adapting to counter the defensive weapons used to protect human beings.

In addition, *The Coming Plague* looks at means by which humans are actually aiding and abetting the microbes through ill-planned development schemes, misguided medicine, errant public health, and shortsighted political action/inaction.

Finally, some solutions are offered. Fear, without potential mitigating solutions, can be very volatile. It has, throughout history, prompted the lifelong imprisonment of the victims of a disease. Perhaps less onerously, it can lead to inappropriate expenditures of money and human resources aimed at staving off a real or imagined enemy.

What is required, overall, is a new paradigm in the way people think about disease. Rather than a view that sees humanity's relationship to the microbes as a historically linear one, tending over the centuries toward ever-decreasing risk to humans, a far more challenging perspective must be sought, allowing for a dynamic, nonlinear state of affairs between *Homo sapiens* and the microbial world, both inside and outside their bodies. As Harvard University's Dick Levins puts it, "we must embrace complexity, seek ways to describe and comprehend an ever-changing ecology we cannot see, but, nonetheless, by which we are constantly affected."

Now in his eighties and retired from the daily practice of medicine, my Uncle Bernard wonders how many American physicians today would recognize a case of malaria, diphtheria, rheumatic fever, tuberculosis, or typhus without needing the guiding advice provided by time-consuming laboratory diagnostic analysis. He doubts whether most physicians in the industrialized world could diagnose an old scourge, like yellow fever or dengue fever, much less spot an entirely new microbe. As he and the rest of the pre-antibiotic era physicians of the developed world retire and age, Bernard asks if doctors of the year 2000 will be better or worse equipped to treat bacterial pneumonia than were physicians in his pre-antibiotic days.

Preparedness demands understanding. To comprehend the interactions between *Homo sapiens* and the vast and diverse microbial world, perspectives must be forged that meld such disparate fields as medicine, environmentalism, public health, basic ecology, primate biology, human behavior, economic development, cultural anthropology, human rights law, entomology, parasitology, virology, bacteriology, evolutionary biology, and epidemiology.

The Coming Plague tells the stories of men and women who struggled to understand and control the microbial threats of the post–World War II era. As these disease vanquishers retire, the college laboratories and medical schools grow full of youthful scientific energy, but it is not focused on the seemingly old-fashioned, passé tasks that were invaluable in humanity's historic ecological struggles with the microbes. As we approach the millennium, few young scientists or doctors anywhere in the world can quickly recognize a tiger mosquito, *Peromyscus maniculatus* mouse, pertussis cough, or diphtherial throat infection.

The skills needed to describe and recognize perturbations in the *Homo sapiens* microecology are disappearing with the passing of the generations, leaving humanity, lulled into a complacency born of proud discoveries and medical triumphs, unprepared for the coming plague.

1

Machupo

BOLIVIAN HEMORRHAGIC FEVER

Any attempt to shape the world and modify human personality in order to create a self-chosen pattern of life involves many unknown consequences. Human destiny is bound to remain a gamble, because at some unpredictable time and in some unforeseeable manner nature will strike back.

—*Mirage of Health*, René Dubos, 1959

■ Karl Johnson fervently hoped that if this disease didn't kill him soon somebody would shoot him and put him out of his misery. The word "agony" wasn't strong enough. He was in hell.

Every nerve ending of his skin was on full alert. He couldn't bear even the pressure of a sheet. When the nurses and doctors at Panama's Gorgas Hospital touched him or tried to draw blood samples, Johnson inwardly screamed or cried out.

He was sweating with fever, and he felt the near-paralytic exhaustion and severe pain he imagined afflicted athletes who pushed their training much too far.

When nurses on the Q ward first looked at Johnson lying beside his two colleagues they recoiled from the sight of his crimson blood-filled eyes. All over Johnson's body the tiny capillaries that acted as tributaries flowing to and from the veins' rivers of blood were leaking. Microscopic holes had appeared, out of which flowed water and blood proteins. His throat hurt so much he could barely speak or drink water, thanks to a raw and bleeding esophageal lining. Word around the hospital was that the three were victims of a strange and contagious new plague that felled them in Bolivia.

In brief moments of lucidity Johnson would ask how many days had passed. When a nurse told him it was Day Five, he groaned.

"If my immune system doesn't kick in fast, I'm a dead man," he thought.

He'd seen it happen plenty of times in San Joaquín. Some of the people died in just four days, but most suffered over a week of this torture.

Over and over he reviewed what he had seen in that isolated village on Bolivia's eastern frontier. He hoped to think of something that could help him recover and solve the San Joaquín mystery.

It had all started exactly a year before—in July 1962. Johnson had just arrived at the Middle America Research Unit (MARU) in the Panama Canal Zone, having had his fill of cataloguing respiratory viruses at the U.S. government's National Institutes of Health in Bethesda, Maryland.

Since 1956, then a young physician fresh from his medical training, Johnson had tediously studied viruses that caused common colds, bronchitis, and pneumonia. The work was getting plenty of praise, but Johnson, who was always impatient, was bored. When word got out that the National Institutes of Health was looking for a virologist to staff its MARU laboratory, he jumped at the chance.

Shortly after Johnson arrived in Panama, his newfound MARU colleague, Ron MacKenzie, volunteered to assist a U.S. Department of Defense (DOD) team that was heading into Bolivia to conduct nutritional surveys.

"A nutrition survey?" Johnson asked snidely.

"Well, I could use the experience, and I've never been to Bolivia. So why not?" MacKenzie said.

When MacKenzie and the DOD team met with Bolivia's Minister of Health in La Paz, the official said he had no problem authorizing their research plans, provided they first take care of a more pressing problem hundreds of miles away.

"I need an expert in mysterious diseases to investigate an epidemic in the eastern part of the country."

All eyes turned to MacKenzie, who, as a pediatrician and trained epidemiologist, came closest to fitting the bill. He shifted uncomfortably in his chair, mumbled something about not being able to speak Spanish, and tried to imagine what eastern Bolivia might look like.

The minister went on to explain that the mysterious epidemic was fairly sizable, and two La Paz physicians had tentatively labeled it *El Typho Negro*—the Black Typhus.

The following morning found the tall, somewhat gawky MacKenzie—dressed in a black suit, starched white shirt, and wing tips—standing on the La Paz tarmac, a briefcase at his feet. He greeted Bolivian physician Hugo Garrón, microbiologist Luis Valverde Chinel, and a local politician, and the quartet boarded an old B-24 bomber bound for the town of Magdalena, in the country's eastern frontier region. MacKenzie looked around for a seat: there were none. The plane had been stripped down for hauling meat, and the only passengers usually on board were sides of beef.

So MacKenzie stood behind the pilot, holding on to the hull for dear life during the long acceleration down the gravel runway. Because La Paz was at an elevation of 13,000 feet, planes had to reach great speeds to gain enough lift for takeoff. After what seemed an extraordinary amount of time, the Bolivian Indian mechanic who was squatting between the pilot and the copilot pulled a lever in the cockpit floor, withdrawing the landing gear, and they took off.

Like a tired old condor, the bomber circled La Paz slowly several times, spiraling upward to 16,000 feet, high enough to fly into a narrow pass

AMAZONIA

between the Andean peaks that towered around La Paz. MacKenzie found himself staring aghast at avalanches of ice cascading off dangerously close cliffs.

When the plane escaped the claustrophobic mountain pass, it was enveloped in a dense fog which forced the pilots to fly by instruments alone: a magnetic compass, stopwatch, map, and notepad.

MacKenzie figured this was already enough adventure. Just three years ago he had been patching broken bones and vaccinating kids in a bucolic town north of San Francisco. This new exploit was a bit more perilous than anything he had gambled on when he left private practice to go into public health.

As the plane descended into the fog, MacKenzie began to feel the heat and humidity increase, sweat built under his stiffly starched shirt, and when the fog cover broke, he watched seemingly endless grassy savannas pass beneath them. These were broken up by outcroppings of low, tree-covered *alturas* hills, and by long winding rivers lined with thin *bandas* strips of rain forest.

"It looks just like Florida," MacKenzie thought. "Kind of like the Everglades."

After two more long hours the plane landed in the little town of Magdalena, and MacKenzie couldn't believe his eyes.

"My gosh," he exclaimed, "there must be two hundred people out there, standing around the plane." The women in the throng were dressed entirely in mourning; the men wore black armbands. The bereaved people of Magdalena had gathered to greet "the experts" who had come to end their epidemic.

"Experts?" MacKenzie muttered, casting an uncomfortable glance at Valverde and Garrón. "Well, I'm it."

With the grim entourage around them, the group dodged lumbering oxcarts as it made its way past a scattering of thatch-roofed adobe houses to the town plaza, a large courtyard surrounded by a circular arcade and the homes and stores of Magdalena. A sad, lethargic feeling pervaded everything.

At Magdalena's tiny clinic MacKenzie found a dozen patients writhing in pain.

"My God!" he exclaimed as he watched one after another vomit blood. MacKenzie shuddered, feeling the tremendous onus of his position and cursing the naïveté with which he had walked into the situation. It seemed that only yesterday he was doling out antibiotics in a clinic in Sausalito to kids whose frolicking was briefly interrupted by sore throats. What MacKenzie saw on the ward forced him to push aside his pediatrics training and, for the moment, draw upon the lessons in courage and horror he had learned during World War II combat.

He was told that most of the sick were outsiders from Orobayaya. The mere name of that distant village sent shivers through the Magdalenistas, who spoke of it with unconcealed fear.

Soon the lanky MacKenzie, who towered over the Bolivians, was crouched in a dugout canoe making its way by moonlight upriver toward the plagued village. As they glided along MacKenzie kept spotting enormous "logs"— far larger than their canoe—sliding down the riverbanks toward them. The hair on the back of his neck stood up when he realized the "logs" were alligators.

The next day the group rode forty kilometers on horseback to Orobayaya.

It was deserted. The six hundred residents had fled days earlier in panic, leaving the village to pigs and chickens that scampered madly about in search of food.

MacKenzie returned to Magdalena, collected some blood samples from local patients, and headed back to Panama, where he tried to convince MARU director Henry Beye and the NIH bosses in Bethesda that the Bolivian situation warranted further investigation.

"It's probably just the flu," was the consensus from NIH officials.

"It's something strange and dangerous," MacKenzie insisted.

Both MacKenzie and Johnson thought the Bolivian villagers' symptoms resembled those brought on by a recently discovered Latin American virus, found near the Junín River in 1953 in Argentina. The Argentine virus was a close cousin of Tacaribe, which caused a disease of bats and rodents in Trinidad, also only recently discovered. While there was no evidence that Tacaribe could infect human beings, Junín was clearly lethal in many cases. In sparsely populated agricultural areas of Argentina's vast pampas, Junín appeared as if out of nowhere among men working the corn harvests. It too was a human killer that disrupted capillaries, causing people to bleed to death. Nobody was sure how the Argentines got Junín; there was speculation the virus might be airborne.

No point in taking stupid chances, Johnson thought. Though the NIH had not approved a MARU investigation of the epidemic, he flew to the U.S. Army's Fort Detrick, in Maryland, to see Al Wieden. A pioneer in laboratory safety, Wieden had turned Fort Detrick into the world's premier center of research on deadly microbes. Johnson wanted something unheard of: a portable box of some sort so he could safely study Junín in the field—or whatever was wiping out the people of San Joaquín.

At Fort Detrick there was a lot of research underway on "germ-free mice"—animals that had such weak immune systems that virtually any microbe could prove lethal to the mutant rodents. To keep the mice alive, scientists housed them inside airtight boxes that were constantly under positive pressure, pushing air past special filters to the mice, and then out again, toward the scientists. In this way, the mice breathed only sterile air. Scientists worked with the mice by inserting their hands into airtight rubber gloves that were built into the sides of the pressure box. The "glove boxes," as the steel contraptions were called, were about the size of large coffins and weighed hundreds of pounds.

Johnson's idea was to convert one of these contraptions from positive to negative pressure so that all air would go inward, toward samples of possibly dangerous animals or microbes. That way, he could work relatively safely in a portable laboratory.

Such a portable laboratory had never before been used, and Wieden wasn't sure how to jury-rig the positive-pressure boxes. But, racing against time, Johnson and Wieden found a new, lighter-weight plastic glove box and surrounded it with a vast rib cage of aluminum poles to prevent the container from imploding when the pressure reversed from inside-out to outside-in. To their mutual delight, it worked.

Meanwhile, MacKenzie still faced tough opposition in Bethesda, as well as at the Centers for Disease Control in Atlanta. Though he was a physician and had public health training, some higher-ups frankly doubted whether the thirty-seven-year-old MacKenzie had enough tropical experience to be able to recognize a new epidemic. They insisted it would be a waste of time and resources to deploy a team to investigate what would probably turn out to be some garden-variety bug such as influenza.

In the fall of 1962 MacKenzie appealed to Bill Reeves, his old mentor from public health graduate studies at the University of California in Berkeley. He described Magdalena to Reeves, who insisted that MacKenzie "stand up to the Bethesda bureaucrats."

"Go for it. You got something there. Don't let 'em discourage you," Reeves urged.

On January 9, 1963, a meeting of the top brass in the NIH's infectious diseases division was held in Bethesda, and MacKenzie persuasively pleaded his case. It was decided that he and a MARU ecologist named Merl Kuns should first undertake a scouting mission to assess the extent of the epidemic, collect blood samples, and define the nature of the local ecology.

The pair made their journey in March, and returned a week later even more firmly convinced that a serious epidemic was underway. Kuns, a University of Wisconsin-trained ecologist, was stunned by the thousands of bats that lived in the thatched roofs of towns like Magdalena, swooping out at night to forage for food. They were small bats, about the size of monarch butterflies, but they clustered in huge flocks that could suddenly fill the village sky. For his part, epidemiologist MacKenzie was convinced that nobody was actually getting infected in Magdalena, and the real epidemic was some fifty miles away in a town called San Joaquín. The pair returned to Panama with more than adequate evidence to gain approval for further investigation.

With his new laboratory contraption in crates, Johnson headed to Bolivia in May 1963, along with MacKenzie and Kuns. After arriving in the capital, the team chartered an old USAF B-17 bomber and flew to the eastern edge of the Andes, then down the eastern Andean foothills to the Iténez River, and from there to the river's Machupo tributary, eventually landing on a field outside San Joaquín. They then hauled their 10,000 pounds of equipment into the tiny town on mules.

Nestled atop a sloping hill just above the Machupo's flood line, San Joaquín was, the flabbergasted Johnson thought, "the last frontier of the New World." Nothing in his scientific career had prepared him for conditions so primitive: no roads, no real health facilities, no fences, no electricity, no telephones, no running water. Cows outnumbered humans roughly two to one and roamed freely about the town. The people of San Joaquín were an evenly divided mix of pure Spanish, pure Indians, and mestizos whose ancestors had built the town in the seventeenth century. The wealthier citizens resided in tile-roofed, whitewashed adobe homes; the rest of the population lived in mud-stick houses with thatch roofs. Six thin strips of marsh formed the "roads" of San Joaquín, which converged in a modest central plaza.

The Spanish people of San Joaquín were descended from cowboys who for a few generations had tended the large herds of a wealthy Brazilian family in control of an Amazon River fleet of refrigerated steamships. The ships took the beef out of San Joaquín, up the river system some 1,400 miles to the northwest, where the Amazon met the sea. From there the beef was shipped to Europe or North America, reaping excellent profits for the Brazilians.

In tiny San Joaquín, however, the cowboys, their families, and the local Indians were entirely dependent on the "benevolence" of the Brazilian ranch owners and on the food and supplies that returning steamships brought to their remote town.

In 1952 a revolution had brought the Movimiento Nacionalista Revolucionario to power. The land reform party stripped old Bolivian and Brazilian oligarchies of vast tracts of land, and the people of San Joaquín suddenly found themselves property owners. Unwilling to buy back from the local people the cattle they once owned, the Brazilians and their steamships left, never to return; and the villagers found themselves isolated, impoverished, and facing severe malnutrition unless they could grow crops to supplement the all too abundant supply of beef.

When Johnson, MacKenzie, and Kuns arrived in San Joaquín they found a modest town of some 2,000 people managing to survive on beef, the yields of home vegetable gardens, and small rice and corn fields scattered throughout the savannas.

A steady stream of travelers passed through the town on their way from even more remote areas in the savannas to larger Bolivian towns via the steamships that moored occasionally at the San Joaquín dock.

Upon arrival, Johnson immediately set up his portable laboratory contraption, and the team set out to assess the extent of the mysterious outbreak. By then the epidemic had already been underway for some fourteen months, the people mourned every day when the church bell tolled another death, and fresh graves filled the cemetery.

With the townspeople's eager help, the team mapped the area and painted numbers on all the adobe houses. Every family was interviewed extensively, and asked the questions most essential to the team: How many people in this house have had the disease? How many have died and how many have recovered? What were they doing in the days before they got sick? Is there any chance one family member gave the disease to another? Have any animals been sick?

It became obvious immediately that nearly half the people had been infected, and, of those, nearly half had died of the disease. That was a terrifying finding because few microbes kill nearly 50 percent of those they infect. One family lost nine of eleven members in 1963.

"That's almost a Roman decimation," Johnson told his colleagues, referring to the great epidemics of ancient Rome's republican era when at least a quarter of the population was felled by a disease now thought to have been smallpox.

The first order of business was to figure out what type of microbe was killing the people of San Joaquín: bacterium, virus, or parasite. Circumstances pointed to a virus, possibly spread by insects, so the team set up two small laboratories located seventy-five yards apart. The first, an existing tile-roofed adobe building, housed Johnson's glove-box contraption and a variety of other equipment and research animals used to isolate microscopic organisms from blood and tissue samples. The second laboratory was built to order by the local people out of lashed poles and thatching. It housed wild insects and animals Kuns and his assistants caught in the San Joaquín area. The team planned to study those animals to determine what species might be carrying the deadly microbes.

The facilities were kept separate to avoid cross-contamination, and the buildings were fitted with window screens and tight doors. Finally, the laboratories were heavily doused with DDT and ringed with rodent traps to protect the scientists from whatever creatures might be carrying the disease.

In June, after days of haggling with the San Joaquín community over the propriety of such things, Valverde convinced the local priest to allow MacKenzie to perform an autopsy on one of the recent victims of the epidemic. A few days later, a two-year-old boy died and from his spleen and brain the team was able to isolate a substance that, when injected into hamsters, produced the disease. Days after the boy died the team completed several more tests that proved the mysterious disease was caused by a virus: they ruled out a parasite or bacteria on the basis of both the minuscule size of filters through which the microbe readily passed and its ability to withstand antibiotics. They also showed the microbe could destroy human cells and cause disease in wild mice.[1]

Midway through the autopsy on the child, Hugo Garrón's scalpel slipped, flew across the autopsy table, and hit MacKenzie's hand. Looking at the blood that instantly filled his punctured glove, MacKenzie looked up at Garrón and predicted the worst.

An anxious week passed without symptoms, and MacKenzie decided he was, indeed, a very lucky person. With greater care, he and Garrón performed several more autopsies and were struck by the level of devastation the mysterious microbe produced. Most alarming were the disease victims' brains: where clear cerebrospinal fluid should have been there was, instead, crimson blood; all of the meningeal protective layers around the brain were blood-soaked. Eerily, most of the hair fell off victims' heads before they died.

Toward the end of June the town had a party, which the scientists used as an opportunity to celebrate their rapid discoveries. The next logical steps in their research would involve characterizing the virus and figuring out exactly how people got infected. Johnson, Kuns, and MacKenzie felt confident all the answers would soon reveal themselves, and enthusiastically joined the celebration, eating and drinking the local specialties. While all three men were in the mood for a fiesta, it was Johnson who, with characteristic gusto, threw himself into the spirit of the event, drinking, dancing, and joining in the local macho sport of telling tall tales. Though not a classically handsome man, Johnson carried himself with a mix of cowboy swagger and charisma that inspired other men and attracted women. MacKenzie too threw himself into the gaiety of the evening, while the shyer, more serious Kuns quietly observed the goings-on.

On July 3 Johnson and MacKenzie were some twenty miles outside of San Joaquín gathering ticks from the bushes around a *chaco*, or small cattle ranch. They suspected the virus might be carried by insects and were collecting samples to take back to their field lab for analysis.

When they began the long trek back to San Joaquín, the shorter Johnson kept having to slow down to avoid outpacing his usually athletic, long-legged colleague.

By the time they reached the river and started to canoe downstream to San Joaquín, Johnson noticed he was pulling most of the weight.

"I feel lousy. Really lousy," MacKenzie said as he staggered off to bed.

The next morning Peace Corps nurse Rose Navarro, who had been sent in to help with translating, took one look at MacKenzie and pronounced his condition serious. She also noticed that Angel Muñoz, a Panamanian lab technician who had recently arrived from MARU to assist Kuns, had similar symptoms.

Johnson and Kuns contacted Panama through a cumbersome radio relay system, and a USAF C-130 flew in that day—the Fourth of July—to evacuate the two ailing researchers.

As he waved goodbye to MacKenzie, Johnson felt a feverish chill come over his body and thought, "Damn! I should have been on that plane too!"

Over the next four days Johnson slowly hitchhiked his way, plane to plane, across Bolivia, Peru, and Colombia, finally reaching the Gorgas Hospital in Panama.

And now here he was, bleeding to death. To his left lay MacKenzie, on the right Muñoz. Johnson could imagine his brief obituary: "promising young research physician, born in Terre Haute, Indiana, 1929. Dead, age 34. Unmarried."

He knew there were two ways the virus could kill him. He'd seen it in San Joaquín. He might soon develop neurological symptoms, getting tremors and losing control of his muscles; eventually, he would have a grand mal seizure and die. Or the sheer volume of blood hemorrhaging from his capillaries would become so devastating that his body would go into shock and he'd die of cardiac arrest. Either death could strike in a matter of hours, or days.

In any event, there was no cure, no antitoxin. There was just lying and waiting.

After several more days of agony all three men showed signs of improvement, thanks largely to the efforts of an Army doctor flown in especially for them from Washington, D.C. Though he had never treated this particular ailment, the doctor had handled dozens of cases of another viral hemorrhagic disease called Seoul Hantaan, which first came to the attention of Westerners when 121 trench-bound American soldiers fighting in the Korean War died bleeding deaths that were not unlike the one threatening Johnson. (Nearly 2,500 U.S. soldiers suffered the Hantaan disease from 1951 to 1955.)[2] Nobody had yet identified the Hantaan virus and it wasn't clear how the disease was transmitted, but U.S. Army doctors had discovered that patients' chances of recovery were greatly enhanced by careful supervision of their electrolytes and fluids. In all hemorrhaging diseases, as the capillaries leaked out precious fluids and proteins, the delicate chemical balances of vital organs such as kidneys, hearts, livers, and spleens were severely disrupted. Long before the immune system had a chance to mount a counterattack against the Hantaan virus, the organs would cease functioning and the patient would either convulse or go into shock.

Also in from Bethesda was Pat Webb, Johnson's petite fiancée. Born in England and trained in both medicine and virology, Webb was doing research at NIH and had planned to move to Panama soon to join Johnson. Short, thin, and prematurely graying, Webb had an often caustic, opinionated style of speech. But for those who persevered, knowing Webb meant experiencing a woman possessed of a profound sense of humanity that infused her medical and research work.

Now she sat beside her future husband and caressed, kissed, or embraced him whenever he could tolerate the pain of being touched. By deliberately touching Johnson to illustrate there was no danger, she hoped to allay the fears of the frightened hospital staff.[3] Webb's fear was not the virus, but that Johnson would die, and a couple of times his condition seemed so grave she was convinced he wasn't going to make it.

The Army physician's efforts, however, paid off. Johnson survived.

As soon as he was up and about, Johnson started studying samples of the San Joaquín virus brought back to Panama with MacKenzie and Muñoz. He was able to confirm in the sophisticated MARU facilities what had tentatively been discovered in his glove-box contraption on the Bolivian frontier: the disease was caused by a virus that was similar to, but not the same as, Junín and Tacaribe.

With Johnson safely recuperating, Webb headed back to Washington in late August. It had been two weeks, the worst was over, and it was time she got back to work. On board the plane she was suddenly seized by a pounding headache, muscle pains, and waves of shaking chills. The symptoms escalated until Webb knew that, despite all her protestations to the Gorgas nursing staff, she had gotten the virus from kissing and embracing her fiancé. She was treated at the NIH hospital and, after ten distressing days, had recovered enough to go home. A few weeks later, Webb moved to Panama and eagerly joined in the detective work.

Though they had no way of knowing whether their painful illnesses had actually made them immune to the virus, Muñoz, MacKenzie, and Johnson made the journey back to San Joaquín in September, traveling now aboard USAF planes. They were quite naturally nervous, even fearful, but they felt compelled to return to the danger zone. The men shared a powerful scientific curiosity that pushed both doubt and fear aside, replacing them with a sleuthing urge every bit as powerful as that of a detective hunting down a serial killer. They needed to find out how the virus was transmitted in order to stop its spread.

On the way, Johnson and MacKenzie reviewed all the possible ways the three of them could have become infected. Since the infections seemed to have been simultaneous, it was unlikely they were due to accidents or carelessness in their primitive field laboratories. The window screens and DDT had probably protected them from any virus-carrying insects that might lurk in San Joaquín. And the fact that many family members who tended to dying relatives were not ill seemed to rule out person-to-person transmission of the virus. Of course, Webb's illness forced an opposite conclusion.

The only experience the three had shared shortly before becoming ill was the town party. But what possible association could there have been between the party and their near-deaths?

In their absence Kuns had remained in San Joaquín, painstakingly capturing samples of every species of insect and mammal—from bedbugs to teeth-baring bats and slithering anacondas—he could get his hands on, all the while aware of the need to handle the animals with extreme care. As the wild rats tried to claw at him, or mosquitoes dove for his vulnerable flesh, Kuns deftly manipulated the creatures.

"I understand the ways of animals," Kuns told his Bolivian assistants. He had a Ph.D. in veterinary sciences, specializing in the study of diseases that affected both humans and animals, and he had minored in wildlife ecology studies. Kuns was a details man; his training reinforced a natural tendency toward searching for answers by tediously sifting through mountains of minutiae. He and Johnson—the impatient man of action—were a study in contrasts. While the rest of the MARU team recuperated in Panama, Kuns organized forty San Joaquín men to assist in the capture of local creatures. All the men were volunteers who believed they had already had the strange disease and their survival presumably rendered them immune. Over a year's time Kuns and his volunteers would collect some 10,000 mammals of dozens of species, all of which had to be identified and studied for viral contamination. Even more insects were amassed, and Kuns pored over microscopes, used field texts to figure out just what species each and every creature represented, and trained his assistants to do the same. When identity was impossible to ascertain, Kuns shipped samples to the Smithsonian in Washington, D.C., or the Field Museum of Natural History in Chicago, where experts made the final judgment calls.

In coming months Kuns would complete one of the most exhaustive ecology surveys ever conducted in South America, all done in the pressure-cooker atmosphere of an epidemic. Having flown bombing missions over Germany in World War II, Kuns had no trouble staying cool, though he never forgot that any insect or animal he held in his hands might carry the deadly virus.

The team went to enormous lengths to find the culprit. A typical rectangular thatched/wattle house was purchased and meticulously deconstructed pole by pole, every insect and animal scraped off and analyzed. By night Kuns and MacKenzie, wearing miner's headlamps, waded into knee-deep water to collect nocturnal animals such as vampire bats and night mosquitoes. Even though they tried to maintain a modicum of nonchalance on such occasions, MacKenzie couldn't help but feel chills when he would turn suddenly and spot beady red eyes staring at him. After a few nights on such forays, Kuns and MacKenzie figured out that the eyes belonged to giant constrictor snakes, anacondas, one of which they measured at over eighteen feet in length.

The townspeople were anxious to help, and Kuns had to warn people not to handle the animals unless they had already survived the disease. One afternoon the leader of Kuns's volunteer ecology army, Einar Dorado, held a big gray mouse in his hands while Merl tried to jab the beast with a needle in order to withdraw a blood sample. The angry animal squirmed, bit Dorado, and urinated over his hands.

Two weeks later Dorado was dead. His prior ailment had probably been the flu and he was not immune to the mysterious virus.

Kuns thought about that large gray mouse—a type of rodent that was pervasive throughout San Joaquín. He remembered visiting a small home in the middle of town and spotting a six-year-old girl asleep atop a cowhide laid across the dirt floor. When the girl awoke, Kuns lifted the hide and dozens of those same big gray mice raced out of a nest in the floor.

As soon as Johnson and MacKenzie arrived in San Joaquín, the team spread word that they would give money to anybody who captured sick wild animals. Despite the strong monetary incentive, villagers were able to bring in only five sick animals over the next ten months, all large gray mice. The rodents were of a wild *Calomys* species normally found in the bush. Three of the animals died, suffering symptoms similar to those seen among the people of San Joaquín. The other two recovered and became virus carriers. The mysterious virus, which the team had dubbed Machupo after the local river, was found in the blood, spleens, or brains of the five animals.

The team hypothesized that the disease was carried the same way as the plague: by insects that inhabited the fur of the rodents. An alternative idea held that mosquitoes or ticks that fed on the mice might, in turn, feed on human blood and thereby spread the virus. In either case, an insect carrier had to be found.

From September 1963 to November 1964, the team, now including Webb, made numerous frustrating trips back and forth between San Joaquín and their Panama laboratories. They captured thousands of insects, ranging from tiny fleas and mites to larger ticks and mosquitoes. The insects were mashed up and their extracts scoured for Machupo virus.

None was found.

Night after late night in the Panama laboratory Johnson complained to Webb, "I can't tell what the hell is going on. I'm running out of ideas here."

After having solved the first parts of the mystery with awesome speed, the team was now mired in confusion. Kuns was convinced that better insect traps and more extensive forays in the foothills around San Joaquín would eventually smoke out the culprit, but Johnson was dubious.[4] Some villagers in San Joaquín were downright fed up with the investigation, and the Americans became targets of wholesale theft of critical supplies, such as the diesel fuel for their generator. Things became so unstable that the Bolivian government declared martial law in the area, flying in fifty-five soldiers to maintain order. Thirty-seven of those soldiers eventually got the disease.

The team's work was further delayed by angry anti-American uprisings in Panama that erupted into full-scale riots, forcing closure of the Canal Zone airport and delaying return trips by Johnson, Webb, MacKenzie, and Kuns.

One hot June day in 1964 Johnson and Webb were going over their laboratory records in Panama and noticed an odd disease pattern among the hamsters they had infected experimentally. If they injected the virus into baby hamsters, the animals almost always died and the adult hamsters would eat the bodies. The adults would then become infected at a low level, but would survive. The adults, however, would pass on lethal infections to previously disease-free baby hamsters. How had these babies become infected?

Johnson and Webb, now married, happily spent long hours together in the laboratory, isolating viral samples from hamster blood. There were no shortcuts, no ways to get around the long tedious hours and days of work required to get a tiny pellet of viruses from thousands of hamster cells. Once the infected animal red blood cells were grown in petri dishes, the virus had to be purified out by a series of fractionization techniques. First, they mixed the hamster cells and blood with ammonium sulfate, which created a salted-out layer of gunk that sat at the bottom of a test tube. The fluid was poured off, and the virus-contaminated mass at the bottom was mixed with an alcohol, creating another layering. The virally infected layer, now smaller and purer, was spun about at low speed in a centrifuge; objects in the test tube migrated under spin to various positions according to their weights. The garbage—extraneous bits of hamster cells—formed a visible band in the tube, which was removed. Then the tube was spun again, this time at a very high speed, adequate to separate objects that differed only slightly in weight.

A pellet would be left, at long last, at the tube's bottom—a nearly pure sample of concentrated, deadly Machupo viruses.

Johnson and Webb discovered that the adult hamsters were shedding virus in their urine. They then bred wild mice from San Joaquín and found the same thing—the animals actively urinated Machupo virus. The baby rodents had become infected because they were caged in an atmosphere of wood chips and sawdust drenched in Machupo virus.

Johnson felt the archetypical cartoon lightbulb flash above his head and heard himself shout, "Aha!"

He returned to San Joaquín and did a very simple experiment: He divided the town in half. On one side he set five-cent mousetraps throughout houses and corn storage areas. The other half of San Joaquín was left alone. One woman living in the trap side of town begged Kuns to give her as many traps as possible; he could only spare three. In a single night she caught twenty-two mice in her home, proudly presenting them all to an astonished Kuns the following morning.

Within two weeks the difference was obvious. While the epidemic continued at the same pace on one side of town, no new Machupo virus cases occurred where mousetraps had been set. Two weeks later, having set traps throughout San Joaquín, Johnson's team stopped the Machupo epidemic.

"This is unbelievable," Johnson proudly said to himself. "Within just eighteen months we isolated the virus, discovered its mechanism of transmission, and stopped it cold."

Between 1962 and 1964 over 40 percent of the residents of the San Joaquín region were sick with Machupo virus; some 10 to 20 percent of the villagers died of the disease. If the region hadn't been so sparsely populated, the impact could have been devastating. As it was, for the people of San Joaquín, Magdalena, and the surrounding area, Machupo virus was a scourge that claimed at least one member of every family and was carried aboard mouse-infested supply carts to remote parts of the eastern frontier. Its impact on people's lives would not soon be forgotten.

Over the next three years the Panama-based researchers would fill in some of the remaining pieces of the Machupo puzzle, and successfully stop a second outbreak of the disease deeper in the Bolivian savannas.[5]

Johnson put together a best-guess history of Machupo virus, and together with MacKenzie, Kuns, and Webb published several scientific papers between 1964 and 1966 describing most aspects of the virus. He decided the epidemic's roots lay in Bolivia's social revolution of 1952, when the people of the San Joaquín area suddenly found themselves without an employer or steady source of food supplies. In their haste to grow corn and other vegetables, they chopped down dense jungle areas of the *alturas* and *bandas* wherever the land naturally formed a relatively flat mesa above the Machupo River flood line. In so doing, they unwittingly disrupted the natural habitat of the *Calomys* field mouse and provided the rodent with a superior new food source: corn.

The mouse population swelled during the 1950s and the rodents literally invaded the town of San Joaquín in the early 1960s.

By the time the first cases of Bolivian hemorrhagic fever (as the disease was now dubbed) surfaced, the mice could be found anyplace the townspeople stored food and grain.

And each night while the mice nibbled away at the humans' food supplies, they urinated.

The virus could be eaten or inhaled or could gain entry through cuts in the skin; in any event, Machupo could be lethal.

Johnson noticed that there was a ritual common to every household in San Joaquín. Before dawn the mothers and grandmothers would awaken and quietly prepare breakfast for the men and children. While pots boiled, the women would sweep their dirt and clay floors.

"And each time they sweep that broom," Johnson realized, "they're sending mouse urine-infected dust and crumbs drifting all about in the air." Every time the families of San Joaquín assembled for breakfast, they shared virus-contaminated air. Johnson also decided that he, Ron, and Angel got sick as a result of eating contaminated food at the San Joaquín party.

Researchers from the Rockefeller Foundation Laboratories in New York City and the University of Buenos Aires eventually reached a similar set of conclusions about the Argentine Junín virus. An Argentine team led by Dr. A. S. Parodi concluded that another species of wild *Calomys* mouse had been flushed out of its pampas habitat by post-World War II changes in local agricultural practices. Farmers had long had difficulty growing profitable crops of corn because short broad-leafed weeds had invaded the fields. After World War II, herbicides effectively eliminated the short grasses and dramatically increased crop yields.[6]

As harvest time approached, however, taller grasses that were not affected by herbicides would grow in the corn fields, thriving just when humans entered the fields to reap the corn. As it turned out, a fairly rare species of field mouse naturally subsisted on the seeds of these tall grasses. As the grasses proliferated, so did the mice, until the once-rare species became the dominant rodent of the region.

The mouse, of course, carried Junín virus, the cause of Argentine hemorrhagic fever.[7]

MacKenzie thought another Bolivian factor also played a role in the San Joaquín epidemic. On all his trips to Magdalena, Orobayaya, and San Joaquín, he was struck by the remarkable absence of cats in the villages. When he asked the people what had happened to the cats, he was told they all died.

The feline die-off coincided with the rise in the mouse population, allowing the *Calomys* to take over the town without battling predatory cats. There were two theories about why the cats died: they were also killed by the virus, or the felines were DDT victims. Bolivia was in the midst of a massive DDT-spraying campaign to eliminate malaria, and the quantities used in the homes of these remote areas were often so great that all furniture and walls were coated with a thin white film of insecticide powder.

MacKenzie did a simple experiment. He injected six cats with the virus and force-fed DDT to six others. The virus-treated cats were completely unaffected by the experiment, but the DDT-poisoned animals all died of symptoms identical to those seen among the domestic pets of San Joaquín.

Valverde was so impressed with MacKenzie's hypothesis that he went on national radio to issue a call for donated cats. In June 1964 hundreds of cats were airlifted into San Joaquín, and the epidemic's halt soon followed.

Kuns, the ecologist of the group, wasn't ready to buy either the cat dieoff idea or Johnson's notion of a mouse invasion.

"It's stupid! Absolutely stupid," Kuns said of the cat/DDT link, noting that felines killed only the weak and sick members of mouse colonies and rarely had an enormous impact on the overall size of a rodent population. He wasn't at all convinced people could get infected as a result of eating food contaminated with mouse urine either. To Johnson's theory that three of the scientists got the disease as a result of feasting at the fiesta he said, with a wink, "You ought to think about what you did *after* the fiesta, boys, not what you ate."[8]

Kuns believed the epidemic began when the virus itself changed and became more virulent. And he felt the *exact* mode of transmission of the disease from mouse to human was still unresolved.

Kuns told the NIH that he wanted to put fluorescent chemicals in mouse food and then use ultraviolet lights to follow the animals' urine trails in San Joaquín. That, he hoped, would answer the question by showing where the animals' urine came in intimate contact with human noses, mouths, and inhalations. His hunch was that the rodents scampered around the villages while people slept, directly infecting the slumbering men, women, and children.

But once the epidemic was abated the NIH withdrew all research funds and Kuns's efforts came to a full stop.

"Hardly anything has ever disappointed me more in all my thirty-nine years than having to pull out of here without finishing the job," Kuns mumbled to himself as he packed up his microscope and thousands of animal samples. When a few cases of the disease appeared in San Joaquín a year later, Kuns told *The Saturday Evening Post*: "You might compare us to firemen. We've discovered the location of the blaze and we've put it out. But we still don't know where or when it could start again."[9]

For his part, however, Johnson was quite pleased.

By the end of 1964 Johnson was able to look over his recent accomplishments with great pride. Together with MacKenzie, Kuns, and Webb, he had solved an intriguing mystery, stopped an epidemic, published in prominent scientific journals, and organized a crackerjack virology laboratory in Panama that was prepared to tackle anything that might surface in the Americas. Furthermore, the NIH promoted him to director of the whole MARU facility.

He had also found love, honor, and a mission in life.

He married Patricia Ann Webb, received the Decorated Order of the Condor from the Bolivian government, and discovered his personal call of the wild. He had survived a near-death experience and then gone on to defeat the enemy on its own turf.

"That's enough for some people for a lifetime," thought the thirty-five-year-old. "But that's just the start for me."

His life dreams now changed forever, Johnson sought ways to combine science, clinical medicine, and good old-fashioned detective work. Wherever he went, whatever epidemics might come his way, he knew these were the skills he would use, and the challenges he would relish.

From then on, Johnson stressed the need for calm in the face of epidemics, for reason, science, sound clinical training, and the ability to work with a team of diverse expertise. These were lessons passed on first to those who worked with Johnson in Panama and later to a whole generation of infectious disease "cowboys." Over the next two decades Johnson and his "cowboys" would fight dozens of skirmishes and a few pitched battles with the microbes, always maintaining a healthy respect for both their microscopic enemies and the human bureaucracies, governments, and institutions whose rules Johnson would regularly defy.

The war was on, and the battlefield was the entire planet.

2

Health Transition

THE AGE OF OPTIMISM—
SETTING OUT TO ERADICATE DISEASE

Germs come by stealth
And ruin health,
So listen, pard,
Just drop a card
To a man who'll clean up your yard
And that will hit the old germs hard.
—Dr. Almus Pickerbaugh,
in *Arrowsmith*, Sinclair Lewis

I

For Western physicians, the 1950s and 1960s were a time of tremendous optimism. Nearly every week the medical establishment declared another "miracle breakthrough" in humanity's war with infectious disease. Antibiotics, first discovered in the early 1940s, were now growing in number and potency. So much so that clinicians and scientists shrugged off bacterial diseases, and in the industrialized world former scourges such as *Staphylococcus* and tuberculosis had been deftly moved from the "extremely dangerous" column to that of "easily managed minor infections." Medicine was viewed as a huge chart depicting disease incidences over time: by the twenty-first century every infectious disease on the chart would have hit zero. Few scientists or physicians of the day doubted that humanity would continue on its linear course of triumphs over the microbes.

Dr. Jonas Salk's 1955 mass experimental polio vaccination campaign was so successful that cases of the disease in Western Europe and North America plummeted from 76,000 in 1955 to less than 1,000 in 1967.[1] The excitement engendered by that drama prompted optimistic declarations that the disease would soon be eradicated from the planet.

Similar optimism enveloped discussion of nearly every infectious disease affecting human beings. In 1948, U.S. Secretary of State George C. Marshall declared at the Washington, D.C., gathering of the Fourth International Congress on Tropical Medicine and Malaria that the conquest of all infectious diseases was imminent.[2] Through a combination of enhanced crop yields to provide adequate food for humanity and scientific break-throughs in microbe control, Marshall predicted, all the earth's microscopic scourges would be eliminated.

By 1951 the World Health Organization was so optimistic that it declared that Asian malaria could soon reach a stage through careful local management wherein "malaria is no longer of major importance."[3] A key reason for the excitement was the discovery of DDT and the class of chemicals known as organochlorines, all of which possessed the remarkable capacity to kill mosquitoes and other insect pests on contact and to go on killing for months, perhaps years, all insects that might alight on pesticide-treated surfaces.

In 1954 the Fourteenth Pan American Sanitary Conference resolved in its Santiago, Chile, meeting to eliminate malaria completely from the Western Hemisphere, and PAHO (Pan American Health Organization) was instructed to draw up an ambitious eradication plan. The following year the World Health Organization decided to eliminate all malaria on the planet. Few doubted that such a lofty goal was possible: nobody at the time could imagine a trend of worsening disease conditions; the arrow of history always pointed toward progress.

Every problem seemed conquerable in the decade following World War II: humanity could reach the moon, bombs too terrifying to ever be used could create a balance of terror that would prevent all further worldwide wars, American and European agriculturalists could "green" the poor nations of the world and eliminate starvation, civil rights legislation could erase the scars of slavery and bring racial justice, democracy could shine in startling contrast to communism and provide a beacon to which the nations of the world would quickly flock, huge, gasoline-hungry cars cruised freshly paved highways, and their passengers dreamed of a New Tomorrow.

From the capitalist world came thousands of zealous public health activists who rolled up their sleeves and dived like budding Dr. Pickerbaughs into amazingly complex health crises. Sinclair Lewis lambasted such zealous health optimism in *Arrowsmith*, creating the character of Almus Pickerbaugh, physician-congressman-poet, whose gems included:

> *You can't get health*
> *By a pussyfoot stealth,*
> *So let's every health-booster*
> *Crow just like a rooster.*

Never mind the seemingly daunting obstacles presented by, for example, cholera control in India; all was possible in the Age of Boosterism.

The notion developed of a Health Transition, as it was called. The concept was simple: as nations moved out of poverty and the basic food and housing needs of the populations were met, scientists could use the pharmaceutical and chemical tools at hand to wipe out parasites, bacteria, and viruses.

What would remain were the slower chronic diseases that primarily struck in old age, particularly cancer and heart disease. Everybody would live longer, disease-free lives.

Such glowing forecasts were not limited to the capitalist world. Soviet and Eastern bloc health officials presented ever-rosier medical statistics each year, suggesting that their societies were also well on their way to conquering infectious diseases. And Mao Zedong, leader of the nearly one-billion-strong Chinese nation, declared in 1963:

> *The Four Seas are rising, clouds and waters raging,*
> *The Five Continents are rocking, wind and thunder roaring.*
> *Away with all pests!*
> *Our force is irresistible.*[4]

Throughout the 1950s and 1960s, the Chinese Communist Party waged a peasant-based war on infectious diseases, mobilizing millions of peasants to walk through irrigation ditches and pluck schistosome-carrying snails from the banks.[5] According to British physician Joshua Horn, who fully embraced the campaign and Maoism, in 1965–66 virtually no new cases of schistosomiasis, a serious liver parasitic disease, occurred in China— a result, he claimed, of the Communist Party campaign.[6]

Though the ideological frameworks differed markedly, both the capitalist and communist worlds were forecasting brighter futures in which there would be a chicken in every pot, a car in every garage, and a long, infectious-disease-free life ahead for every child. Both sides of the Iron Curtain agreed that mass mobilization of the global populace to fight disease would inevitably result in victory. Never mind in what rhetoric public health campaigns might be wrapped, humanity would triumph over the microbes.

In September 1966 the U.S. Centers for Disease Control assessed the status of American health:

The status of diseases may be classified as follows:
1. Diseases eradicated within the United States (bubonic plague, malaria, small-pox, etc.)
2. Diseases almost eradicated (typhoid, infantile paralysis, diphtheria, etc.)
3. Diseases that still are health problems, although technology exists for effective control (syphilis, tuberculosis, uterine cervix cancer, injury, arthritis, breast cancer, gonorrhea, etc.)
4. Diseases where technology is in early developmental stages or nonexistent— and where little capability exists for alleviating or preventing health impairment (leukemia and some other neoplasms, some respiratory diseases and strokes)[7]

As the 1960s opened, the U.S. Department of Health, Education, and Welfare convened a team of medical experts to decide the future mission of the entire government public health effort. Praising the accomplishments of the 1950s, the advisory team declared that "science and technology have completely transformed man's concepts of the universe, of his place in it, and of his own physiological and psychological systems. Man's mastery over nature has been vastly extended, including his capacity to cope with diseases and other threats to human life and health."[8]

By 1967 U.S. Surgeon General William H. Stewart would be so utterly convinced of imminent success that he would tell a White House gathering of state and territorial health officers that it was time to close the book on infectious diseases and shift all national attention (and dollars) to what he termed "the New Dimensions" of health: chronic diseases.[9]

"In the words of the song, 'The fundamental things go by,' polio and measles can be eradicated and should be eradicated," Stewart would tell his exuberant audience. "Venereal disease and tuberculosis can be sharply reduced and should be sharply reduced. These are tasks that no one will perform for us. So long as a preventable disease remains, it must be prevented, and public health must be the primary force for prevention."

Not content to stop with the predicted eradication of all known infectious diseases, the optimists set out in search of rare and remote disease agents. Biology research stations were established throughout the Southern Hemisphere, staffed largely by scientists from the Northern Hemisphere. All sorts of agencies funded and administered these outposts, including the Rockefeller Foundation, agencies of the governments of France, the United States, Germany, and the United Kingdom, as well as a variety of small private interests.

Johnson's Panama Canal Zone laboratory was just such an outpost. The U.S. government alone operated twenty-eight overseas laboratories, and the Rockefeller Foundation's Virus Program operated facilities in eight countries through which over sixty viruses would be discovered between 1951 and 1971.[10]

But much of what these searching scientists were to find proved terrifying. As officials prepared to uncork celebratory champagne, Johnson and his colleagues were unlocking some of nature's nastiest secrets.

Boosters of the 1950s and early 1960s had some basis, born of ignorance, for their optimism: they knew comparatively little about genetics, microbial evolution, the human immune system, or disease ecology. Given the state of knowledge in the public health world of that day, it may have seemed appropriate to view infectious diseases in simple cause-and-effect terms. Seen in such a reductionist manner, problems and solutions appeared obvious and readily conquerable, bravado warranted.

As early as the 1930s scientists guessed that the genetic traits of large creatures, such as plants, animals, and humans, were carried in packages called chromosomes. These structures, which, when examined through a microscope, resembled dark, squiggly worms, were inside the central core, or nucleus, of every cell in a plant or animal. By manipulating chromosomes in test tubes, scientists could change the ways cells looked or grew; exposing chromosomes to radiation, for example, could transform healthy tissue into cancer colonies.

True, Gregor Mendel showed in 1865 that some characteristics were passed on as dominant traits from one generation to another, while other genetic characteristics were recessive. But nobody knew exactly how all this worked, why blue-eyed parents had blue-eyed children, or a bacterium could seem to suddenly develop the ability to withstand higher temperatures than normally tolerated by its species.

Until 1944 nobody knew what was responsible for this neat passage of genetic information, from the tiniest virus to the largest elephant. That year, Oswald Avery and his colleagues at the Rockefeller Institute in New York showed that if they destroyed a specific molecule found inside all living cells, the organisms could no longer pass on their genes.

The molecule was called deoxyribonucleic acid, or DNA.

In 1953 researcher Rosalind Franklin, working at King's College in London, made the first X-ray photographs of DNA, showing that the molecules had a unique helical structure composed of various combinations of the same five key chemicals.

Later that year, America's James Watson and Britain's Francis Crick, working at Cambridge University, figured it all out. One of the chemicals—a sort of carbon chain linked by powerful phosphate chemical bonds—created parallel curved structures similar to the poles of a long, winding ladder. Forming the rungs of the ladder were four other chemicals, called nucleotides. The order of those nucleotide rungs along the carbon/phosphate poles represented a code which, when deciphered properly, revealed the genetic secrets of life.

DNA, then, was the universal code used by one meningococcal bacterium as the basis for making another meningococcal bacterium. It was the material wrapped up inside the chromosomes of higher organisms. Sections of DNA equaled genes; genes created traits. When the chromosomes of one parent combined with those of another parent, the DNA was the key, and which traits appeared in the children (blue versus brown eyes) was a function of the dominant or recessive genes encoded in the parents' DNA.[11]

While government officials were bragging that everything from malaria to influenza would soon disappear from the planet, scientists were just beginning to use their newfound knowledge to study disease-causing viruses, bacteria, and parasites. Scientists like Johnson were of the first generation of public health researchers to know the significance of DNA. Understanding how DNA played a direct role in the emergence of disease would take still another generation.

Starting at nature's most basic level, scientists at Cold Spring Harbor Laboratory on Long Island, New York showed in 1952 that viruses were essentially capsules jam-packed with DNA. Much later, researchers discovered that some other viruses, such as polio, were filled not with DNA but with its sister compound, RNA (ribonucleic acid), which also carries the genetic code hidden in sequences of nucleotides.

When Karl Johnson was virus hunting in Bolivia, scientists had a limited understanding of the vast variety of viruses in the world, the ways these tiniest of organisms mutate and evolve, or how the microbes interact with the human immune system. The state of the art in 1963 was best summarized in Frank Fenner's animal virus textbook, the bible for budding microbiologists of the day:

Suppose that we have isolated a new virus and have managed to produce a suspension of purified particles. How can we classify the virus, and how do we find out about its chemical composition? A lead may be provided by its past history —the species of animal from which it was isolated and whether or not it was related to a disease. This information, in conjunction with that obtained by electron microscope examination of . . . particles, might be enough for us to make a preliminary identification. [12]

Scientists could "see" viruses with the aid of microscopes powerful enough to magnify up to visual level objects that were nearly a million times smaller than a dime. With that power of magnification they could detect clear differences in the appearance of various species of viruses, from the chaotic-looking mumps virus that visually resembles a bowl full of spaghetti to the absolutely symmetrical polio virus that looked as if it were a Buckminster Fuller-designed sphere composed of alternating triangles.

Researchers also understood that viruses had a variety of different types of proteins protruding from their capsules, most of which were used by the tiny microbes to lock on to cells and gain entry for invasion. Some of the most sophisticated viruses, such as influenza, sugarcoated those proteins so that the human immune system might fail to notice the disguised invaders.

In 1963 laboratory scientists knew they could also distinguish one virus species from another by testing immune responses to those proteins protruding from the viral capsules. Humans and higher animals made antibodies against most such viral proteins, and the antibodies—which were themselves large proteins—were very specific. Usually an antibody against parts of the polio virus, for example, would not react against the smallpox virus. Indeed, some antibodies were so picky that they might react against a 1958 Chicago strain of the flu, but not the strain that hit the Windy City the following winter.

Jonas Salk used this response against outer capsule proteins of the polio virus as the basis of his revolutionary vaccine, and by 1963 medical and veterinary pioneers all over the world were finding the pieces of various viruses that could be used most effectively to raise human and animal antibody responses.

Back in the lab, they could also use antibody responses to find out what might be ailing a mysteriously ill person. Blood samples containing the victim's attacking microbe would be dotted across a petri dish full of human or animal cells. Antibodies would also be dotted across the dish, and scientists would wait to see which antibody samples successfully prevented viral kill of the cells in the petri dish.

Of course, if the virus was something never before studied, all the scientists would be able to get was a negative answer: "It's not anything that we know about, none of our antibodies work." So in the face of something new, like Machupo, scientists could only say after a tedious process of antibody elimination, "We don't know what it is."

With bacteria the process of identification was far easier because the organisms were orders of magnitude larger than viruses: whereas a virus might be about one ten-millionth of an inch in size, a bacterium would be a thousandth of an inch long. To see a virus, scientists needed powerful, expensive electron microscopes, but since the days of Dutch lens hobbyist Anton van Leeuwenhoek, who in 1674 invented a microscope, it has been possible for people to see what he called "wee animalcules" with little more than a well-crafted glass lens and candlelight.

The relationship between those "animalcules" and disease was first figured out by France's Louis Pasteur in 1864, and during the following hundred years bacteriologists learned so much about the organisms that young scientists in 1964 considered classic bacteriology a dead field.

In 1928 British scientist Alexander Fleming had discovered that *Penicillium* mold could kill *Staphylococcus* bacteria in petri dishes, and dubbed the lethal antibacterial chemical secreted by the mold "penicillin."[13] In 1944 penicillin was introduced to general clinical practice, causing a worldwide sensation that would be impossible to overstate. The term "miracle drug" entered the common vernacular as parents all over the industrialized world watched their children bounce back immediately from ailments that just months before had been considered serious, even deadly. Strep throat, once a dreaded childhood disease, instantly became trivial, as did skin boils, infected wounds, and tuberculosis with the quick discovery of streptomycin and other classes of antibiotics. By 1965 more than 25,000 different antibiotic products had been developed; physicians and scientists felt that bacterial diseases, and the microbes responsible, were no longer of great concern or of research interest.

Amid the near-fanatic enthusiasm for antibiotics there were reports, from the first days of their clinical use, of the existence of bacteria that were resistant to the chemicals. Doctors soon saw patients who couldn't be healed, and laboratory scientists were able to fill petri dishes to the brim with vast colonies of *Staphylococcus* or *Streptococcus* that thrived in solutions rich in penicillin, tetracycline, or any other antibiotic they chose to study.

In 1952 a young University of Wisconsin microbiologist named Joshua Lederberg and his wife, Esther, proved that these bacteria's ability to outwit antibiotics was due to special characteristics found in their DNA. Some bacteria, they concluded, were genetically resistant to penicillin or other drugs, and had possessed that trait for aeons; certainly well before *Homo sapiens* discovered antibiotics.[14] In years to come, the Lederbergs' hypothesis that resistance to antibiotics was inherent in some bacterial species would prove to be true.

The Lederbergs had stumbled into the world of bacterial evolution. If millions of bacteria must compete among one another in endless turf battles, jockeying for position inside the human gut or on the warm, moist skin of an armpit, it made sense that they would have evolved chemical weapons with which to wipe out competitors. Furthermore, yeast—the molds and soil organisms that were the natural sources of the world's then burgeoning antibiotic pharmaceutics—had evolved the ability to manufacture the same chemicals for similar ecological reasons.

It stood to reason that populations of organisms could survive only if some individual members of the colony possessed genetically coded R (resistance) Factors, conferring the ability to withstand such chemical assaults.

The Lederbergs discovered tests that could identify streptomycin-resistant *Escherichia coli* intestinal bacteria *before* the organisms were exposed to antibiotics. They also showed that the use of antibiotics in colonies of bacteria in which even less than 1 percent of the organisms were genetically resistant could have tragic results. The antibiotics would kill off the 99 percent of the bacteria that were susceptible, leaving a vast nutrient-filled petri dish free of competitors for the surviving resistant bacteria. Like weeds that suddenly invaded an untended open field, the resistant bacteria rapidly multiplied and spread out, filling the petri dish within a matter of days with a uniformly antibiotic-resistant population of bacteria.

Clinically this meant that the wise physician should hit an infected patient hard, with very high doses of antibiotics that would almost immediately kill off the entire susceptible population, leaving the immune system with the relatively minor task of wiping out the remaining resistant bacteria. For particularly dangerous infections, it seemed advisable to initially use two or three different types of antibiotics, on the theory that even if some bacteria had R Factors for one type of antibiotic, it was unlikely a bacterium would have R Factors for several widely divergent antibiotics.

If many young scientists of the mid-1960s considered bacteriology passé—a field commonly referred to as "a science in which all the big questions have been answered"—the study of parasitology was thought to be positively prehistoric.

A parasite, properly defined, is "one who eats beside or at the table of another, a toady; in biology, a plant or animal that lives on or within another organism, from which it derives sustenance or protection without making compensation."[15] Strictly speaking, then, all infectious microbes could be labeled parasites, from viruses to large ringworms.

But historically, the sciences of virology, bacteriology, and parasitology have evolved quite separately, with few scientists—other than "disease cowboys" like Johnson and MacKenzie—trained or even interested in bridging the disciplines. By the time hemorrhagic fever broke out in Bolivia, a very artificial set of distinctions had developed between the fields. Plainly put, larger microbes were considered parasites: protozoa, amoebae, worms. These were the domain of parasitologists.

Their scientific realm had been absorbed by another, equally artificially designated field dubbed tropical medicine, which often had nothing to do with either geographically tropical areas or medicine.

Both distinctions—parasitology and tropical medicine—set off the study of diseases that largely plagued the poorer, less developed countries of the world from those that continued to trouble the industrialized world. The field of tropical medicine did so most blatantly, encompassing not only classically defined parasitic diseases but also viruses (e.g., yellow fever and the various hemorrhagic fever viruses) and bacteria (e.g., plague, yaws, and typhus) that were by the mid-twentieth century extremely rare in developed countries.

In the eighteenth century the only organisms big enough to be studied easily without the aid of powerful microscopes were larger parasites that infected human beings in some stage of the overall life cycle of the creature. Doctors could, without magnification, see ringworms or the eggs of some parasites in patients' stools. Without much magnification (on the order of hundreds-fold versus the thousands-fold necessary to study bacteria) scientists could see the dangerous fungal colonies of *Candida albicans* growing in a woman's vagina, scabies acariasis roundworms in an unfortunate victim's skin, or cysticercosis tapeworms in the stools of individuals fed undercooked pork.

As British and French imperial designs increasingly in the late eighteenth century turned to colonization of areas such as the Indian subcontinent, Africa, and Southeast Asia, tropical medicine became a distinct and powerful science that separated itself from what was then considered a more primitive field, bacteriology. Science historian John Farley concluded that what began as a separation designed to lend parasitology greater resources and esteem—and did so in the early nineteenth century—ended up leaving it science's stepchild.[16]

Ironically, parasites, classically defined, were far more complex than bacteria and their study required a broader range of expertise than was exacted by typical *E. coli* biology. Top parasitologists—or tropical medicine specialists, if you will—were expected in the mid-1960s to have vast knowledge of tropical insects, disease-carrying animals, the complicated life cycles of over a hundred different recognized parasites, human clinical responses to the diseases, and the ways in which all these factors interacted in particular settings to produce epidemics or long periods of endemic, or permanent, disease.

Consider the example of one of the world's most ubiquitous and complicated diseases: malaria. To truly understand and control the disease, scientists in the mid-twentieth century were supposed to have detailed knowledge of the complex life cycle of the malarial parasite, the insect that carried it, the ecology of that insect's highly diverse environment, other animals that could be infected with the parasite, and how all these factors were affected by such things as heavy rainfall, human migrations, changes in monkey populations, and the like.

It was known that several different strains of *Anopheles* mosquitoes could carry the tiny parasites. The female *Anopheles* would suck parasites out of the blood of infected humans or animals when she injected her syringe-like proboscis into a surface capillary to feed. The microscopic male and female sexual stages of the parasites, called gametocytes, would make their way up the proboscis and into the female mosquito's gut, where they would unite sexually and make a tiny sac in the lining of the insect's stomach.

Over a period of one to three weeks the sac would grow, as inside thousands of sporozoite-stage parasites were manufactured. Eventually, the sac would explode, flooding the insect's gut with microscopic one-celled parasites that caused no harm to the cold-blooded insect; their target was a warm-blooded creature, one full of red blood cells.

Some of the sporozoites would make their way into the insect's salivary glands, from which they would be drawn up into the "syringe" when the mosquito went on her nightly sundown feeding frenzy, and be injected into the bloodstream of an unfortunate human host.

At that point the speed and severity of events (from the human host's perspective) would depend on which of four key malarial parasite species had been injected by the mosquito. A good parasitologist in the 1950s knew a great deal about the differences between the four species, two of which were particularly dangerous: *Plasmodium vivax* and *P. falciparum*.

If a human host was most unlucky, the parasites coursing through her bloodstream would be *P. falciparum* and she would have only twelve days to realize she'd been infected and get treatment of some kind before the disease would strike, in the form of either acute blood anemia or searing infection of the brain. In either case, for an individual whose immune system had never before seen *P. falciparum*, the outcome would likely be death.

Scientists knew that injected sporozoites made their way to the liver, where they underwent another transformation, becoming so-called schizonts capable of infecting red blood cells. By the millions the tiny creatures, matured into merozoites, multiplied and grew inside red blood cells, eventually becoming so numerous that the cells exploded. Soon the human body would be severely anemic, its every tissue crying out for oxygen. If the immune system managed to keep the merozoite population down to manageable levels, the results would be prolonged—perhaps chronic lifetime —fatigue and weakness. Unchecked, however, the merozoites would so overwhelm the red blood cell population that the host's brain, heart, and vital organs would fail and death would result.

During the merozoite invasion of the blood supply, a smaller number of male and female gametocyte-stage *P. falciparum* would also be made, and the entire cycle of events would repeat itself when another female *Anopheles* mosquito fed on the blood of the ailing human, sucking those gametocytes up into her proboscis.

Understanding that process of the disease was relatively easy; more difficult was predicting when and why humans and *Anopheles* mosquitoes were likely to come into fatal contact and how the spread of malaria could be stopped.

Several types of monkeys were known to serve as parasite reservoirs, meaning that for long periods of time the disease could lurk in monkey habitats. The *Anopheles* mosquito would happily feed on both the monkeys and the humans that entered such ecospheres, spreading *P. falciparum* between the species.[17]

The size of *Anopheles* mosquito populations could vary drastically in a given area, depending on rainfall, agricultural practices, the nature of human housing and communities, altitude, proximity to forests or jungles, economic development, the nutritional status of the local people, and numerous other factors that could affect mosquito breeding sites and the susceptibility of local human populations.[18]

Almost entirely absent in the mid-twentieth century was an intellectual perspective that wedded the ecological outlook of the classical parasitologist with the burgeoning new science of molecular biology then dominating the study of nontropical bacteria and viruses. Money was shifting away from research on diseases like malaria and schistosomiasis. Young scientists were encouraged to think at the molecular level, concentrating on DNA and the many ways it affected cells.

II

Imbued with profound optimism, coupled with the post-World War II American "can do" attitude, the world's public health community mounted two ambitious campaigns to eradicate microbes from the planet. One effort would succeed, becoming the greatest triumph of modern public health. The other would fail so miserably that the targeted microbes would increase both in numbers and in virulence, and the *Homo sapiens* death toll would soar.

Humanity's great success story would be smallpox.

In 1958 the Soviet Union went before the World Health Assembly—the legislative body of the World Health Organization in Geneva—to request an international campaign for the elimination of smallpox, winning virtually universal support.

Historically, smallpox had proven a particularly vicious killer. It did not, as was typical for most infectious diseases, preferentially attack the most impoverished members of society.[19] In A.D. 165, the Roman Empire was devastated by an epidemic now believed to have been smallpox. The pestilence raged for fifteen years, claiming victims in all social strata in such high numbers that some parts of the Roman Empire lost 25 to 35 percent of their people.[20] It is believed that the virus struck a completely nonimmune population, having first appeared in Asia some one hundred years earlier.[21]

Over subsequent centuries equally devastating pandemics of the viral disease claimed millions of lives in China, Japan, the Roman Empire, Europe, and the Americas.[22] According to historian William McNeill, Cortez's capture of Mexico City with just a small army of exhausted Spanish irregulars under his command was possible only because the Europeans had unknowingly spread smallpox throughout the land. When Cortez launched his final assault on the capital, few Aztec soldiers were alive and well. Smallpox, together with measles, tuberculosis, and influenza, claimed an estimated 56 million Amerindian lives during the initial years of the Spanish conquest.[23]

By 1958, when the Soviets called for global eradication, smallpox was killing 2 million people a year, and cases could be found in thirty-three countries.

The virus could be spread by touch or respiration, and scientists carefully calculated the infectious dose necessary to produce disease—the numbers of viruses present in a tiny droplet of human exhalant and other details of transmission. It turned out that a single milliliter droplet of human lung exhalant contained 1,000 more viruses than were necessary to produce infection in the unfortunate soul who inhaled that minuscule bit of moisture.[24]

Both the historic devastation and the widespread rates of contemporary infection seemed to argue for skepticism about smallpox eradication.

On the other hand, several aspects of the biology of smallpox gave cause for optimism. Foremost was the existence of an extremely effective vaccine that, in various forms, had been in use since 1796.[25] In modern times the cowpox vaccine, made from the bovine form of the virus, had been perfected, making it over 99 percent effective, with virtually no side effects. Smallpox was also easy to diagnose, and cases of the disease could readily be spotted by people with no professional training. During severe illness, grotesque bulbous inflammations formed on the individual's face and skin. The distinct poxing, once healed, left visible scars anybody could recognize.

Because the virus was spread directly from person to person there were no troublesome vectors to control, such as mosquitoes, rats, ticks, or fleas. And the very thing that made smallpox so terrifying—its rapid lethality— also rendered it controllable because the viruses multiplied and spread so quickly that most people were infectious for only four or five days and their debilitation was so great that they could not walk about and infect large numbers of people.

Though eradication would require over 250 million vaccine doses per year and a worldwide effort to reach all citizens at risk for smallpox—even those in the midst of wars, social tyranny, famine, or disaster—the program began amid optimism in 1967 under the leadership of American physician Donald "D. A." Henderson.[26]

Nations of the Northern Hemisphere and Latin America were already well on their way to smallpox eradication in 1967, but the disease was firmly in place in many parts of Africa and Asia, where religion often proved a major barrier to vaccination.

Before the campaign began, researchers scoured pilot project areas to see how accurately smallpox cases were reported. Their conclusion was that an astonishing 95 percent of all cases of the disease were never reported to national or international public health authorities. The reasons for such dramatic underreporting were numerous: local authorities feared being penalized if higher-ups learned that epidemics had occurred in their jurisdictions; in some areas the disease was simply accepted as a fact of life; outbreaks tended to occur in confined areas and could easily be missed by quick national surveys; during centuries of colonialism, people's homes were often burned if smallpox was found in a family member, so people in former colonies naturally concluded it was best not to inform authorities.[27]

Ultimately, Henderson's team at WHO devised a smallpox plan of attack that boldly confronted these issues by dispersing dozens of skilled tropical disease experts all over the world in search of small outbreaks of the virus. Once such an outbreak was identified, local government was mobilized and residents of the area were vaccinated. Inoculation was occasionally carried out forcibly; in some instances people's homes were invaded and local police assisted the inoculators.

Because both superpowers wholeheartedly supported the campaign, few governments resisted public health efforts that often took on military overtones. The WHO teams braved civil wars, floods, religious battles, and a variety of geographic and logistic problems to accomplish their task in eleven years.

In Bangladesh, for example, where the worldwide campaign faced its toughest battle due to the great population density and ancient smallpox endemicity, French physician Daniel Tarantola braved confrontation with an infamous murderer thought to be a smallpox carrier. Without police protection, Tarantola approached the murderer and his outlaw gang in their hideout and faced down guns to immunize them. The word from villagers was that members of the gang had classic pockmarks on their faces and the robbers were spreading the epidemic throughout the countryside. The village intelligence proved accurate, and the immunizations prevented a local epidemic. The gang leader, however, died of smallpox two days after Tarantola's courageous confrontation.

During the late 1960s, Tarantola, then a Paris medical student, volunteered his services to Médecins Sans Frontières, an idealistic organization that sent European medical volunteers into war-torn areas to provide care to civilian populations. In the midst of civil war, the twenty-something Tarantola ran a pediatric ward in Biafra. Two years later, with some course work yet to be completed at the University of Paris Medical School, Tarantola signed on for a two-year stint in Africa in a small hospital in newly decolonized Burkina Faso.

Tarantola was a product of his times. While he studied the intricate workings of human kidneys, riots raged in the streets of Paris. Students formed alliances with factory workers and, inspired by such heroes of the day as Mao Zedong, Che Guevara, Ho Chi Minh, Herbert Marcuse, and Kwame Nkrumah, challenged the very existence of the De Gaulle government. Such bold, youthful actions were reflected all over the world, from Washington to Jakarta, as college-age young adults challenged the established order of things. A mood of activism and boldness infected the usually staid halls of medical schools internationally, inspiring would-be physicians like Tarantola to dream of a world in which villagers in Burkina Faso had as much a right to expect an eighty-year life span as did *les parisiennes bourgeois*.

When young doctors like Tarantola looked around the world for inspiration in the 1960s, they saw people nearly their own age leading revolutions against the old European colonial powers, taking control of governments and debating the creation of new types of social orders. Like many European and American idealists, Tarantola thought that with enough energy and Western money, just about anything was possible "if there is political will."

It was with that zeal that he approached his work in the Fada N'Gourma Rural Hospital in Burkina Faso, developing a grass-roots primary health care system that radically reduced infectious disease problems in the area. For his efforts, Tarantola was given the 1973 Albert Schweitzer Award.

The ink was barely dry on his medical degree when Tarantola next signed on with another French charitable group, Brothers to All Men, to do primary health care in northern Bangladesh. Because he had no command of English, the second language of Bangladesh, Tarantola promptly taught himself Bengali.

He had been in the country only six months when he was recruited to work with the smallpox campaign. Like Tarantola, most of the smallpox investigators were young (under thirty-five), Caucasian, idealistic males from Europe and North America. At the time, this cultural and gender homogeneity made some team members uncomfortable, but the grander goal of eliminating one of the planet's most notorious scourges outweighed concerns about neocolonialistic appearances.

In 1972 Don Francis had just barely completed his pediatrics residency at Los Angeles County Hospital and signed on with the CDC when smallpox broke out in Kosovo, Yugoslavia. The young doctor was just setting up a

CDC disease surveillance office in Oregon when Atlanta called, ordering him to go to Belgrade. Francis raced home, grabbed a few changes of clothing, a shaving kit, and his passport, and headed for the airport. Seven hours later he was in Washington, D.C., getting a briefing and cases of vaccination equipment. Before midnight he was asleep on a jet flying somewhere over the Atlantic, and in the morning he hit the ground running in Belgrade.

A few weeks later, the Yugoslavian outbreak safely contained, Francis was in Khartoum, the capital of Sudan, hunting down smallpox cases. From there, he went on to India and Bangladesh.

By the time Francis's obligations to the smallpox campaign were fulfilled, nearly three years had elapsed since he answered that phone call one morning in Oregon.

Another young American physician, David Heymann, saw a one-shot CDC assignment turn into two years of Indian smallpox hunting in Bihar and Calcutta. When Heymann's group vaccinated somebody, they always showed pictures of smallpox-stricken Indians and asked for names of people suffering from the disease. In some areas they offered rewards to people who could steer the team to active smallpox cases. If they found a case, the ailing person was quarantined and everybody in the region vaccinated—some against their will.

Despite the coercive nature of their activities, few of the fieldworkers doubted that, in the greater scheme of things, what they were doing was just: if 2 million people a year could survive because of momentary inconveniences visited upon a few, then how could there be any doubt about the righteousness of their campaign?

The one concern that did constantly haunt D. A. Henderson and his team was the cost of failure. In those brief moments when they allowed themselves to entertain the notion that smallpox might not be eradicated, the scientists knew the world might never again be willing to mobilize across political, national, cultural, racial, and religious boundaries to share a common battle against disease. The stakes, clearly, were high.

By the summer of 1974, the WHO team was about to declare victory in Bangladesh, the last stubborn holdout of the virulent variola major form of smallpox. Officials had even gone so far as to publicly predict that complete elimination of the virus would occur before that November.

But then the rains came, and came, and came. By August, Bangladesh was besieged by water, as dikes and dams burst under the monsoon's force. Refugees poured by the tens of thousands into Dhaka. Famine spread throughout the land, and the country seemed to be coming apart at the seams. Shortly before the floods, the Prime Minister, Sheik Mujibur Rahman, was assassinated, and a series of riots, civil violence, and coups followed, lasting well into 1975.

After being so close to victory, the task of eliminating—indeed, amid the chaos, *finding*—the remaining cases seemed so daunting that most of the eradication staff gave up. They were exhausted, burnt out.

But Tarantola told his staff, "Look, this just means we have to get down to micromanagement. We must look at the trees now, not the forest. Take it day by day."

Slowly staff confidence and morale were rebuilt, smallpox cases were found, and the enthusiasm returned. Within a year, victory once again seemed within grasp. Heymann and Francis were reporting total success in India, and no new outbreaks had occurred in Africa in months. With all eyes on Bangladesh, excitement rose.

One of the last infected villages was outside the city of Chittagong, which was under the command of an Army general. Not knowing what side of the civil war that general was on, or how he felt about foreigners, Tarantola confronted the general and requested permission to vaccinate the villagers. Permission was initially denied, and the disease spread. Once again, it seemed the gigantic obstacles of Bangladesh would force the WHO team to snatch defeat from the jaws of victory.

But the general finally relented, the last outbreak was stifled, and champagne was poured in Dhaka. Tarantola greedily guzzled the champagne, exultant after years of round-the-clock viral pursuit.

The next morning victory once again disappeared when word came that smallpox had surfaced on Bhola, an island off Bangladesh. For the third time the team was forced to remobilize after having been convinced the war was over. This time, when all the affected islanders had been vaccinated, Henderson held his breath a bit before announcing success.

On November 1975, D. A. Henderson was able to announce to the world that a three-year-old Bangladeshi girl named Rahima Banu had been cured, and represented the last case of wild variola major in human history. Two years later, on October 26, 1977, the last case of the less virulent variola minor would be found in Merka, Somalia.

By then Dr. Isao Arita had been in charge of the international effort for ten months, Henderson having retired. The Japanese physician ran the program with at least as much energy as the tall, bombastic American, but with a personal style that was more low-key and witty. In times of tension, Arita told jokes.

His humor was put to the test in the Horn of Africa in early 1977, just weeks after military leader Mengistu Haile Mariam seized power in Ethiopia, installing a Soviet-backed, communist government. The military government in Somalia laid claim to Ogaden, a region then part of eastern Ethiopia, and full-scale war was underway. Ethiopia, backed by Soviet arms and Cuban troops, mounted a powerful defense of Ogaden. But Somalia, despite its left-wing leanings, successfully obtained Cold War countersupport from the United States. As war raged, more than a million civilian refugees fled the area, pouring into nearby Somali and Ethiopian rural provinces that were in the throes of their second and third years of drought and starvation.

It was in that area—Ogaden—that the world's last known cases of smallpox variola minor could be found, primarily among Somali Muslims.

Arita knew that UN flags and WHO credentials would offer little protection to his scientific team members in such a volatile situation, yet he also felt time was running out. It was February 1977, and the Hajj was just ten months away. During the Hajj thousands of devout Somali Muslims would make their pilgrimage to Mecca, where they would eat, sleep, and pray for several days with some 2 million other followers of Islam from all over the world. If infected pilgrims were part of the hajj, all efforts to eradicate smallpox would end in failure.

For months the multinational team struggled against the elements, avoided the war's front, and tracked down smallpox cases among the refugees and villagers of Ogaden. By October the numbers of rumored cases seemed small, and despite the onset of the rainy season, Arita ordered the team to push on. Half the team got mired in mud floes during those rainy days, and one American scientist, Joe McCormick, spent three days alone in Ogaden, stranded in a Land-Rover stuck in a three-foot-deep wall of mud.

Finally, in Merka, Somalia, the team found the world's last case of variola minor.

Ali Maow Maalin would be cured, and all forms of smallpox disappeared. Smallpox had been conquered.[28]

Their jobs completed, smallpox team members dispersed to public health jobs all over the world. Surprisingly, they were not eagerly snatched up by WHO, or congratulated individually for their magnificent efforts. On the contrary, the brash young smallpox scientists were considered arrogant and thoughtless. They violated too many WHO bureaucratic guidelines. And they operated with a single goal in mind—a perspective quite unlike that of those who usually drove the vaguely pro-health WHO and national ministries of health worldwide.

"Science really suffers from bureaucracy," Arita would later declare, adding, "If we hadn't broken every single WHO rule many times over, we would never have defeated smallpox. Never."

Even Arita and Henderson, the heroic leaders of the smallpox eradication effort, were criticized for crossing too many WHO lines. Arita shrugged it off and returned to his hometown, Kumamoto, to run Japan's National Hospital.

After a period of being the target of animosity in the Geneva headquarters of WHO, Tarantola was thrilled to be assigned to an overseas job, running childhood vaccination campaigns in Indonesia. Francis was burned out, so he followed his girlfriend to Harvard, where he planned to do virology research. Heymann moved to Atlanta, signing on with the CDC. All three men would work together again, over a decade later, to combat another global epidemic.

Eradication took eleven years, involving about a hundred highly trained professionals and thousands of local health workers and staff worldwide. It was achieved at a cost of $300 million.[29]

On May 8, 1980, the World Health Assembly formally declared that "the World and all its peoples have won freedom from smallpox, which was a most devastating disease sweeping in epidemic form through many countries since earliest times, leaving death, blindness, and disfigurement in its wake and which only a decade ago was rampant in Africa, Asia, and South America."[30]

A very different outcome awaited those who fought to eradicate malaria worldwide. Between 1958 and 1963 alone, $430 million was spent on a series of failed attempts to eliminate malaria. In 1991 dollars that constituted an expenditure of over $1.914 billion.[31] Between 1964 and 1981, the United States spent an additional $793 million.[32]

When the international effort began, there were millions of cases of malaria every year, largely concentrated in Southeast Asia and Africa. Though reliable numbers were not available for most parts of the world, it was estimated that, for example, about 1 million people had malaria that year in Sri Lanka, some 100 million in India, and untold numbers, roughly estimated in the "hundreds of millions," in Africa.[33]

On the other hand, humanity possessed powerful weapons. Chloroquine and quinine were effective treatments that, when properly and speedily used, stopped most cases of the disease in a matter of days. And chlorinated hydrocarbon insecticides—notably DDT[34]—could kill not only adult mosquitoes that carried malarial parasites but their progeny as well, because the chemicals were virtually nonbiodegradable and their insect toxicity could be expected to eliminate pests that landed on sprayed surfaces for months, even years, after chemical treatment.

Even without such chemical weapons, some countries had dramatically reduced their malaria problem; chief among them was the United States. There was disagreement in research circles about the origins of malaria in the Americas,[35] but whatever the case, by the eighteenth century malaria was a serious endemic disease from Montreal to southern Chile.

For the U.S. military, malaria was an enormous problem—in some theaters of combat, its chief hardship—since the creation of George Washington's Continental Army in 1776. At least a million soldiers suffered malaria during the Civil War, and the disease was a major killer in America's southern states well into the 1930s. U.S. troops also suffered serious malaria problems overseas; 19,000 doughboys contracted the disease during World War I, and over 500,000 GIs had malaria during World War II.[36]

But during the construction of the Panama Canal (1904–14), General William C. Gorgas directed the U.S. Army Medical Corps on a successful campaign to drain swamps, treat local people with quinine, and kill mosquito larvae floating atop pools of water. He did not have DDT at his disposal in those early years, yet the result was virtual eradication of Panamanian malaria. Similar drainage efforts throughout the U.S. sunbelt were already bringing malaria down to negligible levels by 1947.[37]

Then Egypt embarked on the first successful campaign using DDT to eliminate the *Anopheles gambiae* mosquito. Its initial results were so dramatic that the U.S. Congress allocated $7 million in 1947 for a DDT-based program to eradicate malaria within the forty-eight states. Five years later, the program was abandoned when not one case of malaria could be found within the U.S. borders.

Similar successes were reported across the European continent, though malaria stubbornly hung on in parts of Italy, Spain, and Greece well into the 1950s. Buoyed by Brazil's success a decade earlier controlling the *A. gambiae*[38] mosquito population, the 1954 Pan American Sanitary Conference held in Santiago, Chile, resolved to eradicate malaria-carrying mosquitoes from all countries in the Americas, from the Arctic to Antarctica.

In 1956, malariologist Paul Russell, then at Harvard University's School of Public Health, authored a report for the International Development Advisory Board (IDAB)[39] recommending immediate global eradication of malaria. In the report Russell reflected mainstream scientific views of the day when he argued that DDT was such a powerful tool that a multimillion-dollar worldwide campaign could eliminate all malaria-carrying mosquitoes on the planet within less than a decade.

Generally, it takes four years of spraying and four years of surveillance to make sure of three consecutive years of no mosquito transmission in an area. After that, normal health department activities can be depended upon to deal with occasional introduced cases. . . . Eradication can be pushed through in a community in a period of eight to ten years, with not more than four to six years of actual spraying, without much danger of resistance. But if countries, due to lack of funds, have to proceed slowly, resistance is almost certain to appear and eradication will become economically impossible. *Time is of the essence* [his emphasis] because DDT resistance has appeared in six or seven years.[40]

Lest anyone in Congress miss the IDAB report's point, Russell added the following strong words:

This is a completely unique moment in the history of man's attack on one of his oldest and most powerful disease enemies. Failure to proceed energetically might postpone malaria eradication indefinitely.

Russell's plan caught the imagination of several key figures in the American political arena of the late 1950s: Secretary of State George Marshall, Senators John F. Kennedy and Hubert H. Humphrey, and President Dwight

D. Eisenhower. Though malaria no longer existed in the United States, America was, in 1957, the center of virtually all cash reserves on earth. Europe, Japan, and the U.S.S.R. were still smarting from World War II devastation, and what is now called the developing world was largely in the yoke of colonialism or severe underdevelopment. Having won World War II, Americans were of a mind to "fix things up": it just seemed fitting and proper in those days that Americans should use their seemingly unique skills and common sense to mend all the ailments of the planet.

Thus, in 1958 Russell's battle for malaria eradication began, backed directly by $23.3 million a year from Congress.[41] Because Russell had been so adamant about the time frame, Congress stipulated that the funds would stop flowing in 1963. In addition to the $23.3 million to be spent annually by IDAB, Congress shelled out funds generously between 1958 and 1963 for WHO (contributing 31 percent of its overall budget and more than 95 percent of its malaria budget), the Pan American Health Organization (PAHO, which got 66 percent of its funds directly from the U.S. Congress), and UNICEF (underwriting 40 percent of the UN Children's Fund budget).[42] It was a staggering economic commitment, the equivalent of billions of dollars in 1990. Remarkably, American politicians didn't complain about spending so much money to control diseases few U.S. citizens ever contracted, and the effort enjoyed bipartisan support. President Eisenhower called for the "unconditional surrender" of the microbes, George Marshall foresaw the "imminent conquest of disease," and Senator Kennedy predicted that children born in the next decade would no longer face the ancient scourges of pestilence.

The stage was set. The scientists had everything going for them: political support, money, DDT, and chloroquine. So certain were they of victory that malaria research came to a virtual halt. Why research something that will no longer exist?

Yet when Andy Spielman had started graduate school just five years earlier at Johns Hopkins School of Medicine, the budding young Colorado scientist was convinced he would have a lifetime's worth of parasite research puzzles to solve. Socially awkward because of a stuttering speech impediment, Spielman delighted in the introspective world of science. Baltimore colleagues quickly admired his wit, warmth, and ready intelligence. Spielman anticipated decades of studying insects and the parasites they carried.

He had, however, been in Baltimore less than two months when his mentor, Lloyd Rozeboom, grabbed Spielman by the collar and said, "Let's get a beer."

The downcast Rozeboom bought Spielman a pint and after a few quaffs said, "Look, I've got to get this off my chest. I'm conscience-stricken."

"What's the problem?" Spielman asked.

"I should never have accepted you into graduate school. I should never have encouraged you to pursue medical entomology. It's a dead field. DDT is killing it," Rozeboom said.

Spielman argued it was too early in the game to call the score. But Rozeboom was adamant.

"It's all over. There will be no career for you. By the time you've finished your thesis all the insect-borne disease problems will be solved," Rozeboom insisted.

Undaunted, Spielman pursued his Ph.D. despite Rozeboom's warning. He was a firm believer in evolution—he had practically memorized his favorite text on the subject—and he told Rozeboom that "DDT isn't the final answer."

While Congress reviewed Russell's IDAB proposal, Spielman took some courses at the Woods Hole Oceanographic Institute in Massachusetts. There he met a middle-aged marine biologist who was quietly rethinking the whole DDT question. She told Spielman that evolution would come between DDT and the dream of malaria eradication. DDT-resistant strains of *Anopheles* mosquitoes were turning up all over the world, she said.

Her name was Rachel Carson, and the same year the United States and WHO embarked on their ambitious campaign to eliminate malarial mosquitoes, Carson started writing *Silent Spring*.[43] Carson never completely opposed pesticide use; rather, she favored their rational and limited application. Prophetically, she worried that widespread agricultural use of insecticides would endanger efforts to control malaria, typhus, African sleeping sickness, yellow fever, and encephalitis. She wrote:

No responsible person contends that insect-borne disease should be ignored. The question that has now urgently presented itself is whether it is either wise or responsible to attack the problem by methods that are rapidly making it worse. The world has heard much of the triumphant war against disease through the control of insect vectors of infection, but it has heard little of the other side of the story—the defeats, the short-lived triumphs that now strongly support the alarming view that the insect enemy has been made stronger by our efforts. Even worse, we may have destroyed our very means of fighting.[44]

She noted that the first public health use of DDT occurred in 1943. Allied troops sprayed the chemical liberally to eliminate typhus-carrying lice in Italy. The lice were, in fact, killed, and typhus halted, but a year later DDT-resistant *Culex* mosquitoes and houseflies stepped into the vacuum. By 1951, mosquitoes and flies in the region were resistant to DDT, methoxychlor, chlordane, heptachlor, and benzene hexachloride, and Italians had returned to time-honored tactics for insect control: screened windows, flypaper, and flyswatters.

In 1959 Spielman joined the faculty of the Harvard School of Public Health and discovered that no courses on malaria or *Anopheles* mosquitoes were on the curriculum. With the leader of the world's malaria eradication campaign on the faculty, it was considered distasteful to offer such courses. Training young scientists in techniques of mosquito control implied that

Paul Russell's efforts would fail and such knowledge would actually be necessary for future practitioners of public health.

Russell, an ex-missionary, was a kindly, elderly gentleman, and although Spielman had never believed the campaign would succeed, it broke his heart to see the dejection Russell felt when 1963 arrived.

Malaria had, indeed, reached its nadir. But it had not been eliminated. In some countries success was so close that people were already celebrating. Sri Lanka, for example, had 1 million malaria cases in 1955; just eighteen in 1963.

But a deal's a deal. Russell promised success by 1963, and Congress was in no mood to entertain extending funds for another year, or two. As far as Congress was concerned, failure to reach eradication by 1963 simply meant it couldn't be done, in any time frame. And at the time virtually all the spare cash was American; without steady infusions of U.S. dollars, the effort died abruptly.

In 1963 Harvard put malaria control back on its curriculum.

Spielman shook his head and wondered out loud, "How can they just abandon all these people?" He knew that, thanks to the near-eradication effort, hundreds of millions of people now lacked immunity to the disease, but lived in areas where the *Anopheles* would undoubtedly return. Pulling the plug abruptly on their control programs virtually guaranteed future surges in malaria deaths, particularly in poor countries lacking their own disease control infrastructures. As malaria relentlessly increased again after 1963, developing countries were forced to commit ever-larger amounts of scarce public health dollars to the problem. India, for example, dedicated over a third of its entire health budget in 1965 to malaria control.[45]

Everything started to unravel. The Green Revolution—a World Bank-backed scheme to improve Third World economies through large-scale cash crop production—got underway. Turning thousands of acres of formerly diversely planted and fallow land into monocultured farms for export production of coffee, rice, sorghum, wheat, pineapples, or other cash crops necessitated ever-increasing pesticide use. When an area had very diverse plant life, its insect population was also diverse and no single pest species generally had an opportunity to so dominate that it could destroy a crop. As plant diversity decreased, however, competition and predation among insects also declined. As a result, croplands could be overwhelmed rapidly by destructive insects. Farmers responded during the 1960s with heavy pesticide use, which often worked in the short term. But in the long run pesticides usually killed off beneficial insects, while the crop-attacking pests became resistant to chemicals. A vicious cycle set in, forcing use of a wider assortment of insecticides to protect crops.

At the very time malaria control efforts were splintering or collapsing, the agricultural use of DDT and its sister compounds was soaring. Almost overnight[46] resistant mosquito populations appeared all over the world.

As Russell kept a worried eye on the pesticide resistance problem, a new crisis appeared: two people who were taking chloroquine developed malaria in South America.[47] Almost simultaneously, chloroquine-resistant malaria turned up in Colombia,[48] Thailand,[49] Venezuela,[50] and Brazil.[51] The drug had been in use for only fifteen years; widespread use spanned less than a decade's time.[52] By 1950 a second drug, primaquine, was available, and many countries returned to the use of the ancient antimalarial, quinine. But resistance soon developed to those and other drugs introduced in the 1960s.[53] By 1963 U.S. forces fighting in Vietnam encountered chloroquine-resistant malaria, and the Army began a major effort to research and develop new antimalarial drugs.[54]

The drug-resistance problem could only have been aggravated by government decisions in some countries—notably New Guinea—to add chloroquine to all table salt.[55]

By the time the smallpox campaign was approaching victory in 1975, parasite resistance to chloroquine and mosquito resistance to DDT and other pesticides were both so widespread that nobody spoke of eliminating malaria. Increasingly, experts saw the grand smallpox success as an aberration, rather than a goal that could easily be replicated with other diseases.[56]

In 1975 the worldwide incidence of malaria was about 2.5 times what it had been in 1961, midway through Paul Russell's campaign. In some countries the disease was claiming horrendous numbers of people. China, for example, had an estimated 9 million cases in 1975, compared to about 1 million in 1961. India jumped in that time period from 1 million to over 6 million cases.[57]

A new global iatrogenic form of malaria was emerging—"iatrogenic" meaning created as a result of medical treatment. In its well-meaning zeal to treat the world's malaria scourge, humanity had created a new epidemic.

3

Monkey Kidneys
and the Ebbing Tides

MARBURG VIRUS, YELLOW FEVER,
AND THE BRAZILIAN MENINGITIS
EPIDEMIC

*When the tide is receding from the beach it is easy to have the illusion
that one can empty the ocean by removing water with a pail.*
—René Dubos

■ The failures of malaria eradication were overshadowed by the tremendous triumphs of polio control and the campaign for the elimination of smallpox. Western scientists in the late 1960s saw the history of disease as an arrow shooting straight toward a *Homo sapiens* victory over the microbes. Machupo was considered a distant anomaly, news of which hadn't reached most Western physicians or scientists. And more than another decade would pass before the global public health community would stop using the term "eradication" when referring to malaria.

But other "anomalies" soon followed.

I

In August 1967 three factory workers in Marburg, Germany, reported in sick, suffering from muscle aches and mild fevers. The three men were employed at Behringwerke AG, the vaccine-producing subsidiary of pharmaceutical giant Hoechst AG, and though their ailments looked like nothing more than the flu, it was quite unusual for influenza to appear during Germany's hot summer months. The men were referred to the Marburg University Hospital.

The following day the three became nauseated, their spleens enlarged and were tender to the touch, and their eyes became increasingly bloodshot. The attending physicians noted that "the patients had a sullen, slightly aggressive or negativistic behavior."[1]

Day by day more workers from the pharmaceutical plant fell ill, as did a doctor and a nurse who tended the patients. By September, twenty-three patients lay in agony in the Marburg University Hospital wards. Some fifty miles away in Frankfurt, six more individuals contracted the same mysterious disease at the German government's Paul Ehrlich Institute. Four were also workers employed in pharmaceutical research, one was their treating physician, and the sixth was a pathologist who performed laboratory analysis of the cases.

At the same time a third outbreak occurred in Belgrade, Yugoslavia, involving a veterinarian and his wife.

The thirty-one cases struck terror in European research circles because of the ferocity of the disease and its spread from patients to their health care providers. Nobody knew what caused the ailment, how it was spread, what treatments might be effective, and/or how many more people might eventually be stricken.

Because the wives of the Yugoslavian veterinarian and one of the Marburg workers had become ill, there was fear the disease could be passed in the air. Nobody knew how the husbands had originally become infected.

They did know the disease was terrible; the adjective most commonly used to describe it was "agonizing." Each patient suffered the same excruciating chain of events. After a couple of days of flu-like muscle aches and fever, they began to show classic symptoms of acute viremia (physical responses to a flood of newly made viruses into their bloodstreams): large tender lymph nodes along their throats, inflamed spleens, a marked drop in the number of disease-fighting white blood cells, and a sudden shortage of blood platelets and other factors that are necessary to stop bleeding.

By the sixth day the patients were covered with red rashes that made their skin too sensitive to be touched. Their throats were so raw that they couldn't eat and had to be fed intravenous fluids, sugars, and vitamins. Within a week, all were suffering acute diarrhea.

By the eighth day, the rashes gave way to a far more painful and alarming reddening of the entire body caused by microscopic blockages throughout the thousands of tiny capillary networks immediately under the surface of the skin. Because the capillaries were blocked, red blood cells backed up, giving the patients a crimson glow. With the red blood cells immobilized, the oxygen those cells normally carried throughout the body was not reaching its destinations. Nerves responded by causing searing pain.

By the tenth day patients were vomiting blood.

At the three-week mark their skin peeled off, as oxygen- and nutrient-starved cells died by the millions. Most painful was the peeling along the patients' genitals; the testes of some of the men were damaged, shrinking in size.

The doctors noticed a remarkable similarity between their patients' symptoms and those of the acute phase of hemophilia. In both diseases blood loses its ability to coagulate properly, and some larger blood particles such as platelets get blocked up in peripheral capillaries, while the smaller blood-clotting molecules that normally prevent uncontrolled bleeding simply disappear.

The patients were bleeding to death. As the Frankfurt doctors put it: "Blood is pouring from all apertures."[2]

But this wasn't hemophilia—it was a contagious disease; one of the Marburg medical team led by Drs. Rudolf Siegert and Gustav Adolf Martini felt justified in labeling it "a new and up to now unknown disease."[3]

By December 1967 seven of the patients had died. Most succumbed within sixteen days of their first symptoms. Some had clearly suffered a brain-related stage of the disease, becoming severely confused, even demented, in the second week of their illness and then falling into deep comas from which they never recovered. Two of the patients' hearts simply couldn't bear the burden of pumping so much thick blood, and eventually gave out; they died of massive heart attacks.

For the survivors, the long-term effects of the disease were often serious. Several had permanent damage to their livers, leaving them in a lifelong state of chronic hepatitis. All had lost large amounts of weight. One man became psychotic and never recovered from the psychiatric impact of the ailment. A few of the men were left impotent and with no sex drive.

While physicians in Frankfurt, Marburg, and Belgrade tended to their patients, the World Health Organization assembled a multinational search for the cause of the disease. It seemed obvious the patients had a viral hemorrhagic disease, but attempts to extract the virus based on probes used for other known hemorrhagic diseases (including Machupo and Junín) failed. It seemed this was something altogether new.

All of the original cases in Germany and Yugoslavia involved men who worked with monkeys. Furthermore, investigators discovered, each of the men had handled animals, or the tissue of animals, from the East African nation of Uganda. The investigation narrowed when it was learned the monkeys were all of the same species: *Cercopithecus aethiops,* a type of vervet monkey common throughout Africa.

The investigators hit pay dirt when they determined that all the monkeys came from three shipments of wild animals transported from Uganda to Belgrade, then on to Marburg and Frankfurt. When the first shipment of animals arrived in Belgrade, 49 of 99 monkeys were dead, and the survivors were placed under quarantine. The Yugoslavian veterinarian who autopsied the dead animals contracted the disease a week later. Shortly after that, his wife, having nursed her husband at home, also developed what would eventually be dubbed Marburg disease. The veterinarian's autopsies of the dead monkeys revealed that the animals also had suffered massive hemorrhages. Two subsequent shipments of Ugandan monkeys contained large numbers of dead animals.

Martini, Siegert, and their colleagues discovered strange viruses in the blood and tissue of the monkeys. When samples were injected into guinea pigs, the laboratory animals died in a matter of days. But when mice were injected, nothing happened; the mice were somehow able to withstand the virus.

Microscopic studies revealed that the Marburg virus could be found in two different forms. The first looked like a caterpillar, with its long, thin, tubular shape coated with "fuzz." Inside the tube was RNA (ribonucleic acid), the genetic blueprint of the virus. The "fuzz" along the outside of the virus's protein tube was a constellation of extruding protein receptors the virus used to gain entry into target cells.

In its more mature and dangerous form, the viral tube was rolled up into a tight round coil that appeared virtually invulnerable to assaults from the cells and antibodies of an ailing creature's immune system.

In late August, Jordi Casals answered his phone at the Rockefeller Foundation laboratory in New Haven, Connecticut, to hear an operator say, "Now listen carefully, this is Germany talking to you." The Barcelona-born scientist waited patiently, listening to the roar of the transatlantic telephone cable.

"Dr. Casals?" the caller shouted over the noisy line. "This is Dr. Lehmann-Gruber calling from Marburg, Germany. We need your help."

Casals had the largest collection in the world of insect-carried and hemorrhagic viruses, stored under careful security in deep freezers inside the Yale arbovirus laboratory. A few hours after Lehmann-Gruber's entreaty, a similar call for Casals's expert assistance came from the Frankfurt group, which told the Rockefeller scientist they were terrified. Both groups, the callers said, had been doing research using monkey kidney cells, and in just three weeks' time sixteen people had come down with severe hemorrhagic disease and seven had died.

"The people go into shock, they hemorrhage from their noses, anuses, stomachs, mouths," a very anxious Lehmann-Gruber said. "We are at our wit's end. We need your help."

As the Germans described the frightening symptoms, Casals thought of his friend Karl Johnson's near-death bout with Machupo. Since the Bolivian incident, Casals and Johnson had become the world's experts on hemorrhagic diseases, Johnson operating out of Panama and Casals at Yale University, where he ran the Rockefeller Foundation's arbovirus laboratory.

The Germans desperately wanted to know what was killing their laboratory workers, and they begged Casals to screen patient blood samples against all the viruses in his Rockefeller facility. Casals agreed, with the stipulation that they only send serum samples from patients who had survived the disease. That way, he reasoned, the samples would be free of lethal viruses but would contain antibodies that should react against some virus in his vast Yale collection.

For several weeks Casals and his staff tested the German blood samples against hundreds of viruses, but none gave a positive reaction.

Casals called Lehmann-Gruber, telling him, "It's not anything we have here in the laboratory. It might be something distinct, something completely new."

In September 1967 a WHO team was sent into Uganda to find out where exactly the monkey virus originated. They tested monkey serum samples taken from wild animals that had been processed for shipments to zoos and research laboratories all over the world. As early as 1961 some monkeys captured near Entebbe and Kidera were infected with Marburg, and the numbers of infected animals, as evidenced by their stored blood samples, increased each year thereafter until late 1967, when a third of some monkey groups carried the virus.

All the actively infected monkeys were of two species: the vervets (African green monkeys) and red-tailed monkeys. Some other animals captured for the study had antibodies to Marburg virus, indicating they had been exposed to the microbe: chimpanzees, baboons, talapoins, and gorillas. It seemed an epidemic had broken out among Ugandan monkeys sometime around 1961, reaching a serious level by 1967. In laboratory studies, it was possible to infect rhesus macaques and a variety of other primates from the Old World (African and Asian animals), but it was not possible to infect New World monkeys or apes (those species from the American continents).

Experimental infection of Old World primates proved, alarmingly, to be 100 percent fatal. Yet it was clear that many, if not most, monkeys that were infected in the wild survived. The paradox was, indeed, puzzling.

In years to come researchers would make many forays into East African wilderness areas in pursuit of solutions to that paradox, as well as a larger one: where did the virus come from? As with most viruses, it was assumed Marburg had a reservoir, a species of some insect or warm-blooded animal in which the virus harmlessly resided and quietly reproduced. The relationship between such viruses and their reservoirs was commensal; neither organism was harmed, over decade after decade of coexistence. But if that reservoir species came in contact with a vulnerable animal, such as a human being, the virus might jump its peaceful ship for the new, highly susceptible vessel, producing an epidemic. When that happened a disease that had existed unnoticed for centuries in other species might suddenly appear "new" when it attacked human beings.

For three years researchers from the United States, Europe, and East Africa scoured Uganda and Kenya in search of a Marburg reservoir. They tested every monkey, ape, rodent, mosquito, tick, hyena, canine, feline, and bovine they could get their hands on. But no reservoir of the virus was ever found.

In the face of this mystery, WHO could not anticipate when or where Marburg might reappear. The agency could only remark on two facets of the German/Yugoslavian outbreak that were responsible for the spread of the virus from monkeys to humans.[4]

First, said the agency, it was clear that quarantine and export procedures for wild primates were inadequate. All wild animals should be quarantined in the country in which they were captured for a minimum of three weeks, and once quarantine was completed, transport outside the country should be very rapid, preferably by air. The numbers of human beings exposed to the animals during quarantine and transport should be kept at a strict minimum. And during transport, animals, particularly primates, should be separately caged and kept a sufficient distance from one another to ensure there is no clawing or biting during the stressful voyage.

Once animals reached their destination, WHO said, "it is recommended that national veterinary authorities should supervise import and quarantine" for a minimum of six weeks. During the lengthy quarantine period, the animals should, again, be separately caged to avoid spread of disease within the colony, and the numbers of human handlers ought to be minimized. It was considered too obvious to mention that animal caretakers every step of the way should wear appropriate gloves and protective garb, take steps to ensure that they are not bitten by the animals, and remain ever vigilant against allowing animal fluids or tissues to come in contact with any skin cuts or their mouths.[5]

Regrettably, these instructions would be violated repeatedly in years to come; sometimes with tragic consequences.

Nine years later, in February 1975, two young Australian students on walkabout in Southern Africa would unintentionally serve as "canaries" and prove that Marburg disease hadn't disappeared from the planet when the 1967 epidemics apparently ceased among vaccine researchers in Europe and in Uganda's monkey populations.

A twenty-year-old Australian draftsman and his nineteen-year-old girl-friend spent the Southern Hemisphere summer of 1975 hitchhiking around Rhodesia[6] and South Africa. Sitting on the roadside outside the Rhodesian Gwaai River town of Wankie one morning, the young man felt a sudden sharp pain in his right leg. Looking down, he saw a red swelling and concluded he'd been bitten.

Six days later the couple was enjoying the Natal beaches near Margate, South Africa, when the young man broke out in a sudden sweat and felt a wave of fatigue that totally sapped him. After four days of escalating muscle pain, exhaustion, fever, and headaches he was admitted to Johannesburg Hospital on February 15, 1975. Four days later, he died, after suffering internal hemorrhaging so severe that the alveolar air sacs of his lungs were filled with blood.

During his four days at Johannesburg Hospital, the Australian was monitored by fifteen doctors and scientists and ten nurses, one of whom, a twenty-year-old nurse, fell ill with the disease nine days after the man died.

Two days after the Australian man's death, his girlfriend also got the disease. Dr. Margaretha Isaacson, of the South African Institute for Medical Research in Johannesburg, treated the patients with the anticoagulant heparin, saving their lives. No doubt, Isaacson concluded, the vast hemorrhaging seen in the Australian man had been prevented in the other two Marburg victims because heparin stopped mini-clot formation throughout their vascular systems.[7]

Though the Johannesburg team was convinced the Marburg infection began with the bite on the young man's leg, they had no idea what sort of creature—rodent, insect, monkey—had attacked him. It was also possible the young man's mysterious bite had nothing to do with the Marburg infection. Within the ten days before he arrived at Natal, the Australian engaged in several other activities that could have put him in contact with a Marburg-carrying creature. In Rhodesia he slept outdoors on a field that was zebra grazing land; he handled raw meat in Bulawayo; he touched monkeys near the Great Zimbabwe ruins, and he hand-fed monkeys caged in a hotel lobby in Natal.[8] The Johannesburg scientists were no closer than their German and Ugandan colleagues had been in 1967 to solving a key mystery: why do such deadly diseases suddenly appear, then just as suddenly disappear?

And so a mystery remained. It began with two teams of German scientists engaged in one of the most optimistic and potentially beneficial pursuits in health care: development of vaccines. It arose from monkey cells, and it ended somewhere in a geographic space so large and varied, spanning thousands of miles from Nairobi to Cape Town, that no one could begin to sift the clues.

At a time when scientists were talking about artificial hearts and advanced brain surgery it seemed almost inconceivable that twenty years later the Marburg mystery would remain utterly opaque.

But it would.

II

Joe McCormick walked briskly down the long emergency ward corridor of the Emilio Ribas Hospital, trying hard to ignore the terrified faces that stared at him from stretchers and chairs. Hospital staff scurried all about him, fighting against time to get acutely ill children and teenagers under treatment before they died.

"We've got to do everything we can to avoid panic," McCormick thought, reiterating the sentence as much for his own edification as for a larger public health concern.

He stepped out into São Paulo's winter air and stood on the ambulance off-loading dock, watching as another dangerously sick child was transferred from a stretcher to a gurney. Almost in a daze, he glanced at his watch, noted the time, and started counting. In the next thirty minutes, thirteen ambulances arrived, each carrying a Brazilian child or young adult in the grip of meningitis. By day's end, over 200 patients passed through the emergency doors, and the 1,000-bed hospital was filled well beyond capacity.

It was August 1974, and Brazil's burgeoning megacity—then boasting a greater metropolitan population of 20 million—was in the throes of an epidemic of meningococcal meningitis, a severe bacterial disease that could kill an untreated human being in less than twenty-four hours. Caused by *Neisseria meningitides* bacteria, meningococcal infection was spread directly from person to person, carried in the mucoid droplets of a sneeze. Even under the very best of circumstances, meningococcus killed 10 percent of those infected, and these were hardly the best of circumstances. Close to 15 percent of the cases admitted to Emilio Ribas Hospital were dying, and this was the premier treatment facility in São Paulo; death rates in smaller hospitals were as high as 77 percent in infants and 60 percent in adults.[9]

McCormick was flabbergasted by what he had already seen in his few weeks in São Paulo. He was on loan from the CDC's Special Pathogens and Bacteria Branch to the Pan American Health Organization (PAHO), with the task of assisting a team of Brazilian doctors in their efforts to stop the epidemic. Though he was relatively new to the CDC, McCormick's unique background put him in good stead for handling the unfolding Brazilian crisis.

While German public health authorities were panicking about Marburg virus in the summer of 1967, Joe McCormick had been impatiently cooling his heels in a remote northern outpost of Zaire. The former Indiana farm boy had been teaching elementary school in Zaire for two years. He had finished college in 1964, graduating with a stellar record in chemistry. The National Science Foundation offered him a full fellowship for graduate studies in physics, but he had turned it down—for Zaire, and adventure.

Shortly before the first workers fell ill in Marburg, civil war broke out in Zaire, pitting the two-year-old government of Mobutu Sese Seko against Katangan rebel forces led by white mercenaries. Mobutu, who himself came to power through military action, cracked down hard. Among the many measures taken in the summer of 1967 to quell rebellion was mandatory house arrest of all white people residing in Zaire.

This was tough luck for McCormick, who was anxiously pacing about his quarters in Wembo Nyambo. As he had surveyed his cinder-block rooms, which were under a constant state of siege by invading tropical foliage, McCormick felt that, on balance, house arrest or no house arrest, he had made the right decision in turning down the generally coveted NSF fellowship. Like hundreds of other bright members of his generation, McCormick deeply admired John F. Kennedy, was devastated by his assassination, and had planned, in the spirit of "ask what you can do for your country," to join the Peace Corps.

There was a catch, however; the Peace Corps wouldn't let him teach in a foreign language, and McCormick desperately wanted to master at least one language other than his native English. So, twenty-two years old and full of vinegar and wanderlust, he had signed on with a Methodist program that was sending teachers to Zaire.

Given recent events in Zaire, nobody much cared that McCormick lacked teaching credentials. Virtually all European- and American-trained professionals had fled the country over the last few years due to a chain of events that began in June 1960 with the overthrow of Belgian colonialism. That brought the country a new name (Zaire in place of the Belgian Congo) and its first independence leader, Prime Minister Patrice Lumumba. During the violent transition from colonialism to independence, an American missionary was kidnapped, taken to Stanleyville, and publicly assassinated, putting a pall over efforts to recruit foreign teachers, physicians, and other professionals.

Lumumba, an ardent African nationalist admired throughout the continent, ruled for only a few months before he was deposed by elements of the military and assassinated in what was later acknowledged by Congress to be a CIA operation. Four years of civil unrest and United Nations intervention followed, culminating in an Army takeover on November 24, 1965; General Mobutu Sese Seko proclaimed himself President of Zaire. Not all Zairians accepted Mobutu's leadership, and armed rebellion persisted throughout the country when Methodist recruiters scoured the United States and Western Europe in search of schoolteachers.

Into that tension stepped McCormick, a midwestern white kid with a head for science and a knack for repairing anything mechanical. His only knowledge of French was a cram course in Parisian dialect given to him by the missionary program just before his departure for Zaire. But McCormick discovered during his stay in Wemba Nyambo that he had a real knack for languages. He quickly began chatting with students and villagers not only in French but also in Lingala and Otetela, languages spoken by very few non-Zairians, as well as in the universal patois of African trade, Kiswahili.

It wasn't long before McCormick, in his youthful arrogance, felt he had mastered teaching, and started searching impatiently for new challenges. Impressed by the physicians at the local hospital, McCormick decided to become a doctor.

In late 1966 McCormick took the Medical College Admissions Test via correspondence examination, having boned up using whatever texts his newfound physician comrades could provide. To the surprise of no one who knew him, he aced the test and was well positioned to gain admission to a top American medical school. Just days before medical school was to begin, house arrest was lifted, and McCormick made his way to Duke University.

Over the next seven years, the always restless McCormick obtained an M.D. from Duke and a master's degree in tropical medicine, both financed by a federal training plan, activated in 1965 and designed to alleviate what was then considered a severe shortage in the number of American doctors. Under the system, U.S. medical students were subsidized by the federal government in exchange for postgraduate duty in the Public Health Service.[10] In McCormick's case, it meant putting in a few years' service at the federal Centers for Disease Control in Atlanta following completion of his medical studies. Since McCormick's plan was to devote himself to the study of tropical diseases, some time at CDC was precisely what he most desired.

In 1972 McCormick joined a two-year program under the Epidemic Intelligence Service, gaining training at CDC's Atlanta headquarters and being deployed to investigate outbreaks of diseases in the United States. His first assignment was to an American Indian reservation in Parker, Arizona, where people were falling ill from foods contaminated with *Streptococcus* bacteria.

By the time McCormick was ready to join CDC as a full employee of the agency's Special Pathogens and Bacteria Branch, he had already made a name for himself among the "cowboys" of the organization. Karl Johnson, who left Panama in 1971, was back at CDC, and he heard stories of the promising young EIS officer. He decided to keep an eye on the fellow— his Africa experience might one day come in handy.

But Africa would have to wait, for now McCormick was consumed by the crisis in Brazil.

Like most post-World War II physicians, McCormick had assumed that antibiotics would cure all bacterial infections, but it was clear this microbe could kill even children who were injected with massive doses of penicillin or ampicillin, the preferred drugs for meningococcus control.[11]

McCormick was worried. It looked like this epidemic involved a particularly virulent strain of the bacteria; possibly one that was resistant to the sulfa-based antibiotics. Despite drug treatment, the bacteria savagely attacked the membranes—the meninges—that enveloped victims' brains and spinal cords and caused excruciating pain and neurological damage. That apparent drug failure could force doctors to switch from treating patients with the cheap accessible penicillin-class drugs to using more expensive and less predictable antibiotics like rifampin and chloramphenicol.

When he reviewed the medical charts and laboratory findings on the epidemic cases, a few anxiety-provoking facts leapt out. First, the director of Emilio Ribas Hospital, Dr. Carlos de Oliveira Bastos, had noticed that the numbers of meningitis cases admitted to his facility had increased slowly by 21 percent between 1962 and 1971, and then more than doubled in 1972.[12] During that time, the majority of the cases involved infants, and the leading pathogen was meningococcus Type C.

But between January and August of 1974 over 11,000 meningitis cases were reported to Ribas Hospital alone; the patients were older and even included a few elderly individuals, and the dominant pathogen switched from Type C to Type A.[13] The typing of meningococci was based on tiny molecules that protruded from the surface of their bacterial membranes. When people were infected, they produced disease-fighting antibodies against these typing molecules. Antibodies against Type C could not recognize and attack Type A markers. Since Type A infections were previously almost unheard of in Brazil, virtually no citizen in São Paulo was naturally immune, which would explain why some adults were coming down with what typically was a disease of immune-naïve children.

Worse yet, though there was a Type C vaccine available, research on development of a Type A vaccine was only just beginning. Furthermore, Augusto Taunay, director of the Adolfo Lutz Institute in São Paulo, was confirming that virtually all the Type C meningitis bacteria examined at his laboratory were resistant to sulfa-based antibiotics, such as penicillin.

In August, Taunay concluded that most Type C sufferers were under nine years of age, and the increase in mortality in that group was due to sulfa resistance. But Type A primarily attacked teenagers and adults under twenty-five years of age. It appeared that two simultaneous meningitis epidemics were underway.

This looked at first blush like a lose-lose situation. Brazilians generally had some natural immunity for Type C, but for those who were susceptible the bacteria had a new trick—antibiotic resistance. On the other hand, the apparently new Type A strain was vulnerable to antibiotic treatment, but few Brazilians were naturally immune; by the time they reached hospitals their meningitis cases were often too far gone to be cured with the antibacterial drugs.

Studying the patients, McCormick noticed that most had the classic acute form of the disease known as Waterhouse-Friderichsen syndrome; they went from well to severely sick in a matter of minutes, feeling sudden fevers, neck stiffness, and dizziness. Within hours their bodies were covered with tiny red dots—sites of pinpoint hemorrhages of capillaries under the skin. Within twelve to twenty hours they descended into comas, their kidneys hemorrhaged, and death soon followed.

"If you don't pick it up the first time you see it in little kids, they will be dead the next time you see them," McCormick thought with a shudder.

Surviving the disease could result in serious lifelong disabilities. The bacteria attacked the middle of three protective meningeal layers enveloping the brain and spinal cord, and survivors were often left with a range of different types of brain damage. Or the bacteria might attack the kidneys and outer limbs, causing victims to lose their fingers, toes, even feet.

Epidemics of the disease were rare in South America, but relatively common in parts of West Africa. In Chad, for example, 1950s epidemics had attacked at rates of 11,000 cases per 100,000 people. By August 1974 São Paulo's meningitis rate was comparatively low, reaching 100 cases per 100,000 people. But McCormick had done the math, and he knew that the situation could quickly reach West African proportions. It troubled McCormick that nobody knew where the antibiotic-resistant Type C or the

Type A strain came from. Not knowing exactly when and where things started made it harder to forecast the future scope of the epidemic.

There were many possible explanations for the origin of this sudden epidemic, and at various times during his stay in São Paulo, McCormick mulled them all over. It was, for example, possible that a Brazilian had traveled to Africa, become infected, and brought the Type A bacteria to São Paulo. By 1974, over 75 million passengers were flying each year across national boundaries somewhere in the world[14]—perhaps someone flew in, already infected, from West Africa. On the other hand, the bacteria could have arisen from a hospital or clinical setting locally, the result of improper antibiotic treatment. It galled McCormick that he couldn't figure out a way to track the origins of this epidemic.

In September, a WHO investigation reached a painful conclusion: although the Type C vaccine had been tried in only a couple of field settings and the Type A vaccine had to be considered completely experimental, nothing else could possibly stop the burgeoning epidemic.

Furthermore, WHO felt strongly that a vaccine combining immunization against Types A and C was essential, and warned that "the future of any vaccination programme will be in jeopardy" if the public loses faith in health efforts because people continue to die after receiving only the partial vaccination protection; the expected result of using only one vaccine.

The WHO decision was actually pro forma: the military government of Brazil had already decided in an August 5 meeting in Brasília to vaccinate every citizen living in the epidemic area.[15] While panic mounted, France's Institut Mérieux agreed to manufacture the combination A/C vaccine for Brazil, and hurriedly built a new factory outside Lyons, France, for the purpose. Within four months, Mérieux would have attained the ability to manufacture and ship to Brazil 500,000 vaccine doses a week.[16] The vaccine was composed of the sugary pieces—the polysaccharides—of the bacteria that gave them their A and C immunological statuses.[17]

Between August 1974 and mid-January 1975 while waiting for the vaccine, however, McCormick and Brazilian officials had few tools at hand. Joe decided to focus on public education, and immediately taught himself Portuguese. By October he was giving lengthy interviews and holding press conferences, calling for calm and reason.

"I've got to be clear about what I say," McCormick thought, "and give the people confidence."

As an outsider, McCormick had a special role to play—and a delicate one. For ten years Brazil had been ruled by a military junta noted for its brutality. Countless students, labor leaders, religious activists, and representatives of the country's underclass had "disappeared" by late 1974. "Disappearance" was a euphemism for death, preceded usually by kidnapping and torture. Rumors and fear ruled public opinion, and few of the nation's poor trusted government pronouncements.

On the whole, however, the Brazilian Ministry of Health's meningococcal announcements were accurate. It was part of McCormick's diplomatic mission to publicly support the government's statements about the epidemic while clearly maintaining a critical distance from the junta itself. Such dicey politics weren't taught at Duke University Medical School. Or at the CDC training program.

McCormick had to wing it. Over the months he learned how to gently point out that nearly all meningitis sufferers came from communities of acute poverty, such as the massive favelas of São Paulo and Rio de Janeiro, without directly attacking the government policies responsible for that impoverishment. [18] He noted that the disease spread most rapidly in conditions of dense housing and poor hygiene, where people who lacked access to clean tap water rarely bathed or washed their clothing. The bacteria survived in such conditions, and could be passed from one family member to another via shared towels, clothing, cleaning rags, or kerchiefs.

By the time Mérieux had manufactured enough vaccine to immunize the population of São Paulo, the epidemic had claimed over 11,000 lives and caused serious illness in more than 150,000 people in at least six Brazilian states. About 30 percent of the survivors were reportedly suffering long-term neurological disorders of one kind or another.

By New Year's the attack rate of the disease in Rio was 205 cases per 100,000 people and authorities feared the upcoming Carnival would increase its spread. The vision of millions of Brazilians and tourists dancing together for days on end in crowded streets presented the very real possibility that the epidemic would be carried all over the planet by worn-out revelers returning home.

Though they had no idea how effective the Mérieux vaccine might be, or how dangerous, Rio officials felt they had no option: on January 13 they began a twelve-day vaccination campaign with the announced goal of immunizing 80 percent of the population of greater Rio.

Within five days, over 3 million Cariocas were vaccinated and the incidence of meningitis instantly plummeted. During the dreaded Carnival week only ten people contracted the disease.

Encouraged by the Rio experience, the military government organized the largest vaccination campaign in world history. From April 21 to 24, nearly 11 million residents of São Paulo were vaccinated, representing 90 percent of the city's population. This was accomplished by cordoning off all the commuter-intensive areas of the city and lining up as many as half a million people at a time for their shots. The entire mass media was mobilized as a huge propaganda tool for the campaign and military vehicles blasted pro-immunization announcements from roof-mounted loudspeakers.

Throughout Brazil similar militarily precise operations were soon conducted, eventually bringing the epidemic to a halt.

By early 1976, when McCormick returned to CDC headquarters in Atlanta, meningitis had ceased to be a serious problem in Brazil. But the fundamental question of where this virulent Type A strain had come from remained unsolved. At a PAHO meeting in Washington in February 1976, McCormick pushed for inclusion of the following statement in the official summary of the Brazilian episode:

It is not possible at the present time to predict when and where an epidemic of meningococcal meningitis will occur. Therefore it is not clear when and where preventive immunization should be carried out.[19]

Overlooked entirely in the final PAHO report was the significance—possible harbinger—of the bacteria's ability to resist common antibiotics.

III

As was the case with malaria, polio, smallpox, and all bacterial diseases, the 1960s mood surrounding yellow fever control was one of extreme optimism. The tools were at hand: DDT and other pesticides to kill the *Aedes aegypti* mosquitoes that carried the yellow fever virus and an effective vaccine. Since 1937, yellow fever vaccines had been in use, and refined forms of immunization proved so powerful that virtually every vaccinated person was protected for life by a single shot.[20] Beginning with the period of the construction of the Panama Canal at the turn of the century, a variety of successful means had been used to eliminate the *A. aegypti* mosquitoes from the Americas.

Since the seventeenth century yellow fever had been a major and terrifying scourge in the Americas, causing endemic disease in jungle and swamp areas from Canada to Chile and claiming tens of thousands of lives in periodic urban epidemics. It would begin with a headache, fever, and a vague sense of uneasiness, within hours progressing to chills, muscle pains, and vomiting. After five days internal bleeding would commence, the liver would malfunction, and the individual would become jaundiced. If never previously exposed to the virus, the human then had a 50-50 chance of dying. A 1793 yellow fever epidemic in Philadelphia killed 15 percent of the city's population and sent one out of three residents fleeing into the countryside.[21]

In West Africa, yellow fever was so ubiquitous that most surviving adults were immune to the disease. Many historians have noted that their acute vulnerability to yellow fever prevented British and French colonialists from attaining full control over West Africa.[22] So obvious was this deterrence in some areas of Africa that it was celebrated in song and verse by people from the Sudan to Senegal. Well into the 1980s schoolchildren in Ibo areas of Nigeria still sang the praises of mosquitoes and the diseases they gave to French and British colonialists.[23]

It was generally believed the *A. aegypti* mosquito originated in West Africa and was brought to the New World aboard slave ships.[24] The mosquito quickly adapted and thrived in the moist tropical regions of the Caribbean and the Amazon. The first epidemics occurred in Mexico's Yucatán and Havana, Cuba, in 1648. In less than fifty years, the *A. aegypti* population had blanketed the Americas, and yellow fever epidemics were cropping up everywhere.

In 1901 American Army physician Walter Reed and Cuban doctor Carlos Finley figured out the link between the *A. aegypti* mosquito, the virus, and the importance of uncovered pools of clear water, and started a hemisphere-wide effort to eradicate the mosquito. The mosquito, they discovered, could only leave its eggs in clear, clean water, so it thrived around people, lived in human homes, and left its larvae in jugs of drinking water. The insect was also constrained by temperatures below 60°F and only thrived in humid climates over 72°F. It seemed immediately obvious, then, that the entire yellow fever problem could be greatly reduced by simply covering all clean water supplies during warm months.

In 1927 a vaccine was developed and the first official global disease eradication effort began, endorsed by the governments of the Americas.[25]

The language of yellow fever efforts shifted from "eradication" to "control" and "conquest" following Fred Soper's 1932 discovery that some monkeys harmlessly harbored the virus.[26] In subsequent years, scientists discovered that several species of monkeys and apes could carry the virus, both in Africa and in South America. In the Americas, capuchin monkeys were unharmed by the virus, but carried yellow fever and could be a source of the microbe for feeding mosquitoes. In contrast, when yellow fever hit Central America, epidemics virtually exterminated the nonimmune *Ateles* and *Alouatta* monkey populations.[27]

In short order it was also discovered that *A. aegypti* wasn't the only mosquito that could carry yellow fever: *A. africanus*, *A. simpsoni*, and *A. albopictus*, to name a few, could carry the virus. Furthermore, the virus could be passed from one mosquito generation to the next in the insect's eggs; this allowed for long periods of time—several insect generations—when the disease seemed to disappear. But the virus was actually silently residing in generations of monkeys and mosquitoes, ready to reappear in human epidemic form under the proper conditions.[28]

The harsh significance of this jungle/monkey form of yellow fever hit home in 1949 when the disease broke out again in Panama, reversing more than forty years of successful eradication begun in the days of Walter Reed. From there it spread northward through Costa Rica, Guatemala, and Mexico, forcing U.S. military and PAHO intervention for control. By 1959 cases of yellow fever were cropping up in areas all over South America where authorities thought eradication had been successful. In most outbreaks, the first cases involved men who worked in agriculture or timbering on the edges of tropical rain forests; there, they came in contact with wild mosquitoes that fed on monkey carriers.[29]

By the late 1950s scientists realized that there were two types of yellow fever: the urban form associated with *A. aegypti* and the forest or sylvan form that could be found in a variety of monkeys and wild mosquitoes. Eradication of the urban form might be possible through vaccination, covering all water sources, and DDT spraying of insect breeding sites. But jungle yellow fever could not be eliminated without vaccinating all wild monkeys in Africa and South America, a clearly impossible task. Despite these hurdles of nature, WHO and PAHO remained optimistic about eliminating all human yellow fever disease because the vaccine protected people against both forms: if all children living in endemic areas were routinely vaccinated, they reasoned, the disease would only remain a threat to nonimmunized foreigners traveling through jungle areas. Mass vaccination campaigns of the 1940s and 1950s drastically reduced human disease in both South America and West Africa.

In the Americas, PAHO officials decided that the disease could further be prevented by eliminating the *A. aegypti* mosquito from the hemisphere, and from 1947 to 1960, the organization conducted a second massive campaign of mosquito control. In some countries, such as Argentina, Chile, Panama, Venezuela, and Colombia, DDT spraying and systematic covering of water sources radically reduced *A. aegypti* and public health officials were confident the insect could be wiped out of the Americas by the mid-1960s. But the U.S. Congress was never convinced such an effort was important for residents of the Northern Hemisphere, and, despite having formally committed itself to the PAHO campaign, never allocated funds for such an effort inside U.S. territory.

Yet Congress, recognizing the diplomatic importance of appearing to comply with a PAHO edict for which the United States had voted, did order the CDC to attempt eradication. The effort was doomed from the start by hundreds of protesting property owners who threatened to sue if chemicals were sprayed in their yards or homes.

In 1964, Dr. Donald Schleissman, who led the largely unfunded U.S. *A. aegypti* elimination effort, said of U.S. congressional commitment, "The mandate to eradicate *aegypti* with the funds available was equivalent to instructions to fly across the Atlantic with half a tank of gas."

Though its numbers were reduced temporarily, *A. aegypti* was never driven out of the Americas.

Similar campaigns were carried out throughout equatorial Africa, but five yellow fever outbreaks occurred in the 1950s. A 1959 outbreak in Zaire only came to a halt when hundreds of thousands of people had been vaccinated and more than twenty tons of DDT were sprayed over a relatively confined area.[30] In 1960 an enormous yellow fever epidemic broke out in western Ethiopia. By the time the epidemic died down in 1962, over 100,000 people had suffered the disease; yellow fever killed one out of three infected Ethiopians.

A subtle change followed the Ethiopian epidemic. Without really discussing the matter, international experts slowly switched their tactics from the bold eradication ventures aimed at wiping out the disease to fire fighting. Research outposts were set up in yellow fever hot spots worldwide by the Rockefeller Foundation and a variety of government-associated agencies.[31]

It was in such an outpost that Tom Monath, a CDC entomologist, worked at the University of Ibadan in Nigeria. Before he would leave Nigeria in 1972, Monath would travel all over the country trying to figure out where the virus hid between human epidemics. He would discover that a mosquito called *Masoni africana* could carry the virus throughout its habitat: the higher treetop levels of the Nupeko tropical forest lining the banks of Nigeria's rivers.[32]

Monath's commitment to conquering yellow fever was solidified in late 1970, when he was part of a U.S.-Nigerian team that investigated an epidemic in Nigeria's savanna plains of the Okwoga District. Over the Christmas holidays, Monath and his Nigerian colleagues made house-to-house surveys of Okwoga villages and medical clinics, searching for yellow fever cases and assisting Nigerian efforts to control the epidemic.

"Has anybody here been sick lately?" Monath, a white Bostonian sporting a crew cut and smile, would ask when he arrived in a village. Time after time the scenario repeated itself: a villager would nod somberly and lead Monath into a thatched hut. There, a dead man would be sitting up straight in a chair, his eyes staring ahead, cotton plugs stuffed up his nostrils.

The first time Monath beheld such a sight it scared the hell out of him, but after a while it was not the individual cases that troubled him, but the overall level of destruction inflicted by both the disease and the treatments used in some areas to allegedly cure yellow fever.

He was also impressed by the fact that no original source for the 1970 epidemic could be found. Monkeys were scarce, there were no rain forests, yet in some villages one-third to half the residents showed blood-test evidence of recent infection. The overall infection rate in Okwoga was 14 percent, yet the most common yellow fever carrier, the *A. aegypti* mosquito, was virtually nonexistent in the area.[33]

Monath and his colleagues were forced to conclude:

The origin of the epidemic is not known. Two possibilities exist: (1) the Okwoga outbreak . . . resulted from the introduction of Yellow Fever Virus from a distant source at a time favourable for interhuman transmission in an immunologically susceptible population or (2) Yellow Fever is endemic . . . in or near Okwoga District.[34]

In other words, either the disease was brought into the area by a traveler or it was there all along, hiding somehow for decades. There was an enormous biological gap between those two possibilities. Monath realized that not knowing which explanation was correct meant there was no way to determine how best to prevent future outbreaks in the area. Or *any* area.

Hopes for controlling the insects that acted as vectors for yellow fever dimmed further still, as everywhere scientists looked another insect vector for yellow fever turned up. Karl Johnson found other virus-carrier species in Panama, Brazilian physician-scientist Francisco Pinheiro identified still more insect vectors in his country's jungle interior, and U.S. Army researchers discovered that horse mosquitoes in Brazil and common ticks in West Africa could spread the virus.

In 1972, convinced that it was fruitless to try to eliminate yellow fever, the Rockefeller Foundation shut down Monath's lab in Ibadan and the other field stations. Years later Monath would still refer to the decision with bitterness. "A great opportunity has been blown," he told colleagues, noting that between 1947 and 1972, A. aegypti had been eliminated from three-quarters of its pre-World War II habitat worldwide. Nineteen countries had eliminated the yellow-fever-carrying insect entirely by 1972, prompting the Washington consulting firm of Arthur D. Little to do a cost-benefit analysis of mounting a full-scale campaign to rid the Americas of the insect. The Little study determined that such an effort would clearly be desirable, even though the virus had a sylvatic cycle that allowed it to hide for long periods of time in wild monkeys and several other insect species. It reasoned that spending $326 million in the early 1970s to wipe out A. aegypti would bring human incidence of the disease down to near-zero levels in most Latin countries, because only that mosquito was infecting urban residents. Furthermore, A. aegypti was a far more efficient virus spreader; every major yellow fever epidemic of the nineteenth and twentieth centuries was spread by that mosquito. So, the consulting firm reasoned, a global campaign to eradicate A. aegypti could limit the yellow fever problem to levels entirely controllable through routine vaccination of people living or working in jungle areas.

But the CDC disagreed; director David Sencer commissioned a rival study that concluded that A. aegypti eradication within the United States and Puerto Rico and the Virgin Islands alone would cost more than $200 million and speculated that an Americas-wide eradication would exceed $1 billion.[35] The inflated cost was primarily due to private citizens' refusal to allow spraying on their properties and widespread threats of lawsuits. Though poorer nations to the south spent enormous sums, successfully eliminating A. aegypti from much of the Americas, the wealthiest country in the hemisphere refused to get rid of its own mosquitoes.

The effort died.

Frustrated and disappointed, Monath packed his Ibadan laboratory into crates and bade his Nigerian colleagues farewell. But he wouldn't be leaving Africa, not just yet. Something even deadlier than yellow fever awaited him.

4

Into the Woods

LASSA FEVER

I'm not afraid when we have to bring a risky, hazardous virus into the lab. I just hope and pray, and I don't think about it.
—Dr. Akinyele Fabiyi, Lagos, 1993

■ Uwe Brinkmann's head suddenly jerked up, he looked out the car window desperately searching for a familiar landmark along the back roads of northern Germany, and panic fell over him, the true deep panic that comes only when something taps into one's innermost fears.

"Now they've got me," Brinkmann cried out in his mind. "Now I'm going to be gassed. They're taking me to the concentration camp."

The sides of the van seemed to be closing in and Brinkmann had no idea where he was being taken. As he looked out the window in the summer of 1974, the city and suburbs of Hamburg gave way to countryside and then, he noted with fear, the woods.

Sealed off from the anonymous driver and the entire outside world, Brinkmann and his companions stared ahead in shared terror. Brinkmann's patient, German surgeon Bernhard Mandrella, lay in subdued delirium on a stretcher between them. To his side sat British physician Adam Cargill, who had treated Mandrella in Nigeria. With the men were three Nigerian women; a nurse/nun and two nurse assistants.

"Nobody wanted us here to begin with," physician-scientist Brinkmann thought. "So now they are going to get rid of us. All of us."

As claustrophobia overwhelmed him in their tightly sealed mobile unit, Brinkmann later said he was thinking, "How odd. I have spent the last seven days exposed to a lethal microbe, feeling no fear. Now it is people —*my* people—that terrify me."

But Brinkmann had never felt completely at home among his fellow Germans. His part-English family had carefully hidden the Jewish identity of Brinkmann's grandmother throughout the years of the Third Reich, and young Uwe had made a career during the counterculture days of the 1960s out of crafty troublemaking. Similarly, his patient was the son of one of the military officers who tried to assassinate Adolf Hitler on July 20, 1944. Mandrella's father was executed, and his mother was billed by the Third Reich for the cost of the hanging.

"But this time it's really too much, even for me," Brinkmann thought. Taking stock, he could visualize how this group looked to German officials: three African women whose own government had sent them into isolation, an English physician who was suffering a suspicious case of diarrhea, a man apparently dying of a lethal contagious disease, and himself—a hippie troublemaker. He considered his black shoulder-length hair, thick unkempt mustache, tie-dyed T-shirt, and bell-bottoms. And he recalled headlines in German publications just weeks ago that denounced his famine-relief efforts in Ethiopia; Brinkmann was, the German press declared, creating communist communes in the deserts of the Horn of Africa. Though Ethiopian Emperor Haile Selassie had awarded Brinkmann his nation's highest commendations and requested that the young hippie doctor remain indefinitely in the country, the German government recalled Uwe. It seemed that Brinkmann's solution to Ethiopia's ongoing food crisis—establishing village-based communal farms and produce-marketing apparatuses—was a little too left-wing for the conservative West Germans.

"Yes," he thought in those seconds of panic, "it makes sense. They will simply eliminate us and tell the world we died of the disease. That will take care of everything."

There was little to comfort Brinkmann when the caravan reached its destination. Just outside the village of Ebstorf, in the woods some fifty miles south of Hamburg, was an abandoned medieval monastery, recently converted to a high-security facility for smallpox containment. A series of three automatic air-lock doors opened for the anxious group, quickly shutting behind to seal them off from the rest of humanity.

Inside were several sleeping rooms, an autopsy laboratory, and research facilities. Sophisticated research devices rested atop aseptic surfaces.

Though there was an autopsy facility, there was no place for patient treatment.

The group settled in as best they could, but tensions were high.[1] The Nigerian women had never set foot outside of the Jos region in which they were born before Mandrella came down with the dreaded disease. Then, having tended to the physician in their Jos hospital, the nurses accompanied the patient to the University of Ibadan Hospital, one of Nigeria's premier medical facilities.

Mandrella's problems began two weeks earlier when his colleague, Dr. Egon Sauerwald, was treating a patient from the old colonial city of Enugu in their St. Charles Mission Hospital, miles away in Borromeo. The patient had high fevers, chills, muscle aches, and a sore throat. Despite Sauerwald's efforts, the Enugu man died. Days later, the twenty-nine-year-old doctor developed the same symptoms, quickly descending into acute disease.

Mandrella did everything possible to save his colleague from the mysterious ailment, but Sauerwald continued to deteriorate. He sent Sauerwald's blood samples to Ibadan, from where they made their way to CDC laboratories in Atlanta. Word eventually got back that Sauerwald was infected with the recently discovered Lassa virus, a microbe that the U.S. Centers for Disease Control said was "thought to have a unique proclivity for killing doctors and nurses."[2]

The news was too late for Mandrella. By the time word got back that Sauerwald was infected with an exotic lethal virus, the thirty-three-year-old Mandrella had already performed a last-ditch bloody procedure to save his friend. The virus had so devastated Sauerwald's throat that the doctor couldn't breathe, so Mandrella made an incision into his friend's trachea, creating an air hole in his neck. Mandrella was unprepared for the sudden gust of mucus that flew from his friend's throat. He was instantly spattered with Sauerwald's blood. Though he pulled away quickly, Mandrella's face had been very close to Sauerwald's neck as he made the incision, and the surgeon inhaled microscopic bits of blood and mucus.

Mandrella was infected, and in a couple of days he too was shivering with Lassa fever. Still unaware of the CDC laboratory findings, Mandrella saw Dr. Hal White, an American physician who ran the missionary hospital in Jos. White examined Mandrella and warned the young doctor that the symptoms looked suspiciously like those of Lassa fever. As a precaution, White injected Mandrella with a unit of sera donated years earlier by nurse Lily ("Penny") Pinneo. At White's advice, Mandrella immediately drove to the metropolis of Ibadan, where he came under the care of British physician Adam Cargill of the University Hospital.

Nigerian health officials reacted with considerable alarm. They had already had their fill of this terrible disease, which was named after the village of Lassa, located southeast of Jos, in the Yedseram River valley that runs along Nigeria's eastern border with Cameroon. In 1969 an outbreak of the disease in Lassa had brought it sharply to Western attention for the first time when American nurses fell ill in the town's Church of the Brethren Mission Hospital.

It was a long chain of events, stretching back five years, that brought Mandrella, Brinkmann, and their colleagues to this moment of panic in Germany.

On January 12, 1969, a sixty-nine-year-old mission nurse began complaining of a sharp backache. Laura Wine told her colleagues the pain was increasing as days went by, but assumed she'd done something to wrench her spine. Perhaps the daily rounds of bed changing and turning patients were the cause, she thought.[3]

After a week, however, the nurse also had a throat so sore she couldn't swallow, and her colleagues saw ulcers lining her pharynx. Assuming she was suffering from some bacterial infection, such as streptococcus, the hospital staff gave Wine penicillin.

But the antibiotics did no good. Wine's state escalated; fevers of 101°F, acute dehydration, unusual blood-clotting activity, a complete lack of proteins in her urine—these and other symptoms signaled that the woman was

WEST AFRICA

suffering from something wholly unlike the multitude of tropical diseases tolerated by residents of the grassland Yedseram River valley.

Over the next four days Wine began to swell, her skin showed signs of hemorrhaging, her heartbeats became irregular, she grew disoriented and was unable to speak properly.

On January 25 volunteer pilots flew Wine to Jos, rising from the hot grasslands at sea level up to the 4,000-foot-high town of Jos. As they made their journey the air cooled, the humidity dropped, and the tin mines around Jos came into view.

Jos itself was inhabited by some 12,000 people, a large percentage of whom were European expatriates seeking refuge from the heat and malarial mosquitoes of Nigeria's lowlands. Members of all three of Nigeria's leading tribes—Hausa, Ibo, and Yoruba—lived in Jos, and the community had come through the tragic Biafran war fairly unscathed. Though tens of thousands of Nigerians died in the civil warfare of 1967–68 and thousands more were uprooted from their homes, Jos suffered only twenty-four hours of rioting and killing during that period.

Dr. Jeanette Troup and nurse Lily Pinneo greeted Wine at the Jos landing field. Because radioed descriptions of Wine's illness seemed to indicate cardiac problems, the pair immediately strapped an oxygen mask on the ailing nurse and rushed her to their Bingham Memorial Hospital emergency ward. There, Troup and her staff did everything possible to save Wine's life.

They failed. A day after arriving in Jos, Wine went into horrible convulsions and died.

Three days later, a Jos hospital nurse who had tended to Wine felt chills, a headache, and dull pains in her back and legs. Forty-five-year-old Charlotte Shaw had gently dabbed Wine's bleeding mouth with a gauze pad. When she too fell ill, Shaw remembered she had a tiny rose thorn cut on her finger—the very finger she had used to push the gauze along Wine's mouth.

Soon Shaw was experiencing the same symptoms that had claimed her patient: fevers, rashes, hemorrhaging, pains, swellings, heart irregularities. After eleven days of illness, she died.

That night Dr. Jeanette, as she was called, performed an autopsy, assisted by her head nurse, fifty-two-year-old Pinneo.

Pinneo, a Presbyterian missionary, had followed Shaw's progress carefully, monitoring her lab results every day. Shaw and Pinneo had been close friends. As she donned her gown, gloves, and mask to assist in the autopsy, Pinneo thought, "How can I do this? How can I possibly face opening her up?"

Troup and Pinneo gasped when they saw the devastation; every organ of Shaw's body was seriously damaged. The heart was stopped up, with loads of blood cells and platelets piled well into the arteries and veins. Fluids and blood filled the lungs. Dead cells and fat droplets clogged the liver and spleen. The kidneys were so congested with dead cells and proteins that they had failed to function. When the team cut open Shaw's lymph nodes they discovered with some shock that absolutely no lymphocytes—disease-fighting white blood cells—were inside. The nodes had been completely emptied.

A week after assisting in the autopsy, nurse Pinneo also fell ill. This time the medical staff took the case seriously, admitting their colleague to the hospital with the first signs of fever.

It was February 21, 1969, and panic began sweeping over the Jos hospital as Pinneo's colleagues stood by helplessly and watched their friend deteriorate from her early symptoms of a mild fever, reddened tonsils, and some liver tenderness.

Hoping Pinneo had a bacterial infection, Dr. Troup gave the nurse huge injections of penicillin. But the antibiotic proved useless, and on February 26 Troup contacted Dr. John Frame at Columbia University in New York. Frame was a tropical disease expert and director of medical services for Sudan Interior Mission (SIM), which operated a chain of Christian hospitals in West Africa.

Frame saw no alternative: it was imperative that Pinneo, along with blood and tissue samples from her dead colleagues, be flown to New York immediately. While Pinneo was en route from Jos to the Nigerian capital of Lagos, Frame contacted laboratory scientist Jordi Casals at Yale University.

As early as 1955 Frame had been hearing reports of strange illnesses among the mission hospital staff and members of their families. That year eight children of Nigerian missionaries suffered high fevers and convulsions. Though the children survived, they all had some degree of permanent brain damage.

With odd disease reports continuing in subsequent years, Frame hatched the idea of using missionaries as epidemic early-warning systems.

Though most of his colleagues espoused the 1960s concept of a health transition, Frame wasn't at all convinced it was time to close the book on infectious diseases; he had reviewed too many strange medical charts forwarded from nurses and doctors in the field.

In the mid-1960s, Frame met with Wilbur Downs, director of the Yale arbovirus laboratory and Jordi Casals's boss. They decided to test the blood of all missionaries who had recently suffered unexplained prolonged fevers.

Casals screened the blood of sixty-five such cases and was able to ascribe illness in half the missionaries to one or another virus he had in his vast Yale collection.

But the inability to find an explanation for the illnesses of thirty-two cases seemed to prove Frame's point: there were a lot of microbes yet to be discovered.

In 1968 a screening system was put in place, and Troup and other mission hospital directors were told to send blood specimens from all mysterious cases to Columbia University. After a superficial perusal, Frame would pass the blood samples on to Casals for detailed analysis. The 1969 illness of Pinneo was the first real test of the system.

As had been the case two years earlier with Marburg, Casals's expertise in identifying mysterious viruses was needed. Casals agreed to accept the blood and tissue samples from the nurses, and told Frame, "It will just be a matter of routine to identify this agent."

When she arrived in Lagos, Pinneo was dangerously weak. Nigerian and American officials could not reach agreement on how best to transport the ailing nurse to New York, so she was placed in isolation for four days in a small shed near the Methodist Hospital.

Pinneo was tended to by close friend Dorothy Davis, also a nurse. Authorities in Lagos placed the women in the pesthouse, a shed so full of mosquitoes that Davis was forever waving her arms frantically about in hopes of keeping the insects from biting her friend. Both Pinneo and Davis had already noticed that even the tiniest of mosquito bites seemed to bleed for several minutes: for some reason, Pinneo's blood wasn't clotting.

During their first night in the pesthouse temperatures in Lagos soared into the nineties—hotter still inside their tin-roofed hut. Beside Pinneo lay a tiny measles-ridden baby who was also fighting for its life. During the night the baby died, and its mother rocked back and forth for hours, wailing and sobbing.

Fitfully, Pinneo went in and out of delirious sleep.

For four long days Davis fretted over her friend and, knowing what had happened to Shaw after tending to Wine, couldn't help but worry some about her own health. Both devout Christians, the nurses prayed for a miracle.

Nigerian and American officials finally agreed on a way to transport Pinneo to New York after CDC investigator Lyle Conrad, who was coincidentally in Nigeria, intervened. He negotiated air passage for Pinneo aboard a Pan Am commercial jet, placing her, along with Davis and himself, in the otherwise empty first-class section. In the seat beside Pinneo was a box containing blood and brain tissue samples removed from the bodies of her friends Laura Wine and Charlotte Shaw.

During her long flight Pinneo languished apathetically on her stretcher. Though she was fighting for her life, Pinneo's outward appearance was that of an exhausted traveler, just a tad more wearied by jet lag than were most passengers.

But inwardly Pinneo was reeling. Every defensive weapon her immune system could muster against the rapidly growing virus population was coursing through her bloodstream to pockets of confrontation. From her lymph nodes to her liver the battle raged.

Meanwhile, Pinneo's lassitude was striking—the result of having suffered a 101°F fever for six days. She stared blankly ahead and thought, "I don't have time to be sick. I have so much to do. I need to fulfill the Lord's will."

But she also trusted the Lord. "If He wants to take me away, it would be all right."

Just after midnight Pinneo was admitted to Columbia-Presbyterian Hospital, where she was placed in isolation inside a glass-walled room under twenty-four-hour direct observation by intensive-care nurses.

When Frame arrived the following morning, he found the medical staff highly agitated, some clearly fearful. He tried to reason with Pinneo's physician, asking, "What would you do if this were a case of pneumonic plague?"[4]

"I'd die of fright!" was the answer.

Frame, searching for a middle ground between the terror of the Black Death and the casualness inspired in hospital staff by routine bacterial infections, suggested that the physicians and nurses take precautions appropriate for handling scarlet fever. All personnel who attended Pinneo wore protective gloves, masks, gowns, and foot coverings and operated under strictest disease control procedures.[5]

Pinneo was exhausted, severely dehydrated, and feverish. Her temperature was 101.2°F, her muscles ached, and her abdomen was tender. But she was often alert and able to assist the Columbia team with useful answers to their medical queries. Some members of the team were, in fact, so impressed by Pinneo's lucidity and normal heart functions that they expressed the hopeful belief she had already seen the worst and would soon recover from the mysterious ailment.

All hopes were dashed hours later when Pinneo's fever skyrocketed to 107°F and her throat filled with lymphatic fluids. By March 6, Pinneo could no longer eat or swallow because her throat—the esophageal lining—was aflame with florid infection. The worried medical staff noted that Pinneo's face and neck were swelling. Her lungs and chest were also filling with fluids, and X rays showed evidence that some organisms had invaded the linings of her lungs.

Pinneo grew weaker. She became completely apathetic, no longer evidencing a will to fight the disease. On top of everything else, Pinneo developed malaria on March 7, undoubtedly due to a latent *Plasmodium falciparum* infection that was activated by the devastation of her immune system.

Samples of the fluids from Pinneo's throat were sent off to Casals in New Haven, along with the brain and blood samples from her dead colleagues.

Meanwhile, Pinneo deteriorated further. By April 1, she lost control of her eye muscles, and her eyes began jiggling about uncontrollably in their sockets. Muscles all over her body were similarly subject to tics and trembling. A brain scan indicated the mystery virus was attacking her central nervous system.

Ironically, whenever Pinneo was momentarily lucid she seemed more concerned about the fate of the nurses around her than about her own sorry state. "Oh, look at the poor dears, fumbling with those big rubber gloves.

It's so hard to make a bed with those things on," she thought. Each time the nurses had to administer a local anesthetic to Pinneo's ravaged throat before giving her a pill, Penny apologized. She was similarly contrite when day after day she remained unable to swallow food, and had to be maintained on intravenous fluids that required close attention from the hospital staff.

Casals had no immediate answers. Whatever virus was tormenting Pinneo did not match any hemorrhagic agents in his Yale collection. He did know one thing: this organism was unusually tough.

Miraculously, Pinneo began to recover by mid-April, and was well enough to walk—albeit at an odd rightward slant—out of her hospital room. On May 3 she was discharged from the hospital, but she continued to suffer severe headaches, dizziness, and vertigo until the end of the month.

The doctors, in both Nigeria and New York, were at a loss to explain what had happened. The best guess was that Marburg disease was responsible, but Casals could find no evidence of Marburg in any of the nurses' blood samples.

Back in New Haven, Casals continued his search, working under state-of-the-art precautionary conditions. During the four days it took Pinneo to travel from Jos to New York, the dry ice in which the samples were packed had completely sublimated, and the blood was exposed to the hot, humid Nigerian climate. Nevertheless, the hardy viruses arrived intact. To protect his staff, Casals insisted that only he would attend to the mice that had been experimentally injected with the mystery agent. The rodents were kept in a special airtight room, which Casals never entered without first donning a mask, goggles, and gloves.

Day after day, Casals injected samples of the nurse's mystery microbe into test animals, searching for clues to the identity of the agent. His lab also grew cells from African green monkeys, called Vero cells, in petri dishes, poured in microbe-contaminated fluids, and watched the results. The final quick test they used mixed antibodies against Marburg and other viruses into test tubes full of Pinneo's blood.

When they studied their results under powerful microscopes, the mystery for Casals only deepened. None of his vast range of antivirus antibodies —including those that usually attacked the Marburg virus—seemed to lock on to the mysterious microbe. Antibody molecules are very specific; for scientists like Casals, an antibody/virus complementary relationship was as reliable a clue as a detective's discovery of a culprit's possession of the sole key to a safe-deposit box full of the victim's diamonds.

But none of the antibody "keys" in Casals's vast collection fit the "lock" of Pinneo's mysterious virus. Casals and colleague Robert Shope screened over 200 viral antibody types against the microbe before concluding it was "something new."[6]

Equally disturbing were the microscopic clues provided by studies of infected Vero and rodent cells; the enigmatic agent just didn't look like

Marburg, or any other pathogen with which Casals was familiar. Working with ace electron microscope expert Sonja Buckley, Casals searched for recognizable attributes of the microbe. They magnified their samples over 100,000 times to be able to visualize the tiny killers: what they saw were perfectly round balls or spheres, from which projected dark spikes.[7] Inside the balls was the genetic material of the virus. Marburg, by contrast, was a long, thin, fuzzy virus that often coiled up into a tight spiral.

The two viruses just didn't look alike.

Worse yet, Casals concluded that this mystery microbe was lethal as hell. When Buckley diluted samples to a ratio of one drop of Pinneo's blood to ten million drops of benign fluid, the concoction still killed half the Vero monkey cells she grew in her petri dishes within eight or nine days. Even Marburg didn't do that.

In late May, as Pinneo was recovering, Casals began working on his presentation for the upcoming WHO conference on Marburg disease. He was living in an apartment on Manhattan's Upper West Side and commuting daily to New Haven. Preferring to do his writing at home, Casals gathered up papers around his office on June 3 and prepared to head into New York. But suddenly Casals felt a dramatic shivering sensation all over his body that lasted over an hour. He took two aspirin and it went away. The following morning he awoke to make his journey back to New Haven and felt lousy.

On Saturday, Casals awoke to pain in his thigh muscles that was startling in its severity. "I never knew a muscle could hurt that much," he thought, contemplating the scientific significance of the matter.

He tried to get out of bed and walk, figuring he'd shake it out, but was startled to find the pain increased and his legs were almost too weak to hold up the rest of his body.

Dazed, Casals went back to bed, and soon found himself watching the hours slip by without the slightest concern. He felt oddly lethargic and apathetic.

"Probably just the flu," he stubbornly concluded, despite symptoms that he knew indicated something else.

Casals's family had been away for the weekend, and when they returned on Sunday evening, mother and daughter were aghast. Jordi Casals had eaten nothing. *The New York Times* lay unread on the floor beside him. The usually energetic scientist seemed absolutely apathetic, as if he had decided to abandon himself to fate. He was extremely confused and kept mumbling something about the flu.

"This is no damned flu!" his wife declared, quickly calling their family physician, Edgar Leifer. The doctor arrived swiftly, ruled out influenza immediately, and whisked Casals off to Columbia University's Presbyterian Hospital.

The journey proved painful for Casals, as did several hours of waiting in hospital hallways for X-rays and other tests. While physicians worried about how best to limit contagion to others when using cumbersome medical equipment on Casals, the glass-booth isolation unit was prepared.

For the remainder of June and much of July, Casals lived in a glass room specially ventilated with negative pressure and air-lock doors. Only essential medical personnel, adorned in protective gowns, gloves, goggles, and masks, were allowed inside Casals's little world.

Casals was acutely ill. His temperature was 104°F, his blood pressure high, pulse weak, skin flushed, and red and white blood cells were draining at an alarming rate out of his cardiovascular system and into his urine.

The virus was attacking his heart, throat, and veins.

Though Casals continued to mumble that he was probably okay, it was just a little cold, the hospital staff knew he was fighting for his life, in the grip of Lassa fever.[8]

It was a long shot, but Leifer solicited the help of the recovered Penny Pinneo.

"We need your blood," the doctor said, explaining that he hoped Pinneo's blood contained antibodies that could destroy the Lassa viruses then attacking Casals. Pinneo readily agreed. Meanwhile, Casals's boss, Wilbur Downs, called Karl Johnson to ask what had been the Bolivian experience with trying sera from a Machupo survivor as treatment for a victim of the disease.

"Works," Johnson said in his usually abrupt way. "But you better hurry. The longer a patient is sick, the less helpful the immune sera gets to be."[9]

On the fourth day of his hospitalization, having deteriorated quite dangerously, Casals was injected with 500 milliliters of Pinneo's plasma.

"It's miraculous," Casals said the following day as his fever dropped to 101°F and his mind began to clear. With each passing day thereafter Casals gained strength, attained a sense of will while shedding his prior apathy, and felt the muscle pains diminish. Within a week his temperature and cardiovascular signs were normal.

After thirty days of hospitalization, a sadder but wiser Casals went home.

The sadness stemmed from Casals's inability to explain how he had become infected. Certainly he had taken every possible precaution in the Yale laboratory—precautions that had proven quite adequate for the hundreds of other bizarre viruses he studied. It was clear the Lassa agent was especially dangerous.

Over and over again Casals scanned his lab notes trying to pinpoint an error, a moment when he carelessly allowed the occurrence of contact between himself and the virus. Only two possibilities were open. Because Casals insisted that only he have contact with infected mice, the scientist thought it possible that the animals had urinated viruses on the sawdust and wood chips that lined their cages, and somehow these wood particles had been kicked up into Casals's breathing space by the agitated rodents. But he always wore a mask: for such a scenario to work, the mask that had protected him against over a hundred other viruses would have failed with Lassa.

An alternative explanation lay in the dried cracked skin of Casals's hands, full of microscopic holes that might have served as portals of entry for the tough virus. But Casals always wore thick latex gloves and could not remember noticing any leakages. Could the virus get through rubber? Had he unknowingly worn a faulty product one day that was full of microscopic leaks?

The frustrating puzzle remained unsolved, even twenty-five years later, much to the consternation of Yale officials who were less than enthusiastic about having research on mysterious lethal viruses conducted on campus.

Casals reminded inquiring university and Rockefeller Foundation officials that he had, without prior incident, conducted years of successful and safe research in the Yale facility and, earlier, at Rockefeller laboratories in New York City. Ever since the young Catalonian doctor had been stranded in the United States by the Spanish Civil War in the 1930s, Casals had devoted most of his waking hours to studying a succession of deadly viruses: polio, Japanese encephalitis, rabies, St. Louis encephalitis, Junín, Machupo, LCM (lymphocytic choriomeningitis in mice), dozens of hemorrhagic viruses found in people and monkeys, and a host of mosquito-carried South American agents he had discovered.[10]

Indeed, ever since he and Karl Johnson had traveled all over the Soviet Union investigating strange hemorrhagic diseases, the pair had discovered that the real danger was not the viruses, but politics. In the spring of 1965 for example, they joined Soviet colleagues and a handful of other top American researchers for a monthlong investigation of four different types of viral bleeding syndromes, at least three of which were found exclusively within the borders of the Soviet nation: Omsk hemorrhagic fever, Crimean-type fever, and Central Asian hemorrhagic fever. The trip proved immensely useful for all the scientists involved, and several more exchanges followed over the years.[11]

But every time Casals and Johnson returned from the Soviet Union, they were hounded by CIA agents who expected the scientists to reveal all that they had seen and discussed in the communist nation. Casals always obliged, but by 1969, as he argued with Yale officials about the continuance of his Lassa work, he was growing impatient with all the inquiry.

In the summer of 1969 a recovering Casals pursued his Lassa research vigorously, aware that university authorities were debating whether to shut it down. In short order he verified that the disease that had nearly killed both him and Pinneo was caused by an apparently new virus.[12] He further showed that the virus's genetic material was in the form of RNA, rather than the DNA present in human cells. Using techniques similar to those Johnson and Webb had followed for purifying Machupo viruses, he was able to isolate Lassa microbes from all the Nigerian samples, as well as his own taken during hospitalization, including throat swabs, blood, and urine. His lab ruled out the possibility that Lassa was carried by a common African mosquito because they were unable to experimentally infect such insects. On the other hand, they pointed an accusatory finger at the rodent world when they showed that experimental mice could be infected with Lassa and none became ill; rather, the rodents served as lethal carriers of the disease.

The strange virus failed to cross-react with any of the hundreds of agents in Casals's viral library, which ranged from the nonpathogenic Tacaribe to 100 percent lethal strains of rabies viruses, and when he tested his own blood against all those viruses he found that his anti-Lassa antibodies reacted only against the new virus. In other words, Casals was immune to Lassa, but the immunity did not overlap to protect him against any other virus. That was clear evidence that Lassa was, indeed, unique.[13]

Most disturbing, Casals concluded that the virus could be spread in four ways: by inhaling viral particles from an infected human or animal, by contact with contaminated urine, by direct blood-to-blood contact with an infected person, or by some less clear method involving laboratory mice.

In the fall of 1969 Casals was forced to accede to the Yale authorities' concern about the safety of Lassa research, and all samples of the virus were shipped to the Atlanta headquarters of the Centers for Disease Control, where they were studied and maintained in a uniquely designed high-security facility.[14] What prompted Casals to agree that it was wise to cease Lassa studies was the tragedy of Juan Roman, a fifty-five-year-old technician in the Yale laboratory. Roman had decided to spend the Thanksgiving holiday that year with cousins in York, Pennsylvania. The Puerto Rico-born assistant had never been involved in Casals's Lassa research and was strictly forbidden (as were all people at the Yale facility, save Buckley and Casals) to touch anything labeled "Lassa."

On Wednesday evening, Roman left New Haven, apparently feeling fine, and made the drive to York. By Friday he was severely ill, suffering all the classic Lassa symptoms: fevers, shivering, muscle pains, severe fatigue, and lethargy. He was admitted to a local hospital, where he was treated without special contagion precautions for a week before his stymied physician called the Yale arbovirus laboratory to ask whether Roman had handled any strange viruses.

Casals made a worried visit to York that Saturday and found his technician desperately ill. After warning the hospital staff about the need for high-security contagion precautions, he returned immediately to Yale and began preparing samples of his own blood to use as antiserum for the dying laboratory technician. By the time he had arranged to have Roman transferred to Columbia-Presbyterian Hospital in New York, it was too late.

Roman died Monday morning, after just ten days of illness. He never had a chance to try Casals's antiserum.

Though Casals and Yale authorities went over Roman's notes and activities for hours, searched every inch of the laboratory for an improperly labeled tube or dish, and hunted for leaks in the ventilation systems, they were never able to explain how the technician got Lassa. Concerned that panic would spread through New Haven, particularly amid the Vietnam War protests and student suspicions that biological warfare research was being conducted at all the nation's high-security laboratories, Yale and Casals agreed it was time to get rid of the Lassa samples.

Christmas 1969 found Pinneo and Casals happy to be alive, sharing the company of their families. It was possible to look at the glittering lights strung along Manhattan's avenues, feel the crisp night air filled with the promise of winter snowfall, and completely forget the tropical menace that so nearly claimed their lives. Casals blessed his hearty constitution, and Pinneo mused often during that season's church services about her great luck. Her ears still rang all the time due to some damage left by the virus, but she was alive and her energy was slowly returning.

The season's sense of joy quickly dissipated after the New Year when patients started pouring into Pinneo's former hospital in Jos, now renamed Evangel Hospital. In three weeks' time, Dr. Jeannette Troup treated seventeen cases of what looked like Lassa fever.[15] Panic quickly spread among the Evangel staff, and at the weekly prayer meeting of January 21 the reading was from Psalm 91:

> *Thou shall not be afraid for the terror by night;*
> *nor for the arrow that flieth by day;*
> *nor for the pestilence that walketh in darkness.*

Though she suspected Lassa was the culprit, Dr. Jeannette decided to perform one autopsy to confirm the diagnosis. On January 25, 1970, the petite Troup, who sported a Lady Bird Johnson hairdo, glasses, and the conservative cotton dresses suitable for the tropics, stepped up to the autopsy table. Acutely aware of the risks, she took a deep breath and made her first incision.

Minutes later, she cut herself, drawing blood through her protective gloves. Though she insisted to others at the time that it was "just a nick," Troup was terrified.

For good reason, as it turns out. Ten days later Dr. Jeannette told colleagues she had the flu. On February 10 she was admitted to the hospital with a fever of 103.8°F.

As Troup's condition worsened, fear spread in the Jos medical community and her colleagues notified Frame at Columbia University. He, in turn, set into motion plans to fly Pinneo and Casals to Jos. Though both were still suffering Lassa virus-related symptoms, Frame was confident they were now immune to the disease. That made them ideal Lassa investigators.

Unfortunately, civil war once again raged in Biafra, and the Nigerian government—which had never been pleased about Frame's decision to name the deadly disease after one of its towns—delayed granting visas to

Casals and Pinneo. As days dragged by, filled with cables describing Troup's deterioration, the two Lassa survivors became increasingly anxious.

In a frantic effort, Frame shipped Pinneo's antiserum via U.S. diplomatic channels, but it was mistakenly routed to Ibadan, miles from Troup's Jos deathbed. The antiserum reached Ibadan on February 15, and was carried from there to Jos by Pinneo. She reached Jos on February 20.

Jeannette Troup, however, died on February 18.[16]

On March 3 Casals arrived, and a team of five researchers—including Troup's assistant Dr. Harold (Hal) White and Pinneo—was assembled. Pinneo's fluency in Hausa, as well as the great respect she had garnered over the years from the people of Jos, proved invaluable.

After weeks of investigation the team was unable to say from where, exactly, the virus had come, but it could explain the dramatic spread of Lassa inside Evangel Hospital and nearby Vom Christian Hospital.[17]

Tracing back the cases, the team decided that it began with a woman who traveled from Lagos, Nigeria's huge metropolis, to her home village of Bassa during September 1969. There, she gave birth.

Forty days later, on Christmas Day, the woman developed symptoms of acute viremia and was put in a general ward of Evangel Hospital. Throughout her hospitalization the woman's newborn and three-and-a-half-year-old child stayed beside her, and she was closely tended to by her mother and brother-in-law.

By mid-January the woman was well enough to return to Bassa, but shortly after the family reached their home the older child died and the woman's mother fell ill.[18] Altogether, twenty-eight people suffered Lassa fever in the two hospitals between Christmas 1969 and February 26; thirteen died. With the exception of Jeannette Troup, all were Nigerian.

At least sixteen people got Lassa from the first Bassa woman, though most never had physical contact with the patient. Much to the embarrassment of Evangel physicians, it seemed most infections actually occurred on the A Ward of their hospital. There, the Bassa woman struggled with her fever for two weeks, lying on a corner bed beside an open window. The prevailing breezes carried her exhaled viruses down the ward, past the noses and mouths of four patients, six visitors to the ward, and four hospital employees, all of whom developed Lassa. The infected, in turn, passed the virus on to family members after leaving the hospital, indicating it was possible for Lassa survivors to carry the virus for two or more weeks.

Searches throughout Bassa failed to find the source of the epidemic.

For Frame, who was responsible for the health and safety of SIM missionaries in West Africa, the second Lassa outbreak was particularly worrisome for three reasons: it occurred primarily among Nigerians, indicating there might not be natural immunity in the population; the outbreak clearly spread as a result of hospital procedures; and the source of the disease remained elusive.

In 1970 Frame collected blood samples from 712 current and recent

American missionaries working in West Africa and had them tested for evidence of past Lassa virus infection.[19] Five tested positive, four of whom recalled having suffered long, unexplained fevers. Only one member of the group, Harry Elyea, had been ill while in Nigeria. In 1952, Elyea spent a month in Rahama, Nigeria, severely ill. As a result, he suffered a lifelong hearing deficit.[20] Hearing loss would prove a common side effect of Lassa infection. Twenty-five years after her brush with death from Lassa fever, Pinneo would still have a constant ringing in her ears.

The other four cases were missionaries who fell ill between August 1965 and February 1968 in far-off Telehoro, Guinea. Sixty-one-year-old missionary Carrie Moore was rendered stone-deaf by her illness, having suffered total auditory nerve destruction.

Frame's group also tested blood samples that had been collected in 1965–66 from villagers in northern Nigeria as part of a parasite survey; 2 percent showed signs of previous Lassa infection. The scientists suspected Lassa was hiding inside some ubiquitous species of African animal and might well be infecting human beings all over West Africa. With so many fever-producing diseases to worry about in the tropics, it was to be expected that the occasional Lassa case would go unnoticed.

They warned that Lassa would undoubtedly strike again.

It was just months before their prediction proved sadly accurate. Not in Nigeria, nor Guinea, but Liberia.

Garbazu, a Liberian peasant, was four months pregnant when she started bleeding profusely. Though she had been nauseated and sick for a week, she tried to treat herself in her village of Zigida, using traditional antisorcery methods of herbs and incantations. The uterine bleeding, however, scared her enough to prompt a twenty-seven-mile journey to the Curran Lutheran mission hospital in Zorzor.[21]

On March 3, 1972, Esther Bacon, a missionary nurse-midwife from Colorado, performed an emergency dilation and curettage after Garbazu spontaneously aborted twins. Throughout the procedure, Garbazu bled an inordinate amount, hemorrhaging so severely that Bacon's nursing gown soaked through, as did her cotton dress beneath, leaving her torso drenched in Garbazu's blood.

After a few days, Garbazu recovered and went home to Zigida, but two other women with whom she had shared the ob-gyn ward developed similar symptoms: nausea, mouth sores, fevers. By mid-March, Bacon was ailing, as were other members of the hospital staff. At month's end, five patients (including an ob-gyn patient's newborn baby) and eight hospital staff members were sick,[22] and, as rumors spread of Lassa fever, panic erupted in Zorzor. Travelers zoomed through the town, their car windows shut tight against imagined plague. And residents of neighboring towns became dangerously agitated.

The fear was compounded by a very special concern for Bacon. Since her arrival in Zorzor in 1941, the energetic midwife had personally revolutionized health care in the region, and eventually throughout Liberia, by creating a vast infrastructure of trained midwife assistants and prenatal screenings. Twice she was awarded Liberian presidential medals for her years of walking hundreds of miles to far-flung villages to convince women of the wisdom of delivering their babies in clean hygienic settings, assisted by trained personnel.

Located in the high-country tropical forestland of eastern Liberia, close to the Guinea border, Zorzor was a remote area where sorcerers usually resented the intrusion of all forms of competing providers of medicine, yet Bacon managed to gain respect even in villages located so deep in the rain forest that they could only be reached by horseback or on foot.

When word spread of Bacon's illness, prayer meetings occurred spontaneously throughout the region and the fear of Lassa fever grew. People wondered, "What can be so terrible that it can kill nurse Bacon?"

On Good Friday, March 30, Bacon was carried from her home to a small landing strip, where she would board a prop plane bound for the somewhat more sophisticated Phebe Hospital. As her stretcher-bearers wended their way to the airstrip, the people of Zorzor lined the path, some crying out, others sobbing.

Samples of blood drawn from Bacon and other ailing hospital staff members[23] were forwarded to the CDC in Atlanta, and health officials in Liberia, as well as WHO headquarters in Geneva, were alerted. Once again, Penny Pinneo, who had remained in Jos, was asked to donate her services and her blood. Now convinced that God had intended her to survive Lassa for just such a purpose, Pinneo enthusiastically complied.

Though Pinneo brought two units of antibody-containing plasma, the blood proved useless for Bacon. As Karl Johnson had predicted three years before, antisera was effective for viral hemorrhagic diseases only when given early in the illness.

On April 4, Esther Bacon died.

When Esther Bacon fell ill, Tom Monath was just finishing up his research fellowship in virology at the University of Ibadan in Nigeria. The thirty-year-old Bostonian had developed a fascination with insect-carried viral diseases during his medical studies at Harvard, and went to Nigeria to study yellow fever. By March 1972, his work was done, crates were packed, and Monath was already imagining sinking his teeth into a nice juicy all-American cheeseburger when a cable arrived from the CDC.

"Go to Liberia," it read, noting there were reasons to suspect Lassa fever had broken out in a tiny mission hospital. He was instructed to link up with Penny Pinneo and get to Zorzor as fast as possible. His job was to find out where Lassa came from and how it was spread.

For a moment Monath just stared at the cable, struck by a flash of fear. "This is pretty terrifying," he thought. "Nobody really knows anything about this. It's highly contagious, and half of the people who got it in Jos died."

But hours later, as he unpacked gear he had already prepared for shipping to the United States, Monath deliberately downplayed the assignment, telling his wife it was "no big deal, probably routine."

Pinneo and Monath made their way from Lagos to the Liberian capital of Monrovia, and then took a tiny light-wing plane to Zorzor. The moment they landed the pair could feel the pall over the community. Nobody was walking along the roads, and travelers refused to make their usual stopovers for gasoline and food in Zorzor.

"The air smells like fear," Monath thought.

Bacon was still alive—barely. Several hospital beds were occupied by other Lassa patients, and the staff was frankly stupefied by it all. Pinneo and Monath met with the hospital director, Dr. Paul Mertens, and Tom laid out a battle plan.

"First, we go to Zigida, track down Garbazu, and try to figure out where she got the virus," Monath said. While Pinneo and Monath searched the village, Mertens would try to determine how the virus spread within his hospital. Because she was immune, Pinneo agreed to be in charge of drawing and handling blood samples.

Before he left Zorzor, Monath grabbed every mousetrap and net he could find. Though all his training was in insects, Monath knew that Junín and Machupo were rodent-carried diseases, and circumstances just didn't point to insects in either Jos or Zorzor; if insects had carried the disease the cases would not have all occurred indoors or been primarily among adults. As a rule, insect-carried diseases attacked children more than adults because youngsters tended to play outdoors in watery or wooded areas where they came in contact with mosquitoes, mites, spiders, and such.

But in Zorzor only one child had Lassa—a newborn who undoubtedly got infected as a result of blood-to-blood contact with its dying Lassa-infected mother.

As they headed for Zigida, Monath reviewed what he had heard about rodent disease carriers and tried to imagine what a real CDC rodent expert might do in the situation.

Neither Pinneo nor Monath spoke the Liberian languages of Loma or Kpelle, but the government conducted its affairs in English because the nation was founded by freed American slaves in the nineteenth century, and even in remote villages it was possible to get by with basic English. They found Garbazu and gained village approval to take blood samples from her friends and relatives, set mousetraps, and collect local animals. While Pinneo patiently drew blood from anxious villagers unaccustomed to such procedures, Monath hunted.

Night after night Monath crouched by candlelight, a net grasped tightly in one well-gloved hand, the other constantly adjusting his respirator. In this manner he captured dozens of bats, always aware that the fangs of the animals that were thrashing about in his nets might carry the deadly virus. With only a flickering candle to illuminate his actions, Monath carefully placed each captured bat in a thermos full of liquid nitrogen, freezing their bodies for future study at CDC high-security laboratories in Atlanta.

Monath failed to find the animal culprit responsible for the original Zigida cases, but blood tests of 133 villagers showed that, in addition to Garbazu, four people had survived bouts of Lassa fever.

Back at the hospital, Mertens and Monath were joined by Jordi Casals, who was sent by the CDC. Together they scoured the building for pests, but eventually were forced to conclude that the mini-epidemic constituted a classic case of nosocomial transmission: spread of disease between patients and medical staff.[24] Nurse Esther Bacon, they decided, clearly became infected during the dilation and curettage performed on Garbazu, who presumably somehow contracted Lassa fever in Zigida.

Garbazu was on the Curran Lutheran Hospital ob-gyn ward from March 1 to 19. During that time Nessie, who was recuperating from a kidney infection during pregnancy, lay in the bed that was beside Garbazu's for a few days, sharing her newfound friend's food and water. Nessie recovered nicely from her pyelonephritis and was discharged. Five days later she returned, suffering a soaring fever. Nessie died of Lassa.

Liberian midwives Jetty Ziegler and Phebe Hollwanger, having tended to both Nessie and Garbazu, fell ill in mid-March, but recovered fully from Lassa after a few weeks' time.

Sarah wasn't so lucky. Bedded just twenty feet downwind from Garbazu, the Kpaiyean villager was recuperating from an emergency caesarean section and caring for her newborn baby. The very day Garbazu left Curran Lutheran Hospital, Sarah developed a sudden searing headache, her fever spiked at 103°F, and she was unable to sit up. On April 4 her birth canal began hemorrhaging so severely that Sarah went into shock and died. Four days later her baby succumbed.

In all, eleven women got Lassa in Zorzor in 1970, all of whom had been in the Lutheran hospital; seven were members of the staff. Four people died: Esther Bacon, a Liberian obstetrics patient named Sarah, her newborn baby, and Juanita Akoi, a Liberian nursing assistant.[25] Two of the survivors were rendered hearing-impaired, one was completely deaf.

The team tested fifty-nine other patients who had been in the hospital during March—six tested positive for Lassa. Among the fifty-seven hospital staff members, in addition to the seven known to have had the disease, two more tested positive. Both had worked on the obstetric ward tending to Sarah, Garbazu, and Nessie.

Nessie's case particularly troubled the researchers because it implied Lassa could have a long latency time (nineteen days), and could even recur. The prospect of Lassa relapses among his staff was particularly unnerving for Mertens.

Further evidence of relapses emerged when the team studied hospital records going back five years. In the end, they were convinced many past cases of unexplained feverish ailments could be considered Lassa; one such former patient was tracked down, tested, and found to have antibodies against Lassa, showing he had, indeed, once been infected with the virus.

Because the scientists couldn't pin down the original source(s) of the Liberian epidemic, Mertens knew he should expect additional cases of the disease in the future.[26] Though he could do nothing to prevent village outbreaks, Mertens was determined that the disease would never again be spread within his hospital. The entire staff was trained in proper disease control measures, hygiene, instrument sterilization, and other classic practices that have been used with general success to block the hospital spread of microbes since the days of Baron Joseph Lister.[27]

Monath had barely caught his breath and was, once again, preparing to return to the United States when Peace Corps physician Michael Gregg, then working in Sierra Leone, contacted the CDC. The volunteer doctor was convinced Lassa had struck. Once again, Monath gathered up Casals and Pinneo. The trio made their way in September 1972 to Freetown, capital of Sierra Leone, and were joined by CDC investigators David Fraser, Paul Goff, and Carlos ("Kent") Campbell. Together, they solved the Lassa mystery, though their efforts went largely unnoticed back home in the wake of the Watergate break-in and heated U.S. presidential elections.

About two hundred miles east of Freetown, not far from the borders of Guinea and Liberia, Lassa fever struck among villagers and diamond mine workers. At first glance the epidemic seemed a repeat of those in Jos and Zorzor: hospital-based. But it didn't take long for Monath and Casals to recognize that most of the people then suffering Lassa in Panguma Hospital acquired their infections somewhere else. A search through the medical records of six hospitals in the region revealed sixty-three cases of what looked suspiciously like Lassa, occurring between October 1, 1970, and October 1, 1972, with the numbers of sick having increased steadily over the two years.

Once again, Monath donned thick gloves and a respirator to hunt wild animals in the villages and mining camps around Panguma. Casals and Pinneo took blood samples from hospital staff members. The hunt was on. Cats and dogs were grabbed by the villagers, who held the animals still while Campbell, Casals, and Monath drew blood samples. Hundreds of traps for rats and mice were set, bats were netted in the night. Again, the wild animals bared their fangs at their captors, who carefully killed the beasts with gloved hands.

"This is fantastic!" Monath thought, sensing the possibility that here, in these Sierra Leone villages, the animal that carries Lassa viruses would finally be found. So great was the excitement that the team members kept their fears private, never voicing anxieties about getting the dreaded disease. The shorter, older meticulous Monath kept a watchful eye on Kent, however. Towering over everyone, Campbell had the lanky strong build of a basketball player and the impulsive swift movements of an athlete.

Indeed, twenty-six-year-old Campbell got so wrapped up in the quest that he suggested the team do something they would all later agree was "really, really dumb": take phlegm samples from deep inside the lungs of Lassa patients lying on the wards of Panguma Hospital. Temperatures in Sierra Leone were hitting 110°F and humidity topped 90 percent, so Monath and Campbell often found their protective gear (consisting of latex surgical gloves, facial respirator masks, surgical cotton gowns and foot coverings) intolerable and "fudged a little," as Campbell put it, creating spots along their masks and gowns that allowed air circulation. Young Campbell, who had just signed on for a two-year hitch with the CDC to avoid the Vietnam War draft, enjoyed working with the older, more experienced Monath. Though Monath was an urbane Bostonian and Campbell hailed from eastern Tennessee, both men had spent formative years at Harvard studying medicine and public health. Campbell planned to return to Harvard when his CDC hitch was over, to complete his pediatric residency.

Monath, Campbell, and the rest of the group collected more than 640 animals: mice, rats, shrews, bats, and house pets. The animals' lungs, hearts, spleens, kidneys, and blood were carefully removed with sharp dissection scalpels wielded cautiously by gloved hands. All were placed in liquid nitrogen, meticulously labeled, and prepared for overseas air shipment to the maximum-security laboratory at the CDC in Atlanta.[28]

While the team anxiously awaited results, they studied the villages carefully, trying to see what was unique about those that had Lassa cases. In all the eastern Sierra Leone villages around Panguma and nearby Tongo, people lived in large extended families that resided in houses made of mud coated with cement. Their roofs were of iron sheets or thatching, the floors packed mud. Harvested grains were stored in sacks and baskets inside the homes.

The villages were clusters of homes encircling a clearing. Outside the village lay some croplands and rain forest; at times it was hard to tell where one stopped and the other began. Because it was the rainy season, people—and animals—tended to spend their time in shelters.

When the team captured animals they noticed three types of mice and rats scurrying about the villages, and one type—the *Mastomys natalensis* rat—was present in greater numbers in villages stricken with Lassa.

To the joy of all the research team members, CDC laboratory analysis confirmed their hunch. Of 350 animal species initially tested, only *Mastomys natalensis* turned up positive for Lassa virus infection. Better yet, the infected rats came from the same villages where humans had the disease.

Though the major puzzle was finally solved,[29] two questions remained: why had the *Mastomys* suddenly become a problem in key villages; and how did the rats pass Lassa virus on to the people?

Monath's group noticed that *Mastomys* had tough turf competition in the form of the larger, more aggressive black rat, *Rattus rattus*. In some villages, the people had driven out or eaten the big black rats, leaving smaller brown

Mastomys virtually unopposed on the playing field. *Mastomys* often came out of neighboring fields and took shelter from the rains inside homes.

Less clear was how the rats gave people Lassa. Few of the humans who were infected could recall being bitten, and the team was unable to prove one way or another that *Mastomys* could pass the virus in its urine, as had already been seen with Junín and Machupo.

Back at the hospitals in Tongo and Panguma, Pinneo's lifesaving antibodies once again were used in hopes they would help in the recovery of two Lassa patients. But the team discovered in laboratory studies that antibodies from the original Nigerian strain of Lassa virus (now dubbed Pinneo) reacted poorly with the Sierra Leone virus. Weaker still was the reaction to Monath's Liberian strain. This meant there were at least two widely divergent strains of Lassa viruses in West Africa, and Pinneo's antiserum could not be counted on to save patients—and scientists—who got Lassa outside of the Jos region of Nigeria.

It was possible, the group concluded, that Pinneo's antiserum had no real effect on the two Sierra Leone patients to whom it had been given, as there were indications that both women were already recovering.[30]

Any comfort Pinneo's units of blood carefully stored at the CDC might have provided to doctors, nurses, and researchers working in West Africa in the 1970s clearly had to be muted.[31]

Because *Mastomys* was a common African rat, found in fields and villages from Sudan to South Africa, it seemed possible dozens of additional strains of Lassa lurked on the continent—strains Pinneo's antiserum might not be able to combat.

Back at Columbia University, John Frame was convinced Lassa fever could be found throughout West Africa if a scientific search was carried out. With a paltry budget of only $5,000, Frame and Casals screened the blood of missionary workers from countries all over West Africa. They found evidence of Lassa infection in people stationed in Mali, Upper Volta,[32] Ivory Coast, Zaire, and possibly the Central African Republic. That meant Lassa existed in at least eight countries.[33]

Kent Campbell had a similar idea. Mischievously, he thought he could combine some smart science with a CDC-paid extended trip through Ireland by offering to screen nuns who had in the past worked at Panguma Hospital. Since its opening in the 1950s Panguma had been staffed by the Sisters of the Holy Rosary, an Irish Roman Catholic order. The nuns tended to rotate through the African hospital, returning to Ireland after a year or two, so several dozen women who could have been exposed to Lassa now lived in Ireland. Campbell told the CDC that testing those women might reveal the answer to a key question: had Lassa been around Sierra Leone for decades but gone unnoticed amid the plethora of diseases people suffered, or was the virus new?

He argued persuasively to the Atlanta bosses: "If you weren't paying close attention, you wouldn't be able to distinguish Lassa from malaria.

They look exactly the same until the tail end of Lassa when the hemorrhaging starts."

Campbell got the okay, hopped a commercial jet to London, connected with his wife, Liz, and the two of them happily embarked for the Emerald Isle. For four days Kent and Liz traveled all over Ireland, from convent to convent, testing the nuns and sightseeing. For Kent, it was a welcome relief from the hard work in Sierra Leone; for Liz it was a break from pacing about their Atlanta home worrying about her husband's safety.

One afternoon two of the Sisters took the Campbells to Blarney Castle, where they, like thousands of Americans before them, bent to kiss the Blarney stone. When they returned to their car, Kent suddenly reeled, feeling as if he'd been hit hard on the back of the head. Within seconds, sweat poured out of his skin and he became terribly feverish.

By the time the Sisters got the Campbells back to their hotel, Kent was delirious and running a 107°F fever. Liz was hysterical. The Sisters called London authorities, who ordered Kent transferred immediately to the hospital of the London School of Hygiene and Tropical Disease.

Later that day, the Campbells boarded a commercial Aer Lingus jet, and flew without special precautions, amidst hundreds of tourists. No one had instructed Liz to do otherwise, and Kent was in no condition to do more than follow Liz's orders. On arrival in London, again without extraordinary precautions, they grabbed a taxi to the London School. And once inside the hospital, the ailing Kent was placed on an ordinary isolation ward and treated by doctors and nurses who had no idea what should be done.

After a day and a half of delirium, Campbell was given a pint of Jordi Casals's blood antiserum. It was midnight and Campbell barely realized he was being transfused.

Five hours later he opened his eyes to see his friend Tom Monath worriedly hovering over him.

"What're you doin' in London?" Campbell drawled in his gentle Knoxville accent.

"We're getting you out of here," Monath responded abruptly.

Campbell had no idea how much anxious negotiation had surrounded his case over the previous thirty-six hours. State Department and White House officials had been in discussions with 10 Downing Street and Whitehall; CDC bosses had kept an open line to their counterparts at the London School; the decision was made to get Campbell onto American soil as quickly as possible.

That night the Campbells were driven to Heathrow Airport, this time wearing respirators to protect others, and transported in a special ambulance driven by volunteers. Awaiting the couple on the tarmac was a USAF C-141 transfer jet, inside of which was an Apollo space capsule that had been flown from a U.S. military warehouse in Frankfurt, Germany. Sealed off from the outside world, the couple rested in seats designed for astronauts orbiting deep in space. Monath and four USAF medical corpsmen monitored the Campbells during their transatlantic flight.

When the plane landed at New York's Kennedy Airport, another special ambulance greeted the group on the runway, taking the Campbells off to Columbia-Presbyterian Hospital.

For four weeks Kent was treated in the same room in which Casals had once languished. Liz was monitored closely, and remained well. After thirty days, Campbell recovered and was ready to return to his job at the CDC, but officials politely asked that the young doctor "take a little time off": it seemed many employees of the world's most prestigious disease control agency were afraid Campbell might still harbor contagious infection.

During his time off, Campbell got a bill from the U.S. Air Force: $17,000, payable immediately, for medevac airlift services. Kent shrugged and passed the bill on to CDC director David Sencer, who gruffly sent it back to the Defense Department.

In recognition of his recent hardship, CDC officials next gave Campbell a choice assignment in Hawaii, where he and Liz spent several weeks during a rubella outbreak. Upon returning to Atlanta, his obligatory conscientious objector stint at CDC nearly completed, Campbell spotted a help-wanted notice on an agency bulletin board: "Chief Malaria Control Officer: El Salvador."

Kent Campbell re-upped with the CDC, and in 1973 he, Liz, and their two small children moved to San Salvador for what was supposed to be a two-year assignment.

It eventually became a four-year assignment that completely changed Campbell's life, giving him a newfound concern for malaria control and the health problems of people in developing countries.

While the Campbells settled into their new lives in El Salvador, Uwe Brinkmann paced like a lab rat inside the Ebstorf smallpox containment facility, and pondered the stories of Casals, Pinneo, Campbell, and Roman. He knew his predicament in Germany stemmed from all those past incidents, and the high death toll the virus had claimed among Americans and Europeans working in Africa. He thought of all the mysteries surrounding Lassa, and wondered if he had become infected while tending to Mandrella.

It felt as if a lifetime had passed, but it was just days ago that Brinkmann, the controversial "hippie doctor," had met Casals in Ibadan. The CDC now made it a practice to send the sixty-three-year-old Casals to investigate all reported Lassa outbreaks. In five years he had seen two serious outbreaks, and arrived in Ibadan in 1974 to witness his third.

For Casals, the Ibadan case was a tiny episode, highly exaggerated by international press attention and government panic on three continents.

When Casals arrived, Brinkmann stepped out of the group's Nigerian isolation house to greet the famous scientist, who, characteristically, brushed aside the young German, barged into the building, and went straight to Mandrella's bedside.

Casals carefully examined Mandrella, surprised to discover the patient was recovering. Dr. Hal White, who had attended to Jeannette Troup when she succumbed to Lassa four years earlier, had provided Mandrella with a unit of Penny Pinneo's antiserum. And English physician Adam Cargill was looking after the ailing man. Cargill, then thirty-four, was on the faculty of the University of Ibadan Medical School.

"What a ridiculous international brouhaha!" Casals thought. In Lagos, Nigerian government officials had informed him, "There is no Lassa fever in this country. Period. So the German must have brought it here."[34]

Angrily Casals thought, "So nice. They eliminated the disease by saying it didn't exist."

Meanwhile, the Catholic Church, which ran St. Charles Mission Hospital outside Enugu and had employed both Sauerwald and Mandrella, alerted the German Foreign Ministry and the Tropical Disease Institute (Tropeninstitut) in Hamburg, raising concern in Germany. On March 15, 1974, Ibadan's Catholic bishop, Richard Finn, officially requested, on behalf of Mandrella, help from the German government.

The Nigerians insisted that the German must take "his disease" back to Germany, along with all the Nigerians and others he may have touched. Panic escalated, and no nation between Nigeria and Germany would grant permission for a suitably small aircraft to land en route for refueling.

"Ideally, the patient should remain exactly where he is, tended to by the apparently able nurses and these Brinkmann and Cargill fellows," Casals thought, knowing that the stress of traveling could aggravate a condition that otherwise appeared to be improving.

The Lassa expert then focused on Brinkmann, briefly conveying his view that the patient's condition was no longer acute and it would be best for all if he remained exactly where he was. After all, Casals said, transport would only aggravate the patient's pain and increase the number of people worldwide who could be exposed to Mandrella's breath and bodily fluids.

Brinkmann agreed wholeheartedly, and waved goodbye to Casals, hoping the highly respected scientist would succeed in convincing Nigerian and German authorities of the folly of their transport plans.

Unbeknownst to Brinkmann, Casals flew immediately to WHO headquarters in Geneva, where he tried, in vain, to argue for calm. He knew his was a lost cause when he scanned the German and French newspapers on sale at the kiosk located in the organization's lobby.

For days the German press and, to a lesser degree, British and French media were spellbound by the saga of Lassa and the German doctor.

"Who Will Save This Physician?" screamed a *Diepahtzer Nachrichter* headline, adding, "A scandal!"[35]

The *Bild Zeitung* ran huge portraits of Brinkmann, depicted as a heroic figure who was braving death to rescue a colleague. Alongside the front-page story was a sidebar describing practice sessions held by the Hamburg fire brigade and police department, in preparation for Mandrella's top security arrival and transport to hospital facilities.[36]

Such flattering articles were common, and of no help to Brinkmann, whose colleagues considered his journey to Nigeria distasteful. Germany in 1974 was, like the United States, experiencing an enormous generation gap that affected every aspect of society, including science. Brinkmann, with his tie-dyed T-shirts, long hair, and Indian sandals, became a lightning rod for the resentments of older, more traditional tropical disease experts. When the German Foreign Ministry contacted the Tropeninstitut in Hamburg on March 16 asking for information about Lassa, Brinkmann eagerly offered his services. But some older scientists, notably virologists Godske Nielsen at the Tropeninstitut and Fritz Lehmann-Gruber, acting director of the Virology Institute of Hamburg University Hospital, felt it was foolhardy to bring Mandrella back to Germany. They argued that such an effort was overly dangerous for everybody, including Mandrella.

Lehmann-Gruber went further, telling the German press that bringing Mandrella into the country could lead to a Teutonic Lassa epidemic.[37]

"We don't know whether the virus may not find an ideal vector," Lehmann-Gruber told *Bild-Hamburg*.[38]

"Are you talking about an insect?" the reporter asked.

"Yes, a fly. A mosquito. All is possible," Lehmann-Gruber responded, adding, "The danger [of bringing Mandrella to Germany] is still incalculable."

For Brinkmann such talk seemed utterly absurd. Having spent years in Ethiopia with his scientist-wife, Agnes, Uwe knew Europeans tended to exaggerate the dangers of African diseases. And in a manner typical among peers of his generation, Brinkmann favored immediate action. Brinkmann told the institute director he'd gladly take vacation leave immediately and fly at his own expense, if necessary, to aid Mandrella. Brinkmann's offer was made in a room full of Tropeninstitut staff in response to the director's request for a volunteer.

"No, Uwe, you can't do it," the director said. "You have two little children."

Brinkmann shuddered for an instant, thinking of his sons, Patrick, age two, and four-year-old John Vincent. But no other hands were raised.

The director took Brinkmann up on the offer, handing the young scientist cash—drawn from his own wallet—for the airfare.

After Brinkmann arrived in Lagos, the German press started making trouble, fueling the debate between Uwe and his opponents, by building him up as a hero. On March 18, a day after Uwe left Germany, a national television correspondent visited Brinkmann's mother.

"How does a mother feel whose son is flying off to death?" the reporter asked.

Unaware of Uwe's departure, Mrs. Brinkmann invited the TV crew to take tea in her living room, and answered their question with one of her own: "Who is so crazy to ask him to do something like that?"

Meanwhile, the German government, now convinced Lassa was akin to the Andromeda Strain, was frantically trying to find some form of airtight container in which to transport Mandrella. On March 19 the answer seemed at hand when headlines declared, "Danke! Kissinger Will Save the Fever Physician."[39] U.S. Secretary of State Henry Kissinger offered the use of an American military transport plane to fly an Apollo space capsule to Nigeria. Mandrella and Brinkmann would then board the airtight capsule, as had Kent Campbell eighteen months earlier, breathe filtered air, and remain so encapsulated for several hours' flight time from Ibadan to Hamburg. It was a precaution considered brilliant by the German press and political hierarchy, but paranoid by experts like Casals.

While the German, American, and Nigerian governments debated the relative merits of using the Apollo spacecraft, Mandrella and his caregivers grew increasingly anxious in Ibadan. Mandrella had been sick for nearly a month (his symptoms began February 22); and the three men, the nun, and the two nurse assistants had been under virtual house arrest for days. For the three chaste Nigerian women, being co-housed with men was particularly disgraceful, and they greatly feared their reputations would be damaged.

Dr. Cargill was especially agitated. He feared the Nigerian government would mount a campaign of blame against him, primarily because he— the Lassa victim's physician—was a citizen of Nigeria's former colonial power. He paced the small house anxiously, thinking of his wife, Alice, and their two small children—all safely tucked away in Sussex, England.

When he developed serious diarrhea and a mild fever, Cargill was convinced that he too had Lassa. Fear within the group escalated. Brinkmann tried to calm everyone down, saying, "I know in my gut we're all going to survive."[40]

Finally, on March 21, a specially outfitted Lufthansa Condor jet landed at the Ibadan airstrip. The three foreigners and the three Nigerian women were driven to the tarmac, and Mandrella, languid on a stretcher, was placed on a forklift and loaded in the cargo entry of the jet. The others climbed stairs to the plane, finding no comfort inside. To protect the flight crew, German engineers had gutted the passenger section of the aircraft, placing a huge airtight barrier between the plane's tail section and the forward crew compartment. In addition, special air circulators were installed, providing the two halves of the craft with separate oxygen supplies. No flight attendants greeted the group; just a forbidding, barren compartment.[41]

Brinkmann, the last to board, took his seat before realizing there was no crew in their section.

"Close the door immediately!" the captain shouted over the plane's public-address system. Brinkmann jumped out of his seat and stared at the pressurized door, upon which were printed two pages of instructions on proper methods of closure and opening.

"We are leaving immediately. Close the door now!" the pilot said.

"Well, here goes nothing," Brinkmann thought, as he grabbed the door, pulled and twisted the handles, and hoped he had safely sealed them in their strange cabin to Deutschland.

When the plane landed hours later at the Hamburg airport the door was opened by a man dressed in a white head-to-toe outfit reminiscent of astronaut's gear over which he wore a huge clear plastic bubble that enclosed his legs from the knees up and all of his torso and head. From the bizarre inflated bubble protruded his arms, which flopped about almost helplessly. A long plastic hose connected to the back of the bubble provided the man with a germ-free atmosphere.

"This is like a scene from a bad science fiction movie," Brinkmann told his fellow travelers.

The fantastic bubble creature waved his arms clumsily, beckoning the group out of the plane. Three other bubble men ushered them into a waiting van and carried Mandrella's stretcher.

In the process, one bubble man fainted for lack of oxygen, and shouts of "The virus! The virus!" went up among the security entourage. For a few moments the operation was seized with panic.[42]

When they were on their way into the woods, and Brinkmann was struggling to suppress his concentration camp fantasies, he shared an anxious glance with Mandrella. Only later, during their long days of captivity in Ebstorf, would Mandrella tell Uwe that he too had momentarily thought of the Third Reich, remembering his father's execution and the almost unbelievably cruel "bill for hanging" his mother had received.

Until their release from Ebstorf on April 20, the sextet had only occasional telephone contact with the outside world. All papers, food, garbage, clothing, and medicines were sterilized or destroyed when passed out of the facility through specially designed airlocks.[43]

Every day Agnes would bring Patrick and John Vincent to Ebstorf to wave at their father from behind a chain-link fence some twenty yards from the containment facility. Groups of nuns and Catholic parishioners would also gather at the fence to pray for the three Nigerian women and their missionary doctor. Pictures of the chain-fence gatherings graced the pages of German newspapers for over three weeks.

Inside, Brinkmann tried to keep the sorry spirits of the group buoyed with jokes. His sense of humor tending to sarcasm, Uwe told the group, "We could go out right now and become millionaires. It's true! We could walk right out of here, rob the biggest bank in Germany, hijack an airplane, and spend the rest of our lives in luxury on some tropical island. No one would dare stop us, they're all so afraid of the virus."

On March 28, *Der Stern*, one of Germany's two most popular news magazines, published a lengthy article praising Brinkmann. Describing his hippie attire and disheveled appearance, the magazine declared Brinkmann a far greater physician than, for example, Dr. Ernest Fromm, then head of the German Physicians' Association. Fromm was under investigation for allegedly embezzling funds.[44]

The day the article was released a public relations manager for the federal Health Ministry called Brinkmann in the Ebstorf facility, accusing the young doctor of planting the article as a deliberate smear against Fromm.

"You better never come back to Hamburg!" the PR man said. At that moment Brinkmann knew his efforts on Mandrella's behalf were going to demand a high career price.

Cargill feared he too would pay for his actions. A slim, nervous man, Cargill anticipated the worst, and carefully monitored news from Nigeria. Indeed, he was fired in absentia from his hospital job, and the Lagos press accused him of being responsible for the mini-epidemic.

"An expatriate doctor . . . almost caused an epidemic of the disease in Ibadan by arbitrarily getting in contact with a patient of the killer disease," said Lagos press accounts.[45]

When CDC laboratory tests finally confirmed that all six people in the Ebstorf containment facility were free of the Lassa virus, Mandrella having recovered and the others never having been infected in the first place, the group was released. A department store gave the Nigerian women an all-day free shopping spree as compensation for their long captivity. Mandrella quietly retreated to the company of German friends for several more months of recuperation, Cargill joined his family in Sussex, and Brinkmann—despite the warnings of some—returned to the Tropeninstitut.

He found an atmosphere deeply polarized by the Lassa virus events. On one extreme, the federal government offered Brinkmann one of the country's highest medals, which he discreetly declined. On the other, many old-guard scientists bitterly denounced Brinkmann's actions and demanded his resignation. At the director's insistence, Brinkmann took his family on a vacation to allow time for things to cool down in Hamburg.

Two weeks into their vacation, John Vincent went to play in a friend's flat. While adult eyes were briefly turned away, the energetic four-year-old jumped and leapt about, misjudged his footing, and fell out of the apartment window.

News of his son's death crushed Brinkmann. He lost his will to fight the Hamburg old guard, to swagger as the hippie doctor, or to take bold steps to confront tropical diseases. Nearly twenty years later he would find it impossible to discuss Lassa fever without recalling the emotional traumas of the political battles, the long quarantine at Ebstorf, the group's fears, criticisms from fellow scientists, and—most tragically—his son's death.

In August 1974, Dr. Bernhard Mandrella quietly returned to Nigeria, continuing his missionary work at the Borromeo Hospital in Onitsha.

5

Yambuku

EBOLA

Men who never have had the experience of trying, in the midst of an epidemic, to remain calm and keep experimental conditions, do not realize in the security of their laboratories what one has to contend with.

—Dr. Martin Arrowsmith, from *Arrowsmith*, Sinclair Lewis

I

Mabalo Lokela was in a great mood. Sure, he had a fever, but it was undoubtedly just the malaria again. He was sure of that. The important thing was that he was back from a great vacation—one of the few he'd had in his forty-four years.

While he waited for one of the Sisters to give him malaria medicine, Mabalo shared with colleagues at the Yambuku mission stories of his recent travels. From August 10 to August 22 he and six other mission employees had driven around the far north of Zaire, visiting towns all over the Mobaye-Bongo Zone, sampling local delicacies and enjoying the sort of sightseeing that was rare for people in the Bumba Zone. It was possible to travel such distances—it must have been hundreds of miles!—only because Father Augustin was with them: his presence allowed the use of the mission's Land-Rover.

"We got all the way up to Badolité, and we would have crossed over into the Central African Republic, but the bridge was down," he told friends at the mission. When he got back to Yambuku four days ago, Mabalo (whom friends called Antoine) was so happy to be home that he spent a good bit of his schoolteacher income buying fresh antelope meat in the market—something to please his wife, Mbuzu Sophie. Sophie, who was eight months pregnant, dried the meat and made a stew for the family celebration of Antoine's return.[1]

Antoine watched as one of the Belgian Sisters prepared a syringe, and gritted his teeth when the needle punctured his skin. "Chloroquine," she told him as he rubbed his arm, "will cure your malaria." He nodded, confident that all good cures come from needles.

Two days later, on August 28, 1976, a thirty-year-old man came to the

Yambuku Mission Hospital complaining of terrible diarrhea. Though nobody at the mission recognized the man, he told the Sisters that he came from the nearby village of Yandongi. Well, his origins were no matter; the Sisters treated any needy soul who crossed their threshold, sometimes 400 a day, many of whom walked and hitched rides distances of fifty or sixty miles to reach the mission. Most of the sick got injections of one kind or another: antibiotics, chloroquine, vitamins—whatever supplies might be on hand in the modestly funded remote Catholic hospital. And usually that was enough for the people, who would, in any event, supplement whatever the Belgian nurses gave them with potions, incantations, and injections from local sorcerers.

But the case of the man from Yandongi was odd, and Sisters Béata, Edmonda, and Myriam weren't quite sure what was the source of his illness. They put the man in one of the 120 beds in the hospital and, for two days, debated his diagnosis, finally writing in his medical chart a vague "dysentery and epistaxis."[2]

After two days the man left the hospital against the Sisters' wishes, his diarrhea and epistaxis, or severe nosebleed, still unresolved. He was never seen again, though events days after his disappearance would prompt dozens of investigators from all over the world to scour villages throughout the Bumba Zone in search of the elusive patient.

The Bumba Zone lay in Zaire's northern frontier, spanning savanna and dense rain forest lands between the Ubangi and Zaire (formerly Congo) rivers. Some 275,000 people lived in the Zone, most in villages of fewer than 500 people. They earned their living growing cash crops for export to the Zairian capital, Kinshasa, and by hunting. The equatorial jungles and grasslands were rife with game that included such marketable delicacies, pelts, and riches as green monkeys, baboons, black-and-white colobus tree monkeys, chimpanzees, spotted-necked otters, mongooses, civets, elephants, hippopotamuses, bushpigs, buffaloes, bongos, sitatunga antelopes, bushbucks, reedbucks, and oribi.[3]

Since 1935 the major hospital and dispensary for some 60,000 villagers living in the central Bumba Zone was that operated by Belgian Catholic missionaries in the village of Yambuku. A staff of seventeen "nurses"—so designated, though none of the Sisters had attended a certified nursing school—and medical assistants tended to the health needs of the community out of a rather modest set of cinder-block buildings. As one entered the front of the hospital, administrative offices were in a room on the right, followed by a pharmacy, and a surgical block comprised of an operating theater, scrub room, and facilities for "sterilizing" instruments: a thirty-liter autoclave and a Primus stove atop which water boiled.

Outside the surgical block one entered a long alleyway. To one side of the alley was a pavilion bisected by a hall, off of which were large hospital rooms: one common ward with eighteen beds, four eighteen-bed men's wards, and three larger women's wards. As was common throughout Central

Africa, the beds were flat metal ones made tolerably comfortable with thin mattresses and ancient linens. Additional comforts and foods to supplement the basic rice or mealie-meal menu were provided by patients' relatives.

Further along the outer alleyway was an outpatient clinic, through which flowed dozens of people every day seeking prenatal care, injections for a variety of ailments, vaccines for their children, and advice from the Sisters about all sorts of health problems.

There was no doctor in Yambuku. Patients were treated by the staff of four Belgian nuns who had received a modicum of training in nurse-midwifery, a priest, one Zairian female nurse, and seven Zairian men.[4]

This small team of hardworking health providers also monitored patients in another building housing a large ob-gyn ward and two more general medicine wards. The hospital was part of a larger mission complex that included a school where Antoine worked, a church, a variety of other service buildings, and the living and dining quarters of the missionaries. In addition to those working in the hospitals, the missionaries included several more Belgian nuns and priests who staffed the schools, the church, and other facilities.

Though his home was in the village of Yalikonde, about a mile from Yambuku, Antoine spent days on end at the mission, as did two of his older teenaged children. So it was natural that he returned to the Sisters on September 1 when, despite the chloroquine injection, his fever soared over 100°F. They checked his vital signs and told Antoine to rest for a few days. Antoine returned to Yalikonde, where Sophie tended to him.

At about the same time as Antoine was regaling friends with tales of his recent travels while awaiting his chloroquine shot, sixteen-year-old Yombe Ngongo lay in Yambuku Hospital undergoing transfusions to counter her severe anemia. Nearby, twenty-five-year-old Lizenge Embale was recuperating from what seemed to be malaria, tended to by her husband, Ekombe Mongwa.

And over on the men's ward Angi Dobola was recovering from hernia surgery. The sixty-year-old villager from Yalaloa was watched closely by his wife, Sebo Dombe, who complained to the Sisters of exhaustion. Sebo was given vitamin injections, which helped her find the energy to cope with long, tense nights by her husband's post-op bedside.

On September 5 Antoine returned to the mission critically ill. He was vomiting and had acute diarrhea, leaving him so dehydrated that he had "ghost eyes," as the missionaries called them: deeply recessed, dark, glazed eyes surrounded by pale, parchmentlike skin stretched tightly over pronounced facial bones. His chest hurt, he had a terrible headache, fevers continued, he was deeply agitated and confused.

And he was bleeding. His nose bled, his gums bled, and there was blood in his diarrhea and vomitus.

The Sisters had no idea what was wrong with Antoine, nor did they realize that he was not an isolated case. Yombe Ngongo had checked out of Yambuku Hospital on August 30, and was now fighting for her life at home, in the village of Yamisakolo. At the sixteen-year-old's side was her anxious nine-year-old sister, Euza, feeling her own first symptoms of headaches and fever.

And Sebo Dombe's exhaustion now exceeded the benefits of vitamin injections. Though her husband was recovering nicely from his hernia operation and the pair had returned home, Sebo was semi-delirious. She too was hemorrhaging blood. As was Lizenge Embale, who had returned to her home in Yaekenga in the beginning of September but was now struggling to stay alive. At her side, vomiting blood and bleeding from his eyes, was her husband, Ekombe Mongwa.

The Sisters knew only of Antoine's case, and they did everything they could to save their friend. The hospital had no sophisticated laboratory facilities to aid in diagnosis, so they could only guess what might be causing such horrendous things to happen to a human body—perhaps yellow fever, or typhus. They pumped Antoine full of antibiotics, chloroquine, vitamins, and intravenous fluid to offset his dehydration.

Nothing worked. On September 8, Mabalo ("Antoine") Lokela died. Unbeknownst to the Sisters, Yombe Ngongo died the day before in her village home. On September 9, her little sister, Euza, succumbed. That week Lizenge Embale and her husband, Ekombe, died in the hut in Yaekenga—again, the Sisters didn't know.

Antoine's funeral was well attended and, as was customary, his body was carefully prepared before burial by Sophie, his mother, Sophie's sister Gizi, and other women friends. By tradition readying a body for burial required evacuating all food and excreta, a procedure that was generally performed by bare-handed women.

In a matter of days Antoine's mother, Gizi, and Sophie were suffering the same ghastly disease; Sophie and Gizi survived, but Antoine's mother died on September 20, as did his mother-in-law, Ngbua, who had assisted in the funeral preparations. And though Sophie survived those hellish September days, her baby was stillborn—another hemorrhagic victim.

In all, twenty-one of Antoine's friends and family members got the disease; eighteen died.

Soon the hospital was full of people suffering with the new symptoms. Panic spread as village elders spoke of an illness, unlike anything ever seen before, that made people bleed to death. In Yambuku the Sisters were already close to the breaking point, not knowing the why, what, or how of the new disease.

The horror was magnified by the behavior of the many patients whose minds seemed to snap. Some tore off their clothing and ran out of the hospital, screaming incoherently. Others cried out to unseen visitors, or stared out of ghost eyes without recognizing wives, husbands, or children at their sides. Word, and the disease, spread quickly to villages throughout the Bumba Zone. In some, the huts of the infected were burned by hysterical neighbors.

On September 12, Sister Béata developed the sudden fever, muscle aches, nausea, diarrhea, and bleeding gums that she and her fellow nurses now recognized only too well. Sisters Myriam and Edmonda prayed for a miracle and radioed urgent pleas for assistance.

Bumba Zone medical director Dr. Ngoi Mushola scoured the city of Bumba for petrol, finally arranging transport across the roughly fifty miles to Yambuku on September 15. What greeted Ngoi upon arrival was a horror that shook the provincial physician to his very soul. The Sisters and priests beseeched him to tell them what disease was claiming the lives and spirits of their parishioners. In desperation they begged him to help cure Sister Béata.

But Ngoi was every bit as helpless as the hapless clerics. With great care he gathered as much clinical information as possible, and on September 17 rushed back to Bumba in order to cable his report to authorities in Kinshasa.

Republique du Zaire—Region of the Equator -S/Region of Mongala—Bumba Zone—Bumba Medical Service
Inquiry into alarming cases in the community of Yandongi, Bumba Zone, 15–17 September 1976.

I received an urgent call from Yambuku on September 15 from the medical assistant Masangaya Alola Nzanzu of Yambuku Hospital because of alarming cases in the community since September 5, 1976; I went to determine the reality of the situation.

Findings. The affliction is characterized by a high temperature around 39°C; frequent vomiting of black, digested blood, but of red blood in a few cases; diarrheal emissions initially sprinkled with blood, with only red blood near death; epistaxis [nosebleeds] now and then; retrosternal and abdominal pain and a state of stupor; prostration with heaviness in the joints; rapid evolution toward death after a period of about three days, from a state of general health.

Ngoi's report described the first case, that of Mabalo Lokela, and then listed twenty-six cases of the strange illness, giving the names of the patients, noting that fourteen had died, ten were still sick, and four had fled the hospital in terror, their whereabouts now unknown.

Eerily, Ngoi corrected his report just before sending it to Kinshasa to note that two individuals on his "ailing" list had died by the time he reached Bumba. He listed the treatments tried, without success, at Yambuku Hospital: aspirin, chloroquine, nivaquine, blood coagulants, calcium, cardiac stimulants, caffeine, camphor. And he noted that the hospital had used up all its antibiotic supplies.

Nothing helpful had been discovered in the Yambuku Hospital group's microscopic studies of blood, urine, and stool samples, Ngoi noted. And he tactfully added that protective measures by the hospital to isolate patients with the disease "are not strict."

Warning that "there is already panic" in all the villages, Ngoi requested assistance from Kinshasa authorities.

He left Yambuku having recommended that the Sisters take three measures immediately: "(1) Hospitalize the cases. (2) Use public cemeteries.[5] (3) Boil potable water."

What Ngoi had written, though he did not know it at the time, was the first historic description of a new disease. In clear, succinct, and, as time would show, largely accurate terms, Ngoi had described what would prove to be the second most lethal disease of the twentieth century.[6]

At five o'clock in the afternoon of September 19, Sister Béata died. The same day reports came into the mission of illnesses and deaths from the bizarre bleeding disease in over forty villages. By now, there was real danger of a mass exodus of hysterical villagers fleeing to nearby zones— and taking the disease with them. Through the missionary radio relay system, the Sisters sent more urgent pleas for assistance.

Federal authorities dispatched two professors from the National University of Zaire to Yambuku: microbiologist Muyembe Tamfum Lintak and epidemiologist Omombo. They reached the mission on September 23, intending to conduct a six-day study of the problem, but cut their visit short and beat a hasty retreat from Yambuku after just twenty-four hours.

When they arrived at Yambuku Hospital, Muyembe and Omombo saw despair and horror everywhere they turned. Just hours before they arrived, twenty-six-year-old mission nurse Amane Ehumba had died of the disease, and anxieties among the Zairian hospital employees were at near-panic levels.

The professors first focused on a small child who was writhing in agony in a hospital crib. While they discussed what might be done, the child died before their eyes. The academics were shaken from their intellectualizing, and immediately set to work collecting blood and tissue samples from patients and cadavers, interviewing ailing patients and reviewing their medical charts.

As the professors commenced their research, Sister Myriam, who had nursed Sister Béata, was suddenly overcome by piercing headaches and fever. The fear among the mission staff was contagious.

Unfortunately, the academics hadn't taken Ngoi's field report seriously, and brought no protective gloves, masks, or gowns for their use during procedures that put them in contact with infected blood. Still, they worked around the clock, examining five blood samples for signs of malaria, parasites, or bacteria. They found nothing. When they performed autopsies, Muyembe and Omombo were aghast at the extensive damage inflicted by the disease, and removed liver samples to send to sophisticated laboratories for further analysis.

Sister Romana arrived during the day, having traveled all morning from the Lisala Mission, located in the zone to the southwest of Bumba. "I have come," she told the other Belgians, "to replace Sister Béata." The visiting nun set to work immediately, looking after the latest victims.

Among them was Sophie, still severely ill at that point, groaning in agony in her hospital bed. While the professors inspected the wards, their guide, nurse Sukato Manzomba, progressed from being mildly feverish to a life-threatening state. She began vomiting blood and passed into delirium. The stunned professors acceded to the missionaries' pleas and agreed to take Sister Myriam, Father Augustin (who had traveled with Antoine in northern Zaire and was running a high fever), and Sister Edmonda (as an accompanying nurse) back to Kinshasa for treatment.

The group traveled the muddy, bumpy road from Yambuku to Bumba in a Land-Rover, passing several villages along the way, and were airlifted the following day to Kinshasa aboard a Zairian Air Force transport jet. Left to their own devices at Kinshasa's N'djili Airport, inexplicably abandoned by the professors, the missionaries were forced to take a taxi to Ngaliema Hospital—Zaire's premier teaching facility.

From the moment she arrived it was obvious to the Ngaliema staff that Sister Myriam needed not a hospital bed, but a deathbed.

Because they had no idea what pathogen was producing the Sister's illness, the Ngaliema staff didn't know what precautions they should take. Sister Edmonda described the rapid spread of the disease inside Yambuku Hospital and volunteered to do the bulk of Sister Myriam's care. The ailing nun was placed in an isolated ward. A pretty young student nurse, Mayinga N'Seka, offered her assistance and Dr. Lusakumuna took charge of the case. Collectively they did what they could to ease Sister Myriam's suffering.

On September 30, despite their efforts, Sister Myriam died in the Kinshasa hospital.

II

Dr. William Close was in Wyoming at the time, negotiating the purchase of a ranch. For sixteen years he had lived in Kinshasa, serving as personal physician to President Mobutu Sese Seko and directing a nongovernmental medical development group called Coopération Médicale Belge. The American physician and his family[7] had arrived in Zaire when Mobutu seemed a heroic, towering figure on the African landscape, a leader of postcolonial black Africa, and an inspiration to young idealists worldwide. But over the years Close witnessed Mobutu's transformation from a sort of Zairian George Washington to a tyrannical and corrupt despot enamored of the works of Machiavelli and surrounded by family and associates who treated Zaire's national bank as their personal cash register.

Grown cynical, Close was preparing for a new life in Wyoming when Dr. Ngwété Kikhela, Zaire's Minister of Health, called to ask Close to notify American authorities, requesting assistance. Close immediately contacted the Centers for Disease Control in Atlanta, apprising the agency of the situation and formally requesting laboratory support to determine the cause of the Yambuku outbreak.

Back at the mission, more of the hospital staff contracted the disease. Now ten of the seventeen employees were either dead or too sick to continue tending to patients. Following Muyembe's parting recommendations, Sister Genoveva closed the hospital to all but the remaining dying victims of the mysterious disease. Though she had no medical training and was one of the mission's teaching nuns, Sister Genoveva was forced to carry the onus: none of the Belgian medical personnel remained well enough to shoulder such responsibilities.

Sister Romana lay in one bed, vomiting blood, bleeding from her gums, suffering acute diarrhea, and groaning in delirium. The elderly Father Germain Lootens was similarly stricken, and none of the remaining Zairian nurses felt up to staffing the hospital without their supervision.

Lacking medical skills, Sisters Genoveva, Marcella, and Mariette turned to the only weapon in their armamentarium: prayer. For hours on end the grief-stricken nuns and the three remaining priests prayed over the sickbeds of their friends and colleagues, hoping their devout entreaties would bring a miracle.

Despite their prayers, Sister Romana died at noon on October 2. Word of her death, radioed by the Yambuku staff to the Lisala Mission, produced both tremendous grief and justifiable fear among her old friends. Just six hours later, Father Lootens also passed away and this threw the surviving Belgian missionaries into such despair and terror that a visiting team of Kinshasa scientists found the group virtually paralyzed by anxiety.

At Minister Ngwété's request, a team of medical experts had been assembled and flown to Bumba by the Zairian Air Force. From there they drove to Yambuku. The three-man team arrived shortly after the deaths of Sister Romana and Father Lootens. Ministry officials, notified of the deaths by relayed radio messages, ordered the area placed under strict quarantine and "cordons sanitaires" established around Yambuku.

Having no experience in such matters, Sister Genoveva took the order literally. She gathered up rolls of bandage gauze and strung them around the periphery of the mission and suspended signs from the "cordons" warning visitors to stay away. A large bell was hung at the mission entry, with a sign telling visitors to ring, leave their messages or food donations, and quickly withdraw.

Close explained the crisis to President Mobutu, who expressed concern about containing the epidemic, and put his personal Hercules C-130 transport jet at the disposal of the medical effort. He also ordered the entire Bumba Zone placed under strict isolation. All roadways, waterways, and airfields in the region were placed under martial law, and the transport of goods and people in and out of the area came to a full stop within a week. The village elders of the Bumba Zone, recalling the smallpox epidemic of the 1960s, advised their people to remain housebound until the epidemic passed. Overnight all commerce, social life, schooling, and ritual gatherings ceased and the villages surrounding Yambuku looked like ghost towns.

Close helped gather medical supplies, rudimentary lab equipment, and other hospital essentials from warehouses and hospitals around Kinshasa, and these were loaded aboard Mobutu's jet and flown to Bumba.

Meanwhile, the three-man team of Kinshasa-based investigators, composed of Zairian health official Dr. Krubwa, Belgian medical mission director Dr. Jean-François Ruppol, and French medical mission chief Dr. Gilbert Raffier, did their best to comfort the extremely upset Yambuku missionaries. They gathered more blood and tissue samples, examined medical records, and toured local villages. Though the scientists still had little solace to offer, the missionaries were greatly relieved, and radioed gratitude to Bumba for the supplies and physicians.

At about the same time, Paul Brès received word that another strange epidemic was unfolding in a town called Maridi in the grasslands of southern Sudan. Information was scarce, and authorities in the Sudanese capital of Khartoum had no radio contact with that impoverished and distant region to the south. Still, Brès and other experts in the virus branch of WHO thought—from their distant Geneva vantage point—that the Sudanese accounts bore a remarkable resemblance to those from Yambuku. He urged Khartoum to immediately send blood and tissue samples from Maridi patients.

But it was no simple matter for a doctor in Khartoum to make his way to Sudan's southernmost provinces, gather blood samples, store the precious fluids in containers that would protect their contents from the intense desert heat, and make his way back to Khartoum. In addition to the usual—and monumental—logistic obstacles to such a trek, whoever went faced the even more towering blockade of politics.

But the mysterious epidemic was occurring in one of the country's three most southerly provinces, where the people lived and believed as they had since before the Nubians were enslaved by Egypt's Pharaohs. Speaking a variety of ancient Bantu languages, the southern Sudanese were animists who believed all living things, as well as the sun, water, wind, and weather, had a spiritual character. The manipulation of these often unpredictable and fickle spirits and gods was the province of fate: wise sorcerers knew how best to cajole the spirits to support their ends or repel evil spirits that produced illness, death, and misfortune. The southerners lived in small, temporary villages, were often nomadic, had a high rate of illiteracy, and could not be expected to be found in any particular locale at any specific time.

In 1969 Sudan had a military coup d'état. A Muslim-led civilian government backed by the military was installed, and the nation teetered on the edge of civil war, splitting the Muslim north and the Christian, animist south until 1972. Then a semblance of peace took hold when a constitutional agreement was reached, providing the three southern provinces with a fair degree of self-rule. The autonomous region was only nominally connected to the Khartoum-based infrastructure, and it was rare indeed that a Ministry of Health official from the north would be asked, or would agree, to intercede in medical problems to the south.

Still, Brès and other Geneva officials insisted on pushing past the political obstacles to discover what was going on in Maridi. Their greatest fear was that the epidemics of Yambuku and Maridi were one and the same, representing a vast super-lethal disaster spanning an area of about 1,000 square miles in at least two nations.

Blood samples, collected in Maridi and shipped over several days' time to Khartoum, finally reached Geneva. They were in poor condition, but WHO immediately sent them on for analysis in laboratories in the United States and the U.K.[8]

III

WHO enlisted high-security laboratories all over the world in the effort. It wasn't hard, really: everybody wanted a piece of the action. Though the best guess was that the disease was caused by the yellow fever virus, the outbreaks were something new, intellectually exciting. Throughout October and November blood and tissue samples from disease victims in Yambuku, Kinshasa, and Sudan were sent to laboratories in the United States (Centers for Disease Control, Atlanta), the U.K. (the Microbiological Research Establishment, Porton Down, Salisbury), Belgium (the University of Anvers and the Prince Leopold Institute of Tropical Medicine), West Germany (Bernard Nocht Institute for Naval and Tropical Diseases), and France (special pathogens branch of the Pasteur Institute).

On October 11 the Pasteur Institute's director of overseas research, Claude Hannon, told Pierre Sureau to go to Roissy Airport to retrieve a package containing blood samples from Kinshasa, adding that he should "consider the packet's contents dangerous." The perilous shipment was, however, misrouted to Paris's Percy Hospital, passing through many hands before Sureau was able to track it down.

When hours later he obtained the curious box and opened it at his lab bench, Sureau found a thermos flask containing several Vacutainer tubes of blood surrounded by dry ice—a commonly used freezing protective layer. Tucked among the tubes was a note from Dr. G. Raffier of the French Embassy in Kinshasa, dated October 10, 1976:

Sir, the enclosed tubes contain blood samples collected at a mission October 4 to 9 on patients and illness contacts at the hospital of the Catholic Mission of Yambuku, Bumba Zone, Equatorial Region of the Republic of Zaire. This village of Yambuku and another close neighbor, Yandongi, are currently seized by a deadly epidemic of indeterminant nature. It began September 5. It is now in regression (10-9-76). . . . The first assumptions were that the region was hit by yellow fever (but four of the dead Belgian missionaries were vaccinated) or typhoid fever. The first analysis done at the Institute of Tropical Medicine (IMT) of Anvers eliminated yellow fever and typhoid; a virus not seen before was isolated at Anvers.[9] We have not yet received results of a liver biopsy sent to Dakar. A diagnostic assumption of Lassa has been advanced, but not proven to date. The fresh blood samples have been preserved on dry ice.[10]

Sureau knew Lassa could be terribly dangerous—he'd certainly heard of Jordi Casals's near-fatal infection. But he had no reason to believe the suspected virus could be airborne. He placed the nine tubes in a rack atop a sterile lab table, opened the first, and dabbed a sample on filter paper.

The implications of such casual behavior would be obvious a few weeks later. One of the tubes contained Sister Edmonda's blood.

But as Sureau looked at the tubes his only thought was: "What shall I do first? Electron microscopy? Antibody complementarity assays?"

He was smoking a cigarette, mulling it over, when the phone rang. Paul Brès, chief of the Viral Diseases Branch of WHO, was calling from Geneva.

"Pierre, have you received the suspected blood samples from Zaire?" he asked.

"Yes, Paul, I got them this morning."

In an urgent tone Brès stressed that the samples were "highly infectious and must be studied in a maximum-security laboratory. They must be sent on immediately to the CDC in Atlanta. Don't open them!"

"Too late, Paul, I already did," Sureau said, anxiously glancing at the nine neatly lined-up tubes.

Brès instructed Sureau to repackage the tubes immediately and ship them by overnight plane to Atlanta. Then Brès asked Sureau whether he would serve as the official WHO consultant for the mysterious epidemic. Sureau agreed without hesitation and left the following day for a briefing in Geneva. He would be in Kinshasa within thirty-six hours.

As requested, Sureau sent the nine test tubes to Karl Johnson at the CDC, and enclosed his own note summarizing the contents of Raffier's letter and information from Paul Brès, noting that he had repacked the samples in more secure containers.

"I am leaving this evening for Kinshasa on a mission for WHO," Sureau concluded, "to participate on the ground in research. My instructions are to send to the CDC clinical samples I collect."

A week earlier, Peter Piot, then only twenty-seven years old, was completing his virology postdoctoral research at Anvers when the first mysterious blood samples had arrived from Zaire. With Piot were Flemish biochemist Guido van der Gröen, Bolivian physician René Delgadillo, and their boss, Stefan Pattyn. The group looked at the odd blue thermos that reached them via Brazzaville and discussed rumors they'd read about in the Dutch press of, as van der Gröen put it, "something weird in Zaire, involving Belgian missionaries."

An accompanying note from WHO authorities in Brazzaville indicated that yellow fever was suspected.

"Well, that's not very dangerous. Not in the lab anyway," Piot reasoned. He blithely pulled on a pair of latex gloves and, without further precautions, opened the thermos. Inside he found a soup of melted ice, an illegible, water-soaked note from somebody in Zaire, an intact test tube, and another one, broken into pieces, its contents mixed into the watery soup. Piot, under the watchful eyes of his colleagues, removed the intact tube, setting it out on the tabletop in their lab inside a mundane research facility in the city of Antwerp.

Years later, while eating a luncheon salad of *jambon* and *fromage* in a noisy Rive Gauche café in Paris, Piot would explain that he had been "young, foolish, and fearless" and that it wasn't until well after Christmas in 1976 that he stopped to reflect on the tremendous dangers he had faced. Only then did he allow himself to finally experience fear.

But in the first week of October all the ambitious young Belgian saw when he looked at the samples was a wondrous mystery. He and van der Gröen first prepared samples for standard yellow fever antibody tests, using antibodies that would react with the contents if the virus was present. Negative. He repeated the yellow fever test. Still negative. Then he tried typhoid antibody. Also negative.

But van der Gröen confirmed that whatever was in that odd blue thermos from Zaire was quite deadly by putting droplets from the intact test tube into larger tubes containing so-called Vero monkey cells. Within eleven days, the Vero cells were dead, and when van der Gröen withdrew liquid from the dead Vero tubes and put it in tubes full of fresh Vero cells, they too died within ten to eleven days.

The laboratory in which this work was done had no special security or containment facilities, no fancy hoods to draw dangerous bugs up into ducts, away from scientists' mouths. Indeed, the Belgians labored under conditions no more sophisticated or secure than might be found in a typical high school biology lab.

Their folly would prove striking in retrospect, and all concerned would later express astonishment that they suffered no ill consequences from such frivolous disregard of the potential hazards of the microbes.

Indeed, three days into their research, the much older Pattyn removed a rack full of incubating infected Vero cells for examination. He tilted the rack to get a clearer look, and a tube slid out, crashing to the laboratory floor.

Delgadillo and van der Gröen stared in panic at the wet floor, the Bolivian noting that liquid had splashed on his shoes. Van der Gröen, spotting his Bolivian colleague's anxious glances at his shoes, looked at his feet as well: fluid splattered his wing tips in deadly little beads. Delgadillo and van der Gröen exchanged worried glances.

After a few moments, Pattyn suggested that van der Gröen "clean it up," and left the laboratory. With gloved hands, van der Gröen and Delgadillo gingerly wiped up the floor and their shoes, then liberally spread disinfectant around the facility.

Shortly after the Belgian group's Vero cell studies confirmed the dangers of the mysterious Zairian microbes, their government began questioning the wisdom of continuing the Antwerp research effort. They were instructed to pass the samples on to higher-security laboratories outside Belgium. Van der Gröen convinced Pattyn to save one small sample, reasoning that it should be used as a backup, in case the primary samples were damaged or lost in shipment to Porton Down.

Having ruled out the easy answers, Piot eagerly prepared the sample for analysis under an electron microscope. He gasped as he stared at the strange viruses; they were shaped like question marks.

"This is a new virus! It's something we have never seen before," he exclaimed, feeling the thrill of discovery. The virus was a long wormlike tube that coiled at one end and left the other extended. Piot imagined that when he asked, "What is this?" the viruses simply answered back: "????"

Thoroughly committed to solving the mystery of the "???? viruses," Piot was disappointed when WHO telexed on October 7 that the group should cease all research immediately, saying, "Investigations indicate this may be Marburg." Piot packed the last sample, wrote up his findings, and, as per WHO instructions, shipped the lot off to Karl Johnson at CDC. He was intrigued by the diagnosis and wanted to go to the scene of the epidemic to see for himself.

The usually shy Piot uncharacteristically marched over to the Belgian Ministry of Development Cooperation and argued his case. "We have to be there," he said. "There are missionaries, Belgian missionaries who died."

He didn't need to underscore Belgium's unique relationship with Zaire. In 1876 the European power had begun to colonize and brutalize the Congo, as it was called. Now, almost exactly a hundred years since King Leopold II declared the Congo a part of the Belgian Empire, authorities in Brussels were at pains to rid their country of its guilty legacy. On the other hand, the Belgian government was also acutely aware of the risks inherent in offending Mobutu or his government. It was an extremely delicate situation to place in the hands of a twenty-seven-year-old, politically naïve postdoctoral student.

"All right," Piot was told, "You can go. We will only fund one week. And you're representing the Belgian government."

Carrying the only suit he owned, he may have been prepared to meet officials in Kinshasa and travel around Zaire for a week. But he was woefully ill equipped for what would become a three-month stay in a tropical rain forest during the Zairian summer.

Dr. Stefan Pattyn, before sending his samples on to England's maximum-security laboratory in Porton Down, had completed studies in laboratory mice, which showed that the virus was quite lethal to rodents. He had also compared the mystery virus to Lassa, concluding that "it was probably some other arbovirus," not the West African killer. Now he too departed for Zaire, leaving van der Gröen behind to monitor the health of the accident-exposed members of the Antwerp laboratory.

On October 14, Patricia Webb and Fred Murphy completed their first round of studies of the mystery virus, working in the CDC's maximum-security laboratory. In 1976 the lab was designated a P3 facility. A P1 facility was a basic laboratory such as could be found lining the hallways of university science departments; a P2 facility had a slightly higher level of security with entry limited to trained, authorized personnel and actual research work performed under hoods that sucked air away from the experiment, up a ventilator duct, and past scrubbers that disinfected the air with ultraviolet light and microscopically gridded filters; a P3 lab was state of the art in high-security research. For Webb, working in a P3 lab meant passing through a series of guarded locked doors, presenting her security pass for entry. She would then shower with disinfectant soap and don a set of head-to-toe protective clothing, gauze face mask, double latex gloves, and radiation badge to monitor her levels of exposure to isotopes occasionally used in such research. She would then pass through two more air locks lined with microbe-killing ultraviolet lights.

Once inside the inner core, Webb might enter either the laboratory or the animal room. Both rooms were pressurized; all air was forced in past microscopic filters and sucked back out rapidly through several additional layers of filters, ultraviolet lights, high heat sources, and chemical scrubbers.

A further layer of protection was provided by glove boxes: more sophisticated versions of the portable box Karl Johnson jury-rigged for his studies of the Machupo virus in Bolivia. All Webb's samples from Zaire were stored in deep freezers overnight; small amounts were thawed during the day and analyzed inside the boxes. Webb would thrust her already double-gloved hands into a larger set of thick rubber gloves that were permanently installed in the clear-plastic front wall of the hooded box. She would then try, with three cumbersome layers of rubber over her hands, to manipulate test tubes, pipettes, petri dishes, and the like. It was slow-going, arduous work that often proved physically exhausting.

Harder still was the animal work. To find a mysterious microbe, it was necessary to inject samples into mice, guinea pigs, hamsters, and monkeys, all of which were also kept in large glove boxes. The animals didn't sit still in the grasp of bulky gloved hands, and injections were often a test of wills between scientist and guinea pig.

In such a setting the greatest risks to the scientists were accidents, such as cutting oneself with a broken contaminated test tube or receiving an animal bite. Webb had never cut herself, but she had been bitten several times by monkeys that attacked her approaching gloved hands. Fortunately, those monkeys were part of Webb's Machupo research, and, having already suffered the disease, she was immune.

These Zairian samples, however, tested negative for Machupo, and Webb was acutely aware of the need to work with slow, cautious deliberation. It was not her style, really. When Patricia Webb graduated in 1950 from Tulane University Medical School in New Orleans, only eight other women were in her class. In those days a handful of women were given the opportunity to matriculate into a field dominated by males. Unlike most of her fellow students, Webb never planned to spend her life in a profitable practice giving middle-class kids antibiotics for strep throat and monitoring the blood pressures of obese patients.

Since childhood in England Webb had been fascinated with stories of India, Pakistan, and China and saw medicine as a sort of universal passport.

It hadn't gotten her to India yet, but through medicine and research virology she had already seen Malaysia, Panama, Bolivia, California, Louisiana, Georgia, and the Washington-Baltimore area. But now she found herself locked inside an artificial environment day after day.

The further Webb got into her research, however, the more obvious it became that the CDC needed to deploy a team immediately for fieldwork on the ground in Yambuku. With the approval of her CDC seniors, Webb began amassing further information and planning her fieldwork.

She asked the CDC's personnel office to find a staff scientist with three key qualifications: fluency in French, strong African experience, and training in epidemiology. The name Joel Breman popped up.

Breman had spent six years in Africa since completing his medical studies—two years in Guinea and four in Burkina Faso. He had been part of D. A. Henderson's successful campaign to eradicate smallpox, and he was fluent in African-dialect French. But Webb was a little anxious when she noted he was technically an EIS (the CDC's Epidemic Intelligence Service) trainee.

In late September, when the CDC's Lyle Conrad contacted Breman in Michigan, the epidemiologist was knee-deep in another investigation—of Swine Flu. Conrad asked if the EIS trainee would like "one hell of an assignment. It's in Africa, it's a little frightening. Something has killed just about every villager in the area. You'd be gone about a week."

Having spent six years in tropical Africa, Breman knew nothing got done in one week. And he didn't like the sound of this particular mystery. Nevertheless, over the next three weeks the tall, bearded scientist talked almost daily over the phone with Webb, getting a sense of the excitement and fear surrounding the Zaire outbreak. For her part, Webb soon grew used to Breman's long-winded, often cliché-packed ramblings. Beneath his occasionally incoherent conversational style lay a sharp intellect that Webb recognized and planned to push to its limits in Yambuku.

On October 10, Webb and her co-worker, Fred Murphy, officially informed WHO that "the illness is caused by a virus that resembles Marburg (Marburg-like), that the epidemics are probably caused in Zaire and Sudan by an etiological agent that is similar but represents a new immunotype that is in the family of Marburg."[11]

Webb's Marburg speculation prompted an international escalation in scientific security. Thereafter the CDC and Porton Down—the world's most secure labs in 1976—received virtually all samples of the mystery agent.

At Porton Down it was Geoffrey Platt who handled most of the mystery virus research. His lab wasn't exactly an American-standard P3 facility; rather it was a uniquely English mix of P3 and P2 elements. Because the British antivivisection movement was quite militant in its opposition to the use of laboratory animals, security in the form of controlling access to Porton Down was very high. Indeed, most British citizens had no idea where the lab was located or what it did.

Since 1964 Platt had worked at Porton Down with dangerous viruses, particularly Lassa, taking precautions to protect himself, though the microbes were not kept safely inside glove boxes, as was done at the CDC. The rooms were, indeed, pressurized, and the air was decontaminated before being released into the English countryside, but Platt's personal protection was limited to a cloth surgical gown, a double layer of latex gloves, and an old World War II-era gas mask. Though the respirator had been thoroughly tested for its effectiveness in protecting British soldiers from combat gases, it had never been proven that the mask filtered out viruses. Nevertheless, the handful of Porton Down scientists and technicians who worked with super-lethal microbes were limited to using the cumbersome, often hot masks that always seemed to cloud up in the midst of delicate procedures, usually leaving researchers with headaches by the end of the day.

Every night after work, Platt would scrub his mask with Lysol and spray it with formalin disinfectant.

Though mindful of the risks, and very careful in his work, Platt knew there were dangers, especially when working with an unknown, Marburg-like killer.

"Care is absolutely essential," Platt told his colleagues, warning that nobody should enter his lab or animal care area unless absolutely necessary—at least, not until Platt knew what lurked in those test tubes. "You've got to realize you're working in some danger and be able to accept that. It's not good if you're going to go home at night and not be able to sleep."

Platt had no way of knowing that in just three weeks he himself would lose a great deal of sleep worrying about his own chances of survival.

Platt's work on the Sudan samples prompted WHO to release, on October 15, the following urgent bulletin:

Haemorrhagic Fever of Viral Origin. Between July and September 1976, it was observed in the region spanning N'zara to Maridi, in southern Sudan, sporadic cases of fever with haemorrhagic manifestations. It is thought that the first cases occurred among agricultural families. During the last week of September, the situation worsened considerably, 30 of 42 cases occurred in Maridi hospital among members of the staff; it is thought the disease was spread directly from one person to another. By October 9, 137 cases, 59 deaths, were reported for the region comprising N'zara, Maridi and Lirangu. The epidemic has caused panic on the local level. . . .

The report closed with these words: "Samples from Sudan and Zaire have revealed the presence of a new virus, morphologically similar to Marburg, but antigenically different."

Well before WHO officially released that report, the agency had confirmed from three labs (CDC, Anvers, Porton Down) that a deadly new virus had been discovered, and had initiated an international effort to try to stop the epidemics in Zaire and Sudan, identify the virus, and determine how and why it had appeared. In a matter of days, what began as a problem in a missionary hospital would involve investigators and military personnel from eight countries, several international organizations, the foreign ministries of at least ten nations and the entire health apparatus of Zaire. Almost overnight, events would snowball into an effort necessitating over 500 skilled investigators, and mobilizing the resources of numerous European and American institutions, all at an indirect cost of over $10 million.

Direct costs for the Yambuku investigation alone would exceed $1 million.

IV

The snowball effect began modestly enough on October 13, with Pierre Sureau's arrival in Kinshasa. The Pasteur Institute virologist represented WHO for the duration of the epidemic, and had the task of assisting Zairian authorities in any way possible. Sureau's first meeting was with Minister Ngwété Kikhela, who informed the French scientist that it would be several days before transport to Yambuku could be arranged. Such delays were to become a major component of this investigation, one that was constantly plagued not only by the mysterious virus but also by logistical nightmares aggravated by national panic. All commercial flights to Bumba had ceased as a result of the regional quarantine. That left only Zairian Air Force transport to the region, but terrified pilots were rebelling against orders to enter the Bumba Zone.

Though his hopes of getting an immediate look at the Yambuku epidemic were thwarted, the spry, middle-aged French doctor was able to see a case of the disease on his first day in Zaire. Having nursed her dear friend Sister Myriam, Sister Edmonda now lay dying in Ngaliema Hospital's Pavilion 5 isolation ward. Sureau found her semi-delirious, severely dehydrated from days of diarrhea, feeble, anorexic, feverish, completely drained of energy; yet, surprisingly, unafraid.

"She knows what is coming. She has seen it before with Sister Myriam and all the cases in Yambuku. Yet she is calm," Sureau noted, with considerable amazement.

Sister Edmonda thanked the doctor for his attention and "the good conversation," and clutched the hand of an elderly Kinshasa nun, Sister Donatienne. Sureau took a blood sample and departed.

That night Sister Edmonda died.

"My God!" Sureau exclaimed. "That virus is fast!"

The following morning, October 14, Sureau returned to Ngaliema and discovered that a new patient had arrived. Student nurse Mayinga N'Seka, who had tended to both Sister Myriam and Sister Edmonda, was developing the first symptoms of the mysterious disease at about the time Sister Edmonda died. Two days earlier, Mayinga had spent hours in a general administrative office awaiting transit papers for overseas study, where she had contact with numerous strangers and officials. She then took a taxi to Mama Yemo Hospital, where she sat in a crowded waiting room, waiting for someone to treat her fever, headache, and muscle pains.

Sureau and Ngaliema doctors quickly determined that Mayinga had the Yambuku disease, and transferred her to Ngaliema's Pavilion 5 isolation ward. Concern and rumors started to spread through the streets of Kinshasa.

Meanwhile, WHO remained convinced the culprit could still be a strain of Marburg disease, so Sureau and Close contacted the South African team that had treated the Australian tourists a year earlier, asking for antiserum. The politics of such a request were dicey, and necessarily involved notifying the Mobutu government, South Africa's apartheid leaders, and the embassies of France and the United States. Though it violated Zaire's ban on relations with South Africa, all government representatives eventually agreed, for the sake of young Mayinga and the people of Yambuku, to allow Dr. Margaretha Isaacson to fly up from Johannesburg, Marburg antiserum in hand.

"It's our only hope," Sureau told Zairian officials.

Talking incessantly, Isaacson hit the ground running and approached medicine like a field commander, ordering the Ngaliema medical troops about and bringing instant order to a scene that had been dangerously close to chaos. She and Sureau gave Mayinga the Marburg antiserum, and then the South African sat down with Zairian doctors to plan the transformation of Pavilion 5 into a bona fide isolation ward. The Zairian medical staff, which had been in a state of extreme agitation ever since their colleague fell ill, was thrilled to see the "space suits" Isaacson brought from South Africa. Soon the entire staff of Pavilion 5 worked in head-to-toe white suits that had clear-plastic face coverings and respirators. The suits proved horrendously uncomfortable in the steamy Kinshasa heat, but the Ngaliema staff was enthusiastic about the protection.

They were far less enthusiastic about Isaacson's recommendation, supported by the Zairian Health Ministry, that the entire Pavilion 5 staff be placed under quarantine. Health Minister Ngwété made it clear his greatest concern was the possible spread of the Yambuku virus from Ngaliema Hospital into the streets of Kinshasa, endangering the 2 million residents of the capital. For nearly a month, a half dozen staff members would be confined to Pavilions 5 and 2 of Ngaliema Hospital, forbidden to leave the confines of the area to see their families.

Officials tracked down 37 people with whom Mayinga had shared meals or close contact in the days prior to her illness, placing all the unfortunate men, women, and children inside Pavilion 2 for twenty-one days of quarantine. One quarantined woman would give birth during her stay, and all the staff and isolated civilians would fight day-to-day personal battles against boredom, fear, and fatigue. In addition, 274 people who had had recent contact with the Pavilion-bound individuals were found, blood-tested, and kept under close surveillance.

Fortunately, no further cases of the Yambuku disease would develop in Kinshasa.

Years later, reflecting on the extreme precautions taken at Ngaliema Hospital, Isaacson would say that "perhaps we were overdoing things a little bit," but "we could not afford doing less than the maximum precautions that were available. We could not do it ethically, we could not do it scientifically."

Constantly abandoning all precautions—much to Isaacson's consternation—Sureau never wore a mask, and often spent long periods of time at Mayinga's bedside, chain-smoking cigarettes and dispensing calming conversation. Despite huge cultural and generational gaps, the student nurse and the physician became close, and Sureau often voiced his increasingly urgent hope that the Marburg antiserum would save his new friend. Mayinga herself was far from optimistic. Having seen the agony the Sisters had endured, she was frankly terrified.

"Dr. Isaacson is here," Sureau told Mayinga gently. "She is one of the greatest experts in the world on Marburg. You are in very good hands. Have faith."

Later, as he carefully prepared samples of Mayinga's blood for shipment to Pat Webb's CDC laboratory, Sureau could barely contain his excitement about the coming trip to Yambuku.

"For the community of arbovirologists, this is one of the greatest events in contemporary epidemiology," he noted in his diary. "No one of us would pass up such an opportunity for passionate study. Personally, I am delighted to be in this place, and to participate in such an adventure."

Sureau's enthusiasm was tempered the following day, however, when Mayinga's condition deteriorated. Isaacson decided to try a second dose of the precious antiserum, and Sureau again comforted Mayinga by telling her that Isaacson was an expert. But by then the French and South African physicians knew the truth: whatever was infecting Mayinga was *not* the Marburg virus.

On October 18, six weeks after the Yambuku epidemic began, the core of what would that day be dubbed the International Commission arrived. Loaded down with enormous crates of sophisticated laboratory equipment, a plastic isolator for research, state-of-the-art microscopes and protective gear came the Americans: Karl Johnson and Joel Breman of the CDC's Special Pathogens Branch. Still in her maximum-security Atlanta laboratory, Patricia Webb was steaming mad. Just days before her planned departure, CDC director David Sencer had decided the job was "too big," and leadership of the mission was awarded not to the woman, but to her husband.

The Machupo legends had preceded Johnson to Kinshasa, and Sureau's admiration for the man who discovered and survived Bolivian hemorrhagic fever was undisguised. Now a middle-aged veteran of dozens of CDC investigations, Johnson carried himself with a reasoned calm that inspired confidence in the men around him. He would be the foreman for an often adventuristic bunch of disease cowboys. Johnson, Breman, and Sureau became instant friends, and everyone, including the Zairois, deferred to Johnson's leadership.

By the end of the day the core of the International Commission was in place, and its first meeting convened (tensions eased by ample quantities of French wine) at five o'clock in the evening, October 18, chaired by Minister Ngwété. Present were six Zairois, including Omombo, whose twenty-four-hour visit to the Yambuku Mission had dramatically raised levels of anxiety in Kinshasa government circles. Representing WHO were Sureau, smallpox expert René Collas, and two Zaire-based European physicians. The five-man Belgian contingent included Stefan Pattyn and Peter Piot. One South African (Isaacson) and one official French representative (Gilbert Raffier) were present. And Americans Johnson, Close, and Breman were joined by Dr. John Kennedy of the U.S. Embassy. In coming weeks this core group would guide nearly all Yambuku-related activities, operating in several languages, crossing often difficult political and cultural boundaries, each professional adhering to his or her designated responsibilities and all answerable to the acerbic, often flat-out outrageous Johnson.

Niceties and introductions taken care of, Johnson swiftly guided the multilingual group through its marching orders, delegating responsibilities and laying out a strategy for attack that drew heavily from experiences with

Machupo and Lassa. Breman was put in charge of epidemiology investigations: doing the detective work necessary to determine who was spreading the disease, how, and with what clinical results. Together with Belgian Jean-François Ruppol, Piot, Zairian scientist André Koth, and Sureau, Breman was told to prepare for immediate departure for Yambuku.

Johnson reminded the group that the virus they were dealing with was extraordinarily dangerous, and using colorful language peppered with obscenities, ordered everyone to take their temperature twice a day, follow to the letter Isaacson's recommendations for protection, and always work in teams.

That night Piot, Sureau, and Breman prepared, each in his own way, for the next day's journey to Bumba. Young Piot, who had never before set foot outside Europe, was anxious to get out of the wedding suit his government had instructed him to wear, and see the infamous nightlife of Kinshasa. All night long the Belgian doctor strolled the streets of the city, chattering incessantly with the friendly Zairois, listening to the *ramba* rhythms in nightclubs, and sampling local drinks and delicacies.

"This is wonderful!" Piot exclaimed to local team members who showed him the town. "What an exciting place."

He didn't want to sleep, or think about the epidemic. Piot arrived at the airport the following morning groggy and caffeine-sobered. Though sleepy, he grew increasingly alert as the time for the team's departure drew closer.

Sureau was also excited as he sat in the President's plane awaiting takeoff.

He had to admit, however, that he was "a little scared of the unknown," and had been in no mood to party all night with the young Belgian. Instead, Sureau had paid another visit to Mayinga, finding her condition further deteriorated. The young student nurse was emotionally overwrought. He reviewed virus containment and protection procedures with Isaacson. And from her got a copy of *Marburg Virus Disease* by Martini and Siegert,[12] which he was now trying to read aboard the Hercules jet.

Breman, feeling the dull disorientation of jet lag, had spent the evening working out logistics with Johnson and Ruppol and making sure the proper equipment found its way into the plane's cargo bay.

After a three-hour flight the jet landed on Bumba's tiny airstrip. The terrified Air Force pilots left the engines running and ordered the scientists to get out as quickly as possible. Piot drove their Land-Rover, packed with supplies, out of the cargo hold, and hadn't even parked it before the panicky pilots began to taxi for takeoff.

Piot could feel the eyes of hundreds of people upon them as he inspected the Land-Rover. The airstrip was lined many rows deep with anxious-looking people.

"My God, the entire town must be here," he whispered to Sureau.

"They've been under quarantine for days," Breman reminded his colleagues. "They're fed up, and they're scared. I imagine they think we're going to perform some kind of miracle."

That night the scientific team shared the hospitality of Catholic missionaries in Bumba, who brought them up to date on radio messages from Yambuku. Bumba physicians Ngoi Mushola, Zayemba Tshiama, and Makuta briefed the foreigners on their clinical observations, noting that the epidemic had spread to several villages around Yambuku.

Sureau passed the final hours before retiring that night in quiet discussion with the grief-stricken priests of Bumba. Joel and Peter, however, were too high on adrenaline and curiosity to sip vermouth with aging clerics, so they went to a folk music festival at the local cathedral.

The following morning the group looked in on a handful of mystery disease patients at the Bumba hospital, and, fortuitously, met Dr. Massamba Matondo, chief physician for the neighboring Lisala province. Massamba, a careful doctor with an instinct for epidemiology, had already toured the epidemic area and he told Sureau the disease was claiming residents of at least forty-four villages in a fifty-mile radius around Yambuku.

With Massamba and Bumba missionary Father Germain Moke, the scientists made their journey to Yambuku later that morning in two Land-Rovers. The fifty-mile drive took all afternoon. Rarely could the frustrated drivers gain enough speed on the bumpy muddy roads to get out of second gear.

More than three hours later, they reached Yambuku.

They turned off the engines and immediately felt the sad silence of the place. Gone were the noise and activity of typical Zairian villages, the long lines of chattering women and children waiting for vaccinations, the vendors selling their wares. Indeed, gone were the people, altogether.

Piot spotted Sister Genoveva's odd white gauze "cordons sanitaires" strung about the mission, and a sign in French that said: "Do not enter; to call the Sisters ring the bell." As he approached the bell three nuns came running out of one of the buildings, calling, "Don't come near! You're going to die! You will die! Stay away!"

Recognizing their Flemish accents, Piot jumped over the "cordons" and shouted greetings in their shared native tongue. Overwhelmed at hearing Flemish, the nuns broke down, sobbing. Sureau, Ruppol, and Breman quickly joined Piot in his efforts to comfort the women, and the Sisters were pleased that Jean-François had, as promised, returned to their devastated outpost. As tensions and emotions eased, the scientists unloaded their equipment and followed the Sisters to the school. Closed since the fourth week of the epidemic, the barren classrooms became their temporary home.

Over dinner and plenty of wine the Catholic teachers and clerics poured out their stories for hours on end, while the visitors patiently listened, asked gentle questions, and occasionally jotted down notes. Sister Marcella, who had been keeping logs of the dead, presented her grim lists to Sureau.

Speaking in a deliberate monotone, which seemed to help her keep her emotions in check, Sister Marcella explained that in the past month 38 of the 300 residents and employees of Yambuku had died, including all the missionary nurses, four out of six Zairian nurses, one of the three padres, and one of two hospital laundry workers. Then she gave the scientists a sobering list of villagers afflicted. The visitors realized they would have to go to every single village, conducting a house-to-house investigation. No other approach would do.

Sister Marcella also volunteered that the first unusual medical problems at the hospital may have occurred in August, when three women died in close succession on the obstetrics ward. They had bled to death after giving birth. The Sister had checked hospital records for the same time periods in 1975 and found no such cases, and she was unable to tease out of the general hospital records cases of anything similar to the strange new disease prior to August 1976.

"It is new," she told them. "It is definitely something new."

V

While the exhausted scientists slept on the hard floors of the Yambuku school, Mayinga lost consciousness in Ngaliema's Pavilion 5. And the commission members argued late into the Kinshasa night about contingency procedures for handling infected team members.

Joe McCormick had just started unpacking his hundreds of crates of laboratory supplies for Lassa research when he received a cable from the CDC in Atlanta, instructing him to temporarily abandon the lab outside Kenema, Sierra Leone, and make his way as quickly as possible to Kinshasa. The cable stated that his familiarity with northern Zaire, coupled with his epidemiology skills, made him indispensable. He was instructed to bring with him the portable glove-box lab he and Johnson had rigged up in Atlanta just weeks earlier, and other equipment that was needed at Ngaliema Hospital for testing and screening blood samples and preparing antisera against the mysterious disease.

Just a few months earlier, having heard of McCormick's exploits in Brazil, Johnson had snagged Joe one day in the CDC hallway.

"I'd like to send you to Sierra Leone," Johnson said, "to figure out just how widespread Lassa really is."

McCormick hadn't been in West Africa, and the Lassa puzzle sounded "damned interesting," so in March 1976 he packed and prepared to set up a one-man Lassa research station in Sierra Leone. Just before he left, McCormick and Johnson rigged a glove-box contraption similar to the one Karl had used in Bolivia, and Joe gathered enough equipment to study the virus safely under even the most primitive conditions.

Within a week he had his Lassa research station: a small building 200 miles from the capital, outside the town of Kenema. It contained two chairs and his cases from Atlanta. He'd just uncrated Johnson's portable laboratory when the cable arrived from Kinshasa.

McCormick knew there was no easy way to get from Freetown, Sierra Leone, to Kinshasa: virtually all flights between African countries were both notoriously expensive and unreliable; nearly all went from one African country to another via the formerly colonial European countries.

For three days McCormick bluffed, bullied, and bribed his way onto airplanes and through customs in Freetown, Abidjan, and, finally, Kinshasa. He completed the 2,000-mile journey with all equipment crates, remarkably, intact. At Kinshasa's N'djili Airport he sprinkled a little Kiswahili and Otetela in with his French, and eventually convinced customs and immigration officials to let his crates into the country, unopened, undamaged, and unexamined.

Meanwhile, the scientific team awoke with the Yambuku dawn, relayed an abbreviated version of Sister Marcella's reports by radio to Bumba (from where they were ultimately relayed to Kinshasa), breakfasted, and set out in different directions in four teams to inspect the villages. Piot and Sureau were teamed up, and Sister Marcella, ecstatic to be outside the mission after so many days of quarantine, acted as their guide.

"We must limit the numbers of us who are exposed to this virus until we determine how infectious it is," Sureau told the group, instructing that only he and Piot should draw blood.

The trio first arrived in Yalikonde, close by Yambuku, where they quickly learned how to gain the trust of the fearful villagers. A working pattern developed that was repeated in ten villages that day. It would begin with an amble about the middle of the village, during which time the leading elder of the community would introduce himself. The group would discuss the weather for a while, until the elder invited them to share some *arak*.

"This stuff is pure methanol," Piot whispered.

"Drink!" Sureau commanded.

After the *arak* burn had made its way down their throats and into their stomachs, the Yalikonde elder introduced the white men to Lisangi Mobago, a twenty-five-year-old man who had been struggling with the disease for six days. The visitors examined Lisangi, who was far too weak to protest, and drew a blood sample.

Everywhere the group went they noticed the people had taken remarkably wise measures to stop the epidemic's spread. Roadblocks were staffed around the clock near village entries, virtually all traffic on the Ebola and Zaire rivers had come to a halt, the ailing villagers and their families were kept under quarantine, bodies were buried some distance away from the houses, and there was little movement of people between communities.

"These people have really got their act together," Piot told Sureau, who was also impressed by the steps taken.

In one village about ten miles from Yambuku, Piot and Sureau found a husband and wife lying side by side in their hut, both in the final throes of the disease. Pierre took blood from the husband while Peter prepared the wife's arm.

Sureau shifted his weight to face the wife, found a vein, and inserted his syringe. As he released the tourniquet and watched blood slowly fill the tube, the husband let out a deep groan and died. The wife cried out, Sureau quickly withdrew his needle, and she rolled over to embrace her dead husband.

Shaken, they stepped out into the sweltering sun and whispered anxiously. If the husband had died while Sureau was taking his blood, the villagers might have attacked the men, accusing them of responsibility or, worse, of homicide. As it was, many villagers were taken aback when the tall white men—especially Breman and Piot, who both stood over six feet—donned goggles, rubber gloves, and surgical masks before entering the homes of the infected.

For their part, the Europeans and the American had no idea whether these modest precautions were adequate to protect them from what they now understood firsthand was a particularly horrible disease. Breman was a bit anxious that villagers might be offended by his protective gear, but he was also, as the blunt American put it, "scared shitless."

"I'm no Marlboro Man, and I don't mind admitting I'm really frightened. As, I think, we all should be," Breman told his colleagues, who vigorously nodded their concurrence with his sentiments.

When the team members reunited at the mission after their first long and emotionally exhausting day in the villages, they compared notes and agreed that the epidemic had taken a terrible toll—in some cases claiming entire families—but the worst of it seemed to have passed. Breman, in particular, was relieved to see that initial accounts saying entire villages had been exterminated by the virus were gross exaggerations. Without knowing its cause or cure, the people had wisely taken many proper measures to slow the virus's spread. The scientists humbly agreed that their scientific expertise had not been necessary to arrest the epidemic.

But it would require the best their collective talents could muster to solve the mysteries of where the strange virus came from, how it was spread, and how best to prevent its resurgence.

That evening as they relaxed with more mission wine, layered atop several rounds of village *arak*, a day-old radio communication reached them from Kinshasa, via Bumba.

"Mayinga died late the night of October 20," it stated flatly.

Sureau was devastated, as were the Sisters, who felt profound gratitude for the student nurse's courage in tending to Sisters Edmonda and Myriam.

"What we are dealing with is a virus like Marburg, but *more* pathogenic. A super-Marburg. I don't feel alarmed, but I do feel a sense of disagreeable uncertainty. Who will be the next victim among the caregivers? Sister Donatienne? . . . Margaretha or me? The incubation time is usually eight days! How many more victims will there be in the villages? What can be done to stop this epidemic?" Sureau asked.

The others looked at Sureau in sad agreement, for he was voicing thoughts shared by all.

Though Sureau had clung to the increasingly dubious hypothesis that the epidemic was caused by some sort of Marburg virus—probably in a spirit of hopefulness on Mayinga's behalf—Breman had no such illusions. Breman had been on the phone with Pat Webb at least twice a day for the three weeks prior to his arrival in Kinshasa. He knew precisely what Webb had discovered, and he carried with him eight-by-ten microscope photos of the enemy. As he wandered about the villages, Joel would hold up the "????? virus" pictures of fuzzy, curled, wormlike microbes and explain to the Zairois that this new entity was the cause of their suffering.

The trained epidemiologist of the group, Breman laid out a symptomatic definition of the Yambuku disease that the four teams should use as they scoured villages for cases.

That night, Sureau radioed Bumba to tell Kinshasa that first surveys showed 46 villages were affected, with over 350 deaths.

For the next few days the scientists worked in Yambuku, Bumba, and the villages in between, having no way to communicate either with Karl Johnson in Kinshasa or with one another once they split off daily to investigate separate villages. Sureau and Breman occasionally received garbled messages about helicopters due to arrive with more experts and better equipment. When the copters failed to arrive, they simply assumed the messages were mixed up.

It broke Breman's heart to watch the nuns "communicate" with their ancient ham radio equipment. Every day at a designated time a Sister would put her ear to the decades-old speaker, turn on the radio, and listen through horrendous static for the voice of the monsignor in Lisala. One by one, he would call out the name of each mission in northern Zaire and in this crude manner the network of missionaries would order supplies and share important news.

Though he had hooked up newer equipment, the overall system was so primitive that the American-made side-band radios made little difference. There remained no way to communicate directly with Kinshasa, and all communications were subject to the problems inherent in the child's game of Telephone, in which one person passes a sentence on to another, and another, until after a tenfold relay the message bears little resemblance to the original. U.S. Embassy officials told Johnson that setting up a sophisticated communications system connecting Yambuku, Bumba, Kinshasa, and Atlanta "would entail several million dollars and a twenty-four-hour aircraft relay system." In his usual insouciant fashion, Johnson had words with the officials, but Breman, having had his fill of U.S. State Department types elsewhere in Africa, suggested Karl not waste his energy on the bureaucrats.

Meanwhile, Karl Johnson was trying his best to outmaneuver other logistic nightmares created by the Zairian Armed Forces (which didn't want to fly anywhere near Yambuku), the embassies of the United States and France, and a host of international political issues. He needed a top entomologist or ecologist—somebody who could search, as Merl Kuns had in Bolivia, for the insects or animals that carried the disease. After casting its net far and wide, WHO decided to send French Dr. Max Germain, who worked in the agency's Brazzaville office.

Finally, Johnson needed to get a team further out, way up to Sudan, to figure out how the epidemics of Yambuku and Maridi were connected. For that job Johnson knew exactly whom he wanted: "This one's for Joe," he said, anticipating McCormick's imminent arrival.

McCormick landed in Kinshasa on October 23, the same day the CDC's David Heymann cabled word that a NASA space capsule had successfully been transported from Houston, where it was staffed and ready to receive any WHO team members unlucky enough to catch the virus. Also ready at the Frankfurt Airport in Germany was a USAF C-131 transport jet with an Apollo space capsule aboard—the same space capsule that Henry Kissinger had offered two years earlier for use in airlifting Lassa-infected Mandrella from Nigeria to Hamburg.

That day Johnson got an update from Geneva on the Sudanese epidemic. A team of investigators had been selected, comprised of ten Sudanese doctors, Irishman David Simpson, France's Paul Brès, and the CDC's Don Francis. The investigators were instructed to rendezvous immediately in Khartoum and from there make their way south to Maridi.

Johnson also wanted a top-flight lab worker in Kinshasa right away—someone who could improvise and create a diagnostic laboratory out of the meager facilities available. Pattyn recommended van der Gröen, who immediately flew to Zaire loaded down with essential supplies. Key among his gear were microscope slides he had carefully coated with infected Vero cells. Though they were fixed with acetone, van der Gröen had no way of knowing whether or not those slides were covered with contagious organisms.

"It doesn't matter," he told himself. "I'll make sure I'm the only one exposed to these things. But I must have them. There is no other way to diagnose infection."

His plan was to diagnose infections by putting patients' blood samples on the microscope slides, waiting a while, then rinsing the slides. If patients were infected, they would make antibodies against the mysterious microbe that would latch on to the infected Vero cells. He then planned to mix fluorescein—a molecule that glows under ultraviolet light—with monkey antihuman antibodies. When the fluorescent antibodies were coated onto the microscope slides, they would cling to the human antibody-attached Vero cells. And van der Gröen planned to simply flash ultraviolet light at his slides to see which people had infected blood. Though the method had been in use for Lassa since Jordi Casals's days at the Rockefeller Foundation, van der Gröen had never tried the immunofluorescence technique, and hoped that he would be able to perform professionally in the epidemic pressure cooker.

On arrival at Kinshasa's airport, van der Gröen was greeted by Belgian diplomats who ushered his supplies through customs and whisked the bewildered scientist off into the night. Van der Gröen's initiation into the Zairian investigation began with a harrowing midnight drive on the highway to Kinshasa, which, as was typical, was marked by several near-collisions with cars and trucks that careened at high speeds without using their headlights. The Zairois believed that they saved fuel by shunning headlight use. As Ruppol briefed van der Gröen on the Yambuku epidemic, the terrified Belgian stared agape out the windshield as one oncoming vehicle after another suddenly loomed out of the darkness, coming within inches of their car, which was driven by a Zairois who deftly dodged all the nearly invisible cars and trucks.

At that moment nobody could have convinced van der Gröen that within a few days such sights would seem blasé.

After a night of restless sleep, van der Gröen arose early to meet his new boss, Karl Johnson. Watching Johnson shout commands, always laced with florid language, van der Gröen thought the American "a very peculiar man." But during his first of the daily morning epidemic meetings, van der Gröen witnessed Johnson's uncanny ability to coordinate a multilingual, multicultural team of individual egos, creating a single, well-oiled antimicrobe machine. As had all those who arrived before, van der Gröen instantly admired Johnson and accepted his leadership without question. Johnson ordered the young Belgian to go to Ngaliema Hospital and create a modern laboratory—immediately.

Late that night Johnson came by, holding up some blood samples that Sureau and Piot had sent down from Yambuku. The men uncrated Johnson's glove box, placed it atop the wooden table, and set to work. Van der Gröen had never worked with his hands inserted inside heavy stationary gloves, and he found the process terribly cumbersome and frustrating. Johnson taught the Belgian tricks that he had learned nearly two decades earlier in Bolivia, and the pair were soon toiling smoothly as a team.

Sweat poured off van der Gröen, and every step of the procedure performed in the glove box seemed to take ten times longer than it would atop his Antwerp lab bench. But by two in the morning, he had completed each step of the immunofluorescence process. All that remained was the ultraviolet light microscope examination to see whether or not the serum from Yambuku was infected with the mystery microbe. For that, the pair needed a completely dark room.

Van der Gröen hauled a small table into the bathroom, used the toilet as his seat, and turned off the electric lights.

He switched on his ultraviolet microscope, and with Johnson holding his breath in anticipation, van der Gröen put his eye to the lens.

"*Comme les étoiles!*" he exclaimed. "Like twinkling stars . . . in a dark night, surrounded by red cells. Look, Karl, the cells containing the virus are bright, glowing, fluorescent masses."

Johnson took the toilet seat and peered down the lens. It was three-thirty in the morning, but the men were too excited to sleep. The serum they were looking at came from Sophie in Yambuku, who had survived the disease. Finally they had a way to test who was infected, and to find people who had been infected but successfully fought off the microbe without developing detectable disease. They also now had a way to test whether a particular person's blood contained potentially lifesaving antibodies.

That morning Pierre Sureau awoke in Bumba feeling feverish, unsettled. He and Piot spent the day arguing with the Zairian Armed Forces for transport of equipment from Kinshasa.

As he sat on a veranda overlooking the Bumba Mission, Sureau contemplated what he had learned so far: The disease was clearly deadly, and most victims died within a week. It had taken a huge toll at Yambuku Hospital, and left the surviving nuns and priests nearly incapacitated with grief.

Sipping a cocktail, Sureau contemplated the most awkward finding: the condition of Yambuku Mission Hospital. The previous day he, Breman, and Piot had examined the now empty facility, its medical records, and equipment. When they entered the compound, Sureau had been appalled. The sterilization facilities were abominable, the surgical equipment positively antique, and the linen—though washed—was often covered with old blood stains.

Yet when he examined the medical records, he found no telltale link between people who had undergone surgery in the antiquated facility and those who got the disease. He wiped fever sweat off his brow, and couldn't shake the uneasy feeling that the starting point for the epidemic was— somehow—that well-intended but fatally primitive hospital.

With the contemplative Sureau now in Bumba, awaiting an armed forces plane to Kinshasa, were Piot, other team members, Sophie, and Sukato. As the surviving wife of Yambuku's first victim—the index case, in epidemiology parlance—Antoine, Sophie had antibodies of incalculable value. Yambuku nurse Sukato was the only member of the mission's medical staff who had survived infection and illness. These two in coming days would donate many units of blood, from which would be derived a tiny vial of antiserum of inestimable value.

By October 24 Sureau's secret fever disappeared. He told no one about his illness: Johnson had insisted from the beginning that team members take their temperature twice a day and immediately report fevers to Kinshasa headquarters. Sureau chose to disobey, convinced that whatever he had was not the Yambuku disease.

Fortunately, he was correct.

Sureau was still feeling hot under the collar, however, because the long-awaited plane from Kinshasa hadn't arrived. Johnson sent word that President Mobutu was out of the country, in Switzerland, and in his absence the government was nearly paralyzed. No one else dared order the armed forces to fly into the Bumba Zone.

Sureau and Piot were worried that their samples would go bad in the tropical heat if these delays continued. They made more dry ice for their storage thermos and kept their fingers crossed.

At last on October 27 the Zairian Air Force arrived in the form of a C-130 transport, loaded with lumber and supplies for the construction of a villa for Captain General Bumba, a powerful commandant. The pilots refused to turn off their engines and ordered the group to board immediately.

The pilots grew positively enraged when they spotted the unexpected passengers, Sophie and Sukato.

Ruppol explained the two were very important people: survivors of the epidemic. The appalled pilots swore that nobody infected with the disease could board their aircraft, nor would they allow samples of contaminated blood and tissue on the plane.

A truce was reached: the group could board the plane, but none of them could enter the cabin for any reason. Squeezed in among the commandant's building materials, the passengers did their best to get comfortable during the two-hour flight to Kinshasa. For Sophie and Sukato, both novice air passengers, it was a terrifying experience.

The following morning, October 28, Piot nervously strolled the Kinshasa streets. Like Sureau before him, Piot was running a high fever. Worse yet, he had uncontrollable diarrhea and felt distinctly nauseated. Fearing Johnson would order him airlifted out of the country, Piot hid his ailment from the others, deliberately shunned team members, and searched on his own for dysentery treatments.

Meanwhile, both Sureau and Breman told Johnson that they privately believed the source of most of the fatal cases in Yambuku's epidemic was the hospital. Breman described in detail his eerie stroll through the hospital, and showed Karl two syringes he had delicately removed from a pan of water in the outpatient clinic.

"I'll bet these are infected," Breman said, noting that the clinic issued only five syringes to its nurses each morning. They were used and reused on the 300 to 600 patients who required medical attention each day.

When Johnson suggested it might be a bit dicey to point an accusatory finger at four now deceased Catholic missionaries, Breman said, "The villagers clearly understood the hospital was the source. Long before it was closed, the people voted with their feet. They ran away. That place was almost empty when it closed."

Johnson decided to put Breman in charge of a second survey team, responsible for designing a way to test Joel's hypothesis. The plan was to conduct a major epidemiological investigation with nearly all the International Commission directly involved. The first step would require returning a small team to Yambuku to recruit and train a staff of local Zairois, particularly those who had survived the disease and were presumably immune. Johnson warned Breman, however, that one of the locally recruited Machupo investigators in San Joaquín had mistakenly thought a past ailment was Bolivian hemorrhagic fever. It was not, and Einar Dorado had paid for the error with his life.

"Be damned careful," Johnson said.

On October 27 the commission released its first official account of events in Yambuku to foreign embassies in Kinshasa; the following day embassy officials passed the information on to the international press corps. It was a bland statement, conservatively designed to cast a sense of routine around a crisis that had rendered some team members patently terrified.[13] Still, two scientists, a Zairois and an American, who read the report, requested permission to drop out of the investigation. The American had made it as far as Geneva before turning back.[14]

VI

On October 31, Sureau, Germain, McCormick, and recently arrived Belgian researcher Simon van Nieuwenhove gathered on the military strip at N'djili Airport at four-thirty in the morning. Now accustomed to the anxieties of the Zairian Air Force, the group had triple-confirmed the flight plans. Their cargo was impressive: two fully equipped Land-Rovers complete with a three-week supply of diesel fuel, food (C rations), and water. The plan was for the group to fly up to Bumba, where they would leave off Sureau and Germain, then proceed further north, to Isiro.

For McCormick and van Nieuwenhove, Isiro would be just a first stop on long separate journeys. McCormick was destined for Maridi, van Nieuwenhove was assigned to search the remote southernmost expanse between Zaire and Sudan for additional pockets of the epidemic.

Of course, despite all their prior confirmations, the scientists were told the planes weren't quite ready for takeoff, and at five-thirty in the morning the pilots were scrambling to cannibalize parts from other planes. Finally, five hours later, the reluctant pilots—having exhausted all reasonable dodges—were forced to concede defeat to the scientists, and the C-130 took off for Bumba.

The pilots landed in Bumba and, as always, kept their engines running while Max and Pierre unloaded the cargo and waved goodbye to their friends. Then they took McCormick and van Nieuwenhove to Isiro some 300 miles further north near the Sudanese frontier. The two men drove their well-stocked Land-Rovers off the C-130, shook hands, and headed in opposite directions.

In Isiro (known in colonial days as Old Stanleyville), McCormick hunted for information and some additional supplies, quickly discovering there was little of either to be had. After years of central government corruption, Zaire's most remote areas were bereft of all but local trade, and such "luxury" items as toilet paper, matches, and batteries had long since disappeared.

Information was equally scarce, and McCormick found that the lack of trade activity had slowed the flow of traffic and communication between Zaire's various northern zones to a mere trickle. Surprisingly few people in the region seemed aware of the Yambuku epidemic, and nobody could recall a case of anything resembling that hemorrhagic disease.

As McCormick headed northeast toward the Sudanese border his language skills began to fail him; the more remote the area, the fewer people spoke French, or Otetela, or any other Bantu tongue of which McCormick had a passing knowledge. Soon he found himself in villages not visited by a motorized vehicle in months, even years. And he, a bearded white man, would use an assortment of hand signals and languages to try to find out if anybody in the community had recently suffered an unusual disease. It was hard going, and all too often Joe found himself trying to make sense of conversations that, for example, began in Azandi, were translated into Lingala, and then conveyed to him in French. Information was, at best, muddled.

The closer he got to the Sudanese frontier, the less obvious were the roads. Several times he bailed his Land-Rover out of a river, or plowed through yards of six-foot-tall elephant grass praying the road would reappear on the other side. He was making his way through an area that experienced torrential rains nine months out of the year, and was perpetually mud-laden.

At the border he discovered an Italian Catholic mission so removed from its Roman headquarters that the priests were living on five-year-old flour and the "protein" provided by the insects that infested their meager supplies. Elated at seeing a visitor from so far away, the priests were eager to assist McCormick and insisted he share in their sparse food reserves.

The priests told McCormick there were rumors of an epidemic around the Sudanese village of N'zara, located some sixty miles further northeast. Joe told the priests he had no visa or travel papers.

"That is no problem," they said. "We will take care of it."

After a night's rest, McCormick was introduced to the chief of a Zairian village adjacent to the border. At the priests' request, the chief signed a letter that formally asked his counterpart on the Sudanese side of the border to admit McCormick into the country. Arriving at the much-anticipated border, Joe found two posts, atop which rested a long stick. Between the border posts lay the road, now narrowed to a mud footpath and bearing no signs of recent vehicular traffic. A handful of obviously hungry soldiers sat on their haunches around the site; they greeted McCormick's arrival as a source of grand gossip and entertainment, breaking up an otherwise miserably monotonous day.

When Joe finally reached N'zara he sent a relayed radio signal to Karl Johnson to assure his friend in Kinshasa that he had arrived safely and could confirm there was an epidemic afoot. To accomplish such a seemingly simple task: McCormick first sent a ham radio signal to the Italian missionaries back at the Zaire border. They relayed the message to a pilot flying a missionary twin-engine plane. He ascended to sufficient altitude to be able to send an unblocked signal down the length of Zaire, where it came out of the speaker of Johnson's single side-band radio. With such a complex system it was obvious that the message had to be short and sweet: the details would have to await Joe's return to Bumba.

For three weeks McCormick slept in the Land-Rover by night and interviewed epidemic victims and survivors by day. It became obvious that few people traveled between the N'zara area and Yambuku, a distance of over 400 miles.

By the time Joe arrived the worst of the N'zara outbreak was past, and there were no more active cases in the clinic. For several days he questioned residents of N'zara and the outlying villages, and collected blood samples. Satisfied that the epidemic was under control, his supplies dangerously low, McCormick prepared to return to Zaire. But first, with a hint of mischief, Joe wrote a note to his CDC colleague Don Francis. McCormick knew Francis was heading up an official WHO team that was trying to make its way to N'zara from Khartoum.

Before he left, McCormick put the note in a box and left it with a town leader instructed to "give the container to the white man who will come soon from Khartoum."

McCormick had no idea that Francis and his team were trapped in Khartoum, hostage to terrified pilots who were refusing to fly and government bureaucrats uncertain about providing open access to the Europeans and Americans. It would be several days before the WHO team would reach the area. In the meantime, McCormick's container, pregnant with information, waited in the hot N'zara sun.

In American holiday terms, McCormick left Isiro on Halloween and returned to the Bumba Zone the day before Thanksgiving, having been virtually incommunicado the entire time.

Much had happened in his absence. A full-scale epidemiological survey of all the villages surrounding the Yambuku Mission had been conducted, involving most of the International Commission members. For nearly two weeks, the team, augmented by dozens of trained local volunteers, surveyed over 550 villages, interviewed 34,000 families, and took blood samples from 442 people in the hardest-hit communities. In addition, team members gathered a sampling of local insects and animals to test for viral infection.[15]

And on November 6, Zaire's Minister of Health, Ngwété, issued an international report summarizing findings to date: 358 cases of the viral disease had occurred, 325 were fatal. That was an astonishing lethality rate of 90.7 percent.

Ngwété said all tests in labs throughout the world proved that "this agent is a new virus."

"The name 'Ebola,' after a little river in the region where the disease first appeared, is proposed for this virus," Ngwété concluded.[16]

Somehow, having a name for the culprit had brought new energy and focus to the Yambuku investigation, and the fears of the scientists receded with repetition of the word "Ebola." After a while, "Ebola" sounded almost routine, like "measles" or "polio."

That sense of relative calm evaporated when, several days later, the International Commission learned that Geoffrey Platt had contracted Ebola disease in England.

For nearly a month Platt had toiled with caution and deliberation in his laboratory at Porton Down, trying to learn quickly as much as possible about the Sudan strain of the Ebola virus. On the morning of November 5 he was working in the Toxic Animals Wing of Porton Down, passing Ebola samples from one guinea pig to another to see if the virulence of the virus was diminished as it went through successive generations of animals. As always, he was wearing a respirator, protective lab clothes, and three layers of latex gloves.

His hand slipped.

The syringe containing Ebola-infected guinea pig blood jammed into the tip of his thumb, just above the nail.

Horrified, Platt was seized by panic, and for some time—he had no idea whether it was seconds or minutes—he stared at the thumb and saw his mortality.

"Hurry, get a grip on yourself," he said, ripping off the three sets of gloves and squeezing hard at his punctured thumb tip.

"Bleed, damnit! Bleed," he muttered, but no blood appeared. He dashed out to the next chamber and shoved his hand into a disinfectant tank. For two minutes he held his digit submerged, praying against all biological probability that no virus had actually passed into his thumb; or the disinfectant was getting drawn up into the microns-wide pore created by the needle, killing the virus; or the accident hadn't actually happened at all. He could feel his heart pounding hard against his chest, and feared the adrenaline-propelled organ was all too efficiently pumping Ebola virus throughout his body.[17]

He slowly withdrew his hand from the vat, daubed it with a towel, and used a magnifying lens to search for the needle puncture site. He saw no sign of it.

Carefully following lab exiting procedures, Platt left the Toxic Animals Wing and reported to the Laboratory Safety Office, where he was examined briefly, given a thermometer, and sent home with instructions to report any sudden rise in his temperature.

For six days Platt paced the floors of his lab and house, losing sleep for the first time in his many years of working with lethal viruses. His wife, Eileen, did her best to shield their two preadolescent children from the growing anxiety shared by their parents.

At midnight on November 11, Platt's temperature suddenly jumped, and he felt the chills of a fever. The following morning he reported to the Porton Down safety office. By then his fever was over 100°F, and the staff was very worried, not only on Platt's behalf but also for everyone at the laboratory with whom he'd had contact. They immediately took a blood sample from Platt, examining a droplet under an electron microscope.

The dreaded "???? virus" was there.

Platt donned a respirator to protect others from his virus, and a special ambulance staffed by volunteer drivers and guided by a police escort took the English scientist to North London's Coppetts Wood Hospital. While Platt was placed inside a new Trexler negative-pressure plastic isolator, the 160 other patients then in the hospital were hastily packed off to alternative medical facilities.

For forty-nine days Platt languished inside his plastic environment, which was, in turn, inside an otherwise empty hospital. The large medical staff that tended to him, led by Dr. Ronald Emond, was placed under quarantine. And throughout the first week of Platt's life-struggle, he was entirely cut off from family and friends.

Meanwhile, Eileen and the kids were under house quarantine, forced to constantly check their own temperatures, and terrified that Geoffrey would die.

The British government's reaction to Platt's illness was rapid and severe. Porton Down was immediately shut down, all its employees sent home and placed under surveillance. Several friends of the Platt family were also put under home quarantine. Over a month's time some £200,000 was spent by the U.K. government to compensate employees for lost work time, relocate Coppetts Wood hospital patients, and monitor over 300 people for possible Ebola infection.

Meanwhile, Platt suffered most of the symptoms seen among Ebola victims in Zaire and Sudan. His care, however, bore no resemblance to that available to the people of Yambuku.

As Platt's fever climbed to over 104°F, his hair fell out and he passed blood in his stools and vomitus. Dr. Emond's team attacked the virus with every weapon available. Recently isolated human interferon—a crucial chemical of the immune system—was injected into Platt twice a day in large doses (3 million units). The ailing scientist was placed on intravenous feedings, carefully selected to balance his diarrhea-disrupted electrolytes. When *Candida* fungal infections appeared in his throat, Platt got amphotericin B lozenges. Every fluctuation in his vital signs and blood and urine chemistry were monitored closely.

And forty-seven hours after his fever began, Platt was infused with Sophie's plasma, flown in from Kinshasa.

Following the Ebola plasma treatment Platt's condition worsened; his fever spiked again, he was extremely nauseated, his bowel was incontinent, his joints all ached, and he was very weak. Most alarming to Emond was Platt's mental state. The bright scientist was losing his memory, couldn't concentrate long enough to finish reading a sentence and seemed disoriented.

Platt was indeed very confused.

"Why am I in this plastic tent?" he wondered. "Who are these people looking at me? Where am I? Why can't I read? Did I used to be able to read?"

By November 20, nine days into his illness, Platt began to shed his confusion (along with dead skin and hair), and shortly before Christmas the British government was pleased to conclude that nobody else at Porton Down or in the Platt family had become infected.

England, it seemed, was spared.

News of Platt's illness came to Yambuku on November 12, hitting the commission members very hard. Morale plummeted and collective fear rose. Johnson sensed that the anxiety could be impairing the team's efforts. Certainly Max Germain, whose job was collecting wild, possibly infected animals, was on the verge of panic, and Breman had warned Karl that several team members had asked about the reliability of emergency evacuation plans for flying infected scientists to Johannesburg. Johnson tried to reassure the researchers, but he knew every movement the commission had made since the day it formed had been slowed or impaired by logistic problems.

By November 9, Sureau, having personally searched 21 villages, identified 136 fatal Ebola cases, and mapped the complex relationships of all the dead, recovered, and well people in those villages, was ready to leave. Of all the foreigners flown into Zaire during the epidemic, Sureau had been at it the longest. He was burned out, and both he and Johnson felt the epidemic was over. Sureau began his long journey home.

But the mystery of Ebola was far from solved.

On November 16, McCormick cruised into Yambuku, greeting the commission members with startling news.

"Guys," he said, "what we have here is two totally separate outbreaks. There is no relationship between what's going on here and what's happening in N'zara, except they both happen to be Ebola virus."

Johnson looked at Joe as if he had just watched his protégé's mind snap. Breman shook his head in disbelief. And young Piot grinned, thinking, "Jeez, this guy's got balls!"

McCormick explained to his disbelieving colleagues that travel between the two areas was so arduous, and the cultural gaps so great, that people simply didn't go back and forth.

"There's no way the Yambuku epidemic could get to N'zara or vice versa unless some infected person traveled those roads. And I can tell you, guys, my Land-Rover was the first vehicle on those so-called roads in months . . . maybe years."

Furthermore, he argued, there were no Ebola cases in the villages between N'zara and the Bumba Zone, and the Sudanese epidemic seemed less severe; more people appeared to survive Ebola in N'zara. McCormick's theories were dismissed out of hand by most commission members, and official WHO accounts of the 1976 outbreaks implied there was some as yet undiscovered link between the two epidemics.[18]

Joe stubbornly insisted, however, that despite what seemed coincidence on an unnaturally profound scale, the two epidemics were entirely separate events. He would not abandon that belief with the passage of time, and years later would provide irrefutable proof that Nature had, indeed, rolled an incredibly bizarre set of snake eyes.[19]

VII

While McCormick wrote up his Sudanese findings for the commission, epidemiology investigations continued in the Yambuku area. Piot was left alone full-time in Yambuku, while other commission members combed neighboring areas. A few days after Joe's return, Piot got a radio message from Johnson, telling him a Zairian Air Force helicopter would arrive shortly to bring him back to Bumba for a meeting with "U.S. Embassy officials." Piot protested: why should he, a Belgian, fly back to brief a group of Americans?

"Look, Peter, they want to see firsthand what's going on. Don't argue with me. Just get your butt on that copter," Johnson said.

Piot got off the radio grumbling about the "sick tourism" of the U.S. Embassy and CIA interests, but reluctantly prepared to meet the Zairian helicopter.

As he paced about the mission, the skies suddenly darkened and he could tell a storm was coming. Out of the blackened sky came the Puma helicopter. Without shutting off its engines, the pilot opened his cockpit window and called out to Piot. When Piot asked the pilot about the safety of flying such a large cumbersome helicopter in a storm, he smelled the familiar scent of Zairian beer on the pilot's answering breath.

"*Pas de problème*," the pilot insisted.

Piot asked a few more questions, studied the pilots, and concluded the two of them were drunk.

"The hell with it," Piot said. "I'm not going to that meeting." As he waved off the pilots, a Yambuku villager ran up crying, "Doctor, please! I've never been in the air before. If you are not going, may I take your place?"

Piot shrugged, helped the young man into the helicopter, and waved the aircraft on its way.

Two days later a somber Johnson radioed Piot with bad news, telling him that the drunken pilots had crashed the helicopter. Everybody on board died, and a hunter found the wreckage in the jungle southwest of Yambuku.

Piot listened in disbelief as Johnson went on to explain that the Zairian Air Force was holding Piot personally responsible for the deaths.

"They're saying you sabotaged the helicopter because you're some kind of Belgian colonialist," Johnson continued. "And they're insisting you have to go out there, get those bodies and perform autopsies. There's no ifs, ands, or buts on this one, Peter. You have to do it. The entire research effort could be shut down in an instant if the Zairian military tells Mobutu we're a bunch of CIA agents or something."

Shaken and angry, Piot jumped in his Land-Rover and drove as fast as roads would allow to Bumba. There, he was assigned a detail of prisoners from the local jail, who worked all night under Piot's direction, making three coffins. The next morning Piot and the prisoners were flown by bitterly angry Zairian Air Force pilots to a plantation on the edge of the jungle area in which the hunter had spotted the wreckage. With the prisoners in a line behind him hauling the coffins and supplies, Piot cut a path through the rain forest. Whatever image the sight of Piot, the coffins, and the prisoners conjured for local villagers, it was obviously one of great interest. As the grim group cut its way deeper into the jungle, it was joined by clusters of the curious. Eventually, over a hundred people trailed the coffin bearers.

The wind first told them when they had reached their destination, for it carried the stench of three human bodies that had literally cooked for four sweltering days in the equatorial jungle. Piot, standing a foot taller than most of his companions, peered ahead, trying to catch a glimpse of the helicopter. The jungle canopy was so dense that little sunlight penetrated it. Still unable to see the copter, Piot paused and pulled a respirator out of his knapsack.

At the ghastly sight of the wreckage, all the prisoners screamed in horror and ran away. When he turned to look straight at the wreckage, Piot had to struggle hard to hold back a wave of nausea. The bodies had bloated in the humid heat, their eyes bulged, insects crawled over their taut skin, and the stench was overpowering. Fighting back his disgust, Piot forced himself to walk up to the first body, formaldehyde sprayer in hand, to ready it for the coffin.

It was the young villager. Piot swayed, feeling suddenly dizzy. "This should have been me," he thought. "I should have been in that seat, instead of this poor fellow."

He looked at the villagers, at the bodies, and called out, "The shoes! The shoes! Whoever helps me gets their shoes!"

A cluster of boys ran forward, helped Piot stuff the bodies into their tight coffins, removed the shoes, and then carried the horrendous burdens—unlidded to compensate for the swelling—out of the jungle.

When they reached the plantation Piot found the military pilots busily pursuing the business of alcohol consumption. They had refused to assist in the removal of the bodies, and looked on Piot with undisguised contempt.

"Here are your colleagues," Piot said, pointing at the gruesome coffins.

The pilots gulped down more beer and *arak*, ordered Piot to put the bodies in the aircraft, and made it clear that they were in no mood to argue with a white man from Belgium. For half an hour Piot sat white-knuckled, barely able to breathe, clutching the armrests of his helicopter seat as the belligerent, inebriated pilots maneuvered their copter and its macabre cargo to Bumba. When they landed, Piot was beside himself with rage and fear; he called the bluff of Bumba military officers, refused to perform the autopsies they had demanded, and declared, "You have your bodies, I've done my part, the hell with you!"

Peter Piot staggered off into Bumba, feeling more emotionally wretched than he had previously imagined was possible. For the first time in his life, Piot set off, with determination, to get drunk.

After a couple of beers, he felt tears pressing against his eyelids, and thought again of the poor villager who had perished in his place. He bought drinks for anybody in the bar who would hear his story, and soon the modest establishment was packed with thirsty ears.

After a couple of rounds, he heard someone greet him in Flemish, looked up, and saw a white man covered in road dust. Simon van Nieuwenhove introduced himself, explained that he had just returned from a four-week tour of the wilds in search of Ebola cases, and asked if he might join in the revelry.

The two men shared sagas, beers, and emotions, and developed an instant friendship that would bond them like brothers for the rest of their lives.

In the following days Piot and van Nieuwenhove talked for hours, trying to make sense both of the strange epidemic and of its impact on their lives. Piot's backgrounds in medicine and virology had served him well, but the twenty-seven-year-old Belgian had enough humility to recognize that he knew nothing about developing countries, and even less about epidemiology. He had developed a strong admiration for the multifaceted skills of the Americans—Breman, Johnson, and McCormick. And he decided to ask Johnson to recommend him for epidemiology studies at the CDC.

Like so many other members of the International Commission, Piot was discovering that the relatively brief Yambuku experience was completely changing his life. It would be some time before he would discover the effect his African encounter was having back home on his wife, Margarethe. And on van der Gröen's wife, Dina. Unbeknownst to the men, the Belgian government had informed Dina and Margarethe that "there had been a deadly helicopter crash involving Belgian members of the International Commission." It would be several days before the women would learn that their husbands were alive.

Since the ghastly incident with the Zairian helicopter, Piot was gaining a healthy respect for danger, among other things. But most of the other survey team members had settled into routines, staying in the more comfortable town of Bumba, driving their Land-Rovers out to the villages, and going house to house completing huge questionnaires on detailed information considered vital to understanding the epidemic. With routine comes complacency, a lowering of both guard and fear.

VIII

On November 26, U.S. Peace Corps volunteer Del Conn told team members in the Yambuku area that his head and back were killing him. The pain came on suddenly, and then hung on relentlessly. Conn, who had previously worked in a small hospital outside Kinshasa, had joined the Yambuku survey effort ten days earlier and was assisting Piot in collecting blood samples and village data. He had also helped van der Gröen prepare microscopic samples of Ebola-infected tissues for study in a field lab the Belgian had recently constructed in the mission. A month later researchers would learn that some of Conn's samples, despite ultraviolet radiation exposure and acetone treatment, still contained live Ebola virus.

Though Conn's temperature was only slightly above normal, team members were worried. They notified commission headquarters that it might be necessary to activate the complex system of medical evacuation that had been worked out in detail after days of negotiations with the governments of Zaire, South Africa, the United States, and France. Those procedures required that Conn be placed under strict quarantine for thirty-six hours and airlifted out of the region if his condition worsened.

While Conn lay inside a room of the mission facility, tended to by Karl Johnson and Margaretha Isaacson, the team tried to continue their survey work.

"But there's no question about it," Breman radioed to Johnson, "this is a major downer for everybody." Morale plummeted, fear rose.

A Canadian military officer had, coincidentally, arrived a day earlier in Kinshasa with a newly designed portable plastic isolator unit, intended to allow safe transport of contagious individuals.

By November 29, Conn's condition had worsened. His fever was up slightly, blood chemistry showed classic signs of viral infection, back pains were severe, and he was nauseated. In Yambuku, Johnson, Dr. Dennis Courtois, and Isaacson tried to prepare Ebola antiserum from recovered patients' blood, but power failures shut down their centrifuge and other equipment necessary to ensure safe plasma preparation.

"You can't imagine the fear here," Johnson radioed to Bumba.

Under contingency plans, a military helicopter was supposed to fly immediately to Yambuku, pick up Conn, and bring him to Bumba. Meanwhile, a C-130 was supposed to fly from Kinshasa to Bumba, load Conn into the Canadian isolator unit, and transfer him to Johannesburg, after a refueling stop in Kinshasa.

But the Zairian Air Force's pilots balked again. Fearing Conn might give them the disease, the pilots refused to fly their helicopter to Yambuku. All other options closed, Johnson, Isaacson, and Courtois loaded Conn into the back of a Land-Rover and drove the bumpy road to Bumba, their passenger groaning in pain all the way. All three scientists wore disposable protective clothing and surgical masks throughout the journey, which lasted four and a half hours because Conn could not tolerate the sudden jarring produced by hitting ruts at speeds greater than ten miles per hour.

When they reached Bumba, continued Air Force fear was obvious: no plane awaited them. And panic among the townspeople was so great that the Land-Rover was not permitted to leave the center of the Bumba landing strip. Unprotected from the tropical sun and forced to wear a tight rubber respirator mask to allay the fears of the populace, Conn was miserable. Johnson and Isaacson sedated the Peace Corps volunteer and gave him analgesics to ease his pain.

As night fell, there was still no word on air transport for the ailing man, so Johnson and Isaacson were forced to make do with available plasma and equipment. Convinced their colleague had Ebola, they hand-administered a unit of Sophie's antiserum into Conn while the young man lay in the back of their Land-Rover.

At dawn an Air Zaire "Fokker Friendship" airplane landed, the Canadian respirator on board.

The two doctors studied the isolator for a moment. It consisted of a plastic pipe frame that outlined a space some seven feet long, four feet high, and four feet wide. Suspended from the frame was a box tent of thick, clear, pliable plastic. From the sides of the box tent hung attached gloves, into which attending physicians would insert their hands and arms when they needed to "touch" the patient.

The doctors carefully slid Conn into the isolator, attached an intravenous feeding tube to a device installed in the box tent, shut their patient inside, and switched on the pressurized air device. It seemed to work, but the intravenous feeding device was poorly designed and the feed rate fluctuated wildly.

An assortment of drugs and medical supplies were also on board the aircraft, and the doctors decided to administer strong painkillers to Conn before takeoff. Commission members in Kinshasa failed, however, to provide a file with which the vials could be opened, forcing yet another delay while the physicians sought an alternative sterile means to unseal the ampules.

Once the Keystone Kops-like operation was in the air, another failing of the Canadian device was noted: it did not adjust well to altitude-induced air pressure differences. Conn grew anxious as the box tent slowly caved in on him, making his space ever more claustrophobic.

When the plane landed in Kinshasa another snag appeared in the commission's grandiose emergency evacuation plan: no plane was "available" to take the patient to Johannesburg.

Johnson, now fuming mad, contacted U.S. Embassy officials, who relayed an air support request to the USAF. A C-141 Starlifter was dispatched from Madrid, arriving in Kinshasa six hours later. During the long wait, fear of contagion once again forced the group to stay at the airport, this time inside an abandoned hangar. The afternoon heat was so great that the isolator steamed up and was soon creating its own internal rainfall.

Although he received a variety of analgesics, Conn's pain was acute, he was running a fever of over 102°F, and the hours inside the wet, coffin-sized plastic cocoon were driving him crazy. His anxiety reached a zenith when the doctors noticed blood oozing out of the tiny puncture hole through which his intravenous feeder was inserted.

Uncontrolled bleeding, Conn knew, was hemorrhaging; and hemorrhaging was the key symptom of Ebola. Conn had to be heavily sedated.

That night Conn was transferred into the USAF jet and flown to Johannesburg. Because of a storm front, the flight was diverted on a huge loop out over the Atlantic Ocean. The doctors felt Conn would be unable to tolerate turbulence, but the diversion added several hours to their flight time.

The plane landed in Pretoria and Conn was transferred to a South African Air Force plane for his final leg to Johannesburg.

When he was finally removed from his nightmarish cocoon, Conn's entire body was covered with a florid measleslike rash that was not usually seen with Ebola but had been noted in some Machupo and Marburg cases. He had been severely ill for six days before reaching a hospital.

Clearly, the commission's contingency plans had failed completely when put to the test. Johnson was enraged, and scientists still deployed in the field were extremely distressed.

Behind the scenes still more misadventures occurred. The CDC sent a massive hospital containment bed isolator by air from Atlanta, but when the contraption arrived in Johannesburg two crucial components were missing: instructions for assembly and an electrical converter that would allow the American-made device (designed for 110 volt, 60 Hz electricity) to function in South Africa (which uses 220 volt, 50 Hz electricity).

Furthermore, early difficulties in transporting Conn prompted CDC officials to prepare an Apollo space capsule for use in South Africa. That forced a major South African Army mobilization of ground transport capable of maneuvering the eighty-ton capsule. At the last minute space capsule airlift plans were scrubbed.

All in all, the planned thirty-four-hour evacuation actually took over seventy-two hours, at an inestimable cost to the governments of Zaire, the United States, and South Africa.

And when Conn's blood was submitted to repeated examinations, no Ebola viruses could be found. Nor could the South African team find evidence of any other known human pathogen.

Twenty years later, the cause of Conn's bona fide illness would remain a complete mystery.

Conn, it seemed, had "discovered" another new virus.

IX

Don Francis was burned out before he ever got involved in the Sudan episode. Hell, he was burned out before he even got to Harvard.

After two years of chasing down smallpox cases all over Sudan, India, and Bangladesh he was ready for a break. September 1975 found Francis at Harvard University, working on a Ph.D. in virology. With the CDC's permission, Francis was studying in Max Essex's laboratory when the Ebola mystery started unfolding some ten thousand miles away.

When CDC officers called him in October 1976, Francis was a bit flattered at first.

Francis got off the phone and searched out his mentor, Max Essex. He found the Rhode Island-born Yankee, as usual, poring over data, and requested a two-week leave from doctoral studies.

Essex agreed to let Francis take two weeks off; indeed, he later had to talk Don into going when the younger scientist's ego was bruised by learning that, far from being indispensable, he was the CDC's last resort. Every other person on the agency's list had turned down the assignment out of fear.

Word from Zaire had, by early November, been exaggerated in the gossip mills of international virology and finding eager volunteers for the Sudan investigation proved exceedingly difficult. Eventually WHO's Paul Brès gave up his search, bought a Geneva–Nairobi ticket, and assigned himself to the investigation. The WHO team in Maridi would be composed of David Smith (of the Kenyan Ministry of Health), Don Francis, Brès, Irishman David Simpson (of the London School of Hygiene and Tropical Medicine), animal expert Barney Highton (also of the Kenyan Ministry of Health), and Sudanese medical experts Babiker El Tahir, Isaiah Mayom Deng, and Pacifico Lolik. Most would join Francis and Simpson in the south, having made their own way to Maridi via Nairobi or Juba . . . after several days' delay.

Because of the ancient rift between Khartoum and the southern Sudanese provinces, the federal government decided to stop the epidemic by completely cutting the south off from the rest of the country. It was sort of a damage control approach: many might die in the remote south, but the disease would not reach the more densely populated Muslim north. Absolutely no airplanes, trucks, or other vehicles were allowed in or out of the southern section of the country.

For four days Francis and Simpson begged, cajoled, and bribed their way around Khartoum, searching for a way to get themselves—with a couple of tons of supplies—past the quarantine lines, all the way south to Maridi and N'zara. Simpson, El Tahir, and Francis visited all the Western embassies, pleading for assistance. Much was promised, little materialized. Chartering a private airplane was ruled out: Khartoum and Kenyan officials insisted the entire aircraft would be burned, as a protection against contagion, upon return from the quarantine area.

At long last two large British trucks were found, loaded up, and filled with extra tanks of gas. Unbeknownst to the hapless WHO crew, McCormick had already left N'zara by the time Francis finally got behind the wheel of a truck bound for Maridi. It was the rainy season, and what passed for roads had become muddy rivulets. For twelve hours the WHO team kept their trucks in four-wheel drive and their accelerators floored and endured a battering, crashing ride. It was two in the morning when the exhausted group pulled into the town of Maridi, population 2,000.

They were greeted by the Maridi hospital's night watchman, who awoke the town's two public health doctors and installed the tired team in an old British missionary complex.

The following day further impediments to their investigation mounted, and Francis, Simpson, and El Tahir were frankly stunned by the scale of their problems. The national quarantine of the south was bringing on near-famine conditions in the region. Because the rolling elephant grass savanna was often wet and marshy, it was insect-infested. Tsetse flies, in particular, swarmed about, infecting livestock and people with the trypanosomes that cause sleeping sickness. The problem was so severe that most people had years earlier ceased raising animals, and the entire region was dependent upon shipments of meat and protein from the north. No shipments had come through since September 30, when the quarantine was imposed. El Tahir, who had made the first official visit to the epidemic area on September 26, could clearly see the enormous difficulties imposed on the people by six weeks of quarantine.

The three men also found the distances between villages in the region lengthy and untraversable in four-wheel-drive vehicles. Some of the nomadic villages were virtually invisible, hidden in tall stands of elephant grass, reachable only by nearly imperceptible footpaths.

The district's headquarters, Maridi, was a sparsely supplied government town whose sole significant employer was a UNICEF-funded teaching hospital. Constructed of wattle, the hospital was staffed by two poorly paid public health doctors and 120 nurses, most of whom were trainees. Their shared skills and supplies pretty much limited the Maridi staff to tender loving care in their constant war against sleeping sickness, malaria, bacterial meningitis, septicemic plague, relapsing fever, and a host of other tropical diseases. Long cut off from the rest of the world, Maridi had no telephones, so a ham radio was used to relay signals to Juba, where a French scientist remained throughout the Sudan investigation, serving as a communications officer, relaying messages to and from Khartoum. There was no communication with International Commission members in Zaire.

When Francis, Simpson, and El Tahir arrived, the two Maridi doctors were already in the process of closing their hospital, most of the nursing staff having either died of the new hemorrhagic fever or run away in fear, carrying the virus and panic with them back to the villages. The handful of nurses who remained were in the process of closing down the regular hospital facilities and tending to Ebola cases in a specially constructed wattle quarantine building.

Wearing respirators, protective gowns, and gloves, Simpson and Francis inspected the hospital and were horrified by their first sight of Ebola. Neither Francis nor the more experienced Irish physician, Simpson, had ever seen anything even approaching its devastation. Weak, emaciated men and women lay about the mud-and-stick chamber, staring out of ghost eyes at the white men. The virus was so toxic that it caused their hair, fingernails, and skin to fall off. Those who healed grew new skin.

Over the following days Francis, the epidemiologist of the group, questioned hundreds of people in the Maridi area, using local schoolteachers as translators. He drew many blood samples and mapped how the epidemic had spread. Barney Highton led efforts to capture animals and insects, hoping to discover the natural reservoir of the Ebola virus, and El Tahir set up a laboratory inside the abandoned Maridi hospital.

They soon discovered that the major sources of the continuing spread of the virus were the funerals; more specifically, the procedures—not unlike those practiced in Yambuku—used to cleanse the bodies before burial. Francis ordered a halt to all the funerals of Ebola victims, promising that his team would cleanse the bodies according to tribal customs.

The people were outraged, and their collective anger nearly destroyed the entire WHO effort.

"I think they're going to kill us," Francis told his colleagues. "I mean it. Watch your backs."

Fortunately, one of Maridi's public health doctors was the son of a powerful local chief, and with the leader's support the people were eventually coaxed into bringing their dead to Maridi. Francis, Simpson, El Tahir, and Brès would take the bodies a discreet distance away from public view, put on their protective clothing, gloves, and respirators, and remove all undigested food and excreta from the cadavers, as prescribed by tribal custom, which entailed hand removal and manipulation of wastes without evisceration. They would also carefully remove tissue and organ samples for laboratory analysis.

Stopping the funeral cleansings and closing the hospital brought the Maridi epidemic to a halt, so Francis and El Tahir made their way to the even more remote town of N'zara. There Don found Joe McCormick's boxed note, guiding them through the sequence of original Ebola cases.

"Hi Don," the note read. "Found your index case." After providing details, it was signed simply, "Joe."

N'zara was the hub for a population of about 20,000 people, most of whom lived in village clusters of mud huts scattered throughout the surrounding savannas and jungle. The economic center of N'zara was a cotton factory, where some 2,000 men made fabric from locally grown cotton using nineteenth-century machines. Inside the factory conditions were harsh; the tin roof magnified the excruciating equatorial heat, lung-damaging cotton fibers filled the air, bats swarmed out of the roofing periodically, filling parts of the factory with their malodorous guano, and the poorly paid men worked long, exhausting shifts.

McCormick's note explained who had been the first case in the mysterious epidemic and traced the order of subsequent infections. On June 27, well before the apparent onset of the Yambuku epidemic, a man who worked in the N'zara cotton mill fell ill and died on July 6 of hemorrhaging. His death was soon followed by those of two co-workers whose jobs were in the factory's cloth room, the same site where the first man worked. By July about two factory workers each week contracted the virus. By September several workers and their friends and family members had contracted Ebola, and at least thirty-five had died.

Two-thirds of the subsequent Ebola cases in N'zara involved a man named Ugawa, who was comparatively wealthy because he ran N'zara's cultural hub, a jazz club. The factory workers would spend much of their earnings in Ugawa's club, eating, drinking, and buying the sexual favors of the barmaids. Most of N'zara's epidemic evolved from those liaisons.

And it was Ugawa who had enough money to travel to the Maridi hospital when he came down with the disease. Once his virus got into the Maridi hospital, it spread like wildfire.

By the time the WHO team arrived in N'zara in mid-November, the epidemic was on its way out, having sickened over a third of the Maridi hospital staff, forty-one of whom died. It threw the hospital into chaos, from which many staffers fled. Nearly all the staff that got the disease were infected on the job, primarily through exposure to sick patients' fluids.

From the staff, the epidemic spread into the community through several generations of transmission. Later investigations would reveal that the N'zara virus was highly contagious, spreading more than eight generations from the index case, as the scientists put it. The Yambuku strain, in contrast, never spread more than four generations. On the other hand, the Yambuku virus was far more likely to kill those it infected.

By November 20 it seemed the epidemic was over, the spread having halted as a result of the hospital closures and changes in funeral practices. Francis totaled up his case list: 284 Ebola cases, 151 deaths, all but four cases occurring in either N'zara or Maridi. As McCormick had suggested in his report to the commission (which Francis could not see in the Sudan), the Sudan virus seemed less deadly. While upward of 90 percent of those infected in the Yambuku outbreak died, only about half (53 percent) of the Sudanese cases were fatal.

The center of Maridi's epidemic was the hospital, where nearly half of the people hospitalized for other reasons got the disease (93 of 213 patients) and the toll among the medical staff was high.

In N'zara, however, the virus seemed to come, somehow, from the cotton factory, and the WHO team devoted a great deal of time and attention to that building, where nearly a thousand men worked at any given time. Freshly picked cotton came in one end of the structure and was processed room by room into bolts of finished cloth.

Blood tests showed the highest infection and death rates were among the twenty-four men employed in the cloth room: four deaths and five nonlethal cases, for an overall infection rate of 38 percent. Francis and Highton combed the room in search of an animal or insect that carried the Ebola virus. They had no way to test the animals in N'zara, so they were working blind, capturing anything that moved, removing vital organs, and placing them in liquid nitrogen. Eventually, the organs would reach Pat Webb's lab in Atlanta, where she would perform the tests necessary to determine whether any were Ebola-infected.

They found the cloth room heavily infested with bats, rats, cotton boll weevils, spiders, and numerous other insects. By December, Webb would give the WHO team disturbing news: *none* of the animal samples contained Ebola virus.

Thus, the origin of N'zara's epidemic remained a mystery.

Having already spent well over the requested two weeks in the epidemic zone, Francis was anxious to get back to Harvard. The CDC, however, cabled Khartoum to instruct Don to remain in Sudan. It wasn't until Christmastime that Francis, imbued with a bitterness toward CDC leaders that would color his future activities with the agency, returned to Boston, renegotiated an extension on his CDC leave, and set to work completing his Ph.D. research.

By then, Joe McCormick was back in Sierra Leone, setting up his primitive Lassa laboratory. Karl Johnson had returned to Atlanta. Months later, Pat Webb would get her long-desired taste for exotica fulfilled, when she volunteered to join McCormick's Lassa studies in Sierra Leone.

Joel Breman did not return to his Michigan Swine Flu work. Instead, he devoted two more years to African research on behalf of the CDC and WHO. He joined efforts to search for cases of monkeypox, a type of virus similar to smallpox that produced illness, but rarely death, in human beings.

WHO wanted to be sure that it was safe to cease smallpox immunization efforts; it was essential that Breman find out whether wild monkeys carried forms of pox viruses deadly to humans.[20]

Throughout the late 1970s and early 1980s human monkeypox case reports would increase steadily, from zero prior to 1970 to 35 in 1983, most occurring in the rain forest regions of Zaire.[21] In 1984 some 214 cases would be found in Zaire alone.[22]

It would turn out that most monkeypox cases occurred in—yes—the Bumba Zone of Zaire in the villages surrounding Yambuku.[23] All sorts of animals living in the Bumba-area jungles would be shown to carry monkeypox: tree squirrels, forest monkeys, chimpanzees, and antelopes. But in the end scientists would conclude that the rain forest virus was not genetically close enough to the smallpox virus to pose a threat to human populations, and the monkeypox virus spread so inefficiently from person to person that *Homo sapiens* epidemics never occurred.[24]

Breman insisted that the animals and people studied in equatorial Africa during the monkeypox surveys—particularly those surveyed in the Bumba Zone—also be tested for both Ebola and Marburg virus infections. After nearly ten years of testing, no infected animals would be found, although a handful of bats captured in faraway Cameroon would test antibody-positive for prior exposure to Ebola.

The mystery of where Ebola came from would haunt most of those who had been involved in the Yambuku and N'zara investigations for years to come. Guido van der Gröen would spend years working patiently in the highest-security laboratories in the United States, the Soviet Union, and Europe, searching for clues to the origin of Ebola in the virus itself. He was determined to crack the mystery of the organism that he and Karl Johnson had dubbed the Andromeda Strain.

He would participate in two expeditions back to the Yambuku area in 1979, piggybacking on Bremen's monkeypox searches, and would test countless animals in search of the natural reservoir of the deadly virus.

In 1980, David Heymann, who was also fixated on the Ebola mystery, would discover that pygmies living in the dense rain forests of Cameroon had antibodies to Ebola, indicating that they had once been infected with the virus. He would corral support from Pat Webb and Guido van der Gröen, and the trio would spend two months living among the Cameroonian pygmies.

The tall white foreigners would find their African counterparts remarkably receptive to the pursuit, willing to use their awesome hunting skills to capture all sorts of creatures for the scientists to test. Van der Gröen would run immunofluorescence tests on over 3,000 animals of 100 different species, ranging from one-meter-long poisonous snakes to chimpanzees.

Webb and Heymann would eventually discover that 15 percent of the pygmies had antibodies to Ebola, proving that whatever animal served as the reservoir of the deadly virus lurked in the dense rain forests of that region. But none of van der Gröen's meticulously preserved animal samples would be infected.

Still further into the future, Joe McCormick would continue his search, testing animals in the western Ghanaian rain forests. Because it turned out that the natural reservoir of monkeypox was flying tree squirrels, McCormick would capture and test squirrels. And he would find one tree squirrel that had antibodies to Ebola. But it would not be carrying the virus.

The source of both horribly lethal viruses—Marburg and Ebola—remains a complete mystery.

"There is a strong suspicion that the disease is a zoonosis. Monkeys did not seem to play a role in these epidemics and rodents, or bats, may perhaps be the animal reservoir," stated one of the International Commission's reports.[25] A later WHO official report would bemoan that "since the natural reservoirs of Marburg and Ebola viruses are unknown, no control activities can be carried out in Africa."[26]

Perhaps the bluntest statement appeared in the commission's second report: "As in the case of Marburg virus, the source of Ebola virus is completely unknown beyond the simple fact that it is African in origin."[27] But even the assumption that all cases would originate in Africa would prove naïve in years to come.

The commission was, however, able to explain how the apparently extremely rare disease spread quickly throughout the Bumba Zone and Maridi. Knowing why a disease spread could allow local authorities to limit future epidemics to a handful of primary cases, preventing hundreds of deaths. El Tahir put it best: "The hospital must be viewed as an epidemic amplifier." Both in Maridi and in Yambuku the poorly supplied clinics reused syringes hundreds of times a day, injecting drugs from one person to another without sterilizing the needles. McCormick calculated that during the months of September and October 1976, an individual's odds of getting Ebola virus from a single injection at the Yambuku and Maridi hospitals *exceeded 90 percent.* Seventy-two of the primary cases in Yambuku (out of 103) were caused by unsterile needles used in the mission hospital. Sureau calculated that 43 percent of the Yambuku-area Ebola victims who got the disease from another person survived the ailment, but only 7.5 percent of those who were injected with contaminated syringes survived.

At the Yambuku Mission Hospital, for example, the commission eventually figured out that the majority of the early Ebola cases involved women who came to see the Sisters for pregnancy-related checkups. When women were questioned, it turned out the real draw to the mission was a miraculous injection that made pregnant women feel energetic and content.

It was vitamin B complex.

The commission determined that injected Ebola infections were far more likely to result in terminal disease than were secondary exposures to sick friends and family members.

The Sisters did not appreciate this information. Still grieving the loss of more than half their staff and colleagues, the missionaries would not countenance accusations that the very individuals who had given their lives in a saintly struggle against an unknown horror should now be labeled agents of epidemic spread.[28]

X

As Christmas approached, Peter Piot prepared to leave the place that had over two and a half months' time come to feel like something of a home. He had long since sold the wedding suit and wing tips he wore to Kinshasa. Gone too was his naïve arrogance. In its place was a new sense of confidence coupled with a healthy respect for the microbial world.

"I have seen things which most Europeans only read about in books or see in adventure movies," he told Sister Genoveva. "My mother, a typical Flemish woman, always taught me, 'Speaking is silver, silence is gold.' But I have seen too much to keep my mouth shut."

As the Belgian packed crates for his departure, another young adventurer was sitting in Kinshasa, eagerly awaiting his opportunity to go to the Bumba Zone. American CDC scientist David Heymann had volunteered without hesitation to be the last foreign scientist in Yambuku, charged with cleanup epidemiology and, perhaps most important, giving the rest of the crew an opportunity to head home for Christmas.

At Bumba's airport Heymann and Piot met for the first time, shook hands, and headed off in opposite directions. Years later the pair would work side by side, trying to control another, far larger, deadly epidemic. Piot recognized the excited look in Heymann's eyes: it was the same look that had filled the now world-weary Belgian's face when he first arrived so long ago.

As Heymann drove Piot's Land-Rover along the road to Yambuku, he spotted boys along the way playing with homemade toys. Throughout Southern Africa, boys made clever sculptures of cars and trucks from cast-off wire, and rolled their toy vehicles along the roadside in imitations of the real ones. But Heymann saw these Bumba-area boys all had made something very unusual: helicopters. Nowhere else in Africa had Heymann seen children playing with helicopters. One boy, seeing Heymann's white face coming down the road, merrily held his helicopter up in the air and then dropped it to the ground, laughing hysterically.

"Wonder what that was about," Heymann thought.

Back in Bumba, Piot was unknowingly preparing for one more undesired adventure. He glanced angrily at the military pilots who were laughing and guzzling beers with fellow officers while the huge C-130 was loaded. The Bumba quarantine having finally been lifted, hundreds of local traders and still nervous families were clamoring for spots on the huge plane, along with their goats, pigs, monkeys, chickens, and sacks of worldly goods. The task of organizing their boarding was left to Piot, who felt no joy in anticipating another airborne excursion with drunken pilots.

Piot and a few other passengers loaded dozens of crates into the cargo hold, having no idea where it was best to place heavy versus lightweight objects. The anxious pilots left the engines running and occasionally shouted for Piot's group to hurry. The men placed most of the lighter objects at the front of the plane, heavier crates of laboratory equipment to the rear, leaving the center open for passengers. With the few nets and ropes provided by the pilots the group did their best to secure all the cargo in place.

By the time all the passengers were on board, crammed shoulder to shoulder without benefit of seat belts, or even actual seats, a storm was brewing. The pilots taxied their huge, heavily laden aircraft to the end of Bumba's tiny tarmac, revved the engines, and roared down the runway. The plane lumbered, groaned, and bounced, unable to gain height with such a heavy load.

"Oh my God!" Piot cried out, seeing the tree line directly ahead. The pilots pulled the throttle sharply, attaining just inches of advantage over the tops of the palm trees. The plane climbed steadily for several minutes until, hitting a pocket of storm turbulence, it suddenly dove a few hundred feet.

The heavy crates to the rear of the aircraft broke loose of their nets, slamming down on the screaming passengers. Blood spattered in all directions, people screamed in pain, and the inebriated pilots responded by jerking the plane up, causing the front-loaded cargo to snap loose. Piot and his bleeding and battered fellow passengers were sandwiched between heavy crates of cargo, some of which carried thousands of samples of lethal Ebola-infected animal and human tissue and blood samples.

Convinced he was going to die, Piot found himself thinking not of his wife or his past life, but of the epidemic.

"Shit," he muttered, "all that work for nothing. Nobody will ever know the answers."

Piot's fellow travelers became nauseated; some had suffered contusions and broken bones. For the rest of their relatively uneventful two-hour flight to Kinshasa the only sound heard above the engines' roar was the sobbing of terrified and injured passengers.

When Peter Piot staggered off the last of a series of planes into the Christmas chill of Antwerp, he found Margarethe obviously pregnant. And suddenly the full weight of what he had been through since September, of his many brushes with death—some foolhardy—hit him like a bolt.

Still, he had tasted adventure, and Piot would never again be satisfied for long with the seemingly mundane life of laboratory science.

Both van der Gröen and Piot were deeply affected by their Zaire adventures, so much so that Guido, whose emotional fuse was normally so long that few had ever witnessed an outburst from the Belgian virologist, discovered rage. Dragging Piot along with him shortly after Christmas, van der Gröen marched into the headquarters of the Sisters of the Holy Heart of Maria.

"Our objective here is education," the seething van der Gröen told Piot as they entered the office of the order's Mother Superior.

The meeting began calmly enough, with the two scientists applauding the Catholic education of children in the Yambuku area—an assignation that dated back to 1935. The men also noted the well-intentioned origins of the order's medical effort, which stemmed from its relatively recent recognition that some 50 percent of the schoolchildren were chronically absent from classes in Yambuku due to illness. The order sought to improve school attendance by maintaining child health.

In the early 1970s members of the order had attended several days of basic medical training at the Tropical Institute in Antwerp. That was the full extent of their nursing training before venturing into the field.

"They're not nurses!" van der Gröen uncharacteristically shouted, realizing he was criticizing deceased nuns. Still, he pushed on. He applauded the holiness and devotion of the Sisters.

"But no one was thinking that if you start such a medical business, and the people of the region are receiving no support from the government of Zaire, and you give out free health care, then you must be prepared to be deluged. You must be ready to safely give 300 shots a day. If you build something you call a hospital, then you must do the logistic planning, provide the resources, and train your personnel accordingly."

Van der Gröen's coup de grace was an accusation: "The price for your lack of planning was high"; half the dead got Ebola in the mission hospital.

Piot insisted that the mission hospital should either be closed or be staffed by a certified physician. And both men warned the Mother Superior that the source of Ebola was never found: it could return, and spread again inside the mission if their instructions weren't followed.

Though their advice was heeded for the Yambuku Mission, Piot and van der Gröen left the religious order wondering just how many missionary health facilities of all denominations operated in developing countries with a similarly imbalanced mixture of hopeful devotion and tragically poor medical training and logistics. The two men, who would remain lifelong friends forever bound by their shared Zairian experience, stepped out into the icy Antwerp January morning, their minds and conversation filled with concern for the far-off tropical villages, the women in their *kangas*, the babies tied on their backs, and husbands earning hard livings selling wild animals they captured in the steamy rain forest—a jungle that hid Ebola.

The following Christmas, Pierre Sureau received a letter from Sisters Marcella, Genoveva, and Mariette. He sat in his comfortable Paris apartment, trying to recall the sweltering heat and primitive mission of Yambuku as he read the Sisters' greetings:

Dear Doctor, we wish you a good New Year 1978. These days we talk a great deal about the events of the past year and you, who have left us with good memories. We would like to thank you again, sincerely, for coming to our aid when others dared not to come. At the moment life has returned to normal. A Zairian doctor is here: he also works with all his heart. Four Belgian volunteers and a Sister are here to rebuild the hospital. The students are back in the school, making plenty of noise. Are you well? We send you our affection. Sister Marcella, Sister Genoveva, Sister Mariette.

Sureau affectionately refolded the letter, put the pages back inside their envelope, and put the missive into a carton marked "Yambuku." He carefully shut the box and placed it in the back of a closet.

Souvenirs of a plague.

6

The American Bicentennial

SWINE FLU AND LEGIONNAIRES' DISEASE

There is evidence there will be a major flu epidemic this coming fall. The indication is that we will see a return of the 1918 flu virus that is the most virulent form of the flu. In 1918 a half million people died. The projections are that this virus will kill one million Americans in 1976.

—F. David Mathews, Secretary of Health, Education, and Welfare

I

The great Ebola drama went almost unnoticed in the United States in 1976, even in the hallways of the Centers for Disease Control. The nation was preoccupied. And Africa was, in the American consciousness, far away.

In its 200th year, the United States of America was busily celebrating patriotism with a mix of red-white-and-blue entrepreneurial souvenir sales, Hollywood extravaganzas reenacting Great Moments of History, tall-ship regattas down New York's East River, and a good deal of boasting about the brilliance contained in the Declaration of Independence and the Constitution. Compounding the national sense of distraction was a striking new political atmosphere in which President Gerald R. Ford was struggling to defend himself in national elections against a virtually unknown southern politician named Jimmy Carter. A national soul-searching was underway, as Americans contemplated the significance of the U.S. defeat in Vietnam and the Nixon administration's Watergate scandals.

Even if Americans hadn't been in an isolationist mood in 1976, their limited attention spans would still have been unable to absorb the events of Yambuku, for they had more than enough disease news upon which to focus. After all, 1976 was the year of two of the most exhaustive and expensive investigations in the history of the U.S. Public Health Service: the Swine Flu affair and Legionnaires' Disease.[1]

Overall, Swine Flu and Legionnaires' Disease boiled down to the same set of troubling perceptions for the American public, and, to a lesser extent, the Canadian, Mexican, Australian, New Zealand, and European publics: something new and very scary was coming; nobody was sure what it was, but the experts were certain it was dangerous; the federal government seemed quite distressed about the matters, but the experts and authorities didn't seem to agree as to what, if anything, should be done to protect the public; and it was all costing taxpayers a pretty penny. In both cases, public apprehension would eventually yield to impatience and allegations of incompetence, even scandal. Each step of the investigations would take place under the full glare of television lights and public scrutiny.

Ultimately, one disease would emerge, the other would not.

While Karl Johnson's team combed the villages of the Bumba Zone in search of Ebola cases, the American public health establishment, from low-level municipal officials all the way up the ladder to the President of the United States, anxiously monitored hospital records and physicians' reports for hints of the emergence of the so-called Swine Flu.

It began in January 1976 at Fort Dix, a U.S. Army training center in New Jersey. A young, highly motivated recruit, Private David Lewis, felt the dizziness, nausea, fatigue, fever, and muscle aches that are the hallmarks of influenza. Several of his fellow recruits were similarly stricken during that cold, wet week following the New Year, and some sought comfort in the base dispensary.

Eighteen-year-old Lewis, however, was determined to excel in basic training. Though he had been assigned by a medical officer to remain in quarters for forty-eight hours, Lewis loaded his fifty-pound pack and joined his platoon for an all-night hike in the bitter New Jersey winter.[2] Overcome by fever, the teenager forced himself to keep marching, though he lagged far behind his fellow soldiers. After a few hours he collapsed.

Lewis died just hours after reaching the base hospital.[3]

Nearly two decades later scientists would still be asking, "Did Private Lewis die because he was infected with a particularly lethal, virulent strain of influenza, or did the young man die because he went on an overnight winter full-pack forced march while in peak viremia with a modestly dangerous influenza strain?"

Knowing the answer to that fundamental question would make all the difference in interpreting the events of 1976. Lewis would be the only American whose influenza death in the 1976–77 flu season seemed out of the ordinary, based on his youth and physical fitness. Typically, influenza sickened thousands of people every year, but claimed only the lives of the very old or those weakened by other ailments that stress the human immune system. It is rare, indeed, for a healthy teenager to die of influenza. A hallmark of the great 1918–19 influenza pandemic was the virus's ability to kill young adults and children.

By the end of January, Fort Dix medical commander Colonel Joseph Bartley had a widespread flu problem on his hands, with some 300 recruits hospitalized or confined to quarters. At his request, New Jersey State Health Department laboratory director Martin Goldfield received nineteen throat-wash specimens from ailing recruits, including a sample taken during autopsy from Private Lewis. Goldfield's lab put droplets from the nineteen sample tubes onto nutrient-rich petri dishes, allowing the cultures to grow. Once the viral colonies were large enough to be studied, Goldfield's team ran a series of antibody tests aimed at determining what strain of influenza was attacking the Fort Dix recruits.

At the time scientists like Goldfield knew that when the human immune system successfully overcame influenza infection, antibodies were made against two proteins that protruded from the outer envelope of the spherical virus: hemagglutinin and neuraminidase. The influenza virus was otherwise well protected by a tough protein-and-fat armor made of two layers of viral enveloping: one layer was almost entirely composed of the human heart's nemesis, cholesterol. But the virus was caught in something of a Catch-22: it could not infect and destroy cells without the use of its neuraminidase and hemagglutinin proteins, yet these very compounds were what attracted the usually successful attack of the immune system.

Over 700 of these proteins protruded from the surface of each virus. The long rod-shaped hemagglutinin proteins performed the job of grabbing on to red blood cells, connecting one cell to another and causing formation of clumps of cells in the bloodstream. Neuraminidase in turn pinched off pieces of the cellular membrane that were wrapped around newly formed viruses, allowing the microbes to flood out into the bloodstream.

In 1976 scientists believed that the relative danger and virulence of a particular influenza strain were a function of three things: the efficiency of its hemagglutination ability, the functional abilities of its particular neuraminidase proteins, and the immunity of the animal or human host it infected. The first two factors were controlled by viral genetics; the last was under host regulation.

Unlike such relatively simple viruses as Lassa and Marburg, influenza proved to have a complex genetic organization. Long, single strands of RNA genetic material were entwined around themselves, forming a spiral structure. Five such RNA spirals were further entwined with protective proteins, forming genetic packages similar to those seen in human and animal cells, called chromosomes. When the virus reproduced itself, the chromosomes had to unwind and make duplicate sets of their proteins and RNA. In the process, parts of one chromosome might overlap with another, extraneous bits of RNA from the cell in which the virus resided might get copied as well, and the whole mess would be reassorted and reassembled to yield an intact parent virus and its packaged, somewhat different, offspring.

At the heart of such complexity lay many opportunities for genetic change, some of which might be lethal for the viruses, others of which might prove fatal for the targeted human hosts.

By 1976 virus specialists were beginning to appreciate that influenza was a sort of microbial chameleon that had thrived over the millennia by rigorously adhering to a single maxim: Adapt or die. If this constant process of genetic shuffling didn't frequently yield new types of hemagglutinin and neuraminidase, all target humans could eventually be immune to influenza and the virus species might die out. While the chances of the planet's entire human population becoming immune to a rare virus such as Ebola were nil, it was possible that an easily transmitted, ubiquitous respiratory virus like influenza would infect billions of human beings in less than five years' time, kill off all the susceptibles, and leave the world's survivors completely immune.

Global pandemics were, in fact, a hallmark of influenza that spanned recorded human history. Charlemagne's conquest of Europe was slowed by an A.D. 876 flu epidemic that spread across the continent and claimed much of his army. Many suspected influenza epidemics followed, though history can only vaguely discriminate between ancient accounts of influenza and other respiratory diseases. In 1580, however, the world was clearly hit by a major pandemic that followed trade and early colonial routes across Africa, Europe, and the Americas. So devastating was the epidemic that "some Spanish cities were said to be nearly dispopulated."[4]

By examining the more clearly recorded histories of influenza epidemics of the eighteenth and nineteenth centuries, scientists were able to recognize some patterns. First, the virus seemed to successfully change itself often enough so that at least once in every human generation a significantly new strain appeared that could elude the human immune system. Usually, after a great pandemic that killed hundreds of thousands—or millions—of people, survivors would have made antibodies that recognized and quickly neutralized the neuraminidase and hemagglutinin proteins of that strain. For several years thereafter the influenza viral population would undergo incremental changes that would result in modest alterations in its two crucial proteins, but not enough to render most immunized people vulnerable to infection.

The hemagglutinin and neuraminidase proteins constituted antigens, or targets for the human immune system's antibodies. As influenza made small genetic changes, the antigens would "drift" a bit from their standard form, and antibodies might not be able to lock on to the new types with quite as snug a fit over time. Nevertheless, most otherwise healthy people could successfully overcome antigen drift swiftly, making suitably adapted antibodies that would obliterate the viruses after a relatively mild bout of flu. Only people whose immune systems were weak, such as the elderly or malnourished, would die of influenza during such antigenic drift epidemics.

But sometimes something far more serious would happen. The antigens would do more than drift incrementally from their original genetic blueprint; a serious mutation event would occur and the hemagglutinin and neuraminidase proteins would suddenly be so different as to render human antibodies utterly useless. If the new forms of the proteins were also highly efficient in their tasks of clumping up blood cells and punching holes in cell membranes, a devastating global epidemic could result.

Though historians disagreed about details, there was an emerging consensus about when and where great flu pandemics had occurred in the previous 276 years, with what levels of human devastation.[5] Maps were drawn, illustrating the directions of influenza's spread around the world during the great pandemics of 1729, 1732, 1781, 1830, 1833, and 1889.

But it was the devastating pandemic of 1918–19 that most concerned President Gerald Ford and his advisers in 1976: the specter of another 500,000 dead Americans, 21 million dead worldwide, and over 10 percent of the U.S. workforce bedridden during the upcoming winter. That was the scale of what remains the twentieth century's worst pandemic.[6] It occurred when the world's population was far smaller, and human mobility limited to slow forms of transport, such as steamships and locomotives. Nevertheless, the epidemic moved completely around the globe in less than five months. The Ford administration knew that such an epidemic could be far worse in the age of jet travel and overpopulation.

The world was at war in 1918, fighting a largely ground struggle with millions of troops holed up in muddy trenches from the English Channel to the Crimea. The virus appears to have swept the world in three waves over less than two years' time, gaining virulence with each new assault. By October 1918 its strength was so great that people died with spectacular speed. There were reports of women boarding a New York subway in Coney Island feeling little else than mild fatigue, and being found dead when the train pulled into Columbus Circle, some forty-five minutes later.

In New York City alone, over 20,000 people died during the fall of 1918. The virus spread so extensively that travelers later discovered that entire Inuit villages in remote parts of Alaska were obliterated by influenza. Autopsies performed by London coroners revealed huge hemorrhages in the lungs, unlike anything the physicians had seen in the influenza epidemics of 1873 and 1889.[7]

The epidemic was by no means restricted to the war-torn Northern Hemisphere; influenza found its way from Europe to every nation on the planet. One out of every twenty citizens of Ghana died of the flu between September 1 and November 1, 1918.[8] The population of Western Samoa was overwhelmed by the virus. During the months of November and December 1918, nearly all of the 38,000 residents of Western Samoa contracted the flu, and 7,500—nearly 20 percent of the population—died.[9]

Medical science was then at a loss to explain the epidemic or provide sound advice to the terrorized population of any country. A Virginia State Department of Health pamphlet told the public that the disease was caused by "a tiny living plant called the germ of influenza." The less erudite *New York Post* told its readers that "epidemics are the punishment which nature inflicts for the violation of her laws and ordinances." Well-known Chicago physician Albert J. Croft said influenza was not a contagious microbe, but "small amounts of a depressing, highly irritating, high-density gas, present in the atmosphere, especially at night."[10]

Among the factors said by prominent American physicians to be responsible for influenza in 1918 were nakedness, fish contaminated by Germans, dirt, dust, unclean pajamas, Chinese people, open windows, closed windows, old books ("stay out of libraries"), and "some cosmic influence."

Sadly, nobody saved blood or tissue samples from victims of the disease. Such scientific forward thinking simply wasn't commonplace in those days. In 1976, more than a few U.S. health officials would curse the oversight, regretting there was no historic sample with which Private Lewis's influenza killer could be compared.

The 1918–19 epidemic did, however, spark a wave of aggressive research, and in 1932 Richard Shope (father of the Yale University researcher Robert Shope, who three decades later worked with Jordi Casals) did the experiment that would result in the moniker "swine flu": he removed nasal secretions from influenza-ailing domestic pigs and successfully infected other animals by rubbing the swine secretions on their noses or mouths. The following year, the British team of Wilson Smith, Christopher Howard Andrewes, and Sir Patrick Playfair Laidlaw isolated the influenza virus, for the first time giving the world an identity for its constant enemy. Two years later, Shope showed that people who were alive during the 1918–19 epidemic had antibodies against his pig virus, but children born after 1920 lacked such antibodies.

Shope's conclusion, which would remain the dominant hypothesis six decades later, was that the great pandemic was caused by a swine type of flu virus. Shope argued that the virus came originally from some other animal, went on to infect people, and then was transmitted from people to pigs. There, the virus found a safe haven, where it remained for years.[11] Nobody knew in 1976 whether the deadly strain had remained stable— and lethal—in the pig population for six decades, though it seemed unlikely that such a lethal virus could have failed for sixty years to cause disease in at least some pig farmer.

When Goldfield and his team of New Jersey scientists tested the Fort Dix samples they didn't have on hand a test tube full of the 1918 influenza virus—nobody did. But they did have a sample of Shope's swine flu, and were able to show that some of the Fort Dix recruits had antibodies that neutralized the Shope swine flu. On the basis of those two layers of hypothetical thinking, Goldfield suggested the Fort Dix strain might be the same as, or similar to, the virus that sickened over one billion people worldwide in 1918–19, killing more than 21 million.

The Centers for Disease Control quickly repeated the New Jersey studies, confirming Goldfield's findings. Hypothetically, then, it seemed they had found a relationship between Shope's 1935 pig virus, a massive human epidemic in 1918, and some of the soldiers at Fort Dix who had antibodies that could neutralize the swine strain. Furthermore, the influenzas extracted from all the infected Fort Dix soldiers were A-type flus, the one of three influenza serotypes that was most often responsible for large-scale pandemics. When the CDC completed all their "fingerprinting" of the Fort Dix virus it turned out to be influenza A, H1 (hemagglutinin 1) N1 (neuraminidase 1). Shope's swine virus was also an influenza A, H1N1.

In contrast, the most prevalent flu strain in the world during the early spring of 1976 was influenza A/H3N2. Dubbed the A/Victoria/75 strain, it first appeared in Victoria, Australia, a year earlier, causing relatively mild flu outbreaks from Johannesburg to Minneapolis.[12]

The appearance of the Fort Dix virus, dubbed A/New Jersey/H1N1, caused considerable anxiety inside the U.S. Public Health Service.

"By every available scientific measure, the Shope strain was indistinguishable from the 1918 strain, and also indistinguishable from the Fort Dix strain," Dr. June Osborn said nearly two decades later. In 1976, she was one of seven members of a U.S. Food and Drug Administration committee responsible for reviewing all American vaccine policies. Then a professor of medicine at the University of Wisconsin, Osborn was certain, and would remain so convinced, that the dreaded 1918 swine flu recycled back through the animal population, killing Private Lewis in the winter of 1976.

But investigations underway at Fort Dix revealed that only Private Lewis had died, and most of the illnesses at the base were due to the A/Victoria/75 strain. Furthermore, Lewis's sergeant attempted to revive the teen soldier when he collapsed, giving him mouth-to-mouth resuscitation. A month later the sergeant remained well and showed no signs of influenza A/New Jersey/H1N1 infection. That certainly argued for caution in ascribing great powers of virulence and transmissibility to the new virus.

However, several dozen Fort Dix recruits tested positive for A/New Jersey/H1N1 infection, and the stakes were awfully high. As health officials would later explain, the consequences of being wrong, of denying the possibility that Private Lewis's death was a harbinger of an epidemic akin to that of 1918–19, and then having such a devastating incident catch the nation unprepared, were simply too dreadful. Though some signs in February 1976 already pointed in the direction of skepticism, they were overshadowed by the fear of having hundreds of thousands of deaths ascribed to health officials who had chosen to take a wait-and-see approach.

A final element that tipped the balance at CDC in favor of acting on the assumption that a dreaded epidemic was imminent came in the form of a widely held scientific theory. One of the most prominent virologists in the world in the mid-1970s was Dr. Edwin Kilbourne of the Mount Sinai School of Medicine in New York City. Kilbourne had shown a decade earlier that influenza viruses unusually rich in neuraminidase proteins were more easily spread from person to person. As viruses were formed in mass quantities inside a human cell, their packaged chromosomes migrated to the outer membrane of the invaded cell. Scientists could visualize this with an electron microscope, which would reveal long rows of dark spheres pushing the membrane edges, creating bulges. Eventually, the viruses would push hard enough to pull a glob of cell membrane around their inner envelope and chromosomes, creating an outer protective coating. In this process, called budding, the new viruses would protrude sharply from the cell, but remain tethered by a final strand of host membrane. Kilbourne showed that the viruses' neuraminidase proteins would snip the tethers, freeing the newly formed microbes to enter the lungs, nasal fluids or tears of an ailing human, from there going on to infect another person. The greater the number of neuraminidase molecules, Kilbourne argued, the more rapidly viruses could complete their budding process and spread.

In essence, Kilbourne had found a possible key to both high transmissibility and virulence, explaining why some epidemics produced viruses that rapidly flooded the bloodstreams of infected people and readily became global pandemics, while others caused fairly localized mild outbreaks. [13]

He proved his point by quantifying the density of neuraminidase proteins on the surface of the influenza strain responsible for the 1957 flu pandemic, a fairly severe wave that swept the world and claimed an estimated 60,000 American lives. That strain had the highest neuraminidase concentration of any influenza discovered since the 1930s.

But with the arrival of the Hong Kong A flu in 1968, public health experts worldwide were taken by surprise. Though most had predicted the winter of 1968–69 would have only mild flu, the Hong Kong strain caused a huge global epidemic that proved less deadly than the 1957 pandemic, but caused far more widespread illness. Whereas the Hong Kong strain had undergone sharp antigen shift in its hemagglutinin proteins, its neuraminidase component was unchanged from the previous year's mild influenza. Epidemiologists, most of whom were dumbstruck by the Hong Kong pandemic, were sobered by the virus's ability to outsmart their collective human intelligence.

"The epidemiologist must recognize that prediction of future epidemics remains a hazardous business," warned noted Harvard epidemiologist Alexander Langmuir following the Hong Kong pandemic. "There does seem to be a periodicity that must relate to the balance of immunes and susceptibles and to the mutations of the virus. In a way, influenza predictions are like weather forecasts. As with hurricanes, pandemics can be identified and their probable course projected so that warnings can be issued. Epidemics, however, are more variable and the best that can be done is to estimate probabilities."[14]

Still, most influenza experts believed in the early 1970s that flu epidemics appeared in predictable cycles, with B-type and A-type viruses undergoing antigenic shifts in separate but fairly regular time periods. This was argued most persuasively by Kilbourne, who noted that major antigenic shifts had recently occurred in roughly ten-year cycles: 1947 (H1N1), 1957 (H2N2), and 1968 (H3N2). [15]

In February 1976, Kilbourne wrote an opinion piece for *The New York*

Times in which he predicted that a major pandemic was coming soon, based on the theory of ten-to-eleven-year influenza cycles. He warned that "those concerned with public health had best plan without further delay for an imminent natural disaster."[16]

Although Kilbourne spoke for the dominant tendency in influenza scientific thinking of the day, there were dissidents who felt that influenza cycles were longer, shorter, varied in length by subtype, or were entirely random and unpredictable. Several scientists argued that swine strains, in particular, appeared in 90-100-year cycles and forecast a repeat of the 1918–19 disaster for sometime in the 1990s.[17] Blood tests of Americans during the 1968 Hong Kong flu pandemic showed that elderly people who had lived through the 1889 flu epidemic were immune to the 1968 strain. That seemed to indicate a cycle, at least for that A strain, of about eighty years.[18] Finally, Australia's leading influenza expert, W. I. B. Beveridge, argued that influenza was a "capricious virus that is not possible to predict," noting that a long-term view of the historical record showed pandemics of one kind or another had surfaced after intervals of ten to forty-nine years, a range too great to represent a basis for forecasting future outbreaks.[19]

While influenza was spreading around Fort Dix, most of the world's flu experts were gathered in Rougemont, Switzerland, for an international influenza meeting.[20] Completely unaware of events then unfolding in New Jersey, the scientists devoted their collective energy from January 26 to 28 to the task of deciding how humanity should best respond to a 1918-like flu pandemic.

In 1947 the World Health Organization, shortly after its inception, had created a network of laboratories throughout the world that agreed to collaborate in efforts to monitor changes in influenza patterns. By the time scientists gathered in Rougemont twenty-nine years later, there were nearly a hundred laboratories in the WHO influenza network, crossing most Cold War and economic boundaries of the day. These laboratories regularly collected samples of influenza from human flu patients, and sporadically from ailing birds and livestock, hoping to be able to detect dangerous perturbations in global influenza before millions of people were infected.

Based on the work of the CDC's Dr. Walter Dowdle and Robert G. Webster of St. Jude's Children's Hospital, Memphis, the gathered scientists knew that ducks and other wild birds carried influenza around the world along their migratory routes, passing it on to other animals via fecal droppings. The bird droppings were an ideal ecological environment for the viruses. Influenza could survive over three days outdoors or in the milieu of fecal-contaminated water.

"It appears that influenza viruses are constantly circulating among many avian species without causing panzootics [cross-species epidemics]. This suggests that influenza is a natural avian infection and may have been so for thousands of years," Webster told the Rougemont gathering.[21] "The conclusion can be stated simply. All the genes of the influenza viruses of the world are being maintained in the aquatic bird population, in gulls and ducks, and periodically they are transmitted to other species, including humans, usually after reassorting."

Perhaps more important, in light of subsequent immediate events, was the Rougemont discussion of vaccination policies. Clearly, it was difficult to motivate voluntary compliance with vaccination campaigns. For example, between 1968 and 1974 the best turnout for flu immunization in the United States occurred during the 1968 Hong Kong epidemic, when a mere 10.7 percent of the population got their shots, despite the epidemic's severity. By 1974 U.S. flu vaccination rates had plummeted to 8 percent of the general population and a poor 17.4 percent showing among the elderly, who were considered at special risk for flu. Even in states such as California, where flu shots were offered gratis to citizens over sixty-five, much of the vaccine supplies rotted in warehouses in 1975 for lack of physician and public interest in immunization. And in the U.K., fewer than 12 percent of National Health Service physicians promoted the use of flu vaccine to their elderly or hospitalized patients. Most striking, less than 6 percent of London nurses agreed to have themselves vaccinated against flu in any given year from 1968 to 1975.

Were a major epidemic on the horizon, the Rougemont gathering concluded, special efforts would have to be made to find the virus soon enough to allow large-scale manufacture of vaccine. And extraordinary steps would have to be taken to mobilize massive public compliance—on the order of 80 percent of the elderly and at-risk populations—in vaccination.

"The most important outcome of the next epidemic may well be the lessons it will provide that can help in controlling both pandemic and epidemic influenza," Walter Dowdle told the Rougemont gathering. He had no idea his words would prove immediately prophetic. Two weeks later the health establishment would seriously entertain the possibility that the Fort Dix virus was the same as, or a close cousin of, the 1918–19 killer microbe, and events would snowball down an icy policy slope. The perceived threat would grow larger every week as scientists were forced to yield decision making to politicians, and the highly hypothetical basis of all Swine Flu conjectures would recede further into the background. While early scientific and policy reports carefully alluded to the tentative basis of such dire forecasts, official government prognostications presented only certainties and absolutes. This contrast between a very tentative hypothetical outlook in the immediate wake of Private Lewis's death, and the federal mobilization weeks later was so striking that Dr. Joseph Califano, upon taking office a year later as Secretary of Health, Education, and Welfare in the new Carter administration, would commission an outside investigation. Two Harvard University policy analysts would be given a simple question to answer in the winter of 1977: "What went wrong?" Richard E. Neustadt had served in government during the Democratic administrations of Truman, Kennedy, and Johnson, and was on the faculty of the John F. Kennedy School of

Government at Harvard during the Swine Flu affair. Dr. Harvey Fineberg wore two hats, holding degrees in both medicine and public policy. Together in 1977 they would review the events of the previous year, trying to pin down exactly when, and why, the federal health establishment started to suspend disbelief and began its journey down the Swine Flu slippery slope.

"From one case—Private Lewis—you learn nothing," Fineberg, seventeen years later as dean of the Harvard School of Public Health, would say. "It was in the great quiet afterwards that more genuine information was to be found," he said, noting that the entire health establishment held its breath during the winter of 1976, waiting for signs of Swine Flu spread. Signs that never appeared.

"The issue was not the overinterpretation of early cases [at Fort Dix], but the subsequent foreclosure of doubt" about the possibility of fulfillment of a worst-case scenario, Fineberg said. "My personal view is that often in such cases it's hard to separate likelihoods from consequences. In this case the consequence of being wrong about an epidemic were so devastating in people's minds that it wasn't possible to focus properly on the issue of likelihood. Nobody could really estimate likelihood then, or now. The challenge in such circumstances is to be able to distinguish things so you can rationally talk about it. In 1976, some policy makers were simply overwhelmed by the consequences of being wrong. And at a higher level [in the Ford administration] the two—likelihood and consequence—got meshed."

Osborn, known for her blunt and often eloquent expressions, shared no such perspective in 1976, or nearly twenty years later. She told fellow FDA advisers that there had been a long spring-to-summer silence following the first flu outbreaks of 1918—a silence that was followed in September by the greatest pandemic of the early twentieth century.

"To decide not to do something, to decide to go on pause because the virus went on pause," Osborn argued in long conference calls to fellow scientists, "would be utterly irresponsible."

From his perspective as director of the CDC, David Sencer also saw a very different picture in 1976—which would remain at odds with the Neustadt-Fineberg report two decades later. Jovial, self-effacing, and well liked by most CDC staffers during his reign (1966–77), Sencer was most persuaded to action by what he believed were clear cases of soldier-to-soldier spread of the apparent swine virus at Fort Dix.[22]

"The fact of transmission is the key," he told his staff.

Dr. Walter Dowdle, a soft-spoken influenza virologist who viewed the world in a serious and cautious fashion, telephoned Sencer late the first Tuesday night of February to tell the director that the New Jersey state labs claimed to have found five cases of Swine Flu. Dowdle rarely raised red flags of alarm around the CDC. Forty-eight hours later, however, Dowdle told Sencer that CDC scientists had confirmed the Swine Flu findings. Sencer immediately called a meeting of top government scientists from all over the country, and Saturday morning in Atlanta, Dowdle presented the Fort Dix case report.

Dowdle carefully laid out what had transpired, leaving for last evidence that Private Lewis died of a virus which cross-reacted with the Shope swine strain. When Dowdle concluded his remarks with the words "The isolates were swine," General Phillip Russell's jaw dropped, and the tall, muscular Army physician abruptly sat forward. As head of all military medical research in the U.S. armed forces, Russell had complete responsibility for decisions concerning vaccination of armed forces personnel. Well versed in the history of the 1918–19 epidemic and its rampant spread among World War I military personnel, Russell saw no choice in the matter: the United States should immediately develop a Swine Flu vaccine.

Sencer agreed wholeheartedly. It wasn't a question of probabilities, he argued, but of disease prevention. Even if the likelihood of a 1918–19 type of virus reappearing in the fall of 1976 was immeasurably small, he said, it would be wrong to avoid taking steps to prevent it.

"We have the technology, we have the evidence of transmission," he told the group. "It would be irresponsible to do anything else except develop a vaccine."

In its first official pronouncement the CDC's words were crafted with careful attention to the ambiguities and tentativeness inherent in interpretation of the Fort Dix cases. After hours of meetings with top CDC and Washington Public Health Service officials, the agency published its first Swine Flu notice on February 14, in its regular weekly publication.[23] It explained that a small influenza outbreak had occurred at Fort Dix during the previous month, involving one death. Eleven blood samples had been tested, seven proving to be the relatively harmless A/Victoria strain; four resulted from an A/H1N1 strain "similar to swine influenza."

The report also noted, "There is some evidence from antibody prevalence studies that occasional infections with swine influenza virus might have occurred in more recent [since 1970] years among persons in frequent contact with swine."

In 1974, a sixteen-year-old Minnesota boy suffering from Hodgkin's disease (a type of blood cancer that produces severe immunodeficiency) died of what appeared to be Swine Flu. A year later, an otherwise healthy eight-year-old Wisconsin boy contracted the disease, surviving thanks to his body's production of antibodies that cross-reacted with Shope's 1930s Swine Flu virus. Both boys lived on farms and handled pigs. More important, in both cases the infection never spread to other schoolmates, and though most of the Wisconsin boy's immediate family tested antibody-positive for exposure, none had developed the flu.

In February, then, the agency readily acknowledged that there might be some low-level background rate of Swine Flu infections among people who lived around domestic pigs, and the presence of antibodies to swine antigens did not, in and of itself, indicate that a particularly lethal or highly transmissible form of influenza was afoot in America.

Many years later CDC influenza expert Nancy Cox, who was not directly involved in the events of 1976, would summarize a large body of evidence indicating that people who lived and worked around domestic livestock were routinely exposed to the viruses those animals harbor, including swine strains of influenza. The great 1889 pandemic, for example, began as an epidemic of "the cough" among European horses (probably Russian) sometime in the early 1880s. Nearly ninety years later the 1968 Hong Kong flu also proved capable experimentally of producing "the cough" in horses.

Swine influenzas, Cox would later explain, were particularly worrisome because pigs were highly permissive hosts, capable of harboring influenzas from a wide range of animals, birds, and humans. Inside the swine, various influenza strains shared genes, and recombined, resulting in major antigen shifts.

"We do see in hindsight that the farm Swine Flu cases in 1976 were separate and isolated events from what occurred at Fort Dix," Cox explained.

A week after the CDC's first 1976 publication, the agency noted discovery of six more Fort Dix soldiers with Swine Flu, bringing the total to ten (including the deceased Lewis). The remainder of the base's epidemic appeared to result from A/Victoria flu. An additional agricultural Swine Flu case was reported, involving a young man from Mississippi, who also suffered from Hodgkin's disease and worked in a pig slaughterhouse.[24]

Large-scale blood testing at Fort Dix soon revealed a total of 273 individuals who may have had swine antibodies, thirteen of whom had actually contracted influenza. Unclear, and never clarified in any subsequent CDC publications, was how many of the influenza-ailing soldiers were co-infected with A/New Jersey/76 and A/Victoria/75. Even thirty years later no technology could tell which strain was responsible for disease in an individual who was co-infected, although it is generally assumed that whichever strain is present in largest numbers is the pathogenic culprit.

The CDC interviewed the Fort Dix recruits to see which might have had direct contact with swine, and found twenty-two men who had been around pigs and had antibodies to the Shope swine influenza. Investigators then tested the families of those twenty-two soldiers: one family tested antibody-positive. Four out of eleven members of that family tested positive; none were sick. They were not farmers, and the flu-exposed members were all under twenty-five years old. When 200 classmates of the children were tested, no further spread of the apparent infection could be found.

Policy decisions and actions moved forward rapidly, though investigations at Fort Dix were far from complete. Army and CDC investigators would spend several more weeks combing the base for clues to the origin and spread of the apparent Swine Flu, eventually concluding that no more than 155 recruits were definitely infected with the virus. Another 300 Fort Dix soldiers were infected with the A/Victoria/75 strain.

More important, the investigators concluded that the only time or place shared by all soldiers infected with Swine Flu prior to their illnesses was the Fort Dix reception area. The mini-epidemic began, they concluded, sometime in the first or second week of January, when hundreds of fresh recruits were processed onto the base following the Christmas holidays, and assigned to basic training. In the reception center the new recruits were given physical examinations, vaccinations, and basic military instructions.

The first of the new recruits subsequently shown to have Swine Flu arrived at the reception center on January 5. Designated only as V4, he complained of illness on January 28.

Private Lewis came through the reception center the following day. All Fort Dix Swine Flu illnesses occurred between January 12 and February 8, the time span of high reception center activity. The infections probably incubated between the initial transmissions in the reception center in early January, and the flu illnesses appeared two to three weeks later.

The only other possible shared source of infection for the thirteen soldiers struck with Swine Flu was the base medical system. All the men made visits to the base dispensary for a variety of health problems prior to developing the flu. Under General Russell's personal command, Army investigators searched for a source of viral contamination at both the reception area and the dispensary: none was found. It is not likely, however, that weeks after the events any evidence of viral contamination of equipment or medical instruments would persist, available for discovery. Thus, the possibility remained that America's Swine Flu outbreak of 1976 was iatrogenic.

By mid-March influenza of all types was on a sharp decline worldwide, even at Fort Dix, and the agency's virus branch director, Dr. Walter Dowdle, said, "Influenza in the United States has decreased markedly, and there is no longer evidence of nationwide epidemic activity."[25]

"By the beginning of March," Dowdle would write six years later, "the only signs of the Swine Flu epidemic *in the world* [his emphasis] were at Fort Dix. But the possibility of a Swine Flu outbreak in the future could not be disproved. What could not be disproved must be allowed for. Most of the scientists were well aware of the professional risks they incurred if they mounted a national immunization program and the virus did not appear. Most were equally aware of their responsibility for the public's safety in the event of an epidemic. Something had to be done."[26]

On March 13, the CDC director, David Sencer, completed a special memorandum for his superiors in Washington, detailing the evidence for a Swine Flu outbreak and requesting a $134 million congressional allocation for development and distribution of vaccines. Within less than a week, word of Swine Flu was all over Capitol Hill. By March 18, Sencer's memo had been signed by Assistant Secretary for Health Dr. Theodore Cooper, and lay upon the desk of HEW Secretary F. David Mathews awaiting his urgent attention.

Stated as certainties, rather than hypothetical conjectures, were the following points listed under the memo's heading *"Facts"*: The virus found at Fort Dix is "antigenically related to the influenza virus which has been implicated as the cause of the 1918–19 pandemic which killed 450,000 [American] people; every American under the age of fifty "is probably susceptible to this new strain"; severe flu epidemics "occur at approximately ten-year intervals."

After laying out four different plans of suggested action, the Sencer memo suggested mass vaccination, sponsored by the federal government, conducted by local authorities and supported publicly at the highest possible level.

Within two weeks the snowball was roaring down an Alpine slope, gathering size as most sectors of the federal government, from congressional aides to the White House Office of Management and Budget, signed on.

March 24 found an extraordinary group of scientists gathered in the White House at President Ford's request. Edwin Kilbourne, polio vaccine inventors Jonas Salk and Albert Sabin, and a host of CDC and other federal researchers were asked point-blank by Ford, "Do you agree that the nation is facing a Swine Flu epidemic, and mass vaccination is necessary?"

There were no voices of dissent in that room.

That night, President Ford went on national television in a press conference that found him flanked by the icons of immunization, Sabin and Salk.

"I have just concluded a meeting on a subject of vast importance to all Americans," the President said. "I have been advised that there is a very real possibility that unless we take counteractions, there could be an epidemic of this dangerous disease next fall and winter here in the United States. . . . Accordingly, I am asking the Congress to appropriate $135 million, prior to the April recess, for the production of sufficient vaccine to inoculate every man, woman, and child in the United States."

Congress had no choice but to support the President. The politicians were nearly unanimous in their shared apprehension about being responsible for massive numbers of influenza deaths should they balk. Former Senator Edward Kennedy staffer Arthur Silverstein said there "was an almost unseemly race" on Capitol Hill to approve the President's $135 million vaccination appropriations request. Senator Kennedy said, "There is nothing more frightening to a society than an epidemic," throwing his liberal weight behind the Republican President's request.

Only two members of Congress were sharply vocal in their criticism of the program. California Democrat Representative Henry Waxman and his New Jersey colleague Andrew Maguire denounced the program as a "rip-off" that was guaranteed to generate profits for vaccine manufacturers. Consumer advocate Ralph Nader accused the government's health establishment of crying wolf, wasting taxpayer dollars.

Recognizing Ford's position, some members of Congress decided to exploit the President's absolute support of the flu campaign by attaching a long list of liberal riders to the immunization bill, adding $1.8 billion worth of social service spending and environmental protection funds to a bill they knew Ford could not possibly veto.

Meanwhile, when Osborn saw President Ford's press conference on television, she was outraged. Forced to work by telephone from Madison, Wisconsin, rather than in direct conference in Washington because her twin daughters were just seven years old, Osborn couldn't believe that the sober, cautious approach that all her colleagues had tried to follow was suddenly shoved aside.

"Everybody knows Salk and Sabin detest one another, and that they're the two most famous vaccinologists in the world," Osborn told fellow FDA advisers she reached that night by phone. "Neither of them has been involved in this in any way. Putting them together with the President like that spells disaster."

Years later, Osborn and Sencer would both argue that Ford's March 24 press conference marked the turning point, bringing healthy skepticism to an end and putting politicians in the Swine Flu driver's seat.

As support built in Washington, the pharmaceutical manufacturers played their trump card, telling Ford directly that their insurance carriers would not indemnify such hastily produced vaccines. Unless the government absorbed liability for all possible ill effects from the vaccines, drug companies could not possibly cooperate in the $135 million effort. Well before Congress approved, and the President signed Public Law 94-266 allocating funds for the flu campaign, word was out that the real price tag might exceed by millions of dollars the requested sum. Some liberal members of the House of Representatives accused the pharmaceutical industry of trying to pull off a major scam, milking taxpayers for hundreds of millions of dollars and refusing to accept any responsibility for vaccine product quality.

But on April 15, 1976, PL 94-266 was signed by President Ford in a televised ceremony. As he placed his signature on the bill, Ford dropped all pretense of doubt or conjecture, saying the Fort Dix virus "*was* the cause of a pandemic in 1918 and 1919 that resulted in over half a million deaths in the United States."

Though dissent and controversy would increase in political and scientific circles over subsequent months, a seemingly intransigent White House and public health establishment would speak with ever-greater certainty about the likelihood of a catastrophe, and all semblance of theorizing and guesswork would disappear from official pronouncements. Anger built during the late spring in response to pharmaceutical manufacturers' insistence that no vaccine could or would be produced before the federal government agreed to absorb full liability. Some politicians accused the industry of casting off all vestiges of public responsibility, while corporate representatives reminded members of Congress that they were working in a tough, highly competitive free market in which profit making (or, at the very least, breaking even) was essential to survival.

With a good deal more dissent than was engendered by the PL 94-266 enactment, Congress would eventually pass a law that officially waived corporate liability for Swine Flu vaccines, placing all legal culpability squarely on the shoulders of the U.S. taxpayers. It would be signed on August 12 and designated the National Swine Flu Immunization Program of 1976 (Public Law 94-380), scheduled to go into effect October 1, the same day the CDC planned to kick off the national flu vaccination program.

The nation would then be irrevocably committed.

It might not have made that leap from April's bill to the August open-ended liability price tag had it not been for a unique and entirely unexpected set of events in July. Throughout the spring and early summer of 1976, opposition to the very concept of a mass epidemic was building in both scientific and political circles.

Several leading physicians, notably consumer advocate Sidney Wolfe, vocally protested the government's dire forecasts of a million dead Americans, noting that the CDC had projected those numbers from a base of 500,000 dead in 1918–19, multiplied by the increase in the U.S. population size since that time, and factoring for other changes, such as air travel and urbanization, which were thought to speed the spread of airborne microbes. The dissident doctors attacked the projection, noting that medical science had advanced considerably in its ability to diagnose and treat influenza, and it was highly unlikely that even a super-virulent strain could kill 21 million people worldwide in 1976. After all, they said, most influenza deaths were usually produced not by the virus but secondarily by bacterial infections that took advantage of the weakened immune defenses of influenza-infected lungs. Bacterial pneumonia was easily treated in 1976 with a number of readily available antibiotics. Though the CDC insisted (and still would nearly twenty years later) that the 1918–19 virus killed massive numbers of people directly, without secondary bacterial infections, many vocal physicians maintained that some, perhaps most, of the lung hemorrhages and fatal heart attacks reported in 1918 might be treatable in intensive-care units in 1976.

There was also increasing skepticism about the basic assumption that the Fort Dix strain was equivalent to the deadly 1918–19 Swine Flu. There was no evidence of mass spread, and Fort Dix medical officer Bartley told *Science* magazine's Philip Boffey that Private Lewis might well have lived if he hadn't gone on the long winter march. Some Army physicians quietly told their civilian colleagues that it was possible even the dreaded virulence of the 1918–19 strain was more an environmental than genetic issue. Rather than ascribing the rampant spread and quick die-offs to some unique characteristics of the virus, these researchers discreetly insisted it was World War I trench warfare conditions, horrendous overcrowding in military encampments, and the movement of hundreds of thousands of troops in jam-packed ships, submarines, and train cars all over the world that spread the disease.

Dr. E. Russell Alexander, chair of the University of Washington's School of Public Health, was a dissident member of the CDC's flu advisory committee. From the beginning he had advised that the government hold off on a mass immunization campaign and instead stockpile vaccines for possible use should an epidemic appear. As months rolled by without additional cases, Alexander's position drew support from many circles.

If the CDC was facing obstacles due to controversies in the United States, its ability to win over public health counterparts in other countries was firmly blocked by strong scientific skepticism. Though the World Health Organization lent official support to a global Swine Flu campaign in a special meeting convened in Geneva April 7–8, the backing was lukewarm, and fell short of recommending widespread vaccination. Instead, WHO suggested that national health ministries worldwide be on the alert for unusual flu outbreaks, consider adding Swine Flu vaccines to the list of immunizations offered to elderly citizens of richer nations, and stockpile supplies of vaccine once it was available. It seemed Russell Alexander's position was garnering support outside the United States, with the exception of Canada, where the CDC's arguments held sway.

On July 3 the prestigious British medical journal *The Lancet* published three articles critical of the American campaign. In the first, physicians from England's Harvard Hospital in Salisbury compared the A/New Jersey flu strain with two other varieties found in American pigs and a strain then common in England. Six human volunteers were injected with viral samples. The researchers concluded that the Fort Dix virus was "evidently intermediate in its virulence for man between a human virus and a swine virus. . . . The conclusion, therefore, was that in its present form, A/NJ/8/76 was less virulent in man than an established human influenza-A virus, but a good deal more infectious and virulent than two swine pathogens . . . tested previously."[27]

In an accompanying essay, University of Sheffield Medical School professor Charles Stuart-Harris argued it was "a time for a continual reappraisal of all possibilities rather than for a change of tactics." It simply wasn't time for a mass immunization campaign, he said, though it might be wise to stockpile vaccine just in case.

As had been stated previously by dissident American colleagues, Stuart-Harris insisted it was unwise to compare the 1918 pandemic to any 1976 possibility. The current virus simply wasn't as virulent, he said, and nowhere were human conditions as severe as those seen in the battlefields of World War I.[28]

Finally, the publishers of England's leading medical journal dismissed the dangers of the Fort Dix strain specifically, but not generally of influenza, and ominously concluded that "the whole exercise will be valuable practice for doing it when a really new influenza A does finally appear."[29]

Throughout May, June, and early July, the arguments in U.S. government circles centered not on whether or not to vaccinate but on how best to accomplish the task of making 200 million doses—for Americans alone— and mobilize local health authorities and the public before the fall. On June 22, despite serious difficulties, both the CDC and FDA vaccine advisory committees voted to proceed with the Swine Flu campaign.

"We have no choice," Osborn told her colleagues.

But the vaccine trials hadn't gone at all well, and Albert Sabin did a 180-degree turnaround, becoming a vocal opponent of the campaign. None of the products seemed to work at all in children; the vaccines performed so poorly in young adults that even some campaign proponents openly worried that an acceptable formulation might not be found before the fall; nobody was sure how much flu antigen, or human antibody, was necessary to protect against a super-virulent virus, and one company, Parke-Davis, made 2 million doses of vaccine against the wrong flu strain.[30]

Meanwhile, the Pharmaceutical Manufacturers Association continued its pressure on Congress and the White House, saying that no vaccine would be made unless something could be done about the liability issue. Two bills were stalled in Congress (HR105050 and S3785) that aimed in different ways to clear vaccine manufacturers of liability by shifting the burden to the federal government.

Debate on the matter was feverish both on the editorial pages of America's leading newspapers and in the halls of Congress, and pharmaceutical industry lobbyists were clearly concerned that their liability protection would be defeated.

Until August 2, 1976.

On that day, newspapers across America carried the headline news that several men had succumbed as a result of sudden severe respiratory ailments contracted, apparently, following attendance at an American Legion convention in Philadelphia during the week of July 21–24.

II

Few groups in the United States took patriotism as seriously as the American Legion, and in the country's bicentennial year it was more than appropriate that an organization dominated by World War II veterans should convene in Philadelphia, the cradle of the country's Declaration of Independence and Constitution. For four days in July several hundred members of the Pennsylvania Legionnaires division held meetings, sat at banquets, danced, and sipped cocktails in four Philadelphia hotels.

Liquor flowed most freely in the hospitality suites of thirteen candidates for Legionnaire offices. Scattered throughout the luxurious old Bellevue-Stratford Hotel, these suites were sites of energetic handshaking and free cocktails.

On the second night of the meeting, two of the Legionnaires fell ill with symptoms that included fevers, muscle aches, and pneumonia. Because they were older men, the first cases raised no alarms.[31]

Within a week, however, the Pennsylvania Department of Health was flooded with reports of acute pneumonia illnesses and deaths among people who had been inside Philadelphia hotels during the latter half of July. The count would eventually reach 182 cases (78 percent males); 29 deaths. Some 82 percent of the cases, when final numbers were tallied, would turn out to be American Legionnaires.

On August 2, with about 150 cases and 20 deaths then reported, the Pennsylvania health authorities issued a statement that was instantly front-page news worldwide. Given the media moniker "Legionnaires' Disease," the mysterious Philadelphia epidemic caused an intense escalation in Swine Flu fears.

Word hit Congress and the White House like a jolt of electricity, shocking the argumentative politicians into action. Various bills that sought to break the long deadlock on vaccine liability were hastily approved by House and Senate subcommittees and rapidly made their ways toward the floor for full legislative debate and possible approval. Fearing that the dreaded Swine Flu epidemic had arrived, the country's political leaders acted with atypical speed.

By August 5, when Sencer testified before the Senate's Health Subcommittee, the stage was set for almost immediate approval of liability-waiving legislation.

Proud of the rapid and thorough investigation responses of his CDC staff, Sencer could barely contain himself. His office had first learned of the Legionnaires' outbreak on Monday, and by Tuesday, CDC staff were able to positively rule out influenza as the cause of the mysterious deaths. Of course, they didn't know what *was* the cause of the Philadelphia ailments. Sencer was delighted to sit before Congress on Thursday, just four days after the outbreak was reported, able to allay the public's fears that the dreaded Swine Flu had finally arrived.

But Congress, far from applauding the news, grit its teeth and immediately reverted to its prior argumentative stance. Within minutes, all hope of rapid passage of a vaccine liability law was lost.

The following night, President Ford lashed out, saying he was "very dumbfounded" by Congress's actions. He came right out and told the Capitol Hill politicians that they would personally bear responsibility for millions of American deaths should a Swine Flu epidemic materialize.

Congress caved in.

Six days later, President Ford signed the National Swine Flu Immunization Program of 1976 (Public Law 94-380). The die was cast.

Within the bill's tediously argued language lay the seeds of the Swine Flu vaccine program's destruction. Though the politicians, the pharmaceutical industry, and insurance companies generally felt that the bill masterfully eliminated all remaining stumbling blocks to mass immunization, it actually created dangerous new obstacles. As would be seen in years to come, the public health problems created by Public Law 94-380 resonated perilously not only throughout the 1976–77 flu season but in all future new American vaccine campaigns. In a very real sense, the Swine Flu vaccine campaign of 1976 would eventually work to the advantage of the planet's microbes. The bill stated:

The United States shall be liable with respect to claims submitted after September 30, 1976, for personal injury or death arising out of the administration of swine flu vaccine under the swine flu program and based upon the act or omission of a program participant in the same manner and to the same extent as the United States would be liable in any other action brought against it. [32]

The bill guaranteed that the U.S. government would respond financially to claims of damages associated with Swine Flu vaccination; put the legal defense job in the hands of the Attorney General; prohibited pharmaceutical companies from making "any reasonable profit" off Swine Flu vaccines; and ordered all individual immunizations to be accompanied by a signed informed consent form that "fully explained" the risks and benefits of the product. To mollify vaccine producers who did not take kindly to being ordered to forgo Swine Flu profits, the bill allowed "reasonable profit" for A/Victoria flu vaccines—"reasonable" was never defined. In the end, worried that Congress might later attach some arbitrary level of profit as unreasonable, the pharmaceutical companies donated A/Victoria and bivalent (A/Victoria plus A/New Jersey) vaccines to the federal government.

All of this was precedent-setting in the United States. Though vaccination campaigns, from point of purchase to distribution, were entirely government-operated in most of the world, they were a rarity in the United States. Typically, Americans were vaccinated by a personal physician, school nurse, or public health clinic nurse. The federal government's role was usually limited to identifying what types of vaccines were needed (through the CDC) and regulating their purity and safe manufacturing (through the Food and Drug Administration).

But now the U.S. government was up to its neck in the vaccine business, and the Pharmaceutical Manufacturers Association sat back comfortably, awaiting its instructions, its fifty-some members freed of liability concerns.

Though the CDC laboratory had before August 5 successfully ruled out Swine Flu as a cause of the Legionnaires' deaths, by the time the influenza vaccine liability law had been signed by President Ford it was obvious to the lab boys that solving the Legionnaires' puzzle was going to be terribly difficult. Though they worked around the clock, using every obvious scientific trick they knew to tease out the culprit, it remained utterly elusive. Joe McDade and Charles "Shep" Shepard were stumped.

By August 31, their labs had scanned hundreds of tissue samples at the subcellular level of magnification with electron microscopes. They had used fluorescent antibodies against over a dozen different microbes to see if any of the patients were infected with the agents, including chlamydia, rickettsia, typhoid, pertussis, tularemia, plague, coccidioidomycosis, histoplasmosis, Marburg, Lassa, influenza, and choriomeningitis. They searched for fifteen different types of yeast and two types of mycoplasma. The team had also tried isolating viruses by infecting chicken eggs, monkey cells, human cells, guinea pigs, and mice with blood samples from the patients.

Microbiologist McDade and physician-scientist Shepard conducted standard tests on blood tissue samples aimed at finding out what sort of microbe was responsible. They dumped various antibiotics in cultures, but got no consistent effects. They had used standard stains for visualizing bacteria under microscopes and saw nothing.

So they switched tactics. They put blood samples in test tubes with antibodies against all sorts of different microbes, and looked for the clumping reactions that meant a positive reaction had occurred. With this method, they had already tested for over twenty-six different microbes by the end of August, including various influenzas, Q fever, mumps, measles, adenoviruses, and a host of extremely rare diseases.

They tried another tactic.

Lung, liver, and kidney samples from the deceased Legionnaires were subjected to radioactive assays for heavy-metal poisoning (by such things as mercury, arsenic, thallium, nickel, cobalt—a total of twenty-three potentially toxic metals). And in hopes of finding a pesticide or other toxic chemical product, the team had already done 300 gas chromatography and mass spectral analyses by August 31. If any of these tests had found a contaminant, huge spikes would have appeared on the graph paper that spewed from the analytical machines. There were no unusual spikes.[33]

This really annoyed McDade, who had been with the CDC just one year but had ten years' experience in microbiology detective work under his belt. Meticulous in both his work and his personal style, McDade routinely took every imaginable precaution against contamination or error in the lab. The bespectacled, blue-eyed scientist liked everything around him to be neat, trim, and predictable.

In his detective work, McDade spoke of the "algorithm of the investigation," assigning each clue to its "dot position in the matrix." Depending on which clues first fell into place, the algorithm would move across the matrix in a certain direction. Normally, the careful scientist had a set of clear epidemiological dots to assign to the matrix before he even received blood or tissue samples to study in his laboratory. And it usually wasn't long before McDade discovered several more matrix dots in the lab and the algorithm quickly flowed to completion.

But this Legionnaires' Disease paradox wasn't yielding any useful clues for the matrix. The epidemiologists couldn't figure out how, whatever the culprit was, people got sick. They couldn't even narrow the clues enough to tell whether the killer was a chemical or a microbe.

And absolutely nothing was working in the laboratory.

The CDC lab boys then began to realize the enormity of their task.

"Something is lurking in those hundreds of tissue and blood samples amassing in the CDC refrigerators," Shep told Sencer, "but it isn't possible to narrow the search down by broad categories. We can't tell at this point whether we should be looking for a virus, bacterium, fungus, parasite, toxic chemical, or whatever. We need more information from the field."

Sencer deployed two large teams of CDC investigations to Philadelphia. One concentrated on the Philadelphia hotels, the other studied the surviving patients and their families.

Fresh out of his residency training in clinical medicine and just four months away from his extraordinary journey to Yambuku to assist in the Ebola investigation, Dr. David Heymann was part of the Legionnaires' investigation. Lean, dark-eyed, and shy, he would look back on 1976 years later, expressing astonishment at his "unbelievable" experiences. The twelve months from August 1976 to the summer of 1977 would find the twenty-something Heymann in the eye of microbial storms in Philadelphia, Yambuku, and Cameroon.

When he left Atlanta for Philadelphia in August 1977, however, the junior EIS officer had no inkling of what was to come. Nor was he prepared for the public panic in Philadelphia, the extraordinary complexity of the Legionnaires' mystery, or the personal safety concerns that would ebb and flow discreetly through the CDC team.

Panic does not always go hand in hand with epidemics, nor does its scale correlate with the genuine gravity of the situation. Indeed, history demonstrates that population responses to disease are rarely predictable, often peculiar, and always key features of frustration for disease detectives who must sift through public accounts to find clues to the origin and cause of the epidemic. Where a hefty dose of public concern was warranted, as in the case of the 1918–19 pandemic, an oddly common feature was nonchalance. The usually vigorous New York press, for example, reported virtually no flu news that year until the week of November, by which time some 20,000 residents of the city had already died, victims of influenza.[34]

Similarly, the specter of a latter-day global epidemic provoked little more than a shrug from most Europeans and North Americans. A Gallup poll in September 1976 revealed that while 93 percent of the adult population of the United States knew what Swine Flu was and were aware that an apparent strain of the disease had stricken Fort Dix, fewer than 53 percent said they were willing to be vaccinated. The abstract possibility of a million American flu deaths seems to have caused no collective or individual panic in the United States, except, perhaps, in some corners of government.

In contrast, public reaction to the twenty-nine deaths in Philadelphia was extraordinary. Heymann, completely uninitiated in matters of public relations and the media, found himself working in a massive fishbowl, every move made by CDC team members under the constant, often hostile, scrutiny of the nation's citizenry. The perlustration was compounded by widespread fear of contagion in Philadelphia.

Phrases like "explosive outbreak," "mysterious and terrifying disease," "Legionnaire killer," and "killer pneumonia" filled press accounts, as well as the on-camera statements of Philadelphians and politicians.[35] Because American Legion members were stricken, and the nation had been polarized during its war with Vietnam, some members of the highly conservative organization insisted the deaths and ailments were tied to an unidentified left-wing sabotage. From the perspective of the left, events in Philadelphia fit neatly with the then vogue view that an unregulated chemical industry was raining toxic compounds upon the American people.

The CDC investigation, led by veteran disease detective Dr. David Fraser, searched on all fronts, abandoning few theories, no matter how tenuous. All the disease survivors and their available relatives, and over 4,400 Legionnaires and families, were questioned repeatedly, cadavers were autopsied at the microscopic level, and hospital staff that had treated the ailing hotel guests were grilled for hours on end. Most of the victims were Legionnaires, or spouses who had accompanied their veteran husbands to the convention's cocktail parties and banquets. Some hotel staff got sick, but none of their families appeared to be affected. The only clue that tied all the cases together was the victims' presence in the five Philadelphia hotels that hosted the Legionnaires' convention.

By September, the focus shifted completely to the hotels in which the Legionnaires and their spouses had stayed during the convention. Heymann and half the CDC team were housed in the Bellevue-Stratford Hotel. As the investigation dragged on, they were just about the only patrons of Philadelphia's revered landmark hotel. Less than a year later, unable to counter the torrent of negative publicity, the Bellevue-Stratford's management would be forced to close the seventy-two-year-old hotel.

Like the lab boys, the CDC field team was stumped. Members collected air, water, soil, dirt, and materials samples from every room that had been occupied in Philadelphia hotels during July. They questioned hotel staff and reviewed pneumonia records from all local hospitals. None of the collected samples seemed to contain a questionable microbe or toxic chemical.

Grabbing at straws, Fraser sent Heymann out to track down a magician. Just prior to the Legionnaires' gathering, hundreds of magicians had held their annual convention in the Bellevue-Stratford. Heymann's task was to discern whether the conjurers had used any unusual devices or chemicals in the creation of their illusions. The sorcerers had not used anything, however, that hadn't been part of the magicians' standard bag of tricks for centuries.

Back in Atlanta, Sencer was taking phone calls twenty-four hours a day from anyone who had a helpful theory. Most came from "well-intentioned nuts," as Sencer kindly called them, but some represented plausible explanations that the CDC director passed on to his crew in Philadelphia. Though it meant weeks of interrupted sleep, Sencer felt it necessary to make himself available at all hours to public suggestions in order to quell panic and head off accusations of cover-up or stonewalling at the CDC.

But accusations came anyway. From Congress.

On October 27 the congressional Subcommittee for Consumer Protection, part of the House Interstate and Foreign Commerce Committee, released a report that not only condemned the CDC's efforts in Philadelphia but accused the agency of sabotaging its own inquiry.[36] Chaired by Congressman John M. Murphy, a Democrat from Staten Island, New York, the committee lashed out with unusual venom. The CDC spent too much time trying to prove Swine Flu had struck Philadelphia, the report charged, and in the course of its overly virus-focused investigation, lost tissue and urine samples that might have demonstrated that a toxin or chemical caused the ailments.

"It appears to be the consensus of opinion that the failure to save, take, and keep free from contamination the tissue of the victims of the epidemic is clearly the reason that ultimate resolution of the cause of the Legionnaires' disease may never be found," the report accused. At the top of the list of potential chemical killers, the report concluded, was nickel carbonyl, an odorless, unstable compound that can produce some of the same symptoms seen among the Legionnaires.

Dr. William F. Sunderman, Jr., of the University of Connecticut School of Medicine, studied some Legionnaires' lung tissue samples in mid-September, concluding that unusually high levels of nickel were present. But by the time the congressional report was released, an embarrassed Sunderman was at pains to explain that he had never proven that the chemical caused deaths among the Legionnaires and he had reason to believe some of the nickel measured in the tissues actually came off surgical instruments used in the original autopsies.[37]

Congressman Murphy's attack on the CDC was not waylaid by the demise of the nickel carbonyl hypothesis, however, and by December he had escalated his assault, charging it was outrageous that in a country "supposedly with the most advanced technology in the world, we find ourselves in a position of not knowing what happened in Philadelphia."[38]

For CDC director Sencer, Congressman Murphy became a haunting nemesis, whose name and actions provoked unusual outbursts of anger. The Staten Island representative was, in Sencer's mind, living proof that politicians shouldn't be allowed to meddle in ongoing scientific investigations. Over a decade later, Sencer would speak with a happy gleam in his eyes and a lightened tone of voice about Congressman John M. Murphy's 1981 indictment and later conviction for conspiracy and accepting bribes in the so-called Abscam scandal.

"What goes around comes around," Sencer would say.

III

As the CDC toiled over its very full load of enterprises in the fall of 1976, the public became increasingly skeptical of the nation's premier public health agency. And in a year that was dedicated to the celebration of Americanism, the public came to question the "can do" outlook that had dominated the culture ever since World War II.

Here it was, 1976, and the leaders of the scientific community appeared to many members of the public and the political arena to be in absolute disarray. Record amounts of taxpayers' money, including a then breathtaking sum of $135 million for the Swine Flu campaign, were being expended for biomedical investigation and public health, yet the nation felt besieged by confusing medical threats. A dizzying array of common chemicals seemed to be implicated in cancer, everything Americans loved to eat seemed to play a role in heart disease, their long romance with cigarettes appeared destined to send millions of people to early graves, the Pentagon's Vietnam War chemical weapons may have also harmed American troops —even their offspring.

And now the scientists couldn't shoot straight. They were pointing their guns, it seemed, incorrectly at Swine Flu, and couldn't figure out what had killed the Legionnaires.

For many members of Congress and the press, the operative terms now used to describe the government's Swine Flu effort were "fiasco," "debacle," "farce," "travesty," and "waste of taxpayers' dollars."

At the CDC this sudden public revulsion was hard to stomach. Not accustomed to such controversy, or to juggling so many epidemic crises simultaneously, the agency was overwhelmed. Its four-man public relations effort became entirely defensive, reacting day by day to a torrent of accusations raining down from hundreds of sources. And every one of the roughly three hundred scientists at the agency, including junior EIS officers like Heymann, was up to his or her neck in work.

By early October, Karl Johnson and Joel Breman had been dispatched to Zaire to quell Ebola. Heymann was pulled back to Atlanta to coordinate emergency rescue operations and backup for the multinational team in Yambuku. The Legionnaires' Disease problem dragged on, mysteriously defying McDade and Shepard's resolution. And the Swine Flu vaccination campaign, which officially commenced on October 1, continued to draw fire.

On October 11 the night edition of the *Pittsburgh Post-Gazette* broke the story that two elderly people died shortly after getting their Swine Flu shots from the Allegheny County Health Department. Within a couple of hours, newspapers all over America were hurriedly remaking their final morning editions, having grabbed the report off the United Press International wire service.[39]

Within forty-eight hours the story had escalated radically, though the actual number of dead never exceeded three, all over the age of seventy. Reporters led the investigation. The CDC was strangely silent for several days, giving the public the impression that journalists were the only professionals doing any detective work on the cases. Or that the CDC was covering up something insidious. The allegations poured forth: all the dead had received vaccine manufactured by Parke-Davis, which had two months earlier admitted making an entire batch of vaccine against the wrong flu antigen; the vaccine was made by "recombining genetic elements," a point of great confusion at the time; deaths seemed also to be occurring in other cities all over America.

"The Scene at a PA Death Clinic" was a *New York Post* headline. United Press International ran a daily tally of additional suspected vaccine-caused deaths.

Science enthusiast Walter Cronkite, whose anchorman status with the American public was so prominent that insiders at the CBS-TV network referred to him as "VOG" or "Voice of God," brought President Ford before the cameras on October 13, giving the political leader a massive evening platform from which to argue that 215 million Americans should ignore the scare stories and get their shots—just as he and First Lady Betty Ford would do the following day during a photo opportunity.

But two days later, the national immunization tally had reached only 40 million. While that was the best two-week immunization turnout in U.S. history, it could hardly be considered cause for joy among those who still believed a repeat of the 1918–19 pandemic was likely. After all, by October 16, 1918, influenza had already killed millions. The gathering of international influenza experts in Switzerland just ten months earlier had concluded that a minimum of 85 percent of high-risk populations would have to be vaccinated to ensure society's protection against an analogous epidemic.

Though the agency fought hard to put a positive spin on the campaign's progress, many CDC insiders quietly despaired and military officials openly declared that the less than satisfactory turnout was evidence that the American people would fail to respond in an appropriate and timely fashion should an enemy hurl biological weapons at the country.

The CDC tried to counter the impact of the Pittsburgh deaths with an analysis of the statistical relationship between public campaigns of any kind and cardiac deaths among senior citizens. Their conclusion: the three deaths were statistical anomalies, not events caused by the vaccine.

"Persons vaccinated in the [Pittsburgh] clinic died at a rate of 5/100,000/ day in contrast to the expected rate of 17/100,000/day for persons 65 years and older in Pennsylvania," CDC scientists argued.[40] The FDA tested the Parke-Davis vaccine immediately following the Pittsburgh deaths and proclaimed it free of contamination.

By mid-December the number of deaths and illnesses allegedly linked to Swine Flu vaccines would reach 283, more than half limited to headaches or mild fevers.[41] Two months later, the agency would report that the fall of 1976 marked an unusually disease-free time, with the numbers of pneumonia and influenza-associated deaths at their lowest points since 1972. This terrific record would not be ascribable to the vaccine campaign; rather, it would be the result of a virtual absence of influenza virus in North America.[42]

So it was that long-haired hippies and close-cropped businessmen found common ground on something: the Swine Flu shot was not to be trusted.

A year later, President Jimmy Carter's tactless brother Billy would best capture the mood by averring that if he "had to get stoned to death," he'd rather do it with booze than a Swine Flu shot.[43]

On November 2, Gerald Ford lost his bid for the presidency, and Georgia governor Jimmy Carter, a liberal Democrat, was elected. The already demoralized federal public health establishment now had a lame-duck champion for its efforts. Fewer than 5 million Americans would voluntarily get vaccinated after November 1; virtually none after word got out of Guillain-Barré syndrome.

IV

The first case of Guillain-Barré syndrome appeared in Minnesota, during the third week of November. Days after he got his flu shot, a man's arms and legs grew increasingly weak; his reflexes worsened, eventually ceasing altogether; he lost feeling in his hands and feet. For all intents and purposes, he was paralyzed. His physician correctly diagnosed the ailment as Guillain-Barré syndrome and, suspecting an association with the flu vaccine, reported the ailment to officials at the CDC.

First identified in the 1920s by French neurologists Jean Alexander Barré and Georges Guillain, the syndrome was rare, usually reversible, occasionally lethal, and normally occurred in the absence of any other associated illness. No cause or treatment for Guillain-Barré was known, nor could anybody explain why some individuals recovered completely after about a month of paralysis, a few were permanently paralyzed, and still fewer died of respiratory distress when neurological symptoms affected their lungs, hearts, or diaphragms.

The first Guillain-Barré report was quickly followed by others, and the

CDC ordered nationwide active surveillance for syndrome cases in all fifty states.

On December 14 the CDC issued a press release announcing that thirty people had developed the syndrome within a month after their Swine Flu vaccinations; an additional twenty-four Guillain-Barré cases had occurred in people following a lapse of more than thirty days after immunization.

Two days later Sencer called a halt to the Swine Flu campaign, pending further investigation of the Guillain-Barré cases.

On Christmas Eve the CDC revealed that 172 Guillain-Barré cases had turned up in twenty-four states. Ninety-nine cases involved flu vaccines— six of whom had died. The cases spanned all age groups, genders, and races, and no geographic clustering of cases could be seen.[44] Something was going on.

By New Year's Eve the reported number of cases had soared to 526; of these, 257 had received flu shots.

Though CDC officials tried to argue that, like the three Pittsburgh cardiac arrest cases, these Guillain-Barré episodes might represent a normal background rate of the syndrome, the American people—and their politicians —were appalled. Ralph Nader and his consumer action group called for Sencer's immediate resignation. In congressional hearings during December, Senator Edward Kennedy declared the Swine Flu campaign dead.

The CDC continued to downplay the association between the vaccine and the syndrome, though agency insiders had already concluded that the Guillain-Barré rate among those vaccinated against Swine Flu was at least four times that in the unvaccinated population. As further syndrome reports poured in during the early weeks of 1977, some agency representatives suggested that the publicity had created hysteria, prompting cases nationwide of psychologically induced paralysis and limb weakness. But studies in various communities showed no such panic, and found that most cases had been diagnosed by qualified neurologists.[45]

By the time Jimmy Carter had been inaugurated and named Joseph Califano as his nominee for Secretary of Health, Education, and Welfare, the CDC's Guillain-Barré total had topped 1,100, half of whom had received Swine Flu shots, with cases reported in all fifty states. Fifty-eight deaths had occurred. Agency analysis showed a clear clustering of cases around the months of November and December, coming on the heels of the Swine Flu campaign's peak. The average lag time between vaccination and developing the syndrome was six weeks. Over 5 percent of cases were lethal, and nearly a quarter of the Guillain-Barré sufferers had to be placed on respirators.

The researchers concluded that America's normal, inexplicable Guillain-Barré rate was about one case in every million people per year, for an expected 1976 total of some 215 cases. But among Swine Flu vaccine recipients, the attack rate was about ten times greater, at one case in every 100,000 Americans.[46]

Overnight, lawyers filed claims with the U.S. Attorney General's office on behalf of clients alleging they had suffered ailments of various kinds due to the Swine Flu vaccine. So great was the deluge that the White House Office of Management and Budget was forced to approve an emergency allocation on January 28, 1977, of $1.2 million for immediate processing of claims. Justice Department attorney Jeffrey Axelrod and his staff of ten lawyers worked in twenty-four-hour shifts for several months, rushing to meet the letter of Congress's August 1976 liability law to determine which claims should receive immediate settlement, which appeared bogus, and which should go to court.

In the end, 4,181 claims would be filed seeking a total of $3.2 billion. For over sixteen years, the cases would wend their way through the legal system: by 1993, three cases would still be pending. Axelrod's team would end up deciding that bona fide cases of Guillain-Barré syndrome among vaccine recipients represented vaccine-caused ailments and would settle those without court proceedings.

After a decade and a half of legal proceedings, the U.S. government would settle 393 claims for $37,789,000. Another 1,605 cases would end up in the courts, with 53 resulting in judgments against the federal government (for a total of $17 million) and another 56 cases lost in litigation (for another $30,683,000).

By 1993 the U.S. government had paid out nearly $93 million in taxpayer dollars to Swine Flu claimants.[47] Though the final liability to the U.S. government was well below early dire forecasts of "hundreds of millions of dollars," it would prove to be great enough in the eyes of Congress to have a long-lasting impact on global immunization programs. A gun-shy Congress, afraid for years to come of approving any federal immunization efforts for fear of repeating the Swine Flu fiasco, would prove only part of the vaccination problem.

Swine Flu also left its mark in the judiciary system, setting precedents for government culpability in cases of large-scale public health efforts. In years to come, lawyers would file suits on behalf of clients claiming damages from all sorts of vaccines. Even the sacred cow of public health—the polio vaccine—would come under fire. In the 1970s and 1980s, individuals would sue the U.S. government claiming they got polio from the 1962 Sabin oral vaccine. In 1993, the federal district court for Maryland would rule that individuals had a legal right to sue the government even though the 1962 polio campaign was originally considered above the law because it was deemed an extraordinary humanitarian effort.

From the perspective of the microbes, no immunized *individuals* mattered: the issue for microbial survival was the overall rate of immunity in an entire population of potential human targets. If a critical percentage of the human population was rendered immune by virtue of vaccination, the microbes, unable to thrive and reproduce, would either retreat into an animal reservoir or vanish.

Nobody knew in 1962 in the case of polio, in 1976 in the case of influenza, or in 1993 for the majority of the world's diseases precisely what percentage of a human population had to be vaccinated in order to defeat the microbes; the assumption was, and would continue to be, that greater than 80 percent of a population had to be immune to stop most communicable microbes.

Public health advocates worldwide welcomed the dawn of mass immunization in the 1950s and 1960s, feeling that any possible risk to a few individuals far outweighed the extraordinary benefit to society as a whole that would follow defeat of such terrible microbial scourges as polio, pandemic influenza, diphtheria, whooping cough, and typhus. And for two decades in the United States—the center of more than 80 percent of the world's vaccine production—insurance carriers, politicians, drug companies, and the judicial system adhered to the basic principle that the rights of an immunized society superseded those of small numbers of individuals. Indeed, the courts in some states ruled that officials had the right to overrule the objections of family members in some cases, forcing vaccination of elderly residents of nursing homes and children entering the public school system.

Swine Flu threw ice water on the previously warm relationship between public health and individual rights. It set a precedent that would haunt *all* vaccine efforts inside the United States for decades, and shatter the confidence of vaccine manufacturers (and their insurers), not only vis-à-vis their U.S. markets but globally. Many in coming years would abandon vaccine production entirely; by 1993, only four U.S. pharmaceutical companies—Connaught Laboratories, Inc., Lederle-Praxis Biologicals, Merck & Co., Inc., and Wyeth-Ayerst—would remain committed to production of vaccines.[48]

In early 1993, as a result of the courts' decisions, the U.S. Department of Justice would settle five 1960s polio vaccine cases, each individual receiving "seven-figure sums," according to Axelrod. Precise figures were sealed by the courts.

V

By the second week of January, David Sencer could clearly see the writing on the wall: somebody was going to have to take the blame for the entire failed Swine Flu effort, and he was the most likely fall guy. Nobody in Washington political circles cared much about the disease victories that had occurred on his watch: smallpox, Ebola, dramatic decreases in all U.S. childhood diseases.

"Somebody's head is going to be on a stake," Sencer told his staff. As the Guillain-Barré toll mounted and members of Congress called for their sacrificial lamb, Sencer tried hard to keep a business-as-usual profile in Atlanta.

On Friday afternoon, January 14, an exhausted Sencer sat in his office sorting through a stack of messages from the likes of his nemesis, Congressman John Murphy. He was in no hurry to return the calls. Three of his key men, Walter Dowdle, Joe McDade, and Shep Shepard poked their heads in his office, asking if the director could spare a moment.

"What's up?" Sencer asked as the men gathered around his desk.

Dowdle smiled and said, "Shep and Joe have isolated an organism that causes Legionnaires'."

"What!" shouted Sencer as he leapt to his feet, searching the faces of Shep and Joe for corroboration.

The pair nodded, and Shep told the director, "It's a bacterium."

Sencer grabbed his phone and called in members of the CDC's top brass, including public relations director Don Berreth.

"Now we're all here. Let's go over this very carefully," Sencer said, looking around at the cluster of his most trusted scientists.

In painstaking detail, Shepard and McDade explained what caused the Philadelphia deaths, and how the mysterious bug had eluded laboratory discovery for six months. The bacteria would not grow in the laboratory, for reasons the two scientists had yet to determine, but they had succeeded in obtaining evidence of its existence and pathogenicity through a series of experiments.

First, McDade removed lung samples from one of the deceased Legionnaires, mashed up the cells, and injected samples into chicken eggs. After incubating the eggs for some time, he cracked the shells and extracted the yolk sacs. The sacs were then also mashed up, and extracts were injected into the foot pads of guinea pigs; the animals developed symptoms similar to those seen in the Legionnaires'.

The scientists then took blood samples from thirty-three disease survivors—samples that presumably contained antibodies against the causative agent—and mixed them with the yolk sac isolates. Yes, they reacted, confirming that whatever was in the yolk sacs was the same agent that caused illness in those thirty-three people. Conversely, the yolk sac isolates did not prompt antibody reactions with blood samples from people who hadn't had the disease.

McDade explained that he had been stumped for months by several unusual characteristics of the bacteria. First of all, it wouldn't grow under typical laboratory conditions. They had tried putting samples from the Legionnaires' blood and tissues in petri dishes filled with standard fluid media used to grow hundreds of other bacterial varieties—nothing happened. "At that point, we figured we were dealing with a virus," McDade explained. "So we tried to culture it in medium with antibiotics."

Virologists always filled their culture media with antibiotics in order to eliminate any chance of bacterial contamination. By using virus-appropriate media, McDade said, they had wiped out the very organism for which they were searching. It wasn't until they injected samples into eggs not treated with antibiotics—using the eggs as substitutes for petri dishes and growth media—that they saw clear evidence that living organisms inhabited the bodies of the deceased human beings.

Another stumbling block had been mice. Specially bred varieties of mice were the most commonly used laboratory animals, and throughout the fall of 1976, the CDC scientists had tried in vain to produce Legionnaires' Disease in the rodents. It wasn't until they switched to guinea pigs that their efforts bore fruit, and Shepard and McDade were now certain that mice were immune to the elusive bacteria.

The men told Sencer that they still didn't know why the bugs wouldn't grow in petri dishes. And they hadn't been able to get a look at the bacteria, though they were certain it was present, swimming about on the fluid surfaces of the microscope slides. In the matrix, McDade said, they clearly had enough clues to set the algorithms in motion. It seemed to him that even lacking such matrix points as visualization of the culprit organism, the lab boys had enough dots on the matrix to point their fingers at a bacterial source for the infections.

Sencer wasn't sure what the always precise but cerebral McDade meant. "It's just not showing up," Shepard explained, "but I'm sure it's there."

"Shep, how sure are you?" Sencer asked, leaning toward the laboratory scientist.

"Better than ninety-five percent," Shepard said, "but I'd like to run a few more experiments before we go public on this."

"No way!" Sencer shouted. He reminded the lab men that in the real world, far from their isolated laboratories, there was genuine fear abroad, rage directed at the CDC, congressional probes, hourly media inquiry, and a serious mandate to get reasonably reliable information to the public as quickly as possible. Berreth chimed in, describing the range of angry press queries his office was fielding.

Shepard objected to haste, and argued that their findings should be written up and submitted to a scientific journal for publication. Yes, he said, it was potentially a six- to nine-month process, but necessary to maintain scientific credibility.

"I am not going to have Joe McDade made fun of by his peers," Shepard declared.

Sencer chewed the idea over for a moment, turned to Berreth, and asked, "How fast could you generate a special issue of *Morbidity and Mortality Weekly Report?*"

Berreth assured his boss that such a CDC product could be generated and posted by Tuesday. A press conference, he said, could be held Tuesday afternoon.

"That satisfy you, Joe? If we publish all your data in *MMWR*, can you abide by release next week?"

McDade agreed, and dashed off to, as was his wont, triple-confirm every possible detail before Tuesday. He called his wife and warned her that neither she nor their two children would see much of him for the next four days, as he expected to pull all-nighters right up to Tuesday.

Sensing the CDC staff's acute need for a morale boost, Sencer arranged a most unusual press conference for Tuesday, January 18, 1977. Every CDC staffer was invited to attend, as were the Surgeon General and members of the Washington, D.C., health hierarchy.

Just hours before the expected gathering, Shepard stormed into Sencer's office and breathlessly announced, "The same organism caused the St. Elizabeth's epidemic!"

On July 27, 1965, sixty-two mental patients living in St. Elizabeth's psychiatric hospital in Washington, D.C., had fallen ill with pneumonia. Within a month's time, nineteen more patients fell ill, fourteen had died. Overall, 1.3 percent of the hospital population had been ill, nearly all of whom had lived in the same wing of the facility. At the time, authorities scoured the hospital for clues, and tested hundreds of blood and tissue samples, but no cause was found.[49]

Wisely, somebody put blood and tissue samples from St. Elizabeth's in the CDC deep freeze, and there they had remained for eleven years. Until McDade, anxious to make as strong a case as possible in the Tuesday presentation, recalled the unsolved mystery and injected the old samples into chicken eggs, afterward running antibody tests on the extracts.

"Did you get reactions?" Sencer asked.

"Yes. Definitely," Shepard effused.

"Well, write it up and we'll add it to today's *MMWR*."

At three o'clock that afternoon, most Atlanta employees of the CDC, from janitors to Ph.D.s, were assembled in the agency auditorium, along with a sizable press corps. As Shepard and an exhausted McDade presented their data and the hurriedly produced *MMWR* was distributed,[50] nobody so much as whispered. When Shepard said, "Slide projector off, please," and finished his closing statement, a hushed moment followed.

Then revered virologist Alex Langmuir leapt to his feet and exclaimed, "Shep, that was great!" Applause and a deluge of press questions followed, all observed by a grinning Sencer, satisfied that he had successfully fulfilled his mission—namely, shielding his staff from politicians and the public while they went about the business of doing science. Sencer now felt like gloating.

VI

Sencer's sense of sweet revenge wouldn't last long, however.

Two weeks later David Sencer would have the dubious distinction of being the first federal official fired on national television, when President Carter's new Secretary of Health, Education, and Welfare, Joseph Califano, led him out into an HEW hallway.

"I want to say something nice about you here today," Califano said, asking Sencer to name a few of CDC's triumphs of late that he could mention to soften the blow. But then Califano added, "I'm going in there in a few minutes and announce your resignation."

That evening, a disheartened Sencer turned on a network TV newscast and saw film footage of Califano's whispered conversation in the HEW hall. In the voice-over to the footage, the telejournalist told viewers that they were witnessing Califano's firing of Sencer.

For years, the Swine Flu events of 1976 would be debated, analyzed, and scrutinized for lessons that might guide future public health officials, politicians, and microbe hunters faced with potential pandemics. After more than fifteen years, little controversy would be shed, and consensus on what went awry would elude the American public health community. A year after the events, Dr. Arthur Viseltear wrote:

The short and not so very happy life history of the national swine influenza program has already become a classic health policy case study because the elements of policy and politics are so illustratively intertwined. . . . If one wishes to find heroes and villains in the piece, they will certainly be found; if one wishes to view the Congress as a moribund or trifling institution, there is abundant evidence to do so; if one wishes to interpret the actions of the administration and its scientists as being politically motivated and self-serving, he will find circumstantial evidence to support the theory; and if one wishes to view the President's decision as being based upon some real or imagined bicentennial or electoral bonanza, he will also find evidence to support that thesis. But he will also find, as others have found, men and institutions muddling through, making their decisions hastily and under conditions of chronic obscurity, where chance, accident, confusion, and stupidity play a larger role than certitude or calculation.[51]

On an ominous note, Viseltear concluded: "If the events of 1976 are not to recur, then the Congress and the administration had better ensure that these issues are addressed now."

Echoing that sentiment, Neustadt and Fineberg, while conceding that improper decisions were made in 1976, wrote:

We find no villains in the Federal government's officials and advisors then and think that anyone (ourselves included) might have done as they did—but we hope not twice. . . . These remain our sentiments. The opposite danger, of course, is that the lessons of the crash program are learned too well—too literally—producing stalemate in the face of the next out-of-routine threat from influenza. Someday there will be one.[52]

Former congressional staffer Silverstein insisted there were no human culprits in 1976, no grave errors in political or public health judgment. Rather, he said, the only correct site upon which to fix blame was the

Swine Flu virus itself, "which failed to appear and 'justify' the program of preventive medicine."

Among the most haunting questions reviewed repeatedly by historians and participants in the Swine Flu campaign are: Where did the Fort Dix flu go? Would the vaccine have protected Americans if a pandemic had materialized? What caused Guillain-Barré syndrome, and could it have been prevented? If a major pandemic of any kind were on the horizon, would the American people respond to the public health authorities?

The dominant theory, reached incrementally over subsequent years, explaining the disappearance of the Fort Dix flu, was one of competition. Numerous virologists put forward the view that in any given ecological setting, two very closely related viruses would be forced to compete for hosts, and the virus with the greater transmission capabilities would be victorious. As a rule, viruses could carry only a finite amount of genetic baggage, and many species of the tiny microbes sacrificed one set of genetic capabilities for another. Thus, a highly transmissible virus might carry loads of genes that conferred the ability to remain alive while suspended in the air or resting atop steel tables, but sacrificed genes that conferred the ability to outwit certain elements of the human immune system or reproduce rapidly inside human cells.

The A/Victoria strain appears to have had the advantage on the transmission side, as it rapidly spread around the planet in several cycles, each lasting roughly one year. The A/New Jersey virus may have been virulent, if it was the cause of Private Lewis's death, but it clearly was not particularly transmissible. So, the argument goes, the two viruses were both on base at Fort Dix during January 1976, seeking human hosts to infect. In such a setting, competition would favor the more transmissible A/Victoria virus.

"The failure to detect spread of influenza A/New Jersey virus to civilian populations, however, also suggested the possibility of deficient transmissibility, as did its failure to thrive in a military population," wrote Martin Goldfield, the New Jersey State Health Department scientist who first isolated the Fort Dix virus.[53]

Nobody would ever figure out which soldier or recruit was first exposed to the pig virus, or why Private Lewis, in particular, succumbed. Not knowing how and why the virus spread inside the military setting also heightened difficulties in ascertaining why, conversely, it did not spread to the civilian population. Even the relative virulence of the A/New Jersey strain continued to be a matter of debate years later, not only because the British study in the spring of 1976 failed to find the virus particularly dangerous to human volunteers but also because the U.S. Army discovered that several of the most acutely ill Swine Flu patients at Fort Dix were simultaneously infected with *Haemophilus influenzae*, a bacterial disease that can produce pneumonia.[54] CDC studies in 1977 would show that the A/New Jersey virus replicated fairly slowly under laboratory conditions.

Throughout 1976 the National Institute of Allergy and Infectious Diseases and FDA tried in vain to come up with a vaccine combination or dose that would raise protective anti-Swine Flu responses in youthful adults—those around the same age as Private Lewis. Though it was never the subject of widespread public attention at the time, the government never succeeded in developing a vaccine for young adults that raised much confidence in FDA or CDC circles.

Eventually, the CDC figured out that the Swine Flu vaccine worked best for adults born before 1957, a year marked by a massive global influenza A epidemic. Those who survived that epidemic seemed to respond to the Swine Flu vaccine nineteen years later as if it were a booster shot. For those born after 1957, however, the vaccine was never particularly effective, and some scientists—notably Dr. Anthony Morris of the FDA's Bureau of Biologics—openly speculated that those people would have been vulnerable if the A/New Jersey strain had spread to the civilian population.

"If we go back through time to January 1976 and have a second chance at decision making, given the same information, what would we do?" Walter Dowdle asked.[55] "We know that antigenic shifts do not necessarily lead to pandemics and that the vaccine has some risks.

"Knowing this, would the vaccine have been made at all? If so, would it have been stockpiled or given to the entire population, all those who wanted it, or only to certain target groups in the population? With the benefit of hindsight, we can now make the right decision for 1976. But what about our decision next time? It is highly unlikely that the circumstances surrounding the next potential pandemic will be precisely like those surrounding any other."

Though errors were clearly made in the handling of a possible emergence of Swine Flu in 1976, no such flaws could be seen in the Legionnaires' Disease effort. After their dramatic January 18, 1977, announcement, Shepard, McDade, and the rest of the CDC team swiftly determined why the elusive bug was so hard to isolate and grow in the laboratory.

The *Legionella* bacterium, as it was dubbed, had peculiar dietary needs. Standard laboratory culture media wouldn't support growth of the persnickety microbe: it needed supplements of the amino acid cysteine, vitamins, and minerals, particularly iron. Accustomed to living in what is politely referred to as pond scum, *Legionella* preferred dark, nutrient-rich, almost anoxic environments. It also enjoyed living inside the cytoplasms of larger one-celled organisms.[56]

These conditions rendered the organism impossible to see through microscopes with standard techniques. When treated with silver, however, the organism clearly revealed itself, and Shepard and McDade saw the long, round rods of *Legionella* squirming on their slides.

The epidemiology team in Philadelphia meanwhile noticed that most of the Legionnaires' Disease sufferers had spent time schmoozing in the five cocktail suites run by the candidates for leadership of the veterans' group. Further analysis revealed that the bacteria thrived in the Bellevue-Stratford

Hotel's cooling tower. From that water supply, the hotel derived its air conditioning. The *Legionella* organisms were hidden in biofilm "scums" along the edges of the cooling tower, and were actively pumped into the hotel's hospitality suites during the hot month of July.

It wasn't long before similar cases of Legionnaires' Disease surfaced all over the world. First, the CDC spotted isolated cases in eleven different states.[57] By September 1977, the federal agency was busily tracking three hospital outbreaks in Ohio,[58] one in Vermont,[59] and one in Tennessee.[60] The combined fatalities in the Ohio, Vermont, and Tennessee outbreaks and sporadic isolated cases in 1977 reached thirty-two by December, about 25 percent of all reported Legionnaires' cases.[61]

In the fall of 1977 a small epidemic of Legionnaires' broke out in a hospital in Nottingham, England, leaving three patients dead.[62]

In the summer of 1977 Legionnaires' struck a brand-new hospital located in one of the wealthiest parts of Los Angeles. The Wadsworth Medical Center, a veterans hospital, situated between the posh communities of Bel Air and Brentwood, was the site of a yearlong outbreak of the disease that infected about 3 percent of all patients who passed through the facility, caused disease in both staff and patients, and claimed sixteen lives.[63]

By late 1978 scientists had discovered the *Legionella* bacterium in soil samples, ponds, cooling towers, water-driven condensers, slow-flowing creeks, mud, polluted and silty water, at construction sites, and in steam turbines. In coming years, they would find the dangerous organism in shower heads, grocery store vegetable counter misters, hot tubs, fountains, and a wide variety of humidifiers and other devices that aerosolized water.

Clinically, it was soon apparent that the organism was most dangerous to cigarette smokers, people recovering from surgery, and individuals who were suffering some type of immunosuppression. It seemed the bacteria were inhaled from the environment; never were they transmitted from person to person. Once somebody was infected, the *Legionella* were tough to defeat because they were resistant to a wide spectrum of antibiotics.

Air-conditioning standards changed after 1976, with federal agencies all over the world requiring far more stringent cleaning and hygiene provisions for cooling towers and large-scale air-conditioning systems.

In the case of *Legionella*, a new human disease had emerged in 1976, brought from ancient obscurity by the modern invention of air conditioning.

At the CDC's International Legionnaires' Disease meeting in 1978, several particularly ominous facets of the bug were scrutinized. CDC scientists revealed that the organism could be found in tap water, shower nozzles, and other allegedly clean water sources. One tap water study showed *Legionella* could survive over a year inside pipe biofilms, emerging in wholly infectious form once the faucet was turned on full force. It thrived in temperatures from ice cold to steamy hot. Even distilled water samples occasionally contained small numbers of *Legionella* organisms.

A team of scientists from the Denver Veterans Administration Medical

Center was particularly prescient, predicting the bacteria might survive chlorine purification efforts. "The residual amount of chlorine recommended (0.2 ppm) for standard water purification may not be sufficient for killing the LD bacterium when it is present in high concentrations," the group wrote.[64]

Bacteriologist Mortimer Stall, of the University of California, Davis, warned that the soils and waters of the earth were replete with organisms not yet identified, many of which might, like *Legionella*, one day be provided the proper circumstances for their emergence as human pathogens. Plant bacteria such as *Serratia* and *Pseudomonas* were known to cause human disease, he noted, and it would be arrogant for humanity to assume it had identified all of its flora, marine, and soil microbial enemies.

"The existence of plant-animal ambilateral harmfulness is generally unrecognized, even though I have assembled more than 200 bacteriological and mycological examples, mainly in the 'questionable' category," Stall told the international gathering. "It seems, then, that ambilateral harmfulness may have a significant bearing on the 'emergence' of 'new' infectious diseases . . . because the ability of a plant microbe to harm an animal (or vice versa) in any manner whatsoever would seem to indicate that the 'emergence' of a 'new' pathogen is not far off!"

The CDC estimated that somewhere between 2,000 and 6,000 people had been dying every year of Legionnaires' Disease, probably for decades, certainly since the advent of air-conditioning technology and, long before that, indoor plumbing. Prior to the dramatic Philadelphia outbreak, these cases had simply been dumped into the category of "pneumonia of unknown etiology."

Armed with such observations, medical historian Robert Hudson of the University of Kansas closed the international gathering on a particularly frightening note. After describing the Black Death plague of medieval Europe, Hudson warned that "when we grant that our knowledge of existing microscopic pathogens is deficient, we necessarily grant the possibility at least of a return of the great epidemics of the past. . . . the possibility exists that a deadly and common organism could emerge that is easily spread from person to person and that might be aloof to all available therapeutic and preventive methods."

"The Philadelphia event remains unsettling because it shows the very real limitations of our tools for investigating an apparently new microbial disease," Hudson concluded. "If we are to retain public confidence in the face of some future serious epidemic, it is important that our limitations be widely understood. As a medical community, there is no cause to feel humiliated by the Legionnaires' affair, but it is altogether proper that we be humbled."[65]

Chagrined by events of 1976, the U.S. public health community looked to the future, for the first time in the late twentieth century, with a vague sense of unease.

7

N'zara

LASSA, EBOLA, AND THE DEVELOPING WORLD'S ECONOMIC AND SOCIAL POLICIES

Improvement in health is likely to come, in the future as in the past, from modification of the conditions which lead to disease, rather than from intervention into the mechanisms of disease after it has occurred.
—Thomas McKeown, 1976

The microbe is nothing; the terrain everything.
—Louis Pasteur

■ While his colleagues in Atlanta anguished over Swine Flu damage control, Joe McCormick was content to finally have a chance to uncrate several thousand pounds of laboratory equipment and build his remote Lassa Fever Research Unit in Sierra Leone. It hadn't been easy getting all the gear by ship from Atlanta to Freetown and by assorted trucks along the sporadically paved roads to Segbwema.

But here he was, at long last, on his own, in charge, doing what he most loved: science. Months earlier, long before he was detoured off to the Sudan to chase down the Ebola virus, McCormick sat down with Karl Johnson and mapped out a strategy for his Lassa research.

There was so much to do.

He wanted to find out just how widespread Lassa virus infection was in the West African *Mastomys* rat population. McCormick planned to do antibody tests on thousands of Sierra Leonians to see how many had ever been infected with the virus.

"While I'm at it, might as well check for antibodies to Marburg and Ebola. Wouldn't it be a kick if it turned out those things were up there," McCormick told Johnson just before he left Zaire.

Pat Webb would later join Joe, and that pleased him. She was one of the best field lab workers he'd ever seen, and McCormick knew he could count on Webb's data: it would always be reliable. He appreciated her caustic, opinionated way of looking at the world, always coupled with a great sense of humanity. She could be counted on for ample hilarity.

Not long after getting settled in and starting up laboratory operations, McCormick received a cable from the U.S. Embassy in Freetown summoning him for a top secret meeting. After traversing the backbreaking roads from Segbwema to Freetown, McCormick was informed that his presence was required immediately at the U.S. Embassy in nearby Monrovia, Liberia. It was an official diplomatic summons, passed from the Liberian government of President William R. Tolbert, Jr., to the Carter administration in Washington. The U.S. State Department cabled word to its embassy in Freetown that Liberia had requested McCormick, by name. And he had better go.

When McCormick finally reached the U.S. Embassy in Monrovia, he was informed that the Tolbert government, which was aligned with that of the United States in Africa's crazy-quilt mixture of Cold War allegiances, was concerned about the recent arrival in the country of four Soviet scientists. The Soviets, who were interested in Lassa research, had arrived in Monrovia a few days earlier, unannounced, requesting unusual and basically unobtainable supplies, such as tanks of liquid nitrogen and various compressed gases.

The U.S. Embassy set up a formal meeting, attended by McCormick, the four Soviets, and representatives of the U.S. and Liberian governments. One of the Soviets was a bona fide scientist: Sasha Kachenko had collaborated years earlier with Karl Johnson on studies of hemorrhagic fevers in Russia and the Ukraine. The true identity of the other three was less clear, though Joe was certain at least one was a KGB agent. The fellow certainly didn't know anything about basic biology. It was obvious after twenty minutes of vague conversation that the Soviets had no lucid plan for studying the Lassa virus.

"Hell," McCormick told embassy officials, "it sounds like they're just going to wander around collecting rodents."

Once outside the embassy, far from electronic surveillance, the Soviet team pressed McCormick for assistance. They wanted Lassa antibodies, reagents, and, most important, samples of the virus. They also wanted to know how to do Lassa research.

Convinced it would be dangerous to turn over samples of such a lethal virus to men he was certain worked directly or circuitously for Soviet intelligence, McCormick simply smiled and told his luncheon companions that all such requests would have to be submitted in writing to the director of the CDC in Atlanta.

Throughout 1977 and 1978 the Soviet researchers continued importuning McCormick and the CDC for Lassa virus samples. And the CIA grilled McCormick after every contact with the Russians. McCormick and then former CDC director David Sencer were convinced that both sides in the

Cold War feared the other was developing Lassa as an instrument of biological warfare. As a weapon, Lassa certainly had many desirable characteristics: better than 90 percent lethality in nonimmune populations, extraordinary virulence that needed only minute doses for infection, apparent viral tolerance of a variety of hostile environments, and, most important, no clear treatment or antidote. Furthermore, because of the circumstances surrounding Mandrella's infection, military researchers on both sides of the Iron Curtain were convinced that lethal infection could result from inhalation of the virus.[1]

The Soviets had first approached the U.S. government five years earlier, requesting a sample of the Lassa virus during a Moscow visit by U.S. Secretary of State Henry Kissinger. Because he had publicly vowed complete openness in United States/U.S.S.R. information exchange as part of ongoing nuclear and biological arms negotiations between the Nixon and Brezhnev administrations, Kissinger ordered the CDC in 1972 to accede to the Soviet request.[2] Sencer later did so by personally hand-carrying a vial of the virus to Moscow.

It was, therefore, unclear to the CDC in 1977–78 why the Soviet researchers in Liberia needed further samples of the virus, and though the superpowers had signed a treaty forbidding biological weapons use and development, the U.S. agency remained suspicious of Soviet intentions.

Though the Soviet effort in Africa over the coming years was far larger than anything McCormick and Webb could muster in Sierra Leone on their meager CDC funds, it accomplished nothing. Four separate teams of Soviets "were wandering around in a thick fog, with no sense of direction," McCormick told Webb. Both the CDC scientists felt that a golden opportunity to do stellar collaborative work was lost. The Americans couldn't conduct research in Guinea because the United States did not have diplomatic relations with President Sékou Touré's left-wing government. Similarly, the Soviets couldn't freely roam about the countryside in search of Lassa in Senegal, Liberia, Sierra Leone, or Nigeria.

For years separate and often isolated research was conducted, and both superpowers would eventually shut down their West African Lassa laboratories, leaving the Africans the ultimate losers. At McCormick's insistence the CDC would maintain a collaborative Lassa research effort with Sierra Leone and occasionally Nigeria well into the 1990s. But the Soviet operations would be abandoned in 1984 following a Guinean coup d'état, leaving behind no obvious scientific legacy.[3]

As time went on in the late 1970s, McCormick wasn't even sure his efforts in Sierra Leone constituted a lasting legacy for the people of that nation. Though he, Webb, and other CDC colleagues amassed an impressive store of information, they had tremendous difficulty in translating the newfound knowledge into meaningful action. The virus was carried by the *Mastomys* rats, which were ubiquitous throughout the villages, swamps, and forests of West Africa. The rats lived inside huts and homes, people tolerated—even ate—the rodents, and the rats urinated on stored grains and the dusty dirt floors. Once people became ill, they went to hospitals that failed to follow hygienic practices aimed at preventing spread of the virus from patient to patient.

McCormick drew blood samples from hospitalized patients in Sierra Leone's eastern province, discovering that on any given day some 5 to 15 percent of the adult patients were infected with the deadly virus. McCormick and his CDC colleagues surveyed the remote villages of the country's northern savanna area, finding that up to 40 percent of the adult residents of some locales had antibodies against Lassa, which proved that they had been previously exposed to the virus. Nationwide, nearly 9 percent of Sierra Leone's citizens tested positive for antibodies against the virus.[4]

Together with Guido van der Gröen, then with the Institute of Tropical Medicine in Antwerp, McCormick collected all available blood studies and data on Lassa from all over Africa for the World Health Organization. In 1977 a second type of Lassa virus was discovered in *Mastomys* rats in Mozambique, and eventually one or the other Lassa strain was found in the blood of those rodent species in every country in which they were able to look, from Mozambique and Zimbabwe in the southeastern corner of Africa all the way up to Senegal and Mali in the northwest.

McCormick and van der Gröen told WHO that the scope of rodent and human infection in Africa clearly showed that what had seemed an extraordinarily rare and mysterious disease in 1969 when Penny Pinneo took ill was actually a highly endemic problem in scattered villages across the continent.[5]

When an infectious vector-borne disease was that thoroughly entrenched, and its carrier—the rat—that perfectly adapted to cohabitation with human beings, classic public health training dictated just three options for limiting the further spread of the disease: a vaccine, a cure, or elimination of the vector.

But there was no Lassa vaccine, nor did one appear likely in view of the lack of interest in developing one at the laboratories and pharmaceutical companies of the wealthy nations. McCormick's initial hopes that stockpiling plasma from people who had recovered from Lassa would result in a storehouse of curative antiserum were quickly dashed. He soon discovered that few people made sufficient antibodies to be useful in preventing full-blown disease in other infected individuals. Indeed, he found most people didn't even muster a strong enough immune response to prevent their own reinfection, and repeated episodes of Lassa fever were fairly common.

In the villages of Sierra Leone, fevers were a constant presence and had been for millennia. The people assumed most fevers were caused by mosquito-borne diseases—malaria and yellow fever—or by sorcery and evil spirits. McCormick discovered that about one out of every ten cases of high fever in the villages was caused by the Lassa virus.

Most of the time—perhaps over 98 percent of the time—people recovered from Lassa fever, but their illnesses lasted many days, even weeks. During that time they were unable to work, and McCormick could see the enormous economic toll Lassa took on the villages. In contrast, if people got infected in a hospital, through blood-to-blood contamination of medical equipment, their chances of getting sick and dying were far greater: 16 percent of such infections were fatal.

McCormick and Webb experimentally treated Lassa with a new injectable drug called ribavirin. The antiviral drug had already proven effective in treating some other viral infections by blocking the ability of the viruses to reproduce. It worked on Lassa, McCormick and Webb found, provided it was administered before full-blown symptoms developed.

From the beginning, however, McCormick knew that impoverished societies like Sierra Leone could never afford to buy enough ribavirin, build enough hospitals, and train adequate personnel to curtail Lassa fever deaths. As he struggled to find a solution, including searching for ways to eliminate *Mastomys* rats, McCormick began to appreciate the scale of the problem. As had many Europeans and North Americans before him, and as would others afterward, McCormick was acquiring a deep sense of the "infrastructure" problem. Every day his midwestern can-do mettle was put to the test as his time was wasted repairing an electrical generator, rebuilding a washed-out bridge, sewing up holes in mosquito nets, negotiating receipt of illegible carbon copies of documents from self-important bureaucrats, training bright unskilled people in basic hospital practices, and traveling from one incredibly remote location to another.

"It does sometimes matter whether you know how many logs it takes to float a Land-Rover," McCormick would tell his colleagues back in Atlanta.

In the late 1970s Sierra Leone had a population of 4 million people, representing a polyglot mixture of more than ten tribes, at least five distinct language groups, and three mutually hostile religions. Most Sierra Leonians survived on marginal or subsistence agriculture. What wealth existed in the country was concentrated in the very few hands that had a role in the management of the nation's diamond or bauxite mining and exportation industries.

The average baby born in Sierra Leone in 1977 had about a one-in-ten chance of surviving a host of infectious diseases and chronic malnutrition and reaching adulthood; having approached that milestone, men could expect to live to the ripe old age of forty-one, women six years longer. Infant mortality was high: 157 of every 1,000 babies died before their first birthday. And for those older children and adults who fell ill, scant curative facilities were available. Fewer than 150 doctors, many of them foreigners, treated the 4 million citizens of Sierra Leone in a patchwork of hospitals and clinics nationwide that could provide only about 4,000 hospital beds. Not surprisingly, most of the population sought medical help from traditional herbalists and sorcerers, rather than what was offered in these meager Western-style facilities.

Though the British prided themselves on leaving the stamp of English civilization upon all their colonies, less than 10 percent of the Sierra Leone population was literate when the country gained its independence in 1961. In 1787 Britain had founded the nation of Sierra Leone where no such country had previously existed, carving out boundaries for a slave-free state. Though the British continued to play an active role in the slave trade well into the nineteenth century, the government was compelled by domestic English dissent in the late eighteenth century to provide a safe haven for escaped slaves and the descendants of interracial couples. Thus, Freetown was created.

Nearly two hundred years later, the creole descendants of those freed slaves were a distinct but tiny minority population of some 60,000 people, representing the bulk of the well-educated elite of the country. In the first decade of self-rule, Sierra Leone's creole-dominated government spun wildly out of control, with corruption, graft, and mismanagement rife throughout state-operated sectors of the society. Roads, schools, and hospitals deteriorated, new construction was concentrated in the country's three main cities—Freetown, Bo, and Kenema—and subsistence existence in the villages became even more difficult.

By the time McCormick and Webb set up their remote Lassa laboratory, Sierra Leone was coming out of a ten-year period of political instability and violence, had established a one-party republic, and was so far in debt to the International Monetary Fund, the World Bank, and other, largely British, creditors that annual national revenues were diverted from the people and projects in desperate need nationwide to pay interest to lenders in London, Geneva, New York, and Paris.

Unfortunately, there was nothing unique about Sierra Leone. The lack of basic infrastructures, such as roads, schools, hospitals, shipping and supply routes, electricity, and telephone systems, was hobbling African development. Political instability and the corruption that seemed to go hand in hand with militarism and elitist oligarchic government were draining the lifeblood of once proud agrarian societies from Casablanca to Cape Town.

While the wealthy nations made large commitments to infrastructural development in Latin America and parts of Asia, no such obligation seemed to be felt toward Africa. The continent most ravaged by colonialism, resource exploitation, slavery, and cultural destruction was, as a result, now starving and dying of so many different infectious diseases that even sophisticated physicians often found it impossible to assign specific causes of death to their patients. Twenty-six of the world's seventy most impoverished nations were in Africa. In most of these countries, the daily caloric intake of the average person was below that considered essential to support health.[6]

Several factors well outside the control of even the best-managed underdeveloped countries were suffocating the economies of the world's poorest nations. Western scientists like Karl Johnson, Joe McCormick, Pat Webb,

Pierre Sureau, and Uwe Brinkmann constantly felt the frustrations of working in what was then termed the Third World. Though the plight of the poor nations was hardly a revelation to their own populations, the genuine causes and effects were often surprising and disturbing to well-intentioned foreigners.

This paradigm of perpetual poverty became obvious to McCormick and others like him the moment some piece of essential machinery broke down: a car, generator, centrifuge, microscope, autoclave, or respirator unit. Rarely could someone be found to service the broken equipment because such wealthy-nation gadgetry was simply too scarce to support a domestic service economy. So scientists like McCormick often spent hours under the hoods of their Land-Rovers trying to identify the culprit in malfunctioning engines.

Once the faulty transmission, for example, was identified, the next step was finding a replacement. If no other Land-Rover was available to cannibalize for spare parts, McCormick would have to order a new transmission, shipped from London at enormous cost. Because he had U.S. dollars, McCormick could pay the British exporters for the needed transmission: a Sierra Leone resident, whose leones were worth less than U.S. $0.08, had no currency that the British exporter would accept.

Even with the advantage of valued foreign exchange—"4-X," as it was colloquially called—in the form of U.S. dollars, McCormick's difficulties in obtaining a new transmission would only have begun. Once the desired part arrived at the Freetown docks or airport, already having cost McCormick an enormous amount of money in purchasing and shipping, it would remain locked up for days or weeks in a government warehouse while the American negotiated a maze of bureaucratic paperwork and duty fees. If any single piece of paperwork was deemed improper, McCormick might never be allowed his transmission.

And during that time, the precious transmission, whose value in any African country in 1977 was extraordinary, would rest in a loosely guarded warehouse, ripe for pilfering.

This scenario was not unique to Sierra Leone, or to Africa. Rather, it was the state of affairs in nearly all of the world's poorest nations in the 1970s, and would remain so well into the 1990s.

While their populations exploded in size, national debts mounted, and political instability increased, the world's poorest countries searched for ways to raise foreign exchange capital that would enable them to purchase essential goods for infrastructural development, such as generators, highway construction materials and equipment, and hospitals. Those nations that possessed mineral resources of value to the West mined their bauxite, copper, diamonds, gold, silver, and other ores and gems at a furious pace, selling the materials in exchange for strong foreign currencies or gold. If no such prized goods could be gleaned from the soils or waters of the country, governments sought ways to exploit their agricultural, forestry, or fishing resources for the highly prized foreign dollar, pound, franc, yen, or mark.

But they soon discovered that the buyers for all their goods were far better organized than were the scattered competing sellers. The buyers set the prices, and throughout the 1970s global pricing for most resources fluctuated wildly. Corn, rice, coffee, cocoa, wheat, sugar, bananas—all the classic export crops raised in developing countries—sold at radically variable prices year by year.[7] The variation made it almost impossible for these countries to plan domestic economic development.

Despite such market irregularities, the World Bank, the International Monetary Fund, and major foreign aid spenders on both sides of the Iron Curtain continued to fund and promote investments in large-scale projects such as enormous hydroelectric dams, international airports, and container-ized shipping ports. Such projects, which would often be named after the receiving nation's head of state or a recent political hero, appealed to national pride and the prestige of both donors and recipient political leaders.

But they usually had no ameliorating impact on the health of average citizens, and all too often worsened conditions, giving further advantages to the microbes.

For example, malnutrition was a widespread and increasingly severe problem throughout the least developed parts of the world in the 1970s, and would continue to be serious, occasionally reaching famine conditions, as the millennium approached. Among the cells of the human body most dependent upon a steady source of nutrients are those of the immune system, most of which live, even under ideal conditions, for only days at a time. As nutritional input declines, these vital cells literally run out of fuel, fail to perform their crucial disease-fighting tasks, or, in worst cases, die off. The body may also lack nutritional resources to make replacement cells, and eventually the immune deficiency can become so acute that virtually any pathogenic microbe can cause lethal disease.

Yet the primary economic change in most of the world's poor countries in the 1960s and 1970s involved the creation of export crop systems. Regardless of their political tendencies, governments allocated even more prime agricultural land to production of crops intended for export sale, all in pursuit of foreign exchange. The result was a decline in domestic food production and higher local market prices for grains, vegetables, dairy products, and meat.

Noting that five corporations controlled 90 percent of all international grain sales, four corporations monopolized 90 percent of the world's banana trade, and one multinational had cornered 80 percent of the global markets in corn, soy oil, and peanut oil, American critics Frances Moore Lappé and Joseph Collins warned that "multinational agribusiness corporations are now creating a single global agricultural system in which they would exercise integrated control over all stages of production from farm to con-sumer. If they succeed, they—like the oil companies—will be able to effectively manipulate supply and prices on a worldwide basis through monopoly practices."[8]

In the early 1970s the world's poorest countries formed a voting bloc in the United Nations, dubbed the Group of 77. They sought to force an open discussion of world economic reform issues and use their UN voting leverage to create a "strategic solidarity" against the multinational corporate interests of wealthier nations.

Though the Group of 77 effort quite effectively disrupted United Nations activities for years and resulted in dramatic personnel changes throughout the entire system, it did not fundamentally alter the course of events in crop exportation and food distribution. The Western capitalist governments generally ignored the Group of 77's demands when possible, and effectively counterargued when necessary. The two primary counterarguments were that food scarcity was a function of swelling population sizes rather than of global food distribution patterns. And second, that restricting the activities of multinational corporations was not only unfair to those companies and their stockholders but also counterproductive. In the face of hostile restrictions, they argued, corporate investors would simply abandon the poorest nations altogether.

Further, in the tense Cold War atmosphere of the 1970s all debate about the wisdom and fairness of various policies of development reform was sharply polarized, and it was nearly impossible for countries to navigate independent nonaligned pathways toward advancement. In capitalist circles and at the world's leading lending agencies, the general view was that nations had to modernize first, developing industrial capacities and sizable consumer classes. The benefits of economic modernization would eventually trickle down throughout society, resulting in improvements in education, transportation, housing, and health.

The staunchest advocates of modernization pointed to the Marshall Plan recovery of post-World War II Europe and the MacArthur Plan's efficacy in rebuilding Japan. They argued that a concerted path toward free market capitalist industrialism was the ideal way to raise the standards of life and health of a nation's people.

Stalinist modernists also promoted the notion of industrial development first, social advancement second. Throughout the Soviet Union and the Eastern bloc, massive steel and iron production foundries were glorified, the workers depicted as strong, healthy human beings. According to official Soviet statistics submitted to the World Health Organization in the 1970s, virtually every imaginable infectious disease was on the decline or had disappeared, thanks to communist policies. It was widely believed in international health circles at the time that these statistics were wholly fabricated.[9]

Both superpowers and their allies favored funding projects that had high propaganda value, and most funding was strategically directed. For example, in 1978, half of all World Bank lending went to Brazil, Indonesia,

Mexico, India, the Philippines, Egypt, Colombia, and South Korea.[10] Half of all U.S. nonmilitary foreign aid went to ten strategic nations, five of which were also on the World Bank's list of key beneficiaries: Egypt, Israel, India, Indonesia, Bangladesh, Pakistan, Syria, the Philippines, Jordan, and South Korea.[11] In 1952 none of the U.S. foreign aid budget went to Africa. By 1968 U.S. nonmilitary aid to the continent had increased only slightly: excluding Egypt (which was considered of Middle East strategic interest), 8 percent of all nonmilitary foreign aid went to African countries.

Though details of Soviet nonmilitary foreign aid policies were rarely disclosed, the bulk of its donated largesse went to Cuba, Vietnam, Laos, the Eastern bloc nations, and key strategic points of mixed Cold War allegiance, notably Egypt and India.

From both sides of the Iron Curtain, donors' monetary contributions to poor nations were all too often linked to prestigious showpieces: hydro-electric dams, international airports, university complexes, tertiary care hospitals. Usually ignored were community-based projects, such as schools, medical clinics, skills training programs, or public health campaigns. Worse yet, donors preferred one-shot investments, and disappeared for the long-haul maintenance of their high-profile efforts; even the dams, airports, and massive construction projects soon took on a shoddy, potentially dangerous reality under their previously polished veneers. Lacking the foreign exchange to purchase replacement parts, hire expertise, or carry out routine maintenance, the poor countries had no choice but to let cracks go unchecked in their dams, watch helplessly as the tarmacs of their runways deteriorated, and use staircases when the elevators of their fancy office buildings broke down. Over a third of typical developing country budgets was eaten up by recurrent costs, while donors insisted on funding only new, prestigious programs.

Nongovernmental investment in developing countries came exclusively from the capitalist and social democratic states of North America and Europe, and was heavily targeted toward the acquisition of vital resources. In Africa, in 1977, 56 percent of U.S. private investment was in petroleum, 26 percent in mining, and 6 percent in manufacturing.[12]

From the socialist and nationalist movements and intellectual circles of South America and Africa emerged the dependency theory of development. Overall, the dependency theorists provided cogent criticisms of Western modernization strategies and investment policies, avoided issues related to Soviet activities, and had no consensus on an alternative approach to raising the standards of living and health of the people of the Third World. They represented more a force of opposition than an alternative scheme for development.

Most of these critics (notably such intellectuals as André Gunder Frank, Theotonio dos Santos, Fernando Henrique Cardoso, and Enzo Faletto) argued that acceptance of loans and aid from multinational corporations and lending agencies led to cycles of ever-greater dependency and debt. For example, the poor country that wishes to build a hospital turns to a wealthy nation for donations and loans. Once granted, the hospital's construction leads to a new dependency on Western-style medicine, drugs, and machines. Purchasing replacement parts for American X-ray machines or French autoclaves exhausts the country's small foreign exchange resources. Eventually, the hospital becomes a drain rather than a boon to the society, adding a budget line to the Ministry of Health's already overdrawn accounts. The dependency theorists argued that poor nations lost out in two ways: they were compelled to purchase all equipment and expertise from the richer countries, and whatever products they, in turn, produced had to be sold back to those same wealthy-nation interests at prices set by the purchasers. This, they insisted, represented a lose-lose situation.

By the late 1970s even Western investors were beginning to recognize that modernization wouldn't inevitably bring twentieth-century European standards of living and health to the Third World. In the early 1970s the U.S. Agency for International Development had focused on gross national product (GNP) growth as the crucial measure of success for Third World countries. By 1977 the agency's administrator, John J. Gilligan, was compelled to reverse that policy.

"This approach had certain notable successes," he said in a key policy address:

During the last quarter century, per capita GNP growth in developing countries has averaged three percent—nearly the same growth rate as the rich countries. Average life expectancy in developing countries has increased from 35 years to 50 years—the level attained in Western Europe only at the beginning of the twentieth century. . . . Some developing countries have achieved such high rates of growth that our grant aid to them has ended, and our principal form of economic interaction with them is now largely in trade and private investment.

These overall gains, however, have masked a crucial fact: that while some developing countries have achieved dramatic per capita GNP growth—some at rates of over 7 percent—many others have made very little progress. These averages also conceal wide differences in the extent to which various groups within the poor countries have benefitted from development. For in most less developed countries, the so-called modern sector of urban areas and large farms have been the major beneficiaries of growth, while the urban and rural poor—whose numbers have been rapidly increasing and who form the majority in most developing countries —have generally been left behind.[13]

The World Bank didn't begin to view health care as a specific part of its mission until 1975, when its Health Sector successfully argued that trickle-down modernization would never adequately remedy the acute needs of the poorest of the poor. Between 1975 and 1978 the World Bank gave loans or provided technical assistance for seventy health-related projects in forty-four countries, ending up the world's biggest health lender. During that three-year period, the World Bank loaned poor countries $400 million for primary health care facilities and mosquito control; $160 million for family planning and nutrition projects; and $3.9 billion for water sanitation efforts.[14]

At the close of the decade, the World Bank again assessed its efforts, deciding to shift policy further toward provision of financing the development of primary health care infrastructures for, among other things, "promotion of proper nutrition, provision of maternal and child health care, including family planning, prevention and control of endemic and epidemic diseases."[15]

As the twentieth century drew to a close, the majority of the world's population still suffered and died from diseases due to unclean water.[16] During the 1970s one out of every four people on earth suffered diseases due to roundworms, acquired from polluted waters or foods. A World Bank study found that 85 percent of the residents of Java had hookworm. Some 1.7 billion people annually suffered some additional parasitic infection acquired from polluted water, according to WHO.[17]

Sometimes a major water development project could directly increase the incidence of disease by changing the local ecology in ways that were advantageous to the microbes. The most often cited example of this was the Aswan High Dam, with its apparent association with an increased incidence of schistosomiasis.[18]

Schistosomes are parasitic organisms with a complex life cycle in which, at different stages of the organism's development, the creature grows inside snails, on the surface of freshwater plants, and inside human beings. Its eggs are excreted via human waste into water supplies and are taken up by riverbank and lakeside snails. Inside the snails the eggs hatch and the organisms advance into the larval stage. Those larvae are excreted by the snails back into the lake or river, where they come to rest on the stems and leaves of underwater plants, usually along banks. People who bathe, play, or work in the watery area brush against these plants, and the larvae readily pass through their skin into the bloodstream.

Depending which species of schistosome is involved (*Schistosoma japonicum, S. haematobium, S. mekongi, S. mansoni,* or *S. intercalatum*), the larvae make their way into the human liver, spleen, urinary tract, kidney, rectum, or colon, where they grow into worms. The worms may remain indefinitely, secreting their eggs, which the human then passes on into water supplies, repeating the cycle.

The worms can produce an enormous range of illnesses in people, from minor local skin infections and virtually unnoticeable mild fatigue to life-threatening heart disease, epilepsy, kidney failure, and malignant cancer in the organs in which they reside. Because the range of symptoms is so vast, it is virtually impossible to say with certainty how many people in an endemic area have schistosomiasis: indeed, the definition of schistosomiasis has always been a matter of dispute.

Given the uncertainties inherent in schistosomiasis diagnosis, it was always difficult to prove specific trends in the incidence of the disease. Nevertheless, there was scientific agreement that the enormous Aswan High Dam radically changed the ecology of the Nile, slowing the flow of the once uncontrolled river, preventing annual floods, and creating the huge Lake Nasser. And those changes prompted shifts in the schistosome population.

For millennia nearly every Egyptian had lived in close proximity to the Nile, the rest of the country being largely desert, so the potential of human exposure to any changed disease risk along the river was very high. Yet at no stage of the 1950s planning or construction of the Aswan High Dam was the ecology of human disease taken into consideration, by Egyptian authorities, Western financial interests that initiated the project, or the Soviet government, which, with much fanfare, completed the dam.

The slowing of the Nile flow rates caused a marked shift in the types of schistosome species prevalent in Egypt, from S. haematobium to S. mansoni. For the Egyptian people this meant a shift from organisms that primarily attacked young children, mostly producing urinary tract disorders, to organisms that targeted young adults, causing often severe disorders of the spleen, liver, circulatory system, colon, and central nervous system.[19] Similar shifts in schistosome populations and human disease followed construction of the Sennar Dam in Sudan and the Akosombo Dam in Ghana.

The Aswan High Dam's impact on schistosomiasis was questioned by some because there was a lack of sound comparative data on the incidence of the disease in Egypt prior to construction. But there was an additional reason to challenge the wisdom of building massive water projects without first assessing their potential health impact: Rift Valley fever.

Carried by mosquitoes (Aedes pseudoscutellaris), the Rift Valley fever virus was, prior to 1977, considered largely a veterinary disease that primarily attacked bovine and ovine livestock, though sporadic cases among ranchers were seen. First noticed in 1930, when an outbreak of spontaneous abortions, stillbirths, and adult die-off occurred among sheep and cattle in Kenya,[20] Rift Valley fever epidemics occurred throughout Africa wherever European livestock species, which had no immunity to the virus, were introduced to the continent.[21]

The virus produced hemorrhagic disease similar to yellow fever, with marked lethal effects on developing fetuses and newborns. In nonimmune animals its impact could be devastating: intravenous injections of minute quantities of the virus into laboratory mice produced death within less than six hours in 100 percent of the test animals.[22]

In 1977, six years after completion of the Aswan Dam, James Meegan and his colleagues with the U.S. Navy Medical Research Unit based in Egypt proved that a widespread human epidemic in the Aswan area was due to Rift Valley fever. Over 200,000 people fell ill, 598 died of hemorrhagic disease, and livestock losses were so great that the country experienced severe meat shortages.[23] The scientists concluded that the epidemic began as an isolated outbreak among livestock in northern Sudan, but spread—either via human migration or wind-carried mosquitoes—to Aswan. Once in Aswan, the infected mosquitoes thrived in the 800,000 hectares of dam-created floodlands. The disease had never previously been seen in Egypt.

Similar dam-related epidemics of Rift Valley fever would occur during the 1980s in Mauritania, Senegal, and Madagascar, and in the 1990s the disease would revisit Aswan, causing a severe epidemic.[24]

By the mid-1980s major donor groups, particularly the World Bank, would acknowledge the health care downside to dam construction and instruct applicants for major water project funding to submit disease impact studies as part of their project proposal. In all cases, however, it would be decided that the benefits to society of hydroelectricity and flood control far outweighed the disease potential, particularly if steps were taken to improve local primary health infrastructures.

By 1980 the World Bank would conclude, belatedly, that the worldwide malaria eradication campaign had failed, noting that cases of the disease had increased an astonishing 230 percent on the Indian subcontinent over a mere four years' time (1972–76). Most other vector-borne diseases, just a decade earlier considered easy to eliminate, had experienced "a startling increase in their incidence over the last decade."[25] Sleeping sickness (trypanosomiasis), bilharzia (schistosomiasis), river blindness (onchocerciasis), and Chagas' disease were all increasing in frequency, often in the very countries that had been recipients over the period of billions of donated and loaned U.S. dollars.

Something was clearly amiss. The world's leading agencies were forced to retreat from the grand optimism of the fifties and sixties. Explanations had to be found, blame fixed, solutions suggested.

By the end of the 1970s the World Bank's solution was to urge poor nations to spend more on primary health care and disease prevention. This was done mostly through persuasion, such as the World Bank implying that "because of the emotional appeal of health issues, it may be politically attractive to redistribute welfare through government provision of health care."[26]

Reaching U.S. health care expenditure levels, even as a function of per capita annual spending, would, however, represent an extraordinary feat for most of the world's poor nations. According to the Carter administration, in 1976 in the United States there was a 1:600 ratio of physicians to the general population; virtually 100 percent of drinking water supplies were considered free from infectious disease; people consumed, on average, 133 percent of their minimum caloric need every day; 99 percent of adults were literate; 3.3 percent of the federal GNP was directed toward health care spending for a per capita spending rate of $259.

In contrast, Tanzania, for example, had one physician for every 18,490 citizens; safe drinking water was available to less than 40 percent of the population; the average citizen consumed only 86 percent of the minimum daily caloric need; 34 percent of the adult population was illiterate; and the government spent 1.9 percent of its GNP on health care for a total of $3 annually per capita. Even if Tanzania doubled the percentage of its GNP devoted to health care, reaching U.S. percentage levels, it would still be spending less than $10 a year on each of its citizens. To reach U.S. annual expenditure rates of $259 per citizen, the Tanzanian government would have to rob nearly every other program in the government.[27]

"It is stupid to rely on money as the major instrument of development when we know only too well that our country is poor," Tanzania's one-party state proclaimed in its historic Arusha Declaration of 1967. "It is equally stupid, indeed it is even more stupid, for us to imagine that we shall rid ourselves of our poverty through foreign financial assistance rather than our own financial resources."

Tanzania sought to create an infrastructure of modestly trained paramedics who worked out of tiny concrete or wattle clinics dispersed throughout the villages inhabited by most of the nation's ten million citizens. Between 1967 and 1976, the Tanzanian *Mtu ni Atya Chakula ni Uhai* village health campaigns increased the numbers of maternal/child health clinics by 610 percent, rural paramedics by 470 percent, and built 110 new medical facilities (for a total of 152 clinic structures nationwide by 1976). Life expectancy over that time increased seven years, reaching 47 (compared to 70 in Europe in 1976). Infant mortality also showed modest improvement, decreasing to 152:1,000 babies, compared to a 1967 level of 161:1,000 (with 1976 European infant mortality at 20:1,000).[28]

Recognizing its acute need for physicians, the government built Muhimbili Medical School in Dar es Salaam and sent many bright young Tanzanians overseas for medical training, hoping to increase its national physician population by about 65 doctors a year. By 1975 the paramedic-to-patient ratio was 1:454, but the physician-to-patient ratio had actually worsened, in part due to anti-Asian bigotry. Many of East Africa's best-educated residents were Indians, brought decades earlier as indentured labor by British colonialists in need of a literate bureaucratic class. In 1972 Uganda's dictator, Idi Amin (whose proclaimed hero was Adolf Hitler), ordered all Asians, numbering some 50,000 to 80,000, to leave the country immediately or face execution. No hue and cry of protest was raised by any other African government. Thousands of Indians, most of whom had spent all their lives in East Africa, fled not only Uganda but the continent as a whole.[29]

Though such problems plagued all the poor nations on the planet, they were particularly acute in Africa because of its severe political and military instability. Nowhere else in the world were governments so recently freed from centuries of European colonialism. The Portuguese colonies of Guinea-Bissau, Angola, Mozambique, and Cape Verde only gained independence in the mid-1970s, after more than a decade of bloody civil war. In the

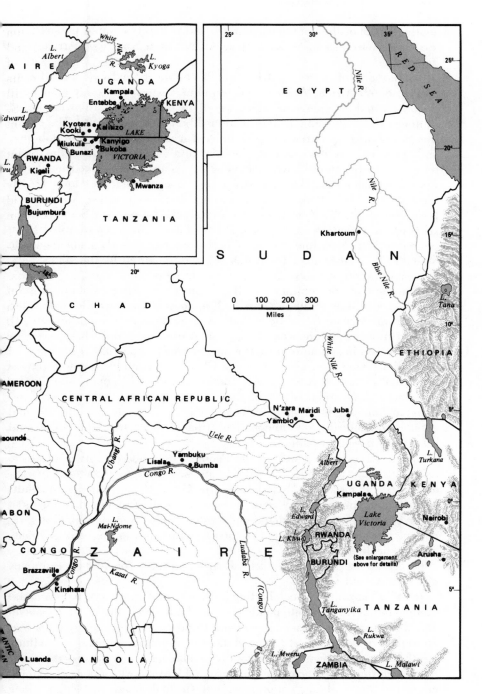

CENTRAL EAST AFRICA

southern part of the continent, warfare and instability would persist until the fates of Rhodesia, South Africa, Angola, and Southwest Africa were decided.

To the north of those countries (which would eventually be named Zimbabwe, South Africa, Angola, and Namibia, respectively), lay a string of majority-ruled independent states sworn to boycott the still white-ruled southern states and support their various liberation movements. The Frontline States, as they were called, included Tanzania, Zambia, Mozambique, and, to a less militant degree, Lesotho and Botswana. Guerrilla troops representing the future governments of the region freely moved inside the Frontline States, and Lusaka was a sort of command post for SWAPO (South-West Africa People's Organization), ZAPU (Zimbabwe African People's Union), ZANU (Zimbabwe African National Union), and South Africa's ANC (African National Congress). Political exiles from the troubled south poured into the Frontline States, exacerbating their already acute economic difficulties. Furthermore, trade was severely impaired by the states' self-imposed boycott of South African ports and markets.

Elsewhere on the continent, civil instability was legion. Mobutu brutally smashed all dissent within Zaire. Self-appointed Emperor Bokassa ruled the Central African Republic with such brutality that he would eventually be overthrown by French paratroopers and tried for cannibalism and genocide. In an alleged anti-corruption cleanup campaign, junior elements of the military violently seized power in Ghana. Civil unrest due to religious and tribal disputes raged through Sudan, Morocco, Ethiopia, Mauritania, Angola, and Rwanda. Much of the warfare stemmed from the artificial national boundaries created by colonial powers in the seventeenth and eighteenth centuries, dividing ancient tribal lands, extended families, and traditional power structures.

The superpowers, as well as the People's Republic of China, sought to manipulate these seemingly endless battles, hoping to align African governments with either the United States, the U.S.S.R., or China. As a result, obscene amounts of money were spent on the military and police forces of impoverished countries, squandered by dictators who made "gifts" to their nation's power elites in exchange for support, wired to the bank accounts of arms dealers worldwide.

Clearly, those funds were not spent on health care. Consider the examples of Tanzania and Uganda.

In 1979 Tanzania was celebrating its recent military victory over Uganda. Though the world's seventh pandemic[30] of cholera had struck Dar es Salaam and the lethal *Vibrio* bacteria coursed through the open sewer lines that crisscrossed the streets of the capital, little attention was paid to anything but the war. Pretty young girls proudly proclaimed victory across their rear ends, wearing *kangas* made from fabric emblazoned with the news. Young men wore their military uniforms as they strutted, heads held high, along Independence Avenue or past ANC headquarters on Nkrumah Street.

On his way to the Dar es Salaam airport in April 1979, Yusufu Lule anxiously cast his eyes about, looking for what he suspected was his last time at the city's street scene. After years of exile, he was about to take the reins of government in Uganda. Though he had agitated for Idi Amin's overthrow for years, the prospect of returning was frightening.

"It is chaos. We have a whole generation who don't know right from wrong. For years they have seen such brutality—rape, murder, theft, torture. I am going to a place where morality has no meaning," Lule said with apparent dread.

Sixty-eight days later, Lule would be overthrown and Uganda would spin into a cycle of short-lived and vengeful governments.

It all began in 1971, when the Ugandan military overthrew the elected government of Milton Obote, putting a semi-literate, temperamentally violent man named Idi Amin in charge of the nation of some 18 million. Ten years earlier, Uganda had been considered one of the finest jewels in the British Empire's crown; a rich cornucopia of agricultural wealth with a well-established infrastructure of colonial and missionary schools, hospitals, roads, and trade. But Obote's government was also marked by corruption that fueled unrest and the 1971 military coup.

Amin destroyed the nation's prosperity and drove his country into a state of hellishness unlike anything it had previously experienced.

In 1975 Tanzanian President Julius Nyerere denounced Amin as "an oppressor, a black fascist, and a self-confessed admirer of fascism." A few months later, Amin declared that, by ancient tribal rights, parts of Sudan, Kenya, and Tanzania belonged to Uganda. To drive home his point, Amin publicly executed a group of Kenyan students studying at universities in Entebbe and Kampala.

By 1977 Amin's government had committed so many atrocities both domestically and against its neighbors that the Western powers and Soviet Union had terminated diplomatic and trade relations. In response to British condemnation of Amin's human rights practices, said to include wholesale rape of women nationwide, as well as summary executions of tens of thousands of citizens of all ages, the dictator personally executed Anglican archbishop Luwum in front of hundreds of witnesses and television cameras.

"Thousands of innocent Ugandans have been floating in the river Nile in what the dictator and butcher Amin calls accidents," charged Radio Tanzania on the day of Archbishop Luwum's execution. "If black African states condemn white minority rule [in South Africa and Rhodesia], they must also condemn atrocities committed in black-ruled states."

By early 1978, according to the International Commission of Jurists in Geneva, Amin had summarily executed some 100,000 of his citizens, the trade agreement of the East African Community of Uganda, Kenya, and Tanzania was formally dissolved, and both Kenya and Tanzania were placed on war-readiness status.

In October 1978, Amin's troops invaded the Kagera District in northern

Tanzania. A pastoral area lining the western shores of Lake Victoria, Kagera had no industry, only one small city (Bukoba), a scattering of hamlets, and no ability to defend itself against Amin's marauding forces. The Ugandan Air Force softened up the rolling verdant hillsides of Kagera with bombing raids. Troops followed, laying waste to the thatched huts and wattle structures of villages from one end of the district to another. For two months Amin's troops occupied a 700-square-mile area of Tanzania, killing hundreds of peasants, practicing deliberate rape of the women that was intended to humiliate their men, slaughtering most of the region's livestock, and driving some 40,000 peasants into exile.

Nyerere appealed for support from the Organization of African Unity (OAU) and the United Nations. None was forthcoming.

In December 1978, Tanzanian troops went to war with Uganda, fighting over the Kagera region for two months. Having beaten back Amin's troops, the Tanzanians pushed on toward the capital, Kampala.

On April 11, 1979, the Amin government was toppled. Idi Amin went into exile in Libya, and Tanzania put Lule in power.

The five-month war between Tanzania and Uganda—which was puny by international standards—devastated the infrastructures of Uganda and northern Tanzania, and left the economies of both nations in a shambles. The combined impact of war and previous years of Amin's wantonness left Uganda in need of $2.3 billion in emergency reconstruction aid. It hurt Kenya's coffee trade, which had relied in part on Ugandan beans. And for the tiny, landlocked nations of Burundi and Rwanda it brought all trade to a standstill.[31]

When Lule's staff took over the national bank, they discovered that Uganda was $250 million in debt to foreign interests, and less than $200,000 could be found in the nation's coffers. During his six-year reign, Amin simply printed more money whenever resources dwindled, causing annual inflation to run at 200 percent a year. Prior to the war gasoline sold in Kampala for $39 a gallon, housing rents increased 41 percent in a single year, while per capita income plummeted.[32]

Well before the war erupted, most health professionals who could manage to do so had fled the country, and the severe economic difficulties created by the Amin government prompted wholesale looting of all undefended facilities.

Widespread famine followed the end of the war, claiming at least 50,000 lives. Wildlife conservation groups throughout the world protested as starving Ugandans slaughtered and consumed elephants, hippos, elands, giraffes, monkeys, and other animals by the thousands.

Between 1975 and 1980, Uganda, its entire health infrastructure devastated, experienced epidemics of malaria, leprosy, tuberculosis, cholera, visceral leishmaniasis (kala-azar), and virtually every vector-borne ailment known to the continent.[33] A French team found evidence of more exotic diseases as well, when they took blood surveys of villagers in western Uganda. Ebola, Marburg, Lassa, West Nile fever, Crimean-Congo hemorrhagic fever, and Chikungunya were among the viruses found in the blood of the region's populace.[34]

Between 1971 and 1977, Uganda had its worst measles epidemic in over forty years, with high death rates among children seen all over the country. So great was the country's chaos that no agency kept count of the death toll. Gonorrhea soared during the Amin years, particularly among soldiers. Because the country was bereft of antibiotics, most cases went untreated. Routine vaccination for such diseases as whooping cough and tetanus came to a halt, and the incidence of these diseases rose dramatically.

Starving, sick refugees poured by the tens of thousands across borders to Zaire and Sudan, taking their diseases with them.

Makerere University, which had been the primary medical training center for East Africa's doctors, was looted right down to its electrical sockets and bathroom tiles. By the end of the 1970s, the nation of Uganda would be completely out of toilet paper, antibiotics, aspirin, sterilizers, cotton wool, bed linens, soap, clean water, light bulbs, suturing equipment, and surgical gowns.[35] Rumors of strange disease outbreaks were rampant, but there was nobody left to investigate these claims.

Such tragic events, with the resultant epidemics and health crises, were mirrored all over the world. From Pol Pot's reign of terror in Cambodia to the Cold War-manipulated battlefields of Central America, the world's poorest countries spent extraordinary amounts of money on domestic military operations and warfare. And the microbes exploited the war-ravaged ecologies, surging into periodic epidemics.

The World Health Organization, with a staff of only 1,300 people and a budget smaller than that spent on street cleaning every year by the city of New York, tried to combat such seemingly intractable public health problems with donated vaccines, technical assistance, and policy statements.[36]

On September 12, 1978, WHO convened a meeting of ministers of health from over 130 nations in Alma-Ata[37] in the U.S.S.R. The conference issued what would be hailed years later as a pivotal document in the international public health movement: the Declaration of Alma-Ata. Inspired in part by U.S. Surgeon General Julius Richmond's Health Goals 1990, which in 1975 systematically outlined the status of Americans' health and set goals for improvement, the Alma-Ata Declaration called for "the attainment by all peoples of the world by the year 2000 of a level of health that will permit them to lead a socially and economically productive life."

The ten-point Alma-Ata Declaration defined health as "a state of complete physical, mental, and social well-being, not merely the absence of disease or infirmity," and declared it "a fundamental human right." It decried health care inequities, linked human health to economic development, and called upon the governments of the world to develop financially and geographically accessible primary health care facilities for all their people.

Declaring health a human right forced issues of disease control onto the newly powerful agenda of global civil liberties. In 1976 the UN General Assembly voted to enter into force the International Covenant on Civil and Political Rights.[38] It was the strongest vilification of tyranny, discrimination, violations of basic freedoms, and injustice ever passed by the UN. Also that year the UN passed the International Covenant on Economic, Social, and Cultural Rights,[39] which specifically recognized "the right of everyone to the enjoyment of the highest attainable standard of physical and mental health."

John Evans of the World Bank elucidated three key demarcations in health problems that he felt were tied to the economic development and status of each nation: infectious disease stage; mixed phase; and chronic disease state. In the poorest, least developed nations of the world, the majority of the population suffered illness and death due to communicable and vector-borne diseases. With improvements in economic development, Evans said, came a painful period of mixing, in which the poorer members of society succumbed to infectious diseases while the wealthier urban residents lived longer, disease-free lives that were eventually cut short by chronic ailments such as cancer and heart disease.

In the most developed nations, Evans argued, infectious diseases ceased being life-threatening, some disappeared entirely, and the population generally lived into its seventh decade, succumbing to cancer or heart disease. The bottom line, from Evans's perspective, was that infectious diseases would no longer pose a significant threat to postindustrial societies.

"We must never cease being vigilant," Richmond said, "but it is altogether proper to shift resources towards prevention of chronic diseases. With political will, tremendous strides can be made."

Though the World Bank perspective informed most long-term planning, there were voices within the academic public health community who loudly questioned the three-phase assumptions. While not disputing that curative medicine had made genuine strides, particularly since the 1940s, and agreeing that control of disease was linked to societal wealth, they refuted the idea that there might be a direct correlation between stages of national development and individual disease. In their view, the ecology of disease was far more complex, and waves of microbial pestilence could easily occur in countries with enormous gross national products. Conversely, well-managed poor countries could well control pestilence in their populations.

The debate centered on a two-part question: when and why did most infectious diseases disappear from Western Europe, and what relevance did that set of events have for improving health in the poorest nations in the last quarter of the twentieth century?

University of Chicago historian William H. McNeill spent the early 1970s studying the impact epidemics had on human history since the beginning of recorded time, and then reversed his query to ask which human activities had prompted the emergence of the microbes. In 1976, his book *Plagues and Peoples*[40] created a sensation in academic circles because it argued with the force of centuries of historical evidence that human beings had always had a dramatic reciprocal relationship with microbes. In a sense, McNeill challenged fellow humans to view themselves as smart animals swimming in a microbial sea—an ecology they could not see, but one that most assuredly influenced the course of human events.

Like Evans, McNeill saw stages over time in human relations with the microbes, but he linked them not so much to economic development as to the nature at any given moment of the ecology of a society. He argued that waterborne parasitic diseases dominated the human ecology when people invented irrigation farming. Global trade routes facilitated the spread of bacterial diseases, such as plague. The creation of cities led to an enormous increase in human-to-human contact, allowing for the spread of sexually transmitted diseases and respiratory viruses.

Over the long course of history, McNeill said, pathogenic microbes sought stability in their relationships with hosts. It was not to their advantage to wipe out millions of nonimmune human beings in a single decade, as happened to Amerindians following the arrival of Columbus and Cortez. With the Europeans came microbes to which the residents of the Americas had no natural immunity, and McNeill estimated, "Overall, the disaster to Amerindian populations assumed a scale that is hard for us to imagine. Ratios of 20:1 or even 25:1 between pre-Columbian populations and the bottoming-out point in Amerindian population curves seem more or less correct."[41]

This was not an ideal state for the microbes, he argued, because such massive death left few hosts to parasitize. After centuries of doing battle with one another, humans and most parasites had settled into a coexistence that, if not comfortable for humanity, he argued, was rarely a cause of mass destruction. Still, he sternly warned, "no enduring and stable pattern has emerged that will insure the world against locally if not globally destructive macroparasitic excesses."

Other historians of disease had tried to link the emergence of epidemics to the social and ecological conditions of human beings,[42] but none had presented as lucid an argument as McNeill's, and it promoted widespread reappraisal of both historic events and contemporary public health policy.

Nobel laureate Sir MacFarlane Burnet was moved from his perspective as an immunologist to issue similar warnings about humanity's overconfidence. True, he said, vaccines and antibiotics had rendered most infectious diseases of the Northern Hemisphere controllable. But, he cautioned, "it is almost an axiom that action for short-term human benefit will sooner or later bring long-term ecological or social problems which demand unacceptable effort and expense for their solution. Nature has always seemed to be working for a climax state, a provisionally stable ecosystem, reached by natural forces, and when we attempt to remold any such ecosystem, we must remember that Nature is working against us."[43]

The policy implications were clear, Burnet said. Start by looking at the ecological setting of disease transmission. If the ecology could be manipulated without creating some untoward secondary environmental impact, the microbe could be controlled, even eradicated.

René Dubos, who served in the 1970s as a sort of elderly patron saint of disease ecology because of his vast contributions to research on antibiotics and tuberculosis during the pre-World War II period, also favored an ecological perspective of disease emergence, but laid most of the blame for epidemics on *Homo sapiens* rather than on the microbes. In Dubos's view, most contagious disease grew out of conditions of social despair inflicted by one class of human beings upon another. Dubos believed tuberculosis, in particular, arose from the social conditions of the poor during Europe's Industrial Revolution: urban crowding, undernutrition, long work hours, child labor, and lack of fresh air and sunshine.

"Tuberculosis was, in effect, the social disease of the nineteenth century, perhaps the first penalty that capitalistic society had to pay for the ruthless exploitation of labor," Dubos argued.[44]

For Dubos, unbridled modernization could be the enemy of the poor, bringing development and freedom from disease to the elites of societies, but consigning their impoverished citizens—particularly those living in urban squalor—to lives of microbial torture.

"The greatest strides in health improvement have been achieved in the field of disease that responded to social and economic reforms after industrialization," he wrote.[45] He strongly felt that infectious diseases remained a major threat to humanity, even in the wealthy nations, and warned physicians not to be fooled into complacency by what he termed "the mirage of health."

At the University of Birmingham in England, Thomas McKeown led a team of researchers who reached the conclusion that rapid urbanization, coupled with malnutrition, was the key factor responsible for the great epidemics of England and Wales from medieval times to the beginning of the twentieth century. Conversely, McKeown credited improvements in access to nutritious food for England's lower classes with at least half the reduction in premature mortality in the country between 1901 and 1971, and insisted that the bulk of all improvements in survival preceded the advent of modern curative medicine.[46] McKeown based his assertions on a meticulous scanning of English and Welsh government medical records maintained over the period, which indicated that premature mortality rates decreased radically before the age of antibiotics.

Joe McCormick had heard it all, argued one position or another over beers with CDC colleagues, and recognized grains of truth scattered through each position, from the World Bank to the angry socialist dependency theorists. But all the hand-wringing and theorizing wasn't going to provide the resources needed to get rid of Lassa.

For nearly three years he had been tramping around West African villages testing residents and rats for Lassa virus infection. By 1979 McCormick had reached the conclusion that Lassa was an entrenched endemic disease, causing thousands of cases of illness of varying degrees of severity each year. The only way to rid Sierra Leone of human Lassa cases would be to eliminate contact between the rats and humans—an option he considered doable if millions of dollars were spent improving the country's rural housing and hospitals.

The alternative was mass education about rat avoidance and ribavirin therapy for those who suffered Lassa fever. That prospect was also orders of magnitude too expensive for the impoverished state.

In late June 1979, McCormick returned to CDC headquarters to take over Karl Johnson's job as chief of the Special Pathogens Branch, leaving Webb in charge of the Sierra Leone laboratory. For many years to come, he would return to the West African country to further study the Lassa virus, hoping to find ways to limit the impact of the disease on the developing countries of West Africa.

Shortly after his return to Atlanta, the World Health Organization called to formally request McCormick's assistance in investigating a suspect epidemic in Sudan. It was believed that Ebola was the culprit.

According to Sudanese epidemiologist Osman Zubeir, the outbreak began sometime in early August in N'zara, spread quickly, and was still raging when he notified WHO in mid-September. He placed the area under quarantine, and Zubeir was preparing a surveillance effort.

McCormick hastily gathered supplies and the first assistant he could get his hands on—a new EIS officer, Dr. Roy Baron. Within a matter of hours, the pair were on board a flight to Khartoum, and McCormick was giving Baron a rapid-fire lesson on Ebola, Sudan, field operations, and self-protection.

Joe tugged at his dark brown goatee with anticipatory excitement, relishing a second chance to crack the mysteries of Ebola. McCormick showed Baron the only available maps of the region, made in 1955. He described the difficulty of finding villages, which were deliberately hidden in the ten-foot-tall Sudan grass and swamps.

And he gave Baron a quick sketch of the political and social situation. Tensions between the Muslim Nubian and Arab north and animist and Christian south remained high, though no civil war had flared since 1972, when Colonel Jaafar Nimeiri granted the southern area some degree of autonomous power. The Nimeiri government in Khartoum continued to maintain highly centralized government oversight over all matters deemed of national interest, including disease control.

Since McCormick's last visit to the region during the 1976 Ebola outbreak, the relationship between Sudan's north and south had grown more strained, and the Nimeiri government had been chipping away at the scope of autonomous rule in Juba. Though few realized it in 1979, the country was on the brink of a civil war that would flare up when Nimeiri rescinded southern self-rule in 1983 and would persist for over a decade. In 1979 the most visible signs of a rift between Khartoum and Juba could be seen in government-financed facilities in the south: schoolrooms without desks, books, or teachers; unpaved, often impassable roads; almost nonexistent postal services; no central electricity outside Juba; and hospitals lacking any furniture save the steel-framed, unmattressed beds.

Twenty-four hours after he got the call from WHO, McCormick, with Baron in tow, was in Khartoum getting a quick briefing from federal health authorities. Anxious to get to N'zara as soon as possible, McCormick left Baron behind in Khartoum to arrange methods of air-shipping tissue and blood samples to Atlanta, set up some sort of communications system, and smooth over relations in the capital.

The next morning, McCormick had breakfast in Juba and began what would be a maddening three-day search for transport to N'zara. The area, which had been under strict quarantine for several days, was almost completely lacking in gasoline for the handful of functioning Land-Rovers and ancient British vehicles that comprised the local government fleet. Furthermore, the people of Juba were terrified: nobody was willing to guide McCormick into the epidemic zone.

Using every means of persuasion at his disposal, including offers of cash, McCormick finally garnered a police airplane and a pilot who agreed to fly him into N'zara on September 22. En route, the pilot told McCormick that he would land but not stay. For the entire flight, Joe tried cajoling, threatening, bribing, and instilling guilt in the frightened police pilot, hoping to persuade him to remain in N'zara long enough for the scientist to gather samples that the pilot could fly up to Baron in Khartoum.

Ultimately, promises of great financial reward upon arrival in Khartoum triumphed over the pilot's anxieties, and a deal was struck: the pilot would remain inside his plane, having no contact with anybody in N'zara, until dawn.

The plane touched down at 5:30 p.m. on the grassy field outside N'zara, McCormick hurriedly reminded the trembling pilot of the rewards awaiting him in Khartoum, swiftly gathered his supplies, and set out to find his designated translator.

"I've only got a few hours, and it's going to be dark soon," McCormick told the young schoolteacher who translated his English into the local dialect. "I need to see the sick people immediately."

The translator nodded and led McCormick through the hamlet of mud-and-wattle structures to a round hut on the periphery. Lighting his kerosene lamp and shouldering his medical bag, McCormick stepped inside.

When his eyes adjusted to the darkness only dimly lit by his upheld lamp, McCormick saw what he would later describe as a vision from hell. Some twenty men and women lay upon grass mats, crammed one against another in a small, dark atmosphere of overpowering heat and stench. Most were in agonizing pain, horribly ill, groaning aloud or crying out in demented visions. Some, their skin in excruciating pain, had torn off their clothing and lay in naked terror.

McCormick took a deep breath, stepped over to the first ailing man, and resolved to draw blood samples and gather vital statistics on every person in the room before dawn.

All night long McCormick, wearing only latex gloves and a constantly steamed-up respirator for protection, knelt beside the Ebola victims, giving them thorough physical examinations, painstakingly noting all information on a pad, and taking blood samples.

Shortly after midnight, McCormick reached the midway point in his work: an elderly woman who was delirious, burning with fever. She seemed to be hallucinating, and McCormick assumed that his masked visage was disturbing. He carefully set the lantern on the dirt floor next to his bent knees, placed a tourniquet on the woman's upper arm, and uncapped the needle on a fresh syringe. McCormick paused, waiting for calm to come over the woman, then deftly inserted the needle in the ailing woman's arm, simultaneously releasing her tourniquet.

The instant the needle hit her vein, the woman thrashed wildly, the syringe popped out and landed in McCormick's thumb. Horrified, he swiftly recapped the needle, squeezed his thumb, and applied disinfectant to the invisible microscopic wound site. He looked at the frenzied woman, so ill she probably hadn't even realized what transpired. And then he held his watch up to the lantern.

"Only five hours until dawn's first light," he thought. Mustering composure and deliberately shoving aside all thoughts of being stuck by the needle, McCormick scrupulously completed his rounds, prepared all the samples for shipment, putting them inside a small tank of liquid nitrogen and placing that in a case of dry ice. He then raced out to find the pilot.

Once the plane was aloft, McCormick let himself feel his exhaustion—and fear. He asked his translator for a place to sleep, and followed zombielike, carrying his supplies to a thatched-roof hut. He reached into his bag and withdrew a supply of precious Ebola antiserum that Joel Breman and Peter Piot had collected in Yambuku three years earlier.

"Nobody really knows if this stuff works," McCormick mumbled as he injected himself with two units of plasma. He then grabbed the radio he'd brought from Atlanta, called Juba, and got the signal relayed to Khartoum. Once Baron was on the line, he told the young doctor that samples were on their way and the pilot should be rewarded.

"Oh, and by the way, I seem to have stuck myself. I just mainlined some plasma, and I think I'm going to pass out for a while," he told Baron.

He then downed the only medicine in which he had any genuine confidence, given the situation: a bottle of Scotch.

Twelve hours later, he awoke to a N'zara late afternoon, stepped out into the 100-degree heat, and took stock of the situation. It was clear that Ebola had struck again. It was also obvious that he had been exposed to the virus. Recalling his previous experience with Ebola, McCormick figured the virus would incubate for five to seven days before he got sick, leaving him plenty of time to get to the bottom of this epidemic.

"I was not in a panic," McCormick would explain years later. "If I got a fever, my plan was I'd get on the horn, get a plane in there, and evacuate to Europe. I'd faced the possibility of dying before, and I just didn't see the point of going to some hospital and getting everybody in a stew, sitting and waiting to get sick, and thinking all the while about the work I should have been doing in N'zara."

With a shrug, he added, "I'm a fairly stoic Midwesterner."

Nevertheless, McCormick had no death wish. Waiting for him in Atlanta were his wife and three children, aged two to nine years.

Over the next week, McCormick, joined by Baron and Zubeir, reconstructed the history of the N'zara epidemic, collected more samples, and took steps to stop the outbreak. All the while, he kept a constant eye on that "death hut," as he called it, keeping track as the dead were removed by relatives for burial.

Meanwhile, he was determined to observe enough patients in Yambuku in sufficient detail to formulate a copious description of the signs and symptoms of the disease. The disease seemed to strike people very suddenly: one moment an individual might be laughing and sharing local moonshine with a friend, the next he would have a searing headache, be drenched in sweat, and feel too weak to stand. It was this seemingly instantaneous illness that the Sudanese people found especially terrifying. Over the next three days, things would escalate rapidly. Patients would tremble with chills, fevers would soar over 105°F, and every joint and muscle would ache with pain so severe that the Ebola victim could find no position in which to comfortably lie down or sit. Their throats would become so sore that most couldn't tolerate swallowing their own saliva. Eating was out of the question.

By the fourth day, hemorrhaging would begin. Ebola victims would vomit blood, excrete blood, bleed profusely from their gums, and stare at McCormick through bloodshot eyes.

McCormick soon realized that the frequently used expression "the patients bleed to death" wasn't accurate in the case of Ebola—or for Lassa for that matter. It wasn't the bleeding that killed the Sudanese Ebola victims, he concluded, but shock due to fluid loss. Somehow the virus was causing the endothelial linings of the patients' veins to break down, giving rise to leakage of water from the bloodstream into neighboring tissue. As vascular volume decreased, the patients went into shock. If fluids were pumped into the patients' bloodstreams, the result was death by pulmonary edema, because leaking veins in the lungs flooded the airways with fluids.

McCormick treated patients with the Yambuku plasma, with mixed results: some improved, whereas others showed no response to the putative antiserum. He wasn't convinced it was useful, which added to whatever concern he allowed himself to feel about his own status.

One afternoon, he spotted the old woman from the death hut strolling through N'zara, a jug of water on her head, clearly full of energy. McCormick was ecstatic. CDC blood test results cabled from Atlanta shortly thereafter indicated that she alone among those in the death hut was uninfected. Whatever her ailment, it wasn't Ebola.

And Joe McCormick had never been infected with the deadly virus.

With the passing of days the people of N'zara concluded that the odd-looking white man with the weird face mask did seem to have special powers. They observed his actions and followed those orders that seemed to make sense. The key order they wouldn't follow, however, was "bring your sick and dead to N'zara Hospital." And for good reasons.

It seemed the hospital was, as it had been three years earlier during the first Ebola outbreak, the focus of the epidemic. Shortly before McCormick, Baron, and Zubeir arrived, two nurses in the hospital died of Ebola, apparently contracted from patients. The people knew that many checked into N'zara Hospital; few checked out.

The second and far more difficult problem was the fate of the dead. In 1976, Don Francis had allayed the people's spiritual concerns by performing the ritual burial practices himself, but the epidemic in 1979 was too big for McCormick to handle the evacuation of wastes from all the bodies—especially since those wastes undoubtedly contained viral contaminants.

So he settled on a novel idea: let those attending funerals wear respirators, gloves, and surgical garb, and conduct the burial preparations themselves. In exchange for such protection, the relatives usually allowed McCormick to remove tissue and blood samples from the bodies.

Day after day Zubeir and McCormick scoured the high Sudan grass in search of hidden communities, and negotiated for the sick and dead. It was a process fraught with cultural difficulties for both sides, but they usually succeeded.

Within a month, the team had the disease under control and was able to recommend that Khartoum lift the area's quarantine: not a moment too soon, as the entire province was out of gasoline, most food supplies, and even medical supplies. Famine would have quickly followed.

When the team reconstructed the events of the summer of 1979, they discovered many parallels with the 1976 outbreak, but were still unable to say where the virus came from. Once again, the first case involved a man who worked in the run-down colonial-era cotton factory that was filled with huge swarms of bats and a vast array of insects. He fell ill on August 2 and died of the disease in N'zara Hospital three days later.

Three of the man's family members who cared for him fell ill. So did a man who lay in the bed alongside him in N'zara Hospital. A woman who frequented the hospital ward, tending to her ailing husband a few beds down from the first man, got Ebola. And the two nurses on the ward got the virus.

Every additional illness involved members of the families of those first five cases or close friends who tended to their illnesses or burial preparations. All infections could be tied to some direct blood or fluid contact between an ailing Ebola victim and another individual. The best nurses—those who provided the closest care for the patients—were five times more likely to be infected than their more aloof colleagues.

The team was able to find fifty-six Ebola cases, many hidden in the tall grasses. Sixty-five percent of those who got infected died.

Though it seemed obvious to McCormick that some Ebola-carrying animal or insect lurked inside the cotton factory, none of the fauna samples he sent to the CDC were Ebola-positive. Their inability to pinpoint the reservoir for Ebola would bother McCormick for years, nagging constantly at the back of his mind whenever he had reason to recall the events in N'zara. He would always tell anyone who asked, "It's probably the bats. We just have to get in there and capture a few more of them and we'll find the virus."

Barring identifying the reservoir for Ebola and eliminating the culprits from the human ecology of the N'zara area, McCormick suspected isolated cases of the disease would always crop up.

"Because the cultural and social structure in Sudan tends to limit contact with severely ill persons to a few adults in a relatively secluded compound, sporadic cases of Ebola virus disease may have little impact on the community at large," the team wrote, summarizing their findings.[47] "In this outbreak, however, the hospital appeared to be the focal point for dissemination of infection to several family units after the admission of the index patient."

As was the case with Lassa, poorly run hospitals operating under conditions of extreme deprivation were the amplifiers of microbial invasions. What might have otherwise been individual illness, limited to one or two cases of Ebola, was magnified in a hospital setting in which unsterile equipment and needles were used repeatedly on numerous patients. N'zara Hospital couldn't afford mattresses for its steel bedframes or penicillin— it could hardly be expected to throw away every single plastic syringe simply because it had previously been used.

McCormick was at a loss for a solution. Once again, elimination of a disease threat seemed inextricably bound to economics and development. The poverty of southern Sudan exceeded anything he had seen before, and McCormick had little reason to hope that some government or agency with the wherewithal to do so would deem it politically expedient to assist such godforsaken parts of the planet.[48] Yet McCormick felt certain that Ebola and other dangerous diseases would continue to haunt the most impoverished communities on earth, constantly threatening to explode into epidemics, some of which might one day lap at the shores of the planet's richest nations.

Out of such poverty, from the African Serengeti to the burned-out tenements of the Bronx, would soon come microbial invasions that would bear out McCormick's prophesy.

8

Revolution

GENETIC ENGINEERING AND THE
DISCOVERY OF ONCOGENES

*Man is embedded in nature. The biologic science of recent years has
been making this a more urgent fact of life. The new, hard problem
will be to cope with the dawning, intensifying realization of just how
interlocked we are. The old, clung-to notions most of us have held
about our special lordship are being deeply undermined.*
— Lewis Thomas, 1975

*Scientists at the Massachusetts Institute of Technology have com-
pleted the synthesis of the first man-made gene that is fully functional
in a living cell.*
— Massachusetts Institute of Technology press release, August 30, 1976

■ The revolution happened with such astonishing speed that few participants fully appreciated what had transpired. The collective consciousness
of science and medicine changed in the blink of a historic eye, rendering
those who failed to adapt obsolete overnight. In less than five years every
aspect of biology and medicine was so thoroughly shaken to its core that
science students trained afterward thought it had always been so. The
excitement could be felt from the floors of the world's stock exchanges to
the halls of parliaments worldwide.

Just as the hopeful spirits of the post-World War II scientific conquest
of the microbes seemed to be flagging, humanity discovered genetic engineering.

When science learned how to manipulate the genetic material of plants,
animals, and microbes—the DNA and RNA—an entirely new world revealed itself. Suddenly it seemed possible to understand the secrets of the
microbes, appreciate at the molecular level how the human immune system
destroyed (or failed to destroy) its microscopic challengers, and invent
radically new weapons to use in waging war on disease.

Once again, optimism pervaded biological research. Once again, scientists predicted bold victories over everything from cancer to malaria. The
speed of discovery from the early 1970s into the 1980s was dizzying, even
for those who started it.

"I wasn't surprised about much of anything until 1966," declared Sir Francis Crick in a 1983 interview. "But after that, well, the last ten years have surprised us enormously. We had no idea. No idea."

The English scientist turned and nodded to his American colleague, James Watson, who readily agreed with Crick's assessment. Together in 1953 at Oxford University, with the unwitting "assistance" of X-ray crystallographer Rosalind Franklin, they discovered the relationship between an enormous and strange molecule, deoxyribonucleic acid, and human genetics. They proved that DNA contained the genetic code of life, a discovery for which the men shared the Nobel Prize in Medicine.[1]

Reflecting thirty years later on the revolution that had transpired since, Watson said, "None of this could have been predicted. Now it's hard to imagine things going any faster. But it will be faster. Mysteries will tumble. All is now open to experimental attack, and problems we can't even foresee today will be identified and solved within less than a decade."

His prophesy would prove remarkably accurate, as massive global computer interconnections and fax machines would become the preferred method of sharing the excitement of biological discovery, the pace becoming so furious that by the mid-1980s most researchers would consider journal publication of their findings a matter of historic obligation, rather than a primary method for informing their colleagues. By the time results were published, most molecular biologists would already be two or three experiments further along in their laboratories.

Since the early 1970s, biologists had been working on ways to chop up DNA and RNA in order to figure out what various pieces of the genetic code actually controlled. It had been determined that nearly all living systems had repair mechanisms to fix damaged DNA. In addition, they knew that something in the DNA regulated when the genes for, say, growing fingers were turned on and when they were switched off. There was also a sense that the malfunctioning of such genetic signals lay at the core of the causation of cancer, because tumor cells seemed to behave as if all their internal policing mechanisms were out of control.

Scientists soon realized that the world of DNA was replete with specialized proteins that busily moved up and down the vital molecule's lengthy sequence performing a myriad of tasks, ranging from snipping out a single defective nucleotide to making a copy of an entire DNA molecule, or chromosome. These proteins, which were themselves made according to instructions inside the DNA, were the key to regulation of the massive genetic code. Like switching signals in a vast computer data base that ensured desired bits of information were displayed on the VDT screen when—and only when—the human user wanted to see them, these proteins, particularly a group known as restriction enzymes, made sure that genes were expressed only when necessary, cut out if troublesome, inserted if needed, and remained silent information stored in the DNA data banks at all other times.

The world's top molecular biologists concluded that the best way to decipher DNA was to manipulate these regulatory proteins and see what effect removal of this or that piece of DNA might have on the virus, bacteria, or cell it controlled. Scientists like Stanley Cohen and Paul Berg at Stanford University made batches of these proteins, mixed them with DNA, and watched the results. Berg and Cohen soon figured out how to excise minute, discrete pieces of DNA with the precision of molecular surgeons. They also learned how to insert genes into DNA sequences by opening up the DNA, attaching the desired segments, and then allowing the DNA to recombine, the new gene now having been incorporated.[2]

In late 1973, Berg hit on an idea: put genes into harmless viruses, let the viruses infect cells, thereby carrying the genes inside. The genes might then be recombined into the cellular DNA. This was especially easy to achieve by using bacteriophages—viruses that infect bacteria—to carry experimental genes into well-understood, simple organisms, such as the *Escherichia coli* bacterium. Berg envisioned treating genetic deficiency diseases one day through just such a mechanism.

Berg's idea not only worked but caused an international upheaval in biology. Within a year every molecular biologist who could get her or his hands on the proper chemicals and viruses was using the genetic engineering technique to study life in a test tube. But Berg worried that his own experiments, using monkey virus SV40 to carry genes inside *E. coli* bacteria, could be dangerous, and in 1974 convened a meeting of the world's preeminent biologists to establish safety rules for their enterprise.

While some critics would attack genetic engineering research as ungodly or risky, the field was as unstoppable as a speeding locomotive. What was considered experimental in 1976 was routine by 1979. The SV40 experiments Berg and Cohen fretted over in 1974 were graduate student training exercises by 1980. And the term "genetic engineering" was transformed from an almost whimsical description of a handful of experiments performed in a few select laboratories during the mid-1970s to a label applied to a global multibillion-dollar industry in the 1980s.

For the disease detectives the revolution was a mixed blessing: on one hand it offered new tools for solving microbial mysteries, but it was also immediately obvious that funding—never generously available to parasitologists or infectious disease researchers—was becoming even scarcer as resources shifted toward molecular pursuits.

Bright, young scientists followed the excitement—and the money. And why not? Clearly, in 1976 the opportunity to work, for example, as one of twenty-four postdoctoral fellows in the MIT laboratory of Nobel laureate Har Gobind Khorana manufacturing the first fully functional man-made gene was a great deal more seductive pursuit than joining a team that used old-fashioned light microscopes to count the number of malarial sporozoites in a mosquito.

Yet what the microbe hunters learned when they applied their newly honed genetic manipulation skills to the task only heightened their sense of concern. They soon discovered that microbes could share genes with one another that made them more formidable human enemies; many viruses not previously thought to do so could cause cancer; some microbes possessed the ability to chemically manipulate the human immune system to their advantage; and there were viruses that could hide for years on end inside human DNA.

It was Barbara McClintock who first suggested that genetic signals could move about, be mobile, producing changes in the fated appearance of an organism. During the 1940s and 1950s, well before Watson and Crick discovered the link between genes and the structure of DNA, McClintock studied maize plants at the Cold Spring Harbor Laboratory on Long Island, New York. She showed that genes could move from one position to another on maize chromosomes, causing radical changes in the appearance of corn kernels. The cause of these differences would not be inherited genes per se, but the movement or transpositioning of those genes. The movable genes were dubbed transposons.[3] Only years later would the full impact of her pioneering efforts finally be evident, and McClintock would be awarded the 1990 Nobel Prize in Medicine.

A decade after McClintock discovered transposons in maize, Joshua Lederberg showed that bacteria had movable bits of DNA that conferred the ability to resist antibiotics. And by the 1970s, when Berg and Cohen invented the techniques of genetic manipulation, scientists all over the world realized that certain bacterial genetic traits commonly jumped about from place to place within a cell's chromosome, or between bacteria. These were not rare events. In fact, it seemed that at the bacterial level, genetics, far from being the rigid blueprint envisioned less than a decade earlier, was more akin to a game of Scrabble in which each organism came into existence with a finite set of letter tiles, or genes, but jumbled those tiles around according to a set of rules creating a vast variety of different words.[4]

These Scrabble tiles of movable genes could be in the form of discrete packages of DNA that moved about along the bacterial genome—McClintock's transposons. They could be singular genes that appeared to leap about almost at random, designated "jumping genes." Or they could be highly stable rings of DNA, called plasmids, that sat silently in the bacterial cytoplasm waiting to be stimulated into biochemical action.

It became alarmingly obvious that microbes used this constant game of genetic Scrabble to their advantage in a variety of ways. Bacteria could occasionally undergo a process called sexual conjugation, stretching out portions of their membranes to meet one another and passing plasmids, transposons, or jumping genes—including genes that conferred resistance to antibiotics.

Naturally, if humans could manipulate the Scrabble game to their advantage in the laboratory, so could the microbes in the real world. It wasn't a long intellectual leap from jumping bacterial genes, for example, to viewing viruses as well-packaged transposons capable of corralling the genetic resources of the bacterial, or even human, cells they invaded.

The quintessential example of Lederberg's notion of genetic entanglement was discovered by Howard Temin at the University of Wisconsin in Madison and David Baltimore at Massachusetts Institute of Technology: retroviruses. These tiny RNA viruses were unique in that they gained entry into cells and made reverse mirror-image copies of their RNA (running backward compared with the normal course of events) to produce a DNA version of their genes. And then they exploited vulnerable locations along the host's DNA to insert themselves, like a transposon, into the cell's genetic material. The retroviruses accomplished this feat through the use of a unique enzyme called reverse transcriptase, which performed the mirror-image flip of viral RNA genes into DNA.

Shortly after the discovery of retroviruses, National Cancer Institute scientists Robert Huebner and George Todaro proposed a theory to explain the ability of these viruses to cause cancer. They suggested that there were places along animal chromosomes where transposons rarely went, and into which a viral insertion could spell cellular disaster. According to their hypothesis, if a retrovirus inserted itself near certain host genes, those cellular segments of DNA would be switched on, and they, in turn, would cause wild cell growth and misbehavior—the hallmarks of cancer. Driving their theoretical point home, Huebner and Todaro named these cellular DNA sites of special viral vulnerability "oncogenes."

Baltimore believed in oncogenes. He also believed that retroviruses were capable of inserting themselves permanently in animal germ line DNA, right alongside these oncogenes, and being passed on in that form via sperm or eggs to the next animal generation. In this way, he reasoned, virally induced cancers could be inherited. Baltimore cautiously predicted that human retroviruses would be found that, as theorized by Huebner and Todaro, triggered cellular oncogenes.

Having shared the 1975 Nobel Prize with Howard Temin and another leading microbiologist, Renato Dulbecco, Baltimore turned his attention broadly to the role of retroviruses and the more traditional RNA viruses in cancer.

"What is cancer?" he asked in 1978.[5] "This question is at the heart of present efforts to control this disease, and the most manipulable model systems for studying it have been virus-induced cancers. That viruses cause cancer in animals is a certainty; that they do so in humans is less certain but probable."

Temin and Baltimore, working independently, had already shown that two retroviruses caused cancer in animals: Rous sarcoma (in chickens) and Rauscher mouse leukemia viruses. Other animal retroviruses, by virtue of their ability to get inside and disrupt cellular DNA, were shown to be associated with cancer: avian leukosis virus (leukemia in chickens), Moloney leukemia virus (in mice), Kirsten sarcoma virus (in mice), Gibbon ape leukemia virus, cow and feline leukemia viruses, visna virus (in sheep), mammary tumor virus (in mice), and a host of so-called foamy viruses (found in monkeys, cats, and cattle).

Faced with these discoveries, Joshua Lederberg said that the only reasonable way to look at viruses was to recognize that "the very essence of the virus is its fundamental entanglement with the genetic and metabolic machinery of the host."[6]

During the early 1980s, the genetic engineers discovered that those genetic entanglements could be deliberately manipulated in hundreds of different ways, allowing scientists to learn what tasks a given gene sequence normally performed by moving, switching off, turning on, or mutating that sequence. This could be done by inserting artificially constructed plasmids into cells, or by attaching genes to bacteriophages—minuscule viruses that infect bacteria.[7]

In California, Michael Bishop and Harold Varmus were in pursuit of oncogenes. In their laboratories at the University of California, San Francisco, long-haired, bearded Michael Bishop and his taller, leaner bespectacled counterpart Harold Varmus formed a Mutt-and-Jeff team that zeroed in on the Rous sarcoma virus. It was such a potent cancer-causing agent that all chicken muscle cells in petri dishes could be transformed to cancer cells within twenty-four hours of infection. Researchers at Rockefeller University had previously discovered that the virus contained a gene they called *src* (for "sarcoma") that seemed to cause the tumor transformation of infected cells.

Between 1976 and 1983, Bishop and Varmus discovered that *src* was, indeed, a potent cancer-causing virus product that was a near-duplicate of a gene normally present in chickens. To differentiate between the two, Bishop and Varmus designated the viral oncogene *v-src* and the normal cellular oncogene *c-src*.[8] The pair of energetic young researchers then asked just how widespread was the *c-src* oncogene in the animal world. To the surprise of many, they quickly discovered *c-src* in the DNA of other birds, animals, insects, and humans.[9]

Why would humans and chickens share a common gene—one that caused cancer, no less? Varmus and Bishop quickly discovered that *c-src* was the genetic blueprint for the manufacture of a protein that ended up nestling on the inner lining of the cell membrane. There, it acted as a kinase, chemically altering passing proteins by adding phosphate ions to specific amino acids. This radically changed the biochemical reactivity of the proteins, and the impact was so profound that nearly every aspect of cell structure and activity was adversely affected. The discovery "sent the thrill of recognition down the spines of biochemists," Bishop said,[10] because they had long recognized that nearly every essential activity inside a human or animal cell was affected by phosphorylation.

Other researchers quickly discovered that the same pattern held true for a variety of cancer-causing retroviruses: the viruses carried genes that mimicked oncogenes that were commonly found in the DNA of all animals, humans, even insects. And those oncogenes controlled very powerful enzymes that could alter hundreds of different essential proteins inside cells, causing the cells to transform into cancer.

"The genes of retroviruses assume principles that are very similar to what we call jumping genes," Varmus explained. "And they, too, have evolved mechanisms for getting around, for picking up new genes, for making mutations. And carrying out evolutionary changes."

The retroviral genes "jumped" better along the cellular genome than did the "garden-variety oncogenes" inside the cell, Varmus asserted. And they had the ability to insert themselves into host DNA, reproduce right along with the host cell, and, as Varmus put it, "carry out God-knows-what."

Scientists hypothesized that normally oncogenes were switched on only at given times in an animal's development. For example, as a fetus grew, such wild cellular activity might be key to its development from a fertilized egg to a baby.[11]

Bishop hypothesized that these oncogenes acted "as a keyboard on which many different carcinogens can play, whether they be chemicals, x-rays, the ravages of aging, even viruses themselves. With the revelation that there were a limited number of genes in cells that were affected, it became natural to see them as a keyboard on which many different causes of cancer play. It's not an endless keyboard—it's a keyboard of perhaps less keys than a standard piano keyboard. And out of this comes the manifestation of cancer—the melody, if you wish. An enemy has been found—it is part of us—and we have begun to understand its lines of attack."

The discovery of oncogenes would cause a shift in thinking among cancer experts worldwide, prompting many to wonder for the first time just how many human tumors were started by microbes.

And, sure enough, in 1979 researchers at the U.S. National Cancer Institute, the Tokyo Cancer Institute, and Kyoto University discovered a retrovirus that caused cancer in human beings. Dr. Robert Gallo and his NCI colleagues found evidence of a virus inside the T cells (disease-fighting white blood cells) of a twenty-eight-year-old African-American man who had come to Bethesda, Maryland, in 1979 from his Alabama home for experimental cancer treatment. The NCI group quickly found two other individuals who suffered T-cell lymphomas and seemed to be infected with a virus: an immigrant woman from the Caribbean and a Caucasian man who had traveled extensively in the Caribbean and Asia.

Two years earlier Kiyoshi Takatsuki, an epidemiologist with the Tokyo

Cancer Institute, had discovered groups of people living on outer Japanese islands who apparently had cancer involving their immune systems' T cells.[12] The Japanese researchers dubbed the disease adult T-cell leukemia or ATL. Gallo's laboratory isolated their virus and named it HTLV, or human T-cell leukemia virus.[13] The Gallo group also identified the existence of an oncogene in the HTLV virus that gave the microbe the ability to produce leukemia.[14] Attempts at collaboration between the Japanese and American researchers went awry and Yorio Hinuma and Mitsuaki Yoshida of Kyoto University announced discovery of a different virus in the Japanese leukemia patients, named ATLV, or adult T-cell leukemia virus.[15]

Ultimately, Mitsuaki Yoshida led a Tokyo Cancer Institute study in 1980 that compared ATLV and HTLV and found them identical. They furthermore showed that Japanese monkeys (*Macaca fuscata*), Indonesian rhesus monkeys, and African green monkeys captured in Kenya and held in captivity in Germany had antibodies to ATLV/HTLV, and that the virus—or a monkey version of the human virus—could be transmitted from one co-caged animal to another.[16] The finding posed several questions, the researchers wrote, including "Are monkeys the natural reservoir of ATLV? Is ATLV transmissible from monkeys to humans through a certain vector? What is the mode of infectious transmission of ATLV in monkeys?"[17] The finding, and questions it posed, would be echoed with other diseases in coming years.

The following year, 1981, David Golde at UCLA found a patient who was suffering from a particularly aggressive type of blood cancer, hairy-cell leukemia, so named because the damaged white blood cells appeared "hairy" under the microscope. Golde discovered that something in the blood of this patient was capable of producing the "hairy" effect on human T-cell lymphocytes grown in the laboratory. Golde named the patient's cell line MO.[18]

Several scientists wondered why Golde's cell line grew so well in test tubes, since heretofore it had been nearly impossible to raise human T cells in the laboratory. Robert Gallo and UCLA's Irvin Chen both thought the lab growth capability plus evidence that "something" from the MO cells could transform other human lymphocytes indicated that an infectious cancer-causing agent was involved.

The hunt was on.

Chen discovered a second cancer-causing retrovirus in the MO cells, which was dubbed HTLV-II (Gallo's first virus was then redesignated HTLV-I).[19] Chen concluded that HTLV-II had no oncogene, however, and the cancerous behavior of the MO cells seemed to be caused by a defective form of the virus that had emerged in the laboratory as a result of culturing conditions. The finding was confirmed within weeks in three other laboratories.[20]

The impact of these findings was striking. The U.S. National Cancer Institute, for example, would quickly shift resources toward cancer virology, encouraging scientists to search for other cancer-causing human viruses and to further elucidate the link between oncogenes and microbes.

"We have found oncogenes. We have sequenced those oncogenes. And we have learned that we have these genes in our human genomes normally. That's both frightening and exciting," National Cancer Institute director Dr. Vincent De Vita said in 1981. "We've put one billion dollars into viral oncology research. Jim Watson asked me to say was it useful or not. What value would I place on it? Every nickel we've spent or committed so far has been worth it. We've had dividends beyond imagination."

That year De Vita ordered all the work in the NIH's Frederick Laboratory facility switched to the pursuit of links between viruses, oncogenes, and cancer. Thanks to the new molecular biology technologies, it was now possible to conduct such searches with a reasonable degree of speed and efficiency. One segment of known DNA or RNA from, for example, HTLV-I could be used as a probe to search quickly for the presence of its genetic mates in all sorts of animal and human cells.

The notion that cancer could result from a contagious process was extraordinary, particularly in view of how hard cancer patients and scientists had fought for centuries to dispel precisely that notion. Ever since medical science had learned to differentially diagnose cancer, people had feared the disease's victims. Prejudice and shame often went hand in hand with the biological horrors of cancer.

That cultural perspective had begun to shift in the 1960s when the public recognized the link between cancer and a host of chemical toxins, particularly those contained in smoked tobacco. While fears of contagion were erased, they were replaced by apprehension and a considerable amount of anger directed at the sources of chemical carcinogens.[21] During the mid-1970s, most Western countries had erected government infrastructures devoted to the regulation and control of human exposure to such chemical carcinogens, monitoring food, water, air pollution, pesticides, auto emissions, industrial waste, housing materials, and so on.

By the time molecular biologists zeroed in on oncogenes and retroviruses, the political and consumer power of the environmental movement was quite considerable, particularly in North America and the Scandinavian countries. That explains Michael Bishop's hesitancy to overemphasize the role of viruses in causation of human cancers. His "keyboard" metaphor for triggering oncogenes with a variety of carcinogens—hormones, chemicals, and microbes—was an important way to reconcile the previous emphasis on chemical origins of cancer with the new insights into viral mechanisms of pathogenesis.

In years to come epidemiologists would strive to understand how such viruses were spread, who gave HTLV-I, for example, to whom. Japanese and German researchers would discover antibodies to HTLV-I in African monkeys and chimpanzees, as well as hunters in Kenya.[22] It would quickly be apparent that HTLV-I infections of human beings were clustered in populations not only in Japan[23] and the Caribbean[24] but also in Surinam[25] and Italy.[26]

Harvard Medical School virologist Bernard Fields, who tried to perform studies of viral disease agents simultaneously on the micro, laboratory level and the macro, clinical scale, wondered just how relevant all this gene jumping was to human health. He likened viruses to spaceships on a voyage in to a hostile environment. Their payload—the viral DNA or RNA, replete with all its jumping gene and transposing potential—was hidden inside a delivery system that, in Fields's analogy, included propulsion and navigation systems and a protective capsule capable of withstanding the viral equivalent of an Apollo spacecraft's heated reentry into Earth's atmosphere. To gain access to its target cells in the human bloodstream, liver, brain, or whatever organ it was designed to infect, the virus had first to pass through significant hostile territory: the skin, intestinal lining, mucosal barriers in the reproductive tract, protective linings of the nose, mouth, and lungs, and the blood/brain barrier that barred entry to the central nervous system. Fields sought to keep concerns about the newly discovered viruses in check, insisting that the tiniest of microbes would die out if they mutated radically because they would, in the process of mutation, damage their vital payload and delivery systems.

A few scientists focused their attention on the origins of such viruses. Gallo, for example, forwarded the hypothesis that the HTLV-I and HTLV-II viruses made their way around the world along the shipping routes pioneered by Magellan and slave traders. He and Yamamoto asserted that the viruses originated in African monkeys, spread somehow into the human population, and then found their way around the world via sexual transmission between slaves.[27] An alternative theory was that the virus originated in Africa and was carried by Portuguese sailors to the port city of Kyushu during the sixteenth century.[28]

A few months before the discovery of HTLV-II was publicly announced, the U.S. National Cancer Institute and the Tokyo Cancer Institute held their 1982 annual Japanese-American cancer meeting, focusing on HTLV-I. The Japanese had trouble narrowing the pool of HTLV-I researchers down to the limit of seven participants who could attend the elite gathering. Research on HTLV-I had exploded in Japan, with over a dozen large laboratories attacking the scientific problem.

For the Americans, the reverse was the case. Besides Gallo and his staff at the National Cancer Institute, nobody was devoting much attention to HTLV-I, and many leading cancer experts in January 1982 pooh-poohed the significance of the virus. Since the meeting was to be chaired by the Americans, the National Cancer Institute was at pains to find a leader for the gathering who was generally familiar with cancer-causing retroviruses. The institute selected Harvard virologist Myron "Max" Essex, who was one of the world's experts on the feline leukemia virus (FeLV), a retrovirus that caused cancer in cats.[29] The Rhode Island-born Yankee was both a trained veterinarian and microbiologist. At the time the Japanese-American meeting took place, Essex was the newly-appointed Chair of the Department of Cancer Biology at the Harvard School of Public Health.

As the meeting unfolded it was clear that the Japanese were studying HTLV-I at a feverish pace, and had made impressive strides in elucidating the relationship between the virus and genesis of hairy-cell leukemia.

Gallo was not happy.

"I can't even get people in my own lab interested in working on this virus," Gallo told Essex. "Nobody in the U.S. is taking this thing seriously, but the Japanese are working like crazy on it. They're pulling ahead of us."

Essex acknowledged that the range and quality of data that various Japanese scientists had presented were quite good. Gallo leaned over and looked earnestly into Essex's eyes.

"Max, you've got to get involved."

Essex protested that his lab was already overwhelmed with work on other cancer-causing viruses, notably FeLV and hepatitis B,[30] which produced liver tumors. But Gallo's insistence won him over.

Essex applied the tools his lab had developed for studying T-lymphocyte responses to the cat virus to answer questions about how the human immune system reacted to HTLV-I. He soon demonstrated that, as was the case with FeLV in cats, humans infected with HTLV-I had aberrant immune systems. In particular, their T cells were suppressed or deficient in number, leading to an overall inadequacy of the entire immune system.[31]

Researchers from the Tokyo Cancer Institute showed that 100 percent of the Japanese islanders who had hairy-cell leukemias were infected with HTLV-I. But about 12 to 15 percent of the adult residents of the area were also infected, without having cancer: they did, however, suffer a range of immune system disorders.

Essex was convinced that these striking similarities between HTLV-I and FeLV pointed to some distant time when the viruses moved between host species. Similarly, he was convinced that hepatitis B viruses in various animal species all evolved from a common ancestor: the genes of the human virus were over 40 percent identical to those of a liver-cancer-causing virus found in, of all things, woodchucks.[32]

In both species, it would later be shown, the virus caused nearly all hepatocellular carcinomas; perhaps 100 percent of such tumors in the woodchucks and about 90 percent of those in human beings. Worldwide surveillance would eventually reveal that millions of people were infected with the hepatitis B virus, about 15 percent were chronically ill, and as a result, perhaps five million developed liver cancer every year.[33] Hepatitis B was not a retrovirus, of course, but a large virus whose genetic material was in organized segments of DNA. Scientists had no idea how the virus caused cancer, and there was no clear link between hepatitis B and any known oncogenes.

By 1980 there was also strong evidence linking some other DNA viruses to human cancer. As early as the 1960s, Denis Burkitt, a British physician working in Uganda, had noticed that a certain type of lymphoma was extremely common in East Africa, and that its distribution in the human population seemed to follow a clustering pattern: whole families or villages might be afflicted in one area, while virtually no cases of the cancer could be found in a nearby village. He hypothesized that the disease was caused by a transmissible virus.[34] British researchers Michael Epstein and Y. M. Barr discovered a new type of herpes virus in cells from Burkitt's lymphoma patients.[35] The tumor was dubbed Burkitt's lymphoma, the virus Epstein-Barr virus or EBV. Like hepatitis B, EBV was a fairly large DNA virus and scientists could find no immediate explanation for how it caused the lymphomas. Similarly, human papillomavirus was linked to genital cancers, particularly cervical carcinoma.[36]

Though much about the connection between viruses and cancer remained obscure, it was an accepted tenet of biology by 1982 that viruses could directly, or perhaps through intermediary chemicals or host genes, cause the changes in cells that were the hallmarks of cancer. It was also generally accepted that such viruses might take years to produce clinically noticeable symptoms in those humans or animals who were infected. Thus, the concept of *slow* viruses had emerged—an idea epidemiologists found extremely challenging because of the difficulty of showing that a population of people have cancer today due to a virus they were exposed to ten or twenty years ago.

The remarkable genetic similarities between oncogenes found in all animals, humans, even insects seemed to signal a commonly shared point of vulnerability in a huge range of the planet's fauna. If a virus adapted to infect, for example, a monkey and deftly switch on the simian's oncogene, could it not also, with some evolutionary or rapid mutation, gain the ability to enter human cells and switch on the nearly identical *Homo sapiens* oncogene?

Given the slow pace of the disease process produced by such viruses, and the ability of some to hide inside animal or human DNA, these microbes were extremely difficult to detect.

How many more might exist in nature?

How many types of cancer might prove to be caused by such viruses? Were there other diseases slow viruses might be causing right under the medical establishment's nose?

In a very short time scientists would unearth frightening answers to their collective inquiry.

9

Microbe Magnets

URBAN CENTERS OF DISEASE

*When one comes into a city to which he is a stranger, he ought
to consider its situation, how it lies as to the winds and the rising of the sun; for its influence is not the same whether it lies
to the north or to the south, to the rising or to the setting sun.
These things one ought to consider most attentively, and concerning the waters which the inhabitants use, whether they be marshy
and soft, or hard and running from elevated and rocky situations,
and then if saltish and unfit for cooking; and the ground, whether it
be naked and deficient in water, or wooded and well-watered, and
whether it lies in a hollow, confined situation, or is elevated and
cold . . .*

*From these things he must proceed to investigate everything
else. For if one knows all these things well, or at least the greater
part of them, he cannot miss knowing, when he comes into a
strange city, either the diseases peculiar to the place, or the particular nature of the common diseases, so that he will not be
in doubt as to the treatment of the diseases, or commit mis-
takes, as is likely to be the case provided one had not previously
considered these matters. And in particular, as the season and
year advances, he can tell what epidemic disease will attack
the city, either in the summer or the winter, and what each individual will be in danger of experiencing from the change of
regimen.*

—Hippocrates, *On Airs, Waters, and Places, c.* 400 B.C.[1]

I

In 6000 B.C. there were fewer humans on earth than now occupy New York
and Tokyo. Earth's roughly 30 million prehistoric residents were scattered
over vast expanses of the warmer parts of the planet, and few of them ever

ventured far from their birthplace. According to what little archaeological information and scientific conjecture is available, their microbial threats came primarily from parasites in their food and water or were carried by local insects.

Over the next 4,000 years the human population slowly increased and people congregated around rivers, ocean ports, and sites of rich food resources. Trade routes emerged, connecting the nascent urban centers, and the city's residents thrived off their merchants' exploits and the taxes they levied on their poorer rural subjects.

By the time the Egyptians ceased building pyramids, around 2000 B.C., there were several cities with thousands of inhabitants each: Memphis, Thebes, Ur—the religious or political capitals of empires. And by 60 B.C. the vast empires of Rome and China boasted urban centers of tens of thousands of people, which functioned as the hubs of trade and culture for the planet's 300 million residents.

By 5 B.C. Rome's 1 million residents consumed 6,000 tons of grains a week. After the fall of the Roman Empire, no city would again attain such a size for 1,800 years, when London would become the largest metropolis in history up to that time.[2]

Cities afforded the microorganisms a range of opportunities unavailable in rural settings. The more *Homo sapiens* per square mile, the more ways a microorganism could pass from one hapless human to another. People would pass the agent to other people in hundreds of ways every minute of every day as they touched or breathed upon one another, prepared food, defecated or urinated into bodies of water with multiple uses, traveled to distant places taking the microbes with them, built centers for sexual activity that allowed microbes to exploit another method of transmission, produced prodigious quantities of waste that could serve as food for rodent and insect vectors, dammed rivers and unwittingly left cisterns of rain water about to create breeding pools for disease-carrying mosquitoes, and often responded to epidemics in hysterical ways that ended up assisting the persistent microbes.

Cities, in short, were microbe heavens, or, as British biochemist John Cairns put it, "graveyards of mankind."[3] The most devastating scourges of the past attained horrific proportions only when the microbes reached urban centers, where population density instantaneously magnified any minor contagion that might have originated in the provinces. And microbes successfully exploited the new urban ecologies to create altogether novel disease threats.

Warfare, trade, the occasional need to put down local peasant uprisings during times of elevated taxation or famine, religious pilgrimages, and the seductive lure of the city for adventurous youth guaranteed that continuous cycles of new microbial invasions would beset urban populations which generally lacked protective immunity.

The microbes' transmissive success was guaranteed among a city's poor, and every urban center had its marginalized neighborhoods where malnourished, immunodeficient people lived in high-density squalor. Urban poverty and disease went hand in hand not only because insufficient diets weakened people's immune systems but also because of their living conditions. If the Roman patricians occasionally suffered dysentery because of bacteria in the aqueducts, the plebeians downstream were guaranteed a doubled exposure due to the additional bacterial burden of the patrician's contaminated waste.

The life expectancy of ancient Rome's populace was far shorter than that of the Empire's citizens in rural Mediterranean or North African areas. Only about one of every three Roman residents saw the ripe old age of thirty, compared with 70 percent of their rural counterparts. Virtually nobody in the city lived to eighty, whereas about 15 percent of the pastoral citizens attained that goal.[4]

Ancient urbanites recognized some of these special hazards. Accounts going back 2,000 to 4,000 years tell of scourges carried by lice, bedbugs, and ticks—all disease-associated insects that the writers noted were more abundant in the dense housing conditions of the cities. Though their understanding of the relationship between these insects and specific diseases was muddy, writers in ancient Egypt, Greece, Rome, India, and China all drew attention to the insect problem. Similarly, Galen in Athens and Herodotus in Rome recognized a connection between the expansion of their cities into marshy areas and the increase in malaria.[5] Chinese records dating back to 243 B.C. also noted massive epidemics—claiming millions of lives—which arose constantly from the cities of China's far-flung empire.[6]

On the basis of historical accounts from Greece, Rome, Europe, and the post-Columbian Americas, twentieth-century scholars have tried to interpret which diseases plagued ancient urban centers. For example, during the Peloponnesian War of 430 B.C. a devastating epidemic hit Athens, probably imported by returning soldiers. Thucydides said of it, "No scourge so destructive of human life is anywhere on record. The physicians had to treat it without knowing its nature, and it was among them that the greatest mortality occurred."

It was later thought that the epidemic, which Thucydides said caused illness in every Athenian and killed up to half the population, was either typhus, the plague, or smallpox.[7] Hundreds of great global pandemics followed. Four diseases that seemed to William McNeill and other medical historians of the 1970s to have gained particular benefit from the urban ecology over the previous 2,000 years were pneumonic plague, leprosy (Hansen's disease), tuberculosis, and syphilis. As far as could be discerned from historical records, these were rarely—if ever—seen prior to the establishment of urban societies, and all four exploited to their advantage human conditions unique to cities.

The world has experienced at least two great pandemics of bubonic/pneumonic plague, a disease caused by the *Yersinia pestis* bacterium—carried by fleas which resided on rodents, particularly rats. Though the bacterium has never been eradicated, ideal ecological conditions for its rapid spread among *Homo sapiens* occurred only a handful of times in recorded human history. Once *Y. pestis* got into the human bloodstream, either via a flea or rat bite or by inhalation of the bacterium, it quickly made its way into the lymphatic system. There, the bacterium killed massive numbers of cells, giving rise to formation of often grotesque pustules and pus-filled boils. Bacteria produced in these infected sites then migrated to the liver, spleen, and brain, causing hemorrhagic destruction of the organs and demented behavior that during the Middle Ages was interpreted as intervention by Satan.

The occasional case of plague was seen during the twentieth century,[8] but well before humanity had invented antibiotics to treat it, the disease had ceased to threaten further pandemics.

Sometime around 1346 the Black Death began on the steppes of Mongolia: infected fleas infested millions of rodents which, in turn, raided human dwellings in search of food. Why the disease emerged that particular year was never clear, though scientists in the 1980s speculated that the weather may have favored a rodent population explosion. The disease made its way rapidly across Asia, carried by fleas that hid in the pelts of fur traders, the blankets and clothing of travelers, and the fur of rodents that stowed away aboard caravans and barges crisscrossing the continent. Rumors of the Asian scourge preceded its arrival in Europe, and it was said that India, China, and Asia Minor were literally covered with dead bodies.[9] The Chinese population plummeted from 123 million in 1200 to 65 million in 1393, probably due to the plague and the famine that followed.

It reached the prosperous European trading port of Messina, Sicily, in the fall of 1347 aboard an Italian ship returning from the Crimea, and immediately exploited the city's ecology. Rats from the plagued ship joined the abundant local rodent population. Ailing men from the ship passed the bacteria on to the Messina citizenry directly, exhaling lethal microbes with their dying gasps.

As the plague made its way across Europe and North Africa, each city anticipated its arrival and tried by a variety of means to protect itself. Travelers were barred entry, drawbridges were raised to seal the wealthy urbanites off from their less fortunate peasantry, great purges and outright slaughter of tens of thousands of Jews and alleged devil worshippers were staged. The city of Strasbourg alone savagely slew 16,000 of its Jewish residents, blaming them for spreading the Black Death.[10]

Some who had no scapegoats blamed the plague on their own lack of piety. The Brotherhood of Flagellants were Christian men who daily beat themselves to the edge of death to purge the sins that were responsible for their disease. All over Europe, these men, encouraged by crowds of crazed aristocrats and peasants alike, would thrash themselves with leather whips embedded with small iron spikes.

The terrorized European population did everything save what might have spared them: ridding their cities of rodents and fleas. The cities fell not only because of rat infestations but also due to both human population density and hygienic conditions. Bathing was thought to be dangerous, and few Europeans ever washed, making them fertile ground for flea and lice infestation.

The pneumonic form of the plague, which rarely spread in less populated rural areas, was easily transmitted inside the densely populated medieval cities. Once a rat-driven bubonic form took hold, pneumonic cases in humans soon appeared, spreading the disease with terrifying rapidity.

Each city would be in the grip of the disease for four or five months, until the susceptible rats and humans had died. The survivors would then face famine and economic collapse, caused by the sharp reduction in workforces.

The daily death rates were staggering: 400 in Avignon; 800 in Paris; for Pisa, 500; Vienna buried or burned 600 bodies per day; and Givry, France, 1,500 daily. By the end, London, with a pre-plague population of 60,000, had lost 35,000. Half of Hamburg's and two-thirds of Bremen's populations perished. Most historians believe that at least one-third of Europe's total human population (20 to 30 million people) died of the plague between 1346 and 1350.[11] The highest per capita losses were consistently in the cities.

Over subsequent centuries, there were numerous outbreaks of urban plague in Europe, Asia, and the Middle East, though few spread far beyond the cities due to quarantines and to slow improvements in hygienic conditions.

In 1665, London suffered the Great Plague, a *Yersinia* epidemic that claimed over 100,000 lives in a year's time. The epidemic began a year earlier, probably in Turkey, and was carried by trading ships to Amsterdam and Rotterdam and on to London during the winter of 1664–65. By that summer as many as 3,000 of the city's residents perished each day.

The royal family and the aristocracy fled at the first sign of pestilence, taking up residence in the English countryside. The residents of London, the world's largest and most densely populated city, were left to fend for themselves. They lived in thatch-roofed, brick, and wood row houses: an ecology made in heaven for rats.

In considering the pestilence a generation later, Daniel Defoe recommended that city authorities in the future

. . . not let such a contagion as this, which is indeed chiefly dangerous to collected bodies of people, find a million of people in a body together, as was very near the case before. . . . The plague, like a great fire, if a few houses only are contiguous where it happens, can only burn a few houses; or if it begins in a single, or, as we call it, a lone house, can only burn that lone house where it begins. But if it begins in a close-built town or city and gets a head, there its fury increases: it rages over the whole place, and consumes all it can reach.

Shortly after the plague subsided, in 1666, London was overcome by a real fire that engulfed most of the city. McNeill believed it was the Great Fire which stopped the Great Plague, burning off the thatch roofs, which were replaced with tile and slate.

Leprosy, as it was then called, claimed only a fraction of the lives felled by the plague, but was no less feared. Throughout history leprosy was dreaded more for its disfiguring and crippling effects on the human body than for its slow capacity to kill.

By the 1970s leprosy would be referred to as Hansen's disease (named after Armauer Hansen, who in 1873 described the first definitive differential diagnosis of the disease) by those who wished to separate the bacterial ailment from the centuries of horror and prejudice that went with the word "leper."

There was great debate in the latter half of the twentieth century about the age of the *Mycobacterium leprae* organism and how long it had been producing significant disease in human beings. Though the Bible referred to ancient Hebrews suffering disfiguring diseases often translated as leprosy, the usually meticulous records of Egyptian scribes bore no hint of it. Searching for evidence of bone damage produced by the gnawing bacteria, studies of skeletons revealed no sign anywhere in the world prior to A.D. 500, when apparently leprotic bones were buried in the graveyards of Cairo, Alexandria, and parts of England and France.

A previously unrecognized disease did, however, sweep over Europe.[12] Leprosy seemed to follow the rise of European cities during the medieval period, reaching a peak sometime around 1200. Nobody was certain then, or now, exactly how the fussy, slow organism was passed from one person to another. It obviously required close contact, but may have originally been more easily transmitted among the then totally nonimmune *Homo sapiens*. Once in a person's body, however, *M. leprae* attacked the nerves and skin cells of cooler, peripheral parts of the body, causing them to go numb, weaken, and often be destroyed as a result of unfelt injuries. The disfigurement that resulted from loss of fingers, toes, ears, noses, and other external body parts marked "lepers" as targets for stigmatization and fear.

By 1980 most of the world's five billion humans had antibodies to *M. leprae*, proving they'd been exposed without apparent harm.

But in medieval Europe leprosy took a high toll and seemed to spread rapidly through the congested cities. Some biologists in the 1980s theorized that factors unique to medieval urban life helped promote the mycobacterium's spread, including the lifetime avoidance of bathing, always wearing wool rather than cotton clothing, and the practice of sharing bedding to stay warm.

Whatever the case, European leprosy died out with the Black Death of 1346. Nobody was certain why this was so, but it was generally suspected that the Black Death decreased the human density of urban areas, thus reducing human-to-human contact. It may also have been possible that plague survivors' immune systems were less susceptible to a broad range of bacteria, including both *Yersinia* and *M. leprae*. Or conversely, those who were vulnerable to leprosy may have also been less able to respond to a range of other bacterial assaults.

In leprosy's place, exploiting the post-plague urban chaos, came tuberculosis. Unlike the leprosy bacterium, *M. tuberculin* was truly ancient, and clear evidence of its affliction of *Homo sapiens* dated back to at least 5000 B.C.[13] The disease was described by all ancient literate cultures, except those of the Americas, and archaeological evidence of bone damage predated the written descriptions of "consumption," "phthisis," or "tuberculosis," as it was variously labeled. But the true impact of the disease wasn't felt until after the Black Death when, according to theories popular in the 1980s, the tuberculosis organism exploited human ecological niches vacated by *M. leprae*. An urban environment was not required for its transmission, but it was clearly advantageous.

The rise of European tuberculosis was not sudden. Like leprosy, the organism was fastidious and slow-growing, producing overt and highly contagious illness only after months, or years, of infection. While the fast-growing plague bacteria could kill a human in a matter of hours, few *Homo sapiens* were felled by *M. tuberculin* without prior years of debilitating illness.

On the other hand, the bacteria could be spread by airborne transmission, assuring that humans sharing close quarters with an afflicted individual would be exposed. By the 1980s scientists knew that infection did not guarantee illness or death: about one out of ten infected individuals eventually developed the disease, and without twentieth-century treatments about half would die.[14]

But conditions in European cities of the fifteenth to the seventeenth century were ideal for transmission of *M. tuberculin*, especially during the winter, when the practice was to shut all windows and huddle around a heat source. The microscopic droplets exhaled by a tuberculosis victim would drift continuously about the home.

The household might take steps to avoid exposure to visible droplets of coughed or sneezed tubercular material, but these were actually harmless. To take hold in the human body the bacteria had to be carried inside droplets small enough to pass through the barriers of the upper respiratory tract. Such tiny droplets could remain suspended in the air, drifting on currents, for days, containing live, infectious tuberculosis germs.[15]

There was only one thing seventeenth-century Europeans could have done to decrease their exposure to household tuberculosis: open the windows. One good flushing of the air could have purged 63 percent of the suspended infectious particles exhaled each day by an ailing resident, and the sun's ultraviolet light would kill those organisms it reached.[16]

Medieval Europeans had no such options during the winter, however, particularly in northern latitudes. For poorer city dwellers especially, it was inconceivable to open windows during the winter, as fuel of any kind was scarce and extremely expensive. Europe's wood had been used to build her cities.

The rates of tuberculosis slowly but steadily increased. The hardest-hit cities were also the largest, London particularly. By the time London was devastated by the Great Plague and subsequent Great Fire, one out of every five of its citizens had active tuberculosis. And this time the plague had no purging effect on a mycobacterial epidemic: the rates of TB continued to climb long after the 1665 plague passed.

As European explorers and colonialists made their way to the Americas, they carried the deadly mycobacterium with them, adding to the disease burden of tuberculosis, which had already for centuries plagued the Amerindian population.[17] By the time the United States was torn asunder by the Civil War, tuberculosis was firmly entrenched in its northern cities, particularly Boston and New York City.

Between 1830 and the eve of the Civil War, Americans' life expectancy and death rates fell to the levels that existed in London. In 1830, with a population of 52,000 citizens, Boston's crude annual death rate was 21 per 1,000—half that of London at the time. By 1850, Boston's crude death rate nearly equaled London's, hitting 38 per 1,000. Tuberculosis wasn't the only responsible factor, but it was a major contributor. Cases of consumption, as it was called, increased every year in Massachusetts, rising 40 percent between 1834 and 1853.[18]

The old families of New York City, Philadelphia, and Boston groaned in disbelief as their cities' populations swelled, filth abounded, and disease ran rampant. Immigration, the Industrial Revolution, crowded slums, no public water supply, moral decay, no sewage systems—these were but a few of the factors that the civic leaders blamed for their crises.

The Western world's urban crises peaked between 1830 and 1896, when Europe and North America suffered four devastating pandemics of cholera that spread primarily via the cities' fetid water and sewage "systems." Though physicians of the day didn't understand why, quarantines didn't work for cholera, so the rich generally fled the cities at the first hint of the dreaded dysenteric disease, leaving the common folk to fend for themselves. It would be decades before scientists could prove that cholera was caused by a bacterium that entered human bodies via contaminated food and water, and got into the water through the fecal waste of infected people.

The death toll from cholera in the nineteenth century due to waves of the disease was astonishing: 10 percent of the population of St. Louis in three months during the 1849 epidemic;[19] 500,000 New York City residents in 1832; 8,605 Hamburg, Germany, residents in three summer months in 1892; 15,000 residents and hajj pilgrims in the city of Mecca, and some 53,000 Londoners, in 1847. The Mecca tragedy was repeated during the hajj of 1865, when 30,000 pilgrims to the city perished.

Though they had no idea what caused cholera, New York City authorities were appalled by the 1832 epidemic and blamed it on municipal filth. Reform followed. The Croton Aqueduct brought in clean drinking water for the first time, muddy streets were cobblestoned, and the squalid slums were slowly upgraded. As a result, subsequent waves of cholera took a minor toll.

Such was not the case in most other cities, however, where the connection between urban filth and disease remained a matter of vociferous debate among civic leaders. The fact that, without exception, cholera and other epidemic diseases—including tuberculosis—took their greatest toll among the most impoverished residents of the world's metropolises seemed only to reinforce the belief by those in power from Moscow to Madrid that lower-class "immorality" was the root of disease.

During London's devastating 1849 cholera epidemic, physician John Snow demonstrated that cholera was transmitted via water by removing the handle of the Broad Street pump, the sole water source for an impoverished and cholera-ridden community. The local epidemic, of course, came to a halt.

Authorities were unconvinced, however, so during London's 1854 epidemic Snow mapped cholera cases and traced their water supplies. He showed that those neighborhoods with little cholera were receiving water drawn from the upper Thames, while cholera-plagued areas drew their water from the lower Thames, which included human waste from upstream.

Snow failed to convince authorities directly of the need to clean up water supplies, but the epidemics of cholera and other devastating diseases spurred improvements in basic urban hygiene all over the industrializing world. Citizens' sanitary action groups cropped up in many cities, garbage- and waste-disposal practices improved dramatically, outhouses were replaced by in-house toilet systems, and "cleanliness" became "next to godliness."

Many urban diseases, including tuberculosis, declined in the cities of the Northern Hemisphere at about the same time as these social reform campaigns emerged. In addition to these changes in physical ecology, urban residents' lives were improved through such political and Christian reform efforts as elimination of child labor, establishment of public school systems, shortening of adult workweeks, creation of public health and hospital systems, and a great deal of boosterism about "sanitation."

At the peak of the Industrial Revolution, before such reforms were widely instituted, life in the cities had become so unhealthy that the birth rates were *lower* than the death rates. For a city like London this meant that the child and adult workforce of nearly one million people could be maintained only by recruiting fresh workers from the countryside. But by 1900 the birth rates soared, the death rates plummeted, and life expectancies improved markedly. Nearly all contagious diseases—including tuberculosis—steadily declined, reaching remarkably low levels well before curative therapies or vaccines were developed. In England and Wales, for example, the tuberculosis death rate dropped from a high of 3,000 per million people in 1848–54 to 1,268 cases per million in 1901 to 684 per million in 1941, just before antibiotic treatment became available.[20]

A similar pattern could be seen for infectious diseases, particularly tuberculosis, in the United States. In 1900, TB killed about 200 of every 100,000 Americans, most of them residents of the nation's largest cities. By 1940, prior to the introduction of antibiotic therapy, tuberculosis had fallen from being the number two cause of death to number seven, claiming barely 60 lives in every 100,000.[21]

By 1970, tuberculosis was no longer viewed as the scourge of the industrialized world's cities.[22] The World Health Organization then estimated that about 3 million people were dying annually of the disease, some 10 to 12 million were active TB cases, and with antibiotic therapy the mortality rate had dropped to about 3.3 deaths in every 100,000 TB cases. Most new infections were then occurring not in the industrial cities of the Northern Hemisphere but in villages and cities of the developing world. The microbe's ecology had changed geographically, but continued to be concentrated in urban areas.[23]

The enormous decline of tuberculosis in the Northern Hemisphere was viewed as a great victory, even though at the time TB raged across Africa, Asia, and South America.

Why this apparent victory over a microbe had occurred—what specific factors could be credited with trouncing tuberculosis—was a matter of furious debate from the 1960s through the 1990s. Resolution of the debate could have been useful in two ways: in helping public health authorities anticipate problems in their cities that might promote the emergence or reemergence of infectious diseases in the future, and in guiding urban development in the Third World by identifying which expenditures—drawn from ever-shrinking national reserves—might have the biggest impact on their public's health.

But the waters were muddy. British researcher Thomas McKeown argued that nutrition was the key—improved diets meant working people could withstand more disease.[24] René Dubos was equally certain that it was elimination of the horrendous working conditions of the men, women, and children of the Industrial Revolution, coupled with improved housing, that merited credit for the decline in TB.[25]

Medical historian and physician Barbara Bates, of the University of Kansas, skillfully asserted that the bold TB control programs of the early twentieth century, sparked by German scientist Robert Koch's 1882 discovery of the *M. tuberculin* bacterium, which led to mandatory quarantines in medical sanitariums, had little or no impact on the decline of the disease.[26]

Bates insisted: "The goal of prevention was frequently compromised. Physicians often discharged still infectious patients, and men and women with communicable disease left institutions against advice. Instead of a system that cured and prevented disease, society had built one that met some needs of sick and dependent people, spared families some of the burdens of care at home, and reduced the public's fear of infection, if not the actual threat. These unanticipated results grew out of political, social, and economic transactions in which medical understanding of tuberculosis played only a subordinate part."[27]

Another way to answer the question of what factors had been key to Europe's TB decline was to study the disease in an area that was making a transition during the twentieth century that was roughly equivalent to Northern Europe's Industrial Revolution a century earlier. South Africa fit the bill, and was a good place to test the hypotheses of McKeown, Dubos, and others: the country's European descendants had a standard of living and disease patterns analogous to their counterparts in the Northern Hemisphere. But the African and Indian residents suffered from marked economic and social deprivation. Their communities bore a striking resemblance to the squalid living conditions endured by London's working classes in the 1850s.

Though antibiotics and curative medicine existed, coupled with scientific knowledge of the modes of transmission of the bacteria, tuberculosis death rates rose 88 percent in South Africa between 1938 and 1945. Cape Town's rate increased 100 percent; Durban's 172 percent; and Johannesburg's 140 percent. In the rural areas, despite poverty and hunger, active TB cases and deaths never exceeded 1.4 percent in any surveyed group. But in the cities, incidence rates as high as 7 percent were commonplace by 1947. Nearly all TB struck the country's black and so-called colored populations.

The key to the increase in TB seemed not to rest with the health care system, for little had changed during those years. Nor were the diets of black South Africans much altered—they had been insufficient for decades.

The answer, it seemed, was housing. From 1935 to 1955, South Africa underwent its own Industrial Revolution, having previously been a largely agrarian society. As had been the case a century earlier in Europe, this required recruitment of a cheap labor force into the largest cities. But South Africa had an additional factor in its socioeconomic paradigm: racial discrimination. The recruited labor was of either the black or the colored race, prescribed by law to live in designated areas and carry identity cards which stipulated their sphere of mobility. The government subsidized an ambitious program of urban housing development for white residents of the burgeoning cities, but government-financed housing construction for black urbanites during the period of metropolitan expansion actually declined by 471 percent.[28]

By the 1970s, South Africa was generating 50,000 new tuberculosis cases a year, and the apartheid-controlled public health agencies were arguing that blacks had some unidentified genetic susceptibility to the disease. In 1977 the government made many of its worst TB statistics disappear by setting new boundaries of residence for blacks. The TB problem "went away" when hundreds of thousands of black residents of the by then overly large cities were forcibly relocated to so-called homelands, or when their urban squatter communities were declared outside the city's jurisdiction and, therefore, its net of health surveillance.[29]

If the South African paradigm could be applied broadly, then, it lent some support to Dubos's theories, underscoring human squalor as the key ecological factor favoring *M. tuberculin* transmission, but it failed to support Dubos's other assertions about the role of working conditions.

One thing that cities of the wealthy industrialized world and the far poorer developing world had in common was an ecology ideal for the emergence of sexually transmitted diseases, particularly syphilis. Certainly people had sex regardless of where they lived, but cities created options. The sheer density of *Homo sapiens* populations, coupled with the anonymity of urban life, guaranteed greater sexual activity and experimentation. Since ancient times urban centers had been hubs of profligacy in the eyes of those living in small towns and villages.

Houses of both male and female prostitution, activities mainstream societies often labeled "deviant" such as homosexuality, orgies, even religiously sanctioned sexual activity, were common in the ancient cities of Egypt, Greece, Rome, China, and the Hindu empires and among the Aztecs and the Mayans. The double standard of chastity in the home and risqué behavior in the anonymity of the urban night dates back as far as the beginning of written history.

It seemed surprising, then, that syphilis did not reach epidemic proportions until 1495, when the disease broke out among soldiers fighting on behalf of France's Charles VIII, then waging war in Naples. Within two years, however, it was known the world over. Syphilis seemed to have hit *Homo sapiens* as something completely new, because in the fifteenth and sixteenth centuries it struck in a far more fulminant and deadly form than it would take by the dawn of the twentieth century.

Syphilis was caused by a spirochete—or spiral-like—bacterium called *Treponema pallidum*, the same organism that caused the childhood skin disease known as yaws. Evidence of the existence of yaws clearly dated back to ancient times, yet of syphilis there was no hint prior to the fifteenth century.

Several twentieth-century theories were offered to explain this puzzle. The most obvious solution was to blame Amerindians, Christopher Columbus, and his crew. Their 1492 voyage to the Americas and subsequent return to Spain coincided with the 1495 emergence of the disease during the Franco-Italian wars. So it seemed circumstantially convenient to conclude that syphilis originated among the Native American peoples, was picked up by Spanish sailors, and carried back to Spain.

There were, however, two problems with that suggestion. First of all, yaws was an ancient disease on both continents, passed by skin contact between people. If yaws had existed on every continent, certainly the potential for syphilis had also always been present worldwide.

Second, syphilis wiped out Amerindians in the fifteenth century with a ferocity equal to the force of its attack on North Africans, Asians, and Europeans. If it had been endemic in the Americas, the Amerindians should have developed at least partial immunity to it.

The most likely explanation for the apparently sudden emergence of syphilis came in the late 1960s from anthropologist-physician Edward Hudson, who argued that syphilis was a disease of "advanced urbanization," whereas yaws was "a disease of villages and the unsophisticated."[30] In Hudson's view the spirochete could best exploit the ecology of the village by taking advantage of the frequent cuts and sores on the legs of children, coupled with the close leg-to-leg contact of young people who slept together in rural hovels and huts.

When the spirochete settled under the skin it produced only a localized infection that eventually healed. Transmission could occur only during the few weeks when the sore was raw and skin-to-skin contact could allow the organism to jump from one person to another. This occurred most easily among children who played or bedded down together.

Sexual transmission of the spirochete, however, required a far more complex human ecology in which many hundreds or thousands of people interacted intimately every day and a large percentage of the population regularly had sexual intercourse with a variety of partners.

Other twentieth-century theorists went further, arguing that the sexually transmitted microbes could *only* emerge in a population of *Homo sapiens* or animals in which a critical mass—perhaps even a definable number—of adults in the population had frequent intercourse with more than one partner. Clearly, they argued, a strict society in which every adult had sex only with their lifetime mate would have an extremely low probability of providing the *Treponema* spirochete with the opportunity to switch from a skin contact yaws producer to sexual syphilis.

Conversely, in cities where social taboos were less enforceable or respected, the possibility of multiple-partnering and, therefore, sexual passage of disease was far greater.

Following the Black Death of the fourteenth century, most of Europe experienced two or three generations of disarray and lawlessness. Death had taken a toll on the cities' power structures and, in many areas, the worst of the survivors—the most avaricious and corrupt—swept in to fill the vacuums.

"The crime rate soared; blasphemy and sacrilege was a commonplace; the rules of sexual morality were flouted; the pursuit of money became the be-all and end-all of people's lives," Philip Ziegler wrote.[31] The world was suddenly full of widows, widowers, and adolescent orphans; none felt bound by the strictures of the recent past. Godliness had failed their dead friends and relatives; indeed, the highest percentage of deaths had occurred among priests. Europe, by all accounts, remained so disrupted for decades.

One could hypothesize the following scenario for the emergence of syphilis: the spirochete was endemic worldwide since ancient times, usually producing yaws in children. But on rare occasions—again, since prehistory—it was passed sexually, causing syphilis.[32] These events were so unusual that they never received a correct diagnosis and may well have been mistaken for other crippling ailments, such as leprosy. But amid the chaos and comparative wantonness that followed the Black Death, that necessary critical mass of multiple-partner sex was reached in European cities, allowing the organism to emerge within two or three human generations on a massive scale in the form of syphilis.

In the late twentieth century similar debates about the emergence of other sexually transmitted diseases would take place—debates that might have been easier to resolve if questions regarding the sudden fifteenth-century appearance of syphilis had been settled.

II

By 1980 there were five billion people on the planet, up from a mere 1.7 billion in 1925.

The cities became hubs for jobs, dreams, money, and glamour, as well as magnets for microbes.

Once entirely agrarian, *Homo sapiens* was becoming an overwhelmingly urbanized species. Overall, the most urbanized cultures of the world were also, by 1980, the richest; and with the notable exception of China, the richest individual citizens usually resided in the largest cities or their immediate suburbs.

Propelled by obvious economic pressures, the global urbanization was irrepressible and breathtakingly rapid.

By 1980 less than 10 percent of France's population was rural; on the eve of World War II it had been 35 percent. The number of French farms plummeted between 1970 and 1985 from nearly 2 million to under 900,000.[33]

In Asia only 270 million people were urbanites in 1955. By 1985 there were 750 million, and that figure was expected to top 1.3 billion by 2000.[34]

Worldwide, the percentage of human beings living in cities showed a steady climb, and from less than 15 percent in 1900, was expected to exceed 50 percent by 2010.[35] About 60 percent of this extraordinary urban growth was due to babies born in the cities; 40 percent of the new urbanites were young adult rural migrants or immigrants moving from poor countries to the large cities of wealthier nations.[36]

The most dramatic rural/urban shifts were occurring in Africa and South Asia, where tidal waves of people poured continuously into the cities throughout the latter half of the twentieth century. Some cities in these regions doubled in size in a single decade.[37]

The bulk of this massive human population surge occurred in a handful of so-called megacities—urban centers inhabited by more than 10 million people. In 1950 there were two megacities: New York and London. Both had attained their awesome size in less than five decades, growing by just under 2 million people each decade. Though the growth was difficult and posed endless problems for city planners, the nations were wealthy, able to finance the necessary expansion of such services as housing, sewage, drinking water, and transport.

By 1980, however, the world had ten megacities: Buenos Aires, Rio de Janeiro, São Paulo, Mexico City, Los Angeles, New York, Beijing, Shanghai, Tokyo, and London. And even wealthy Tokyo found it difficult to accommodate the needs of its new population, which grew from a mere 6.7 million in 1950 to 20 million in 1980.

But this was only the beginning. Continued urban growth was forecast, and it was predicted that by 2000 there would be 3.1 billion *Homo sapiens* living in increasingly crowded cities, with the majority crammed into 24 megacities, most of them located in the world's poorest countries.[38]

Throughout the 1980s a key shift would occur, and most of the nations experiencing the greatest population growth would also rank among the poorest countries in the world. They would be hard pressed to meet the health and service challenges posed by the cities' extraordinary escalation in need.

The World Health Organization concluded that "urban growth, instead of being a sign of economic progress, as in the industrialized country model, may thus become an obstacle to economic progress: the resources needed to meet the increasing demand for facilities and public services are lost to potential productive investment elsewhere in the economy."[39]

According to the World Bank, African cities were increasing in size by 10 percent a year throughout the 1970s and 1980s, which constituted the most rapid proportional urbanization in world history.

In 1970, in the Americas there were three city residents for every rural resident; by 2010 the ratio would be four to one. The same shift was forecast

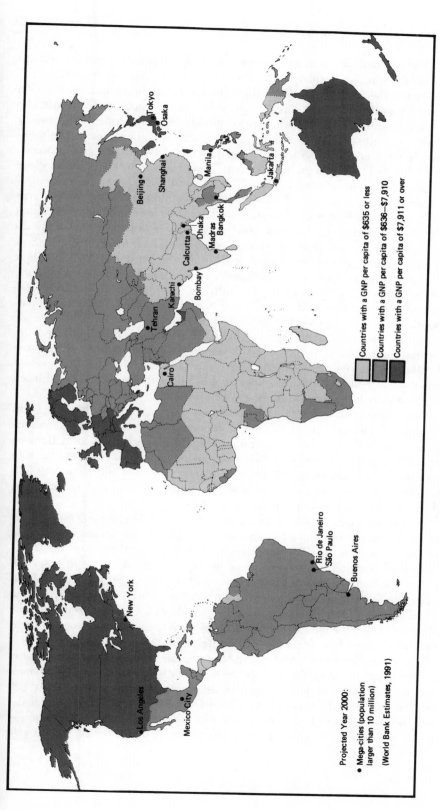

Projected Year 2000:

● Mega-cities (population larger than 10 million)

(World Bank Estimates, 1991)

■ Countries with a GNP per capita of $635 or less

■ Countries with a GNP per capita of $636—$7,910

■ Countries with a GNP per capita of $7,911 or over

GLOBAL PER CAPITA EARNINGS

for Europe, both Western and Eastern. Some Asian countries were predicted to have five urban residents for every one rural individual by 2010.

During the 1970s and 1980s this crush of urban humanity was causing severe growth pains that directly impacted on human health, even in the wealthier nations. Japan, which was quickly becoming one of the two or three wealthiest countries on the planet, was reeling under Tokyo's expanding needs. By 1985 less than 40 percent of the city's housing would be connected to proper sewage systems, and tons of untreated human waste would end up in the ocean.[40]

Hong Kong, a center of wealth for the Chinese-speaking world, was dumping one million tons of unprocessed human waste into the South China Sea daily. Nearby Taiwan had sewage service for only 200,000 of its 20 million people, two-thirds of whom lived in its four largest cities.

But for the poorest developing countries, the burden of making their growing urban ecologies safe for humans, rather than heavens for microbes, proved impossible. Except for a handful of East Asian states which developed strong industrial capacities (i.e., South Korea, Malaysia, Singapore), the developing world simply had no cash in 1980.

In addition to growing national debts, developing countries faced the steady capital drain of paying off development loans obtained during the 1960s and 1970s and investing in newer sources of potential revenue generation. Some countries simply couldn't bear the burden, and shirked or attempted to renegotiate their multibillion-dollar loans.

The capital drain would turn into a hemorrhage. In 1980 the Latin American nations collectively were receiving from their external creditors—major banks, the International Monetary Fund, the World Bank— about $11 billion more than they were losing in capital transfers back to wealthy-nation interests. But by 1985 these nations would be losing $35 billion more a year in capital transfers to North America and Europe than they received in loans and investments.[41]

Africa was also staggering under the burden of debt servicing and capital transfer, though the impact wasn't as profound in dollar terms as that seen in Latin America. In 1979 the recently elected Prime Minister of the U.K., Margaret Thatcher, addressed these concerns in frank terms in a speech before the Commonwealth Conference, convened that year in Lusaka, Zambia. Ministers from the poorest of England's former colonies hoped that Thatcher would extend a pound-filled hand, but she proffered only the sorry news that the once-great Empire was itself feeling the economic pinch. In short, it was time for a global belt tightening.

The cities worsened, some coming to resemble their teeming counterparts in nineteenth-century Europe. By the mid-1980s, 100 million newly homeless adults would roam the streets of developing-world cities; at least 100 million abandoned street children would haunt the urban nights. Half the city dwellers of developing countries who were *not* classified as homeless would live in shantytowns and slums that, among other things, lacked safe drinking water. Forty percent would be without public sanitation or sewage facilities. A third would live in areas in which there were no garbage or solid waste collection services.

As was the case in ancient Rome, it was healthier to remain in the villages and small towns of the developing world—even in times of drought and crop failure—than to live in the filthy, unwieldy metropolises. The average child living in a typical developing-country urban slum was forty times more likely to die before his or her fifth birthday of a preventable infectious disease than was a typical rural child in the same country.[42]

Disasters, and the very real opportunities they afforded the microbes, were everywhere. The streets of Cairo, for example, were flooded in December 1982 with sewage water that in some places was knee-deep. The flooding persisted for day after day, while authorities struggled to identify its cause.[43]

Nearly every Egyptian had been, for over 4,000 years, dependent on a single water supply—the Nile. The river's annual floods would carry away a host of human sins in the form of waste and soil overuse, and leave behind a thick layer of fresh, fertile silt.

But construction of the Aswan Dam, coupled with Egypt's extraordinary human population explosion, had erased the Nile's majesty. Now slow-moving and predictable, the Nile was filling up with silt, fertilizers (which the farmers now needed because they no longer got the topsoil from the annual Nile floods), sewage—both treated and untreated—and industrial waste. Scientists predicted the imminent demise of Alexandria's freshwater delta lagoons, a rise in the level of the Mediterranean Sea, and serious public health risks due to chemical and biological pollution of the Nile. They suggested that, given Cairo's growth rate, there was nothing that could be done to prevent future environmental and public health disasters.[44]

The World Bank rated 79 percent of the housing of Addis Ababa "unfit for human habitation" in 1978. A quarter of the city's houses were without toilet facilities.

A quarter of Bangkok's residents in 1980 had no access to health care, according to the World Bank. In the Dharavi slum of Bombay, then inhabited by over 500,000 people, conditions were so appalling that 75 percent of the women suffered chronic anemia, 60 percent of the population was malnourished, pediatric pneumonia afflicted nearly all children, and most residents contracted gastrointestinal disorders due to parasitic infections. In Jakarta in 1980, the life expectancy was only fifty years, several years less than in the countryside. By 1980, 88 percent of Manila's population lived in squatter settlements constructed of discarded pieces of wood, cardboard, tin, or bamboo. Forty percent of Nairobi's 827,000 people in 1979 lived in housing so poor that their neighborhoods were deliberately omitted from all official maps.[45] The flood of people into the Sudanese capital, Khartoum, led to epidemics of malaria, diarrheal diseases, anemia (presumably produced by malaria), measles, whooping cough, and diphtheria in the early 1980s. In the Ivory Coast, rural tuberculosis rates by 1980 were down to 0.5 percent—a success story. But in the large capital city of Abidjan the TB rate was 3 percent and climbing.[46]

These and hundreds of other examples of urban squalor and its concomitant diseases were compounded by large-scale chronic malnutrition. Except in times of famine, drought, or other natural disasters—or of the man-made disaster of warfare—rural residents of even exceptionally poor countries usually had access to a variety of types of food, including protein. But when they moved to the city, people had to buy foods that were produced and marketed by others. Lacking sufficient earning power to purchase goods, the urban poor were forced to forgo adequate foods. Even in times of food plenty for the nation, most of its urban population might, as a result, be malnourished. This, of course, contributed to weakening their disease-fighting immune systems.[47]

III

By 1980, several traditionally rural parasitic diseases were emerging, for the first time, as urban epidemics.

Uwe Brinkmann, having left Germany following the Lassa episode and settled at the London School of Tropical Medicine and Hygiene, traveled all over West Africa surveying the incidence of onchocerciasis, or river blindness, a disease carried by blackflies. For two years Uwe, his researcher wife Agnes, and their young son went from village to village, primarily in Ghana and Togo, studying the disease and teaching villagers how to avoid it.

From there, Brinkmann moved on to study primary health care systems in Yemen and Sierra Leone, schistosomiasis in Congo and Mali, and onchocerciasis and cysticercosis in Central America.[48]

By 1982, when Brinkmann had joined the faculty at Harvard, he had seen disturbing evidence that the parasitic diseases that he and other scientists worldwide were so successful at limiting in the villages and farmlands of developing countries were invading the cities, often in different forms.

Cysticercosis was usually produced by tapeworms normally found in undercooked pork and some other types of animal flesh. The worms invaded numerous organs of the human body—the worst cases involving infections of the brain. But Brinkmann noted that a change in the human/parasite relationship was occurring in Mexico City—then the world's fastest-growing megacity. People were not getting the worms from uncooked meats, which, as it turned out, they couldn't afford to purchase. Rather, the parasite had taken advantage of the highly favorable ecology provided by the extraordinarily polluted Tula River, the city's primary freshwater source. Tens of thousands of people living in the squalid outskirts of the megacity downstream of the urban center's sewage system were infected with the dangerous *Taenia solium* parasites.[49]

By 1980 the tapeworm had found its way to Los Angeles, carried by human immigrants from Asia and Central America. Some 500 cases of cysticercosis were treated between 1973 and 1983 at four Los Angeles hospitals. Most involved people who were infected in their home countries or while traveling in endemic areas. But at least twelve individuals acquired the disease in Los Angeles, and random stool sample assays revealed that some 0.5 percent of the tested population were infected with the tapeworms.[50]

The *Ascaris* roundworm was another parasite that was invading cities. The *Ascaris* eggs lived in a dormant state in the soil, where they could survive in infectious form for over ten years. Humans and pigs became infected as a result of inhaling contaminated dust, oral contact with dirtied hands, or ingestion of unwashed foods grown in contaminated soils. Once the eggs made their way to the human gastrointestinal tract they would mature into worms that would wreak havoc upon numerous organs, including the entire gastrointestinal tract, liver, appendix, pancreas, heart, and lungs. The human would then excrete more parasite eggs, further contaminating local soils. Prior to the 1970s this cycle was considered an entirely rural, village-based problem.

During the 1970s in Dakar, however, a third of the city's slum residents were infected with the parasite, acquired within the city limits, while less than 3 percent of their rural counterparts carried the disease.[51] At the same time locally acquired ascariasis increased dramatically in the city of Cape Town, South Africa, accounting for 15 percent of all emergency room admissions for acute abdominal disorders.[52]

Schistosomiasis turned up in Dar es Salaam (Tanzania), Harare (Zimbabwe), Kinshasa (Zaire), and São Paulo and Belo Horizonte (Brazil) during the 1970s.

Chagas' disease, caused by *Trypanosoma* protozoa and carried by a variety of insects, was turning up in cities all over Latin America. Capable of causing encephalitis and severe heart disease, the *Trypanosoma* organisms made their way into the continent's burgeoning cities, infecting up to 60 percent of the common household bugs. Eventually the *Trypanosoma* found a more direct way to infect people: bypassing the insect vector, the protozoa entered the blood-bank systems. By the mid-1980s, blood-bank infection rates would be horrendous: 6 percent in Buenos Aires and up to 20 percent in other Argentine cities; 15 percent in Brazil's capital, Brasília; an astonishing 63 percent in Santa Cruz, Bolivia.[53]

For centuries sandflies had been inserting their pointed proboscises into the human epidermis, injecting anticoagulating chemicals and withdrawing enough blood to bloat the insect. In 1824 the flies of Jessore, in Bengal, added something to this process, injecting parasites along with the anticoagulants.

Tiny one-celled *Leishmania donovani* swam into the bloodstreams of Jessore merchants, visiting traders, women, and children. Soon kala-azar (as the disease was called) was attacking the abdominal veins of humans in cities all along the Ganges, causing deadly pneumonia and dysentery. Such illnesses may have been occurring for centuries, isolated and unnoticed. But the 1824 outbreak struck a major trading post, gaining the immediate attention of British colonial interests then controlling the subcontinent.

Another round of sandfly-carried kala-azar struck Assam, India, in 1918, killing more than 200,000 people. Still another hit the area in 1944.[54]

Soon leishmaniasis-producing organisms of various species were turning up in flies that infested cities in Latin America and the Indian subcontinent, producing both kala-azar and the cutaneous, or skin infection, forms of the disease. Various factors seemed to have contributed to the emergence of urban leishmaniasis, including widespread DDT spraying for malaria control. When spraying programs were stopped, either because the mosquito population was effectively controlled or due to government financial restrictions, the fly population would surge, filling the ecological niche vacated by the competitive mosquitoes.[55]

The surge of sandflies in the wake of the malaria eradication campaign was often dramatic. In cities and small towns all over Latin America the insects swarmed in, often hitting communities for the first time in human history. Pioneers searching for wealth in the vast Amazon rain forest often returned to Brazil's eastern cities with little more than a whopping leishmaniasis infection. Even if the Latin American sandflies of the cities hadn't carried the parasites before, they now picked up the microbes as they fed on recently returned Amazon fortune seekers in cities all over the continent.

The Indian kala-azar parasitic strain was able to infect dogs and domestic animals, as well as humans, providing steady reservoirs for the microbe's continued presence in a community. By 1980 scientists in Colombia and Brazil spotted the same phenomenon developing in their cities and towns, primarily among pet dogs and chickens.

Ki denga pepo is Swahili for "it is a sudden overtaking by a spirit." The phrase was used by East Africans to describe a mosquito-carried disease that would abruptly overwhelm human beings, producing horrible headaches, eye pain, and a swelling achiness of the joints.

When the disease swept over Philadelphia in 1780, Dr. Benjamin Rush gave it the moniker "breakbone fever," a reference to the aching joints. By the mid-nineteenth century the disease was endemic throughout the Americas.

And it then had a permanent name—dengue, a Spanish adaptation of the Swahili *denga*. In most cases dengue wasn't a life-threatening ailment, though it was certainly a miserable experience for the afflicted. The disease was caused by four different strains of dengue viruses—cousins of the yellow fever microbe. The dengue viruses were carried by mosquitoes, particularly the female *Aedes aegypti.*

As countries throughout the world conducted *A. aegypti* eradication campaigns during the early twentieth century to rid the earth of yellow fever, dengue outbreaks virtually ceased. A comfortable dengue silence set in during the 1940s.

Then, in 1953, the city of Manila was hit by an apparently new form of dengue that caused hemorrhagic petechial skin rashes—pinpoint-sized red spots, sites of breakthrough bleeding—shock, and soaring fevers. The disease seemed more lethal than any previous dengue outbreaks, and was caused by viral strain dengue-2.[56]

Five years later dengue hemorrhagic fever, as the new disease was called, hit Bangkok, causing 2,297 illnesses—primarily among children—and 240 deaths. Searches of human blood samples from the past revealed that various dengue viruses had infected Bangkok residents harmlessly since 1950, but the population was never infected with dengue-2 prior to World War II. After the initial urban outbreak in 1958, however, the dengue hemorrhagic fever epidemic persisted in Bangkok for five years, eventually sickening 10,367 people and killing 694.[57]

U.S. Army medical researcher Dr. Scott Halstead, who was based at the military's laboratory in Bangkok at the time, teamed up with Thai microbiologist Charas Yamarat to figure out the origin of the apparently new deadly disease. They determined that, as was the case with yellow fever, the *A. aegypti* mosquito that carried the dengue-2 virus was a fully urbanized insect. Lacking the aggressive characteristics of wild jungle mosquitoes, *A. aegypti* only thrived in proximity to human beings, laying its eggs in open containers of fresh water and maturing inside human shelters.

When the men closely examined the medical records of people who suffered acute dengue hemorrhagic fever they discovered that nearly all of the victims had at some recent time been exposed to another, milder dengue strain. Though that first infection caused little or no apparent illness, it sensitized the humans' immune systems for the later arrival of dengue-2.

Usually when people develop strong antibody immune responses against a virus they are protected against future exposure to the microbe. But dengue-2 had evolved an extraordinary ability to exploit human antibodies to its advantage. When the human antibodies attached to the outer envelope of the dengue-2 virus, the microbe played a game of stealth, allowing the antibodies to send their signals to the large immune system macrophage cells. In a process that was usually lethal to the microbes, the macrophages would then engulf the viruses, but instead of dying, the dengue would take control of the immune system's primary killer cells.

Thus, dengue-2 evaded the immune system defenses and gained entry to every organ in the body, carried by macrophages that acted as Trojan horses for the virus. As the immune system struggled to overcome its sneaky invaders, various biochemical reactions were triggered that produced soaring fevers—as high as 107°F—convulsions, classic allergylike shock, and death.[58]

The new dengue disease paradigm spread through South and East Asia, carried by ever-expanding hordes of A. aegypti and another species, A. albopictus, otherwise known as tiger mosquitoes. Unlike A. aegypti, the A. albopictus insects were sturdy creatures adapted to coexistence not only with Homo sapiens but also with a wide range of warm-blooded creatures that thrived in urban environs—even rats.

During the 1950s and 1960s, dengue types 1, 2, and 3 all made sporadic appearances in the Americas,[59] but A. aegypti control programs were strong enough to prevent epidemics. Nevertheless, the viruses were present in the region, particularly in the Caribbean, and the mosquitoes were never fully eradicated.

The stage was set, and dengue invaded the moment a slackening in mosquito abatement programs allowed the A. aegypti population to grow to critical proportions.

In May 1981, the city of Havana experienced the worst dengue hemorrhagic fever epidemic seen up to that time anywhere in the world. The epidemic raged for over six months, causing at least 344,000 illnesses, more than 116,000 hospitalizations, and 158 deaths. At its peak in July some 11,000 Havana residents sickened each day. The epidemic cost the Cuban government $103 million in control efforts and medical care—a large sum for the nation of 10 million people whose per capita annual incomes were less than $1,500 that year.

Havana, with a population of 2 million and fewer than 25,000 hospital beds, was overwhelmed. More than 10,000 health care workers had to be corralled into full-time dengue treatment and control efforts, not only in Havana but eventually nationwide. Nearly 10 percent of the residents of Havana suffered symptomatic dengue-2 infection between May and September 1981.

When researchers tested the city's residents for prior exposure to dengue, they discovered that the Halstead-Yamarat theory of serial infections and immune system deception was correct: 44.5 percent of Cuba's urbanites had antibodies to dengue-1 as a result of a very mild epidemic of the virus that swept imperceptibly over the island in 1977, causing little more than mass natural immunization of the population.

But that was enough.

When dengue-2 hit Havana it found the urban population sensitized and ready to succumb to its immune system trickery.[60]

The Cuban epidemic sent shock waves through the public health communities of the Americas. A year earlier, two residents of Laredo, Texas, had developed dengue hemorrhagic fever, carried by infected A. aegypti mosquitoes found in that city, proving that the viruses had made their way to North America as well.[61]

In October 1982, New Delhi, India, suffered a mass epidemic of dengue hemorrhagic fever that sickened more than 20 percent of its 5.6 million residents. By then the World Health Organization regrettably had to announce that attempts to stifle the spread of dengue-2 since its initial appearance in Manila in 1953 had failed, and the virus was now endemic in and around the major cities of Burma, Thailand, Laos, Vietnam, eastern India, and Sri Lanka.

During the 1980s, Duane Gubler, of the CDC's laboratory in San Juan, Puerto Rico, would with considerable apprehension chronicle the steady rise of dengue, of all types, in the cities and towns of Latin America. Each year the number of hospitalizations would increase, infected mosquitoes would expand their territory, and the specter of a hemispheric urban dengue disaster would become more imposing. By 1990 he would be forced to conclude that dengue was endemic to Latin America.[62]

A key factor in the expansion of dengue threats to the Americas would be the 1985 arrival on the continent of *A. albopictus*. Carried aboard a shipment of water-logged used tires sent from Japan for retreading in Houston, Texas, the extremely aggressive mosquitoes—capable of carrying both dengue and yellow fever—would quickly outcompete more timid domestic mosquito species in the United States. Within two years *A. albopictus* tiger mosquitoes would be seeking human blood in the cities and towns of seventeen U.S. states.[63]

"The presence of *albopictus* dramatically increases the probability that exotic viruses will be brought into the urban human environments of the Americas," Gubler declared. "The tiger mosquito will feed on anything, a rat for example, and then turn right around and feed on a human."

Gubler warned that dengue wasn't the only virus *A. albopictus* threatened to spread: the midguts of the females were capable of harboring a broad spectrum of viruses, including types found in other mammals and not yet known in humans. In contrast, the finicky *A. aegypti* fed almost exclusively on human blood; it could therefore spread only known human diseases.

"We're in crisis management on this, that's all," Gubler would angrily say. "We just wait for crises to occur and then get around to intervening. We could have seen this coming, we could have been vigilant. But the money was never there; the surveillance was never there."

When Tom Monath set out to reconstruct the events that led to the global emergence of urban dengue hemorrhagic fever he concluded that every historical advance of the microbes and their mosquito vectors was a direct result of human activities. Still at the CDC's laboratory in Fort Collins, Colorado, five years after his discovery that rats carried the Lassa virus in West Africa, Monath scoured historical records and contemporary laboratory evidence for clues concerning dengue.

He concluded that World War II was responsible for the emergence of *A. aegypti*-carried dengue in Asia. Massive human migrations, aerial bombing campaigns, densely populated refugee camps, and the wartime disruption of all mosquito control efforts allowed for an unprecedented surge in the *A. aegypti* population: it may very well have numbered more in 1945 than at any time in the planet's previous history. The mosquitoes were able to use bomb craters filled with water as breeding sites and to draw blood from millions of human victims of war whose homes were destroyed and no longer provided nighttime protection from the hungry insects.

Rapid troop movements by air transit, coupled with massive refugee migrations, allowed the various dengue types to get into new ecospheres, carried by humans who were unaware that they were infected. Almost overnight, areas such as the Philippines, which for centuries had only a single dengue strain infecting its human and insect populations, were overrun by all four dengue types. During World War II Japanese, European, and American troops landed on Filipino soil after having been in other dengue-infested parts of Asia, such as Burma, Thailand, Indonesia, islands throughout the Pacific, and China. The soldiers carried various strains of dengue in their blood—strains that were absorbed by local Philippine *A. aegypti* mosquitoes.

After a few years of circulation among humans and mosquitoes in Manila, the immune system cycle necessary for the creation of the acute hemorrhagic and shock syndrome of dengue-2 was in place. Such serial infection of one dengue type after another hadn't been possible before World War II, Monath concluded, because few—if any—areas of Asia had endemic dengue of more than one type.

The Korean and Vietnam wars only created further opportunities for mosquito breeding and dengue cross-fertilization. By the conclusion of the Vietnam conflict in 1975, dengues of all four types were endemic in urban centers throughout the region. The Cuban epidemic of 1981, interestingly, followed a period of intensive cooperative postwar exchange of personnel between the two countries for professional training and Vietnamese reconstruction efforts.

By the time dengue hit Latin America in the 1960s, ecologies that were favorable to *A. aegypti* in Asia were also in place in the Western Hemisphere, the result of enormous tidal waves of migration into the area's largest cities. Conditions in the favelas and slums of cities like São Paulo, Rio de Janeiro, Caracas, and Santiago were similar, from the mosquitoes' perspectives, to those in wartime Asia. The surge in commercial air traffic throughout the world during the 1970s, Monath concluded, facilitated the spread of people whose bodies were incubating as yet asymptomatic infections of dengue-1, -2, -3, and -4. Once they arrived in the cities of Latin America and their dengue illnesses set in, these people transmitted the four dengue viruses to local mosquitoes and, eventually, to other human city dwellers.

When dengue hemorrhagic fever broke out in the streets of Havana the viruses had become so thoroughly entrenched back in Manila that millions of cases of the disease had occurred, and epidemics—particularly among children—had become an annual feature of urban Philippine society. By 1981, as regular as clockwork, dengue arrived each year shortly after the onset of every rainy season in Manila; tens of thousands of children contracted dengue hemorrhagic fever; and 15 percent of those children died.

It hadn't been so before World War II. But by 1980 dengue was one of Asia's most prevalent childhood illnesses. And would remain so.

All things considered, Uwe Brinkmann estimated in 1981 that some 300 million residents of cities located in developing countries suffered debilitating illnesses at any given time due to chronic parasitic infections, over and above periodic viral epidemics, such as dengue. And though the costs of prevention through large-scale housing, sewage systems, potable water, insect control, and improvement in garbage collection might have seemed daunting to the governments of poor countries, Brinkmann argued that the price of doing nothing was far greater. Treatments for parasitic diseases were either extremely expensive by Third World standards—$240 for treatment of leishmaniasis, for example—or nonexistent. Since few such countries could afford to treat their citizens, the true price was an ever-greater trend toward urbanization of previously rural diseases and the tremendous toll they took in human life and productivity.

Tragically, events during the 1980s would prove far worse than Brinkmann had imagined.

10

Distant Thunder

SEXUALLY TRANSMITTED DISEASES
AND INJECTING DRUG USERS

The Snake Pit raid is one more illustration of the ugly games that
straights inflict on gays, driving them underground, to be periodically
chastised by the city's conscience, driving them into self-conscious,
paranoid postures, driving them finally into an up-front struggle for
liberation to establish, once and for all, that gay is neither perversion
nor sissy nor sick nor faggot nor silly. Gay is good.

The Village Voice, 1970

I

Around midnight on Friday, June 27, 1969, Deputy Inspector Seymour
Pine of the New York Police Department reviewed procedures with the men
under his command in the public morals section. Under the pretense of
liquor license violations, they were about to shut down a bar at 53 Christopher Street in Greenwich Village.

It was a homosexual bar, and raids such as this had been a routine part
of New York City "morals" enforcement for decades. Though rarely legal,
the raids succeeded in driving out of business many establishments catering
to gay men and scaring away closeted men and women who feared being
identified as sexual deviants on police rap sheets.

As they had done many times before, Pine's undercover team got out of
their unmarked cars and strolled across Christopher, past the Lion's
Head—an often wild heterosexual bar—to the Stonewall.

With military precision the men surrounded the gay bar, Pine went
inside, the management was presented with papers ordering a shutdown:
routine closure of the Stonewall began. One by one the clientele were
ushered out to Christopher Street, where they hammed it up playfully for
the gathering crowd of Greenwich Village onlookers—hippies, gays, bo-
hemians, the denizens of New York's most notoriously offbeat neighborhood. The mood was calm, even playful.

Until the fifteen paddy wagons appeared.

Within minutes a full-scale riot was underway as the Village's gay men
fought the police, removed colleagues from custody, and declared the

immediate neighborhood "Home of the Queens." Rioting would continue throughout the weekend, often with a joyous giddiness to it.

By Monday morning everyone involved—both rioters and police—knew something dramatic had happened. Overnight, new gay political organizations appeared, not only in New York City but in other cosmopolitan American centers, notably San Francisco and Los Angeles.[1]

"The nights of Friday, June 27, 1969, and Saturday, June 28, 1969, will go down in history as the first time that thousands of homosexual men and women went out into the streets to protest the intolerable situation which has existed in New York City for many years—namely, the Mafia (or Syndicate) control of this city's gay bars in collusion with certain elements of the Police Department of the city of New York," declared a leaflet from a group calling itself the Homophile Youth Movement, which urged the city's gay population to boycott mob-controlled bars and demand an end to police raids.

Within days printed signs appeared all over the Village, stating bluntly for "gays" and "straights" alike: "Do you think homosexuals are revolting? You bet your sweet ass we are."

The gay liberation movement burst like champagne from a highly agitated, just uncorked magnum. The politics of the movement, for many, went no further than open, unabashed displays of their previously closeted gay sexuality. Groups of gays in New York and San Francisco stood up, however, and publicly declared not only their identity but also their right to their sexuality, and the two cities became magnets pulling long-suppressed homosexuals from the small towns of America; indeed, from all over the world.

The cities, it seemed, had more than just economic opportunity to offer.

A year after the "Stonewall Riots," New York's activist homosexuals staged a commemorative parade in Central Park. It was attended by a crowd *The New York Times* estimated at 20,000. The same day, 1,000 gay men marched in Los Angeles and about 100 in San Francisco.

A wiry young Brooklyn activist named Marty Robinson mugged for a TV camera crew that day in Central Park, then changed his mood, stared defiantly into the camera, and said that the parade "serves notice on every politician in the state and nation that homosexuals are not going to hide anymore."

No one could have imagined on that day in 1970 that a mere eight years later June 27 would be commemorated in cities all over the world as Gay Freedom Day, drawing crowds of well over 375,000 to San Francisco and tens of thousands more to the streets of Washington, D.C., Los Angeles, Miami, New York, and Chicago. There would even be small sympathy gatherings in Paris, London, Amsterdam, and Berlin. By 1978, the U.S. gay rights movement mobilized massive protest demonstrations against a former beauty queen turned spokeswoman for the far right, Anita Bryant. The outspoken, "pro-Christian" Bryant had become a leading advocate of both consumption of Florida orange juice and revocation of the hard-won civil rights recently afforded homosexuals in a few cities, notably San Francisco. The leader of San Francisco's gay community, Harvey Milk, called upon homosexuals nationwide to come to the city for the June 27, 1978, Gay Freedom Day parade to "send a message" to Bryant and other opponents of gay rights. And they did.

By 1978 San Francisco's gays were a potent political force. According to the city's noted gay chronicler Randy Shilts,[2] gay immigration to San Francisco between 1969 and 1978 outstripped California's gold rush, adding 30,000 gay men to the population. After 1979, San Francisco would attract an additional 5,000 gay migrants yearly until 1988.

In November 1978, the U.S. gay rights movement attained that dubious notoriety offered to all grass-roots efforts whose leaders are assassinated because of their beliefs. Harvey Milk, by then the city's first openly gay elected official—a member of the Board of Supervisors—was shot dead in his office, along with the mayor, George Moscone. The assassin was another supervisor and former police officer, Dan White, who would later get a light sentence based on his creative plea of temporary insanity, caused by the overconsumption of sweets (Hostess Twinkies). The jury's acceptance of the so-called Twinkie defense would be interpreted by the gay community as an obscene display of homophobia.

Milk's murder placed the political fate of the gay rights movement in the United States solidly in the ranks of other civil rights movements. If African-Americans resented analogies between their civil rights struggles and those of homosexuals—and there were strong protests over comparisons drawn between Martin Luther King, Jr., and Harvey Milk, or between the Stonewall riot and Rosa Parks's refusal to sit at the back of segregated buses—the sentiment had little impact on the youthful exuberance of gay activists.

A party atmosphere pervaded the gay communities of San Francisco, New York, and, to a lesser degree, Montreal, Los Angeles, Washington, D.C., Paris, London, Berlin, and Amsterdam in the late 1970s. Night after night the gay neighborhoods filled with young men determined to make up for lost time, dancing through trysts with such haste that niceties, like partners' names, might be overlooked.

"I was an ecstatic slut," Bobbi Campbell would say later of his days—and nights. A member of the Sisters of Perpetual Indulgence, a group of humorous drag queens who dressed in nun's habits for all major San Francisco public events, Campbell, and thousands of others like him, found plenty of time to indulge in the mass revelry.

Worldwide, the 1970s were a time of sexual liberation and experimentation for young adults—straight as well as gay—who poured into trendy metropolises from Nairobi to Amsterdam in search of the excitement and anonymity of urban nightlife. The birth control pill gave young women freedom from concern about unwanted pregnancy, and, for the first time in history, heterosexual exploration seemed safe. In Europe and North America it was gay men who took greatest advantage of the new climate; in developing countries, particularly in Africa, it was young heterosexuals. From London's posh West End to downtown Abidjan the nexus of all this activity was a new sexual milieu: the disco. In bars all over the world, young adults drank or danced to electronic music, their eyes peeled for potential partners. In the often harsh, alienating atmosphere of big, un-friendly cities, discos provided instantaneous intimacy. If there was po-tential danger in leaving the disco with a stranger, it might only enhance the sexual allure of the adventure. And for millions of women, particularly in developing countries, this new atmosphere provided what was often the only potential source of independent income: prostitution.

Finally, throughout the developing world new patterns of male employ-ment appeared during the late 1970s and the 1980s. Young men, tied by marriage and family to their villages or small towns, commuted to large cities for work. They made their mass exodus every Monday morning, converging on cities like Nairobi, Harare, Bombay, Lima, and Abidjan from the countryside, stayed in flophouses or workers' barracks until Friday night, and returned to their villages for the weekend. For many, a disco cycle set in: on city nights they might pick up a young lady—prostitute or not—but they returned on the weekends to their wives.[3]

Such things had happened in cities before. During the days of Aristotle and Plato, Athens was so replete with (homo and hetero) sexual activities that even the gods had orgies. But the scale of multiple-partnering during the late twentieth century was unprecedented. With over five billion people on the planet, an ever-increasing percentage of whom were urban residents; with air travel and mass transit available to allow people from all over the world to go to the cities of their choice; with mass youth movements at their zenith, advocating, among other things, sexual freedom; with a fem-inist spirit alive in much of the industrialized world, promoting female sexual freedom; and with the entire planet bottom-heavy with people under twenty-five—there could be no doubt that the size and drama of this world-wide urban sexual energy was unparalleled.

"Why do faggots have to fuck so fucking much? It's as if we don't have anything else to do . . . all we do is live in our ghetto and dance and drug and fuck," moaned an exhausted character in *Faggots*, a play by a gay New York author, Larry Kramer.

Though the emotional price of all this anonymous sexuality was obvious to many participants by the close of the 1970s, its microbial toll was apparent only to those few public health authorities who were paying at-tention. It was easy to miss.

By 1980 most Americans and Western Europeans were, on average, remarkably healthy compared with their counterparts of a previous gen-eration, or with their contemporaries living in the Southern Hemisphere.

Nearly 100 percent of U.S. deaths that year were due to chronic diseases, accidents, suicides, and diseases of old age.[4]

Reflecting this, only 34 percent of National Institutes of Health resources in the United States were spent on that gamut of problems that included infectious diseases. The agency's infectious disease prevention and control budget had by 1980 declined 16 percent from 1969–76.[5]

Given the reported mortality statistics, this resource shift seemed wholly appropriate. Sexually transmitted diseases had declined dramatically all over the industrialized world since the discovery of antibiotic treatments for syphilis and gonorrhea. In the 1920s over 9,000 Americans died each year of syphilis, and 60,000 children were born infected with the spirochete. In 1940, just before the introduction of antibiotics, 13,000 Americans died of syphilis. But by 1949, with the availability of antibiotic treatments, fewer than 6,000 Americans died of syphilis, and all signs pointed toward a continuing decline as physicians improved their use of the drugs and more infected people sought treatment. Nobody, therefore, considered it inappropriate to slash venereal disease control budgets from a 1949 commitment of $18 million down to a 1955 U.S. federal expenditure of barely $3 million.

By 1970 fewer than 0.02 of every 10,000 Americans—or two out of every million—succumbed to syphilis. The gonorrhea death rate had also plummeted and most physicians considered both diseases easily curable and, therefore, controllable.[6]

But by 1975 the folly of such overconfidence was apparent: gonorrhea reports in the United States tripled between 1965 and 1975, syphilis reports quadrupled. By the early 1980s over 2.5 million people were getting gonorrhea annually, and syphilis ranked behind gonorrhea and chicken pox as the third most common infectious disease in the United States.[7]

Though few Americans were dying of gonorrhea in the post-antibiotic era, it was not a harmless disease. It clearly contributed to infections in the ovaries and fallopian tubes that comprised pelvic inflammatory disease (PID) in women, with the resultant risks of major surgery and infertility. About 20 percent of the 850,000 women who contracted PID in the United States each year between 1977 and 1980 suffered PID as a result of underlying gonorrhea infection.

A woman who survived a case of PID without obvious lasting effects was ten times more likely to suffer subsequent ectopic pregnancies—which could be life-threatening—due to infectious damage to her reproductive tract. The ectopic pregnancy rate in the United States soared from 19,300 cases in 1971 to 42,000 in 1978. Not only did the numbers of ectopic pregnancies increase, but so did the likelihood that any given pregnancy would be marred by that dangerous complication. In 1970 just over 4 out of every 1,000 U.S. pregnancies was ectopic; a decade later more than 13.5 of every 1,000 pregnancies was ectopic, a fourfold increase.[8]

Finally, about 15 percent of all women who suffered from PID were rendered sterile either as a result of ovarian infections or hysterectomies necessitated by advanced, life-threatening disease. One out of five PID cases required hospitalization. Estimates of the costs—direct and indirect—of PID by 1978 were already starting to approach the billion-dollar mark in the United States. By the mid-1980s the U.S. direct and indirect medical cost of PID would top $2.6 billion per year, and researchers would predict that, given an apparently out-of-control increase in the incidence of the syndrome and its underlying venereal diseases, societal costs could exceed $3.5 billion by 1990.[9]

Other microbes could also produce PID, including the *Chlamydia trachomatis* bacteria, which by 1983 would cause some three million new infections a year among American adults. Like gonorrhea, *Chlamydia* incidence increased steadily in the United States throughout the 1970s and 1980s. The risk of both infections rose in direct proportion to the number of different sexual partners an individual had over a given amount of time.[10]

In 1976 there was a dramatic turn of events, further worsening the sexually transmitted disease situation.

On August 27, 1976, the CDC reported that two individuals—one in Maryland, the other in California—had become infected with an apparently new, mutated strain of gonorrhea that defied penicillin treatment. On closer examination the CDC determined that the *Neisseria gonorrhoeae* made an enzyme that destroyed penicillin; the strain was dubbed Penicillinase-Producing *Neisseria gonorrhoeae*, or PPNG.[11]

By October the CDC had identified ten more cases of the penicillin-resistant gonorrhea, and traced all but one of the U.S. cases to recent travel in East Asia, either by the ailing individual or by his/her sex partner. At the same time public health authorities in the port of Liverpool, England, reported that forty cases of PPNG had surfaced in their city during the previous eight months.[12]

By early 1977, PPNG reports were coming in from all over the United States, and a third of the cases involved U.S. military personnel recently returned from Asia, particularly the Philippines.[13] The U.S. Navy and Air Force both had enormous bases in the Philippines, surrounded by a dense urban sprawl of tens of thousands of people eager to earn U.S. dollars. Notably, prostitution thrived around both bases, and black-market penicillin was sold to the brothels and the hookers.

Surveys in the Philippines revealed that some 40 percent of all gonorrhea cases in cities near U.S. military bases were PPNG. And half of all U.S. military personnel stationed in the Philippines who had gonorrhea were infected with PPNG.

"It seems unlikely that efforts to control emergence of penicillinase-producing gonococci will do more than delay their worldwide spread," a U.S. National Institutes of Health panel concluded in 1977.[14]

The same week the CDC reported the Philippines link, it also reported on a Georgia man suffering from a new type of gonorrhea that was resistant to the two other most commonly used treatments for the disease:

spectinomycin and ampicillin. The CDC scoured over 9,000 gonorrhea isolates collected nationwide prior to 1976 and found no evidence of the spectinomycin-resistant strain in the United States prior to February 1977. It had existed in Denmark, however, where two cases were discovered in 1976.

By May 1977, the penicillin-resistant PPNG strain had been spotted in seventeen countries, and all the North American and European cases traced back to either the Philippines or West Africa. The United States, by that time, had 150 PPNG cases, most were in New York City, and three of the cases involved microbes with triple resistance—penicillin, ampicillin, and spectinomycin. The CDC warned that physicians had to handle antibiotic use in their gonorrhea patients very carefully or "the probability of PPNG acquiring spectinomycin resistance will increase."[15]

Within four years treatment of gonorrhea would become terribly complicated; not only would there be PPNG and spectinomycin-resistant microbes to worry about but also a strain that was resistant to the entire tetracycline family of antibiotics.[16] Soon the microbes were rampant among urban gay men and black and Hispanic heterosexual males in the United States.[17]

Herpes simplex Type I, or HSV-I, had been a ubiquitous pediatric disease for as long as anybody had been able to diagnose the distinctive cold sores it produced. Better than 90 percent of elderly residents of North America and Europe in 1980 had antibodies to HSV-I, indicating that they had been infected with the virus sometime during their lives. But the childhood infection rates declined in the industrialized world during the 1950s and 1960s due to improved personal hygiene standards and an understanding that herpetic sores shed contagious viruses.

As a result of declines in the usually less dangerous HSV-I, herpes simplex Type II (HSV-II) was able to infect a wider range of people. HSV-I offered the infected human an opportunity to make antibodies against it which weakly cross-reacted against HSV-II, offering some protection against the more dangerous virus.

HSV-II was primarily passed sexually between human beings. The virus had several characteristics that allowed its easy passage within a sexually active human population.[18] It could infect nerve cells and hide for years on end inside the relatively quiet host. At any time—perhaps decades after infection—the viruses could emerge from those nerve cells, replicate thousands of copies of themselves, and create painful sores around the genitals, mouth, or anus of the infected human. During such times, these areas would be sites where millions of herpes viruses were shed, and the chances of passage to a human sexual partner were extremely high—approaching 100 percent under certain circumstances.

HSV-II was usually found in teenagers and adults, and prior to the 1970s active cases of the disease were primarily seen among prostitutes and their clients.[19] But a 1981 survey of middle-class young adults in the city of

Toronto found that 15 percent were infected with the genital herpes virus. A Seattle study concluded that nearly half of the city's homosexual population and a quarter of the women living in the community's poorer neighborhoods were also infected.

Between 1966 and 1981 the number of Americans treated by their doctors for genital herpes increased ninefold.

A similar escalation was seen in the U.K., Israel, Thailand, New Zealand, and throughout Western Europe, where, overall, visits to clinics for treatment of HSV-II increased at a rate of 12 percent per year from 1975 to 1982.[20]

Researchers discovered that the virus could lie silently in a woman's uterine lining for years, causing damage only when she became pregnant. Then it might precipitate an abortion, or be passed to the fetus, producing painful infections all over a neonate's body.[21] Treatment of the neonates required intensive care,[22] and, in many cases, the illness was fatal.

Despite public alarm, the incidence of HSV-II infection would continue to rise dramatically, reaching levels in 1986 as high as 60 percent of all adult men living in key U.S. cities.[23]

During the 1970s, researchers reported similar rises in nearly every other microbe known to be sexually transmitted. Cytomegalovirus, or CMV, was increasingly found in the blood or genital tracts of men and women attending STD clinics in the United States, and by 1980 up to 25 percent of women examined in such clinics had active CMV infections of the cervix.[24]

Chancroid, a bacterial disease causing ulcerous sores in the rectum and genitals, was less common than the other major sexually transmitted diseases, but the sores served as "portals of entry," as public health authorities put it, for the passage of other microbes into the human body. Chancroid reached its lowest point in U.S. history during 1975, when fewer than 500 cases would be reported to federal authorities. But almost immediately the trend reversed itself, and outbreaks occurred in Orlando, New York City, Boston, Philadelphia, Dallas, Los Angeles, and other U.S. cities. By 1987 annual U.S. chancroid reports would soar tenfold, topping 5,000 reports a year—the 1950 level.

Like gonorrhea and chlamydia, the chancroid bacterium—*Haemophilus ducreyi*—mutated around antibiotic treatment. On mobile plasmids that could be passed from one *H. ducreyi* to another were genes that made the microbes resistant to ampicillin, sulfonamides, chloramphenicol, and tetracyclines. As a result, treatment of chancroid would, by the mid-1980s, be difficult.[25]

In 1982 H. H. Handsfield focused on why U.S. medicine had been so slow to deal with this rise in all sexually transmitted diseases:

. . . following World War II and the discovery of penicillin, many doctors and public health authorities believed that syphilis and gonorrhea, then the most important known forms of sexually transmitted diseases in the United States, would shortly be all but abolished. It was widely felt, therefore, that the problem could be safely left to public venereal disease clinics. Many private physicians were quite content with this approach, since it more or less absolved them of having to deal with diseases widely considered "not nice" and which confronted the doctor with the difficult and often delicate problem of contact tracing. The public clinics, however, were relegated to second class status, underfunded and understaffed, even within health departments. Simultaneously, there was a radical de-emphasis of STDs in medical schools: Since the problem was "under control," there was obviously little point in training physicians to deal with it. Whole academic divisions of "syphilology" disbanded, and venereology became divorced from both medical research and medical training. Most U.S. medical schools now provide no more than a few hours of lectures on STDs, usually during the preclinical training years, and only a small handful provide clinical training of any kind.[26]

Between the late 1960s and the early 1980s most STDs also climbed in Western European countries, but swift public health action generally prevented U.S.-scale epidemics. In the U.K., for example, syphilis began making a comeback in 1968, but a prompt national control effort brought the incidence down markedly in 1978 among heterosexuals. Homosexual spread of syphilis, however, continued unabated. U.K. control efforts were less successful for gonorrhea, which climbed steadily after 1957 in all sexually active demographic groups.

No country had much luck controlling the spread of genital herpes simplex. Between 1970 and 1984, U.K. herpes cases skyrocketed from 4,000 a year to more than 20,000. On the other hand, chancroid had practically disappeared from most European countries by 1980.[27]

In developing countries the STD crisis was at least as pronounced as in the United States. Pelvic inflammatory disease cases accounted for an ever-increasing percentage of all gynecological visits, particularly in Africa. By 1980, PID was the reason for 30 percent of all gynecological visits in Ugandan cities; 26.5 percent in Zambia; 30 percent in Ethiopia; Nigeria, 30 percent; Kenya, 40 percent; and Zimbabwe, 44 percent.[28] Ectopic pregnancy rates were also rising, and in some countries were responsible for up to a third of all maternal deaths.[29]

Most PID was due to either gonorrhea or chlamydia, both of which were out of control in the majority of cities in developing countries. Gonorrhea had become so widely antibiotic-resistant by the early 1980s that effective doses had to be a hundred times stronger than doses in 1950. In some Asian countries over half of all cases involved PPNG, and strains resistant to more than one antibiotic were on the rise.

The prevalence of gonorrhea among young adults was high by the early 1980s. The greatest reported incidence was in Uganda, where 40 percent of the women attending family-planning clinics in Entebbe and Kampala were infected. In nearby Nairobi, Kenya, 64 percent of the lower-paid street prostitutes were infected, and a quarter of Nairobi's high-class prostitutes carried the bacteria.

Syphilis rates varied among women attending family-planning clinics from a low of 1 percent in Saudi Arabia to 35 percent in Khartoum. Among prostitutes, syphilis was extremely prevalent in most developing countries, and rates of 50 to 75 percent were common.

Chlamydia rates among pregnant women were also alarming. In rural South Africa, for example, only 1 percent of women were infected, but in Johannesburg and Cape Town rates of over 12 percent were seen. Kenya reported chlamydia rates of 29 percent; Fiji led the world, with 45 percent of its tested pregnant women found infected with chlamydia.

The numbers were telling a story, issuing a warning that was largely unheeded.

II

In 1978, Dr. Subhash Hira was looking for a change. Ever since he finished his medical training in Baroda, India, and worked in Bombay, the young physician had felt restless. Opportunities for an honest doctor trained in Western medicine were limited in India, unless he had family money with which to start a private practice. It was particularly hard to find positions that offered a young government physician a chance to escape the Health Ministry's bureaucratic stranglehold.

He was, therefore, easily recruited by the Zambian government to run a new national program to control syphilis, gonorrhea, and other sexually transmitted ailments—the "shame diseases."

When he arrived in Lusaka, Hira immediately noticed that it was overwhelmingly populated by young people who were unmarried or who felt unfettered by their vows on Saturday nights. The city was exploding with new arrivals, and housing was scarce. Men outnumbered women; many were guerrillas fighting to overthrow governments in Rhodesia, South Africa, or Namibia.

In 1978–79 Hira occupied a modest cinder-block office at University Teaching Hospital, Zambia's premier medical school. He designed an STD control program aimed at keeping case records and planned to replace "shame" with prevention and treatment.

He also surveyed small groups of Lusakans to determine the incidence of STDs, something the old Northern Rhodesia colonial government had never done, and postcolonial Zambian health officials hadn't considered a priority given the crises of malaria, malnutrition, and other deadly pediatric diseases, like measles.

Hira had no way of knowing what sort of trend his STD numbers followed, but even taken alone they were startling. Syphilis was rampant. Responsible for 19 percent of all miscarriages, it infected 32 of every 1,000 babies born in Lusaka, about 16 of whom died immediately before or after birth.[30]

Fourteen percent of the Lusaka women tested at family-planning or pregnancy clinics had syphilis; some 11 percent had gonorrhea. Chlamydia and chancroid were also rampant, and Hira suspected that the rates of all four STDs would prove even higher in young men.

Outside of Lusaka and the densely populated area around Zambia's copper mines, these diseases were relatively rare, but Hira noticed that many urban workers returned to their wives in rural villages during holidays. He wondered how long it would be before STDs plagued the villages as well. And how long before the thousands of freedom fighters took Lusaka-acquired microbes to their home countries.

In 1980, Hira set up an STD testing clinic at the University Hospital. It was soon swamped. Men and women, some clutching their children, waited, moved beyond shame to the more familiar ennui of "queuing up." Some waited for hours, as Hira and his assistants desperately tread water before the tidal wave.

Far away in snowbound Michigan, Dr. June Osborn was reviewing grant applications for National Institutes of Health funding to conduct sexually transmitted disease research. When she had started in her NIH advisory role, Osborn (like most American STD researchers) was focused on heterosexually spread herpes simplex, but by 1979–80 she noticed something troubling: all STDs were increasing, at a rate of about 1 percent a year in the general population, but by an incredible 12 percent annually among gay men.

"I fear we're looking at a new ecology here," Osborn told NIH colleagues. Wherever there were large gay communities, there was also a striking disparity in disease rates, particularly for syphilis, gonorrhea, and hepatitis B.

For Osborn one of the most startling findings involved the *Entamoeba histolytica* parasite. Normally found in the food and water of densely populated areas in developing countries, it was rarely seen in the United States. In the late 1970s, however, it turned up in the bowels of gay men in New York, San Francisco, Los Angeles, and a few other cities. In developing countries where *E. histolytica* was endemic, the parasites formed bowel cysts, creating ulcerous sores which shed more organisms, some of which might get into the liver, causing severe damage.

As reports of *E. histolytica* outbreaks among gay men escalated, Osborn and other public health officials became seriously alarmed. By the close of the decade more than 20 percent of the U.S. gay male population was infected with *E. histolytica*—five years before, no locally acquired cases were reported in the United States. Fortunately, the microbial strain rampant in the United States was not particularly virulent and most men had few or no symptoms.

But things were moving quickly—"too fast," Osborn said—for the NIH research planners. The all-heterosexual, mostly middle-aged research advisers simply couldn't fathom what was going on in the U.S. gay community at the time.

"Every time we do an NIH site visit the definition of 'multiple sex partners' has changed. First, it was ten to twenty partners per year. That was 1975," Osborn complained. "Then in 1976 it was fifty partners a year. By 1978 we were talking about a hundred sexual partners a year and now [1980] we're using the term to describe five hundred sexual partners in a single year.

"I am duly in awe. Perhaps somewhat disbelieving, but duly in awe," Osborn concluded.

Preliminary 1980 reviews seen only by Osborn's NIH panel and federal authorities at the CDC revealed that CMV was spreading quickly among gay men. By 1981 the CDC would tell doctors nationwide of another unprecedented new gay male epidemic—of cytomegalovirus. Widespread rectal transmission of CMV among adults had never been seen anywhere before.

Reports of rare diseases among gay men were coming in from public health authorities in Canadian and Western European cities. In Paris, Amsterdam, London, Rome, Madrid, Montreal, Toronto, Copenhagen—wherever researchers looked—the trend was the same.

"We've got to pay attention to this ecology," Osborn warned. "There's something disturbing going on."

To the happy participants in the gay freedom movement, it was the ecology of sexual liberation. A price to pay, so to speak, for newfound freedom.

"I calculated that since becoming sexually active in 1973, I had racked up more than three thousand different sex partners in bathhouses, back rooms, meat racks, and tearooms," gay pop singer Michael Callen wrote. "As a consequence, I also had the following sexually transmitted diseases, many more than once: hepatitis A, hepatitis B, hepatitis non-A/non-B; herpes simplex Types I and II; venereal warts; amebiasis, including giardia lamblia and entamoeba histolytica; shigella flexneri and salmonella; syphilis; gonorrhea; nonspecific urethritis; chlamydia; cytomegalovirus (CMV), and Epstein-Barr virus (EBV) mononucleosis; and eventually cryptosporidiosis."[31]

Another factor in the spread of disease was a change in the culture of homosexuality. In the past, role playing had been a common feature, with some men always being the passive receptors in anal intercourse and others consistently the aggressors.[32] But within the culture of gay liberation such role playing was shunned, even taboo, and more men played both roles. That changed the ecology of anal intercourse for the microbes, and allowed a more rapid epidemic spread of disease. If John, for example, played the passive role one week and got rectal gonorrhea from Sam, his chances of passing the microbe to Charlie the following week were greater if John played the aggressive role. If John remained the passive player, however, his partners might not contract his gonorrhea. So if a man had 500 sexual partners in a year, he might receive an extraordinarily rare microbe from just one of his 250 aggressor partners, then pass it on to 250 men with whom he took the aggressive role. The potential for rapid spread was further enhanced by the lack of a strong local immune system in the anal/rectal area. Thus, the one man could amplify a weak microbe signal 250-fold, creating an epidemic din.

None of these details of homosexual behavior were comfortable intellectual terrain for public health scientists in 1980, however. While they charted the upward curves on the STD graphs in the gay communities of North America and Europe, and discussed the trends at meetings, few scientists wanted to discuss the "new ecology." It was unsettling, and politically volatile.[33]

Having obtained his Harvard Ph.D. in virology, Don Francis was back working for the CDC in Phoenix, Arizona. The hepatitis B virus remained his greatest concern, and he hadn't overlooked its alarming rise among gay men. By 1980 he had established a national cohort of gay men who agreed to be tested periodically for hepatitis B. By following these 6,875 men, most of whom were in San Francisco, Francis hoped to establish the dynamics of the microbe's spread within the homosexual population.

Francis had become one of the world's experts on hepatitis B. In 1979 he helped investigate an outbreak in India, caused by injections of hepatitis-contaminated human immunoglobulin in 325 people.[34] From 1978 to 1983 he participated in three other investigations of nosocomial transmission of the virus: a Baltimore dentist who passed the virus to six patients,[35] a Connecticut oral surgeon who infected over a hundred patients during 1978–79,[36] and a Mississippi gynecologist who infected three women upon whom he performed surgery during 1979–80. In all three cases, transmission ceased with the routine use of surgical gloves.[37]

This demonstrated to Francis that the hepatitis B virus, unlike the food-borne hepatitis A, was primarily transmitted through blood-to-blood contact. And that contact could be entirely prevented by a layer of latex.

To Francis this seemed to be ample reason to recommend that gay men start using condoms, but such proclamations were uncomfortable for the CDC, which still adhered to the old venereal disease paradigm of identifying cases, contacting all their partners, and treating everyone with antibiotics.

But hepatitis B was a virus; it couldn't be effectively treated with any drug. And contact tracing was clearly impossible if an individual had multiple, anonymous sex partners. Francis saw no alternative but prevention to block transmission of the virus, by creating either a physical barrier (condoms) or immunity (vaccination). By 1980 he was actively pushing both angles and, in his usual gruff but earnest manner, making enemies among the more traditional bureaucrats and venereologists at the CDC. Francis, however, was a man with a mission. Prone to impatience and maverick action, he became increasingly outspoken about hepatitis B prevention. On the basis of lab work he had done with Max Essex, and studies he made of Native Alaskans and their parallel epidemics of liver cancer and hepatitis B infections,[38] Francis was convinced that the escalating hepatitis epidemic among gay men presaged a plague of gay liver cancer. He showed that spread could be blocked with condom use.[39] And it was well known that one out of ten adults newly infected with hepatitis B went on to become chronic carriers of the virus, potentially infecting others for decades and putting themselves at risk for liver cancer.[40]

In 1978, federal researchers estimated there were approximately 200,000 hepatitis B carriers in the United States, but as the gay epidemic struck San Francisco, New York, Washington, D.C., Los Angeles, Miami, Paris, London, and other key cities, it became clear that the numbers were mere guesses.[41] By late 1981, San Francisco Health Department officials would estimate that 73 percent of gay men in the city "either have or have had" hepatitis B, and physician Pat McGraw would reckon that at least 1,000 gay men were carriers, or roughly one in fifty of the city's openly gay citizens.[42]

At CDC headquarters Drs. Harold Jaffe and Jim Curran read the field reports coming in on hepatitis B and recognized that all young sexually active Americans—particularly homosexuals—seemed to stand a far greater chance of acquiring a sexually transmitted disease in 1980 than did their counterparts just a decade earlier. The trends, they felt, augured for loss of what little control public health authorities still claimed over the STD microbes, and they tried to argue that case both inside the CDC and at medical conventions.

Jaffe always returned from such meetings sizzling mad. Few physicians or scientists shared his concerns, and many publicly retorted as late as 1980 that "there's really no further need for an infectious disease specialty" in medicine.

Jaffe, thirty-four, answered to Curran, who was just thirty-six. Jaffe had a bit of Northeast passion; Curran was a classic case of Midwest cool. In late 1978 Curran had come to the CDC from Ohio University College of Medicine, where he'd been a professor of preventive medicine. He headed the CDC's research branch for the Venereal Disease Control Division.

Curran's low-key style and extremely fastidious presence led many people to mistakenly conclude that he was a conservative straitlaced sort. But well before he joined the CDC, Curran had concluded that the old-fashioned approaches to venereal disease control—indeed, the very word "venereal"—were outmoded. He favored new approaches to a problem he readily acknowledged was out of control.

Like Jaffe, Francis, and Osborn, Curran recognized that something unique was occurring among gay male residents of some cities with large homosexual populations. He tried to warn the physicians Michael Callen called "the clap doctors," their patients, and gay organizations. But everybody was enjoying the party too much. Besides, after so many decades of maltreatment by every imaginable government agency and medical organization, including officially being labeled mentally ill during the 1950s by the American Psychiatric Association, gay men weren't about to let another bureaucrat tell them to slow down.

III

By 1980 thirty-year-old Greggory Howard had been carrying a heroin addiction for thirteen years. Cut off from his family, Howard was a member of a community of heroin addicts who lived amidst the extraordinary squalor of Newark's burned-out tenements.

It was shortly after the 1967 riots that Greggory Howard, then a high school junior, first shot heroin into his veins. His warm personality and good grades held promise that he might escape New Jersey's notorious slum to a better life.

"Life was good to Greggory," Howard said, always referring to himself in the third person. "Yes, it was. It really was. My parents did everything right, they were very good to me. But Greggory just had . . . just had to drift away."

In 1967 racial tensions in the United States were as high as anyone could remember. The civil rights movement had passed from polite sit-down demonstrations and peaceful marches to unfettered rage when its leaders shifted their foci from the Deep South to the industrialized North. By the mid-1960s racial tensions had reached a tinderbox level.

Newark ignited in 1967; block after block went up in flames amid riots between residents of the city's slums and the police and National Guard. Tanks patrolled the streets.

Frightened, Greggory stayed out of the fray, but it left him with a sense of tragedy and hopelessness. Afterward, he took to walking along Prince Street and Hamilton, staring at the charred structures that had once housed his friends, teachers, and relatives, and he tried heroin for the first time. One of his ex-girlfriends was already shooting up, and she seemed to like getting high. Why not try?

Now it was 1980. Howard's nose was broken, a nasty scar zigzagged across his left cheek, and he walked with a jerk, all thanks to beatings by dealers, crooks, and hoodlums. Those veins that hadn't disappeared were on the verge of collapse or embolism from the thousands of needles he'd jammed into his arms, neck, and thighs. Howard's liver was shot, because of hepatitis.

To avoid being arrested for carrying drug paraphernalia[43] Howard rarely had on his person either heroin or the gear—the cooker, tourniquet, syringe, and needles—that was needed to prepare and inject the drug. Like most street-savvy addicts, Howard always had on hand a supply of minor street drugs, possession of which did not constitute a major crime: Valium,

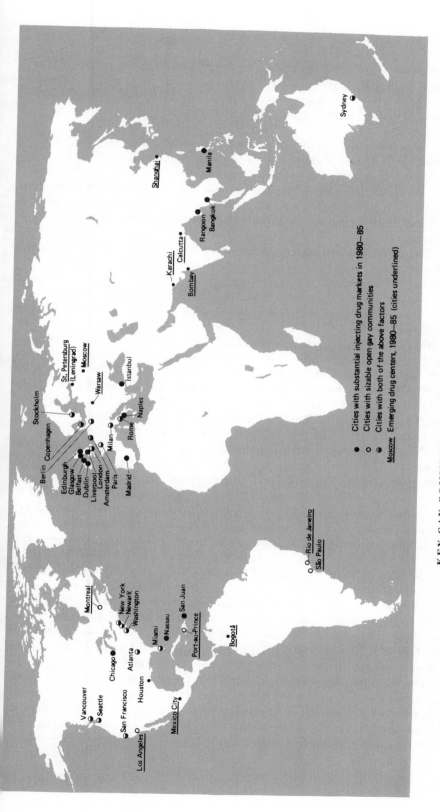

KEY GAY COMMUNITIES AND CENTERS

OF INJECTING DRUG USE, 1980s

- ● Cities with substantial injecting drug markets in 1980—85
- ○ Cities with sizable open gay communities
- ◑ Cities with both of the above factors
- Moscow Emerging drug centers, 1980—85 (cities underlined)

marijuana, a variety of "downers" that could help him stay relatively steady between heroin highs. And when he had the cash, Howard went to dealers at any of a number of apartments, abandoned buildings, alleys, parked cars, or parks and shot up, using works supplied by the dealer or another junkie.

"Someday I'm going to detox my ass," Howard would say, staring off on a high at Newark's hundreds of empty lots—the legacy of 1967.

Heroin, cocaine, amphetamines, and the host of other drugs easily purchased in the full glare of sunlight in most large American and European cities hadn't always been public health disasters. But they clearly did pose a crisis for urban health by 1980—an opportunity, a new ecology, for the microbes.

The anonymity of cities provided cover for illegal activity. The density of the population offered a steady flow of consumers, even for self-destructive products. And the alienation ensured that there would always be people willing to trade their health, wealth, and esteem for something that would take their minds to another place, be it alcohol, Valium, or heroin.

Once it reached Newark, a two-pound bag of pure heroin might be "cut" or "stepped on" with some other chemical by 90 or 95 percent, giving the wholesaler nearly 200 pounds (90 kilos) of street-quality heroin to sell to a retailer. The retailer might further cut the drug to increase his profit potential, so that Greggory Howard's daily high might be on a solution ranging from 2 to 15 percent heroin. In 1974, by the time it reached Newark, one acre's opium could yield more than $40 million.[44]

Profits in the opium/morphine/heroin network were greatest when the risks of police interdiction or interference from competitors was low. Large cities with slums afforded ideal environments, particularly if they were near ports or international airports. If hefty sales could be maintained in the crime-ridden inner-city areas there was no need to risk law enforcement's attention in small towns or tight-knit suburban communities. In slum neighborhoods where most people were hostile to the police, retailers could operate with near-impunity. If kids in the suburbs wanted heroin—and by 1980 many of them very much did—they could come into the city to obtain supplies.[45]

Despite expenditures of billions of federal law enforcement dollars since 1969, when the U.S. Congress voted for a "full-scale attack" on the heroin problem, the number of heroin addicts in the United States would rise from 55,000 in 1955 to 1.5 million in 1987 and they could be found, regardless of race, in any community that offered a steady supply of the drug.[46] In Greggory Howard's state, New Jersey, by 1980 heroin users were of all races, ages, and economic backgrounds, though the majority were white men between twenty-five and thirty-five years of age. About 40 percent of the state's heroin users were holding down jobs. Most had sought treatment several times—and failed several times.[47]

The urban heroin environment was ideal for dozens of different microbes. The drug user generally had an impaired immune system due to the narcotics' effects, to the constant injection of other people's blood cells carried on shared syringes, and to the numerous compounds used to cut the product. On the one hand, they had overexcited antibody responses provoked by all the immune system stimulators, such as other people's cells. Many therefore tested positive for rheumatoid factor and other markers of a system so overstimulated that it might be producing antibodies against itself. This autoimmunity could lead to the body's inability to distinguish genuine microbial threats from vital human cells. [48]

On the other hand, the large phagocytic cells of the immune systems of injecting drug users, which usually bore the responsibility of ingesting and destroying bacteria and other invaders, were alarmingly nonresponsive. And the T-cell system, comprised of cells that usually tipped the rest of the immune system off to the presence of potential threats, was seriously dysfunctional, in part because some lymphocytes bore receptors for opiates, and were dampened directly by heroin. [49]

As a result, microbes found the environment of the body of a heroin user far less hostile than that presented by healthy *Homo sapiens*.

The drug user's basic lifestyle also offered unique opportunities for microbial passage from human to human. Most of the addicts shared one another's injecting equipment. When an addict injected heroin, or any other drug, some of his or her blood might be pulled into the syringe when the equipment was drawn from the vein and the plunger reset. If the injector was infected with, for example, *Staphylococcus* bacteria, the microbes would be withdrawn into the syringe as well.

All the staph bacteria then required was the genetically acquired ability to withstand whatever environment the syringe rested in when not in use. That might mean a few hours hidden outdoors on a subzero Newark night—conditions too tough for most organisms. On a hot, humid August night lying in a watery "cleaning dish," however, the microbes found a highly favorable ecology. The least challenging situation was the shooting gallery, where one person immediately passed his contaminated syringe to another.

Needles also helped microbes bypass the multitudinous barriers humans had in their skin, nostrils, and lungs, and go directly to the bloodstream. Even organisms with weak "delivery systems," as Bernard Fields called them, could thrive in the heroin ecology.

Finally, many heroin addicts lived in acute squalor, ate poorly, worked as prostitutes for drug money, and used a broad range of additional intoxicants, each of which uniquely altered the *Homo sapiens* ecology in ways the microbes might find advantageous.

In 1929 malaria caused by *Plasmodium falciparum* broke out in downtown Cairo, Egypt, due to needle sharing by local drug addicts. By the late 1930s a similar heroin-driven malaria epidemic was spreading through

New York City, reaching such high levels among drug addicts that it was considered endemic. Six percent of New York City jail inmates at the time had signs of malaria infection—all of them injecting drug users. One hundred and thirty-six New Yorkers died of malaria during the period— none of them had been bitten by mosquitoes.[50] The epidemic stopped when the heroin retailers, concerned about losing their customers, started adding quinine to their cut heroin.

Such dealer "benevolence" was more than offset, however, by routine contamination of heroin products, the result of poor chemical processing or the use of microbe-supporting substances to dilute the drug.[51]

Numerous types of microbes managed to successfully exploit the heroin ecology. For example, between 1969 and 1974 physicians at San Francisco General Hospital noticed an increase in endocarditis—life-threatening infections of the heart. In seventeen of the nineteen cases, the individuals were drug addicts. And the organism responsible was the *Serratia marcescens* bacterium. Despite vigorous antibiotic treatment, 68 percent died. Searches of hospital records as far back as 1963 revealed no prior case of *S. marcescens*-induced endocarditis in San Francisco, proving it was a newly emergent microbial threat.[52]

Endocarditis was increasingly a problem worldwide in heroin-plagued cities. Bacteria and fungi entered the bloodstream on dirty needles and colonized the heart valves and other components of the vital organs. In most cases, antibiotic therapy proved fruitless. Prior to 1976 New York City experienced such endocarditis outbreaks among drug addicts, caused by *Staphylococcus, Enterococcus, Candida,* and *Pseudomonas.* Chicago, Helsinki, Seattle, Washington, D.C., San Francisco, and Detroit also witnessed outbreaks driven by those organisms, as well as *S. marcescens.*

Bacterial and fungal infections of all sorts became so prevalent among drug injectors by the mid-1970s that many began to prophylactically medicate themselves with antibiotics. An antibiotics black market, operating in tandem with the heroin trade, developed in many cities in Europe, Asia, and the United States, servicing heroin users with a variety of antimicrobials. But their use of these medicinal drugs was counterproductive, because the black market's supplies were sporadic and rarely offered consistent varieties of antibiotics.

Drug addicts, therefore, became ideal breeding grounds for antibiotic-resistant organisms. From a public health perspective the problem was restricted to the drug users themselves, who in increasing numbers throughout the 1970s suffered and died from antibiotic-resistant bacteria.

But in 1982 injecting drug users in Boston and Detroit were taking black-market methicillin whenever they could to prevent bacterial infections. Strains of the bacteria emerged that possessed two different types of transferable methicillin-resistance genes.[53] When the infected heroin users were hospitalized, a new resistant *Staphylococcus*, dubbed MRSA, spread to the medical staff and other patients.

Tuberculosis also lurked in the heroin ecology. Health authorities in most industrialized countries thought the TB scourge of the pre-antibiotic era was licked, and in absolute numbers of active cases in humans it certainly had declined dramatically by the 1970s. But in New York City hospitals in 1979, Dr. Lee Reichman spotted a trend that had gone unnoticed: TB cases in the city's poor, black neighborhood of Harlem were appearing at a rate of 406.6 per 100,000 residents who eschewed injectable drugs. But among the addicted residents of Harlem an astonishing 3,740 per 100,000 had active tuberculosis.[54]

It seemed reasonable to hypothesize as early as 1979 that TB was spreading among members of the heroin-addicted populations of the wealthy nations, even as the disease was disappearing from their general populations. It might have warranted concern that decades of TB control efforts might be defeated if a subpopulation of actively infected individuals was left alone.

Indeed, Reichman had great difficulty finding a medical journal willing to publish his 1979 findings, and the paper was rejected several times, not because of any inherent flaws in the study, but because the journals simply didn't consider a high level of active TB among junkies terribly important.

Reichman absolutely believed in 1979 that injecting drug users were passing TB infections to one another. Unfortunately, virtually all societies on the planet held injecting drug users in contempt, viewing them as dangerous criminals, pathetically weak individuals, filthy denizens of ghettos, perilously insane characters, or satanically inspired deviants. Microbial threats to such individuals were generally ignored. Nearly every legal system defined some or all drug-related activities as criminal offenses.

Injecting drug abusers were outcasts, at the bottom of the social totem poles of nearly every culture on earth.

Furthermore, physicians generally detested working with addicted patients because the individuals rarely told the truth about activities that might affect their health, often failed to follow doctors' orders, sold their prescription drugs on the streets, and, if given the opportunity, stole needles and drugs from the hospitals and clinics they visited. Physicians who chose to specialize in treating and researching the unique health problems of drug abusers often suffered denigration from their colleagues, and wealthy private hospitals wanted nothing to do with either the drug-using patients or the physicians who cared for them.[55]

As a result, few professionals in the world in 1980 were in a position to notice what was going on in the heroin ecology.

One way drug users legally obtained money with which to buy narcotics was by selling their blood to hospitals and blood banks—a practice that would be outlawed in most industrialized countries by the mid-1980s but would continue in much of the developing world well into the 1990s. Most blood banks worldwide in 1980 didn't test their products for microbial contamination.

Toward the end of the 1970s a new set of players appeared on the international narcotics scene; South American cocaine cartels surfaced that converted the coca leaves of Bolivia, Colombia, and Peru into a potent white powder. Designed to be inhaled rather than injected, cocaine appealed to a different social class. It seemed "clean," its high produced a surge of energy rather than opiated enervation. And it was very expensive.

By 1980 cocaine had supplanted vintage wine in some cities as the drug of choice for the upwardly mobile. Its popularity was so great that icons such as pop stars, society matrons, literary celebrities, and professional athletes were fairly candid about their using it. Stories of pop heroes running quickly through $20,000 to $100,000 to support a cocaine addiction filled the gossip columns.

Few microbes were able to exploit the powder cocaine ecology effectively. The powder was dry and acidic—an environment hostile to most organisms. And few addicts could afford the kinds of long-term habits seen in heroin users that allowed for the slow growth and mutation of microbes over several generations of bacterial, fungal, or viral time. But some people turned to cocaine injections, allowing the microbes to exploit a new ecology that offered most of the benefits of the heroin environment.

In 1980 Don Francis found himself in the midst of an outbreak of a new strain of hepatitis B, spread among injecting cocaine users in New Bern, North Carolina. It seemed to have begun among the teenage sons and daughters of the city's upper-crust families, who had started shooting cocaine as an adolescent fad. Soon their poorer peers were following suit, having discovered that the expensive drug went a lot further when injected, rather than snorted.

When Francis got called to New Bern to head up a CDC investigation, ten of the teenagers had died of fulminant hepatitis B infections of their livers, and many more were sick. The virus was spread, of course, through shared needles. What alarmed Francis was how rapidly these kids got sick and died. These otherwise healthy adolescents were "dropping like flies," Francis told colleagues at the CDC. He suspected that it was what he called "a two-hit phenomenon"; some other bug was in the kids—possibly also passed by the needles—that acted in concert with the hepatitis virus to produce a disease more lethal than either could create on its own.

Francis injected hepatitis extracted from the blood of the infected New Bern teens into chimpanzees, but no disease occurred in the animals. For weeks he tried to make other test animals sick, with no result. In the end Francis was forced to give up. The New Bern teenagers stopped dying as soon as they ceased injecting cocaine, and there was no evidence that the mysterious microbes had found their way beyond the tight-knit cocaine ecology of the North Carolina city.

Not knowing what lurked out there, waiting for an ideal ecological opportunity to pounce, galled Francis no end.

Even as he closed the books on the New Bern case, another microbe was exploiting to its advantage the unique ecologies of cities on three continents.

11

Hatari: Vinidogodogo
(Danger: A Very Little Thing)

THE ORIGINS OF AIDS

Everybody knows that pestilences have a way of recurring in the world; yet somehow we find it hard to believe in ones that crash down on our heads from a blue sky.
—Albert Camus, *The Plague*, 1948

And that was the day that we knew, oh! In the world there is a new disease called AIDS. I thought surely this will be the greatest war we have ever fought. Surely many will die. And surely we will be frustrated, unable to help. But I also thought the Americans will find a treatment soon. This will not be forever.
—Dr. Jayo Kidenya, *Bukoba, Tanzania*, 1985

PRELUDE

Greggory Howard stood across the street from the ugly brick building and watched as junkies went in and reformed addicts came out.

Howard had tried methadone before—who hadn't? It was easy enough to buy on the streets during tough times when the police were busting local dealers or the supply from wherever hadn't made its way to Newark.

But today he was going to walk in that door and sign up for the methadone maintenance program. Last night's hit was the last.

He'd said that before, of course, but this time Howard was fed up with beatings, arrests, and looking up at the stars from a filthy alley. He was sick and tired of being sick. He wanted to "feel good about Greggory again."

Inside the Essex Substance Abuse Center, fluorescent light and iron bars greeted him, and Howard almost fled. But then he spotted the Dixie Cups. Methadone didn't come in Dixie Cups on the streets, but he had heard about this. It was almost impossible to steal a paper cup of neon-pink liquid and sell it on the streets. You had to drink it down right here, under glaring lights with the authorities watching.

Howard's body was trembling with the anticipation of that pink substitute high.

He stepped up to the iron-barred window and announced that he would like to quit his heroin habit.

Three thousand miles away in San Francisco, Bobbi Campbell stood adjusting his nun's habit. Campbell and friends from *Fruit Punch*, a gay men's radio talk show, had formed the Sisters of Perpetual Indulgence. The dozen or so Sisters would don their habits and carouse, given any good public forum. Handsome, black-haired Andy would shed his usual reserve and twirl a rosary while loudly declaring the beauty of gay love. Tall, thin Charlie would dance in circles singing "I Enjoy Being a Girl." Fred, with scraggly beard and wire-rims poking out of his face-framing habit, created endless clever chants, plays on Catholic homilies.

Still in graduate school at U.C. Berkeley, Bobbi, already a nurse, was the baby-faced member of the group. He wanted being gay in 1981 to be playful and joyous. Never mind those serious-politico-homosexual-rights-types who were embarrassed by flamboyant queens. Nurse Campbell, "Soeur en Drag," called himself Sister Florence Nightmare.

Everything about the full-time party that was San Francisco seemed fabulous to Campbell. True, everybody he knew seemed to have more than their share of one bizarre illness after another, but if it was all so joyous, who cared?

In Manhattan, Michael Callen was making music: disco dance tunes, gay love ballads, anthems. He, too, was thoroughly enjoying these days of liberation.

"Promiscuous" was a special word for twenty-six-year-old Callen. By the logic of the day, if it was liberating to openly declare one's right to have sex with a man, "it seemed to follow that *more* sex was *more* liberating," Callen said.

Like many, if not most, of the members of Manhattan's exploding gay community, Callen had left small-town America to escape the claustrophobia of his native Ohio. Raised a strict Methodist, the slender, non-athletic youth sang in the church choir and tried to belong. But he clandestinely devoured literature on homosexuality, most of it written by straight male psychologists. And he reached two conclusions: if homosexuality was a sickness, then he had it; and the best place to be "sick" was New York.

At age seventeen he had arrived in Manhattan, and soon discovered the gay bathhouses and sex palaces. With the exception of a several-months-long affair with a gay police officer, Callen's life from 1972 to 1981 was an endless string of sexual trysts and anonymous encounters—well over a hundred per year.

Thousands of miles and as many cultural leaps away, along the shores of Lake Victoria, Noticia finally had a dignified job as secretary to a Bukoba businessman. True, his tiny business wasn't much and her pay, even by

Tanzanian standards, was rather modest, but the job was honest and covered her bills.

After a year in Mombasa and Nairobi working as a prostitute, secretarial work wasn't at all bad. She had left her village of Nganga in late 1979 when it became obvious that her family would never recover from the shame of her rape by occupying Ugandan soldiers. Now, no man would marry her. Noticia could not have risen above outcast status unless she left Nganga. So she had followed the example of many other Mhaya women of Kagera province and made the long, difficult journey across Lake Victoria by steamship, then overland hundreds of miles to the turquoise Indian Ocean.

In the Kenyan seaside city of Mombasa, Noticia serviced the sexual needs of three or four men a day for very little money. Later, in the Sofia Town slums of Nairobi, she fared a bit better, making more money than she had in Mombasa. She saved enough money to return to Bukoba and start a new, independent life.

Noticia was a shy young woman, and her voice was as soft as silk. Her high cheekbones and dignified carriage attracted the men of Bukoba like bees to honey. They would beg her to go to the disco to dance, drink Safari beer, and listen to flattery.

Noticia felt hopeful about her future.

A thousand miles to the south, Dr. Subhash Hira and his staff at Lusaka University Teaching Hospital went over their medical records in a routine meeting. It was the usual daunting list of sexually transmitted diseases: syphilis, gonorrhea, chlamydia, chancroid, and the like. One of Hira's assistants pointed out that there was a woman on the ward suffering from an unusual case of herpes zoster: tough, perhaps a special kind of herpes.

Hira suggested that everybody keep an eye out for such things, and the meeting moved on.

I

In the fall of 1980, Dr. Michael Gottlieb was in his office at the University of California at Los Angeles Medical Center when a colleague asked if he would look at a particularly unusual respiratory case. A short while later, a frail man of thirty-three waited in one of the outpatient clinic's private rooms.

Gottlieb was startled by the obvious severity of the man's ailment. He appraised the patient carefully: pale, almost ashen; extremely thin, bordering on classic anorexia; a mouth full of the white "cottage cheese" indicative of a fungal infection; coughing uncontrollably, and evincing severe lung pain. It looked like pneumonia, but it was exceedingly rare that Caucasians of this age developed such brutal illness in Los Angeles.

Gottlieb ordered a bronchoscopy, as well as scrapings from the mouth sores, and had the sputum samples sent to the lab. The results astonished him: *Pneumocystis carinii* pneumonia, or PCP, filled the young man's lungs. Caused by a parasitic protozoa, PCP was almost exclusively seen among newborn infants in intensive care, terminally ill cancer victims, and/or elderly individuals living in nursing homes and other group settings. While nearly everyone had some *Pneumocystis* in his or her body, the organism was usually considered harmless because it was effectively kept in check by the immune system. What typical patients with PCP shared were exceptionally weak immune systems and concentrated exposure to other immune-deficient humans.

One thing was certain: it was rare, to the point of inconceivable, that this otherwise healthy man would have PCP.

"This is a red flag for something," Gottlieb told colleagues at UCLA. "This patient has no prior history of illness that should predispose him to *Pneumocystis*. It makes no sense."

The lab also reported that the white sores in the patient's mouth were caused by *Candida albicans* fungi, which could be sexually transmitted. And another sexually transmissible, usually harmless microbe was found in the patient's blood: cytomegalovirus.

Gottlieb took a careful history but learned little to explain his illness. True, the patient was a homosexual, and had had a few sexually transmitted diseases, but *Pneumocystis* wasn't spread sexually, and none of the three infectious agents ravaging him usually caused illness in healthy young adults. It just didn't make sense.

When Gottlieb ran blood tests the mystery deepened: the young man's antibody-producing capacity seemed intact, but his T-cell response was virtually nil. T, or thymus-derived, cells performed a range of crucial functions in response to infection, including identifying an invader and signaling the rest of the immune system to take defensive action against the microbe. Without an intact T-cell system no higher animal—be it mouse, dog, or *Homo sapiens*—could hope to halt the advance of even something as normally benign as *Pneumocystis*.

By March the patient had to be hospitalized. Gottlieb and his UCLA staff tried a variety of experimental and long-shot drugs on him, including the antiparasitic drugs trimethoprim-sulfamethoxazole and pentamidine and the antiviral acyclovir. The patient died on May 3, 1981: the autopsy found *Pneumocystis* throughout his lungs.

The terse litany of a medical report could never capture the drama of this patient's illness and death. For Gottlieb it had been shattering to witness, with uncharacteristic impotence, the patient's entire body fail, one organ after another, seemingly overwhelmed by waves of infection.

Even if this had been Gottlieb's only such case he would have felt compelled to chronicle the mystery for scientific scrutiny in some obscure medical journal.

But it wasn't the only case.

A Los Angeles private practitioner with a sizable gay clientele had, since late 1979, been spotting numerous cases of persistent long-term fatigue, reminiscent of mononucleosis, among his patients. Most of Dr. Joel Weisman's fatigued gay men were infected with the usually harmless cytomegalovirus.

In January 1981 one of Weisman's patients worsened significantly. In a few weeks, the thirty-year-old man's lymph nodes had swollen markedly, he'd lost more than thirty pounds, developed a pronounced *Candida* infection, and was running a daily fever of over 104°F.

By February, when it was clear the man wasn't improving with amphotericin B antifungal therapy, Weisman had him admitted to the UCLA Medical Center. Weisman and Gottlieb discussed the case, as well as other apparently odd infectious diseases seen among local homosexuals. When Weisman's patient also developed PCP in April, the doctors feared they were seeing a pattern.

By then Gottlieb had three other homosexual patients under treatment for PCP, none of whom was responding to treatment.

The similarities were striking: all five men were Caucasian, gay, aged between twenty-nine and thirty-six years at the time of PCP diagnosis, suffered PCP along with *Candida* and cytomegalovirus infections, had abnormal immune responses, reported multiple sex partners, and occasionally used amyl nitrite "poppers" as sexual stimulants.

One admitted to using injectable narcotics.

The "poppers" intrigued Weisman because he knew that use of the cardiovascular stimulants had recently become a fad all over the United States. Men believed the stimulants magnified the orgasmic rush of sex and enhanced their prowess.

Gottlieb wrote up a brief report and sent it to the CDC's Sexually Transmitted Diseases (STD) division, where Dr. Mary Guinan found it interesting enough to bring to Jim Curran's attention. They discussed the coincidences and, knowing that a number of STDs were epidemic in the gay community, speculated whether this might be due to any of several microbes then rampant in that population. Guinan pointed out that orders for pentamidine, an anti-PCP drug that physicians ordered through her office, had jumped from the usual fifteen requests a year to thirty in the first five months of 1981.

Curran decided to put the Gottlieb paper in the CDC's *Morbidity and Mortality Weekly Report*, and on June 5, 1981, U.S. physicians read for the first time of a curious new health problem in homosexual Americans.

The section written by Gottlieb and his Los Angeles colleagues was followed by an editorial, penned by Curran.

The occurrence of pneumocystis in these 5 previously healthy individuals without a clinically apparent underlying immunodeficiency is unsettling. The fact that these patients were all homosexuals suggests an association between some aspect of a homosexual lifestyle or disease acquired through sexual contact and *Pneumocystis* pneumonia in this population. . . .

All of the above observations suggest the possibility of a cellular-immune dysfunction related to a common exposure that predisposes individuals to opportunistic infections such as pneumocystis and candidiasis.[1]

On July 1, 1981, Dr. Paul Volberding opened San Francisco General Hospital's first designated cancer clinic. Not long out of residency, Volberding was pleased to be appointed acting chief of oncology for the city's primary public hospital, which also served as a teaching facility for the University of California at San Francisco Medical School. He selected as his nurse Gayling Gee, an experienced health provider whose staff record displayed a rare mix of administrative and patient care talents.

No sooner had the clinic officially opened than a nurse from another ward handed Gee the charts on an indigent cancer patient who had already been seen by several of the hospital's doctors. All of the physicians were baffled by the case. Gee looked at the diagnosis: Kaposi's sarcoma.

"Never heard of that one," Gee said.

"Well, take a look," the other nurse said. Soon, Gee and Volberding were examining a thin young man with pleading eyes. He had made the rounds of doctors, seen the befuddlement his case prompted, and was frightened.

Volberding studied the purplish-blue splotches on the man's body. These endotheliomas—out-of-control growths of the surface vascular networks on the skin—were a form of cancer extremely rare in the United States, though common in some parts of Africa.

"What do you do for a living?" Volberding asked, wondering if there might be some toxic chemical explanation for the tumors.

"I'm a hooker," the man replied. "Can you help me?"

Volberding had no idea how to respond.

Four days later the CDC published a report linking Kaposi's sarcoma, PCP, and homosexuality.[2] It described twenty-six cases of gay men in California and New York City who, though averaging just thirty-nine years of age, had all contracted the rare skin cancer usually seen in the United States only among elderly men. Eight of the men had died of either the cancers or other infections, most succumbing within a year of diagnosis. All but one of the men were Caucasian; the one exception was black. All were gay; no information about possible injecting drug use was provided.

The CDC also reported that the numbers of PCP cases were up, from five in Gottlieb's report a month earlier to a total of fifteen, all in California.

Credit for seeing a link between the skin cancer and prior PCP reports went to New York City dermatologist Alvin Friedman-Kien, who had documented an additional fifteen Kaposi's sarcoma cases by the time the CDC's report was released. That meant that at least forty-one gay men had Kaposi's sarcoma in New York, Los Angeles, and San Francisco, and some fifteen others had *Pneumocystis* pneumonia.

A review of the medical records at New York City's Bellevue Hospital showed that no men under fifty years of age had been diagnosed with Kaposi's sarcoma during the previous decade. Suddenly, there were thirty-three such cases in New York City.[3] San Francisco had two cases, though records at the city's five largest hospitals revealed no Kaposi's in men under sixty-five during the prior decade.

"Why, at this time, the disease would appear among gay men is unclear," Dr. John Gullet of San Francisco's St. Francis Hospital said. "All over the country scientists are working on this with a sense of urgency. Maybe we have a new virulent strain of CMV [cytomegalovirus]. That would be the most plausible explanation."

He added that the patient he had treated for Kaposi's "had no T cells. Zero. Zip."

Curran, Guinan, and Harold Jaffe were convinced that something serious was going on, but they lacked the resources for a full-scale study. Curran appealed to CDC director Dr. Bill Foege, who was fighting a losing battle with the new budget-cutting administration of Ronald Reagan. Swept into power in November 1980 on the promise of slashing the federal bureaucracy, Reagan vowed to reduce spending in all areas other than the military, domestic law enforcement, the space shuttle program, and a handful of other sectors. He had also promised to cut taxes, and sent a bill for the largest tax reduction in U.S. history to Congress for approval.

When Curran asked for funds for a full-scale investigation of the mysterious outbreak among homosexual men, he was told that massive cuts in the CDC budget were expected. The White House was, at that moment, lobbying hard for its tax reduction plan, which would be passed on July 29. Reagan's budget-axer, David Stockman, was submitting daily memos to federal department directors pointing out areas of alleged fat and duplication in their budgets. Directors such as Foege were meant to take such memos seriously.

To protect Curran's budget Foege took the epidemiology group out of the STD division, which expected severe budget cuts, and hid it in his own discretionary budget under the name Task Force on Kaposi's Sarcoma and Opportunistic Infections. He told Curran that ought to protect the admittedly paltry funds from David Stockman's ax. Nobody in the White House would know what Kaposi's was until they researched it and learned it was a cancer of elderly men—Reagan's constituency.

Curran was discreetly named director of the quietly created task force, overseeing a budget of less than $200,000 and a staff of twenty, most on loan from other programs.[4] The entire CDC budget for 1981 was just $288 million.[5]

Meanwhile, Gayling Gee was having a terrible time dealing with her Kaposi's patient. Homeless, moving from one San Francisco crash pad to another, the young prostitute would scrounge enough change every morning to buy a cup of coffee, a doughnut, and bus fare to the hospital.

"Help me, Gayling," he would plead. "I don't know what to do."

Too weak to work at any trade, he fell way outside the social services safety net of the day. Gee had no idea how to help.

In August, Volberding admitted him to the oncology ward: soon, he was dead.

There was little time to mourn. Volberding and Gee admitted three other gay men with the same strange cancer, and elsewhere in the hospital Dr. Constance Wofsy was handling an ever-increasing load of *Pneumocystis* cases.[6]

By the end of August the CDC had reports of 107 cases of either Kaposi's sarcoma, PCP, or the two combined in ninety-five homosexual men, six heterosexual males, five men of undetermined sexual orientation, and one woman.

"Whatever this is, it's not going to go away by itself. And it isn't an isolated event," Jaffe told fellow CDC task force members. Curran and Don Francis, who was assisting the team from his Phoenix laboratory, felt certain an infectious agent of some sort was responsible. But Jaffe wasn't ready to rule out a role for "poppers" or other factors in the gay scene. At two recent physicians' meetings, he had learned eye-opening facts about sexual practices in the gay community and about the rapidly growing, largely unreported numbers of cases of what appeared to be a radical immunodeficiency disease.

"Something terrible is happening," Jaffe said. "Something really terrible."

When the staid, married, heterosexual physician traveled to San Francisco, Los Angeles, and New York to see things firsthand he discovered what seemed like an unimaginable world. Local physicians who specialized in treating gay clients told him the new disease was related to practices in the bathhouses. From them, Jaffe learned of "fisting," "rimming," and a variety of stimulating drugs, all of which, the physicians said, could play a role in the odd ailment. The doctors assured Jaffe that these were the sexual practices of a clear but very sexually active minority of the gay community—some having upward of 200 partners a year.

The San Francisco Health Department's Dr. Selma Dritz was a key source for Jaffe. Since 1974 she had logged the escalation of sexually transmitted diseases within the city's gay population. Of the roughly 75,000 San Franciscans who entered the city's venereal disease clinics each year during the 1970s, 80 percent were gay men, Dritz said. Between 1974 and 1979 she had seen staggering increases in disease rates among homosexual men: amebiasis had increased by 250 percent; giardia infections jumped from one in 1974 to 85 in 1979; hepatitis A case reports doubled, hepatitis B tripled. Twenty percent of randomly tested gay San Franciscans in 1979 were gonorrhea carriers, perhaps 10 percent carried herpes simplex, and some smaller percentage were infected with syphilis.[7]

Most sexually active gay men living in cities like New York and San Francisco didn't go to straight doctors—they had their own physicians. By the time Dritz's words appeared in a leading scientific journal, the gay medical world had become nearly as separated from the mainstream as had the gay community as a whole. Even venereologists like Jaffe had barely an inkling of the profound biological events taking place in the gay population. And as savvy physicians like Dritz opened his eyes, Jaffe was shaken: what if this new ailment were caused by a sexually transmissible agent?

By August, CDC sociologist Bill Darrow was thoroughly convinced that the strange, lethal ailment was caused by some sexually transmitted microbe. He was also persuaded by the evidence that other factors, such as "poppers" and "fisting," had no direct role in the disease. But he had to prove it.

Toward the end of the summer of 1981 Darrow began to urge fellow epidemiologist Andrew Moss at the University of California at San Francisco to get involved in the investigation. That fall Darrow and Jaffe met with Moss, hoping he would help the CDC gain access to San Francisco research data.

Moss listened, asked a lot of questions, and pondered the implications for San Francisco. In 1983 the city's top gay Democratic Party leaders estimated that their constituency was 70,000 strong in a municipality of 650,000 people. If a sexually transmitted microbe was loose in such a large gay population, the potential for disaster was obvious.

In his characteristically perfunctory manner, the English-born Moss made suggestions and comments, never shying away from sexual matters or, as did most of his scientist colleagues, mincing words.

"Have you done the math, Bill?" Moss later recalled asking.

"Well, what are you driving at?" Darrow replied.

"Look, we've got men in the city [San Francisco] fucking maybe 300 other men every twelve months, okay? So, for the sake of argument, let's say only five percent of the gay community is that promiscuous. That's about 2,750 men, seeing 300 partners a year, for, let's say, five years. That's 4,125,000 sexual encounters in five years. Now, even if only ten percent of those original men—say, 275 of them—were infected with whatever this is, that would still mean 412,500 sexual encounters in five years. Assume an efficiency of transmission of, oh, let's say just one percent to be very conservative. That still means that 4,125 men in San Francisco are infected," Moss concluded.

Darrow succeeded in raising Moss's interest, and within weeks the English epidemiologist was discussing with Volberding the possibility of setting up a disease survey of the gay community.

Though it wasn't something Volberding would ever acknowledge publicly—he'd taken the Hippocratic oath, after all, obligating him to treat patients regardless of their ailments—if this was an infectious disease, he was frightened. He had seen a number of patients by then, witnessed their slow, agonizing deaths, and concluded that "this is the worst disease I can imagine."

He didn't want to get it, or to feel responsible for the safety of Gayling Gee or other staff at San Francisco General Hospital. Procedures such as bronchoscopies to test for *Pneumocystis*, frequent blood tests, and skin biopsies put him and his staff in contact with the patients' body fluids.

"I've got two kids at home," Volberding often thought, never allowing himself to mentally complete the sentence.

Volberding had often faced death among his predominantly elderly oncology clientele. All physicians had tricks for maintaining enough emotional distance from their patients' ordeals to avoid the risk of becoming emotionally paralyzed and unable to practice medicine. It wasn't difficult to accomplish when the patient was fifty years older than the doctor. But, like Volberding, most of the men with this disease were white middle-class guys who had gone through college during the 1960s. The more time Volberding spent with them, the more he found that he had in common with the dying men. It was easy to feel afraid.

In coming months, with no words of comfort from the CDC or the National Institutes of Health, Volberding's fears sometimes prompted a call to a fellow clinician in Boston to say, "Gee, I've got a fever. Do you think I might have it?"

Volberding was far from alone. Most of the physicians caring for the Kaposi's/PCP patients in 1981–82 were very worried about their personal safety, as well as the health of their staff. But the majority pushed on, got past the fear, adhered to their Hippocratic oath, and treated the patients. No study to determine the risks to health providers would be funded until 1984. To allay fears, the CDC would issue a list of recommendations for safe practices by health providers and laboratory personnel on November 5, 1982, suggesting that hepatitis B precautions already in place were adequate. But hepatitis B infection rates were soaring among health providers, and few took comfort in the shared "adequacy" of safety measures taken for the two diseases.[8]

In Antwerp, Peter Piot closely followed the reports about the new Kaposi's/PCP syndrome. He had an insight that gave him a cold chill.

Ever since his rite of passage into global disease research in Yambuku, Piot had maintained close links with Africa and the United States. Unlike most of his Belgian colleagues, Piot didn't find Americans crude and vulgar—in fact, he rather liked them. And he couldn't imagine embracing the neocolonialist attitudes toward Africans still so prevalent in 1981 among Belgians. Whenever the money could be found, Piot returned to America for more training and to Africa for research.

Which was why the CDC's reports in the summer of 1981 struck him with a sense of unease and recognition. Since 1978 he had been involved in STD research in East Africa, and many Africans came through his Belgian facilities for diagnosis when they suffered from an unusual ailment. Gottlieb's report of *Pneumocystis* cases among Los Angeles gay men reminded Piot of the Greek fisherman he had treated in Antwerp in 1978.

The man had commercially fished Lake Tanganyika, from the Zairian side, during the late 1970s. By the time he had reached Antwerp for treatment, he was just moments from death and could give little medical history. The autopsy was so astonishing that years later Piot would recall performing it in vivid detail.

The fisherman appeared to be in his late thirties, an outwardly healthy man. But when Piot opened the body the stench and sight of "pure and complete rot" greeted him. Every organ, each bone, all the tissues were covered with some type of mycobacterium. When Piot cultured samples of it in his laboratory neither he nor any of his colleagues could identify the organism. Whatever it was, this strange mycobacterium was not, in test-tube studies, a killer of human cells, and this fisherman shouldn't have died.

Having learned in Africa of the possible future value of such mysteries, Piot had carefully labeled and frozen samples of the fisherman's blood and tissues.

Piot wondered whether a new, lethal sexually transmitted disease might not be present already in many parts of the world, hidden under layers of neglect, racism, and poverty, and possibly masked by other diseases. He reviewed files on other strange cases that had come through his laboratory since 1978, finding three more bizarre deaths among Africans who sought care in Belgium. Though all three were young adults (one was female), they had, like the fisherman, succumbed to strange fulminant infections of organisms usually known to attack only immunodeficient humans: cryptococcal meningitis, other strange mycobacteria, and *Pneumocystis*.

All three patients, as well as the fisherman, had come to Antwerp from Zaire. And they all died before 1980. Could there be a link, Piot wondered, between whatever was killing homosexuals in California and these Zairian deaths?

By the end of 1981, Michael Callen was feeling lousy. Fatigued, feverish, incontinent, he sought help from a Greenwich Village private practitioner known in gay circles to be a good "clap doctor," Dr. Joseph Sonnabend.

The South African-born Sonnabend had been practicing medicine and conducting clinical experiments in New York for years, and was known for his brusque, outspoken style. In December 1981, Sonnabend told Callen that his illnesses were due to an underlying immunodeficiency of some kind. Unable to explain its cause, Sonnabend decided to aggressively treat all the other organisms that were taking advantage of Callen's beleaguered state, and put him on a prophylactic therapy with trimethoprim to prevent PCP.

It would be six months before Callen would be officially diagnosed as a GRID case—Gay-Related Immunodeficiency Disease.

Sonnabend asked Callen to participate in a study to test his hypothesis that the new disease was directly correlated with promiscuity. Having witnessed the steady rise in infectious diseases among New York's gay men, Sonnabend had a hunch that they had been exposed to ever-greater numbers of microbes, producing a sort of immune system overload, causing it to go haywire and self-destruct.

To test the hypothesis, Sonnabend divided his gay patients according to three tiers of relative promiscuity: monogamy, fewer than fifty partners a year, and men who, like Callen, had hundreds of sexual encounters a year. He sent the blood samples from the men to the University of Nebraska, where Dr. David Purtilo ran them through a fluorescence-activated cell sorter which separated out and counted specific immune system cells.

The study found that in some of the men a special class of T cells, called CD4 or T-helper cells, was virtually absent. These cells normally drew the body's defensive apparatus to the site of an infection and marshaled responses to rid the bloodstream of invading organisms. Without CD4 cells the immune system would be hard pressed to fend off any microbes.

Purtilo's data indicated that the most promiscuous men had the lowest CD4 counts, while the monogamous participants in the study had normal numbers of the T-helper cells.

The finding prompted Sonnabend and Callen to speak out to New York's gay community, warning that continued promiscuity could be lethal. New York gay playwright Larry Kramer echoed their warnings, issuing pleas for a slowdown on the sexual fast lane. All three men were rewarded with cries of outrage, denounced as "anti-gay faggots," homophobes, fearmongers, and fools.

Though vilified, the three were not silenced. Sonnabend flatly told his patients, "You're fucking yourself to death." Callen and Kramer tried to cook up ways to awaken their fellow gay liberationists to reality.[9] Toward the end of summer of 1981 Kramer called a meeting in Manhattan of like-minded gay activists. A handful of men turned up to hear his plea for health action. Money was raised and a name was selected for their new organization: Gay Men's Health Crisis (GMHC). The group's first public approach to their community was via the gay press and brochures distributed on Fire Island, the resort area to which Manhattan's endless gay party moved during the hot, humid summer.

They were roundly ignored.

By fall Bobbi Campbell noticed a few purple blotches on his skin. He had heard of the so-called gay plague. This looked like the ailment about which he had read.

Campbell went to see University of California at San Francisco physician Marcus Conant, who had been the first local doctor to spot Kaposi's sarcoma in a young homosexual. In his slight southern drawl Conant confirmed

Campbell's worst suspicions, and soon the youngest member of the Sisters of Perpetual Indulgence had joined Paul Volberding's expanding clientele at San Francisco General Hospital.

Almost immediately Campbell went public, declaring himself "The KS Poster Boy," sporting a canary-yellow "I Will Survive!" button and giving interviews to any and all media interested in the plight that he and a growing pool of San Franciscans shared.

Like Callen and Kramer in New York, Campbell began preaching caution to fellow gays, though he was less willing to condemn promiscuity, and personally continued visiting the city's bathhouses. To convince gay men of the danger in their midst, all he thought he had to do was point to his disfiguring, yet painless, purple tumors, and say, "See this?"

As the Christmas holidays of 1981 approached, scientists with the CDC and numerous U.S. medical centers reviewed the data on what they had dubbed GRID, Gay-Related Immunodeficiency Disease. It had occurred in 270 known U.S. cases during 1981, most—but not all—of whom were young, homosexual male adults.

Two leading symptoms marked the GRID syndrome: Kaposi's sarcoma and *Pneumocystis carinii* pneumonia. But other odd ailments were also seen: thrush, caused by *Candida* fungal infections; pronounced herpes simplex-II throughout the body; blood contamination of active cytomegalovirus with unknown effect; mononucleosis due to Epstein-Barr virus; marked lymph node swelling; radical infections of the stomach and gastrointestinal tract with *Entamoeba histolytica*; diarrhea and gastric problems caused by the *Cryptosporidium* parasite; similar symptoms caused by, of all things, *Mycobacterium avium*, a tuberculosis bacteria usually found in chickens; galloping infections in many organs of the *Cryptococcus* fungus; out-of-control bacterial infections with common organisms, such as *Staphylococcus aureus*, *Escherichia coli*, and *Klebsiella*.

One or more of these eventually killed many, if not most of the patients. A New York study of gay men with Kaposi's found that half died within twenty months of diagnosis,[10] and there was dire speculation that the strange syndrome might prove almost universally lethal.

Autopsies revealed that the young victims displayed severe organ damage. Vast expanses of tissue were necrotic. Microbes of all types—bacterial, fungal, and viral—had invaded and it seemed every organ showed signs of having been colonized and damaged. Much of the worst damage was caused by microbes that were usually utterly harmless to humans.[11]

The only possible explanation was total collapse of their immune systems.

Gottlieb's group at UCLA carefully studied the immune systems of four gay men with PCP, using techniques that were fairly routine in 1981. First, they measured the abilities of their patients' immune systems to muster antibody responses toward a variety of organisms, and calibrated the levels of antibody-producing B lymphocytes in their bloodstreams. All things considered, the men seemed to have normal antibody and B-cell responses.

That meant the arm of the immune system that produced specialized antibody proteins, tailor-made to recognize and attack very specific targets, was intact. But the T-cell side of the men's immune systems showed total disarray, and the chaos worsened as patients got sicker.

In 1981 immunologists were just beginning to appreciate the extraordinary complexity of the T-cellular immune response, and techniques for separating out various types of T cells were brand-new. For example, Dr. Len Herzenberg, at Stanford University Medical Center, had just a few years earlier invented the fluorescence-activated cell sorter, or FACS, which sorted different types of blood cells and could be used either to give researchers a pure cell population to study or to count how many of some particular type of cell were present in a blood sample.

Different types of T cells—which were white blood cells—had various proteins protruding from their surfaces that served to identify their function and form to other components of the body. Every single cell, from those that comprised a heart muscle to the brain's neurons, had such protein markers on their surfaces, allowing cells to "see" and "recognize" one another. Without such "sight" and "recognition" a collection of billions of cells could not organize itself into the complex entity that is a magnolia, leopard, or human being.

By the early 1970s immunologists had begun identifying various protein markers found on the surfaces of T cells, and understood that these markers distinguished groups of cells that had different jobs to perform in response to microbial invasion. This could be visualized by making antibodies in the laboratory against a given T-cell marker, attaching a fluorescent molecule to the antibodies, and mixing it all up in a test tube. If cells of a particular type were in a blood sample, the fluorescent antibodies would cluster on their surfaces, and scientists could see and count the cells using a fluorescence microscope. It was a tedious process, and it took days to count the marked cells in a patient's blood sample.

The FACS device reduced the counting time to a matter of minutes by dripping the prepared blood sample one drop at a time past a laser beam. The laser bounced off fluorescent cells, deflecting them into a separate test tube and simultaneously taking their count.

This and other pioneering techniques enabled immunologists to distinguish one population of T cells from another, and by 1981 they had come to appreciate the elegant complexity of the immune system. Hundreds of distinctly different types of cells, ranging from tiny, free-floating lymphocytes to huge, relatively stationary macrophages, were necessary to recognize an incoming microbe, latch on to the foe in order to draw the attention of other components of the immune system, signal secondary and tertiary lines of defense, and eventually consume and destroy the invader. Once the enemy was defeated, other immune system cells had to call off the attack, and dampen the response, lest the entire system overreact and destroy human cells.

Most of the job of marshaling immune system forces for microbial attack fell to T-helper cells that bore markers designated CD4. The job of calling off the attack and calming the agitated T-helper cells fell to so-called T-suppressor cells, which bore CD8 markers.

When Gottlieb at UCLA, Henry Masur and his team at New York Hospital in Manhattan, and Frederick Siegal's group at Manhattan's Mount Sinai Hospital scrutinized the cellular immune responses of their GRID patients, they discovered that the CD8-to-CD4 ratios were way off: most of the patients had too many CD8s and too few CD4s. Furthermore, it seemed the slow diminution in CD4 cells paralleled the patients' decline.

As a result, the patients appeared incapable of responding properly to most secondary infections. They were deficient, overall, in white blood cells and had radically diminished abilities to respond to foreign microbes. In some cases the GRID men's reactions to such things as *Candida* or streptococcus toxins were more than 150,000 times below normal. As measured in the laboratory, some patients' immune systems had *no* ability to kill any type of invading microbe.

The discovery of such profound immunodeficiency certainly explained why these men were ravaged by usually rare or benign microorganisms. But the solution to one mystery only deepened another: Why was it happening?

Several leading researchers were convinced that cytomegalovirus, or CMV, was the culprit. They had witnessed the extraordinarily rapid increase in active CMV cases in the gay population, jumping in less than a decade from less than 10 percent to over 94 percent of the nation's homosexuals.[12] But there was nothing special about the CMV running rampant in the gay community, and the virus was a common pediatric infection that never produced such serious immune system devastation in children. Some theorized that it was CMV superinfection—repeated episodes of sexual exposure to the virus—that resulted in the strange, deadly syndrome.

Acknowledging that CMV superinfection might have occurred in some of the cases, Masur warned, "In patients with evidence of cytomegalovirus infection, it is unclear whether the viral process was the precipitating cause of the immune depression or the result of reactivation subsequent to the initial immunosuppressive process. We are not aware of previous data suggesting that immunosuppression has been frequent among homosexuals."[13]

Sonnabend and some other New York physicians favored the multifactorial theory: the notion of microbial overload. They theorized that gay men had simply been exposed to too many microbes, of all kinds.

In the fall of 1981 Bill Darrow and a team of CDC researchers released a survey study of 4,212 gay men who responded to questionnaires distributed for the federal agency by the National Gay Task Force. The survey could not establish how representative the respondents were of the gay community as a whole (an important drawback), but its findings were striking for the STDs about which the men were asked: pediculosis, gonorrhea, urethritis, venereal warts, scabies, herpes, syphilis, and hepatitis B. For all eight diseases, gay men had rates of initial and repeated infections far greater than heterosexual men, and more frequent than had been seen in a CDC study conducted just five years earlier. And when Darrow's team evaluated what seemed to put gay men at such risk, they found that men who always took the anal-receptive role were at a somewhat greater risk of infection.

When Darrow's group charted their findings, a clear picture emerged. For all eight diseases, the incidence shot upward with each increase in the number of reported lifetime sexual partners. For example, a gay man with twelve lifetime sex partners had an 8 percent risk of contracting gonorrhea. But a man with 1,000 gay sex partners in his life ran a 75 percent risk.

A similar chart showed that chances of contracting one of the STDs increased depending on the population density of the individual's town. The CDC concluded that this further bolstered the notion that the number of different sexual partners in a lifetime was key, as small-town residents had less opportunity to form new gay liaisons.[14]

The *lifetime* risk factor seemed to imply a cumulative effect, making the men increasingly susceptible to disease as a result of years of microbe overexposure. So, concluded the STD-overload theorists, GRID appeared when a gay man's lifetime load of disease exceeded some crucial point, beyond which the immune system failed.

A serious problem with this perspective lay in mounting evidence that GRID was transmissible. How could immune system dysfunction be contagious?

"The fact that this illness was first observed in homosexual men is probably not due to coincidence," Gottlieb wrote. "It suggests that a sexually transmitted infectious agent or exposure to a common environment has a critical role in the pathologies of the immunodeficient state."[15]

The only environmental factor under serious consideration was poppers—the amyl nitrites some gay men used in the bathhouses. Though many in the gay community and in physician circles favored the notion that the well-described, mild immunodepression that could be produced by nitrites explained the profound disturbance seen with GRID, the argument was not considered compelling to most scientists in the field.

There was, of course, a fundamental flaw in trying to solve GRID on the basis of factors unique to the large U.S. gay communities: they weren't the only people suffering and dying from the mysterious new disease.

Frederick Siegal determined from a review of his medical records that his first Mount Sinai GRID patient was a thirty-year-old black woman from the Dominican Republic who died of profound immunodeficiency and related pneumonia in 1979. She, clearly, was not a gay man, nor apparently was her husband. She seemed to be a poor housewife with two children and no history of prostitution, drug abuse, or anything else to explain the lethal chaos of her T cells.[16]

In Masur's first New York City pool of eleven GRID patients were three heterosexual heroin or methadone users, one heterosexual cocaine addict, and two homosexual heroin injectors. In other words, for more than half of the men in the first New York group, drug use rather than gay sexual activity might have been the responsible factor. Clearly the phrase "gay-related immunodeficiency" couldn't apply to the four heterosexual drug users; nor could theories of causality that centered on behavior and infections unique to the most promiscuous elements of the gay community.

In Europe a smattering of cases among gay men were also noted in 1981; thirty-six in all, half of them in France. The first French GRID case was spotted by Dr. Willy Rozenbaum of the Claude Bernard Hospital in Paris in July, but Rozenbaum didn't connect the strange symptoms experienced by the man, a gay flight attendant, with those described in the American reports until a month later, when his patient developed PCP.[17]

By early 1982, GRID had devastated the immune systems of at least 310 men and a handful of women in the United States and Europe since 1978, killing 180 of them, and it appeared to be transmissible. Yet it still had aroused little interest or concern (outside a handful of public health circles), even from the populations at the greatest apparent risk. Total fiscal year 1981 U.S. federal expenditures on GRID research at the CDC and, minimally, at the National Institutes of Health came to less than $200,000.

Fiscal year 1982 (begun on October 1, 1981) promised Jim Curran a GRID budget of $2,050,000, though the dollars did not concretely exist on any agency budget line, and twenty-five scientists, most of whom had to be diverted temporarily from other CDC programs. Darrow was tracking down the sexual contacts of known GRID cases, seeking to bolster evidence of GRID's transmissibility. Guinan and Jaffe were running interference with the medical and gay communities, listening to theories and speculation while pushing for hard data.

Though Jaffe thought reports of GRID among injecting drug users just about proved that the syndrome was caused by a transmissible agent, he couldn't be sure. Most of the heroin-using GRID patients were dead by the time local doctors notified the Task Force on Kaposi's Sarcoma and Opportunistic Infections, so Jaffe couldn't interview the men to rule out the possibility that they might also have been homosexuals who were reluctant to reveal their sexuality to the physicians.

Curran was convinced that GRID was an infectious disease, but he was a far more political animal than most of the men and women on his team. He knew that only very solid evidence would persuade the nation to take steps to stop the epidemic, and pushed his team to keep searching.

And in the meantime he fended off seemingly endless countertheories, most emanating from the gay community and its physicians.

"They seem to constantly want to consider other causes," Curran would say. "A lot of people in the gay community are having a hard time accepting the idea that there is a new sexual disease. And a lot of heterosexuals want to think it's some sort of uniquely gay plague."

Curran was also fighting on the financial front.

He, infectious diseases division director Walter Dowdle, and Foege spent hours juggling the CDC's budget numbers and personnel lists in a desperate search for funds and scientists. Among the programs they robbed of funds and personnel during the first eighteen months of the outbreak were hepatitis surveillance, rabies control, studies of the long-term effects of Legionnaires' Disease, flu vaccine efficacy trials, Joe McCormick's Lassa research in Africa, other STD programs, laboratory supplies budgets, and tuberculosis control.[18] At the close of 1981, Curran drafted a bottom-line budget for his team's needs for the coming six months, requesting $833,800. This modest request prompted dissent from some members of the team—notably Don Francis—who felt that far greater resources were necessary. But Foege and Curran thought the figures were rock-bottom reasonable, and the CDC director took the request to Assistant Secretary for Health Dr. Edward Brandt, Jr., four times during December and January.

He was rebuffed.

By early spring 1982, the body count was rising fast, and Curran and Jaffe were convinced a vast iceberg lay below the visible tip of PCP/Kaposi's cases. They knew there had to be an asymptomatic stage to the disease, and they already had received reports of what sounded like a prodrome phase involving swollen lymph nodes and fatigue. Curran had no money to install permanent health advisers in the cities then reporting the greatest numbers of cases: Miami, San Francisco, Los Angeles, and New York. He had no funds for an active surveillance program to discover just how wide-spread the ailment might be in the United States. He couldn't fund case control studies of the various populations of people who seemed to be coming down with the disease. And only through such studies could the agency possibly say with credibility how the disease was transmitted, by whom, to whom.

Curran's tone in memos and letters became increasingly plaintive.

In April 1982, Curran warned the House of Representatives Subcommittee on Energy and Commerce, which controlled the CDC's purse strings, that "this problem is going to get larger . . . and some very, very large studies will probably be necessary in terms of defining the natural history of the syndrome. The role that CDC plays in these studies . . . has yet to be totally defined."

Though Foege supported Curran inside the Public Health Service, and argued repeatedly with higher-ups for funds, he was circumspect before Capitol Hill politicians. In congressional hearings he struggled to protect his agency from larger cuts at the hands of the White House's budget slashers, refusing to air the CDC's behind-the-scenes hostilities.

"As we have in the past when we have a health emergency, we simply mobilize resources from other parts of the Center," Foege told congressional inquirers. "If we reach a point where we cannot do that, of course, then we will come back and ask for additional funds, but at the moment that is the way we intend to handle it."

Virologist Gary Noble was on Curran's team, trying to set up a laboratory in which to search patient blood and tissue samples for evidence of a new virus. He spent most of his time writing memos requesting donations of surplus equipment, tables, and chairs from other CDC labs.[19]

Off in Phoenix, Don Francis was furious. The moment he had heard of the PCP cases in Los Angeles months earlier, Francis had called his former Harvard mentor, Max Essex. Francis thought, as early as June 1981, that the ailment was caused by a virus, though he had no idea what microbe might be blamed. Essex confirmed that it was reasonable to hypothesize a viral causality.

From 1981 to 1983, Francis worked in Phoenix on blood samples shipped from Atlanta and gave Noble a hard time about the essentially nonexistent GRID lab at the CDC.

"You gotta steal resources," Francis would say. "You gotta be an entrepreneur, a Milo Minderbinder type," he said, referring to Joseph Heller's *Catch-22.* "Scrounge!"

Noble offered to relinquish the viral effort to Francis.

"Come on, Gary. I'm in Phoenix, two thousand miles away. That's ridiculous," Francis said. Eventually, however, he agreed to make monthly trips to Atlanta to review progress and assist in the "scrounging."

As 1982 got underway the people most concerned about GRID organized. In Paris a group of physicians, scientists, and gay activists formed the French AIDS Task Force;[20] its goal was to trace the origins of France's cases and determine the cause of GRID. Having already noted that several of the early cases were gay men who had traveled to the United States, the group initially followed the hunch that GRID's cause was a transmissible agent that originated in America's gay community.

In New York City, Kramer's GMHC group was busy preparing a musical benefit, from which they hoped to raise money to care for the growing numbers of sick, publish educational pamphlets to distribute in gay bars and bathhouses, and lobby for research on what they considered a truly terrifying disease.

San Francisco's doctors and patients were also getting organized. Savvy gay political leaders who played prominent roles in the Democratic Party pooled efforts with Dr. Marcus Conant to form an organization which after three name changes would come to be known as the San Francisco AIDS Foundation. And on Wards 5B and 86 of San Francisco General Hospital, Volberding and Gee were creating what would become the first hospital facility in the world dedicated specifically to the care of people with GRID. Volberding was seeking Kaposi's volunteers willing to take experimental drugs and was lobbying the National Institutes of Health for research money.

But there was no outpouring of funds anywhere. GRID researchers world-wide in early 1982 were scrambling for crumbs and robbing other scientific enterprises to pay for the detective efforts they felt compelled to carry out.

Though he had no designated research funds, in response to Darrow's prodding Moss prepared an incidence survey during the winter of 1982 designed to give him a sense of just how many gay San Franciscans might already have GRID, or some sort of "pre-GRID."

Moss, with his University of California at San Francisco colleagues Peter Bacchetti and Michael Gorman, organized Selma Dritz's GRID information into a scientifically accessible form. They indexed the cases by zip code, then overlaid 1980 U.S. Census information on a San Francisco zip code map, rating zones according to numbers of never-married men over fifteen years of age. The zip codes with the most never-married men were in and around San Francisco's Castro District, the hub of the city's gay community. The majority of the GRID cases in Dritz's Health Department files were from the same neighborhoods.

Moss's group drew three startling conclusions: "the incidence of [GRID] in San Francisco is following an epidemic pattern," the incidence among all the city's never-married men was about 102 per 100,000, but the incidence among never-married men in the Castro area was 285 per 100,000.[21]

"Christ, this is big!" Moss told colleagues in an April 1982 seminar presentation of the data. He felt a knot in his stomach as he plotted projections from the incidence data.

"What we see is that about three out of every one thousand gay men in San Francisco already have this disease," Moss explained. "Now, if we assume the disease is caused by a transmissible agent, and we also assume this three-to-one-thousand rate has arisen fairly quickly, from something approaching zero back in 1977, then we can plot ahead. And it looks something like this."

On the graph's vertical axis were percent-of-gay-population-infected numbers. On the horizontal axis were years, from 1977 to 1985.

A line started at near-zero percent infected in 1977, and then climbed upward at a greater-than-45-degree angle, to 5 percent infected in 1978, about 15 percent in 1979, and over 40 percent of the city's gay population *already infected* as the group sat in that room, looking at, but not wanting to believe in, the chart.

By 1985, Moss predicted, three out of every four of San Francisco's gay men would be infected if, as most of the scientists in the room believed, the cause was a sexually transmissible agent. And if nothing was done to prevent the horror from unfolding.

Skeptical questions were asked, but Moss was known as an excellent, careful epidemiologist.

Moss had secretly hoped somebody would find a critical flaw in his study, revealing it to be overly dramatic or exaggerated. When no such mistake was identified, Moss was emotionally shaken, and began having what he called "fits of paranoia" and nightmares. He would lie awake nights trying to shake a vision of ten thousand dying men, some of them undoubtedly his friends and colleagues. The study was a political hot potato, and the new team of eight scientists and doctors working under Volberding's leadership—including Moss—were at odds over how best to release the findings. Feeling that the gay community should see the data as quickly as possible, Moss and Bacchetti discreetly leaked copies of their unpublished charts to key leaders of the San Francisco gay elite: members of the Harvey Milk and Alice B. Toklas Democratic Clubs.

The information would not be formally released for nearly a year, however, when Conant would describe the study's findings in a speech before physicians in New York City. And it wouldn't be published until April 23, 1983.

Moss was politically savvy and cynical enough to recognize that the U.S. President's most avid constituency was composed of right-wing religious moralists, and he suspected that his dire forecasts wouldn't muster much of a response in Washington.

His worst-case scenarios began to come true. Though San Francisco and the California legislature authorized research funds for the unfolding epidemic, pleas to Washington and Bethesda for funds were met with silence.

"This is an actual nightmare!" Moss said. "The sky is falling, we know it. You tell them it's falling, but nobody listens." He likened his funding search to "whacking a brontosaurus on the tail in San Francisco, and praying the neural message finally makes it all the way up the beast, to its pea-sized brain at the NIH and HHS [U.S. Department of Health and Human Services]."

Throughout 1982 Moss would keep on whacking that tail, while going on with his research—with or without research funds.[22]

In New York City, Dr. David Sencer, who had been forced out of his directorship of the CDC in 1977, was Mayor Edward Koch's Health Commissioner. Still in touch with old allies in Atlanta, Sencer knew that their hunch was that GRID was infectious and not solely gay-related. In March 1982, Sencer called a meeting of the physicians in New York who were most involved in GRID research.

"What do you want, and what do you need?" Sencer asked.

The audience wanted answers to puzzles that were expensive to solve. What causes GRID? Who in New York had the disease, and what populations were at risk? How was GRID spread? What treatments should they give their patients? Was there any risk that doctors and nurses could get GRID from their patients—was it seriously contagious?

Sencer agreed to contact the director of the National Institutes of Health, Dr. James Wyngaarden, urgently requesting funds to look for the answers.

And he assured Mayor Koch that the CDC was lobbying hard for increased attention to the problem, obviating the need for large expenditures drawn from the already stretched municipal treasury.

But NIH was not convinced that GRID warranted such urgent, high-priority consideration. Wyngaarden was appointed by the White House, as was his boss, Assistant Secretary Brandt. However persuasive the evidence of a dangerous epidemic might be, they were unlikely to win points with the White House by calling for urgent concern over what appeared to be a gay sexual disease.

Reagan appointees throughout the federal public health structure reflected the administration's concern for extremely conservative interpretations of health policy. Vociferous abortion opponent Dr. C. Everett Koop was named Surgeon General. Assistant Secretary Brandt was a "states' rights" advocate who believed that most sensitive health issues—such as venereal disease prevention—were best handled locally, rather than at the federal level. For Health and Human Services Secretary, Reagan chose a mainstream Republican, Dr. Richard Schweiker. And Schweiker's Deputy Secretary was Dr. Robert Windom, an ultraconservative Florida physician whose *Ask Dr. Bob* radio talk show, wildly popular in conservative circles, had positioned him as a leading celebrity fund raiser for the 1980 Reagan campaign. CDC director Foege, a close ally of former President Jimmy Carter, would soon be replaced by James Mason, a Mormon physician strongly supported by the conservative senator from Utah, Orrin Hatch.

Inside the White House, Reagan surrounded himself with domestic policy advisers who considered even Schweiker and Brandt too liberal: Jack Svahn, Gary Bauer, Nancy Risque, Carl Anderson, Bob Sweet, and Becky Dunlap. These powerful six had their political roots in extremely conservative religious and policy groups.

As Koop would later describe it: "The Reagan revolution brought into positions of power and influence Americans whose politics and personal beliefs predisposed them to antipathy toward the homosexual community."[23] So sensitive was the GRID situation in the eyes of the White House that, far from ignoring the epidemic (as has been alleged by many critics), key insiders sought almost from the beginning of the Reagan era to hold all federal actions on the matter under tight, centralized control. Koop, for example, though he was the Surgeon General and, therefore, logically the spokesperson for federal epidemic control, was flatly forbidden to make *any* public pronouncements about the new disease. More than five years would pass before Koop's gag would be untied. A CDC budget outline and description of funding needs, written in response to congressional inquiry, was blocked by Secretary Schweiker's office, and Democrats had to threaten congressional subpoena action to obtain the report in late 1982. Similarly, officials such as the NIH's Wyngaarden, the CDC's Mason, and HHS's Brandt knew they were expected to clear all potentially controversial comments on the topic with the Domestic Policy Council inside the White House.[24]

With time, some of Reagan's appointees would surprise observers at both ends of the political spectrum with their independence of thought and action. But there was no visible abundance of it inside the Reagan administration in 1982.

Not surprisingly, David Sencer's letter to Wyngaarden didn't result in the prompt action he and Mayor Koch had expected. It would be more than three months before the NIH would issue its first Request for Applications on GRID research (on August 13, 1982), and a full year before the selection and granting process would be completed. Checks for the first formal research grants to basic scientists wouldn't be cut until May 1, 1983.[25]

Researchers on the front lines warned that precious time was being lost, and the disease was spreading. But the NIH officially deflected most interest in GRID back to the CDC. In response to Sencer's urgent plea, Wyngaarden suggested the New York City Health Commissioner wait a year, for the NIH's next annual grant round. When an internal NIH report, signed by the director of the National Cancer Institute, urgently recommended creation of an emergency joint NIH/CDC task force to study the mysterious disease, Wyngaarden responded coolly: "While NIH does not bear a direct responsibility for controlling the outbreak, it is apparent that an epidemic of this sort may offer significant scientific opportunities. . . . I hope that NIH will not fail to capitalize on any opportunity to contribute. . . ."[26]

At the annual meeting of the American Public Health Association during the summer of 1982, the group's president, Dr. Stanley Matek, charged that the CDC was forced to stoop to "robbing Peter to pay Paul . . . Peter is currently the money for venereal disease and other vital public health problems."

By that time the CDC had spent just under $1 million for some thirteen months of GRID research.

In roughly the same amount of time—or less—the CDC had spent $9 million in pursuit of the cause of death of twenty-nine Legionnaires in 1976–77; more than $1 million on Ebola hemorrhagic fever investigations in Central Africa; at least $135 million on Swine Flu investigation and vaccine development. By the end of 1982, Brandt would defend the Reagan administration by pointing out that between June 1981 and December 1982 a total of 5.5 million federal government dollars were dedicated to the GRID effort, dispensed to the CDC, NIH, and Food and Drug Administration.

This would not appease critics.

"There is no doubt in my mind that if the disease had appeared among Americans of Norwegian descent or among tennis players, rather than gay men, the response of the government and medical community would be different," charged powerful Democratic Party member Congressman Henry

Waxman of California. "I want to be especially blunt about the political aspect of Kaposi's sarcoma. This horrible disease afflicts members of one of the nation's most stigmatized and discriminated against minorities. The victims are not typical mainstream Americans. They are gays mainly from New York, Los Angeles, and San Francisco. Legionnaires' Disease hit a group of predominantly white, heterosexual, middle-aged members of the American Legion. The respectability of the victims brought them a degree of attention and funding for research and treatment far greater than that which has been made available so far to the victims of Kaposi's sarcoma. I want to emphasize the contrast between the 'more popular' Legionnaires' Disease—which affected fewer people and proved less likely to be fatal— and Kaposi's sarcoma. What society judged was not the severity of the disease but the social acceptability of the individuals afflicted with it."[27]

With more than 500 diagnosed GRID cases in America, an apparent death rate of 50 percent, and no sign the epidemic would spontaneously abate, the mysterious ailment had become thoroughly politicized. Battle lines were drawn. Public health scientists and physicians were forced— against the basic natures of most—to choose sides. With time the situation would only worsen, antagonisms would heighten.

Swine Flu and Legionnaires' Disease had certainly been politicized epidemics, but scientists working on the front lines had, for the most part, been shielded from the squabbles and allowed to pursue their investigations. And they never lacked sufficient resources. If GRID had been, for example, a lethal contamination of a commercial food product, there would have been no question of the CDC's public health mandate: order a recall of the product, issue high-profile public warnings, and identify and disinfect the source of the contamination.

But what constituted proper health action in 1982 for GRID?

Curran and Jaffe felt a key part of their job was to warn the gay community. In public forums in New York, San Francisco, and Los Angeles the CDC scientists labeled GRID "an epidemic unprecedented in the history of American medicine" and urged gay Americans to shake themselves out of a state of collective denial. Curran would point to Bill Darrow's data showing that the more sexually active a man was, the greater his risk of contracting GRID.

Meanwhile, Darrow had for months been using the standard sociology techniques he had applied to other disease problems during his twenty-one years with the CDC, to try to disprove etiologic roles for "poppers," "fisting," and other environmental factors, and prove that GRID was caused by an infectious agent. He searched for an irrefutable infectious link between people who had the disease.

A crucial clue came on March 6, 1982, when the Los Angeles Department of Health got a phone call from a gay man who had previously been interviewed by CDC investigators, as had dozens of GRID victims, mostly in California and New York. The man called from an L.A. hospital, where his lover had just succumbed to GRID.

"There are two other guys here in the hospital with the disease right now, and I know they had sex with my lover," the man said.

The call was referred to Dr. David Auerbach, an EIS trainee for the CDC, based in Los Angeles. Auerbach met with the informant hours later, and heard a sexual saga that began in October 1979, when five previously unacquainted gay couples shared a table at a benefit banquet.[28]

The informant and his boyfriend, like the other four couples, had a long-standing but nonmonogamous relationship.

During the summer of 1980 one of the couples threw a backyard barbecue party, inviting a pair they had met at the benefit, who brought with them a gay prostitute. That night all five men had sex with one another. Sometime later, the informant's lover had sex with a member of the barbecue crowd.

Two months later, two members of the barbecue quintet contracted *Pneumocystis* pneumonia. A few weeks earlier the informant's lover had discovered Kaposi's sarcoma splotches on his skin.

The three men died on October 6, 1981; February 6, 1982; and March 6, 1982.

"Six-six-six, you get it?" the informant asked. "Six-six-six!"

Moved by the biblically ominous coincidence of sixes, the man had called the Health Department.

Auerbach telephoned Darrow in Atlanta, who took the next flight to Los Angeles.

For several high-paced days Auerbach and Darrow crisscrossed Los Angeles and Orange counties, interviewing the eight surviving GRID patients of the nineteen cases diagnosed in the two counties prior to April 1982. To gain information on the eleven who had died, the CDC investigators sought out family members, ex-lovers, and friends. Many refused to cooperate, but within two weeks the scientists had solid information sexually linking nine of the men.

By April 7, Darrow and Auerbach had established that two members of the barbecue party group had in 1979 and 1980 had sex with two other Los Angeles GRID cases—individuals who hadn't yet been linked in any way to the rest of the group.

As they crisscrossed Los Angeles that spring day, something eerie happened. Two unacquainted men with GRID independently mentioned a handsome French-Canadian flight attendant with whom they had had sex. The coincidence was striking.

The CDC investigators were further "astounded," they said, when "on the same day the companion of a third case in Los Angeles said that his roommate had had sexual contacts with two friends of this same out-of-California [Canadian] case."

Though Darrow and Auerbach went to great lengths to protect the confidentiality of the men they interviewed during the 1981–82 investigation, even destroying all photographs and identifying material, somebody close to the inquiry leaked the Canadian's name to San Francisco Chronicle reporter Randy Shilts.

And Gaetan Dugas would be, after his demise, vilified and crucified, mistakenly named as the man who personally spread GRID around North America.[29] In 1985 Dugas's photograph would hang in the STD clinic in Lusaka University Teaching Hospital in Zambia, captioned: "The Man Who Started the Epidemic." Because four of their Los Angeles GRID cases named Dugas as their sexual partner, the CDC investigators designated the Canadian as "Patient Zero." This would later be mistakenly interpreted as indicative of a primary, causative role for Dugas.

Darrow flew the following day to New York City to interview Dugas, the man's physician, Alvin Friedman-Kien, having agreed to introduce the two when Gaetan came in for a checkup.

Darrow was struck by Dugas's candor and swagger. Though Dugas had a few Kaposi's lesions, he seemed unconcerned. He said he felt fine. And he anticipated sexual encounters in the dozens of cities he was scheduled to fly into over coming weeks.

Dugas matter-of-factly laid out his sexual/disease history for Darrow. By late 1978—Darrow's study period—Dugas was averaging 250 encounters a year. Between December 1978 and April 1982, Dugas guessed, he'd had sex with about 750 men. He estimated that his lifetime total of sex partners, since he became active in 1972, exceeded 2,500.

During that time Dugas had suffered ailments that weren't diagnosed as GRID until July 1981. In 1979 his lymph nodes swelled appreciably and he felt as if he had a severe case of the flu. A few months later he came down with Pneumocystis pneumonia and was hospitalized in Canada. By early 1981 Dugas had Kaposi's sarcoma, and in July 1981 he came under the care of dermatologist Friedman-Kien.

The CDC's Mary Guinan had interviewed Dugas during the summer of 1981, and his story was already in the agency's files when Darrow asked Dugas to list as many of his sex partners as he could remember. Dugas had never learned the names of most of his bathhouse partners. But he was able to confirm those four Darrow and Auerbach had discovered in Los Angeles, and add sixty-eight more, including four New Yorkers whose cases then led Darrow to a cluster of men who partied on Fire Island during the summers of 1979 and 1980.

By June 1982, Darrow and Auerbach had gathered enough circumstantial evidence to link forty gay GRID victims to a casual sexual network that spanned New York City, Atlanta, Houston, Miami, San Francisco, and Los Angeles. Darrow presented their evidence to Curran, Jaffe, and other members of the Task Force.[30] Curran and Jaffe found it compelling proof that GRID was a sexually transmitted infectious disease, and published the Los Angeles component immediately in the Morbidity and Mortality Weekly Report.

But was it?

Auerbach and Darrow thought they were looking at a disease that rapidly progressed from infection to symptoms, and death. By focusing on the most recent sexual exploits of the men they questioned, the scientists got the impression that, for example, eight members of their study group were infected by Dugas in 1979 or 1980, and developed symptoms within an average of ten months thereafter. They reasoned that GRID was a new disease, and the causative agent had been in the United States only since 1978.

But it would later be clear that the disease's latency period for gay men averaged over ten years, and healthy white, middle-class men in particular almost never developed symptoms as serious as Kaposi's sarcoma or PCP within seven to fourteen months of infection, as Darrow and Auerbach had assumed.[31]

Nevertheless, the cases interconnected so perfectly, with Dugas at the hub, that Darrow and the entire CDC team were absolutely convinced that the disease was due to a sexually transmitted agent. On the basis of those findings, University of Washington clinical researcher Lawrence Corey urged in September that further attention be given to use of condoms to prevent passage of sexually transmitted diseases,[32] but Curran was reluctant to go up against the anti-birth control forces in Reagan's powerful advisory circle without stronger evidence that GRID was caused by a sexually transmitted agent.

Complicating efforts to link GRID cases was increasing evidence that groups of non-gays were coming down with the disease. By mid-1982 the Task Force was convinced that injecting drug users were contracting the disease. Only a minority of them were gay men, and some were heterosexual women. While gay men seemed to be uniquely at risk for Kaposi's sarcoma, the other prime symptom of the new disease—*Pneumocystis*—was striking a broader spectrum of human beings. One out of four men with PCP was heterosexual, the CDC reported in June, and of 152 cases closely scrutinized by Curran's team, 26 were heterosexual men and 8 were women. Twenty-one of the 34 heterosexuals were intravenous drug users.[33]

A month later the CDC reported that a disease that appeared identical to GRID had broken out among Haitians. Thirty-four cases were described among young male and female adults living in Miami and New York. In addition, the report referred to eleven Kaposi's sarcoma cases diagnosed in Port-au-Prince, Haiti. Most of the individuals were heterosexuals with no history of intravenous drug use. And in addition to the opportunistic infections already noted among gay GRID sufferers, the Haitian patients experienced profound tuberculosis and toxoplasmosis infections.[34] All of the Haitian patients in New York and Miami had recently immigrated from Haiti's poverty and severe political oppression. Most feared publicity might result in their deportation.

The CDC had actually been informed of the Haitian cases during the fall of 1981, when Drs. Margaret Fischl and George Hensley spotted cases at Miami's Jackson Memorial Hospital, and Dr. Sheldon Landesman reported treating such individuals at Kings County Hospital in Brooklyn. Jaffe dispatched Belgian CDC physician Alain Roisin, who spoke Creole, to Port-au-Prince, and Roisin was able to confirm that the cases there were identical to those reported by Fischl and Landesman.

An unintended result of the CDC's Haitian report was a new round of blame. Perhaps, National Cancer Institute physician Bruce Chabner publicly speculated, the disease was caused by a "Haitian virus" that was brought back to the United States by homosexuals. It was suggested that the Caribbean resort had become a favorite of gay Americans during the late 1970s and early 1980s. GRID, some scientists said, might even have originated in Haiti.[35]

Some North American researchers familiar with the Haitian situation insisted that most Haitian men denied homosexual behavior because of social stigmas and that all the Haitian cases were due to clandestine homosexuality. The fact that a significant percentage of Haitian GRID patients were female was conveniently ignored.[36]

Darrow knew, from his studies of the gay GRID network of sexual liaisons, that at least one of the New York men was a flight attendant (not Dugas) who frequently flew to Haiti, and other East Coast men acknowledged having vacationed on the island. Dr. Friedman-Kien, who had a large gay clientele, told Jaffe that many New York men vacationed in Haiti because they could buy sex for less than five dollars in the impoverished country, where the average daily wage was less than two dollars.

As for injecting drug users, the CDC knew that several of the 1981 cases had involved men who were both gay and narcotics addicts. These seemingly separate communities had overlaps, and Jaffe privately pictured what he now considered to be an epidemic as a series of circles of varying sizes, overlapping wherever there were people who shared more than one type of behavior that could put them at risk for the disease.[37]

While no causative agent for GRID had yet been discovered, the uniquely high incidence of the disease among members of a specific immigrant group prompted finger-pointing. It would worsen with time, spurred by what Haitians experienced as racist views of their culture, their lifestyles, and themselves as individuals. It would, unfortunately, not be the last time a nation and its people felt unfairly blamed as the source of the new disease; indeed, such blame would remain a hallmark of the epidemic for over a decade.

While top federal authorities pondered the significance of discovering GRID among Haitians, the disease turned up in three men born with the genetic blood-clotting disease known as hemophilia. Because they suffered frequent blood loss, the men had received many injections of Factor VIII blood coagulant concentrate. The three men ranged from twenty-nine to sixty-two years of age and came from parts of the country not yet known to be affected by the epidemic: Denver, Colorado; Westchester, New York; a small town in northeastern Ohio.[38]

Factor VIII was made from the pooled plasma of thousands of donors, so people with hemophilia were particularly vulnerable to contaminants in the blood supply. A typical surgery patient might need six units of transfused blood, donated by, at most, six people. But individuals with hemophilia were exposed to the blood of thousands of people each time they injected Factor VIII. Not surprisingly, they had high rates of blood-borne diseases, such as hepatitis.

The CDC met on July 27 with the National Hemophilia Foundation, the American Red Cross, and the FDA, and mutually agreed to an aggressive surveillance effort.

By December all three of the original hemophilia/AIDS patients were dead.

Yes, AIDS.

In August the CDC had quietly dropped the term GRID, changing the name of the disease to Acquired Immune Deficiency Syndrome to reflect the recognition that it wasn't just a disease of gay men.

Five more Americans with hemophilia contracted AIDS in 1982, one of them a seven-year-old boy. And in the fall of 1982 Curran's team got word that Dr. Arthur Ammann, a pediatrician with the University of California at San Francisco Medical Center, was treating a baby with PCP. The twenty-month-old patient had received multiple transfusions at birth.

AIDS was in the U.S. blood supply. The CDC ended their report with the following statements:

Of the 788 definite AIDS cases among adults reported thus far [December 10, 1982] to CDC, 42 (5.3%) belong to no known risk group (i.e., they are not known to be homosexually active men, intravenous drug abusers, Haitians, or hemophiliacs). Two cases received blood products within 2 years of the onset of their illnesses and are currently under investigation.

This report and continuing reports of AIDS among persons with hemophilia A raise serious questions about the possible transmission of AIDS through blood and blood products.[39]

For physicians it was alarming news. At the time most U.S. blood banks and blood factor manufacturers purchased plasma, and the prime "$10 donors," as they were called (though they might earn $100 per donation), were drug addicts and alcoholics looking for quick cash. New York physician Frederick Siegel immediately called for cessation of blood and plasma purchasing as well as stern advice to gay men that they not donate their blood.

No immediate action followed.

A week after the agency released its blood report, the CDC announced that four other babies and infants definitely had AIDS and eighteen were suffering suspicious immunodeficiencies. None of these children had received blood transfusions, but most had mothers who either had AIDS or fit into an already defined risk group for the disease. Of the thirteen mothers interviewed for the investigation, eight were injecting drug users, one was both a drug user and a prostitute, and two were Haitian. The children all came from areas obviously affected by AIDS: San Francisco, New York, Newark.

"Transmission of an 'AIDS agent' from mother to child, either in utero or shortly after birth, could account for the early onset of immunodeficiency in these infants," the CDC scientists wrote.[40]

In San Francisco, Mrs. Profit, as she was called, gave birth to two children in 1981–82; both had AIDS. A prostitute who worked San Francisco's tough Tenderloin district, Mrs. Profit already had an advanced case of AIDS herself when Dritz's public health team caught up with her in late 1981. By that time Profit was already fairly incoherent, and doctors later concluded that she had AIDS dementia.

Profit wasn't able to be of much help to Moss, Dritz, and other researchers who quizzed her. Though the white woman worked as a prostitute, Moss concluded it was through her injecting drug habit that she had become infected. Or via her extremely secretive and also AIDS-plagued husband, a heterosexual male who refused to provide the scientists with any information.

As the new year opened, the United States had its one thousandth official case of AIDS, and CDC scientists knew the true numbers were far greater.[41]

By the end of 1982 the CDC had nailed down every basic aspect of the epidemiology of AIDS save one: identifying the causative microbe. But they knew it was infectious, was in the nation's blood supply, could be passed by gay men to one another through sexual intercourse, by mothers to their babies, and among drug injectors who shared needles.

And though in this matter the CDC erred, the agency already had evidence of heterosexual transmission of AIDS. All but five of their fifty-five heterosexual cases were improperly given a segregated label as "Haitian." (A nationality designation did not equal a mode of transmission.) The other five suspected heterosexual cases were women whose steady sexual partners were male injecting drug users.[42]

Though puzzles remained, the essential epidemiological outlines were in place, and some decisions about public health action could be taken to stop the epidemic's spread.

Yet a decade later many of the preventive steps that seemed obvious in January 1983 would remain untaken. Delays would be numerous. People would continue to die, and the epidemic would expand.

II

Though microbes know no politics, and *Homo sapiens* of all ideological stripes could be infected with the agent responsible for AIDS, every aspect of AIDS research, control, and treatment was highly politicized by 1983.

When it came to AIDS, every organization and agency seemed to be breaking its own long-standing protocols vis-à-vis infectious diseases. Municipal public health departments, afraid of offending gay voters or civil libertarians, were reluctant to close down bathhouses, despite clear evidence that many—perhaps most—of the homosexuals diagnosed at that point with AIDS acknowledged having frequented the baths. Bathhouse owners, afraid of scaring off customers, only agreed under legal pressure to post signs warning of the risks that might be associated with casual sex. Gay rights organizations all over the United States split politically over what levels of alarm and action seemed appropriate.

Blood bank administrators gave lip service publicly to concern about blood supply safety, but privately told government authorities that no steps could be taken to ensure product safety without incurring prohibitive costs. And the National Heart, Lung, and Blood Institute told Congress that it had no intention of funding research into the safety of the nation's blood supply until the end of 1984. No serious survey of the nation's blood supply would be undertaken until September 20, 1984.[43]

Throughout the fall of 1982 the National Hemophilia Foundation (NHF) met with blood industry representatives and federal scientists in hopes of finding a way to ensure the safety of products used by its membership. They were repeatedly met with calls for proof, for data linking something in the nation's blood and plasma supplies to the ailments suffered by people with hemophilia and transfusion recipients. Only the CDC team wholeheartedly supported the NHF's call for action.

At the behest of the NHF and CDC a meeting was convened in Washington, D.C., on January 4, 1983, with leading blood industry representatives and officials from the Food and Drug Administration to discuss the first handful of transfusion and blood coagulating factor AIDS cases. The CDC's Bruce Evatt, Curran, and Francis hoped to convince the blood industry to take steps to decrease the possibility of passing on whatever caused AIDS via blood, not only to protect Americans who had hemophilia or were undergoing surgical procedures that required transfusion but also the millions of people internationally who relied on blood products from the United States, the world's largest blood product exporter.[44] Conservative estimates valued the U.S. share of the industry at $150 million a year. Four American companies dominated: Baxter Travenol Laboratories, Inc.; Alpha Therapeutic Corporation; Armour Pharmaceutical Co. (a division of the Revlon Cosmetics corporation); and Cutter Laboratories. The domestic blood market, controlled by diverse for-profit and nonprofit interests, was conservatively estimated to be worth roughly $250 million a year.[45]

Less conservative estimates of the global blood industry, compiled by the *Philadelphia Inquirer* Pulitzer Prize-winning journalist Gilbert Gaul, put plasma annual sales revenues at $2 billion, with U.S. companies and blood banks responsible for over 60 percent of worldwide sales of some 6 million liters of plasma every year.[46]

The use of blood products doubled in the United States between 1971 and 1980, due both to an increase in the number of surgical operations and to perfection of procedures for isolation and preparation of Factors VIII and IX, the coagulants most commonly needed, for genetic reasons, by people with hemophilia.[47] New surgical innovations were putting unprecedented pressure on the blood supply. For example, a transplant procedure might require over 150 units of whole blood for a single patient. In 1980 just over 11 million units of blood were collected in the United States. By 1982 that figure had jumped to 12.6 million, and about 4 million Americans received blood products that year.

All but 2 percent of the nation's blood supply was voluntarily donated. Plasma was another matter. Donation of plasma was a three-hour procedure, during which the individual's blood was removed and spun down in a centrifuge to separate the watery plasma from blood cells. The cells were then infused back into the donor's body. Because it was a demanding, uncomfortable procedure, most of the world's plasma came from paid "donors," and the United States bought more plasma in this manner than did any other country in the world. Legally, in the United States, individuals could sell their plasma twice a week, or up to 60 liters a year, which was four times WHO's recommended maximum for individual "donation."

Individuals were typically paid around $25 for donating plasma, and most "donors" were regulars: hard-luck characters who frequently sought quick cash from local hospitals and storefront plasma centers.

With fewer than 240,000 units of whole blood purchased in the United States, and perhaps less than 10 percent, or 24,000, of those units contaminated with any of the dozens of microbes known to be passed among injecting drug users, the odds that any given blood recipient in 1982 could become infected with, for example, hepatitis B, might be calculated to be well below 1:11,997,600.[48] Since nobody knew what microbe caused AIDS, it wasn't possible to calculate the odds in 1983 of becoming infected following transfusion with a single unit of blood.

But the clotting factors were proteins found in extremely minute quantities in plasma, and about 5,000 plasma units[49] were required to manufacture enough Factor VIII or IX concentrate to stop serious bleeding episodes among people with hemophilia. That upped the ante considerably: the odds of getting hepatitis were probably at least 1:3,000. For people with hemophilia, each Factor injection carried a reasonably high chance of infection with some microbe.

The estimate of 1:3,000 was based on an assumed minute level of contamination among the paid donors. Gay men were no less altruistic in donating blood in 1982 than were their heterosexual counterparts. If one conservatively assumed that 4 percent of all blood donors were gay, and 10 percent of them carried some type of blood-transmissible agent, that could mean up to 47,000 units of the American donated—unpaid—blood supplies were contaminated, or 0.003 percent. Again, a minute risk if one were exposed to only one or two transfused units of blood, but a considerable hazard for people with hemophilia.

Retrospective tests of the blood supply would later show that in 1978 at least one batch of Factor VIII was contaminated with the AIDS agent. It was dispensed to up to 2,300 men and boys with hemophilia that year.[50]

The risk for people with the genetic blood-clotting disease was compounded by their need for frequent injections, amounting to between 25,000 to 65,000 international units (IU) of the crystallized protein powder per year. An average treatment ampule was equal to 100 IU. Therefore, the typical person with hemophilia was annually exposed to the blood of 1,250,000 to 3,250,000 people. Severe cases of hemophilia could require use of products derived from 13,555,000 units of blood each year. With that level of exposure, even an extraordinarily rare microbe found in only one out of every 3 million Americans could pose a serious threat.[51]

Widespread, nonemergency use of Factors VIII and IX worldwide did not begin until 1975, when the U.S. Congress created financial incentives for the manufacture and distribution of the product with the passage of the Hemophilia Diagnostic and Treatment Center Program Act. By 1982, 75 percent of all people with hemophilia already had abnormal liver function, due to hepatitis (types A or B) infections, and more than 90 percent of them had been exposed to the virus. Though the blood industry was using a heating technique to eliminate live microbes from blood used in the manufacture of another human protein product, albumin, it did not heat-treat clotting factors. A variety of reasons for this were offered, but they all boiled down to the small size of the hemophilia product market, the added costs of sterilization, and the lack of studies demonstrating its reliability in the case of Factors VIII and IX. In 1980 a single Factor VIII dose cost a patient about $90.

In 1987, after heat treatment and other blood safety techniques were common practice, that same dose of Factor VIII would cost more than $1,000. By 1989 hemophilia patients would, on average, spend over $50,000 per year. Despite the more than tenfold increase in costs, the industry would remain quite healthy, and people with hemophilia in the United States, Europe, and Japan would continue to receive lifesaving supplies of the compound.

An estimated 26,000 Americans—most of them boys and post-adolescent men—had hemophilia in 1983. Their life expectancy had improved radically because of Factors VIII and IX; prior to 1970 few could hope to live past the age of twenty-five, with an average age of 11.2 years. But by 1980, people with hemophilia were averaging thirty-eight years of age.

When Curran, Evatt, and Francis met with blood industry and FDA representatives in January 1983, they had woefully little data on hand to argue for urgency because of AIDS. But they did know how rapidly hepatitis B had entered the blood supply, and statistics on infection among people with hemophilia were known to all in that room.

Several options were discussed, including use of a recently developed hepatitis test that could directly detect the presence of the virus (core antigen) in blood samples. Evatt argued that many people with AIDS—well over 50 percent—had histories of hepatitis B infection, and use of such a test could dramatically decrease the danger of contracting AIDS from blood products.

Blood bank representatives protested that the epidemiological link between hepatitis B and whatever caused AIDS was not as clear as Evatt indicated. Furthermore, they noted, many authoritative scientists were arguing that the disease wasn't due to an infectious agent at all, but to "poppers" or "lifestyles." For those who were willing to accept the notion of a single infectious cause of AIDS, there remained great reluctance to believe that there could be an asymptomatic carrier state of the disease that couldn't be ruled out with a simple symptoms questionnaire. (Such an asymptomatic state existed, of course, for most blood-borne diseases.) Finally, they claimed that such testing would be expensive, adding a cost of two to five dollars per blood unit, which they would be forced to pass on to consumers.

Francis lost his temper. He pounded the table and raised his voice, accusing the blood bank officials of callous disregard of the health of millions of Americans.[52] The meeting deteriorated amid hot tempers and accusations.

In the end, the group agreed to nothing. A course of voluntary action that Dr. Frederick Siegel had recommended six months earlier was adopted by some members of the industry: actively screen out gay, drug-injecting, and Haitian donors.[53]

On March 25, 1983, Assistant Secretary for Health Brandt formally recommended—but did not mandate—"interim measures to protect recipients of plasma, blood and blood products until specific laboratory tests are developed to screen blood for AIDS."

The three recommended measures were: to educate donors about who should refrain from donating, to teach blood drive workers how to recognize medical histories among donors that might be indicative of AIDS, and to establish systems for storing or disposing of suspected AIDS-contaminated blood.

Encouraging educated gay men to refrain from donating blood was relatively simple, and was largely achieved by the gay press. But discouraging injecting drug users from selling their blood and plasma was nearly impossible. As long as someone was willing to buy, they were eager to sell.

Convinced that a hepatitis core antibody test, however imperfect it might be, would offer a vital margin of protection to people with hemophilia, the NHF would continue lobbying the FDA well into 1983. Strangely, at a time when both hepatitis viruses and whatever caused AIDS were in the U.S. blood supply, the Reagan administration radically decreased the overall size of the Food and Drug Administration, reducing by 25 percent, among other things, the size of the staff dedicated to overseeing the blood industry. In the spring of 1983 FDA Commissioner Frank Young announced a 50 percent cut in the number of his agency's blood industry quality control inspections. It would be more than five years before the repercussions of Young's decision would come under review, and during that time the industry would grow dramatically. The American Red Cross alone would expand its blood program by over 150 percent. So by the time in 1988 when the FDA reconsidered its policies, most blood and plasma collection and processing facilities would have been left uninspected for three or four years.

The CDC and FDA would be at odds over the blood supply throughout 1983, and Assistant Secretary Brandt generally would side with the FDA's "wait and see" approach. In November the National Hemophilia Foundation's scientific advisory board decided to demand that the plasma industry use the hepatitis B core antibody assay to screen out as many contaminants as possible.

The FDA's Blood Products Advisory Committee agreed to meet in Bethesda in December 1983 to debate the NHF's demand. But industry representatives gathered in secret on the eve of the FDA session and devised a stalling tactic: they would call for creation of a task force, dominated by their scientists, that would spend several months examining the blood-testing issue and eventually tell the FDA that it was unable to reach a consensus.

And that is exactly what transpired: in May 1984 the FDA's task force told the agency that it simply couldn't reach agreement on use of the hepatitis test to screen out possible carriers of AIDS. Most of the world's blood and plasma supply, therefore, went unsterilized and untested for the first four years of the epidemic.

For drug injectors like Greggory Howard information about the new disease was scarce in 1983. No government agency, at any tier, distributed leaflets or educational materials to the country's most derided population. Drug users had no idea that some scientists wanted them to stop "donating" their blood and plasma. Howard hadn't heard of AIDS. All he, and tens of thousands of addicts like him, knew was that "something else," some additional health hassle, was out there. And there were rumors of fellow junkies who got sick, went into the public hospitals, and disappeared.

The two federal agencies that were supposed to deal with the health of people like Greggory Howard seemed utterly disinterested in the AIDS problem in January 1983. Neither the Alcohol, Drug Abuse, and Mental Health Administration (ADAMHA) nor the National Institute on Drug Abuse (NIDA) would request funds from Congress for AIDS research until mid-1983, and neither agency would undertake any research on the transmission of the disease through the use of syringes until late 1984.

The most conspicuous lack of interest, however, was at the National Institutes of Health in Bethesda. Though a handful of scientists inside the

NIH, particularly at the National Cancer Institute, were using their general research funds to tentatively explore the AIDS problem, the agency demonstrated no immediate enthusiasm for solving the AIDS mystery.

"AIDS is the leading cause of death of men between thirty and forty in San Francisco. And we need more money," said Dr. Donald Abrams. Seated in his tiny office on Ward 86 of San Francisco General, where he and Volberding treated the city's swelling AIDS population, the young oncologist carefully chose his words. He pointed to stacks of files, filled with handwritten and manually typed pages.

"We've collected on our patients reams and reams of data. We don't have a computer to analyze the data, so what is the point of doing all the fancy [T-cell] testing if we don't have it put together and can't publish it?" Abrams asked. "This is a problem that is unique in the history of medicine. People keep telling us, 'The money is coming.' From this city office, or that state or federal office. But it never materializes. And then the reality is that every day we've got more patients out there, waiting for answers in our clinic."

Though only in his early thirties, after eighteen months of working with Volberding on the AIDS problem, Abrams looked exhausted. His voice was weary, his body leaden.

Imbued with something of an activist spirit, Andrew Moss was able to muster a bit more energy from his team, and himself. Now ensconced in shoe-box offices at one end of Ward 86, Moss's group was trying to make epidemiological sense of the epidemic. He felt that the only reasonable approach involved matching AIDS cases with demographically similar non-AIDS gay and straight San Francisco men, and following them over time to see what factors put them at risk for the disease.

But that would be expensive.

There were ten people working with Moss—not one of them was receiving a dime for AIDS research. Some, appalled by the epidemic's toll, were volunteers. Even Moss was, technically, a volunteer, as all his funding was earmarked for brain tumor and testicular cancer research.

"Guerrilla science," Moss called it, only half jokingly. "You see a crisis and you just go do what you have to do, and figure out how to pay for it later."

From the outset Curran had tried to raise interest in the GRID/AIDS problem inside the NIH. In the fall of 1981 he went to Bethesda to sketch an outline of what was then known about the disease, its victims, and the unanswered research questions. Robert Gallo heard Curran's pitch, as did several other key NIH researchers. And though many thought the situation grave for homosexuals, few were persuaded that the outbreak posed any intriguing basic research questions. Traditionally, NIH scientists left epidemic problem solving to the CDC.

It was not until a year later, when Curran returned dangling a new tantalizing basic research problem, that the Bethesda scientists took up the challenge.

"We have evidence that a new infectious agent has entered the blood supply," Curran told them. "And it produces severe immunodeficiency primarily via T-cell changes."

Now that sounded like a terrific puzzle to Gallo, who immediately thought of the virus he had recently discovered, HTLV-I.[54] He knew the virus caused immune system disruptions and cancer, though nothing like what occurred in people with AIDS. Gallo left Curran's talk thinking that the mysterious disease might be caused by some new variety of HTLV.

Later, Gallo spoke on the phone with Max Essex, at Harvard, who was familiar with the AIDS problem via Don Francis. Essex's lab had long since established that the feline leukemia virus altered T-cell activity in cats, and he had tentative evidence that HTLV-I similarly disrupted T cells.[55] Essex worked with Curran and Francis, who sent blood samples to the Harvard laboratory for scrutiny. By June 1982 he had a serious effort underway at Harvard searching for the cause of AIDS.

Gallo also had an AIDS effort underway.

After some soul-searching about the possible contagious peril for his staff, Gallo decided that the epidemiology indicated that the mysterious agent was transmitted by blood, not through the air. In May 1982 he ordered lab personnel to start trying to grow a virus out of blood samples from people with AIDS.

A few months earlier another NIH team had begun searching for a link between Kaposi's sarcoma and "poppers." Jim Goedert, Bill Blattner, and Dean Mann studied fifteen gay New York men, comparing their amyl nitrite uses and immune system status. The study found that five of seven men who didn't use poppers had evidence of immune system dysfunction, compared to five of eight users. They also found that CMV infection histories were identical in the two groups. They concluded that amyl nitrites probably didn't play a role in the disease, though the drugs could alter immune function.[56]

And at the National Institute of Allergy and Infectious Diseases, a division of the NIH, a team of scientists led by Drs. Anthony (Tony) Fauci, Henry Masur, and Cliff Lane were studying the nature of the immune system dysfunction in people with AIDS. The NIAID researchers discovered that, in addition to T-cell abnormalities, gay men with AIDS had severe problems in their B-cell systems: though they had lots of B cells of the highly activated antibody-producing type, other classes of B cells were deficient, even entirely absent. The NIAID team concluded that the B-cell system recognized it was challenged by a microbe, but, due to massive disruption of the T-cell system, was unable to respond with the control and precision customary when both arms of the immune system functioned properly.[57]

With much fanfare the NIH announced on April 25, 1983, that it was soon releasing $240,000 in research funds to four external laboratories. A week later the NIH announced six additional research grants to a variety of institutions. The money—in total less than $2 million—would fund studies of the immunology, treatment, genetics, pediatrics, and cancer of AIDS.

Volberding's group, for example, was awarded the first installment of a five-year $526,229 study of pre-AIDS symptoms and immunologic profiles. Two weeks after the NIH announcement, Volberding and eleven co-workers penned a letter to Margaret Heckler, the newly appointed Secretary of Health and Human Services (Schweiker having resigned on January 1, 1983). The letter thanked Heckler's department for the research funding, but noted that it was less than half the sum the San Francisco team had originally requested.

"This amount of funding is unrealistic if we are to make significant progress in finding the cause of this disease," the group wrote. "In addition, we are unable to use equipment that is generally employed by other laboratory personnel because of the fear of the spread of the AIDS agent. Thus, unless funds are provided to purchase new equipment for this research, our work cannot continue."[58]

By mid-1983 every aspect of the AIDS research situation had become a partisan matter in the United States. Republicans generally defended the pace of research and financial expenditures, while Democrats attacked the Reagan administration on all fronts. The situation polarized irreparably, as the war of words in Congress became increasingly emotional and hostile. The Democrat-dominated House of Representatives repeatedly demanded an emergency posture toward AIDS research. And the Republican-controlled Senate and White House sought to keep AIDS spending down. In congressional hearings throughout the summer and fall of 1983 the two parties traded insults and jockeyed for control of the AIDS agenda.

"The [Democrats] fail to define what would be 'adequate' funding," wrote a group of ten prominent Republican members of Congress.[59] "In addition, temporary diversions of funds to help meet the AIDS problem should not be considered permanent. . . . Finally, the [Democrats'] recommendations that an independent panel be created to develop a comprehensive strategy for responding to AIDS should be rejected as unnecessary."

Leading the counterattack for the Democrats was New York representative Ted Weiss, who denounced "inexcusable and unconscionable gaps in the Federal effort to resolve this crisis" and accused the Reagan administration of deliberately delaying or canceling research funds for what he termed "the Nation's Number One health priority."

Medical research money per se was not usually a partisan matter in the United States. Republican Nixon started the War on Cancer, Democrats Johnson and Carter bolstered funding for cancer and heart disease research, and in emergencies—Legionnaires' Disease, Swine Flu, Ebola fever—resources had been found quickly, regardless of which party controlled the Congress and White House.

But AIDS was unique. It touched every nerve that polarized Americans: sex, homosexuality, race (Haitians), Christian family values, drug addiction, and personal versus collective rights and security.

Of the 1,200 AIDS cases identified in the world by March 1983, all but a handful were in the United States and Haiti.[60] The epidemic's political dimensions would not become obvious in other countries until the sizes of their respective outbreaks were sufficient to push the mysterious disease onto the public agenda.

At the CDC those responsible for finding the cause of AIDS—Gary Noble and Don Francis—still couldn't scrounge enough dollars and equipment to conduct decent laboratory experiments. The most obvious way to prove that an infectious agent was involved would be to inject human patients' blood samples into laboratory monkeys. If AIDS then appeared in the animals it would indicate that an infectious agent was in the patients' blood. However, the reverse was not true: if animals didn't get sick it could be due to an immunity the nonhuman species had to the humanly contagious microbe.

But the CDC had no primate research money. In August 1982, Francis and Noble injected four marmoset monkeys with patients' blood, and waited. And waited. Months went by, and the marmosets thrived. Francis tried reinjecting the animals with blood from a different patient. And then, again, waited.

Francis lobbied for other animals, particularly the rare and expensive chimpanzee, but the CDC didn't even have facilities in which to safely and humanely house large primates. In a joint agreement with Emory University's Yerkes Regional Primate Center outside Atlanta, the CDC's animal research program wouldn't begin until the spring of 1983, albeit moderately, with two chimpanzees and a dozen rhesus macaques. A year and a half later the agency scientists would still be waiting for some physical response in the animals to injections of contaminated human blood.[61]

The NIH had an enormous primate facility in San Antonio, Texas (Southwest Foundation for Biomedical Research), and two chimpanzees there were injected with infected human blood in early 1983, producing rapid T-cell changes and lymphadenopathy.

In Paris, the Groupe de travail français sur le SIDA (French AIDS Task Force) had dismissed all environmental factors, such as "poppers," almost from the outset because the history of the French patients so clearly followed an infectious trail. The first observed case was a flight attendant who got infected, it seemed, during one of his many trips to the United States, and passed the infection on to sexual partners in France. Indeed, frequent travel to the United States was such a striking hallmark of European AIDS cases among gay men[62] that the 1982 appearance of exceptions—of two French homosexuals with no American connections—was cause for note in a leading European medical journal.[63]

One of the most energetic of the French scientists was Jacques Liebowitch, a physician and immunologist who argued his cases with almost as much physicality as language. Gesticulating feverishly, pacing about, jumping in and out of his chairs, the handsome young Liebowitch had a habit of reaching an intellectual conclusion and then holding on to it tenaciously, seeking to convince others along the way, until data either proved him right or proved him wrong. For Liebowitch the most intriguing European AIDS cases were not among gay men—that was simply the American paradigm implanting itself on European soil, he said. Rather, he was moved by the occasional African and Haitian immunodeficient individuals that he and other European doctors had recently seen.[64]

In 1982 Liebowitch put forward the hypothesis that AIDS was a viral disease of African origin that caused illness and death by, as he put it, "completely burning out the immune system." He urged his French medical colleagues to scour recent records for bizarre immunodeficiency cases among African-born residents of France or among French citizens who had traveled or resided in Africa. He further asserted that the Haitian cases represented a Caribbean expression of the African phenomenon, linked somehow by travel between French-speaking African countries and Haiti.

In Belgium, Peter Piot was hard at work on a very similar hypothesis. From the moment he had heard of the first Los Angeles PCP cases, Piot had considered the possibility that AIDS was the culprit in similar ailments among African residents of Belgium. So had Dr. Nathan Clumeck, a low-key physician working in Brussels' St. Pierre University Hospital. In 1982–83 he was treating five upper-class Zairois who either lived in Belgium or had come to the former colonial power for treatment of their profound immunodeficiencies.[65]

Clearly these African cases veered strongly off the course of AIDS events defined by the Americans: none were gay, used injected narcotics, or had visited Haiti. And more than a third were women. All evidence indicated that AIDS had been recently imported to the European continent from Africa, Haiti, or the United States. The American importation centered on blood products and gay sexual transmissions, while the African and Haitian cases seemed to involve importation of a heterosexual pattern of the disease. Regardless of the mode of transmission, however, it all argued strongly for, as the French Groupe de travail put it, "a transmissible agent now present in Europe."[66]

It was a heterosexually transmitted microbe as well as a homosexually transmitted one, the European doctors said. And most members of the Groupe de travail favored the notion that AIDS was a viral disease.

France didn't have an NIH-type financial bureaucracy for dispersal of scientific research funds. Rather, it had a somewhat more elitist system that concentrated the bulk of the country's biomedical research inside the prestigious Pasteur Institute in Paris. Though the Pasteur could hardly compare in size or wealth with the vast American scientific establishment, it had a distinct advantage in an emergency: individual scientists with "tenure" could initiate research in nearly any direction without answering to a bureaucratic superstructure. Though this was less democratic than the American system, the Pasteur system did intentionally allow the nation's scientific elite tremendous creativity and flexibility. The leading Paris hospitals—Claude Bernard, Raymond Poincaré, St. Louis, Pitié-Salpetrière—were free to collaborate with the Pasteur Institute scientists. In 1982, then, a loose AIDS collaboration had developed connecting Dr. Françoise Brun-Vézinet of Claude Bernard Hospital, Willy Rozenbaum of Pitié-Salpetrière, and a Pasteur group headed by virologist Luc Montagnier.[67]

On January 3, 1983, Rozenbaum removed an enlarged lymph node from the neck of AIDS patient Frédéric Brugière, a gay man who had traveled in the United States. The precious tissue was rushed to Montagnier's lab, where virologist Françoise Barré-Sinoussi began analyzing it. On January 25, Barré-Sinoussi told Montagnier she had discovered evidence of reverse-transcriptase activity in Brugière's cells.

Only one entity on the planet was known to use the reverse-transcriptase enzyme: retroviruses. The tiny RNA viruses used the enzyme to make mirror-image copies of their RNA genetic material, creating a DNA version of themselves that could be incorporated into the genes of the animal cells that they infected.

Two human retroviruses were known to exist at that time—HTLV-I and HTLV-II—and the bulk of the world's research on them was done at Robert Gallo's laboratory in the National Cancer Institute. Montagnier initially assumed that the reverse-transcriptase activity indicated that AIDS was caused by one of these two agents, and his laboratory notebook for January describes Barré-Sinoussi's finding under the later-scratched-out heading of "HTLV-I."[68]

Montagnier telephoned Gallo in early February 1983, described Barré-Sinoussi's findings, and a tempestuous collaboration/competition between the two laboratories began. Later that month Gallo's lab also had reverse-transcriptase activity in laboratory isolates of cells extracted from men with AIDS.[69] Gallo was thoroughly convinced that his initial insights, voiced the previous year in his fateful phone conversation with Max Essex, remained accurate: AIDS was caused either by HTLV-I or by one of its close cousins.

"HTLV is endemic in the Caribbean and seems to be relatively common in Africa, and of course AIDS has some link with the Caribbean island of Haiti and with Kaposi's sarcoma found traditionally in Africa," Gallo told the *Journal of the American Medical Association* in August 1983.[70] Admitting that no cases of AIDS had appeared in Japan, where HTLV-I was endemic, Gallo asked, "Could the leukemia-causing virus [HTLV-I] be a variant of the immunosuppressive virus? We don't know. But if it is, it's a very subtle variant with a minor antigenic difference."

While Gallo's and Montagnier's groups raced to find the HTLV link, Jay Levy's tiny team of scientists toiled away in San Francisco, backed by just a few thousand dollars from the recently organized California state-university-wide AIDS Task Force.[71] Despite his meager resources, Levy had a key advantage over the Bethesda and Paris researchers: virtually unlimited access to a large and cooperative AIDS patient population. While Montagnier struggled with samples from one key patient, Gallo with a handful, Levy had blood and tissue samples from more than forty gay San Franciscans. His vast research pool let Levy select patients at random, avoiding any unintended biases that might result from overinterpreting data extrapolated from one or two patients. It would be several years before the importance of Levy's randomly selected samples would be obvious.

Levy and UCSF colleague John Ziegler had a theory that "AIDS is itself an opportunistic infection. It causes disease only in individuals who are already immuno-compromised by hepatitis B, cytomegalovirus, parasites, or other immunosuppressive factors."[72]

They saw the AIDS disease process in fairly complex terms. Probably because their entire patient population in 1983 was composed of gay, sexually active men, they thought AIDS was the final step in a multistaged process that began with an immune system assault by an array of other agents—particularly cytomegalovirus and Epstein-Barr virus—after which an as yet undiscovered "AIDS virus" entered the individual's body.

Levy postulated that "the virus has mutated itself to be such a close imitator of the immune system—of some component of the immune system—that when the system tries to attack the virus, it ends up attacking itself."

The result, Ziegler said, was a profound autoimmunity, or immune system self-destruction, in which the mighty forces of the B- and T-cell systems mistakenly attacked the body's defenses. Levy's guess was that the observed T-cell imbalances in AIDS patients were the direct result of such a process.

But other scientists were looking at the same data on AIDS patients and reaching very different conclusions. Some felt AIDS was simply a new manifestation of the hepatitis B virus,[73] or of some unknown contaminant of the hepatitis vaccine, first experimental trials of which had been on gay Americans.[74]

A variety of other AIDS causality theories floated about the popular and scientific literatures during the early 1980s. Most shared a fundamental flaw: they sought to explain the existence of the disease solely on the basis of observations in the American gay community, ignoring contradictory epidemiological evidence arising from a broader look at all the people who were contracting AIDS. The majority of suggestions put forward by credible scientists were subjected to scrutiny at the laboratory or epidemiologic level, and then withdrawn or amended by the initial proponents when found lacking.

But, as had been the case with so many previous epidemics, there were zealots who denounced their critics in conspiratorial terms and insisted, long after data proved them wrong, that the AIDS causality they had identified was correct. In some cases, their public pronouncements had harsh impacts on the behavior of people who were potentially at risk for AIDS.

Among the early AIDS theories that gained the greatest attention were the notion that the disease was syphilis or that syphilis acted as a co-factor to some other microbe.[75] Many physicians continued to insist, despite contrary data, that unique fast-lane practices in the gay communities, including "poppers," "fisting," and the use of steroid skin creams, were key.[76] A New Zealand team asserted that AIDS was caused by the same tiny odd protein elements then thought to spark scrapie disease in sheep.[77]

Two American scientists became controversial in 1983—and continued to be celebrated into the 1990s in the *New York Native*—for their theory that AIDS was caused by African swine fever.[78] A serious veterinary problem, ASFV infected a variety of different types of pig cells. Human infections with ASFV were rare events, but could produce fevers and immune system disruptions. The researchers who hypothesized the ASFV/AIDS link noted the intersection of events they believed conspired to create an aberrant form of the microbe: a 1978 outbreak among pigs in Cuba and Haiti, which, they argued, was part of a CIA effort to destabilize the Castro government by destroying its livestock; mass movement of refugees from both islands to the United States in 1980; and alleged consumption of undercooked pork by New York homosexuals while on vacation in Haiti.[79]

Still others favored the notion that AIDS was caused by "factors" of some kind in blood products used by people with hemophilia.[80] Conversely, some argued that all the cases allegedly associated with the blood supply were merely misunderstood ailments of other kinds.[81] Thus ignoring evidence that blood product recipients throughout the industrialized world were contracting AIDS and that, when they could be traced, the donors were usually found to have AIDS.[82]

A 1983 Tulane University study of men with hemophilia and their wives proved three key points germane to the causality debate: (1) AIDS in people with hemophilia was precisely the same disease as AIDS in gay men; (2) but the hemophilic patients had no histories of any causes of AIDS proposed for gay men; and (3) some of the men seemed to have passed the disease on to their wives. The researchers concluded that "chronic infection with a blood product-transmissible agent is the most likely source of the abnormalities noted. As hemophiliac patients are not generally exposed to other risk factors previously implicated, future studies . . . as to the cause" of AIDS ought not to focus solely on "persons with nontraditional lifestyles."[83]

The CDC gave out mixed messages in 1983. Concluding their first limited case control study of fifty gay men with AIDS, the agency researchers said that they "cannot exclude the possibility that . . . illicit drug use" and "certain aspects of their lifestyle" were correlated with AIDS. Though the CDC team never described lifestyle issues as *causative*, many members of the gay community read the results as supporting a role for "poppers" and such.[84]

Meanwhile, Francis and Dr. Martha Rogers issued word from the CDC lab that beyond higher-than-average levels of CMV and Epstein-Barr viruses, the gay men with AIDS had nothing in their bodies that could explain their terminal illnesses. "We suggest that future laboratory studies be designed to identify an infectious agent that may circulate freely in the blood or with peripheral blood leukocytes, and that may also be found in rectal secretions, semen, or other secretions of homosexual men."[85]

Amid the confusion, physicist John Maddox, editor of England's most distinguished scientific journal, *Nature*, penned an April editorial entitled "No Need for Panic About AIDS."[86]

"There is now a serious danger that alarm about the disease physicians call acquired immune deficiency syndrome (unhelpful, AIDS for short) will get out of hand," he wrote. "For the characteristics of this previously unrecognized and *perhaps non-existent* [emphasis added] condition are so alarming that the temptation to portray it as a disease invited by a decadent civilization—a kind of latter-day version of the fate of Sodom and Gomorrah—is almost irresistible." Maddox denounced the "pathetic promiscuity of homosexuals," calling it "the most obvious threat to public health."

Dismissing AIDS as a disease that had occurred among fewer than one thousand people, 70 percent of whom were homosexuals, Maddox berated alarmists, adding that "mercifully, the disease—whatever its causation— is neither especially infectious . . . nor certain in its effects."

By contrast, during the spring of 1983, Curran, Francis, and Harvard's Max Essex collaborated on an editorial wake-up call to be published in the *Journal of the National Cancer Institute*. Their intention was to state in no uncertain terms that AIDS was caused by an infectious agent and to suggest that one of the candidates for causation of AIDS was HTLV-I.[87] Essex already had evidence that many AIDS patients were infected with HTLV-I.

"We checked a series of 75 patients with AIDS that were sent to us from CDC," Essex explained in May. "And the patients were classified as either having Kaposi's or *Pneumocystis*. And in that series of 75 . . . between a quarter and a third of the patients had evidence of prior exposure to the HTLV.

"One possibility that I should underline is that the HTLV has nothing to do with this disease, and that the HTLV is opportunistically infecting some of the patients with AIDS, but not all of them," Essex hastily added.

Gallo was ecstatic. Essex had evidence that implicated his personal nominee for the AIDS culprit. Gallo's lab staff had just isolated HTLV-I from the white blood cells of three New York City gay men with AIDS, and in a survey of 33 AIDS patients in New York Hospital, Gallo's group found HTLV-I in the T cells of two men. Together, these findings seemed to argue strongly, in Gallo's view, that HTLV-I, or one of its close cousins, caused AIDS.

The four Essex and Gallo papers were published as a package in the journal *Science*, along with a study from the Pasteur Institute group that an official U.S. Department of Health and Human Services press release that day described as reporting "isolation of an HTLV-related virus from a homosexual patient with persistent, multiple lymphadenopathies and evidence of infections who may be at risk of developing AIDS."[88]

But that wasn't what the French study showed. Not at all.

On February 4, 1983, the Pasteur Institute's Charles Dauguet observed dozens of spherical viruses poking out of Frédéric Brugière's T cells. However, though the mysterious viruses under Dauguet's microscope and HTLV-I were both spherical, they did not appear to the French scientist to be identical. More importantly, Montagnier's group was unable to get strong cross-reactivity between antibodies against Gallo's HTLV-I virus and their AIDS-related microbe. They suggested that the two agents might share some genetic similarities, but were clearly different species of viruses.

Nevertheless, at Gallo's urging Montagnier had inserted the following in his article: "We tentatively conclude that this virus, as well as all previous HTLV isolates, belong to a family of T-lymphotropic retroviruses that are horizontally transmitted in humans and may be involved in several pathological syndromes, including AIDS."

Throughout the summer of 1983 the two competing laboratories toiled to grow the apparent AIDS viruses in cell cultures. But the viruses only grew well inside human T cells, which they also killed. So in a matter of days all the cells in a culture would be dead, along with the elusive viruses. Barré-Sinoussi and Chermann tried a variety of unsuccessful strategies to grow the viruses. Finally, during the dog days of summer Montagnier's team figured out that the trick was continuously, every three days, passing virally infected liquid (supernatant) from cells grown in the presence of T-cell stimulators interleukin-2 and phytohaemagglutinin to fresh T cells, and repeating the process over and over for several weeks. Eventually one would get a culture dish chock-full of viruses.

Meanwhile, panic was growing in North America.

Though the absolute number of reported AIDS cases in Canada and the United States was still below 2,000, the dimensions of the epidemic were expanding. Drs. James Oleske at the New Jersey Medical School in Newark and Arye Rubinstein of the Albert Einstein School of Medicine in the Bronx were treating babies and toddlers who seemed to have contracted AIDS from their parents. Oleske was treating eleven such children, Rubinstein twenty-five.[89] All of the children had a parent who either used injectable narcotics, had recently emigrated from Haiti or the Dominican Republic, or was "promiscuous," as the physicians put it.

"Clearly none of the children that we have seen were sexually abused or given illicit drugs," Oleske said in May 1983. "So the implications are that, if you will, 'normal' people can acquire AIDS."

Rubinstein agreed, saying that it was likely most of the children got the presumed AIDS virus from their mothers during or immediately after pregnancy, but "we find discrete immune deficiencies in other members of the [families his group was studying]. Something that may suggest that the transmissible agent can be acquired in a different mode: not only transplacentally, not only sexually, not only by sharing of needles."

In early 1983, a joint CDC/Montefiore Medical Center study in New York City described two women with AIDS who had no other apparent risk factors save marriage to men who had the disease.[90] By May, Montefiore's Dr. Neil Steigbigel had uncovered five more cases of apparent heterosexual transmission.

"We do feel now that this does show that AIDS should be considered threatening to the health of our general population, not only to male homosexuals, abusers of intravenous drugs, Haitians, or hemophiliacs," Steigbigel said at the time. "Of course, if one is dealing with a potentially fatal disease, that is tremendously frightening. To have a potentially fatal venereal disease, that is . . . present in our general population."

Other studies confirming the sexual passage of the mysterious AIDS agent flooded in during the summer and fall of 1983.

In many ways the most alarming news for the CDC's AIDS Task Force members came from the users of injectable drugs. Curran, Jaffe, Francis, Guinan, Darrow, and the others all cut their public health teeth on sexually transmitted diseases; even so, they were surprised, even shocked, by what they learned about the sexual fast lane in the gay community. Before AIDS, they were similarly ignorant about the drug-using population. They didn't know about all the years that Greggory Howard, and thousands like him, had been stashing their "works" in shared hiding places. They didn't know about the allegedly abandoned buildings filled with the commerce of narcotics.

When New York City and Newark drug researchers brought their familiarity with the desperate details of drug addiction into the growing circle of American AIDS scientists, their insights hit Curran and his colleagues with a jolt: one could debate theoretical probabilities of contracting AIDS through sexual transmission, but injecting it into one's bloodstream seemed to guarantee infection.

Soon the CDC group was learning about shooting galleries where junkies could pay to get injected with just about anything by a dealer who used the same needle and syringe on dozens—even hundreds—of customers a day. Experts like Dr. Don Des Jarlais, who ran a drug rehabilitation program inside Manhattan's Beth Israel Hospital, told the CDC scientists that few addicts in 1983 used just one drug: they were addicted to two, three, or more drugs, often including cocaine, alcohol, amphetamines, barbiturates, Valium, and other benzodiazepines. After years of periodic heroin "famines," due either to police actions or to wholesaler market manipulations, expert narcotics users had adapted by mixing their drugs. One "cocktail" to start the day, another to smooth the rocky edges of coming down off the first, and still another to shoot the user straight to temporary paradise.

It was naïve in the extreme, the CDC scientists learned, to build stereotypes around junkies, or to assume that any single behavior explained the skyrocketing increase in AIDS among users. The array of individual drug-use patterns could range from lethargy to hyperactivity. These people were neither easy to study nor easy to educate.

"We just don't know what to make of all of this," Curran said. "We can't explain why almost all IVDU (intravenous drug user) cases are showing up in New York and New Jersey, while most of the West Coast cases—more than 90 percent—are among gay men. We don't really understand the distribution."

"All you have to do is walk the streets," Howard would claim. "Greggory knows what's going on."

Howard was trying to stick with methadone, but it was tough. The clinic staff treated the junkies like animals, he said, and it was often questionable which was more demeaning: pulling down your pants in front of a hulking clinic guard and struggling to pee into a Dixie Cup while in drug withdrawal so they could test for heroin; or searching frantically for a usable vein to bare to a scowling dealer who jabbed the needle in, shoved down the plunger, released the tourniquet, and turned to the next customer while you swayed off into suspended animation.

Though Howard didn't yet know much about AIDS by the fall of 1983, he was an expert on lifestyles of the stoned and addicted. He could have told the CDC team enlightening and unsettling stories, if they had bothered to ask. But they didn't. Curran knew his team was out of their depth when it came to injecting drug users, and he lobbied hard for research efforts at the agencies that were supposed to be on top of such things, particularly NIDA (National Institute on Drug Abuse). But under the Reagan administration, NIDA was far more concerned with eliminating drugs than with keeping users alive.

If anybody had asked him, Howard would have told the government scientists the same things he said to anyone who asked. "*How much* Greggory uses," Howard would say, "*when* he uses it, *how* he uses it, all depends on *what* he's using. It's as simple as that."

If it boiled down to nothing more than heroin—which it rarely did—one or two injections per day with his personal works would be adequate. If, however, he mixed heroin with downers like alcohol or barbiturates, and "wildness" like injected cocaine or speed, things got more complicated.[91] Heroin might last for hours, but cocaine's rush persisted for only minutes. A heavy injected dose of speed might have the user walking through Newark, barefoot and unfueled by food or sleep, for two or three days unless he smoothed it over with some serious downers, like Valium or barbiturates, both of which would come on faster if the pills were washed down with high-proof alcohol.

So Howard would explain as he strolled the familiar Newark slums, the drugs dictated what he did every day, how many times the needle entered a vein, whose needle it was, and how many other people used it.

The CDC wanted studies done among drug users as soon as an AIDS test of some kind was available. If the drug experts were right, the addicts might have an even greater AIDS incidence than gay men. But gaining access to drug users, especially those who weren't in methadone or reha-bilitation programs, or didn't live in drug ghettos like Howard's Newark niche, would be extremely difficult. The greatest challenge would be finding drug injectors who simultaneously led middle-class existences and clan-destine lives of addiction.

Even without such data, concern was high that AIDS would make its way to the general population via addicted prostitutes or the sexual partners of injecting drug users. The heterosexual transmission reports—excluding those mistakenly labeled as "Haitian cases"—were predominantly among the female sexual partners of male drug addicts.

Another grave concern for the CDC stemmed from reports by two U.S. primate research centers of outbreaks of what looked remarkably like AIDS among their monkeys, in California[92] and Massachusetts.[93] Though these were not of particular concern to the general public, scientists worried that whatever new virus was killing people might have fairly recently arisen from monkeys. If that were so, then few, if any, humans could be expected to have natural immunity to the monkey microbe.

As public awareness of the epidemic's widening scope increased, so did panic. Police officers in San Francisco demanded, and received, specially designed masks and gloves for their protection when performing artificial resuscitation or "handling" of "potentially dangerous" citizens. A New York City garbage collector was terrified that he had become infected as a result of grabbing a trash bag from which protruded a syringe. Also in New York the city's Health Department was swamped with calls from fearful citizens asking whether it was safe to share laundry facilities with gay men, whether the virus could be passed via seats or handrails on the subways, or on public toilet seats.[94]

In Europe, thoroughly respectable, usually conservative scientists were openly comparing AIDS to the plague.[95]

And inside America's gay community a great political-cultural battle was being waged. Many men were duly terrified and were radically altering their behavior. Bathhouses in San Francisco, for example, reported 40 to 60 percent declines in revenues during May 1983.

But as June 27 approached—the anniversary of the Stonewall riot, now celebrated as Gay Freedom Day—shouting matches reverberated through the halls of government in San Francisco and New York. Those most concerned about AIDS feared that the party atmosphere and bathhouse frolicking that had prevailed since 1969 was too dangerous in 1983. Opponents of bathhouse restrictions derided such sentiments as government-inspired paranoia, intended to stifle the gay liberation movement.

San Francisco Public Health Director Mervyn Silverman and New York City Commissioner of Health David Sencer were caught in the middle, forced to decide the fates of local sex parlors and bathhouses while the opposing sides in the gay community issued political threats. The battles would remain heated for over two years, and neither city's leading health officials would survive politically. Eventually the establishments would be closed, though after-hours, semi-clandestine gay sex clubs would continue to exist, albeit illegally, into the 1990s.

"It's sort of depressing," an exhausted Silverman said just days before the Gay Pride Parade and Stonewall remembrance brought over 300,000 celebrants to San Francisco. "You have individuals who are filled with anxiety about AIDS. And because of that anxiety they are going to the bathhouses and indulging in high-risk sex to relieve that stress. It's very, very paradoxical."

Bathhouse owner Hal Slate, proprietor of the Cauldron, corroborated Silverman's observation. "So we're caught in a Catch-22 where we're now dealing with an extraordinary level of stress and anxiety and confrontation with death, all of it surrounding the very mechanism that we see as there for us to help us deal with our anxiety and stress," Slate said.

Though he proselytized about the dangers of AIDS, Bobbi Campbell continued to patronize the baths. He and Michael Callen had created a new self-help and political action group called PWAs—or People With AIDS. He had watched several friends die, and by the spring Bobbi had changed his self-appointed title from KS Poster Boy to AIDS Poster Boy of 1983. He had also suffered more than eight major opportunistic infections, had been in and out of local hospitals several times, and was getting scared.

A few months earlier, Campbell had nursed a close friend named John, who eventually died of AIDS.

"Seeing him in ICU [intensive-care unit] with tubes in his nose was horrifying to me," Campbell said. "More horrifying than where I was at myself. I could deny sometimes that I was sick. But I couldn't deny it anymore, seeing him lying there. And I could see myself, or others I love, lying there.

"John and I talked, I left, and in a week or so he was dead. I cried and I cried, went out and got drunk. And I said, 'I'm alive, goddamnit! I'm alive!' And I am alive, and I want to make that real for myself. I do face death, but until I'm dead, I'm alive, damnit."

The Gay Freedom Day celebrations took place in New York and San

Francisco, and the bathhouses remained open. Though they were the largest such festivals in U.S. history, participants could feel the change. Who could doubt that the party was over when contingents of men with AIDS marched in the parade—or were pushed in wheelchairs?

Warning signs were posted in gay establishments, bowls of free condoms were placed in gay bars and hotels, Health Department pamphlets were distributed advising men to practice safe sex, and the world witnessed it all on international television. Elegant drag queens, Dykes on Bikes, Whores Against Wars, the Sisters of Perpetual Indulgence, the San Diego Gay Softball League, the Harvey Milk Democratic Club, Women Against Imperialism, and a host of other photogenic gay contingents filled television screens as somber announcers remarked on the odd juxtaposition of such frivolity with an epidemic.

The television coverage ignited backlash. The Reverend Billy Graham cried out that "AIDS is a judgment of God." Television evangelist and leader of the Moral Majority Jerry Falwell denounced "perverted lifestyles," saying in a nationally televised sermon, "If the Reagan administration does not put its full weight against this, what is now a gay plague in this country, I feel that a year from now, President Ronald Reagan personally will be blamed for allowing this awful disease to break out among the innocent American public.

"AIDS is God's punishment," Falwell concluded. "The scripture is clear: We do reap it in our flesh when we violate the laws of God."

Bobbi Campbell shortly thereafter denounced the religious leaders before the San Francisco Board of Supervisors.

"I don't consider myself a sinner, and I don't think this is God's will!" the AIDS Poster Boy shouted. "I'm angry at Senator Jeremiah Denton from Alabama, who said, 'Oh, let the faggots die.' I'm angry at the people who fire us and evict us from their homes. It's our crisis. We need support, not hate."

Some scientists complained that a "Plague of Fear" was overwhelming efforts to control the viral plague rationally.[96]

Meanwhile, the scientific competition also heated up. Shielded from most of the public turmoil, the laboratories of Jay Levy, Luc Montagnier, and Robert Gallo generated their own controversy as they raced to discover the virus that caused AIDS. Though no further evidence to support HTLV-I came to light, Gallo continued to publicly proclaim it the most likely suspect, even providing elaborate schemes for the virus's evolution and global spread.[97]

But both Levy and Montagnier were certain by summer's end that HTLV-I was *not* the cause of AIDS.[98]

In September hostilities between the French and Gallo's group escalated from the level of rivalry common to competing laboratories to something markedly worse when Gallo and Montagnier each addressed a virology meeting at the Cold Spring Harbor Laboratories in New York. Gallo presented his arguments in favor of an HTLV-I role in the disease, garnering polite applause from colleagues. Montagnier, however, dropped a bombshell, announcing five crucial accomplishments. First, he said, a new virus (hinted at in his earlier paper in *Science* but now clearly identified) had been discovered and dubbed LAV, lymphadenopathy-associated virus. Second, LAV was successfully cultured from the cells of five pre-AIDS patients who had profoundly enlarged lymph nodes and from three people with AIDS (a gay man, a Haitian woman, and a man with hemophilia). Further, LAV had an affinity for infecting T cells, particularly helper cells that had CD4 receptors on their surfaces. Using a specially made screening test, Montagnier's group had shown that in 63 percent of pre-AIDS cases and 20 percent of full AIDS cases, antibodies against LAV could be found. Montagnier suggested that the lower antibody response in sicker people was due to LAV viral destruction of their immune systems.

Finally, he insisted that all forms of analysis of LAV showed that, far from being a close cousin to HTLV-I, it was a member of the lentivirus family, which included a number of slow-killing veterinary diseases, such as visna in sheep and equine infectious anemia (EIAV) in horses.

An extremely controversial thirty-minute exchange between Montagnier and Gallo followed. It would haunt them, and science as a whole, for over a decade. Gallo asked his French rival eight questions, as was his right in such a meeting. Gallo would insist more than ten years later that his queries were meant to point out inconsistencies or weaknesses that might merit further investigation. Montagnier, however, found Gallo's remarks rude and insulting, and felt the American was throwing down an unmistakable gauntlet. Other scientists in the room were taken aback by the unusually naked competition. Gallo's line of questioning was aimed at disproving the causative role of LAV or, failing that, at undermining Montagnier's assertions that LAV was a lentivirus rather than a sibling in the HTLV-I family.

What ensued over the following twelve months was a race to the finish line, with the three laboratories exchanging niceties—even viral samples —while fighting tooth and nail. Right to the moment he crossed that finish line, well after Montagnier announced the LAV findings, Gallo would continue to insist that HTLV-I was the likely cause of AIDS, and his efforts to prove it would be backed up by the CDC imprimatur in the form of Harold Jaffe, Don Francis, Jim Curran, and other members of the AIDS Task Force.[99]

Essex would continue walking a middle line, finding antibody evidence that 10 to 12 percent of people with AIDS were also infected with HTLV-I, but constantly underscoring that "we certainly don't have any proof that this agent causes AIDS."

Andrew Moss reflected as the year 1984 approached that "there was an enormous PR boom about HTLV, which a lot of people thought would be the AIDS agent, and it's turning out not to be. I think this year has been mostly about PR, funding, and politics. It seems to me that the history of the science of the AIDS epidemic is that there was this wave of science done by people before there was funding. Which was done by scientists mostly by bootlegging and using what they had."

Moss chuckled and rolled his eyes at the new computer on his office desk. Then he added with a sigh, "That's what we did—we scuffled for funds for a whole year. Now that the funds have been gotten, many people will finally be able to do some research."

Though he at last had some research funding of his own, Moss remained angry. He was uncomfortably aware that his forecasts of the AIDS toll on San Francisco's gay population were coming true, and that he, an epidemiologist and statistician educated at Stanford and the London School of Economics, was now cast as a death counter. "Well, if it's numbers you want, San Francisco had the real plague in 1907, and it caused hysteria because sixty people died. Well, we will have more than a hundred AIDS deaths in San Francisco this year, and next year [1984] we will see between two and three hundred AIDS deaths. That's really a very large number. And it is inconceivable to me that we would be facing such a prospect and frankly, as a society, not be alarmed about it if it were not an epidemic of a stigmatized group of people."

In Atlanta, Jim Curran also ended the year counting deaths: 2,042 cumulative AIDS cases reported since May 1981, and 1,283 were already dead. He predicted that the epidemic would continue to expand in 1984, but not at 1983's rapid pace.

On April 7, 1984, the Pasteur group published details on some of the cases Montagnier had described at Cold Spring Harbor seven months earlier, as well as evidence that LAV was in the French blood supply.[100] This revelation sparked no policy action on the part of either French or American blood industry authorities. It did, however, prompt the CDC to send coded blood samples to the Paris laboratory, which quickly returned verdicts: 90 percent of the samples that the CDC later confirmed came from AIDS patients were positive for antibodies for LAV.

"I think it looks very good," declared an obviously excited Don Francis. "The French work is very exciting."

At the National Institute of Allergy and Infectious Diseases, located a stone's throw from Robert Gallo's lab, Dr. Malcolm Martin was even more ebullient, declaring LAV "the best game in town right now."

Gallo and Essex were unconvinced.

"I think HTLV has to be considered the leading candidate at this time," Essex told *The Wall Street Journal*,[101] hinting that proof would soon be forthcoming from Gallo's lab. Gallo declined to comment directly, but "informed sources close to Gallo" told *The Washington Post* that a major announcement was imminent.

And it was. On April 23, 1984, HHS Secretary Margaret Heckler convened a press conference in the agency's offices in Washington, D.C., to announce discovery of the virus that caused AIDS. She was not there to sing the praises of the French effort, but to declare victory for her agency's National Cancer Institute. With Gallo at her side, Heckler announced, "Today we add another miracle to the long honor roll of American medicine and science.

"Today's discovery represents the triumph of science over a dreadful disease," Heckler averred, forecasting development of an AIDS vaccine within five years.

Gallo's group had discovered a retrovirus in people with AIDS, dubbed HTLV-III. And the group had created a cell line that, unlike the system in use at the Pasteur Institute, could grow permanently in the presence of the virus. So his team needn't transfer fluids from one culture dish to another for weeks on end to get a viral sample. The cell line, designated HT, would easily serve as the basis for a rapid AIDS blood test, Gallo said, because it was now possible to make mass quantities of viruses for human antibody screening.

The Gallo group noted the prior French LAV finding and was ambiguous about whether HTLV-III and LAV were different microbes. It was, they said, impossible to say for certain, because the poorly characterized French virus "has not yet been transmitted to a permanently growing cell line for true isolation."[102]

Gallo then predicted that an AIDS vaccine would "be available within two years."

Responding to word that the Pasteur Institute and the French press had not taken Gallo's anointment as "Discoverer of the AIDS Virus" kindly, Gallo told the *Journal of the American Medical Association* that "there was not, is not, and never has been any fight or controversy between us and the French group."[103]

The French and American labs would present mountains of point-counterpoint research papers over coming months supporting an etiological role for LAV or HTLV-III, respectively. Because the National Cancer Institute group had developed the virus-producing HT cell line, they were rapidly able to screen blood samples for infection using a simple antibody test called an ELISA (enzyme-linked immuno-absorbent assay). By November 1984, the ELISA test was being used by researchers in both the United States and Europe to test blood samples for infection and to experimentally screen blood donations.[104]

And by November–December, both LAV and HTLV-III had been cloned in the laboratory and analyzed at the genetic level.[105]

Jay Levy's group would rapidly announce discovery of yet another retrovirus, dubbed ARV (AIDS-Related Virus), in gay men with AIDS.[106] By then they had cloned and characterized that microbe as well.[107]

In December, British researchers who ran a series of immunologic tests on LAV and HTLV-III declared that the pair were "a single species of virus" that infected T cells by attaching itself to the CD4 receptor proteins that protruded from helper cells and some types of macrophages.[108]

By February 1985, all three viruses were completely genetically sequenced and something quite curious was revealed: HTLV-III and LAV differed by less than the usual 1 percent—an amount attributable to human error. In other words, Montagnier said, "they are identical." That implied that during all those months of competition and exchanges of samples something had happened to Gallo's viral cultures. They may have become contaminated with the Pasteur virus.

In contrast, Levy's ARV clearly differed, with 6 percent of its genetic sequence at variance with that in LAV and HTLV-III.[109] And little similarity was found between HTLV-III/LAV/ARV sequences and HTLV-I or HTLV-II. But, as Montagnier had said over a year earlier, that was to be expected: the AIDS virus was a close cousin to well-known lentiviruses, which produced slow-killing immune system disruptions in horses, sheep, and goats.

Working with Gallo, Dr. Flossie Wong-Stahl would soon show that under natural circumstances of infection the AIDS virus mutated rapidly, and it was *impossible* in nature to find two different viruses a continent apart that varied genetically by less than 1 percent.

Eventually it would be shown that Levy's ARV, drawn from randomly sampled San Francisco gay men, was a genuine natural isolate, clearly indicative of what was circulating among unsuspecting human beings. In contrast, HTLV-III and LAV—which were the same agent—had undergone significant genetic changes during all the manipulations it was subjected to in the attempts to culture it in the Paris and Bethesda laboratories. It was not a "natural virus," in that, as time would tell, its key outer envelope sites bore only a partial resemblance to wild viruses.

HTLV-III, LAV, and ARV would all be renamed the human immunodeficiency virus (HIV). Most American laboratory research in the coming decade would be based on HTLV-III, most French on LAV, and all initial vaccine efforts would target the HTLV-III/LAV lab strain: a false target, time would show. Levy's ARV strain would have been a far wiser choice.

Having found HIV, and developed a blood test for antibodies against the virus, scientists optimistically looked forward in January 1985 to quickly solving all the unanswered epidemiological, pathologic, and virology questions about AIDS. A vaccine and an effective treatment couldn't be far away, they thought.[110]

III

In the beginning it was called "Juliana's disease."

It was first noticed in the village of Lukunya, on the Ugandan border sometime in early 1983.

A handsome Ugandan trader had come through selling cloth for women's *kangas* patterned with the name Juliana. A village girl with no money traded the stranger sex for a *kanga*, as did several other women who coveted the beautiful Juliana cloth.

Some months later the first girl became sick; she had no appetite, could hold down no food, and had constant diarrhea, which filled her with shame. Nothing more mortified the Mhaya people of northern Tanzania than being babylike, unable to control the expulsion of their own bodily wastes. In a few weeks she wasted away, grew weak, and had to be carried everywhere. Before she died, two other women, also adorned in Juliana's cloth, came down with the strange disease.

The people of Lukunya decided that the Ugandan was a witch, and that Juliana's cloth had evil powers. To conquer Juliana's disease, healers toiled to lift the stranger's curse.

The people had reason to be suspicious of Ugandans. They were still smarting from their invasion by Idi Amin's Ugandans in 1978 and Tanzania's war with Uganda in 1979. During that time, thousands of Tanzanian soldiers poured into the villages and bivouacked there for weeks on end. As many as 6,000 Tanzanian soldiers were in villages normally occupied by less than a thousand people.

Knowing how bad the blood had been between the Mhaya of Tanzania and the Ganda of Uganda, the people were not surprised when the traditional healers were unable to lift the powerful curse and the death toll continued to rise.

Within a year the curse had spread to the neighboring villages of Kanyigo, Bukwali, Kashenye, and Bunazi. In the village clinics the medical assistants at first dismissed the illnesses as just retribution for having consorted with "unfair dealers from across the border." By 1984, however, Juliana's disease cases were appearing at Dr. Jayo Kidenya's hospital in the Kagera District's capital, Bukoba, and the Ndolage Missionary Hospital on the other side of the district. The Bukoba doctors were convinced the disease was something new. None of their treatments could slow its terrifyingly rapid progress. Kidenya was puzzled because the adults died like children, wasting away as if they were infant measles cases compounded by malnutrition. Some patients had stubborn viral and bacterial infections that could not be treated with Kidenya's small array of antibiotics. Rumors of widespread witchcraft were spreading throughout the Kagera region, and Kidenya felt compelled to solve the Juliana mystery.

He had few resources for medical detective work. The destitute Tanzanian government hospital he ran imposed *unyenyekevu* modesty and lowered the expectations of all who toiled under its tin roof, including Kidenya. Ten years earlier, the soft-spoken, enthusiastic supporter of President Julius Nyerere's Ujamaa program had studied in snowbound Bucharest, thinking a European medical degree would garner a prestigious posting in the capital, at Muhimbili Hospital.

But here he was, forced to augment his meager salary with weekend farming, living by Lake Victoria in an area suffering from the ravages of the war. Ironically, the roughly 10,000 residents of Bukoba found the Ugandan capital of Kampala easier to reach than far-off Dar es Salaam.

Kidenya and his wife, a nurse, were homesick for their homelands in the far south where the soft consonants of Swahili rolled off the tongue. Uncomfortable in this distant outpost, Kidenya waited anxiously for the weekly steamship from Mwanza to bring his mail and hospital supplies.

All too often, however, it arrived nearly empty, the cargo pilfered by its handlers during the 1,000-mile journey from Dar es Salaam. The often reordered electric generator, the refrigerator for vaccines, the long-overdue supply of sterile syringes, the penicillin and surgical equipment never arrived.

Around February or March 1984, Kidenya and his staff noticed that several of their patients had genital ulcer disease that wouldn't respond to normal treatment.

"This is most strange," Kidenya said to medical assistant Justhe Tkimalenka, who readily agreed, and confided that he was afraid of the new patients: it wasn't right for people to die from genital ulcers.

Kidenya admitted that he was also frightened.

"These are very dreadful venereal ulcers," Kidenya later explained to a visitor. "Very deep. Very frightening. We treat these people for chancroid, but they never respond. And they develop long-standing diarrhea and persistent fevers. They all have severe weight loss—the most severe we have ever seen in this hospital in adults."

Kidenya, Tkimalenka, and the hospital's stout surgeon, Clint Nyamuryekunge, scoured medical records for clues. A recent graduate of Muhimbili Medical School, Nyamuryekunge was convinced that the puzzle could be solved using the scientific method. As bombastic and aggressive as Kidenya was discreet, Nyamuryekunge argued that all they needed was a good pool of plague patients to compare with a group of normal venereal disease patients.

Kidenya contacted the directors of other hospitals in the Kagera region—a difficult procedure in an area with no telephone service. He surmised fairly quickly that everyone was seeing cases of strange, lethal venereal diseases, and issued word to bring them to Bukoba.

Shortly, in September 1984, a village paramedic brought nearly two dozen patients to Bukoba in a single day, and Kidenya ordered that blood and stool samples be taken from all. Studies of the samples with microscopes and special stains revealed nothing. The patients all told a similar tale of a young barmaid in their town who gave Juliana's disease to the men, who in turn gave it to their wives.

The three doctors pooled their meager resources and financed a trip to the capital for Nyamuryekunge. Carrying blood samples and all available information on the patients, he traveled southeastward for days by boat, truck, bus, and car, finally reaching Muhimbili Medical School, where his former professors studied the samples, read the reports, and debated the nature of Juliana's disease. In the end they agreed only on one thing: nobody could find a microorganism that could be causing the illness. That meant it must be something new.

Dr. Fred Solomon Mhalu, the university's chief microbiologist, had led the nation's recent fight against cholera and was charged with training Tanzania's future pathologists. He suggested that Nyamuryekunge go to the medical library. "Something about this Juliana's disease reminds me of things I have read from America," he said.

The young surgeon found Gottlieb's description of *Pneumocystis carinii* pneumonia cases among Los Angeles men, the CDC's reports on acquired immunodeficiency syndrome in Haiti, the list of symptoms in the *New England Journal of Medicine*, and evidence in *Science* of a new retrovirus discovery in laboratories in the United States and France. He carefully compared the Kagera patients' symptoms and the Muhimbili laboratory findings from their samples with the observations in the American medical journals.

It was a perfect match, except for the types of patients involved: Kagera had no homosexuals, drug users, or hemophiliacs on Factor VIII. Mhalu suggested that Nyamuryekunge focus on the information about Haitian cases.

Bursting with the information, Nyamuryekunge made the long trek back to Bukoba, and briefed Kidenya and Tkimalenka on his discovery.

"And that was the day," Kidenya later recalled, "that we knew, oh! In the world there is a new disease called AIDS."

The three doctors then realized that their battle against this new virus was hopeless. Even in medically sophisticated New York City all the patients were dying.

Kidenya later remembered that the news hit him like a death sentence for his people.

"I thought surely this will be the greatest war we have ever fought. Surely many will die. And surely we will be frustrated, unable to help. But also I thought the Americans will find a treatment soon. This will not be forever."

It was January 1985.

For more than a year, the three of them struggled to convince other Tanzanian physicians that AIDS had come to the country.

"We told them watch out, there is AIDS in Tanzania," remembered Nyamuryekunge. "But this was causing turmoil. The doctors were refusing to believe us. They were saying, 'How do you know you have AIDS? You don't have a good laboratory! You might be confusing diabetes mellitis! You can't be sure! You are just raising a false alarm.' "

The Bukoba doctors had no way of proving that Juliana's disease was AIDS. The doctors urged the Ministry of Health to get the blood test for AIDS from America so that the debate could be settled.

Following an official request for the CDC's assistance, Nyamuryekunge's samples were shipped to Atlanta, and Jim Curran's team quickly confirmed that they indeed contained the human immunodeficiency virus. Surprised to discover AIDS in an area as rural and remote as Kagera, Curran immediately dispatched CDC investigator Don Forthal to Tanzania.

Forthal's tour of Bukoba Government Hospital began with outpatient clinics, immunization rooms, maternal/child care centers, and the maternity ward, facilities that were minimal but clean. But when they entered a general patient ward the difficulties Kidenya's team faced daily became obvious. Every spartan steel-framed bed was occupied. Thin mattresses and stained sheets were all the comfort most patients were provided. The smell of soiled sheets filled the air, competing for the senses' attention with the din of groans and conversation, amplified as it reverberated off the stark brick walls. The tattered window screens, intended to protect patients from malarial mosquitoes, flapped in the morning breeze.

"I couldn't believe what I was seeing," Forthal later said. "In just seven days I saw maybe twenty-four or twenty-five patients, all sick as hell with AIDS. No doubt about that. All in this one little Bukoba hospital. These patients were so much worse off than American AIDS cases. The disease is different over there. They were just wasting away before your eyes; I could see a difference in these people in seven days' time."[111]

Kidenya later showed an inquiring American journalist the full range of the Kagera crisis. In the general medicine ward, a woman sat stiffly on one bed, holding a small bundle. She looked well fed, but appeared listless and depressed. Her eyes were vacant. Kidenya approached her saying, "*Jambo, mama. Habari gani?*" He carefully lifted the bundle. The woman neither watched nor resisted as Kidenya dramatically unfolded the tiny wad of cloth to reveal a suddenly screaming, emaciated baby girl. Her eyes seemed enormous and her limbs stuck out like afterthoughts, flailing weakly and aimlessly.

"This," Kidenya said, "is malnutrition. This child looks to be a year old, but actually is three years old. This mother is from Uganda. For months she and her child starved. Now she has recovered, but for the child it is much harder.

"As you can see," Kidenya said, "we have many problems here. AIDS is only one." He paused, and as the mother slowly rewrapped the now silent child, he added thoughtfully, "And in truth I do not know—perhaps this, too, is AIDS. We cannot test, so we cannot tell. The child is not responding to treatment."

The visitor was guided to a wing detached from the rest of the hospital: a long, white row of cells, separated from one another like a line of shoe boxes or concrete bunkers.

Joined now by Tkimalenka, Kidenya ushered the visitor into the overpowering stench of one of the cells, where two women lay, one of them on the ground, wrapped in ancient sheets soiled with her own waste. Seeing this, Kidenya and Tkimalenka exchanged words in Swahili and the assistant stepped outside to signal a nurse.

"These two ladies are with HTLV," Kidenya said, using the term HTLV as code in front of the patients. The word AIDS struck such horror that many would kill themselves, usually by ingesting the pesticide Thiodan, if they heard such a diagnosis. Dozens of Kagera's citizens were rumored to have taken the faster Thiodan route to their deaths, and local stores were unable to keep up with the sudden increase in demand for the poison.

Pointing to a young woman who sat listlessly upon her bed, Kidenya said, "You can see her hair is straight, a bit red, and rare. Surely they are wasted. They look very sad."

Kidenya asked the young woman her name.

"Noticia," she replied.

He asked if she would tell her story for the visitor.

"*Jambo. Karibou,*" Noticia said, struggling to muster a welcoming smile. Her long response in Swahili was delivered in a quiet, flat monotone, quite uncharacteristic of Tanzanian use of the language. Kidenya listened patiently to her medical history. She was skeletal. Her lips were ulcerated and small sores dotted her skin. Yet it was obvious that Noticia had once been a remarkably striking woman.

As she spoke to Dr. Kidenya, her eyes glazed over; she knew she was going to die. Birds chirped in the background and the hospital cooks could be heard nearby stirring mashed yams in a cauldron over an open fire. Noticia spoke so softly over the noise that Kidenya bent down and put his ear within inches of her mouth.

Her illness began a few months ago, Noticia explained, when she noted an abscess on her neck. Later similar purplish things appeared all over her body. She poked a bony leg out from under the sheets and barely nodded in its direction, guiding the visitor's eyes to what seemed to be a Kaposi's sarcoma tumor on her calf.

Noticia gulped for air, and continued.

"I became very weak. I developed fevers and chest tightness and coughing." As if to accent that point, Noticia cleared her throat and a fit of coughing ensued.

"Initially she thought she had TB," Kidenya explained while Noticia rested. "We started to treat her for TB, and at first she responded. But then after a month the coughing returned and she had large swellings on her cervix."

Noticia whispered something in Swahili, and Kidenya told the visitor, "She is saying that she knew it wasn't TB. Now, she says, she is twenty-three years old but she feels as if she is a hundred. She is feeling weak, her limbs cannot carry her body. She feels like sleeping all the time."

With obvious reluctance, Noticia related the story of her travels to Kenya, her work in Sofia Town, and the typist job she had recently been forced to quit because she was too weak to push the keys.

As Noticia finished, Tkimalenka spoke softly to the inert form lying on the ground, then replaced her sheet. Noticia watched anxiously, and seemed relieved when Tkimalenka assured her the other patient was still alive.

Outside the cell Kidenya and Tkimalenka agreed that both women would be dead before the end of the week.

The grim picture repeated itself in one chamber after another, and a social pattern emerged from those willing to discuss their lives before AIDS struck: most of the men were former combatants in the Uganda-Tanzania war and/or traveling salesmen and smugglers who regularly crossed the region's borders to trade with counterparts in neighboring countries; most of the women had worked at some recent time as "disco girls," barmaids, or prostitutes, and several, like Noticia, had traveled outside the Kagera District to ply their trades.

Mythologies were instantaneous companions of history in the communities surrounding Lakes Victoria, Albert, Edward, and Kivu. Whether people were by nationality Ugandan, Kenyan, Tanzanian, Burundian, Rwandan, or Zairian, they all told tales in early 1985 of bewitched fabric, Juliana's disease or "slim disease," and the witches, "whores," or visitors who brought it to their people. The Mhaya women, considered particularly beautiful by the men of the region, were also commonly referred to as prostitutes, and in local Swahili dialect the word *Mhaya* was used interchangeably with the word *kahaba*, prostitute.

"One aspect of that notion is quite true," Tkimalenka, a Mhayan, said. "You get prostitutes in Dar from all regions. It is fair to say that the majority of prostitutes are coming from this region. But the people who are dying are not all Mhaya. So we can't accept the notion that AIDS is a disease of the Mhaya."

Still, Bukoba's young men claimed that AIDS was spread by women.

"People are aware of the girls. Afraid of them," twenty-eight-year-old Henry shouted over music in the Bukoba disco. A bachelor, Henry declared that he had "no intention of getting married. I am looking for a girl, but I can't choose which girl has AIDS or doesn't. I am very afraid, because it is death, you know. No medicine. Perhaps I will wait to marry, wait until there is a cure."

And the women, most of whom now boycotted the disco, said with equal certitude that the Juliana disease carriers were all men.

"I know my boyfriend is seeing other women when he travels to this place or that place," a young hotel clerk explained. "But what can I do? When he comes home he is so handsome and I reach out and say, 'Oh, darling, darling,' and all is forgiven."

The Tanzanians of Kagera were adamant that the disease came from Uganda, and with equal certainty the Rakai residents across the border pointed their fingers at the Tanzanians. Most people didn't know what a virus was—there was no word for it in Swahili, the closest approximation being *vinidogodogo*, or very little thing. But they did know that evil existed and could be manipulated by witches and sorcerers to inflict harm on their enemies.

It made sense, then, to assume that the new disease came from old enemies. Kidenya, Nyamuryekunge, and Tkimalenka rejected such superstition and searched for hard facts. They counted the sick and the dead, knowing their total represented a mere fraction of the true AIDS tally. Villagers knew the disease was incurable and therefore wouldn't make arduous journeys on foot to district hospitals. Nevertheless, by the end of 1985 the Kagera District's hospitals had seen 206 AIDS cases, 35 of whom had died in the facilities. Kidenya guessed that they were seeing 5 to 10 percent of the cases.

Blood samples collected by Forthal from local residents were analyzed at the CDC, revealing that antibodies to HTLV-III/LAV were present in 41 percent of the first hundred patients diagnosed symptomatically by the doctors.

Of particular concern to Kidenya and Nyamuryekunge was the discovery that people with AIDS were about five times more likely to have had a series of injections for some reason during the previous two years than were other patients in their hospital.

It worried Kidenya that in America AIDS was spreading via the dirty needles used by people addicted to narcotics. The local practice, born of economic necessity, was to reuse syringes and needles so often that the tips had to be sharpened on whetstones so they could still puncture human skin. Such was the custom in his own hospital, Kidenya said. "But for the time being we believe the problem is not so bad in our clinics. Our people know they must at least try to be clean. But, you see, there is another type of drug supplier, an injectionist. For us, these people are very hard to find. They hide in the fields or whatnot. But they may happen to get hold of a syringe and perhaps some antibiotics. And without any medical knowledge they sell injections. So, you see, the people may go to them. And surely these injectionists do not worry about sterile needles."

The problem had only worsened since the villagers had heard of AIDS. Knowing that the licensed doctors had no cure for the disease, those who suspected they might have Juliana's disease turned to the injectionists, who, like the snake-oil salesmen of America's Wild West, claimed their potions could cure anything.

To show his visitor the true scope of the Kagera District's AIDS problem, Kidenya negotiated privately with the regional party leader for a petrol ration and organized a trip north toward the Ugandan border. The first stop on the tortuous, muddy drive was the Bunazi Rural Clinic, staffed only by medical assistants and midwives. The chief medical assistant gave a tour, anointing each concrete hole with an illustrious title, such as "pediatric ward" or "maternity ward." But few rooms had beds, there was no surgical theater, and the pharmacy had little more than chloroquine and aspirin. In a small side room a woman held pieces of paper, each bearing information about patients. A man was bent over a microscope, studying samples of blood, urine, or stools. No other equipment graced the room.

"This is our pathology laboratory," declared the medical assistant. As he said this, a gust of wind swept all the samples and papers onto the dirt floor. The medical assistant led his visitors on to a general men's ward. As was the case in Bukoba, the term AIDS was never used, but two men were pointed out as suspected HTLV cases. Both were war veterans, coughing from tuberculosis and obviously dying. The assistant explained that usually such cases would be transferred to Bukoba, but there was no petrol for the truck. Asked how many AIDS cases had been sent from Bunazi to Bukoba, the medical assistant said only six.

Asked about syringes, the medical assistant pointed to a small kerosene-fueled autoclave containing several steel syringes and other equipment awaiting sterilization.

An hour further along the muddy road a small village was perched on the Ugandan border. The area was occupied for nearly a year by invading Ugandan troops in 1978. Then Tanzanian soldiers had bivouacked there while battling the Idi Amin government. The village bore the scars of war: bullet holes along the walls, abandoned, rusted vehicles, the complete absence of any valuable supplies.

Tkimalenka parked in front of the only building in the tiny hamlet that didn't sport bullet holes. A bright-faced, energetic young woman stepped out of the modest tin-roofed structure and, recognizing Tkimalenka, grinned and shouted, *"Karibou, Bwana! Jambo!"*

She urged the group inside, where, as eyes adjusted to the dark, three large, barren concrete rooms came into focus. In one there were three beds without mattresses. "They took the mattresses in the war," the village paramedic explained. In another there was no furniture, but a strong cord hung from the ceiling. "This is where we used to weigh the babies to see if they were well. But that was before somebody stole the scale, which hung from the cord." The third room was her office, containing a wooden chair, a small steel desk, and a bare wooden shelf.

"This is my office," she said proudly. "Where I keep track of the health of everybody in the village."

Atop her desk rested a foot-long shiny steel box. She opened the box to reveal two glass syringes, ten needles, and a dead fly resting in fetid water.

"Do you use these syringes?" the visitors asked.

"Yes, when we have something to give. Right now we have nothing. But sometimes we have vaccines for the children, or antibiotics. So then, yes, I use these," the young woman answered.

She described how she sterilized the equipment. Without electricity or kerosene, she couldn't use an autoclave. She had no alcohol with which to swab the needles. Before any round of injections she would hang a steel pot full of water on a tripod over a wood fire outside, boil the equipment, let it cool, and inject whoever needed vaccines or medicines. In such a situation it wasn't usually possible to sterilize the needles between each patient, she explained, but she was able to make sure that the needles were clean from one period of use to another. Proud of her work, and of the polite smile on Kidenya's face, the young woman graciously thanked the group for their visit. Later, when the young woman was out of earshot, Kidenya admitted that the sterilization procedure concerned him.

Back in Bukoba the group discussed the implications of such severe shortages of syringes. If one child in the village became infected with the AIDS virus, all the preschoolers might be infected in a single day's measles immunization campaign.

"Yes, yes, that is very bad. But what about the blood supply?"

Nyamuryekunge shifted his bulk in his chair and reminded the group that he was a surgeon. "It is most difficult for me because we do not have the AIDS blood test." No one could find the prospect of transfusing contaminated blood more alarming than a surgeon, given that virtually all surgical procedures entailed loss of blood that must be replaced to ensure patient recovery. Yet a single blood test cost more than Tanzania's annual per capita medical spending of less than three dollars.

"You see, when I have an elective operation, not an emergency, but the operation itself requires a transfusion, then I'm not very keen to perform that operation," Nyamuryekunge said. "But when the patient must have emergency surgery, then either the patient dies of AIDS in five years or he dies now. So in that case I give the blood now. Save the life now and let's pray the blood is not infected with the virus."

Kidenya sighed and said he hoped that the steps they were taking to educate the people of Kagera about the disease would soon stop the epidemic. Or that the Americans would shortly find a cure.

"It pains me to care for an AIDS patient. It really pains me. Because whatever I give I know it is not helping the patient," Kidenya said. "I don't fear contracting the disease, but it pains me to know that whatever I do, whatever book I turn to, it's useless. Your heart is not settled at all. At times I feel the disease is torturing our patients too much. I would like a disease which kills quicker. This one is too slow in killing. The patient wants to see you, demands your help. The help you cannot give."

Mr. Rutayuge, the hospital's wiry, older administrator, listened. It was his task to order supplies from Dar es Salaam and then fight like hell to see that they reached Bukoba intact. Now the doctors couldn't tell him what to order. Nothing, they said, would help. And so many were dying in his district that Rutayuge began to enter their names in the ledgers where he once itemized supplies and revenues.

"For so long the young people have been running around, not listening to their elders. Even before the war with Uganda some of them were running around. Crazy. They don't listen to the old ways. After the war it was worse. Discos, prostitutes, babies, so many babies!" Rutayuge's frail body shuddered; he seemed to be fighting back tears. "Now some of the elders say to them, 'Look here, we told you! Now you are sick. You are paying for all your running around.' "

Rutayuge appeared to be a practical man, not the sort who normally waxed philosophical or grim. He was a hospital administrator who, day to day, devised ways to replenish medical supplies for a rural clinic that hadn't "officially" received anything in weeks, perhaps months. With no budget, but plenty of ailing patients, Rutayuge negotiated deals with Ugandans, Rwandans, Burundians, even distant Kenyans, exchanging local goods for fuel, bandages, streptomycin, sheets, bedpans, painkillers for the dying, vaccines for the young, and aspirin for the rest.

He looked at his ghastly ledgers and said, "There is no future. It is the end of the world. Without young people how can there be a world?"

Bukoba's plight was little known in the rest of the world. Long after Forthal returned to the CDC, confirming the seemingly odd information that a major rural epidemic was unfolding in Central and East Africa, a preconceived dogma continued to dominate the world's perceptions of AIDS: that it was a disease primarily seen among gay men and injecting drug users, that all the African cases were emerging in major cities, and that the heterosexuality of AIDS in Africa was due to "special cultural factors," such as ritual circumcisions and clitoridectomies.

Some of the misperceptions were the result of the way news of Africa's epidemic unfolded. And some were due to less excusable factors, such as racism.

Before the discovery of HIV and the development of blood test kits, several cases of AIDS among Africans were symptomatically diagnosed in Europe, particularly in Belgium and France.[112] As of November 1983, 22 percent of all European AIDS cases were among people originally from sub-Saharan Africa.

Long before the antibody test was commercially available, the Pasteur group isolated LAV from the blood of a married Zairian man and woman living in Paris, and concluded that "there is strong evidence that AIDS is endemic in central and equatorial Africa."[113]

But it was the Belgians, particularly Peter Piot and Nathan Clumeck, who most aggressively pursued the AIDS/Africa link. Both men were seeing Zairois and Rwandan AIDS patients in Belgium, and they earnestly believed that major epidemics were underway in the two countries.[114] Clumeck and his Belgian colleagues conceived of a quick way to learn what might be transpiring in Rwanda. In October 1983 they mailed questionnaires to all the doctors working in the Centre Hospitalier de Kigali, the capital's main medical facility, describing the symptoms of AIDS and asking if such patients had been seen. Responses in hand, they went to Kigali in January 1984 and ran T-cell tests on twenty-six patients whose symptoms most clearly fit the CDC definition of AIDS: any combination of *Pneumocystis* pneumonia, Kaposi's sarcoma, wasting syndrome, dementia, chronic high fevers and secondary disease due to typically nonvirulent agents, such as cryptococcus and cytomegalovirus.

After four weeks in Kigali, Clumeck and his colleagues returned to Brussels, convinced that "AIDS could be endemic in urban areas of central Africa."[115]

In early 1983, Peter Piot attended a meeting on sexually transmitted disease in Seattle and spotted Jim Curran in the audience. Knowing Curran was in charge of the U.S. AIDS effort, Piot dashed over and asked him to step outside for a moment.

"Look, we have Zaire cases of AIDS in Brussels," Piot told Curran. "And I think they all got the disease in Zaire. I'm looking for money. Nobody in Belgium wants to support such a study."

Piot proposed to return to Zaire and study the nation's possible AIDS problem. Curran was noncommittal, explaining that his office was overwhelmed by efforts to prove to American skeptics that the new infectious disease even existed.

So Piot turned to Dr. Richard Krause, then director of the National Institute of Allergy and Infectious Diseases in Bethesda, and made the same plea. Krause offered a small scientific grant, provided the Belgian took the NIAID's Dr. Thomas Quinn along to Zaire. Filled with a sense of urgency and already having spent over a year searching for research funds, Piot readily agreed.

Krause strongly believed that GRID was an example of the kinds of newly emerging disease problems which he had previously warned the U.S. Congress about, and he made it a point to fly to Antwerp to meet with Piot in September 1983, shortly before Piot and Quinn were to depart for Zaire.

Also at the September meeting in Antwerp were the CDC's head of special pathogens investigations, Joe McCormick, and CDC laboratory expert Sheila Mitchell. Piot was not pleased. After having been ignored by Curran months earlier, he found the agency's apparently newfound interest in African AIDS distasteful and felt McCormick was trying to "horn in" on his study.

McCormick professed to be surprised by Piot's antipathy, and explained that he had been planning a Zaire investigation for months. His presence at the Antwerp meeting was at the bidding of Krause, who realized that Quinn, who had strong experience with AIDS in the United States but had never been to Africa, couldn't possibly handle such an investigation on his own; and Piot, though a veteran of the 1976 Ebola investigation, had no formal connections with the Mobutu government.

Only McCormick had been invited by the government of Zaire—a formal request for an AIDS investigation having been arranged by McCormick's old friend Kalisa Ruti, chief counselor to the Minister of Health. Furthermore, McCormick's African research experiences were extensive, and since his 1979 brush with Ebola in Sudan, Joe had continued investigating hemorrhagic diseases on the continent and in the laboratory. He had, for example, established that people living in the Haut-Ogooué region of Gabon, a rain forest area, were routinely exposed to Ebola, and 6 percent of that population had antibodies to the virus.[116] He and Karl Johnson had completed RNA maps of the Zaire and Sudan 1976 strains of Ebola, proving that McCormick's initial hunches were correct: the microbes represented two different viruses that, in an apparently amazing coincidence, appeared simultaneously in two locales.[117]

On their flight to Kinshasa the American and European scientists argued over who would be in charge of the Zaire investigation. Piot felt that the entire mission had begun in his Antwerp laboratory, and insisted that the effort should therefore be under his leadership. And he noted that Quinn was similarly less than pleased about McCormick's presence. In years to come such tensions between non-Zairian scientists conducting research in that African country would recur with nearly every investigative effort, contributing to difficulties in understanding the depth and nature of the Central African epidemic.

When the scientists reached Kinshasa, they found that the Ministry of Health and physicians from the University of Kinshasa and Mama Yemo Hospital were quite keen to learn whether some of the strange ailments they were seeing in their patient population were due to AIDS. The team, officially headed by Piot, set to work immediately, identifying possible AIDS cases in the hospital, confirming their infections in the laboratory, and determining how the disease was spreading in Kinshasa. The most difficult task—counting T cells one by one on microscope slides—fell to Sheila Mitchell, whose ability to set up a makeshift lab and excel under extremely difficult conditions drew praise from all the Zairian, Belgian, and American men involved in the investigation.

Key among the Zairian physicians was Dr. Kapita Bila Minlangu, who had already recognized the country's AIDS problem. During the first few days of their investigation, Quinn, McCormick, and Piot identified possible AIDS cases on the Mama Yemo wards, most of which had already been pinpointed by Kapita. In addition to the patients present on the wards, Kapita had for several months been saving medical information on odd cases that came through the facility.

Two things were immediately obvious: AIDS was claiming many of the patients in Mama Yemo, Ngaliema, and Kinoise hospitals; and women and men were equally likely to have the disease. Both findings stunned the foreign scientists, whose view of the disease had been shaped by the American and European AIDS model.

Even though the cause of the disease and appropriate blood test kits weren't yet available, Mitchell had little difficulty confirming most of the suspected AIDS cases because many of the patients had *no* T-helper cells.

A total of thirty-eight AIDS cases were identified and confirmed based on T-helper cell counts; 53 percent were men, 47 percent women. An astonishing 26 percent of the patients died during the three-week period of the study, and the foreign scientists observed the same eerie phenomenon Forthal would witness in Bukoba: patients grew sicker almost by the hour, and died before their eyes. The average patient had been symptomatic for only ten months, and all of them had lost more than 10 percent of their total body weight during that brief time.

Comparing their histories with those of controls (patients hospitalized for ailments that clearly could not be AIDS), McCormick found the AIDS patients were more likely to have traveled outside the Kinshasa area, to be either divorced or unmarried, and to have had more than one sex partner during the previous year—the twelve-month median among AIDS cases was seven sex partners.

They found no evidence that any of the cases involved intravenous drug use or homosexuality.

But they did find heterosexual clusters, linked in much the same way as Bill Darrow's Los Angeles gays. They even established that some of the people in Mama Yemo Hospital had had sex with individuals who were on the list of Belgian AIDS cases, demonstrating that the Africa/Europe sexual net could be as complex and far-flung as the gay American one detected by Darrow. Some of the Zairian females were prostitutes; others were the monogamous wives of men who had sex with prostitutes.

When the foreign scientists left Zaire they had no doubt whatsoever that they had witnessed a heterosexual epidemic, and Piot and McCormick, both of whom had studied sexually transmitted diseases in Africa, were deeply concerned. They knew that syphilis, gonorrhea, chancroid, *Chlamydia*, and *Candida* were rampant in most non-Arab African countries, even though none of those microbes was known to exist on the continent prior to Euro-Arab colonialism and the slave trade eras. The two scientists feared that AIDS might follow that pattern of rapid emergence, quickly overrunning the continent.

The Zairian/European/American group wrote up their study and submitted the paper to the *New England Journal of Medicine*. It was rejected because the peer review panel could not believe the disease was heterosexual and insisted that the team had overlooked some other mode of transmission or an unusual African custom that might be spreading the disease. They received similar rejections from a dozen other medical and scientific journals. The Zaire results went unpublished for nearly a year— a year during which Kidenya's group struggled to understand what was killing people in Bukoba, and a year in which AIDS surfaced, unrecognized, all over East, Central, and Southern Africa. Finally, after much revision, the study appeared in the British journal *The Lancet* in July 1984.[118]

Knowing that the existence of AIDS in neighboring Zaire had been proven in October 1983 would certainly have been helpful to Dr. Subhash Hira, whose STD clinic in Lusaka was then filled with mysterious ailments. He had counted a steady increase in particularly aggressive herpes zoster cases since the first had been observed nearly two years earlier.

By late 1983, Hira was seeing patients who were dying of bizarre pneumonias, tuberculosis, and herpes. It rang a bell, and Hira leafed through

French and American AIDS reports in the university library. Though the symptoms he'd seen mirrored those described in San Francisco, New York, and Paris, Hira knew that nobody in Zambia injected narcotics and homosexuality was so rare as to be considered nonexistent among the Bemba, Ndebele, and thirty-five other ethnic groups of the country.

Still, Hira pursued the hypothesis that AIDS was in Zambia. He had his staff tally the numbers of herpes zoster cases seen in the STD clinic since 1980, and the results prompted him to speak to Zambian Minister of Health Dr. Evaristo Njelesani.

What Hira told Njelesani in the Zambian spring of 1983 was that between 1980 and 1982 herpes zoster cases in Lusaka had increased tenfold.

"This all looks like AIDS," Hira told Njelesani, who was both impressed and concerned.

"How can we be sure?" the minister asked. Hira suggested that the Americans might have a way to test his patients, and Njelesani ordered Hira to find the proper groups in the United States with which to collaborate.

But it would be nearly a year before Hira had answers. Only toward the end of 1984 did researchers at the U.S. Army's Walter Reed Hospital in Washington, D.C., complete an HTLV-III search on blood samples from twenty suspected Lusaka AIDS cases: the virus was found in eighteen.

As soon as Hira got the results in the post, he rushed to Njelesani's office. Minister Njelesani studied the Walter Reed paper, refolded it, placed it in his suit pocket, and ordered Hira to immediately set up a national AIDS effort, coordinating all activities directly with his office. Njelesani imposed one strict rule from the outset: tell the press nothing. The Health Minister feared that Zambian AIDS would be exaggerated, affecting tourism and the national economy. And he was upset by rampant speculation in American and European medical journals (though not yet commonly seen in the popular press) that suggested that Africa was the origin of the AIDS virus.

"We have brought to Africa many viruses that were serious for them, and now we get back from them some retroviruses," Luc Montagnier had recently told a visiting journalist in Paris. "It's nothing wrong, just a fact. Also the origin of man is Africa, so it is not surprising to find old viruses in this part of the world. Countries should not hide from it. They cannot escape it. These are facts."

Joe McCormick had no difficulty convincing health authorities in Zaire and Belgium to take AIDS seriously. And the authorities in Kinshasa were enthusiastic about McCormick's suggestion that a joint Belgian/Zairian/ American AIDS research center be established in the country. McCormick's headaches didn't start until he returned to Atlanta, where Curran supported a long-term Zaire AIDS study, and outgoing CDC director Bill Foege was eager to be helpful; but Reagan's newly appointed CDC director, James Mason, seemed lukewarm toward the idea. At Foege's urging, McCormick spoke directly with Assistant Secretary of Health and Human Services Brandt.

"There's a one-to-one sex ratio of AIDS cases in Zaire," McCormick told Brandt, "proving that AIDS can be, and is, a heterosexual disease." Brandt absolutely refused to believe McCormick, maintaining that some overlooked factors had to be involved in Zaire. AIDS, Brandt insisted, simply was not a heterosexual disease.

It would be more than a year before the Reagan administration's health leadership would accept the idea that AIDS in Africa was primarily heterosexual. The administration would never fully acknowledge that the virus might also be heterosexually transmitted in the United States. Indeed, disputes over heterosexual transmissibility of the virus and the applicability of the African (read: black) experience to the Euro-American (read: white) context would rage within the upper echelons of the U.S. government throughout the eight-year-long Reagan administration and well into the term of his successor, George Bush.

The Euro-American scientific community would be similarly divided over interpretations of African AIDS and heterosexual transmission of HIV, and that tension would persist well into the 1990s. Because AIDS had first been noted among gay American men, many scientists and politicians insisted that the modes of transmission of the virus were rigidly limited to those first observed in the United States—anal intercourse, injecting drug use, blood product contamination, and "Haitians."

But, of course, there was heterosexual transmission of AIDS in America, in Europe, in Haiti—in every geographic location on the planet into which HIV had infiltrated. Among the very first cases of AIDS reported in New York City were heterosexually acquired infections.

Some public health officials critical of the Reagan administration quietly argued that there was a racist subtext to the debate: nearly all heterosexual cases reported worldwide by mid-1984 involved people either living in Africa or of African heritage. In Europe and the United States nearly all clearly identified heterosexual transmissions reported to authorities by mid-1984 involved blacks or Hispanics; most were immigrants or visitors from African countries, the Dominican Republic, Haiti, and Puerto Rico.

A well-intentioned effort to gather evidence for heterosexual transmission of the virus began, its focus consciously directed at Africa. In essence, European and North American researchers had domestic agendas that underlay much of their African research.

But Jacques Liebowitch reflected a sentiment more popular among AIDS researchers at the time, saying, "We built a focus on Zaire . . . [to look at] people who didn't fit into any of the other known risk groups, such as being homosexual."[119]

Joe McCormick had no such initial intent for Project SIDA,[120] as the joint Zairian/American/Belgian research effort would be called, nor were the physicians of Kinshasa particularly interested in seeing precious resources wasted on proving what their medical charts already made clear: namely, that AIDS in their country was a heterosexual disease. Curran and McCormick decided that Project SIDA would be a serious African AIDS research center, designed to answer questions important to Africans. Curran immediately began scrambling for funds, carefully avoiding Brandt's office, while McCormick tried to find the right CDC scientist for the job.

The creation of Project SIDA went on quietly in Atlanta and Kinshasa while most of the Euro-American research effort in Africa continued to focus on two issues: heterosexual transmission and the scope of Africa's epidemic. As soon as the Pasteur group had a crude LAV test available, they collaborated with McCormick, Piot, and Quinn on analysis of the blood samples collected in Kinshasa hospitals. They confirmed that 97 percent of the patients Kapita had diagnosed as AIDS cases had antibodies against LAV (HIV). Most troubling: so did many of the controls, which indicated that there was an asymptomatic stage of the disease and that infection was far more prevalent in Zaire than it had initially seemed. Seven percent of the apparently non-AIDS patients hospitalized for noninfectious reasons came up antibody-positive, as did 5 percent of the new mothers who were on the obstetrics ward of Mama Yemo Hospital in 1980. In addition, serum collected from a mysteriously ill woman on the Mama Yemo obstetrics ward in 1977, who died of apparent immune deficiency in 1978, proved positive for antibodies to LAV.

Both the rate of adult infection in Kinshasa and the apparent age of the Zairian epidemic merited serious concern. By contrast, the French overall rate of apparent LAV infection in 1983 seemed to be less than 0.3 percent. [121] The Pasteur group was at the time receiving blood samples from other African countries, and had evidence for similarly alarming rates of LAV (HIV) infection in the general populations of Rwanda and the Central African Republic.

During the 1983 winter holidays Jonathan Mann answered his phone in Albuquerque. Joe McCormick—a scientist Mann admired immensely but had never met—introduced himself and got down to business.

"How would you like to work in Africa?"

Mann was stunned. But the CDC's New Mexico-based epidemiologist and bubonic plague expert listened intently as McCormick described what he had seen in Kinshasa.

Though Jon and Marie-Paule Mann had three young children, and none of them had lived in a developing country, it didn't take much to convince the family to move to Kinshasa. For Parisienne Marie-Paule it meant speaking her native tongue; the kids relished the adventure. And Mann recognized with considerable excitement the scientific importance of such work.

Curran, who had long been impressed with Mann's work, was quite pleased with the choice. Mann had displayed a talent for handling dicey political and press issues during his tenure in New Mexico. This skill, demonstrated from the first day Boston-born Mann had arrived in the state and faced public concern about a case of bubonic plague, would be crucial. The often tense status of relations between the U.S. and Zaire governments and the competing interests of AIDS researchers from all over the world who were eager to investigate the African epidemic would test Mann's mettle.

By March 1984, McCormick and Mann were in Kinshasa, working with Kapita, Drs. Nzila Nzilambi, Ngaly Bosenge, Kalisa Ruti, and other Zairian scientists to establish Project SIDA. McCormick acted as Mann's mentor, passing on in the course of a month as much as he could about Zairian languages, customs, and politics, as well as how to properly play the role of an outside American expert when working in a postcolonial, impoverished country lacking in basic infrastructural support.

Mann learned his lessons well—perhaps too well from the perspective of other foreign scientists and members of the press. He never spoke to outsiders without first clearing his comments with the Zairian Ministry of Health; he fought off foreign researchers who failed to collaborate with Project SIDA on its terms; and primary among those terms was a willingness to collaborate as equals with Zairian scientists and abide by the press and publication limitations set by the Zaire government.

"I'll tell you anything you want," Mann would say to all non-Zairian callers, "if you come here with a letter from the Zaire government. But without that letter, I won't talk to you at all."

Ten years later some rival scientists would still speak bitterly of Mann's policies at Project SIDA, claiming that he froze them out of Zaire and treated the country's AIDS epidemic as his personal "turf." But Zairian scientists would have nothing but praise for Mann, as well as for Piot. Project SIDA would prove to be the most prolific AIDS research effort on the continent from 1984 until its closure, due to civil war in Zaire, in 1991.

While most other African governments either were confused about the extent of their epidemics and still in the rudimentary stages of local research or were deliberately maintaining public silence out of a sense of national pride and economic concern, Zaire was quite open. An unfortunate side effect of the government's candor was a series of false international assumptions that would persist for over a decade: that Zaire had the worst of Africa's epidemics; that AIDS definitely started in Zaire; that all other AIDS outbreaks could be traced back to a Zairian origin.

Though Mann was in charge, Project SIDA included Drs. Henry Francis and Tom Quinn of the National Institute of Allergy and Infectious Diseases. Together with their Zairian colleagues, the team did HIV prevalence studies showing that by 1985 the general population infection rate in Kinshasa was about one-third that seen among gay men in San Francisco, and that multiple heterosexual partners, medical injections with nonsterile needles, and foreign travel were the key risk factors.[122]

As the dimensions of the global AIDS epidemic grew, the CDC organized the first International Conference on AIDS, which convened in April 1985 in Atlanta. About 2,000 scientists and reporters from thirty nations attended the grim gathering, during which the scale of what was by then considered a pandemic became apparent.

Though the Atlanta meeting would, correctly, draw attention to Africa's plight, it would later be established that nearly all the assumptions, and the data upon which they were based, were false. As the scientists assembled in Atlanta, AIDS *was* indeed emerging in Central Africa. But it was not doing so via some of the means described or on the terrifying scales presented.

At the meeting, Luc Montagnier said blood tests on samples drawn in Kinshasa in 1970 showed that one out of every 220 men and women then had antibodies to LAV (HIV); in 1980, he claimed, one out of ten Kinshasa adults was antibody-positive. And, he told the gathered scientists, AIDS was spreading within African households by a variety of nonsexual means. Robert Gallo disputed the household transmission claim, but agreed that AIDS was rampant in Africa, noting that 65 percent of children in Uganda tested positive for antibodies to HTLV-III (HIV).

Nathan Clumeck reported that 88 percent of the female prostitutes tested in Kigali, Rwanda, had antibodies to HIV—up from 70 percent levels of infection in 1982 blood samples drawn from local prostitutes. The general population, Clumeck said, had an infection rate by the end of 1984 of 9 percent.[123]

Dr. Robert Biggar, of the U.S. National Cancer Institute, reported that infection with both the HTLV-I and HTLV-III (HIV) viruses was extremely common all over Kenya, even in remote pastoral areas. On the basis of HTLV-III antibody tests run on blood samples collected by the CDC in Kenya in 1982–84 during various disease studies (not AIDS), Biggar claimed that over half of the Kenyan population had at some time been infected with the AIDS virus and nearly a third had antibodies to HTLV-I. The strongest responses, he said, were among the nomadic Turkana people of northern Kenya, nearly 80 percent of whom were infected with the AIDS virus.[124] Biggar also claimed that up to 15 percent of the children, 25 percent of the elderly, and 20 percent of young adults in the remote Kivu District were infected with HTLV-III.[125] And he told reporters that over half the young women tested on the antenatal ward of Lusaka's University Teaching Hospital—55 percent, to be precise—carried antibodies to HTLV-III (HIV) in 1984.

Similarly terrifying levels of infection in Africa were reported by a team working with Robert Gallo. On the basis of HTLV antibody tests of stored blood samples that had been collected by the National Cancer Institute in 1972 and 1973 from schoolchildren in Uganda as part of a Burkitt's lymphoma study, the team concluded that 66 percent of the children were infected with HTLV-III (HIV) *nearly a decade before anybody realized that AIDS existed.* The blood samples had been collected in the remote West Nile region of Uganda, an area of tiny villages located amid swamps and heavy rainfall.[126]

Finally, Max Essex and his Harvard colleague Phyllis Kanki referred to the recent discovery of a virus in captive rhesus macaques in U.S. primate centers that produced an AIDS-like ailment in the monkeys. The virus was dubbed STLV-III (mac), or simian T-lymphotropic virus type III (macaque). The virus, they said, grew easily in human T cells. A second virus, dubbed STLV-III (agm), was announced at the meeting. It was found, Essex said, in half of all tested wild African green monkeys, or vervets.[127] Essex told the gathering that it was reasonable to assume that AIDS started as an African monkey disease, and only recently, through an unknown means, entered the human population.

Though the essence of nearly every one of these headline-grabbing reports would later prove false, they made their impression: the world was convinced that Africa was witnessing an older, widespread epidemic that originated in monkeys and spread among humans of all ages on the continent via heterosexual transmission and some as yet unclear "household" means.

For the three lone Africans present at the "international" meeting— Project SIDA's Kapita and Nzila and Pangu Kaza Asila of Zaire's Ministry of Health—much of what transpired in Atlanta was deeply offensive. Mann had insisted that the CDC pay to bring the Zairian scientists to the meeting, but he also worried that one of them might unwittingly say something to the aggressive North American press corps that would have dismal repercussions back in Kinshasa. Because none of the Zairois had ever dealt with Western journalists, Peter Piot was asked to stay with them at all times.

Though Kapita, Pangu, and Nzila were upset by allegations that AIDS was Africa's dubious gift to the rest of the world, they managed to keep their anger to themselves until approached by an American journalist who said, "We have all heard what Max Essex said here about AIDS originating as an African monkey disease. Tell me, Doctor, is it true that Africans have sex with monkeys?"

Kapita seethed. The three Zairois pretended not to understand the question, though English was one of the four or five languages they spoke with some degree of facility.

"Peter, *s'il vous plaît, que est-ce-qu'elle a dit?*" Kapita asked Piot, hoping the journalist would give up trying to get an answer. Piot was enraged. He whispered a warning in French: "*Ne répond pas.*" But Kapita told Piot to translate a response to what he considered an exceedingly rude and demeaning query.

"Madam, I don't know what you're talking about," Kapita said. "We don't do those things. But I believe that in Europe they make movies where women have sex with dogs. And I've also heard that in the U.S. there are all these dogs as pets at home, and that they sometimes, well, you know what I mean . . ."

It would not be the last time that distinguished African scientists would be grilled by foreigners—both fellow scientists and journalists—about a variety of alleged sexual and cultural practices that some Westerners believed explained Africa's nonhomosexual AIDS epidemic.

"They just can't seem to accept that you can pass the virus by putting a penis into a vagina," Piot exclaimed at the Atlanta meeting. "I just can't understand this. These people are supposed to be scientists, after all. Would somebody please tell me why a virus would be willing to go from a penis to an anus, but not from a penis to a vagina? These people are disgusting!"

Piot knew better than anyone at the meeting, save perhaps Kapita, Nzila, Pangu, and Mann, that world press coverage of the statements made by Gallo, Essex, Montagnier, Biggar, and other Western scientists would have a chilling impact on AIDS research in Africa. He felt certain that many African governments would react to the finger pointing by shutting down what few research efforts were underway, just as the AIDS epidemics were emerging in their countries.

Sitting in a stairwell of the conference center trying to collect his thoughts, Piot could only shake his head and murmur, "This is a disaster."

Matters worsened following the Atlanta meeting. As Western scientists continued to point at Africa, the continent's leaders—as Piot had predicted—responded in kind.

"African AIDS reports are a new form of hate campaign," decried Kenya's President, Daniel arap Moi.

"If scientists cannot find a home for the virus, Africa is not the solution to their dilemma," declared Kenya's Minister of Health, Peter Nyakiamo, in a speech before his country's Parliament.

"There is no indication whatsoever where the disease started," Dr. Fakhry Assad, director of the World Health Organization's communicable disease program, said. "The disease as we know it appeared here at the same time as in the United States."[128]

The AIDS finger pointing was hitting impoverished Africa at a particularly difficult time. Major wars and insurgencies raged from the Horn of Africa to Cape Town, most fought as Cold War proxy battles fueled by rival industrialized world interests. In addition, several African nations suffered military coups during the early 1980s, prompting additional diversions of scarce resources toward military expenditures, usually at the expense of health and education spending.[129]

In addition, several countries were in the grip of their worst drought of the twentieth century, notably Mali, Mauritania, Mozambique, Zambia, Ethiopia, Somalia, Sudan, and Cape Verde. Scientists argued that the drought, and the famines and massive refugee migrations it produced, were due to structural changes in global meteorological patterns, possibly due to global warming. The Sahel desert belt across the northern part of the continent, they said, was expanding, claiming millions of acres of what had recently been arable land.

Peter Usher, UN adviser to Kenya, said there was a good chance that Africa's drought plight was truly something new, and worsening. "Which could mean that Africa is getting drier, and the future consequences are going to be even more serious than they are now," Usher said.

Bradford Morse, chief administrator of the UN Office for Emergency Operations in Africa, asserted that at least twenty African countries were facing severe drought conditions and food shortages in 1984–85, and at a minimum 30 million Africans were at risk of starvation as a result. In addition, he said, at least 10 million drought refugees were on the move, having abandoned their Sahel belt homes in search of food.

"This is the greatest single phenomenon of this sort in human history," Morse declared.

Ethiopian climatologist Workineh Degefu warned that whether or not fundamental changes were taking place in the planet's atmosphere and weather patterns, history was moving relentlessly toward increasing human need for resources on the continent as populations swelled and demands for farmland and firewood increased. As had happened in the American Midwest during the 1930s, overfarming was producing dust bowls, rendering the once fertile, feral lands nonarable wastelands.[130]

But African leaders knew that the world wasn't much interested in their drought and famine. The crisis began in the late 1970s, but drew little global interest until 1985, when African journalists finally managed to get film of the Ethiopian disaster broadcast on British television and a group of rock-and-roll performers subsequently staged a seventeen-hour benefit concert, called "Live Aid," that was simulcast in 152 countries, raising $70 million for African relief.

African leaders were less than pleased about the attention their AIDS situation was garnering, particularly as they had no idea exactly how serious it really was. Few, if any, of them believed in 1985 that AIDS could possibly match the severity of the drought and famine, or of the region's malaria epidemic, or of its general economic woes.

In Zambia, Njelesani was angry that Robert Biggar told international reporters the results of blood tests done in the country before clearing the data with Lusaka collaborators. Elsewhere on the continent antagonisms were developing against foreign researchers—"safari scientists"—who would dip into a country for a few days, possibly a couple of weeks, leave with cases of blood samples, and write up their results for major medical journals without first clearing the data and interpretations of it with local collaborators.

A chill settled over the nascent African AIDS research community.

Nathan Clumeck and Belgian colleagues decided to convene a meeting on AIDS in Africa during the fall of 1985—in Brussels. By the early summer some African leaders were protesting, saying they wouldn't go to Europe to discuss Africa, particularly if the Americans and Europeans were going to continue blaming Africa for originating the AIDS epidemic. Eventually the governments of Zaire and Burundi pulled all their papers from the Brussels meeting, Project SIDA followed suit, and the CDC took its cue from Zaire, also withdrawing its support and presentations from the conference.

A second, competing meeting was organized by the CDC and WHO to take place in Bangui, capital of the Central African Republic, four weeks before the Brussels gathering.

Shortly before the Bangui meeting Robert Biggar's group published a study that unintentionally provided the first evidence of the serious errors scientists had been making in estimating the size of Africa's epidemic. Biggar's team noted that it seemed strange that the early HTLV-III (HIV) blood tests had discovered the highest incidences of infection in remote areas where nobody seemed to have overt AIDS. So, in May 1984, Biggar's group had journeyed to the Kivu District of eastern Zaire, taken blood samples from 250 hospital patients, and tested them back in the U.S. National Cancer Institute laboratories for antibodies to HTLV-I, HTLV-II, HTLV-III (HIV), and *Plasmodium falciparum* malaria. They found that about 80 percent of the people had antibodies to malaria, far fewer reacted positively to the three HTLV viruses, and the same age groups—even the same individuals—that reacted most strongly against malaria also responded to one or all of the HTLVs.[131]

It was soon apparent that the initial HTLV assays were useless when executed on the blood of people chronically infected with malaria, leishmania, or other parasites, all of which produced what in laboratory lingo was termed "sticky sera." The first HTLV tests involved mixing suspect blood with antibodies to one of the viruses—say, HTLV-III (HIV). If viruses were present in the patient's blood, the antibodies and viruses would form complexes that would stick to the test surface and could be seen following a rinsing step. But parasitically infected blood—particularly malarial blood—formed nonspecific "sticky" complexes that also adhered to test surfaces through rinsings. Thus, the first HTLV-III (HIV) tests produced huge numbers of falsely positive findings.

Given that nearly everyone living below Africa's Sahara Desert chronically carried some malarial parasites in their blood, it was surprising that early AIDS researchers didn't obtain results naming every single African as an AIDS carrier. Instead, they found 50 to 90 percent alleged infection rates. The discovery of the HTLV test flaw meant that all estimates of African AIDS and HTLV-I infection rates made on the basis of that set of assays were thoroughly erroneous.

Some African countries did have serious emerging AIDS epidemics in 1985, but they were certainly not on the orders described at the Atlanta conference. Project SIDA estimates of infection rates of just under 10 percent in some Kinshasa groups were based on LAV antibody tests, which were less vulnerable to the "sticky sera" problem and would prove reasonably accurate.

Amid the exaggerated reports there were several crucial but less dramatic studies that received little immediate attention. Key among them was a joint Anglo/Zambian/Ugandan study of an apparently new disease seen in the Rakai District of Uganda, just across the border from Tanzania's Kagera District. Called "slim disease," the ailment produced dramatic weight loss and overwhelming fatigue, eventually proving universally lethal. The researchers used an improved British-developed HTLV-III (HIV) test on forty-two "slim disease" patients, finding AIDS antibodies in thirty-four. They also discovered that 17 percent of their healthy Ugandan controls were antibody-positive. The implication was that AIDS and "slim" were the same disease.

"Slim," they argued, surfaced in Uganda at about the same time as the "gay plague" appeared in California and New York. There was certainly no evidence that AIDS was endemic in Africa. So, they said, blaming Africa for being the origin of AIDS had no clear basis in available fact.[132]

Speculation arose that the coincident responses in the early HTLV tests to the retroviruses and malaria might indicate that there was mosquito transmission of the virus. This prompted panic not only in Africa but in other parts of the world where *Anopheles* insects were pervasive. Members of Project SIDA and Curran's group at the CDC tried to counter this concern by pointing out that most victims of feeding malarial mosquitoes were small children who took no precautions to protect themselves from the insects and weren't yet immune to the parasites. Yet over 95 percent of all known AIDS cases involved adults.[133]

The epidemiological argument was not enough to quash speculation about mosquitoes, however, and throughout the 1980s concern that insects could transmit the virus would be repeatedly resurrected, particularly by those who were anxious to argue away heterosexually spread outbreaks of AIDS in places such as Belle Glade (Florida), Haiti, Brazil, and India.[134]

By the time African, American, and European scientists gathered in October 1985 in Bangui, there was a fair amount of antagonism in the air. Franco-American tensions were evident, as the Pasteur group and their allies became more vocal in their claims that Gallo's group had stolen not only credit for discovery of the AIDS virus but possibly the virus itself. Some Belgians were angry about the threatened boycotts of their upcoming Brussels meeting. And the Africans shared varying degrees of wrath about the Western portrayals of their epidemic.

Joe McCormick, who engineered the Bangui meeting, made sure that representatives of all points of view were invited, and pushed WHO's Assad to be forceful in his management of the discussion. In McCormick's mind the rivalries and anger were only contributing to the epidemic's spread and its emergence in new areas. He wanted the Bangui gathering to accomplish four things: air everybody's grievances, flush out a true apolitical sense of the dimensions of the pandemic, create a working diagnostic definition of AIDS that could be used in poor countries in the absence of blood-testing capabilities, and set priorities for future research—particularly in Africa.

On McCormick's covert agenda was convincing Assad of the severity of

Africa's AIDS crisis, with the aim of creating a special World Health Organization AIDS program. As far as Joe was concerned, the political fallout of misguided AIDS research and tensions left WHO as the only option for international leadership of pandemic control.

On his way to Bangui, Max Essex passed through Kinshasa, where he met with Jonathan Mann and told him that he had additional evidence for the existence of two different AIDS-like viruses in African monkeys—evidence he felt proved an African origin of the disease.

"Don't talk about that in Bangui," Mann said. "You'll get killed. People will be insulted. It would be disastrous."

International politics, sensitivities to racism, nationalism—all of that was new to Essex. Even years later Essex would say he couldn't understand why his remarks in Atlanta had caused such a furor in Africa, and he wasn't clear why Mann was urging him to censor himself in Bangui. But, recognizing that Mann lived in Zaire and seemed to understand such matters, Essex agreed to save his remarks for the Brussels gathering.

Meanwhile, Essex had set up a long-term collaborative relationship with Dr. Souleymane MBoup of University Cheikh Anta Diop in Dakar, Senegal. The scientists working with Essex and MBoup in the West African country were beginning research on the relationships between various monkey and human AIDS viruses. Essex, who still believed that all the HTLV viruses—including HIV—were closely related, was also looking for evidence of simian T-lymphotropic virus (STLV) and HTLV-I infection in Senegal.[135]

Mann had other concerns. He was deeply upset by what he considered "bad science" done over the previous year by American and European "safari scientists" in Africa. While Project SIDA was at pains to train Zairian technicians and collaborate fully with colleagues in Kinshasa, most other Westerners seemed to give only lip service to collaboration.

"Bad collaboration yields bad science," Mann said. "Suppose a group of foreigners came to some place in the U.S. Midwest, went to a few small hospitals, collected blood samples, then flew home. And then, without consulting with their supposed Midwest collaborators, published a paper in a major international medical journal saying that thirty percent of all adult Midwesterners were HIV-positive," Mann would say. "That would be bad enough—that's bad science on the face of it, extrapolating to a whole population on the basis of isolated, possibly unique, cases. But now suppose you find out your test was all wrong. You goofed. Maybe the real rate of infection in those Midwest hospitals was only two or three percent. Do you honestly think those people in the Midwest would forgive you?

"Why is there no apology? Why hasn't the National Institutes of Health apologized? The U.S. government? When and where will this error be rectified?" Mann asked.

There never would be formal apologies to the affected African governments from either Western governments or scientific institutions. Most of the journals that published the claims of mosquito transmission and rampant AIDS throughout Africa never printed formal retractions or apologies. In a few cases tiny corrections appeared months after initial publication, escaping the notice of the world press and scientific community.

"We can't behave like gods-in-the-sky when we work in developing countries," Mann would say. "And we can't publish without fear of impunity, without a sense of responsibility to the people we study."

These and other grievances were aired at the Bangui meeting, and McCormick's goals for the gathering were met. Western scientists had an opportunity to see abstract concepts like "infrastructural development" or "economics-driven prostitution" come to life when their hotel-room taps produced no water or prostitutes pawed them in elevators in the Central African Republic's best accommodations.

"The most crucial obstacle to comprehending the African AIDS epidemic and bringing it under control is the lack of [local] training and tools of communication and analysis," McCormick said. And the Western scientists, most of whom were on the African continent for the first time in their lives, had a chance to experience firsthand the significance of McCormick's remarks when they tried to telephone their American or European offices or buy batteries for their shortwave radios.

Assad, too, wanted the African government representatives at the conference to get past their resentments and face the reality of AIDS. At one point he demanded that each country representative tell the assembly exactly how many AIDS cases had been diagnosed and what were the suspected infection rates in their nations. On a first pass around the room most African country representatives hedged—some denied any knowledge of AIDS in their nations. Assad then told the group, "You're not being honest. I know, I've been there, I've seen AIDS in your countries."

Assad threatened to cut off WHO shipments of cholera vaccines and other vital supplies to countries that didn't speak up candidly. The following day, most African representatives provided numbers, though everyone knew that no country had epidemiology surveillance systems that could keep track of all its citizenry, and the data greatly underestimated the region's AIDS epidemic.

Rwanda reported that they'd seen 319 AIDS cases since 1983, 86 in small children. Kenya reported ten cases; four were foreigners. Zaire cited the Project SIDA data, which found antenatal clinic infection rates in Kinshasa of about 6 percent. Zambia reported that of 143 women who gave birth at University Teaching Hospital in Lusaka in the days prior to the Bangui meeting, seventeen were infected with HIV, as were fifteen of their babies.

Assad became a convert to the AIDS cause, and readily agreed with McCormick's opinion that a special epidemic effort had to be coordinated out of WHO. McCormick made his private pitch to Assad. He wanted an office established in Geneva that would serve as an international clearinghouse for AIDS information and technical expertise. He wanted it to have enough WHO clout to be able to intervene in multinational scientific disputes.

Assad readily agreed, and asked Joe to do the job, but McCormick had other ideas.

"There's somebody I want you to meet," he said.

Later, McCormick introduced Assad to Mann, and before the Bangui gathering ended, Mann had agreed to become director of a new global AIDS program. For the next six months he would commute between Geneva and Kinshasa, trying to ensure the survival of Project SIDA while giving birth to a new global AIDS effort. There was just enough money in Assad's budget to pay for Mann's plane tickets and a part-time secretary. His salary would still be paid by the CDC.

The Project SIDA group soon found explanations for more of the misleading information that had surfaced in Atlanta and via Western medical journals. The apparent cases of household transmission in Africa turned out to be the results of mother-to-child transmission during pregnancy, the birth process, blood transfusions, and breast feeding.[136] The extraordinarily rapid AIDS deaths that had been observed in the Lake Victoria region did not seem to hold true in Kinshasa, where a years-long asymptomatic infection stage appeared to precede AIDS in Zairian men and women, just as it did in gay American men.[137]

Older children (aged two to fourteen years) were getting infected through nonsterile needles used both in hospitals and by illegal injectionists, and by blood transfusions given to treat anemia episodes brought on by malaria.[138]

Similarly, the 6 percent seropositivity level they discovered among Mama Yemo Hospital personnel was not the result of spread within the facility, but of receipt of contaminated blood transfusions, multiple sex partner heterosexual activity, and nonsterile medical injections.[139]

Biggar, Njelesani, and other American and Zambian scientists coauthored a study reassessing rates of HIV infection in Lusaka's hospital, using the new British-made HTLV-III (HIV) test. They found that rates of infection—far from Biggar's prior estimate of 55 percent for the general Zambian population—actually peaked in 1985 at 29 percent seropositivity, seen among patients in Hira's STD clinic. In the antenatal ward 8.7 percent of the women were infected; hospital employees had a 19 percent infection rate.[140] But even the new, lower figures ranked among the highest infection rates in the world. Furthermore, with no cure or vaccine in sight, Njelesani could foresee a horrendous growth curve for the future of Zambia's epidemic.

Little news of the African debate of 1983–86 or of the spread of AIDS around the continent reached the remote Government Hospital in Bukoba. Drs. Kidenya and Nyamuryekunge were still fighting credibility battles with their Tanzanian colleagues when the continent's health leadership gathered in Bangui. And their hospital administrator continued to enter the names of the dead in his supplies ledger.

As the first January snow of 1986 fell outside his Geneva office, Fakhry Assad reflected on the Bangui meeting.

"The nine [African] countries[141] that officially attended the meeting all said, 'What can we do about it? We are paralyzed. We have no infrastructure, no treatment, no education. We have nothing to give.' This had been completely overlooked by people from the outside," Assad said. "Frankly, just to say, 'Educate your people in regards to sex,' how many would really believe in it? You have to have serious sero-surveys, interviews to assess the problem locally. And even if you find clear evidence of an epidemic, what means will you use to educate? Are you going to do it over the radio: who has transistors? Any country, if he has the means to control a disease, he will. But here [with AIDS] he can only say, 'I have a problem, and I don't know how to solve it.' "

Those most deeply involved in the AIDS fight now realized the microbe had won the first round, successfully emerging in outbreaks on at least three continents in *Homo sapiens* populations ranging from heroin addicts in U.S. ghettos to heterosexual neurosurgeons in Kinshasa; from Michael Callen and Bobbi Campbell to Noticia.

The virus had gone from epidemic to *endemic* status in key population groups around the world. It had defeated the powers of science that just a decade earlier had led public health planners to confidently agree to cut their sexually transmitted disease budgets.

By the time Assad reflected pensively in his Geneva office, Campbell was dead, Callen was battling yet another round of opportunistic infections, Noticia's body was buried in a banana grove in her home village, and Greggory Howard was walking the streets of Newark preaching the AIDS gospel to junkies huddled around trash fires, telling them, "I have the HIV as a result of my drug abuse."

"I'm listening to him, man, cuz he's one of us," said a tall, thin African-American self-described "junkie homosexual." Stabbing the air to drive his points emphatically home, Howard's fan said, "He speaks the truth, man. We all know this AIDS thing is a killer. Especially for us black people. I'm telling you, man. I'm telling you."

IV

Because of the legacy of blame, accusation, and exaggeration concerning AIDS in Africa it was impossible to have an apolitical, "pure science" discussion of the origins of the human immunodeficiency virus during the 1980s. Not until the Sorbonne's Mirko Grmek published his book on the subject in France in 1989[142] would discussion begin to free itself from the fetters of prior blame. Still, there would remain in the 1990s a decided timidity in AIDS academic and policy circles about broaching the subject of the origins of the global pandemic. The official line of the World Health Organization, first enunciated by Assad in 1985, would remain the agency's position in 1994: AIDS emerged simultaneously on at least three continents.

Few scientists accepted that position, recognizing it for what it was—a political compromise. But publicly they went along with WHO's stance because it was too politically dangerous to do otherwise. Far too much finger pointing went on during the 1980s to allow anybody by 1990 to feel that an environment of complete intellectual freedom could surround the question of the origin of AIDS.

"So the origins debate will go on," wrote Canadian analyst Renée Sabatier in 1988.[143] "It is probably optimistic to hope that it will be conducted without continuing imputations of blame, and without a continuing belief by others that blame is being imputed. But scientists, media, and politicians alike would do well to exercise great restraint in this discussion, since feelings of being blamed are already seriously hampering efforts to control AIDS."[144]

Shunning the subject, some scientists would simply say, "Well, it doesn't matter, really. AIDS is here, the pandemic is spreading all over the planet. Let's deal with the here and now. What's past is past."

Or, as Zambian President Kenneth Kaunda put it in 1987: "What is more important than knowing where the disease came from is where it is going."

But few researchers honestly held such a belief. For if AIDS could emerge so successfully worldwide in the age of genetic engineering, antibiotics, sophisticated biochemistry, and global telecommunications, what other microbes might in the future exploit similar conditions? If humanity hoped to prevent its next great plague, it was vital to understand the origins of this one.

Once the HIV antibody tests for screening blood samples had been perfected, the "sticky sera" problems solved, and the Bangui symptomatic definition of AIDS drawn up, it became possible to look backward and ask when and where AIDS had occurred prior to its recognition in California in 1981.

Given the numbers of sexual partners many gay men had prior to becoming infected with HIV, it was considered nearly impossible to trace the epidemic back in time through that population. Researchers could never know who gave the virus to whom, and when.[145]

The clearest tracings could be accomplished by following the AIDS/ hemophilia population, because blood-bank records and stored plasma allowed researchers to match some infections to the HIV-positive donors and to put dates on the times of infection.

Unfortunately, to protect themselves from potential lawsuits brought by people who acquired HIV as a result of transfusions or use of plasma products, many European and American hospitals deliberately destroyed old records and blood samples. Under U.S. law they were required to maintain such records and samples only for five years, and by 1986 hospitals and blood banks all over the country began actively shredding their pre-1982 paper trails and purging computer files. By allowing such wholesale destruction, the U.S. government condoned elimination of a crucial set of clues in the AIDS mystery.

CDC studies of HIV/blood connections in Los Angeles, however, revealed that the earliest date of HIV infection of a person receiving contaminated blood-clotting products was 1978. It was an isolated case, however; the bulk of all blood product infections in the United States occurred in 1983–84.[146]

It would be tempting to conclude that, given the extraordinary numbers of donors' microbes to which people with hemophilia were annually exposed, HIV either didn't exist in North America prior to 1978 or was so rare as to escape chance exposure even for individuals who injected products derived from the pooling of the blood of over 300,000 people a year. Widespread home use of Factors VIII and IX wasn't possible, however, until 1975, so it is conceivable that HIV was present in the U.S. blood donor population for decades prior to 1975, but at such a rare level—say in one out of every million Americans—that the chance passage of blood products wasn't of sufficient likelihood to produce disease that would be noticeable at the population level until 26,000 people with hemophilia started to routinely inject themselves with clotting factors derived from the plasma of tens of thousands of donors per year.

A study by the U.S. National Institute of Drug Abuse found that serum drawn from injecting drug users in 1971–72 tested positive for antibodies to HTLV-III (HIV). Some 1,129 samples obtained from 238 individuals who were surveyed for other reasons at that time were reexamined using the Abbott ELISA test for HTLV-III (the standard test): about 10 percent were positive. The possibly infected samples, which came from all over the United States, were retested using a more precise method—the Western Blot—and fourteen were positive, for an infection rate of 1.2 percent.[147]

Virologist William Haseltine of the Dana Farber Cancer Institute at Harvard ran tests on 1979 blood samples from New York City injecting drug users; 30 percent, he said, were positive for antibodies to HTLV-III (HIV). "It was the druggies, not the gays, who started it," Haseltine declared.[148] The Boston scientist never published his New York City drug users data, which was sharply criticized by researchers who worked closely at the time with the city's heroin- and cocaine-using populations.

Nevertheless, the assumption that the AIDS epidemic of North America began among gay men had to be viewed cautiously; even in Michael Gottlieb's original group of five gay men suffering *Pneumocystis* pneumonia—the study that first alerted the world to the presence of a new disease—one of the men had a history of injecting narcotics. Henry Masur's first report in 1981 of AIDS in New York City described eleven cases, five of whom were injecting drug users; one was both gay and an injecting drug user. And among the original four cases in San Francisco were Mrs. Profit and her husband, both drug injectors. Gay American men in the 1970s were no less likely than other population groups to indulge in such drug use—indeed, some studies found gay men two or three times *more* likely to have injected narcotics, and Harold Jaffe's earliest representations of the epidemic's demographics drew sharp attention to the numbers of men in 1981–83 who had histories of both activities.

Darrow's research showed that the social conditions for emergence and spread of HIV were ideal in the gay communities of the late 1970s in the United States and Europe, particularly because the population was highly mobile and extraordinarily sexually active.

"We found that the earliest cases included gay men involved in international travel," Darrow and his colleagues wrote.[149] "It is impossible to conclude that any of these men is responsible for introducing the virus to the United States. In fact, the virus may have evolved or arrived in some other way. Our purpose is not to pinpoint the source or cast blame, but to show that social conditions in the mid-1970s provided a unique opportunity for the introduction and transmission of an insidious and highly fatal viral disease."

Before the HIV blood test was available, doctors in St. Louis concluded that the bizarre illness and death of a fifteen-year-old under their care in 1968 had been due to AIDS.[150] The teenager was born and raised in St. Louis, had never traveled outside the immediate area, was black, and admitted to "several years" of heterosexual activity. The doctors were unable to cure his medical problems, including galloping *Candida* infections, devastation of his lymphatic system, Kaposi's sarcoma, and fulminant infections of Epstein-Barr virus and cytomegalovirus.

"Although some claim that AIDS is newly imported to the continental United States, the typical features exemplified by our native-born American patient suggest that the syndrome is, at least in part, endemic and appeared more than ten years before the current epidemic," the researchers concluded.

In 1987 scientists presented evidence that the blood of "Robert R.," as the St. Louis case was designated, contained antibodies to HIV, concluding that the virus had been present in the United States in 1968.

"If a virus related to HIV has been present in the United States, Africa, or elsewhere for several decades, its failure to spread in an epidemic fashion earlier may reflect either a recent genetic change in the virus and/or sociocultural factors involving sexual practices or numbers of sexual partners," they wrote.[151]

In 1959 a forty-eight-year-old sailor died of *Pneumocystis* pneumonia and apparent immune deficiency in New York City. The man had traveled widely around the world and was Haitian-born. Though samples of his blood were not available for analysis thirty years later, researchers concluded retrospectively that the sailor died of AIDS.[152]

In Europe several previously unexplained deaths would in the mid-1980s be ascribed to AIDS, among them: Danish surgeon Margrethe Rask, who had long worked in rural Zaire, died in 1977 of acute immune deficiency and *Pneumocystis* pneumonia,[153] and a widely traveled Norwegian sailor, who died in 1966. Over the next decade his wife and one of his three children—born in 1967—also died of immune deficiencies. Later blood tests showed that the three had antibodies to HIV.[154]

Prior to that there were numerous unsolved cases of apparent immunodeficiencies reported in Europe; the most clearly AIDS-associated involved another well-traveled sailor who died in Manchester, England, in September 1959; in 1983 his doctors retrospectively diagnosed the case as AIDS.[155]

All available evidence indicated that the visible AIDS epidemics began simultaneously around 1979 in the United States and Haiti. A review of 1,328 cancer biopsies performed in Port-au-Prince during 1968–77 showed no Kaposi's sarcoma diagnoses. Yet between June 1979 and November 1981 a dozen cases of the rare cancer were diagnosed in the Haitian capital.[156] A French research team tested 211 blood samples collected from adult Haitian immigrants living in Cayenne, French Guiana, in 1983. Using both the standard ELISA HTLV-III (HIV) antibody test and confirmatory Western Blot assays, they discovered that 2.7 percent of the men and 4.9 percent of the women had antibodies to the virus. All of the HIV-positive Haitians had been in Guiana for at least two years, and some since 1974.[157] No Guiana-born individuals tested positive.

Among the original sixty-six AIDS diagnoses of Haitians living either in the eastern United States or Port-au-Prince, only nine definitely fell ill prior to 1981; eight in 1980 and one in 1979.[158]

According to one theory explaining Haiti's relatively high early incidence of AIDS, the country was the unfortunate recipient of the U.S. epidemic, carried there by vacationing gay men who hired local male prostitutes. An opposing argument suggested that the gay epidemic may have originated in Haiti. Again, the putative connection was male prostitution and wealthier North American gay vacationers.[159]

There were then two proffered explanations for HIV's prior presence in Haiti. The first, espoused by Robert Gallo and Harvard's Kennedy School of Government public health professor Yamil Khouri, saw a connection between Zaire and Haiti. Zaire imported nearly 10,000 Haitians a year for short-term contract work between 1960 and 1975. Under the Gallo/Khouri theory, HIV already existed in Zaire at that time, and was carried back to Haiti by returning contract workers.[160]

Always on Peter Piot's mind when he contemplated the origins of the global epidemic was that Greek fisherman he had treated for AIDS in 1978.

When the ELISA test became available, Piot tested the fisherman's blood, confirming that the man who had spent most of his adulthood fishing in Zaire's Lake Tanganyika had, indeed, died of AIDS.

By 1984 Piot and other researchers had determined that 3 to 4 percent of the women who gave birth in Kinshasa hospitals in 1980 carried antibodies to the virus, but none of Nairobi's pregnant women was infected until 1982. By 1984 the infection rate among them was still only 2 percent, Piot said, arguing that "AIDS arrived in Kenya around '82 or '81. In any case, later than in Central Africa."

Between 1981 and 1984 infection rates among Nairobi's poorest prostitutes soared from about 4 percent to over 59 percent, lending further credence to assumptions that Kenya's epidemic was a new, still exploding one. The highest Kenyan infection rates were among recently immigrated Ugandan and Tanzanian prostitutes.[161]

Evidence from countries peripheral to the equatorial center of Africa—Zimbabwe, Zambia, Mozambique, southern Tanzania—paralleled Kenya's: AIDS appeared to have radiated outward from the Lake Victoria region, reaching adjacent areas sometime after 1980.[162]

For example, Dr. Jeff Luande, head of Tanzania's Tumorcentre, located in Dar es Salaam, closely followed Kaposi's sarcoma cases in his country. Working back in time through medical records, Luande said, it was apparent that changes began taking place among the country's cancer patients sometime in 1982. Patients from the country's north, particularly Bukoba, began to appear for cancer treatment that was complicated by numerous infectious diseases, particularly the previously rare *Pneumocystis* pneumonia.

Luande, who had received his medical training at Harvard, had a great deal of experience with cancer treatment, and it was a certainty, he said, that the type of Kaposi's sarcoma he began to see in 1980, first in a trickle and later in torrents, was different from the endemic African form of the skin cancer. The traditional KS, he said, presented with hard, round nodules on the skin of the arms and legs that would enlarge and darken over a period of years. It was undoubtedly a surface skin disease, slow to become malignant, relatively easy to control.

But the new type of KS spread very rapidly and instead of hard nodules, the AIDS KS was splotchy, lighter in color, comparatively soft, and painless to the patients. The lesions were rarely round; rather, they were "spindly," he said. And AIDS KS tumors could be found all over the body—not just on the arms and legs. Luande was particularly intrigued to find so many patients with KS lesions around lymph nodes. This type of Kaposi's sarcoma, he insisted, "is a different, new disease."

Similar shifts in Kaposi's sarcoma were seen elsewhere in Africa. In Kinshasa, the numbers of KS cancers tripled between 1970 and 1984. And the case reports of the new, aggressive type of KS leapt eightfold in 1981 alone. Zambia and Uganda also reported startling jumps in the numbers of aggressive KS cases during 1982. [163]

Based on seroepidemiology—the evidence obtained from blood tests—the highest African infection rates prior to 1984 seemed to have centered on the equator, with latitude ranges of about five degrees northward and ten degrees to the south. Longitudinally the epicenter seemed to range from 15 to 35 degrees. The geographic area, then, encompassed a largely tropical region that included parts of Angola, Zaire, Uganda, Rwanda, Burundi, Tanzania, and Zambia.

The highest infection rates in the region were among female prostitutes, in greatest measure involving women originally from the eastern Lake Victoria region.

Summarizing these findings in a speech before the Second International Conference on AIDS in Paris in 1986, Zaire's Kapita said, "Something dramatic happened in 1975." Prior to that year aggressive Kaposi's sarcoma cases were so rare as to be considered exotic; beginning in 1975, however, the numbers of aggressive KS cases diagnosed in Kinshasa doubled every year. Prior to 1975 cytomegalovirus infections were also rare in Zaire: afterward, they, too, increased dramatically every year, Kapita said.

Kapita could not explain these events. He could only reiterate that "something happened in 1975."

San Francisco's Jay Levy, working in collaboration with Italian and other U.S. scientists, tested a variety of different blood and tissue samples collected in Central Africa between 1964 and 1975, finding *no* evidence of HIV infection. The samples came from Tunisia, Algeria, Uganda, Zaire, Cameroon, and Senegal.

"Our data, as well as epidemiologic studies in Africa, suggest that the AIDS virus was not prevalent and did not spread in that continent until recently," Levy's group concluded. "Thus, HIV appears to have emerged in Africa about the same time as in the United States."[164]

If human factors were the key to the emergence of HIV, there were obvious events in the United States and Europe that could have contributed to sudden viral spread in or around 1975: the gay bathhouse scene, a rapid increase in injecting drug activities, the international expansion of the blood products industry. Less clear were which social factors might have played a role at that time in Central Africa.

The period 1970–75 was marked by guerrilla warfare, civil war, tribal conflicts, mass refugee migrations, and striking dictatorial atrocities in some parts of Central and Southern Africa. Such strife could have affected the historic course of HIV in both direct and indirect ways. Most African military conflict was low-intensity: the weaponry and the strategies were more typical of protracted guerrilla operations than of the Northern Hemisphere's conventional or nuclear warfare. Rather, opposing forces sought to simultaneously cripple one another economically, politically, socially, spiritually, and militarily, often claiming horrendous numbers of civilian lives.

In protracted low-intensity warfare the deeds of war could not be carried out anonymously. The enemy had faces. Soldiers seized villages and imposed their rule on civilians. Brutality and rape easily became companions to more legitimate forms of combat.

The net results were several human activities that were advantageous for sexually transmitted microbes: increased multiple partner sexual behavior (whether voluntary or not), famine or malnutrition that stressed immune systems, large-scale migrations of people from remote areas to central zones of food supplies or safety, increased prostitution, and diminution or devastation of health care services.[165]

During 1970–75 sub-Saharan Africa was the victim of so much strife that it would be difficult for scientists to pinpoint a "worst case" event that could be blamed for the sudden scourge of HIV. It was, for the continent, a time of tremendous instability. The former Portuguese colonies (Angola, Mozambique, Cape Verde, Guinea-Bissau) were given self-rule only in 1975; civil war and revolution raged across Africa's southern belt (South Africa, Namibia, Angola, Mozambique, Zimbabwe); and dangerous despots exerted brutal rule over several nations, and manipulated ethnic conflict, noticeably in the Central African Republic, Uganda, and Zaire. Finally, the entire region was locked in conflict with the only economically powerful nation on the continent, the apartheid state of South Africa.

Uganda's crisis was probably the most acute. Idi Amin's ruthless rule was unchallenged and absolute during the early 1970s and the concomitant massive social and economic disruption is well documented. Over 45,000 Asians were expelled from the country, tens of thousands of black Ugandans sought refuge in neighboring countries, virtually all foreign investors and professionals fled, and Amin, hungry for expanded territorial control in Africa, spent the country into bankruptcy purchasing arms on the world's open market.

As the Amin government printed ever more currency, the official economy became worthless, and the marketing hubs shifted from the old urban centers to remote areas that were conducive to smuggling. Tiny Lake Victoria fishing villages were transformed overnight into busy smuggling ports. As a business, prostitution was second only to the black market.[166] For most women there were only two choices in life: have babies and grow food without assistance from men, livestock, or machinery, or exchange sex for money at black-market rates.

Nowhere was the situation more acute than in the Rakai District, along the Tanzanian border. The area became a vast lattice of mud roads, brothels, and smuggling centers through which flowed a steady stream of truckers carrying cargoes bound for Kenya, Tanzania, Rwanda, Burundi, and Zaire.

The 1979 overthrow of Amin would only continue the crisis, putting it in the hands of President Milton Obote. Famine would strike, particularly in the West Nile region of northern Uganda. An estimated 300,000 had already fled that area as refugees from Amin's brutality. They had taken up unwelcome residence in eastern Zaire, Uganda's Rakai District, and southern Sudan. As famine conditions worsened, the refugee flow would turn into a tidal wave, and the 1983 census would reveal that 57 percent of the region's former residents were either dead or living elsewhere.

The famine would expand in 1982 into southern Uganda—particularly the Rakai and Mbarara districts—and ethnic tensions would become violent. Clashes between local residents and the tens of thousands of refugees in the area (who came not only from northern Uganda but also from Rwanda, fleeing political massacres), would drive even the black-market economy into chaos.

Nearly all of Africa's social and political upheavals had multilateral ramifications, and several served as proxy Cold War confrontations between the United States and the Soviet Union. The region's governments and insurgents were armed to the teeth, and even local tribal warfare became increasingly high-tech and costly in human lives as the decade wore on. The civilian toll, both in direct loss of life and in social disruption, homelessness, famine, and refugee migration, was severe.[167]

Sorting out which, if any, of these upheavals might have played a role in the emergence of Africa's AIDS epidemic seemed a daunting task. In addition to such obvious dramatic events there were the long-term and escalating phenomena of rapid population growth, even more rapid urbanization, and tremendous poverty.

McCormick considered the problem of deciphering what, if anything, happened from the point of view of HIV in 1975, and decided that the easiest first step would be to pull those old vials of blood from the Yambuku and N'zara epidemics out of the CDC's deep freeze and test them for HIV antibodies.

He discovered that 0.8 percent (5 of 659) of the blood samples collected around Yambuku in 1976 were infected with HIV. The infected individuals ranged in age from nine to fifty years; three were female, two male. Similarly, just under 1 percent of the serum samples he had collected in southern Sudan in 1979 had antibodies to HIV.[168] McCormick selected Belgian-born CDC epidemiologist Dr. Kevin De Cock to do the fieldwork, and in early 1985 De Cock made a difficult journey to Yambuku, this time in search not of Ebola but of HIV.

His task was to find the five individuals who had tested positive in 1976 and take fresh blood samples. He also wanted to gather a representative blood sampling of the area for general HIV analysis at the CDC.

De Cock found the local Zairois fed up with all the poking and testing, the unpleasantness of being studied by dozens of foreigners nine years earlier still fresh in the collective memory. There was a haunted, eerie feeling to the place, which still reeled from the terrible, frightening plague of 1976. De Cock was led to graveyards, shown the rows of those family members who were buried one after another as Ebola swept over the population. And every adult spoke on a time scale in which all the world's history was "before Ebola" and events since the fall of 1976 were "after the virus."

He found two of the individuals who had tested HIV-positive in 1976. A middle-aged man and woman were healthy, still tested HIV-positive, and the woman's T-cell count seemed normal. The man had an abnormally low T-cell count.

The other three 1976 HIV-positive individuals had died, all victims of an ailment that could have been AIDS. One of the dead was a woman who lived in Kinshasa from 1972 to 1976 as a "free woman," returning to Yambuku shortly before the Ebola epidemic began. "Free woman," or *femme libre*, was a Zairian euphemism for prostitute.

The overall prevalence of HIV-positive individuals in the Yambuku area in 1985 was the same as in 1976: just under 1 percent.[169] Though the virus was present, there had never been a Yambuku AIDS epidemic.

As McCormick, De Cock, and Zairian colleague Nzila analyzed the data, they reached a set of conclusions: HIV had been present in remote regions of Central Africa for a long time, infecting small numbers of people. The social customs of traditional village life limited the spread of HIV and other sexually transmitted diseases, they argued, as extramarital and premarital trysts were condemned and virtually impossible to conceal in the claustrophobic atmosphere of the tiny communities scattered throughout equatorial Africa.

"The stability of HIV infection in rural Zaire over a long period contrasts sharply with the epidemic spread of the virus in major African cities," they wrote. "Our findings suggest that the traditional village life in the Equateur province carries a low risk of HIV infection. The disruption of traditional life styles and the social and behavioral changes that accompany urbanization may be important factors in the spread of AIDS in Central Africa."

To bolster their conclusion that urbanization and its concomitant erosion of traditional sexual taboos and lifestyles was key to the emergence of HIV in Africa, the CDC and Project SIDA scientists devised a unique experiment: they took blood samples from people living and working along the Congo River, Zaire's equivalent of a superhighway for shipment of goods and travelers. In this way they hoped to track the social pattern of the spread of the virus.

They found a clear pattern of HIV dispersal radiating from river inns, where the male boat workers and traveling salesmen would spend their nights in the company of prostitutes. In the far eastern section of the river, just below the Yambuku area, very few riverside residents, including prostitutes, were infected. But as they progressed southwesterly along the Congo River, drawing ever closer to Kinshasa, the incidence of HIV infection rose steadily among the *femmes libres*, boatmen, salesmen, and local residents. The highest rates of infection were near the river's end, inside Kinshasa.

"In many ways an urban center may be considered an ecosystem that can amplify infectious diseases," they concluded.[170] "This appears to have happened with HIV in various African cities. The conclusion we draw from our study is that AIDS in Central Africa has spread not simply because the virus is present, since in one remote area that prevalence of HIV infection has remained low for over a period of 10 years. A change in the interaction between the agent, the host, and the environment is usually required for an epidemic to develop. In this context, we believe that social change, including the effects of urbanization and population movements, merits consideration in our attempt to understand the changing patterns of disease."

Understanding how human activities were related to the presence and spread of HIV in Central Africa before 1981 still didn't answer questions about when and where the virus first emerged. The 1959 Manchester case argued for HIV's presence somewhere along the sailor's voyages around the planet, going back nearly three decades. But how long had it been in Africa?

Max Essex's group at Harvard, together with scientific teams from Emory University in Atlanta, Duke University in North Carolina, and the University of Washington in Seattle, tested 1,213 plasma samples collected between 1959 and 1982 in Zaire, Congo, South Africa, and Mozambique: one 1959 sample repeatedly tested positive for antibodies to HIV. It came from an individual (gender not known) who resided in colonial Leopoldville in 1959. Leopoldville was renamed Kinshasa when Patrice Lumumba took power; the 1959 sample was designated the "Leopoldville strain."[171]

The debate about where and when the human immunodeficiency virus emerged was radically affected by two other discoveries: that Old World monkeys carried HIV-like viruses, but New World simians did not; and that there was a second species of AIDS virus, dubbed HIV-2, which seemed to exist only in Africa.

The discovery of monkey AIDS viruses dated back to the earliest days of the recognized human epidemic, when scientists with the California Primate Research Center in Davis noted similarities between disease symptoms experienced by gay men and those seen in four strange disease outbreaks among monkeys in their research facilities. The first outbreak occurred in 1969 and lasted six years. During that time, forty-two macaques suffered lymphomas and a host of opportunistic infections related to severe T-cell immune system depression. Two other outbreaks of macaques suffering immune deficiency and disease occurred in the California facility between 1976 and 1978.[172]

The disease, dubbed SAIDS, or simian AIDS, was produced experimentally by injecting the blood of two dying monkeys into four healthy rhesus that had been separately housed. All the injected animals became sick, some developing Kaposi's-like skin patches.[173] The California work indicated two things: the disease was transmissible, it could be experimentally created in susceptible animals, and it had existed—at least among captive macaques—since 1969.[174]

As described earlier in this chapter, in 1985 researchers at Harvard University and the New England Regional Primate Center discovered two simian AIDS viruses infecting their captive animals. The viruses were dubbed SIVmac (previously, STLV-IIImac) and SIVagm (STLV-IIIagm). While SIVmac seemed to be dangerous to macaques, Max Essex's group found that most African green monkeys carried SIVagm without any apparent ill effects.[175]

In March 1986 the Franco-American dispute over discovery of HIV was replayed, as Essex's Harvard group clashed with Montagnier's Pasteur lab over discovery of a second species of human AIDS virus. Dubbed HTLV-IV by Essex and LAV-II by Montagnier, the viruses were found exclusively in West Africa.[176]

For nearly six years the two laboratories would argue over who first discovered the second AIDS virus (eventually named HIV-2), how dangerous the virus was to human beings, and what its relationship was to the monkey viruses.

The Harvard group's virus was found in the blood of healthy Senegalese female prostitutes, and the individuals' immune responses to SIVagm and HTLV-IV were equally strong.[177] Describing his new virus as "the missing link," Essex asserted that it was very close to the monkey virus, and harmless to human beings. Between February 1985 and January 1987, Essex's group analyzed sera obtained from 4,248 West Africans, discovering HIV-2 infection rates that ranged from zero to a high of 19.8 percent among female prostitutes. Few, if any, of the infected individuals were sick, and Essex suggested that HIV-2 might be a "harmless progenitor of HIV-1" that conferred immunity against AIDS upon those carrying the West African virus.[178]

A battle ensued, with Montagnier's group warning that a new lethal virus was rapidly spreading across West Africa, and Essex's laboratory insisting that the microbe was basically harmless.

"We're saying that we're at the dawn of a new epidemic due to a virus that looks like HIV-1, the AIDS virus, but is different, and can induce AIDS," said the Pasteur's François Clavel. "There is an epidemic that is rapidly spreading over West Africa of, if you like, HTLV-IV or HIV-2. And it is accompanied by AIDS."

And Montagnier announced in 1987 that his group had treated thirty individuals who were infected with HIV-2, seventeen of whom had died of AIDS. "This virus is cytopathic for T4 cells," Montagnier said.

It would eventually turn out that both groups were right and wrong. Essex's closest colleagues at Harvard and in Gallo's lab would do a detailed genetic analysis of HTLV-IV and SIVmac, eventually concluding that the viruses were not just close cousins, but were *identical* viruses. Presumably, they argued, contamination occurred in Essex's laboratory, resulting in the mixing of monkey and human samples. [179] Essex and Kanki would eventually publish a concession on the point, acknowledging that their HTLV-IV was essentially identical to a particular macaque strain of SIV found in an animal in the New England Regional Primate Center, samples of which had been in the Harvard lab. But years later Essex would remain personally unconvinced that a contamination error had actually occurred.

"There's no reason whatsoever to consider SIV and HIV-2 different viruses. You don't consider rabies a different virus if it's in bats or dogs or people. You don't consider eastern equine encephalitis a different virus if it's in mosquitoes or birds or horses. But for some reason this one people will forever think of as a totally different virus if it's called SIV in monkeys or HIV-2 in humans," Essex would say.

Montagnier's laboratory would be credited with discovery of HIV-2, but would prove wrong about the lethality of the virus. With time it would become clear that HIV-2 was, as Essex and Kanki claimed, far less virulent and perhaps less infectious than HIV-1. [180] In Senegal, M'Boup would track HIV-2 for nine years, concluding that it was an older, less dangerous virus, found primarily in middle-aged female prostitutes.

As the technology for analyzing genetic material improved during the 1980s, [181] it became possible to compare all the various monkey and human AIDS viruses nucleotide by nucleotide, noting where similarities and differences existed. Using such techniques, scientists would begin to construct family trees for the viruses: lineages of evolution. At the heart of the technique, which was called molecular epidemiology or archeoepidemiology, were a few key assumptions: the more alike two viral genetic sequences were—the higher their degree of homology, as scientists phrased it—the greater was the likelihood that they shared a recent common ancestor, or that one virus was descendant from the other; because genetic divergence required time, degrees of viral variation could be correlated with a timetable of years or centuries; there were certain genetic features that were so essential to the survival of HIVs and SIVs that they would be conserved over generations of viruses; and it was unlikely that evolution progressed from humanly infectious virus to monkey virus, therefore the family tree began with SIV.

Each of these assumptions would be challenged with time, but the basic approach would survive criticism, remaining in use well into the 1990s.

Gallo's lab used such techniques to determine the genetic DNA sequences of SIVagm, HIV-1, and the Pasteur group's strain of HIV-2, discovering that the two human viruses (HIV-1 and HIV-2) shared about 43 percent genetic homology. In other words, they were more different than they were alike. The SIVagm and HIV-1 were also about 43 percent homologous. But SIVagm and HIV-2 shared 72 percent of their genetic sequences. [182]

Vanessa Hirsch's group at Georgetown University in Washington, D.C., found 91.4 percent homology for the envelope genes of SIVagm and SIVmac.

A joint Pasteur Institute/New England Regional Primate Center study of HIV-2, SIVagm, SIVmac, and HIV-1 confirmed the Georgetown findings, showing that SIV and HIV-2 were close, sharing over 75 percent homology. In contrast, HIV-1 had only 40 percent homology with either virus.

As for HTLV-IV, Essex's virus, the Paris/Boston team concluded that it was "a laboratory acquired contaminant": SIVmac.[183] And Beatrice Hahn, then working with Robert Gallo, announced that STLV-III and HTLV-IV were "99% identical and we conclude they are the same virus."[184]

The scientific community recognized that they had a problem on their hands as long as emphasis remained on analysis of monkeys raised or studied in captivity, because the animals were in unnaturally close contact with species of simians they would never see in the wild. Under such conditions disease and contamination were commonplace.

The key lay with the very difficult task of capturing and testing reasonable samplings of wild primates. A Japanese team of scientists did just that, testing enough wild African green monkeys to be able to say definitively that SIVagm was a bona fide wild virus found in about half of all wild African greens on the African continent, but not found in Asian monkeys.[185] The same team sequenced their wild monkey virus, and showed that it was equally similar/dissimilar to both HIV-1 and HIV-2. That meant that neither human virus came recently from SIVagm: rather, they both evolved at some equally distant time from the monkey virus, probably through some intermediaries.

The discovery of other simian AIDS viruses helped clarify the picture. The Japanese team found that wild mandrills carried another virus, SIVmnd. And that virus shared the same percentages of genetic homology with SIVagm, HIV-1, and HIV-2. That put another distant point somewhere on the AIDS family tree.[186]

Eventually, viruses were found in sooty mangabeys (SIVsm1 and SIVsm2), stump-tailed macaques (SIVstm), cynomolgus monkeys (SIVcyn), and chimpanzees (SIVcpz). And careful examination of the genetic sequences of these and various HIV viruses would reveal that some particular strains of HIV-2 and SIVs were so similar that scientists concluded cross-species transmission *had* occurred within the post-World War II period; some were convinced that monkey/human transmission was still occurring, albeit rarely, in the 1990s.[187]

Tragic proof that SIV infection of humans could occur would be found in a July 2, 1992, memo from the office of NIH director Bernardine Healy. Two U.S. lab workers had become infected as a result of bites, needle sticks, and scratches while handling macaques or their tissue. One technician had early symptoms of HIV-2 disease. Genetic analysis of the SIV-2 strain found in one of the workers would show a near-perfect match with a strain found in a sooty mangabey. The scientists who did the genetic analysis would conclude: "Our findings support both the idea that this lentivirus can cause zoonotic infections and the hypothesis that HIV-2 originated from SIV."[188]

Eventually some consensus was reached. The macaque virus (SIVmac) and HIV-2 were so similar that some scientists took to using a new notation system for the two: HIV-2/SIVmac.

In Liberia in 1989, a team of researchers led by Beatrice Hahn of the University of Alabama in Birmingham tested 372 villagers living in the country's remote northern region and 944 employees of the huge Firestone rubber plantation. Three individuals were HIV-1 positive, five carried HIV-2. Detailed genetic analysis of two of the HIV-2 strains found in the Liberian men revealed remarkable homogeneity between the local human virus and two monkey viruses: SIVsm and SIVmac. Capture and testing of wild sooty mangabeys found in Côte d'Ivoire and Liberia revealed that 10 percent of the animals carried SIVsm. And the SIVsm virus had pieces of genetic information otherwise found exclusively in HIV-2.

The researchers concluded that SIVsm, found in wild mangabeys, SIVmac—only seen in *captive* macaques—and HIV-2 were all members of "a single, albeit genetically diverse, group of viruses. Although the evolutionary origins and transmission patterns of this virus group remain to be defined, there is mounting evidence that the sooty mangabey is a natural reservoir and that the human infection probably represents a zoonosis (a disease communicable from animals to man under natural conditions)."[189]

Hahn concluded that SIVsm was probably a sooty mangabey virus that first infected rhesus macaques when the species were co-housed in captivity in a primate research facility or zoo, probably within the previous twenty years. And HIV-2, she averred, was derived from the mangabey virus. She suggested that mangabey-to-man transmission of the SIVsm was an event that had occurred periodically for decades, and still took place in the 1990s, as a result of scratches, bites, or blood exposures people experienced in West Africa while hunting the animals, transporting captured mangabeys, or butchering the monkeys and preparing their meat for human consumption.[190]

If the monkeys had carried SIVsm in days prior to colonialism, human exposure and HIV-2 cases might have occasionally occurred across a vast expanse, from Senegal to Ethiopia. But since the advent of colonialism the rain forest niches of sooty mangabeys had steadily undergone destruction, shrinking the animals' terrain down to a tropical jungle region of Central and West Africa, particularly Congo, Cameroon, Gabon, Liberia, Côte d'Ivoire, Sierra Leone, Guinea, Ghana, Burkina Faso, and Senegal. The postcolonial terrain of sooty mangabeys exactly matched the human HIV-2 region.

In 1993 the Smithsonian Institution's Natural History Museum in Washington, D.C., extracted pieces of DNA from preserved monkey tissue in the museum's archives. They discovered that 57 percent of the wild-caught sooty mangabey samples, dating back to 1896, carried an SIVsm strain that was virtually identical to that found in 1971 and 1981 wild animal samples. That study proved that the monkey virus, which was essentially the same as HIV-2, had been prevalent in Africa for at least a century.

And it begged a critical question: Why wasn't the human disease—HIV-2-caused AIDS—also prevalent prior to 1980?

There was one notable epidemiological exception in the pattern of HIV-2 distribution in human beings: Pygmies. For millennia the Pygmy people had lived in the dense rain forests of Cameroon, Congo, and the Central African Republic, surviving as the continent's most expert jungle hunters. Monkey meat had always been part of the Pygmy diet, and the people—particularly the male hunters—had frequent, often combative, contact with simians.

Yet blood test surveys of Pygmy volunteers revealed no cases of HIV-2 or HIV-1 infection. Both the CDC (David Heymann and Pat Webb) and the Pasteur group (Françoise Brun-Vézinet) screened blood samples extracted from Pygmies during the late 1970s and again in the 1980s, finding no HIV carriers. That seemed to argue that HIV-2/SIVsm zoonotic flux was a relatively recent one, related in some fashion to urban lifestyles.[191] Perhaps, scientists theorized, HIV-2 was a virus that had for decades gone back and forth between humans and monkeys, never evolving particularly well to meet the challenges of infecting either species. In a sense, they argued, HIV-2, SIVsm, SIVmac, and perhaps other simian viruses represented a large fluid genetic pool that shuffled about among a range of primates—including *Homo sapiens*—in West Africa. In contrast, HIV-1 had become such a genetically specialized human killer that scientists were at pains to find ways to infect research monkeys and apes with it, and could not produce clear-cut AIDS in any nonhuman primate.[192]

As evidence mounted supporting Essex and Kanki's original assertions that HIV-2 was less virulent than HIV-1 (though they were wrong to conclude that HIV-2 was harmless), researchers began looking aggressively for evolutionary clues. Natural carriers of the various SIVs were unharmed by the viruses within them, and SIVagm, for example, was dangerous only when it spread from an African green monkey to another simian species.

If HIV-2 were the older, more highly evolved of the two AIDS viruses, then there ought to be many human beings who carried it harmlessly. Essex, Kanki, and MBoup believed that was the case, and their Senegal surveys certainly revealed that well over three-quarters of all HIV-2-positive people in that country were healthy.

In 1989 a German research group discovered a completely healthy woman from Ghana who carried a previously unidentified strain of HIV-2 that bore only 76 percent genetic homology with the classic HIV-2 strains found elsewhere in West Africa, and only 76 percent homology to SIVsm. The group asserted that the Ghanaian HIV-2 strain represented something further back in the evolutionary chain—something close to a common ancestor of other HIV-2s, SIVmac, and SIVsm.

"In our evolutionary tree, HIV-2alt [the Ghanaian strain] is closely related to this common ancestor and branches earlier than SIVsm/SIVmac and the HIV-2 prototypes," said researcher Ursula Dietrich of the Chemotherapeutisches Forschunginstitut in Frankfurt. "It is still unclear whether the host of a common ancestor of the HIV-2/SIVsm/SIVmac group was human or simian. . . . Because captive monkeys were injected with human material [in vaccine studies] back in the 1960s, artificial transmission from a human to a simian host could have occurred. It is possible, therefore, that SIVagm and SIVmac are fundamentally human viruses. In addition, the finding of HIV-2alt, a virus in a human which is evolutionarily older than SIVsm, could indicate that *all subtypes of the HIV-2/SIVsm/SIVmac group are of human origin.*"[193]

In light of such discoveries, several means for human-to-monkey transmission of HIV-2 were suggested, including: tissue culture research in Europe and North America during the 1960s in which monkey and human cells were deliberately mixed, or human cells were injected into captive monkeys; and general export of simians worldwide, and those animals' exposure to human handlers on two or more continents.[194]

Most arguments, accusations, and scientific attention focused, however, on the more lethal HIV-1. And there, the waters were considerably muddier.

It wouldn't be until 1990 that a simian virus bearing significant homology with HIV-1 would be found—in chimpanzees. A Pasteur Institute team, led by Simon Wain-Hobson, discovered SIVcpz in two out of eighty-three wild chimpanzees tested in Gabon. When the Paris group did molecular analysis of the SIVcpz virus they found that it was remarkably similar to several HIV-1 strains, and only distantly related to HIV-2 and all other known SIVs.[195]

In the case of two of the viruses' most important regulatory genes, crucial to the microbes' abilities to get into cells and reproduce (designated *gag* and *nef*), the chimp virus and HIV-1 had about 75 percent homology. Since within the world of all known HIV-1 strains major genetic groups often varied from one another by about 30 percent, the Gabon chimp virus was as similar generally to HIV-1 as the various subtypes of HIV-1 were to one another.

Another SIV chimpanzee strain was found in Cameroon, and it was only 50 percent homologous to the Pasteur group's Gabon strain. That seemed terribly puzzling, until researchers discovered a bizarre HIV-1 strain among Cameroonian people, dubbed ANT70,[196] or Type O. It was highly different from all other HIV-1s, but nearly identical to the new chimp virus.[197]

In 1987, in an attempt to sort out confusion and keep track of the burgeoning genetic information on different AIDS viruses, the U.S. government's Los Alamos National Laboratory decided to dedicate some of its considerable supercomputer space to a special GenBank AIDS project. At its helm, Dr. Gerald Myers kept track of the decoded sequences of every human and monkey AIDS virus in the world. In addition, GenBank became the repository for the genetic sequences of thousands of other species of organisms, as well as the discoveries of the Human Genome Project, an international effort to decipher the entire contents of *Homo sapiens'* twenty-three chromosome pairs.

Using computers to scan sequences for patterns and similarities, the GenBank group was able to construct a family tree, viral bit by viral bit. With PCR and the computerized telecommunications systems that allowed scientists to instantly relay their findings to one another and to GenBank, the AIDS viral files mushroomed in six years to include over 170 sequences, thus reaching proportions that gave Myers and his colleagues information that was statistically highly significant.

When the accumulated HIV-1 data was computer-analyzed, six distinct groups, or "clades," emerged. Within clades the various types of HIV-1 differed by less than 20 percent. The clades were designated A through F, and scientists immediately saw that the various HIV subtypes clustered in distinct geographic areas.

For example, Type A was found in people in Central Africa and India. It was logical that a Central African family of viruses would make its way to India, as tens of thousands of Indians lived in the African region and regularly traveled to the Indian subcontinent. Type B was the only clade of HIV found in North America. Its members could also be found in Peru, Europe, Brazil, southern Thailand, and several parts of Africa.

The most lethal clade—one whose members seemed to kill human beings with terrifying efficiency—was Type D, which was found almost exclusively in Africa's Lake Victoria region, encompassing Rwanda, Uganda, and Tanzania.

Within clades were so-called quasispecies, swarms of HIV types commonly found within individual AIDS patients, varying genetically by less than 10 percent.

And between the six clades, the GenBank group discovered, was a consistent 1992 variation of 30 percent.

When the GenBank group looked at viral strains collected in a given geographic location over a period of years they could see that HIV-1 was evolving—or mutating—at an overall rate of 1 percent per year.[198] Assuming the 1 percent rate had been a consistent feature of the virus since its emergence, that would mean the clades had a common HIV ancestor that existed just thirty years prior, perhaps around 1962. And after ten years of mutating along a single course, the HIV family tree had spread out suddenly, yielding the six distinct clade lineages.

Myers called this "the Big Bang," a deliberate turn of the phrase used by physicists to describe the moment when the density of the universe reached critical mass, causing an inconceivably massive explosion that generated all the nuclear subparticles and mass known to exist. On a humbler scale, Myers suggested that sometime in the early 1970s a biological event occurred that resulted in the sudden and explosive divergence of what had been a virtually linear evolutionary path for HIV-1. He could only speculate as to when and where that event—whatever it might have been—occurred.

Interestingly, the GenBank group's observation, based on viral genetics, that "something explosive" happened in the early 1970s coincided reasonably well with Project SIDA and Kapita's assertion, based on disease epidemiology in Central Africa, that a radical change took place in the region around 1975.

Beatrice Hahn's group, as well as the University of California, San Diego, team headed by Russell Doolittle, were convinced that the human virus was a distant descendant along a lineage that began with SIV in African green monkeys. Because all seven species of African greens carried the virus harmlessly, and some animals located in the wild over 1,000 miles apart had identical SIV strains, Hahn felt certain that the virus originated in the common ancestor of all African green monkeys—a species that theoretically inhabited the continent's rain forests over 10,000 years ago.

One disconcerting finding, however, was that African green monkeys in the Caribbean in the 1990s did *not* carry SIVagm. The animals were all descendant from two vervet monkeys brought from Africa by Spanish sailors to the islands sometime in the sixteenth century. If SIVagm was an ancient virus that had infected half the monkeys in the wild for centuries, it would seem logical that 50 percent of their Caribbean descendants should be infected. Desrosiers argued that the Caribbean situation could be due to a simple progenitor effect: if all the region's African greens were truly descendant from two wild monkeys, pure chance could have resulted in a Spanish sailor's selection of an uninfected pair of African animals.

If SIVagm was the ultimate father of all SIV/HIV viruses, the lineage events between that monkey virus and HIV-1 were sparse and mysterious. The only HIV-1 clade that didn't fit Myers's "Big Bang" theory was Type O, the West African group that included the ANT70 strain, which bore striking similarity to SIVcpz. Hahn and Myers guessed that the Type O clade appeared well before the rest of the HIV-1s, perhaps decades earlier.

Going back even further in time, Doolittle's group compared the HIVs and SIVs to other so-called lentiviruses, showing that, as Doolittle put it, "HIV and visna virus [in sheep] are about as much alike as your average fungus is to a *Homo sapiens.*"

But HIV was more akin to the sheep virus than to HTLV-I and HTLV-II, once considered by Gallo, Essex, and many other American scientists to be the most likely candidates for either the etiology of AIDS or the ancestral origin of the AIDS virus.[199]

The availability by 1990 of advanced techniques for finding and analyzing viral genes hidden in samples of human blood or tissue prompted some scientists to go back and reanalyze the oldest HIV-1 antibody-positive samples to see if the individuals were, in fact, infected with the AIDS virus.

The "Robert R." samples collected in 1968 in St. Louis did *not*, as it turned out, contain HIV viruses. Nor did the 1970 Zaire samples originally tested by the Pasteur Institute. These individuals apparently did not die of AIDS. The 1959 "Leopoldville strain" was lost by American scientists and, therefore, could never be confirmed.

The PCR data seemed to confirm Myers's assertion that HIV-1 underwent some radical event after 1970; before that it was either virtually harmless or nearly noninfectious to human beings.

But the New York City laboratory of Dr. David Ho (at the Aaron Diamond AIDS Research Laboratory) used PCR techniques to study samples taken in 1959 from the Manchester sailor. The sailor was, indeed, infected, and the strain of HIV-1 found in his tissue—extracted painstakingly from paraffin histology blocks made thirty-five years earlier by the Manchester hospital—fit perfectly in the Type B clade. Indeed, Ho concluded that the Manchester sailor was infected with a virus that "looks just like a contemporary European HIV-1 B-type strain."

That seemed to throw a serious monkey wrench in Myers's Big Bang theory. Myers, however, felt the sailor was an aberration; "the preponderance of evidence still argues for an explosive event in the mid-1970s," he said. Furthermore, he insisted that the HIV-1 virus was fairly new, certainly only a few decades old, not centuries.

Ho and a number of other virologists disagreed, arguing that HIV-1 was "an ancient virus" that had existed at a low level in human beings for centuries. If an explosive surge occurred in the 1970s, as all epidemiological evidence indicated, it was due to human events, not to biological changes in the virus.[200]

And what might those human events have been?

First of all, they had to have taken place simultaneously on at least two continents: Africa and North America. Though the Manchester sailor had the AIDS Type B strain in 1959, he could have become infected during travels to the United States or Canada. There is no evidence that he traveled in Central Africa. In the early 1980s all original European AIDS cases directly or indirectly involved visitors from North America and Africa.

If one of the explosive epidemics preceded the other—Africa's or North America's—it could not have been by many years; perhaps one preceded the other only by months.

In light of this duality in occurrence many people sought iatrogenic or conspiratorial explanations for the appearance of HIV-1. One set of theories shared the belief that HIV-1 entered the *Homo sapiens* population via vaccine products. Topping the suspect list was a batch of live polio vaccine that was prepared on African green monkey kidney cells.[201] The vaccines derived from that batch were widely dispensed between 1957 and 1960 in

Zaire, Rwanda, and Burundi. Another polio vaccine distributed by Lederle in 1977 was suspected of containing "C-type particles" that some critics later claimed were AIDS viruses. A scientific panel was assembled in the United States in 1992–93 to review available samples of early polio vaccines, as well as the safety and laboratory techniques used by polio pioneers of the late 1950s. After careful study it was concluded that the polio vaccines were HIV-free.[202]

There were several other reasons to reject the polio vaccine hypothesis. To begin with, HIV-1 and SIVagm were only distantly related at the genetic level, so it seemed unlikely that in less than twenty years' time viruses originally from African green monkey cells could have mutated some 60 percent of their nucleotides to produce HIV-1.[203]

Another theory, that a contaminated polio vaccine batch allegedly used in 1977 by American homosexuals to treat herpes was actually the source of AIDS, could be dismissed both because the vaccine was not AIDS-contaminated and because the timing was wrong. Clearly, 1977 was too late a date. Furthermore, the 1977 vaccine was used widely on populations that did not go on to develop AIDS—Polish schoolchildren, for example.

Two far more elaborate theories of vaccine origin of the AIDS epidemic put the blame on the global campaign to eliminate smallpox. The first version of the smallpox idea was promoted by *The Times* of London in 1987. It claimed that long-latent, ancient HIV infections were activated when people were vaccinated against smallpox. Failing to prove a connection between HIV and the smallpox campaign, five years later the same newspaper would lead an international campaign to discredit HIV's role in AIDS altogether.

The second twist on the smallpox theory, whose chief proponents were an anti-genetic engineering group called the Foundation for Economic Trends and an anti-vivisectionist organization in Los Angeles, claimed smallpox viruses grown on cow cells recombined in test tubes with the bovine leukemia virus, producing HIV.[204] Genetic sequence analysis of the smallpox and BLV viruses indicated that it would be impossible to create, through either deliberate or mistaken recombination of the two, anything that even remotely resembled HIV.

Nevertheless, a retired physician living in London, Dr. John Seale, asserted in 1985 that the AIDS virus was absolutely the result of genetic engineering—the deliberate outcome of biological weaponry experiments conducted at Fort Detrick, Maryland, by the U.S. Army.

"I'm totally convinced it's man-made," Seale said.

In an editorial published by the Royal Society of Medicine,[205] Seale argued that HIV was the result of deliberately mixing bits of the genetic sequences of BLV, visna (from sheep), two other lentiviruses found in horses and goats, and HTLV-I.

"It looks like a recombinant virus to me," Seale said, adding, "We are accusing the retrovirologists as a group of making this virus."

As evidence for his assertions Seale cited the work of Soviet scientist S. Drozdov of the Soviet Academy of Medical Sciences in Moscow. Drozdov and other Soviet scientists were, in turn, influenced by retired East Berlin scientist Jacob Segal, of Humboldt University. Segal wrote a report, read throughout Eastern Europe, that claimed the AIDS virus was made at Fort Detrick in 1977 from a deliberate mixture of visna and HTLV-I. Though it had been the subject of discreet discussion inside the Soviet bloc for over a year, the Segal report was first publicly distributed at the 1986 Summit of the Nonaligned Movement, which convened in September in Harare, Zimbabwe. Over subsequent months the Segal and Seale reports got wide international play, particularly in developing countries.[206]

The seventy-six-year-old Segal claimed to be the victim of CIA harassment. And he said that he possessed, but refused to reveal, documents proving that U.S. prisoners were injected with various experimental combinations of visna and HTLV-I until the perfect lethal form, HIV, was found. All this, he said, took place in 1977.

That such sophisticated forms of cloning hadn't yet been invented in 1977 did not seem to faze Segal. And the Segal/Seale notion that AIDS was the result of a sinister CIA plot found favor in many quarters, particularly African countries that felt unjustly targeted and blamed by American scientists as the origin of AIDS. A popular Soviet cartoon pictured an American scientist who was exchanging a test tube full of swastikas for a wad of cash, proffered by a general. At the characters' feet lay dead bodies.

In a blame-counterblame campaign, the U.S. State Department widely distributed a detailed denial in 1987, charging the KGB had concocted the entire campaign in order to discredit American government credibility in developing countries.[207] Years later, following the fall of the Berlin Wall, the Soviet National Academy of Sciences would formally apologize for the accusation, acknowledging that it had been KGB-inspired.

Another theory of deliberate recombination came from Los Angeles antivivisectionist Dr. Robert Strecker, who gained a large following in 1987 by again claiming, on the basis of a supposed BLV connection, that the CIA made the AIDS virus.

"The AIDS virus was manufactured by crossing BLV and visna virus from animals into man to make the AIDS virus, and growing it in human tissue culture, and that's AIDS. And that's not complicated," Strecker said in a fund-raising speech before wealthy North Hollywood residents. Asked why such a monstrous thing was done, Strecker said the CIA "requested it, to make cancer."

"Why would they want to make cancer?" he was asked.

"You'd have to ask them, I don't know. I don't know! Everybody wants to know why, why, why! I'm just telling you how they did it. I'm not going to tell you why, that's for you to find out," Strecker concluded.

Though Seale, Segal, and Strecker disappeared from the AIDS scene fairly quickly, they were replaced all over the world by others who saw in the apparently sudden appearance of HIV something terribly insidious, deliberate, even conspiratorial. *The New York Native* gay newspaper spent years promoting the notion that AIDS was caused by a CIA attempt to wipe out the agricultural economy of Cuba through release of African swine fever virus—the true cause of AIDS, they said, not HIV. Years later the same newspaper would abandon the African swine fever theory, claiming instead that AIDS and chronic fatigue syndrome were the same disease, both of which were caused by HHV-6, a herpes-type virus. A Vietnam War veteran living in St. Cloud, Minnesota, devoted years of his life to dispersing letters and pamphlets naming dioxin chemicals as the cause of AIDS: again, a conspiracy was afoot, involving a massive cover-up of the worldwide use of Agent Orange and the poisoning of the planet by the petrochemical industry that was destroying humanity's white blood cells. In 1986 the North Korean government charged that AIDS was created in a South Korean laboratory by, of course, the CIA with the goal of wiping out the North Koreans. The fact that virtually no Koreans had AIDS in 1986 was ignored.

Sir Fred Hoyle and Chandra Wickramasinghe, British astronomers, announced in 1986 that the AIDS virus came from outer space.

And sidestepping altogether the issue of the origin of HIV, University of California at Berkeley virologist Peter Duesberg declared that it didn't matter where HIV originated. The virus had nothing to do with AIDS, he said. Duesberg claimed that AIDS was not an infectious disease and had no association with any virus: the disease commonly called AIDS had existed since the beginning of time, but seemed "epidemic" in the 1980s because people were injecting narcotics, snorting nitrites, taking amphetamines, getting parasitic diseases that scientists labeled "AIDS," and leading what he called "a self-destructive gay lifestyle."[208]

"I don't mind to be shot up with it as long as it is a clean virus, without other junk, because I'm fully convinced it's not the cause of AIDS," Duesberg said.

While Duesberg's theories were debunked point by point by scientists all over the world, the public attraction to his ideas was strong, in part because they suggested that such things as consistent condom use might not be necessary. And because blame for having a deadly disease could be leveled straight at the victim—the individual who led a "bad lifestyle" that caused an illness.[209]

Though evidence for HIV as the cause of AIDS, the bona fide existence of a pandemic of infectious immunodeficiency, its evolutionary link to a family of monkey viruses, and its recent large-scale outbreak on earth was overwhelming, collective denial coupled with historically valid feelings of group persecution would continue to support acceptance of dark, conspiratorial theories. The most striking example of this was provided by University of Maryland researchers Stephen Thomas and Sandra Crouse Quinn, who conducted public opinion polls between 1988 and 1990 among working- and middle-class African-American residents of suburban Maryland, Washington, D.C., Atlanta, Charlotte (North Carolina), Detroit, Kansas City (Missouri), and Tuscaloosa (Alabama). Among 999 surveyed members of church congregations, 65 percent either agreed with or were unsure about the statement: "I believe AIDS is a form of genocide against the black race." Nearly 40 percent of African-American Washington, D.C., college students agreed with the statement: "I believe there is some truth in reports that the AIDS virus was produced in a germ-warfare laboratory."[210]

Grasping at straws to explain a spectacularly tragic and explosive disease event that wasn't supposed to happen, that just a decade earlier politicians and physicians proudly proclaimed the stuff of history, humanity was at a loss to look objectively at the rapidly spreading new microbes and reach collective understanding of their origin. Like Marburg, Ebola, Lassa, Machupo, the 1918–19 influenza, and a variety of other viruses described in this book, the HIVs and SIVs that seemed to suddenly appear out of nowhere—dropped from the sky, as Hoyle would have it—actually existed in various forms in nature for decades, or centuries.

"Simian viruses have evolved in simians in parallel with human viruses," Joe McCormick said in 1987, discussing the origin of AIDS. "And the virus in humans has been around for a very long time. For quite a long time, I believe, in Africa. And I believe a whole family of these viruses . . . have co-evolved."

Viruses had generally proven to be remarkably adaptable microbes, capable of altering both their "payload" and "delivery" systems (as Bernard Fields called them) to exploit changes in the animal world around them. If any potential host species underwent significant ecological change, selection pressure would come to bear upon the viruses. Those species that were closest to *Homo sapiens* on the evolutionary tree, chimpanzees and gorillas, seemed to have suffered the most from time's arrow, their populations over the millennia having greatly diminished. Restricted ecologically by their diets, chimps and gorillas had to reside in niches that were inside or near tropical rain forests or forest/savanna junctions. As the forest lands diminished, so did the sizes of their niches. Monkeys, too, had suffered niche encroachment or destruction at the hand of *Homo sapiens*, and many species had lost over half their original territory since the first arrival of European and Arab slave traders.

Two key features of those monkey populations that thrived despite the habitat shrinkage were their ability to adapt to *Homo sapiens'* pressure and their cross-species group behavior. For example, the tough African green monkey species was adept at scavenging human eating and food storage areas and would boldly raid houses.

In the wild, many monkey, and occasionally chimpanzee, species lived and traveled in mixed troops. This worked well when the various species within the mixed troops had different, noncompetitive diets. And the advantage was clear: a larger pack allowed for greater protection from predators and more effective use of a limited ecology.[211]

From the microbial point of view, shrinking primate habitats and mixed-troop behavior opened the possibility for cross-species transmission among three or more monkey/chimp species. In such an environment various SIV strains had ample opportunity to move from immune hosts to vulnerable simian species. And an immune species that thrived alongside *Homo sapiens*, such as vervet African green monkeys, might conceivably serve as an SIV/HIV conduit, carrying viruses back and forth between mixed monkey troops and humans.

It was speculation, of course. No one could be certain how the immunodeficiency viruses zoonotically moved among primates in ancient Africa.

Lentivirus expert Dr. Matthew Gonda, of the National Cancer Institute, argued that biology played no substantial role in the sudden explosion of HIV, which, he said, "has been around for thousands of years." Rather, "the key lies with the demographics of Africa."

In that vein, Dr. Anthony Pinching, of St. Mary's Hospital Medical School in London, maintained that "HIV could have been present, and even causing disease, in a human population in a remote rural region for some time, yet remain undetected. It could then have been transmitted to others following the movements of peoples, especially to the urban areas of Africa. Its subsequent spread would reflect the existing modes of sexual contact in these urban areas. . . . The new seed was thus propagated on the existing soil of human behavior.

"If African countries had had the resources available in the USA during the mid-1970s, we would have seen AIDS emerging [then] as a sexually transmitted disease," Pinching said.[212]

Abraham Karpas, of Cambridge University, felt that human behavior was the key, but put primary blame on widespread use of nonsterile syringes in Africa, which "arrived together with antibiotics. As the early generation of antibiotics came only as injectable medicines, the needle and syringe became inseparable from their therapeutic effect. Even now, injectable medication is the treatment of choice in Africa and in other countries."[213]

In retrospect, social conditions in 1975–80 were clearly ripe for the emergence and spread of even an extremely rare virus. Witness the case of HTLV-I and HTLV-II. Discovered at about the same time the AIDS epidemic was first noted, both were considered extremely rare human microbes, found almost exclusively in remote pockets of the *Homo sapiens* population. In 1980 studies showed fewer than 1 percent of the general populations of Europe, Japan, and North America were infected with HTLV-I. But people with hemophilia were rapidly getting infected in the United States: by 1981, one out of nine Georgians with hemophilia carried HTLV-I, as did one out of six New Yorkers with hemophilia.[214] HTLV-I would prove endemic to pockets of Japan, the Caribbean basin, Melanesia, and Africa, and immigrants from those areas would carry the viruses to new regions. By 1993 the New York City borough of Brooklyn would have an HTLV-I infection rate of 5 percent of the adult population—up from about 0.01 percent a decade earlier.[215]

Like HIV, HTLV-I would be linked to homosexual transmission. In Trinidad, for example, gay men would prove seven times more likely to carry the virus than straight men, and up to 15 percent of gay Trinidadians would test positive for HTLV-I in 1986.[216]

Similarly, HTLV-II would initially be found among Native Americans (in the United States, Panama, Colombia) and be considered an extremely rare event outside those populations. But in 1989 Irvin Chen's UCLA group discovered HTLV-II virus in 21 of 121 injecting drug users in New Orleans, Louisiana.[217] Studies of injecting drug users in Miami and Newark, New Jersey, revealed similar rates of HTLV-II infection, and showed that strains found in the three cities varied genetically by less than 6 percent, implying that their emergence in drug users spanning such distances was an extremely recent event.[218]

The origin and spread of the HTLVs, which was not particularly controversial, could have served as a useful illustration of the principles at work with HIV. The HTLVs were ancient—probably older than HIV—yet they seemed to have spread radically outside isolated human pockets only in the late 1970s or early 1980s. Historical blood sample analysis demonstrated that this rapid spread was not an artifact of discovery. The factors for the viruses' emergence from isolated human groups to larger populations were apparent: injecting drug use and needle sharing, multiple partner sex (both gay and heterosexual), blood products, and transfusions.

It is probably impossible to pinpoint which factor(s) played the greatest role in HIV-1's emergence from an apparently obscure virus that, for example, infected less than 1 percent of the rural Yambuku and N'zara populations in 1976, and perhaps 0.1 percent of isolated populations of Europe or North America during the same period, twenty-four months later exploding into a global pandemic that threatened to kill as many as twenty million adults and over one million children by the year 2000. But Joe McCormick's reminder to "look at human beings" was helpful.

"Human beings have done it to themselves," McCormick said. "And that's not moralistic, it's just a fact."

In Africa, many factors undoubtedly played a role in amplifying the otherwise rare incidence of HIV. The epidemiologic record argues for amplified emergence somewhere around the eastern Lake Victoria/northwestern Zaire region. Crucial data that could help solve that area's AIDS puzzle was never collected, undoubtedly for political and logistic reasons: missing from the equation are representative blood samples from veterans of the Tanzanian/Ugandan war, female victims of rape during that war, and the first wave of female prostitutes that left the war-torn area in search of livings in nearby urban centers.

Nevertheless, it is tempting to conclude, as many of the physicians of the Bukoba and Rakai districts have, that the war played a pivotal role in the emergence of HIV-1 in Central Africa.[219]

Primary truck routes for shipment of goods between the eastern port cities of Dar es Salaam and Mombasa to landlocked Rwanda, Burundi, Uganda, and eastern Zaire all passed through the former war zone. And by 1990 the link between those truck routes and the spread of HIV, via brothels and *femmes libres* along the roadsides, would be thoroughly documented. Finally, the extraordinary incidence of HIV-1 infection in the area, as well as the presence of the particularly lethal D clade of the virus, argued for an especially volatile epidemic.

The postwar dispersal patterns of prostitutes and truckers from the region mirrored the second wave of Central Africa's HIV-1 epidemic.

HIV-1's emergence in North America was almost certainly driven by the overlapping injecting-drug-using/gay male population, but here again the crucial data were missing. It was not possible to work backward in time to the pre-1975 period in New York, San Francisco, Miami, Newark, and Los Angeles to determine which group first had a significant level of HIV-1 infection.

Of note, however, was the fact that gay men of the 1970s were very actively interacting with the U.S. and European medical systems due to their high rates of STDs and comparatively good incomes, which allowed them full access to health care. Several national and local health surveys were underway in the U.S. gay population in the 1970s. And many of America's prominent physicians and nurses were themselves gay. Yet the HIV epidemic wasn't detected in that population until 1981.

In contrast, the injecting-drug-using population was generally outside the medical system, even in countries that had nationalized health care. Drug users interacted primarily with emergency rooms and so-called street clinics. As noted earlier, the medical profession found drug users a difficult, even distasteful, population and few doctors were closely following health trends in that group during the 1970s. Arguably, it would have been easy for isolated early cases of AIDS to go undetected in drug users.

There was a crying need for what Gerald Myers called "fossil viruses," particularly from Western Europe and North America, to help solve the mystery of the 1959 Manchester sailor. Was he, as Myers asserted, an aberration? Or were there European pockets of low-level HIV endemicity in ports of call along his 1950s voyages?

If HIV originated in Africa during the 1970s, scientists must explain why only the Type B clade of the virus had taken hold, after fifteen epidemic years, in Europe and North America. And why HIV-2 had yet to take hold on either continent.

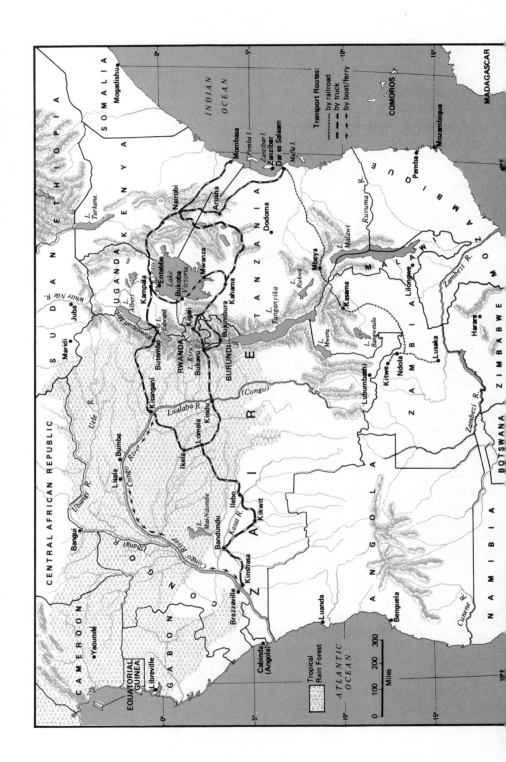

Another piece of missing data concerned the remarkable coincidence of HIV-2 and areas of former Portuguese colonization (Angola, Mozambique, Guinea-Bissau, São Tomé and Principe). The only East African site of HIV-2 was in Mozambique, and West African ex-colonies had among the highest incidences of HIV-2. It would have been helpful if somebody had systematically tested Portuguese and African veterans of the 1965–75 colonial wars to determine whether these soldiers caught, and spread, the virus.

For an obscure blood-borne virus to find its way into large segments of the world's population a crucial amplification step must have taken place. Something new and radical must have occurred that fundamentally altered an ancient homeostatic relationship between humans and the microbe. Ideally, such an amplifier would have provided the microbial population with several key opportunities to spread rapidly outside of its ancient niche.

Between 1970 and 1975 the world offered HIV an awesome list of amplification opportunities: multiple partner sexual activity increased dramatically among gay North American and European men and among African urban heterosexuals; needles were introduced to the African continent on a massive scale for medical purposes, and then resupplied so poorly that their constant reuse on hundreds, even thousands, of people was necessary; heroin use, coupled with amphetamines and cocaine, soared in the industrialized world; waves of other sexually transmitted diseases swept across the same regions, lowering affected individuals' resistance to disease and creating genital and anal portals of entry for the virus; the global blood market exploded into a multibillion-dollar industry; primate research expanded; and governments all over the world turned their backs, convinced, as they were, that the age of plagues and pestilence had passed.

Though it had been the focus of attention of some of the greatest minds in contemporary biomedical science on at least four continents, nobody by 1994 had yet pinpointed a time, place, or key event responsible for the emergence of HIV-1.

But the human factors responsible for amplification of that event, for the rapid expansion of an isolated infection to an outbreak cluster and later epidemic, were very well understood. The World Health Organization was able to delineate those factors repeatedly in pamphlets, and the UN General Assembly would adopt resolutions that cited factors for societal emergence of HIV.

Yet the virus would continually find vulnerable *Homo sapiens* all over the world, for the human factors responsible for spread of the virus would resist change. Governments of countries without AIDS would smugly deny the correlation of such behaviors with the inevitable arrival of the virus. And in nation after nation, when AIDS arrived it would find conditions ideal for rapid spread, and politicians would be unwilling to take unpopular steps to acknowledge the threat, thereby possibly altering the epidemic's course.

Understanding how humanity aids and abets emerging microbes would soon be Jonathan Mann's most important lesson, learned, ironically, in one of the planet's coziest, safest, most sanitized locales.

12

Feminine Hygiene
(As Debated, Mostly, by Men)

TOXIC SHOCK SYNDROME

In general the adaptive relationship between microorganism and host is effective only for the precise circumstances under which adaptation evolved—in circumstances which constitute physiological normalcy for the host concerned. Any departure from this normal state is liable to upset the equilibrium and to bring about a state of disease.
            —René Dubos, 1961

I have an obligation to women to be cautious, to prove Koch's Postulate first, and then—and only then—tell the women of America what they should do.
            —Dr. Michael Osterholm, 1981

■ During the summer of 1982 an angry Dr. Michael Lange took the podium at a GRID conference in New York City and decried the paucity of concern and research funds from the U.S. government. With 177 GRID (AIDS) deaths recorded, he said, the mystery was receiving virtually no attention. In contrast, Toxic Shock Syndrome, which, he claimed, had killed only 85 women in four years, was grabbing headlines, national attention, and federal dollars.

It was not a comparison likely to win over women's health advocates as converts to the GRID cause. Although it was certainly true that the death toll was relatively small, controversy surrounded every aspect of the heated investigation into the cause of Toxic Shock Syndrome, the pathogenesis of the disease, and what steps should be taken to prevent further cases. And for nearly two years American women would be held in the grip of a national anxiety that was fueled by confusion at the highest levels of the public health establishment.

While the cause of AIDS would be determined within two and a half years of recognition of the presence of a new disease, the emergence of Toxic Shock Syndrome, or TSS, would prompt a national debate on its etiology and pathogenesis that would persist in the United States for over a decade. And early steps taken by top public health authorities to limit the spread of TSS would come under sharp fire, attacked as "wrongheaded" and ill-conceived.

So much was certain: *Homo sapiens* females had monthly reproductive cycles, during which time they built up a nutrient-rich uterine endometrium that was prepared to receive a fertilized egg and, once the ovum was implanted, functioned as the blood-rich placenta, feeding the growing fetus. If no egg was implanted, the new endometrium shed out of the uterus, via the vagina, exiting the female's body. The blood-rich expulsion typically lasted for two to six days, during which time the female bled.

It had been so since the beginning of *Homo sapiens* time. And since the beginning of human civilization, women—and men—had sought solutions to the menstrual bleeding problem. Some ancient cultures solved questions of social embarrassment and unsightly bleeding by banning all menstruating women from public view. Hidden away in a designated hut, or in her own home, the female would spend her period away from the males. Such a solution was economically ill-advised for most cultures, however, as women's labor, though rarely valued on a par with males', was still too essential to be easily dispensed with for three or four days a month.

So the females invented clever solutions to the bleeding problem: over forty centuries ago women in Sumer used medicated lint tampon devices, in the early Egyptian dynasties women made diaperlike wraps of papyrus, Roman women inserted woolen balls into their vaginas, medieval Japanese women placed rolled paper tubes in their vaginas. Nineteenth-century American females used rags, cloth diaperlike contraptions, and home-rolled cotton sticks.[1]

None of these ingenious methods created ideal solutions to the bleeding problem, and as women began entering twentieth-century industrial and office workplaces the often embarrassing feminine hygiene issue moved to the forefront. In 1936 a Denver physician named Earle Haas invented a cardboard tube-within-a-tube of compressed cotton that enwrapped a dangling string, and packaged it all so that the outer tube could be inserted into the vagina and the inner tube would act as a plunger, shoving in the compressed cotton plug. The string hung down, allowing for easy removal of the cotton plug.

All in all, it was a clever design that caught on immediately, despite a widespread hue and cry about the immorality of such a device. It was said that tampon insertion stimulated the female excitatory nerves, prompting wanton masturbation. It was also asserted that tampons would puncture the hymen, thus destroying a girl's premarital proof of virginity.

But Haas had no trouble selling his patent to Tampax Incorporated of Palmer, Massachusetts, which promptly proceeded to manufacture the devices. It was a sensation: within a generation most menstruating females in North America used tampons, and 90 percent of all tampon users relied on a single brand—Tampax. A smattering of smaller manufacturers competed for market shares in the United States, Canada, and post-World War II Europe. By the 1960s tampons were in widespread use wherever in the world women could afford to buy them.

Hundreds of millions of women used billions of tampons throughout their reproductive lives. And though history showed that other approaches to the bleeding problem had been associated with elevated risks of some infectious diseases, commercial tampons were sold without any more regulation than hammers or soap. In the United States no federal or state health agencies oversaw tampon production, and the products were never submitted to any required set of safety tests.

During the early 1970s Tampax encountered serious competition as four multinational corporations launched tampon products aimed at grabbing a share of the huge baby boomer market of young women born during the post-World War II American population explosion. The Kimberly-Clark Corporation, Procter & Gamble, Playtex, and Johnson & Johnson entered the market, offering a variety of modifications on Haas's old cardboard-and-cotton design.

The competition turned to a feeding frenzy when the National Association of Broadcasters lifted their long-standing ban on radio and television advertising of tampons, and newsmagazines followed suit, accepting explicit menstrual product ads. By 1975 all five tampon manufacturers were spending millions of dollars on advertising each year. And the key pitches made to women centered on two things: comfort and security. There were few ways companies found to improve upon the relative comfort of the old cardboard tampon design, though some offered plastic tube applicators as an alternative.

Security was Tampax's vulnerable point, for no matter how careful a woman might be, there were those humiliating occasions when the old tampon failed to do its job. Playtex targeted this issue by offering perfumed tampons and the ad slogan: "When you're wearing a tampon you don't worry about odor. But should you?"—implying that small, unseen leaks could still be detected by the sharp olfactory senses of co-workers, friends, and dates.

These new products were allowed on the market without any demonstrated prior proof of safety for either the plastic inserter designs or the perfumes. Following an outcry from Planned Parenthood, Playtex put labels on their perfumed tampon boxes, warning that some women might experience discomfort or irritation from the chemicals.[2]

Meanwhile, competition in the tampon industry escalated radically.

Recent entrants into the field turned to their marketing analysts to determine how better to exploit weaknesses in Tampax's long-standing monopoly, and the unanimous answer was "absorbency." Thanks to feminist challenges to male workplace dominions, American women were filling jobs never, or rarely, before open to their gender. No woman who was among the first of her gender to work as a police detective, firefighter, bank executive, or television news anchor could afford the embarrassment of the bleeding problem.

The first breakthrough in absorbency came in 1974 out of the Procter & Gamble laboratory, where engineers concocted a product based not on cotton and cardboard but on polyester fibers and plastic. Dozens of different types of natural stabilizing fibers had previously been mixed in with cotton to increase absorbency and maintain the tampon's shape inside the vagina. Reportedly among them in the 1950s was asbestos.[3]

Procter & Gamble's use of synthetic fibers, however, changed the entire picture because it allowed engineers an almost unlimited number of ways to vary the shape and relative absorbency of tampons. They could manufacture what amounted to small sponges that ranged from low-density polyester to a very high-density, superabsorbent synthetic.

As was the case with Playtex's introduction of perfumes and plastic inserters into the vaginal ecology, no regulatory agency or medical organization questioned the insertion of petrochemical by-products into the nutrient-rich environment. Again, no safety tests were required. Indeed, with all five competitors quickly putting similar synthetic products on the market, the entire industry declared tampon content to be a matter of trade secrecy.

Among the synthetic materials used in marketed superabsorbent tampons in 1979–81 were polyurethane, polyester, collagen, polyvinyl alcohol, acetyl cellulose, and carboxymethyl cellulose.

In 1979 Procter & Gamble released a tampon comprised of highly compressed beads of, alternately, polyester and carboxymethyl cellulose. The product was, as its name implied, something a woman could Rely upon to prevent embarrassing bleeding accidents, as the synthetic composite was capable of absorbing nearly twenty times its own weight in fluids, and would expand to take the shape of, and fill, the vagina.

With a huge advertising kickoff, Rely hit the North American market and quickly gained enough popularity to radically alter the balance of power among tampon manufacturers. The notion that a tampon could be left in for hours—all night long—without any fear of unsightly failures was extremely attractive to young consumers. Other manufacturers retaliated immediately, marketing Assure!, MaxiSorb, SuperPlus, and other new synthetic superabsorbent products.

The impact on female genital ecology was immediately obvious, as the new tampons were capable of absorbing more fluid than most women actually had in their vaginas at a given time.[4] As the tampons swelled, expanding to touch the vaginal walls, dryness made the usually mucus-coated areas vulnerable. If one of the new tampons was left in the vagina long enough —say, five or six hours—it might adhere to the vaginal walls, and removal would leave behind a residue of synthetic pieces.[5] Some women experienced pain as they removed the new tampons, resulting from the adherent sections of the devices actually tearing cells off the vaginal wall. And still other women required medical assistance to withdraw tampons that had expanded so much that they were too big to come out of the vagina in one piece.

Though nobody officially questioned the introduction of Rely and its competing synthetic products, there were studies that might have served as warnings about the vaginal ecological impact of the new tampons. Tests on rabbits, for example, showed that sterilized tampons made from collagen, polyurethane, polyvinyl alcohol, or acetyl cellulose produced lesions and ulcerations in the vagina's epithelial tissue. In addition, cell regrowth in the epithelium plummeted markedly: by 18–29 percent with collagen and 84–100 percent with other fibers.[6] Another 1979 study, not intended as tampon research, found that carboxymethyl cellulose (CMC) served as an ideal filter for bacterial toxins. In particular, the researchers noted that CMC did a wonderful job of filtering toxins made by *Staphylococcus* bacteria.[7]

From the moment superabsorbent tampons hit the market there were published accounts of vaginal ulcerations, lesions, and lacerations.[8]

In January 1980, Dr. Jeffrey Davis, of the Wisconsin Division of Health, notified the Centers for Disease Control in Atlanta that something potentially dangerous was afoot: he had spotted a sudden surge in Toxic Shock Syndrome cases in the state.[9] On July 15, 1979, a young menstruating woman was admitted to the emergency room of a Madison, Wisconsin, hospital, suffering from shock. Over the subsequent months of 1979, six more TSS victims were admitted to Madison hospitals. All but one of them were menstruating females: the exception was a thirty-six-year-old man. Madison was the state's big college town, heavily populated by postadolescents. All the TSS patients were white, otherwise healthy, and oddly infected with *Staphylococcus aureus*.

Back in 1977, a Denver pediatrician, James Todd, reported having treated seven children, aged eight to seventeen, for an acute life-threatening ailment he called Toxic Shock Syndrome. The children were infected between 1975 and 1977 with bacteria *S. aureus*, which had taken hold in their bodies in an unusual manner, secreting a poison into the youngsters' bloodstreams.[10] The unidentified toxin produced a host of symptoms in the children: fevers of over 102°F, diffuse red rashes all over their bodies, the death and subsequent shedding of skin cells, a marked and dangerous drop in blood pressure, vomiting, diarrhea, muscle aches, kidney dysfunction, liver failure, elevated blood clotting and platelet formation, mental confusion, and loss of consciousness.

One of Todd's patients was a fifteen-year-old girl who had a heavy vaginal discharge and was in a state of shock for two days. Though she eventually survived, the teenager was at death's door for eight days, periodically losing consciousness, and she lost two toes to gangrene. Her vaginal discharge contained a strain of *S. aureus* bacteria that was remarkable for two features: it was genetically resistant to the entire penicillin class of antibiotics, and it appeared to secrete some unique toxin.

Two of Todd's patients were less fortunate than the teenager: one boy died, and another developed "shock lung" that required a laparotomy and resuscitation.

"We suggest that the toxic-shock syndrome is a new staphylococcal-toxin-related disease," Todd's group wrote, adding that "the acute illness which we have described and called the toxic-shock syndrome seems to affect older children."

Todd scoured the medical literature for clues, hoping to find evidence that somebody had previously noted such a severe reaction to *Staphylococcus* infection. He discovered the strange account of a twelve-year-old girl in New York City who developed what looked like scarlet fever in 1927. But it wasn't scarlet fever; it couldn't have been, because the girl was infected with staph bacteria, not the streptococci that caused the bright crimson rashes that were the hallmark of The Fever.[11] Back then, Dr. Franklin Stevens, at the Columbia University College of Medicine and Surgery in Manhattan, treated that ailing girl, whose fever topped 105°F. Her body was covered with "raspberry-like reddened spots," Stevens wrote, and the child complained of pain in her thigh: the result of an unknown injury. When pus was drained from the wound, it was found to be full of staph bacteria.

The New York physician soon saw two more strange *Staphylococcus*-caused scarlet fever cases, in a pair of eight-year-old boys. Much as Todd would do in 1978, Stevens puzzled over the occurrences and marched off to Columbia's medical library in search of clues. He happened upon the 1899 account of rabbit experiments conducted by a German physician, Von Lingelscheim. The German scientist produced scarlet fever in the animals by injecting them with *S. aureus*.

Fourteen years later, in 1941, in Baltimore, Drs. Henry Aranow and W. Barry Wood spent three months struggling to save a fifteen-year-old girl who also suffered from scarlet fever, due, again, to *S. aureus*. The Baltimore girl's symptoms seemed to be a perfect blend of those seen in Stevens's twelve-year-old in New York and Todd's fifteen-year-old Denver case. Like the New York girl in 1927, the Baltimore teenager complained of pain in her thigh, ran a 105°F fever for days, and had "raspberry" formations all over her skin. And as was the case with Todd's ailing teen, the Baltimore girl had discharges from her vagina that were found in the laboratory to be filled with *Staphylococcus*.[12]

During the mid-1970s Japanese pediatrician Tomisaku Kawasaki noticed another odd syndrome in children, involving *Staphylococcus* infection that produced skin shedding and loss of fingers and toes.[13] Todd had no idea whether Kawasaki syndrome, as it was subsequently called, was a manifestation of the same illness he was seeing among Denver children. Certainly, there were differences. The Japanese children were far more likely to suffer heart infections, while Todd's kids seemed to go into shock. In addition, the Japanese children seemed to be much younger than Todd's.

On the other hand, Kawasaki syndrome surfaced in Japan at about the same time Todd first noted TSS cases in Denver.[14]

Meanwhile, Minnesota state epidemiologist Dr. Andrew Dean reported finding five TSS cases during 1979. Both the Wisconsin and Minnesota cases involved teens and adults, about 95 percent of whom were female.

Following the 1980 New Year, the CDC issued an alert to physicians, noting that an apparently new syndrome was surfacing, involving an ancient organism—*S. aureus.*

As calls poured in from around the country during February 1980, the CDC decided to form a task force to investigate the phenomenon, led by the agency's Drs. Bruce Dan, George Schmid, and Kathryn Shands. Dan was in charge of the epidemiological detective work, Shands of laboratory analysis of the staph stains collected from TSS victims. Schmid oversaw group operations.

By May 1980 the federal agency had confirmed forty-three more cases of Toxic Shock Syndrome, and some commonalities were beginning to emerge.[15] The most striking of these were that 95 percent of the cases were female, and 95 percent of the females were menstruating at the time they developed TSS. In most instances, TSS struck on the second or third day of their periods.

The race was on to solve the Toxic Shock mystery, and from the outset the investigation was fraught with scientific backstabbing, rivalries, name-calling, and controversy—most played out in the bright glare of television lights and news photographers' flash bulbs. There would be little interest-free information for public digestion.

One of the first controversies concerned the CDC's definition of Toxic Shock Syndrome, which was drafted by Shands and Todd in February 1980. Though it underwent revisions during the year, the basic case definition remained that of an acute syndrome involving a high fever, scarlet fever-like rash, skin peeling, radically lowered blood pressure, and at least three of the following systemic symptoms: diarrhea and vomiting; muscle aches; vaginal or throat infection; kidney malfunction; liver failure; disorientation or confusion.[16] By focusing on such acute cases, critics charged, the CDC was missing a large pool of people who suffered a milder form of the ailment, and thus underestimating the full extent of the emergence of what might be a new strain of *Staphylococcus.*

"We're using a case definition that is epidemiological, not clinical," Minnesota state epidemiologist Dr. Michael Osterholm said diplomatically. "That means we miss a lot of cases. But it also means that all the cases we name are genuine. The trade-off is that there is no way to answer basic science questions about why these people developed acute shock syndrome, while others who were infected with *Staphylococcus* developed mild or even no symptoms."

Among the first 100 cases reported to the CDC, the agency selected 43 that met the stringent definition of TSS. That meant 57 cases went unexplored—at least some of which might have proven to be milder manifestations of staph infection. As publicity increased, so did the number of ostensible TSS cases that fell outside of the CDC definition.

Throughout the summer of 1980 the number of reported TSS cases rose steadily, reaching 408 reports between January 1975 and October 1980. Of those 408, 14 were in men. The men, of course, had contracted the disease through means other than tampon exposure. They were the anomalies. In five years, 394 cases had occurred in women, 40 of whom had died. Some 95 percent of those women had been menstruating at the time, and 100 percent were tampon users.

The news coverage was terrifying. "Teenager dies of tampon use. Details at eleven!" "Toxic Shock Syndrome survivor tells her story tonight on *Eyewitness News.*" "Centers for Disease Control warning women to beware of tampons. Stay tuned for more!"

Most American women reacted with a sense of helplessness: how could something which had become such an essential part of women's lives turn out to be potentially deadly?

The staphylococcal strain responsible for TSS was genetically resistant to all penicillin-class antibiotics. Many of the acute cases had suffered previous, milder forms of the disease, suggesting that there was some sort of cumulative effect. To the degree that the Minnesota, Wisconsin, and CDC laboratories could be certain, on the basis of currently available technology, the recent TSS sufferers were all infected with the same staph strain. There was absolutely no evidence of person-to-person transmission of the microbes. The outbreaks seemed to cluster in distinct geographic areas of the United States, notably the midwestern states of Wisconsin and Minnesota.

And, according to the CDC, most of the female TSS cases involved superabsorbent tampons. In September 1980 the CDC released its third report,[17] pointing the finger at Rely tampons. In a controlled study of forty-two TSS victims and another pool of non-TSS tampon users, the CDC found that 71 percent of the TSS victims used Rely brand tampons. Other brands came in with markedly lower incidences of TSS: Playtex was used by 19 percent of the cases, Tampax 5 percent, Kotex 2 percent, and OB 2 percent. The CDC pointed out that "consumer use of Rely tampons has increased as the apparent incidence of TSS has increased."[18]

The CDC's investigation also found that a third of the TSS sufferers had had a previous episode of milder menstrually associated symptoms. And the agency suggested that "tampons play a contributing role, perhaps by carrying the organism from the fingers or the introitus into the vagina in the process of insertion, by providing a favorable environment for growth of the organism or elaboration of toxin regardless of the manner in which the organism is introduced, or by traumatizing the vaginal mucosa and thus facilitating local infection with *S. aureus* or absorption of toxin from the vagina."

Though the CDC was convinced that Rely was the bad actor in TSS—and had so informed Procter & Gamble prior to the September 19 release of the agency's findings—the federal scientists were aware that state epidemiologists in Minnesota and Wisconsin had evidence that weakened their case. Osterholm's group surveyed all female TSS cases that had occurred in Minnesota since early 1979, finding that only 35 percent used Rely. Though more TSS sufferers had used Rely compared with matched non-TSS women (35 percent versus 18 percent), the rates were markedly lower than those reported by the CDC.

On September 22, just days after the release of the CDC report, Procter & Gamble voluntarily removed Rely from the marketplace. And they went a step further: together with the Food and Drug Administration, the company designed a massive ad campaign telling women *not* to use their product. The campaign, which began October 6, ran on network television and radio and in over 1,200 newspapers nationwide for four weeks. The FDA, meanwhile, urged women to get rid of their existing supplies of Rely, and recalled inventories of the product from stores nationwide.[19]

Procter & Gamble wouldn't comment on the cost of the campaign, but it clearly was in the tens of millions of dollars. It was unprecedented. As one FDA official privately put it: "We're not used to having such strong company cooperation in a product removal case. And we've never seen a company volunteer to spend millions of dollars to tell people *not* to buy their product." Procter & Gamble, for their part, couldn't be accused of pure altruism. As company representative Marjorie Bradford put it: "Procter & Gamble makes over eighty-eight consumer brands of household and hygiene products. We must maintain a reputation for safe and effective products." Bradford didn't mention the half dozen lawsuits filed by consumers of Rely.

Together with the FDA, the company ran hundreds of market surveys, testing their "do not use our product" ads on women in shopping malls all over America. According to the FDA the ads were 97 percent effective in conveying two messages: don't use Rely, and avoid use of all tampons until the Toxic Shock mystery is solved.

The "don't use Rely" campaign was not applauded by all. In their ads Procter & Gamble quoted the CDC as saying, "Women can almost entirely eliminate their risk of TSS by not using tampons. Women who choose to use tampons can reduce their risk by using them intermittently during each menstrual period." The other manufacturers were outraged, and expected to see their future tampon sales plummet. Tampax took out ads in *The New York Times* and *Washington Star* denouncing Rely, and offering their product as a healthy alternative.

The American College of Gynecologists, representing most of the nation's 23,000 gynecologists, issued a warning to women during the first week of October: avoid using tampons—*all* tampons.

Quietly, some non-CDC scientists involved in investigations of TSS cases were nervous about the agency's position on Rely, and tampons generally. As Osterholm put it: "We had a hepatitis A outbreak in Minnesota a few years ago among people who ate hot dogs. We took tough action and moved quickly on the hot dogs. But it turned out the culprit was the relish. You've got to be very careful about these associative findings."

On October 8 the CDC hastily published the results of a small Utah TSS study that seemed to further implicate Rely.[20] The study compared the tampon use patterns of 29 Utah women who developed TSS during 1979–80 with the behavior of 91 age-matched females who did not have TSS. Sixty percent of the TSS cases involved use of Rely, compared with a Rely use rate among controls of only 23 percent.

Ralph Nader's Health Research Group in Washington, D.C., was not happy. The public advocacy organization, and its lead physician, Dr. Sidney Wolfe, were convinced that the federal government was dragging its feet with the tampon industry, putting the female public at peril. Wolfe attacked part of the message contained in the FDA/Procter & Gamble ads: namely, the statement that "tampons do not cause TSS."

"They most certainly do!" Wolfe asserted. Convinced that the CDC's targeting of Rely was justified, Wolfe charged that the ads produced by Procter & Gamble were unsatisfactory. Though 97 percent of women got the proper messages from the television ads, according to most market surveys, only 89 percent correctly interpreted the company's print advertisements.

Under pressure from the Health Research Group, the FDA held public hearings on TSS in October, and after listening to a range of testimony from women's groups, medical and scientific organizations, and the CDC, issued its first set of tampon regulations.

It was the first time any federal agency had sought to regulate tampon safety.

The FDA ordered tampon manufacturers to put instructions inside their product boxes, describing TSS and explaining the healthy, safe ways to use tampons. Though the agency wasn't certain what "healthy and safe use" might be, there was general agreement that frequent tampon changes—shorter durations of use—were fundamental. The manufacturers were also ordered to list all the ingredients of their tampons and inserters on the product boxes. Also on the outside of the boxes the manufacturers had to list the various available tampon sizes, indicating where on that scale the enclosed devices fit. And the FDA reclassified tampons as Class Three Medical Devices, thus, for the first time, legally requiring premarketing safety tests.

With lawsuits, and the female death toll, mounting, the tampon industry complied without the usual formal corporate protests that typically accompanied escalations in FDA regulation.

Rely tampons had barely been off the market four weeks when Osterholm reported a surge in TSS cases in Minnesota. Reflecting on his earlier study that found that 34 percent of TSS cases had used Rely, Osterholm said the CDC had reached the wrong conclusions in September.

"No other brand showed such a marked difference, so Rely does stand out. But that only accounts for a third of the cases—what about the other two-thirds?" Osterholm said in November. "We say there is risk with *all* tampons."

Describing tampons as an "amplifying factor," Osterholm added that "when we say tampons are guilty, it could be something correlated with wearing tampons."

Todd was also distressed.

"All of our [Denver] cases occurred before Rely got on the market in Colorado," Todd said. "TSS is not caused by tampons. Absolutely not. What we don't know about TSS far outweighs what we do know. You can go to a bar and get twelve theories about where TSS came from. I think TSS is a disease in search of a [bacterial] toxin."

In response, the CDC's Schmid said he and the agency were standing by their Rely study. "The timing for the release of that information was very critical. You act as quickly as you can, based on your findings," he said, noting that Wolfe and other activists were accusing the agency of not moving with adequate haste.

By late November, however, Schmid readily agreed that the focus needed to shift from targeting tampon brands to figuring out what actually caused TSS. And by late December, CDC officials were conceding that "all tampon brands are suspect now," as well as tampon alternatives, such as sea sponges. Two TSS cases had surfaced during the fall among women who, afraid of getting the disease by using tampons, had switched to the then chic alternative, natural sponges.

Operating on the assumption that the tampon acted as "a growth media," FDA officials said that leaving a tampon in the vagina for long periods of time "is not helpful." For the five years prior to the appearance of TSS, the key advertising pitch in the volleys fired by competing manufacturers was aimed at offering ever-longer amounts of "freedom" from concerns about leakage, through the use of higher-absorbency fibers. As a result, women had extended the amount of time they used a single tampon from an estimated range of one to three hours to an average of 6.8 to 7.2 hours, according to the CDC.

Like many local public health officers, Dr. Betty Agee of the Los Angeles County Health Department noted a continuing TSS problem after Rely's withdrawal from the market. She saw no correlation with the brands women used, but did see "a consistent pattern of use of higher-absorbency tampons." Sixty percent of the Los Angeles County cases involved *not* Rely but Playtex superabsorbency tampons, Agee said.

In Connecticut, where Rely was never even sold, there was a sizable incidence of menstrual TSS involving all four of the other tampon brands. And in California, where Rely hit the market in May 1979, the upward surge in TSS cases began in 1977, coincident, local authorities said, *not* with a specific brand but with the introduction by all manufacturers of higher-absorbency products.[21]

There were three possible explanations for the apparent statistical correlation between TSS in women and superabsorbent tampons: like the Rely connection, it could have been a misleading indicator related to some as yet undiscovered underlying mechanism; something inside the new synthetic products could actually promote the growth of *Staphylococcus*;[22] or the superabsorbent products were simply left inside the vagina too long, serving as a sort of petri dish for bacterial growth.[23]

In early October, Dr. Keith White, director of the American College of Obstetrics and Gynecology, linked TSS to the carboxymethyl cellulose fibers used in some superabsorbent tampons. Such fibers, he asserted, were making the vagina abnormally dry and causing tiny lacerations in the vaginal wall through which *Staphylococcus* entered the bloodstream. A parallel theory had it that the carboxymethyl cellulose also bolstered the size of the staph population by serving as a chemical source of sugar for the bacterial colonies.[24]

In the CDC lab, however, Shands was unable to grow staph bacteria on the CMC fiber alone. At the National Institute of Occupational Safety and Health laboratories in Cincinnati scientists showed that the fibers were merely coated with CMC, which acted as a lubricant, probably reducing vaginal irritation. CMC, they noted, was contained in most popular eye drops, and there were no reports of increased conjunctivitis with its use.

In addition, there was a fundamental error in theories that linked fiber-induced lacerations to TSS: the bacteria didn't need to enter the bloodstream to produce TSS. In fact, if the bacteria did get into the bloodstream, a different disease resulted: septicemia. Toxic Shock Syndrome was produced when the bacteria colonized a mucosal area and secreted deadly toxins. The toxins were tiny molecules that could readily make their way into the bloodstream in the absence of cuts, scratches, ulcerations, or other injuries to the vaginal wall. If the fiber-cut/bloodstream theory of the disease were correct, the result would have been a very different disease.

As 1980 came to a close the Wisconsin group surveyed its TSS cases, all but two of which involved menstruating females who took ill after January 1979. They found that *S. aureus* could be cultured from the vaginal discharges of three-quarters of women with TSS,[25] but only from 2.6 percent of randomly selected healthy women who attended a Wisconsin family planning clinic.[26]

"If a new strain of *Staph. aureus* with the potential to produce an unidentified toxin has evolved, it seems likely that conditions in the vaginas of some menstruating women who are using tampons enhance growth of the organism, production of the toxin, or absorption of the toxin," the Wisconsin group wrote.

By January 30, 1981, the CDC was exultant, claiming there had been a marked reduction in TSS cases, from a high of 119 cases reported in August to 37 in December. The agency credited their swift action against Rely for bringing the epidemic under control.

But overall statistics offered a less sanguine picture. Between January 1975 and October 1980 a total of 408 TSS cases, with 40 deaths, were reported. By January 30, 1981, the total had reached 941 cases and 73 deaths. Critics charged that publicity surrounding Rely had contributed to an apparent decline in late November/early December cases, but Rely's withdrawal from the market was just a blip on the social radar screen. The real story, they argued, was that frightened women turned away from tampons—*all* tampons—in record numbers. Tampax reported a 25 percent drop in sales for the eight weeks following the massive Rely/FDA ad campaign, and other tampon manufacturers noted sales declines that ranged from 15 to 25 percent.

"Removal of Rely was clearly wrong," Todd said in January.[27] "If I were Procter & Gamble, I'd sue FDA for $75 million!"

Behind the scenes a bitter feud was raging between scientists in and out of the CDC over what *precisely* was the connection between tampons, *S. aureus*, and Toxic Shock Syndrome. Publicly CDC officials projected confidence in their position on Rely. But privately there was far less certainty.[28]

"The only thing we know is that tampons soaked in blood support bacterial growth," the CDC's Shands said. Privately, the CDC's Dan acknowledged that the strain of *Staphylococcus* responsible for menstrually related TSS appeared to be new. And he admitted that there might well be evidence to support the assertions of a maverick UCLA scientist, Patrick Schlievert, that the new staph strain produced a lethal toxin, unlike any that had previously been discovered in the bacteria.

But publicly the CDC scientists, as well as Jim Todd, denounced immunologist Schlievert, accusing him of weaving bizarre theories in public view, failing to publish his work in refereed scientific journals, and seeking the media spotlight for personal glory. In the Midwest, however, where TSS was taking its greatest toll, Schlievert's theories were considered right on the money, so much so that Minnesota collaborators would soon woo the immunologist away from sunny California, convincing him to take a post at the University of Minnesota Medical School in 1982.

Schlievert got onto TSS research two years before the CDC's January 1980 announcement of the existence of the new ailment. A Pennsylvania physician was handling an odd case of what seemed to be Kawasaki syndrome, or perhaps scarlet fever, and needed help. He sent the patient's blood samples to Schlievert in Los Angeles. The immunologist isolated a previously unidentified streptococcal poison from the patient's blood, dubbing the substance "pyrogenic exotoxin," meaning fever-producing poison secreted by a bacterium.[29]

Convinced that he might have stumbled upon the molecular cause of both scarlet fever and Kawasaki syndrome, Schlievert set to work testing the effect of pyrogenic exotoxin on mice and rabbits.[30]

While the CDC was preparing to move against Procter & Gamble's Rely tampons, Schlievert was busy isolating pyrogenic exotoxin from TSS patients' blood and mucosal samples, sent to him from Minnesota by Osterholm. He quickly confirmed the presence of the toxin, and warned Osterholm that pursuit of a tampon connection, though there probably was one, was diverting attention from the real issue: staphylococcal poison. Using Los Angeles samples sent to his lab by Agee, Schlievert confirmed the presence of pyrogenic exotoxin there as well.

CDC scientists were openly skeptical. While agreeing with Schlievert's assertion that the TSS staph strain was making some new, or usually extremely rare, toxin, Bruce Dan and Kathryn Shands were less than thrilled about discovery of pyrogenic exotoxin.

"More work needs to be done to prove that Schlievert's candidate is the actual toxin," Shands said. There were, she said, several candidates for the TSS toxin.

Schlievert, who was a young, aggressive academic scientist, responded with the kind of certitude that was often misinterpreted as arrogance. "I'm already beyond the stage of trying to figure out what causes the disease," he said. "I don't care what the CDC says."

Having isolated the toxin and shown that it produced disease in animals, Schlievert set out to prove Koch's Postulate. Named after 1905 Nobel laureate Robert Koch, the Postulate was a statement of the experimental evidence required to establish the causal relationship between a given microbe and a particular disease. To prove that an organism or agent actually *caused* a disease, Koch (who discovered the cause of tuberculosis) said a scientist had to identify the presence of the agent in every case of the disease; isolate the organism and grow it (or its toxin) in the laboratory; show that the laboratory-grown sample caused the disease when it was injected into animals; and then re-isolate the organism, or toxin, from the ailing laboratory animals.

Schlievert isolated toxin from TSS patients and injected it into rabbits. Within a matter of hours the rabbits developed classic Toxic Shock Syndrome, complete with high fevers, markedly low blood pressure, and mucosal secretion of *S. aureus*. He was then able to re-isolate the toxin from the rabbits' infected mucosa. Two weeks later Schlievert gave the by then recovered rabbits a subcutaneous second injection of the toxin. Half the animals developed scarlet fever-like rashes.

When Schlievert measured various components of the immune systems of the rabbits, he found that T-cell levels jumped following the first toxin injections. Four days after the injection, with the animals still ailing, their antibody production levels (particularly of IgM) were way down, while their

404] THE COMING PLAGUE

total white blood cell counts were two and a half times above normal. It seemed that suppressor T cells, which were in astonishing abundance, were stifling the rabbits' antibody responses.

By day ten, the toxin had killed off most of the rabbits' T cells, and Schlievert saw a surge in the antibody-producing B-cell population. In particular, IgG antibodies filled the bloodstream, where they sought out their toxin targets. By day twelve the rabbits' blood was full of tightly bound complexes of these antibodies coupled with the toxin molecules—some of which were still attached to the T cells they had invaded.

The immune system then became confused, Schlievert said. It saw the T cells, as well as the toxin, as its enemy and began to autodestruct. The result was autoimmune disease.

On the basis of what he saw in rabbits and mice, Schlievert put forward the following hypothesis of the human disease: a new form of *S. aureus* was in the United States (and, based on case reports, by 1981 in Sweden and Canada); the strain possessed a set of genes that coded for pyrogenic exotoxin A; first-time infection resulted in a mild form of flu-like disease that did not meet the CDC definition of either TSS or Kawasaki syndrome; that first exposure did, however, set in motion a chain of events in the immune system that sensitized the patient; following a second or third round of exposure to the *Staphylococcus* toxin the individual's immune system went into a self-destruct mode, and the unchallenged toxin produced Toxic Shock Syndrome.

Schlievert's model named the menstrual cycle as the ideal vehicle for such a bacterial mechanism. A woman would undergo a sensitizing round of *Staphylococcus* exposure during one menstrual period and would recover, but the bacteria would remain in her vagina. With the following menstrual flow the bacterial population would surge—aided, no doubt, by the provision of an ideal tampon growth surface. And the improperly sensitized immune system would autodestruct.

In Schlievert's hands the model was so clearly demonstrable that he could accurately predict the precise range of symptoms he would produce in a rabbit, based on the quantity of toxin he used and on what dose schedule he injected the animals.

"Tampons are just a passive co-factor in this disease," Schlievert said. "The disease can be produced by your garden variety staph. But five years ago [1975] we got a new staph variety in the U.S."

In addition to being resistant to the penicillin-class antibiotics, the new staph strain was very bad at doing some things classic *Staphylococcus* did, such as kill red blood cells, produce skin boils, and break up fatty acids in the human body. But it seemed that the new strain grew particularly rapidly—100 to 2,500 times faster than normal staph bacteria.

"There was nothing the [tampon] industry could do to put out a safe product once this strain surfaced," Schlievert said. "The disease won't go away without tampons. It just needs a nutrient-rich environment. If you stopped all tampon use in the country today, the organism would adapt. It's just going to pop up somewhere else."

Lending support to his theory of immune system disruption, Schlievert and Osterholm discovered autoimmune diseases in some TSS survivors. Osterholm had a patient in Minnesota who suffered acute Toxic Shock for five weeks; a few weeks after her recovery the patient developed lupus, a classic autoimmune disease. At UCLA another TSS survivor developed such severe lupus that her spleen—the organ that produces B cells—was removed. A survey of twelve other Los Angeles TSS survivors revealed that 75 percent of them were making antibodies against their own cells.

Based on dose studies in rabbits, Schlievert calculated that a single milligram of the pyrogenic exotoxin was enough to kill a 220-pound person. From the blood of one woman who died of TSS, Schlievert extracted over ten milligrams of the toxin.

By the close of 1980 Schlievert was deeply frustrated. He had been sending his data to the CDC, he'd cooperated with public health officials in four states, and he had shared data with Todd's group in Denver, yet the federal scientists continued to discount his findings. Todd accused Schlievert of grandstanding, and told inquiring reporters that "Patrick needs to calm down and publish his findings. Otherwise nobody will take him seriously."

Schlievert was trying to publish his studies, but *Science*, the *New England Journal of Medicine*, and the *Journal of Infectious Diseases* had by November 1980 all either rejected his paper or told Schlievert that they weren't publishing *any* TSS studies.

"Something really needs to be done about this situation," Schlievert said. "I find it very disgusting. I have to defend what I'm doing. It shouldn't be this way. They [CDC] won't tell me what they're doing, but they demand that I tell them every single thing I do."

In January 1981, Todd called for a national scientific forum to settle the dispute. As time went by, the Denver physician was increasingly persuaded by Schlievert's data. He became a convert.

The Institute of Medicine (a division of the prestigious National Academy of Sciences in Washington, D.C.) subsequently convened a special meeting on TSS, and delineated areas of consensus, dispute, and needed additional research.[31] Schlievert and the CDC's Dan finally agreed to a laboratory research protocol that could settle the dispute. Schlievert sent a "cook-book," as he called it, describing the methods he used to isolate the toxin and prove its role in TSS. The CDC, in turn, sent Schlievert a set of coded blood samples, and various *Staphylococcus* strains, not telling him in advance which came from TSS patients. By March it had all checked out, and the CDC team co-published with Schlievert the discovery of TSST-1 —Toxic Shock Syndrome Toxin-1.[32]

Though the CDC team and Schlievert were now "on the same side of the fence," as the UCLA scientist put it, there was bitterness. Schlievert, Todd, and Osterholm all felt that they had suffered for publicly disagreeing with the agency, and resented the CDC's tendency to quash contrary ideas.

Toxic Shock Syndrome gradually fell off the front pages of the nation's newspapers and television news, but the problem did not disappear. The CDC continued to report Toxic Shock Syndrome cases in 1982,[33] 1983,[34] and 1984.[35] The numbers of cases declined, and the epidemiological pattern steadily shifted from the 1980 paradigm that was overwhelmingly an ailment of menstruating women to a more generalized disease that struck a broad spectrum of society, male and female alike.

By April 1984 a total of 2,509 Toxic Shock Syndrome cases had been reported to the CDC; 110 (or 5 percent) were fatal. Of the 2,295 cases in women, 89 percent were menstruating when they fell ill. And 93 percent of the total cases (male and female) that occurred in 1980 involved menstruating women; that dropped to 71 percent in 1983.

Over the years it would become apparent that the highest incidence of TSS was among people of Scandinavian and German extraction, presumably because of a unique genetic susceptibility to staphylococcal infection. That explained the geographic clustering inside the United States in areas such as Wisconsin and Minnesota, to which generations earlier Scandinavians and Germans had immigrated.[36] Outside the United States, the highest incidences of TSS would be seen in Sweden, Denmark, and Germany. TSS cases had by 1984, however, also occurred in every state in the United States, as well as Canada, most Western European countries, Japan, Australia, New Zealand, Israel, and South Africa.

Close CDC examination of the tampon use patterns of 285 women who contracted the disease during 1983–84 revealed that tampon absorbency was strongly correlated with the risk of contracting TSS, though the chemical content of the tampons was not.[37]

And when the VLI corporation of Irvine, California, introduced a contraceptive vaginal sponge to the U.S. market in July 1983 (called Today), the FDA almost immediately received reports of TSS cases associated with the product. With Today, most cases occurred in nonmenstrual days, and the risk of a woman contracting the disease while wearing the sponge was forty times greater than the odds for a nonmenstruating woman who didn't use the product. The sponge was designed to be worn for twenty-four hours, during which it presumably served as a *Staphylococcus* growth site much as did superabsorbent tampons used for shorter durations of time.[38]

The Institute of Medicine report had recommended that women avoid using superabsorbent tampons. And Wolfe's group petitioned the FDA in 1982 to have tampon manufacturers legally required to state product absorbency according to a standardized scale. Wolfe's group wanted the recently developed Syngyna absorbency assay to be used as the gold standard.[39]

The following year the FDA did work out an agreement with industry.

On tampon boxes appeared two TSS messages. The first stated: "ATTEN-
TION: Tampons are associated with Toxic Shock Syndrome (TSS). TSS is
a rare but serious disease that may cause death." In addition, the terms
"Junior absorbency," "Regular absorbency," "Super absorbency," and
"Super Plus absorbency" were standardized, and charts listing their rela-
tionship to blood absorption were put on all boxes.[40]

Toxic Shock Syndrome continued to be a problem in the industrialized
world well into the 1990s. Infection was, by 1990, clearly associated with
surgery, tampon use, and skin injuries.[41]

Clinically, evidence continued to accumulate in support of Schlievert's
original hypothesis of toxin-driven immune system misfunction. For example, during the 1985–86 flu season in Minnesota nine people developed
TSS as a complication of flu; five died. The victims ranged from five to
fifty-six years of age, and none of the cases were associated with tampons
or menstruation.

Physicians were able to culture *S. aureus* from the throats and nasal
passages of the TSS sufferers, and the bacterial strain was a prolific manufacturer of both TSST-1 and another staph poison, enterotoxin B. The
scientists hypothesized that influenza infection caused throat and nasal
irritation. The patients probably became infected with staph by wiping their
contaminated hands over their noses or holding their hands over their
mouths while coughing. Those who developed acute TSS probably had been
exposed to the organism before, which sensitized their immune systems.
Or the influenza had already laid havoc to the immune system, and the
staph toxins simply took advantage of an already dangerous situation.[42]
Physicians in Virginia reported a similar case, involving an athletic, healthy
eighteen-year-old boy who died of TSS following a bout of flu. They warned
that "a newly recognized syndrome of postinfluenza toxic shock syndrome
may be emerging."[43]

Jim Todd, having come around to a complete acceptance of Schlievert's
TSS hypothesis, examined the use of corticosteroids in the treatment of
Toxic Shock. If Schlievert was correct, giving steroids to TSS victims would
be a good idea because the chemicals dampened the immune response.
Todd compared twenty-five TSS patients who received steroids during their
illnesses with twenty who did not. The steroid recipients fared far better,
recovering more quickly from TSS and suffering shorter periods of feverish
hypertension.[44]

Tomisaku Kawasaki, who had been working since 1967 on the syndrome
that bore his name, was not at all convinced that adult Toxic Shock Syndrome was the same thing as the illness he originally observed in Japanese
children. By the mid-1980s Kawasaki syndrome was diagnosed in thousands
of children worldwide every year,[45] primarily among infants. For most
children the ailment was relatively harmless, but in a minority of cases the
syndrome was extremely dangerous, as aneurysms developed in their coronary arteries, causing death due to a heart attack.

Individuals who became infected with *Staphylococcus* through using contaminated needles to inject recreational drugs ran a high risk of endocarditis heart attacks. And while aneurysms of the coronary artery were not seen among such adult patients, brain aneurysms (cerebral or systemic) were observed.[46]

Schlievert and his supporters were nearly certain that Kawasaki syndrome was nothing more than the response of immune-naïve infants to the same, or similar, bacterial toxins that caused TSS in adults. He persuaded Kawasaki to send coded samples of blood from children who suffered Kawasaki syndrome or other ailments, and Schlievert used his assays for TSST-1 to try to determine which samples came from children with the mysterious disease. Schlievert's TSST-1 antibody assays correctly identified half the Kawasaki syndrome cases and did not mistakenly identify any of the controls.

In order to prove Koch's Postulate, however, Schlievert needed to show that injections of TSST-1 could produce Kawasaki syndrome in animals— a feat that, by 1994, he had yet to accomplish.

In collaboration with researchers from the National Jewish Center for Immunology and Respiratory Medicine in Denver and Boston's New England Medical Center, Schlievert demonstrated in 1993 that Kawasaki syndrome in at least some children was related to the TSST-1 toxin. Sixteen children with Kawasaki were compared with fifteen youngsters who were suffering fevers and rashes due to other causes. Bacteria that secreted TSST-1 were found in the blood or mucus swabs of thirteen of sixteen Kawasaki patients, compared with only one of the controls. And the toxin produced proliferation of a specific population of T cells in the kids with Kawasaki (VB2 + T cells); there was no such cellular population expansion seen in the controls.[47]

The 1993 Kawasaki study had provided further evidence for another Schlievert observation: namely, that the strain of *S. aureus* that produced TSST-1 was genetically unique, and had not played a significant role in human disease prior to 1975.

A survey of dozens of samples of TSS-producing bacteria from diverse geographic locales showed that they were all descendants of a single clone of *Staphylococcus*.[48] In addition to producing the killer toxin, the apparently new strain of staph was dependent on its hosts for supplies of the amino acid tryptophan. (Normal *Staphylococcus* make their own tryptophan.) In laboratory cell cultures the new strain actually looked different to the naked eye: normal staph colonies thrived on red blood cells and appeared golden in color, but the new strain seemed unable to digest beef or human blood cells and colonies were white or blanched.

The new strain grew up to 10,000 times faster than normal staphylococcal colonies, churning out massive quantities of the TSST-1 poison, as well as enzymes that rendered it immune to penicillin, ampicillin, and other members of the penicillin-class of antibiotics.

In laboratory tests, TSST-1 and five other toxins extracted from various staph strains proved to be the most potent T-cell stimulators ever found.[49] In the short run, the immune chaos it produced could lead to huge expansions in the CD4 T-cell population (the same population that is destroyed by the AIDS virus). That, in turn, caused secretions of CD4-related chemicals that produced the symptoms of high fevers, shock, and rashes. If lab animals or people were continually reexposed to the toxin, as was the case for many menstruating women who suffered increasingly severe monthly bouts with the bacteria, this CD4 T-cell overstimulation could lead to a wasting syndrome, anorexia, and chronic overproduction of key immune system chemicals.[50]

This dramatic immunological effect resulted in TSST-1's designation as a "superantigen"—an extraordinarily potent immune system stimulator capable of inducing a cascade of activities within the system.

Using sophisticated molecular biology techniques developed in the late 1980s, various scientists were able to show that the unique genetic characteristics for the virulence of the toxic shock staph strain, as well as those responsible for its inability to consume red blood cells and produce septicemia, all clustered together as a continuous segment of bacterial DNA. Furthermore, that segment of DNA was mobile—it could move around along the bacterial chromosome.[51] In most cases, it took up residence alongside either the gene for production of tryptophan or, less frequently, that responsible for making another key amino acid that was occasionally deficient in TSST-1 strains, tyrosine.[52]

The strain's ability to withstand penicillin was due to another transposable DNA segment that caused production of an enzyme, beta-lactamase, which rendered penicillin harmless to the bacteria. This penicillin-resistant genetic segment was first observed in staphylococci shortly after the introduction of penicillin into clinical medicine in North America and Europe, and was known to move about in the microbial world as a plasmid.[53]

The two transposons (TSST-1 and beta-lactamase) appeared to be linked in the new staph strain; TSST-1 was never present in a *Staphylococcus* bacterium without beta-lactamase, and their expression seemed to be simultaneous.

So a tempting conclusion was revealed: perhaps the Toxic Shock Syndrome outbreak followed a unique genetic event in which a plasmid that carried both gene sets was absorbed into an *S. aureus* bacterium sometime in the 1970s under ecologic circumstances that were ideal for that organism's growth and rapid multiplication.

It was tempting to conclude that misuse of penicillin antibiotics was responsible for the event. Because the poison genes and those for antibiotic resistance appeared to be carried together on a plasmid, selection pressure imposed by penicillin use could have caused the mutation event. That was, of course, pure speculation. Proving such an event took place—much less where and when it happened—was impossible.

However the new strain originally emerged, its debut elicited a strong human response. A multimillion-dollar industry was shaken, menstruating women were terrified, scientists feuded, and the credibility of two U.S. federal agencies—the FDA and the CDC—was challenged.

As was the case with HIV and so many other microbes, the new poisonous microbe was victorious. Despite the frenetic (though ineffective) efforts of *Homo sapiens*, the microbe succeeded in carving out a biological niche in the human world and taking permanent hold.

By 1994 Toxic Shock Syndrome was an enduring addition to the list of human pathogens, and though it no longer attracted lawsuits and front-page news, the novel *S. aureus* strain was causing nearly as many infections, ailments, and deaths in the 1990s as it had in 1983. Though Rely had been off the market for over a decade, and tampon boxes were covered with a variety of warnings, menstruating women continued to come down with TSS, particularly those who used superabsorbent products.

One could only take comfort in the fact that the disease, if quickly diagnosed and treated, was curable. Though the TSST-1 bacterium was resistant to penicillin antibiotics, it was vulnerable to other classes of the drugs.

At least, so far.

13

The Revenge of the Germs,
or Just Keep Inventing
New Drugs

DRUG-RESISTANT BACTERIA,
VIRUSES, AND PARASITES

*Consider the difference in size between some of the very tiniest and
the very largest creatures on Earth. A small bacterium weighs as little
as 0.00000000001 gram. A blue whale weighs about 100,000,000
grams. Yet a bacterium can kill a whale. . . . Such is the adaptability
and versatility of microorganisms as compared with humans and other
so-called "higher" organisms, that they will doubtless continue to
colonise and alter the face of the Earth long after we and the rest
of our cohabitants have left the stage forever. Microbes, not ma-
crobes, rule the world.*

—Bernard Dixon, 1994

I

As Toxic Shock Syndrome demonstrated, the bacterial world was in a state
of constant evolution and change. The mutability of bacteria, coupled with
their ability to pass around and share genetic trumps in a microscopic game
of cards, seemed to increasingly leave *Homo sapiens* holding losing hands.

Staphylococcus had plenty of tricks that extended well beyond Toxic
Shock Syndrome. Despite the Age of Antibiotics, staph infections remained
potentially lethal. By 1982 fewer than 10 percent of all clinical staph cases
could be cured with penicillin—a dramatic shift from the almost 100
percent penicillin susceptibility of *Staphylococcus* in 1952. Most strains of
the bacterium accomplished the feat of penicillin resistance in the same
manner as had the TSST-1 strain: by absorbing the beta-lactamase plasmid
into their DNA. Once the plasmid was fully incorporated into the bacterial
genome, and passed from one microbial generation to the next, physicians
witnessed their patients failing to improve with therapy.[1]

Fortunately, alternative drugs existed that did not use the beta-lactam mechanism to neutralize staph, so physicians weren't alarmed. They switched en masse from penicillin to methicillin during the late 1960s, and though a smattering of hospitals in Paris, London, and throughout the United States reported apparent methicillin resistance cases, the overall outcome was positive. Once again, humanity had *Staphylococcus* on the run.

But in the early 1980s, clinically significant strains of *Staphylococcus* emerged that were resistant not only to methicillin but to its antibiotic cousins, such as naficillin. For example, in May 1982 a newborn baby died on the neonatal ward of the University of California at San Francisco's Moffitt Hospital of a strain that was resistant to the penicillins, cephalosporins, and naficillin. The mutant strain had drifted about the hospital and the local community for three years, infected a nurse on the neonatal ward, and then found its way to three babies. The only way the hospital could prevent further cases was to aggressively treat the ward staff and babies with antibiotics to which the bacteria remained susceptible, close the ward off to new patients, retrofit all organic material on which dormant staph might lie (rubber fittings on equipment, curtains, sheets, etc.), and scrub the entire facility with disinfectants.

This was, unfortunately, not an isolated event. Outbreaks of resistant bacteria inside hospitals were commonplace by the early 1980s, particularly on wards that housed the most immune-compromised patients: people who had suffered major burns, prematurely born babies, individuals with end-stage cancer, people who had undergone major surgery, intensive-care patients.

Methicillin-resistant *Staphylococcus aureus* (MRSA) outbreaks increased in size and frequency worldwide throughout the 1980s.[2] By 1990, MRSA would represent a clear economic and health crisis for hospitals all over the globe. The incidence of MRSA infections and deaths would soar steadily, spreading from massive urban medical centers outward, eventually reaching to suburban clinics and rural treatment centers.[3]

In 1992 roughly 15 percent of all *Staphylococcus* strains in the United States were methicillin-resistant; nearly 40 percent of those strains isolated from patients in large American hospitals were MRSA. Significant MRSA problems were soon showing up in far-flung locations, from rural Ethiopia[4] to Perth, Australia.[5] By 1993 only one surefire *Staphylococcus* killer would remain: vancomycin.[6] And even the reliability of vancomycin was in jeopardy, as some physicians reported the existence of MRSA strains that could not readily be cured with the last of the available anti-staph drugs.[7]

Switching from inexpensive penicillins to methicillin increased drug treatment costs for a typical patient approximately tenfold; changing to vancomycin meant turning to one of the most expensive antibiotics on the market. It was a burden in the wealthy countries, but not prohibitive. The increased cost was beyond the reach of poorer nations, however, rendering some staphylococcal infections, practically speaking, untreatable.

Staphylococcus was everywhere: all *Homo sapiens*, as well as some mammalian pets, had staphylococci in their bodies. Most of the time the staph one person passed to another through a handshake, or that a weekend gardener absorbed while turning up soil for a bed of tulips, was rendered harmless by the human immune system. But if the bacteria happened upon a cut, wound, burned skin area, or immune-stressed human, the infection might be extremely advantageous to the organism.

This explained why hospitals and child care centers seemed to be particularly fertile ground for the microbes. Every employee—nurse, doctor, orderly, teacher—could serve as a mobile unit that carried the microbes from one potential human host to another. The vast majority of hospitalized humans had surgical wounds or were suffering ailments that occupied the full attention of their immune systems; similarly, small children in day care centers could be relied upon to have plenty of scrapes, cuts, runny noses, unwashed hands, and dirty faces.

Recognizing the problem, humans living in wealthier nations adopted standardized antibiotic practices, giving the drugs, for example, to all preoperative patients to prevent postsurgical infections. And small children got antibiotics almost as a matter of routine for all manner of infections.

Yet the microbes persevered, resisting the prophylactic and treatment uses of antibiotics. In the United States in 1992 some 23 million Americans underwent surgery, nearly every one of them receiving preoperative antibiotics. Up to 920,000 of them developed postsurgical bacterial infections, the majority of which were due to *Staphylococcus*, particularly MRSA.[8]

Outside day care centers and medical facilities, most dangerous *Staphylococcus* infection was acquired either at random by an ailing individual (one battling cancer, AIDS, heart disease, etc.) or an injecting drug user. In a 1986–89 Danish survey about 7 percent of community-acquired major MRSA infections were the results of sharing contaminated needles: that rate exceeded 10 percent in many inner-city areas of the United States.[9]

Super-strains of staph that were resistant to huge numbers of potential drugs existed naturally by 1990. For example, an Australian research team treated a patient infected with a strain that was resistant to cadmium, penicillin, kanamycin, neomycin, streptomycin, tetracycline, and trimethoprim. Since each of these drugs operated by specific biochemical mechanisms that were used by a host of related drugs, the Australian staph could resist, to varying degrees, some thirty-one different drugs.[10]

In a series of test-tube studies the Australians showed that these various resistance capabilities were carried on different plasmids that could be separately passed from one bacterium to another. The most common mode of passage was conjugation: one bacterium simply stretched out its cytoplasm and passed plasmids to its partner.

Using PCR genetic fingerprinting techniques to trace back in time over 470 MRSA strains, a team of researchers from the New York City Health Department discovered that all of the MRSA bacteria descended from a strain that first emerged in Cairo, Egypt, in 1961. By the end of that decade the strain's descendants could be found in New York, New Jersey, Dublin, Geneva, Copenhagen, London, Kampala, Nairobi, Ontario, Halifax, Winnipeg, and Saskatoon. A decade later they were seen planet-wide.[11]

Fortunately, staph wasn't resistant to vancomycin.

Not yet, anyway.

Staphylococcus wasn't the only bacterial organism that was successfully using plasmids, jumping genes, mobile DNA, mutations, and conjugative sharing of resistance factors to overcome whatever drugs *Homo sapiens* threw at them.[12] In fact, by 1993 nearly every common pathogenic bacterial species had developed some degree of clinically significant drug resistance. And over two dozen of these emergent strains posed life-threatening crises to humanity, having outwitted most commonly available antibiotic treatments.[13]

"The increasing frequency of resistance indicates the need for a stronger partnership between clinical medicine and public health," wrote the CDC's director of bacterial research, Dr. Mitchell Cohen, in 1992.[14] "Unless currently effective antimicrobial agents can be successfully preserved and the transmission of drug-resistant organisms curtailed, the post-antimicrobial era may be rapidly approaching in which infectious disease wards housing untreatable conditions will again be seen."

NIH senior scientist Richard Krause labeled the bacterial situation "an epidemic of microbial resistance." It seemed that new strains of bacteria were emerging everywhere in the world by the late 1980s, and their rates of emergence accelerated every year. In the United States alone, such emergences were adding an estimated $200 million a year to medical bills because of the need to use ever more exotic—and expensive—antibiotics, and longer patient hospitalizations for everything from strep throat to life-threatening bacterial pneumonia. When the costs of extended hospital care were added, the estimated increase due to antibiotic resistant organisms topped $30 billion annually.[15] Though these trends started in huge inner-city hospital complexes, striking elderly and extremely ill patients, they had by the 1990s reached the level of universal, across-the-board threats to *Homo sapiens* of all ages, social classes, and geographic locales.

Jim Henson—famed puppeteer-inventor of the Muppets—died in the spring of 1990 of another common, allegedly curable, bacterial infection. An apparently new mutant strain of *Streptococcus* struck that was resistant to penicillins and possessed genes for a killer toxin very similar to that which Patrick Schlievert had discovered in the Toxic Shock Syndrome strain of *S. aureus*.

Indeed, it was Schlievert who first spotted the new organism in 1989,[16] and dubbed the disease strep A-produced TSLS (Toxic Shock-Like Syn-

drome). By the time Henson succumbed—just a year after its discovery
—lethal human cases of TSLS had been reported from Canada, England,
Scandinavia, Germany, several places in the United States, and New Zealand.[17] In addition, streptococcal strains of all types were showing increasing levels of antibiotic resistance. In the early 1970s these antibiotics,
particularly erythromycin and penicillin, were almost universally effective
against *Streptococcus*, and the appearance of strep-related complications,
such as rheumatic fever and impetigo, were marks of inadequate medical
care, not antibiotic failure.[18]

According to Columbia University antibiotics expert Dr. Harold Neu, a
dose of 10,000 units of penicillin a day for four days was more than enough
to cure strep respiratory infections in 1941. Then, most streptococcal infections in the United States involved bacteria of the strep A type, and the
number one life-threatening complication of strep infection was scarlet
fever.

That strep A strain appeared to be particularly vulnerable to penicillins
and other common antibiotics, and it disappeared entirely from the clinical
scene. American and European medical students of the 1960s had only
picture books to refer to in order to learn what this once-common disease
known as scarlet fever was.

With its ecological competitor out of the way, tough strep B strains
quickly emerged, primarily among newborn babies. By the late 1970s strep
B was the most serious life-threatening disease in neonatal units all over
the industrialized world, and 75 percent of all infections in babies under
two months of age were fatal, despite aggressive antibiotic treatment.[19]

In the late 1980s strep A returned, with the emergence of the hearty
new strain that killed Jim Henson. While strep B continued to dominate
the world's baby wards, strep A struck people of all ages, and did so without
any clear pattern of host vulnerability. But by 1992 the same ailment
required 24 million units of penicillin a day, and might, despite such
radical treatment, still be lethal.[20]

Even more serious was the emergence of virulent, highly antibiotic-resistant strains of *Streptococcus pneumoniae*, or *Pneumococcus*. The bacteria normally inhabited human lungs, and usually did so without causing
undue harm to their *Homo sapiens* hosts. If, however, a person inhaled a
strain of *S. pneumoniae* that differed enough from those to which he or she
had previously been exposed, the individual's immune system might not
be able to keep the organisms in check. And any condition that weakened
a host's immune system could, similarly, allow the pneumococcal population to explode.

Over the years subsequent to the introduction of penicillin, strains
emerged that could resist common antibiotics. For example, parents and
pediatricians noticed during the 1980s that their young children seemed
to suffer increasingly from ear infections, and otitis media-caused hearing
loss became an urgent problem. By 1990 about a third of all ear infections

in young children were due to *Pneumococcus* and nearly half those cases involved strains that were resistant to penicillins.[21]

Initially bacterial resistances were incomplete, meaning that some of the organisms would die off with penicillin treatment, the child's ears would clear up, and both parents and physician would believe the illness had passed. But not all the *Pneumococcus* colony inside the child's ear had, indeed, been killed. With time, the surviving microbes would multiply, and after a few weeks the child's ears would again be in pain. If the parents pulled leftover penicillins out of their medicine cabinets and treated the child again, they would possibly see another apparent recovery in the child. But this time the *S. pneumoniae* colony was more resistant, fewer of the bacteria were killed by the drugs, and otitis media returned quickly with a vengeance.

The old *S. pneumoniae* scourge of rheumatic fever, in which the bacteria colonized human connective tissue, had virtually disappeared from the Western industrialized world by 1970. A dangerous ailment, rheumatic fever usually struck children aged five to fifteen years, causing arthritislike pain in the joints and potentially lethal infections of the heart. In the pre-antibiotic era rheumatic fever survivors often suffered lifelong heart and arthritic problems due to damage wrought by the bacteria.[22]

In 1985 rheumatic fever broke out among white middle-class residents of the Salt Lake City region of Utah. In just three years' time the incidence of the disease skyrocketed eightyfold (between 1982 and 1985), and nearly a quarter of the patients suffered recurrences of the disease despite aggressive antibiotic therapy.[23] The Salt Lake City rheumatic fever outbreak was followed by increasing numbers of cases of the disease occurring all over the United States, and the upward trend would continue into 1994.[24]

At about the same time Salt Lake City physicians were trying to comprehend their sudden surge in rheumatic fever cases, doctors in Oklahoma noted a striking increase in cases of multiply resistant pneumococcal infection. Hardest hit in the Oklahoma outbreak were the state's poor black urban residents—the overall rate of strep pneumonia in blacks was 60 percent higher than that seen in whites. The disease struck with the greatest severity among the state's poorest residents and elderly citizens living in nursing homes. More than 15 percent of those who developed the pneumonia died.[25]

Of course, such ailments as rheumatic fever, strep pneumonia, and general respiratory infections with *Streptococcus* in young children had never disappeared—or even significantly diminished—in the poor countries of the world. Strep infections of the upper respiratory tract and lungs of small children remained, by 1990, major causes of sickness and death in poor countries. The World Health Organization estimated in 1992 that about 2 billion children per year suffered acute respiratory tract infections, 4.3 million of whom died as a direct result. About 800,000 of the deaths each year were due to neonatal bacterial infections, primarily of *S. pneumoniae* or *Haemophilus influenzae.*[26] And overall, 80 percent of the deaths were due to bacterial infection of the children's lungs,[27] the remainder being the result of viral infections (measles, respiratory syncytial virus, influenza, and whooping cough).

In poor countries the prevention and management of pediatric respiratory diseases had to be handled with scarce resources, available antibiotic supplies, and little or no laboratory support to identify the organisms infecting children's lungs. So health professionals defined the disease process not in terms of the organisms involved but according to the parts of the body infected and the severity of those infections. In general, infections of the upper respiratory tract—which were usually viral—were milder, while deep lung involvement signaled potentially lethal bacterial disease.[28]

In 1990 the World Health Organization concluded that the best policy in developing countries was to assume that all pediatric pneumonias were due to bacterial infections, and treat children with penicillins in the absence of laboratory proof of strep or *H. influenzae* infection.[29] Studies done in India, Nepal, and Papua New Guinea showed that presumptive antibiotic treatment of acute respiratory infections reduced the number of child deaths in the test areas by more than a third.[30] Even more striking, there was a 36 percent reduction in child deaths *due to all other causes:* preventing or curing respiratory infections in children stopped not only those lung infections but a host of other secondary pediatric diseases.[31]

That was the good news.

The bad news was that penicillins and other antibiotics offered no more benefit to children with mild, usually viral, respiratory infection than did basic nondrug home care.[32] Antibiotics have no effect on viruses.

"Our results show that there is no justification for use of ampicillin to treat mild ARI [acute respiratory infection] among Indonesian children," wrote a University of Indonesia team. "This practice is both expensive and potentially harmful and is not in the interests of the medical community, the Ministry of Health, or the Indonesian people."[33]

The key danger, of course, was that village paramedics, lacking the training and laboratory support to correctly distinguish viral versus bacterial, and mild versus acute disease, would overuse antibiotics. And that, in turn, would promote the emergence of, among other things, antibiotic resistant *S. pneumoniae.*

Soon, because of drug use policies in both the wealthy and the poor countries, antibiotic-resistant pneumococcal strains turned up all over the world, some able to withstand exposure to six different classes of antibiotics simultaneously.[34] By the 1990s *S. pneumoniae* strains had outwitted all aminoglycoside-type antibiotics, chloramphenicol, erythromycin, and all penicillin-type drugs, leaving physicians with few options, and epidemiologists worrying about when vancomycin resistance would also turn up in that bacterial species.

Genetic analysis of the various new mutant *S. pneumoniae* strains offered some clues as to the origins of these emergences. One multiply resistant strain (dubbed 23F) first appeared in Spain in 1978 in a hospital setting, bearing all its resistance capabilities save invulnerability to erythromycin. That trait was acquired when the organism, carried by an infected human, made its way to Ohio. Subsequent improvements in the bacterium's ability to withstand hostile drug-laden human ecologies came as the organism's descendants made their way to South Africa, Hungary, the U.K., back to Spain, and then again to the American Midwest. By 1992 it was possible to trace every known type of 23F *S. pneumoniae* back to a single mutant clone that arose in the 1970s in Spain.[35]

The nightmare example was *S. pneumoniae* type 19A, which emerged in Durban, South Africa, in May 1977. Five small children came down with the new strain while hospitalized for other reasons at King Edward III Hospital; three died. When the 19A strain was tested in the laboratory it was discovered that it was resistant to a huge list of drugs: penicillin, ampicillin, cephalothin, carbenicillin, streptomycin, methicillin, cloxacillin, erythromycin, clindamycin, gentamicin, fusidic acid, chloramphenicol, and tetracycline.[36] Recognizing the futility of standard antibiotic therapy, the Durban physicians switched to rifampin plus fusidic acid for treatment. Though the organism was somewhat resistant to fusidic acid, it was vulnerable to rifampin.

But the new mutant strain could not be contained. A month after the first baby fell ill in Durban, a three-year-old boy was hospitalized in Johannesburg for heart disease. There, he developed pneumonia due to strep 19A infection, and only recovered after over six weeks of treatments with a variety of antibiotics. Soon it was apparent that the super-strep bug had infected dozens of pediatric patients and hospital personnel, and the entire measles ward was overrun by the mutant microbe. Three of the measles patients died of 19A pneumonia.

Vigorous control measures were taken, including treating all infected hospital personnel with high doses of rifampin and scrubbing down the Johannesburg and Durban pediatric wards. Nevertheless, 19A was never eliminated, and the mutant bacterium resurfaced periodically over the years. In a 1978 survey of Johannesburg's six leading hospitals, over half of all pneumonia patients were found to carry the 19A strain. Fifteen percent of all pneumonia cases in Durban that year also involved strep 19A.[37]

Bacteriologist Alexander Tomasz at Rockefeller University in New York later did genetic analysis of the 19A strain, making what he termed "an astonishing discovery." The Durban strain matched one that surfaced ten years earlier in a little boy living in a remote rural village in Papua New Guinea. By means Tomasz was never able to determine, the bizarre bacterium made its way to South Africa a decade later, and from there to Spain, Hungary, England, the United States, and eventually all over the world.

"But the point is," Tomasz said, "all these bacteria can be traced to a single clone. And it all started with one transformed bacterium."[38]

In response to antibiotic pressure, the microbes altered far more than their ability to withstand the drugs. Tomasz discovered that the strep pneumococci weren't very efficient at absorbing plasmids, as were most other bacteria. But they compensated for that failing by being voracious DNA scavengers. Tomasz actually caught them in the act with his camera and microscope, gobbling up long strings of random DNA. As a result, they changed the biochemical composition of their cell walls so radically, he said, "that we must actually say that these are new species."

Inside their DNA, Tomasz found massive numbers of genes that were just plain wrong—they weren't pneumococci genes at all.

Such emergences of drug resistance usually took place in communities of social and economic deprivation.[39] Poor people all over the world were more likely to self-medicate, purchasing antibiotics on the black market, over the counter in many countries, or borrowing leftovers from relatives. Without consulting often costly physicians, and certainly in the absence of expensive tests that could determine the drug sensitivities of the bacterial strains with which they were infected, the world's poor were compelled to guess what drug might cure the disease that was ravaging their children or themselves.

This state of affairs guaranteed that a sizable percentage of the human population were walking petri dishes, providing ideal conditions for accelerated bacterial mutation, natural selection, and evolution.

Whether one looked in Spain,[40] South Africa, the United States, Romania, Pakistan, Brazil, or anywhere else, the basic principle held true: overuse or misuse of antibiotics, particularly in small children and hospitalized patients, prompted emergence of resistant mutant organisms.[41]

The basic problem with the antibiotic approach to control of pathogenic bacteria was evolution. Long before *Homo sapiens* discovered the chemicals, yeasts, fungi, and rival bacteria had been making antibiotics and spewing the compounds around newly claimed turf to ensure that rival species couldn't invade their niches.

The rivals, of course, had long since evolved ways to rapidly mutate to withstand such chemical attacks. So rivals would make different chemicals, their foes would mutate again, and the cycle repeated itself countless times over the millennia. Humans simply accelerated the natural process by exposing billions of microbes at a time to drugs derived from the natural chemicals, and doing so with less lethal efficiency than had the microbial competitors in their ancient microscopic turf fight.

Often the genetic changes the microbes underwent in order to overcome the antibiotics offered unexpected additional advantages, enhancing the bacteria's ability to withstand wider temperature variations, outwit more elements of the host immune system, or kill host cells with greater certainty.

So the patterns seen with *Staphylococcus* and *Streptococcus* were mimicked with other dangerous microbes.[42] Leprosy, which was caused by *Mycobacterium leprae*, was easily treated prior to 1977 with the antibiotic dapsone. But that year a dapsone-resistant strain of the bacterium surfaced in Ethiopia.[43] Though dapsone remained the drug of choice for treatment of leprosy, resistance increasingly rendered use of the antibiotic problematic. Within ten years the situation had become severe, with high percentages of the *M. leprae* strains from all over the world appearing invulnerable to the drug: 37 percent in Chingleput, India; 39 percent in Dakar, Senegal, and Paris, France; over 30 percent of strains in Guadeloupe, Martinique, and New Caledonia; a quarter of those in Fujian, China; and over half of all *M. leprae* in Shanghai and Jiangsu, China.[44] Subsequently, resistance emerged all over the world to the alternative drug, rifampin, and in Ethiopia a patient was found to have essentially untreatable leprosy, suffering from a strain that was invulnerable.[45]

Gonorrhea was also increasingly difficult to treat, having acquired widespread penicillin resistance during the 1970s and spectinomycin insensitivity by the mid-1980s.[46] The next drugs in line, then, were cefoxitin and tetracycline, and treatment was sufficiently complicated to require special guidelines from the CDC and WHO.[47] In addition to the penicillin-resistance plasmid *N. gonorrhoeae* strains had acquired during the late 1970s, gonorrhea also took on a plasmid around 1985 that gave it the ability to withstand tetracycline.[48]

So the New York Academy of Medicine in 1989 recommended that physicians inject the antibiotic ceftriaxone into their gonorrhea patients and give them oral doxycycline.[49] In addition to being considerably more expensive and available only in injectable form, ceftriaxone was a sulfur drug to which many people (up to 20 percent in the United States) were allergic.

By 1990 physicians all over the world were using ciprofloxacin, ceftriaxone, or another member of the quinolone group of antibiotics to treat gonorrhea, finding the drugs highly effective. But in 1992, Australian physicians reported that the drugs were becoming less effective in treating patients who had recently traveled in Southeast Asia. By mutating changes in its cell wall, making itself less permeable to all the quinolone drugs, *N. gonorrhoeae* was, once again, outmaneuvering another line of human defense. A few resistant cases turned up in England as well—again, among recent travelers to Southeast Asia.[50] Presumably the widespread black-market availability of antibiotics in much of Southeast Asia contributed to selective emergence of quinolone-resistant gonorrhea.

The most dangerous emergences of resistance to antibiotics for people living in poor countries were those in bacteria that caused intestinal disease and diarrhea. In 1991, 80 percent of the people living in the world's poorest countries had no sanitary facilities for the disposal of human wastes. Even in the moderately developed countries—nations with middle-class populations and some industrial capacity—about half the people lacked sanitary toilet/sewage facilities.[51] Under such circumstances it was easy for a water- or food-borne microbe to enter the water supply, be ingested by a human, grow and thrive in the human's gastrointestinal tract, and then be expelled via human feces back into the community water supply.

Not surprisingly, diarrheal diseases were a major cause of death among young children in poor countries. In 1991 the World Health Organization estimated that 3.2 million children annually died before reaching their fifth birthday, victims of diarrheal diseases.

Whether new antibiotic-resistant intestinal pathogens emerged first in the industrialized world or in poor countries made little difference on the net outcome: the microbes' greatest toll was taken among the world's poor- est, weakest children. And as resistant strains pushed up the costs of treatment, forcing the use of more expensive antibiotics, doctors in poor countries had little choice but to ration the drugs, triaging access.

During the early 1960s, *Shigella dysenteriae* became the first diarrheal bacterium to emerge with resistance to penicillins. In the absence of anti- biotic treatment *S. dysenteriae* killed up to 20 percent of the children in whom it caused disease, and fatality rates as high as 15 percent had been seen in adults. Even the less severe types of *Shigella* (*S. flexneri, S. sonnei,* and *S. boydii*) could be lethal diseases in up to 10 percent of those people who fell ill. And natural immunity to the organisms was weak—nearly half the *Shigella* survivors suffered recurring disease.

In September 1983 a middle-aged Hopi woman living on her tribe's national lands in Arizona was hospitalized with *Shigella* dysentery. Doctors soon realized that she suffered from an altogether new mutant strain of the microbe that was resistant to ampicillin, carbenicillin, streptomycin, tri- methoprim, sulfamethoxazole, sulfisoxazole, and tetracycline. It turned out the woman had a long history of urinary tract infections, for which she had taken trimethoprim and sulfamethoxazole off and on for at least three years.

Her intestines had become a breeding colony for resistant bacteria. Subjected repeatedly to antibiotic assaults, the microbes shared resistance plasmids. Colonies of *Escherichia coli* were apparently already in possession of a plasmid bearing genes that conferred resistance to trimethoprim and sulfamethoxazole, and they shared that plasmid with *Shigella* in the Hopi woman's gastrointestinal tract.[52] Though health authorities did what they could to limit the spread of the super-bug, by 1987 up to 21 percent of all *Shigella* infections among the Hopi and nearby Navajo were caused by the mutant strain. Nationwide, 7 percent of all *Shigella* infections in 1986 involved the super-bug, and a third of all cases were also ampicillin- resistant.

In Ontario, Canada, even higher levels of *Shigella* resistance were ap- parent by 1990: eight out of every ten human illnesses with the organism involved resistant strains. And *half* of all *Shigella* infections were caused by bacteria that were resistant to *four or more antibiotics.*[53]

Again, the most devastating impact of such multiply resistant *Shigella* was felt in the world's poorest nations. When a new multiply resistant strain reached the African country of Burundi, for example, the nation's Ministry of Health was unable to come up with enough foreign exchange to purchase alternative drugs from wealthy-nation pharmaceutical companies. So untold numbers of people died of dysentery.[54]

Similarly, between 1960 and 1993 several other enteric bacteria—species that infected the human gastrointestinal tract—acquired profound genetic abilities to resist *Homo sapiens* weaponry. These included *E. coli*, *Klebsiella*, *Proteus*, *Salmonella*, *Serratia marcescens*, *Pseudomonas*, *Enterococcus faecium*, *Enterobacteriaceae*, and cholera. The situation by 1990 was quite grave, particularly in poor countries that lacked sufficient resources or capital to eliminate the unsanitary conditions responsible for the transmission of the microbes from humans to the water supply and from food to humans.[55]

Salmonella, the leading cause of food poisoning, was appearing in the Caesar salads served up in restaurants on Manhattan's posh Upper East Side, or in taco stands along the border *caminos* of Juárez, Mexico.[56] By 1993 it was an essentially untreatable diarrheal disease, as no known antibiotic seemed capable of reducing the three to four days of agony a typical *Salmonella* infection produced in *Homo sapiens*.[57] Fortunately, the microbe rarely caused anything more dangerous in its human hosts than headaches, acute stomach pain, diarrhea, nausea, and dehydration.

One of the most disturbing prospects for physicians worldwide was the emergence around 1988 of vancomycin-resistant *Enterococcus faecium* and *faecalis*. With vancomycin the only remaining reliable treatment for staph and strep infections, there was great concern that resistant enterococcal bacteria could share their resistance genes with the other, otherwise untreatable microbes.

"It hasn't happened yet, but everybody thinks it will," CDC bacteriologist Bill Jarvis said.

Such a bacterial strain, if it did emerge, would be virtually incurable and extremely dangerous, for it would possess not only special drug-resistant genes but also those for heightened virulence.

Physicians and scientists working outside the field of bacteriology in the 1990s generally assumed that, as had been the case before, another class of antibiotics would be developed and the problem would go away.

But they were wrong.

"There's nothing on the shelf. Nothing in the pipeline. If we lose vancomycin we're going to be back to the 1930s with staph," Jarvis said. The same could be true for *Streptococcus*.

"That would be the real nightmare," predicted the CDC's Bill Jarvis.

The nightmare began unfolding in 1988 with first reports of vancomycin-resistant *E. faecium* strains surfacing all over the world, usually appearing first inside hospitals.[58]

For example, a handful of hospitalized patients in New York City hospitals fell ill with vancomycin-resistant strains during 1988: their cases were isolated, and there was no evidence of bacterial spread to other patients or into the community. Between September 1989 and March 1991, however, vancomycin-resistant strains of enterococci emerged in twenty different New York City hospitals. A survey of the first hundred of those New York cases revealed that ninety-eight people became infected while in the hospital; two acquired their infections in the community.

Forty-two of the hundred patients died; incurable enterococci was the direct cause of death for nineteen of them, a contributor in the remaining terminal cases. Most of the dead were elderly individuals.

When laboratory molecular studies were done on bacterial samples from twenty-one of the patients, New York City Health Department researchers found that nineteen were resistant to *all* available drugs. And individuals who were infected with the super-resistant strains also progressed to full-blown blood disease (septicemia) more quickly.[59]

By 1994 all of the Greater New York City large hospitals had cases of vancomycin-resistant enterococci, and infection control had become a major crisis for facilities throughout eastern New Jersey, New York City, and neighboring suburban counties.[60] Similarly grim outbreaks of vancomycin-resistant super-bugs were seen in London,[61] Sheffield, England,[62] and Ancona, Italy.[63]

A CDC survey of key U.S. hospitals found that by 1994 some 7.9 percent of all reported *Enterococcus* infections involved vancomycin-resistant strains. On the nation's intensive-care units, where the risk of infection was highest, vancomycin-resistant strains accounted for just 0.4 percent of all enterococcal infections in 1989, and 13.6 percent by 1993. The highest incidence of the problem was in New York City hospitals, where 8.9 percent of *all* enterococcal infections were of vancomycin-resistant strains.[64]

Inside American hospitals the emergence of super-enterococci was facilitated by practices that allowed the organism to instantly spread from one susceptible human to another: electronic thermometers,[65] catheters and surgical instruments,[66] intravenous lines, mechanical ventilation, and overuse of cephalosporin-type antibiotics.[67] The latter increased the risk of hospital-acquired enterococcal infection because the cephalosporin-type antibiotics had no effect on enterococcal bacteria but did devastate colonies of rival microbes, rendering the treated human especially vulnerable.

By the end of 1993, with vancomycin-resistant enterococci reports coming in from all over the world, CDC and WHO scientists waited anxiously for the seemingly inevitable—exchange of the *vanA* or other resistance plasmids from the *E. faecalis* or *E. faecium* to *Staphylococcus* or *Streptococcus*. It had been done experimentally. European scientists had proven the microbial species capable of such a feat.[68] It only remained for nature to take its course.[69]

The human gastrointestinal tract was an ideal ecology for such microbial events as plasmid exchanges because it was densely populated with dozens of species of both pathogenic and helpful—commensal—organisms. More bacteria lived on a single square inch of the human intestine than there were humans on the entire planet. There were more microbes colonizing a given human's body than there were human tissue cells.

As anybody who had ever taken antibiotics knew, many microbial residents of the gastrointestinal tract performed beneficial functions. As they digested food for their own purposes, these bacteria assisted in the task of breaking down the fats, sugars, proteins, and unwanted chemicals that regularly flowed through the human body. In their absence, human digestion was a difficult, often painful process, as signaled by the constipation, cramps, and gas that often resulted from antibiotic treatment. Antibiotics disrupted the balance of both commensal and pathogenic microbes.

Not long after the advent of the antibiotic medical revolution, an equally radical change in veterinary and livestock practices took place. Expensive livestock lived longer when their ailments were treated with antibiotics. Prophylactic treatment seemed even wiser, as animal husbandry soon included routine antibiotic dosing of chickens, cattle, and dairy cows. The shelf life of meat, poultry, eggs, and dairy products was extended through antibiotic treatment of the animals.

On the face of it this made sense. Why risk *Salmonella* poisoning due to consumption of undercooked meat if it was possible to sterilize the steer's body before slaughter? Why not extend the unrefrigerated shipping distance of eggs by injecting them, or the hens, with antibiotics?

Of course, the microbes were every bit as likely to share genes and mutate around the antibiotics while inside the gut of a cow as they were in a *Homo sapiens* intestinal tract. So in addition to the five billion humans on the planet that might take antibiotics, there were billions of cows, chickens, pigs, cattle, sheep, ducks, and other livestock undergoing prophylactic or treatment exposure to the chemicals, more than doubling the global selective pressure upon bacterial populations.

In the 1970s, Dr. Stuart Levy of Tufts School of Medicine in Boston showed that giving high doses of antibiotics to chickens resulted in the emergence of resistant *Salmonella* strains that could be found in both the meat and the eggs of the animals.[70] Only thorough high-heat cooking could safely destroy the mutant organisms before humans consumed the poultry products.[71]

A Dutch study in 1990 showed that the use of expensive fluoroquinolone-type antibiotics on chickens and eggs led to the emergence of strains of the enteric bacteria *Campylobactr jejeuni* and *C. coli* that were resistant to the drugs in people.[72] Two years later Spanish physicians reported that half of all uncooked chicken in the country contained strains of the bacteria that were resistant to fluoroquinolones. In 1989 the Spanish group had seen virtually no evidence of such resistant strains in randomly tested human stools; by 1993 half of all samples contained the resistant *C. jejeuni/coli*.[73] The same Spanish group had discovered in 1988 that chloramphenicol use in pigs led to the emergence of a strain of resistant *Yersinia enterocolitica* that made its way into humans who ate pork.[74]

In February 1983 an unfortunate landmark was reached when Michael Osterholm in Minnesota discovered that *low*-level livestock use of antibiotics led to mutant bacterial emergence and subsequent human disease.[75] Having recently learned the wily ways of the microbes during their controversial investigations of Toxic Shock Syndrome, the Minnesota State Health Department's Osterholm and the CDC's Mitch Cohen and Scott Holmberg were already on the alert for emergent bacteria when hospitals statewide began reporting an increase in *Salmonella newport* food poisonings.

"This is definitely out of the norm," Holmberg said. "*Newport* is a southern bacterium. You never see it up there."

The *S. newport* bacterium was almost exclusively found in animals and foods common to the near-tropical states along the Gulf of Mexico, yet dozens of people, some acutely ill, were turning up in the ice-cold climes of Minnesota. And the bacterial strain found in their bodies was resistant to penicillin, ampicillin, carbenicillin, and tetracycline.

The patients who took ill in the 1983 Minnesota outbreak were far sicker than was usual: six of them had to be hospitalized for more than a week, several had passed bloody stools, and all of them had suffered at least one of the following symptoms: high fever, diarrhea, stomach cramps, chills, and vomiting.

The investigators first searched victims' medicine cabinets looking for contaminated or aberrant antibiotics, but that proved to be a blind alley. So they turned to the patients' kitchens, hoping to find something unusual in the individuals' eating habits. But what they found were absolutely typical Minnesota diets: lots of meat, potatoes, dairy products, eggs, frozen foods, and prepackaged snack foods.

After several months of frustrated investigation, Osterholm sent out an official memo to his counterparts in neighboring states, asking for clues and suggestions. South Dakota officials called immediately—they had five cases. North Dakota had one. And the common thread between the tristate cases suddenly was obvious: all the sick individuals had consumed hamburger meat shortly before taking ill.

Osterholm, Holmberg, and Cohen traced the hamburger shipments to a herd of cattle that had been grazing in the area where the South Dakota cases were clustered. From there, the 105 cattle were shipped to southern Minnesota and slaughtered on January 8, 1983. Fifty-nine beef carcasses were further processed for retail sale in a Nebraska packaging center on January 10, and shipped to beef brokers in Minnesota and Iowa. From there they were sent to supermarkets across the region. The investigators estimated that 40,000 pounds of contaminated beef, mostly in the form of hamburger, eventually reached Minneapolis, and an untold quantity reached markets in North and South Dakota and Iowa.

It turned out that the herd had been fed antibiotic-laced feed, dosed at levels well below legal standards, thought to be utterly safe. Over time, resistance developed in the *Salmonella* that were infecting the cattle. Most of the humans who fell violently ill were taking antibiotics for other problems, such as sore throats, when they consumed the contaminated hamburger meat. The first antibiotics had cleared their bodies of many other microbes, creating a wide-open, noncompetitive field for *S. newport* colonization.

In a few cases people passed their hamburger-acquired *Salmonella* on to others, via hospital instruments or household exposure.[76]

Levy felt that the Minnesota case proved the folly of continuing to regulate human use of antibiotics in the United States and much of the industrialized world while allowing virtually unregulated sales of the drugs to the agricultural industry, to veterinary medicine, and to most of the world's human population, since few countries required prescriptions for purchase of antibiotics.[77]

Despite these and a host of other examples[78] of the transmission of antibiotic-resistant bacteria from meat, dairy, and poultry products to human consumers, the U.S. Food and Drug Administration, its counterparts in Europe, and the European Community (under the Maastricht Treaty) all failed to take actions that might have limited the use of antibiotics on animals. Government agencies were reluctant to take steps that might impinge upon their country's competitive status in world agricultural markets.

One of the clearest and most troubling examples of animal/human cross-species transmission of mutant bacteria was *Escherichia coli*. The bacteria were ubiquitous, rod-shaped microscopic creatures found in the intestines of all humans and many other mammalian species. Most of the time, in most people, they were harmless. And there was no microscopic organism that was better understood than *E. coli*, as it had been the focus of the majority of the world's molecular and cellular biology research since the 1940s. Scientists liked to work with *E. coli* because all the complex machinery of life was there to study, packaged inside a predictable tubular structure which, almost like clockwork, stretched itself out every 120 minutes, duplicated its DNA, divided down its middle, and—*voilà*—there were two *E. coli*. The hearty bacteria would readily perform this feat of reproduction in the laboratory, always doubling their total population every two hours.

Of course, the bacteria were capable of similar feats of reproduction inside human intestines. If unchecked by the host's immune system, or if of a particularly virulent strain prone to producing tough toxins, the bacteria would cause diarrhea and vomiting. Typically, this occurred in small children whose immune systems weren't yet fully developed, and the ailment was particularly dangerous in malnourished or otherwise seriously ill infants.

In 1982 something new showed up: *E. coli* 0157:H7. It was an apparently novel organism that was capable of causing dangerous hemorrhages of the colon, bowel, and kidneys of human beings of all ages. And it hit suddenly in several U.S. states, as if out of nowhere.[79]

Ten years later the details of 0157:H7 emergence would remain obscure, but its source would not: most cases came from contaminated meat. Like most *E. coli* strains in the 1980s, it was moderately resistant to ampicillin and tetracycline. More important, the mutant bacteria appeared to have acquired the ability to produce *Shigella*-like toxins. Studies of dozens of emergent bacterial species showed that genes for antibiotic resistance and virulence often resided in the same regions of the microbes' DNA, and could move together from one organism to another. Thus, the same selection pressures that led to the emergence of resistance—in this case, use of antibiotics on livestock—also promoted greater virulence.

Because of both agricultural and medical misuse of antibiotics, *E. coli* strains of all kinds were rapidly acquiring broad ranges of resistance during the 1970s and 1980s.[80]

Stuart Levy showed in 1989 that *E. coli* readily spread from pigs and cows to people living and working on a farm. And the resistance factors themselves could move from *E. coli* that were inhabiting a pig, for example, to bacteria that were infecting other higher animals, including humans.[81]

In 1991 in the apple-growing region of Massachusetts there was a small outbreak of *E. coli* 0157:H7 infection, producing serious illness in twenty-seven people, ten of whom required hospitalization. All the cases occurred during the fall apple harvest months. It turned out that the bacteria were in local apple cider. And the cider was made from apples plucked from trees that were fertilized with livestock manure. Presumably, then, the manure was the excreta of 0157:H7-infected animals.[82]

The stage was set for public health disaster.

In January 1993 more than 500 people in Washington State became seriously ill after eating hamburgers prepared in ninety-three Jack-in-the-Box fast-food restaurants. Fifty of the hamburger consumers developed the *E. coli* hemorrhagic syndrome, and four of them—all small children—died. The culprit was *E. coli* 0157:H7, which had arisen in the cattle and was in the hamburger.[83]

Three months later a smaller outbreak occurred in a Sizzlers restaurant in Grants Pass, Oregon. Five diners were hospitalized in that *E. coli* 0157:H7 incident.

Politics immediately entered the picture, as consumer and legal groups demanded that the U.S. government take steps to ensure public safety. They claimed that upward of 25 million Americans suffered food poisoning each year, 6,000 of whom were victims of *E. coli* 0157:H7. The Clinton administration responded by ordering increased meat inspections. But the administration took no steps to get to the source of the problem: the unregulated use of antibiotics on livestock.

II

Many bacteria were capable of using sporulation to their advantage in the face of antibiotics and other threats. Like plant seeds, they would go dormant, toughen their cell walls to a nearly impermeable state, and wait. When conditions were favorable, the bacteria would reactivate, their cell walls once again becoming permeable. Some forms of resistance involved the bacteria's use of genes that triggered sporulation when the microbes were threatened, or created an even less vulnerable cell wall at the time of sporulation.

Under such conditions, microbes could drift about unharmed in solutions designed specifically to kill them. Disinfectants, such as chlorine- and ammonia-based cleansers, soaps, fungicides (yes, fungi could also sporulate), extremely salty or acid solutions, even high heat could all be withstood by hearty sporulation mutants.[84]

By 1992 a number of organisms, including strains of cholera, *E. coli*, and the Legionnaires' Disease bacteria, had developed some resistance, via such sporulation mechanisms, and other means, to chlorine. "Resistance" might have been a misnomer—"partial tolerance" comes closer, because the microbes were able to survive in doses of chlorine that usually killed their species. To ensure safe drinking water in the presence of such bugs, higher doses of chlorine were needed.

Dumping more chlorine in Lima, Peru's water system at the height of its 1991 cholera outbreak provoked little local objection. But in the wealthy United States, where public fear of cancer far outweighed concerns about infectious diseases, thousands of municipalities were lowering their chlorine levels during the late 1980s and the 1990s. Though most evidence indicated that the real chemical carcinogens were chlorinated pollutants, such as polychlorinated biphenyls (PCBs) or dioxins, much public anger was aroused toward *all* uses of chlorine, including sanitation.[85] Greenpeace, the Environmental Defense Fund, and other leading environmental organizations argued that chlorinated compounds accumulated in human body fat and the cancer risk rose over time with each additional chlorine exposure.[86]

That put government—from the municipal to the federal level—in a tight spot, forced to balance the need to limit environmental carcinogens against the threat of infectious diseases. At a time when chlorine-resistant strains were emerging, governments were being pressured to lower sanitation uses of disinfectants.

The first warning shot from the microbes came in January 1987 at West

Georgia College in Carrollton, Georgia. Students in record numbers fell ill with acute gastroenteritis due to *Cryptosporidium*, a one-celled tiny parasite. Not much larger than most bacteria,[87] *Cryptosporidium* caused painful intestinal infections and severe diarrhea.

Rural Carroll County, with a population of only 64,900 people, suffered 13,000 cases of cryptosporidiosis in less than a month. Every household that received water from the central municipal system was struck. The Carrollton water supply met federal standards for water purification, and researchers at the time were uncertain why the outbreak had occurred.[88]

Two U.S. federal agencies whose charters occasionally conflicted on matters of community exposure to chlorine—the Environmental Protection Agency and the Centers for Disease Control—had passive surveillance systems in place. Neither agency conducted active surveillance, aggressively searching for cases of contamination or the emergence of newly resistant strains of microbes. The problem was left to the states and municipalities. And the quality of local surveillance activity varied radically from state to state; some states had no ongoing system in place.

Even this admittedly weak federal data base showed, however, that there was trouble afoot. Between 1991 and 1992, thirty-four outbreaks of disease associated with drinking water were reported to the federal agencies. In 27 percent of the cases the microbial contaminant was identified; 68 percent were unsolved mysteries. Half the cases involved an identified malfunction or deficiency in local water treatment and purification procedures; but in 6 percent of the cases investigators were unable to identify how the water got contaminated.[89]

The key microbes involved were *Giardia*, *Cryptosporidium*, hepatitis A, and *Shigella*. Ten of the outbreaks occurred in communities that used proper chlorine purification. In some of the cases treatment failure occurred when microbially contaminated agricultural wastes got into the water supply. And though *Giardia* had, since the early 1970s, been the dominant microbial contaminant in U.S. drinking water, by 1992 cryptosporidiosis cases equaled those of giardiasis. *Cryptosporidium* were commonly found in cows and their excreta.

In at least three outbreaks the local water treatment facilities were, by all standards, top of the line. They used proper amounts of chlorine, kept the water moving at a rate that could make it impossible for sporulated bacteria to form protective colony clusters adhering to solid surfaces, and passed the water through efficient filtration systems.

Yet the systems failed.

"Evidence suggests that a substantial proportion of non-outbreak-related diarrheal illness may be associated with consumption of water that meets all current water quality standards," the CDC concluded. The agency was also forced to conclude that "*Cryptosporidium* oocysts are resistant to disinfection by chlorine."

The clearest evidence of chlorine failure could be seen in the sudden surge of Legionnaires' Disease, cryptosporidiosis, and giardia among people who used chlorinated hot tubs, swimming pools, and public spas.[90]

In April 1993 some 400,000 residents of Milwaukee, Wisconsin, fell ill with cryptosporidiosis, and the city's AIDS population faced a mortal threat in their drinking water, as their immune systems couldn't control the microbe. The problem was blamed on a combination of chlorine-resistant *Cryptosporidium* and a decrease in filtration efficiency due to a drop in water levels that left the liquid unusually high in particulate levels. U.S. Environmental Protection Agency laboratory studies later showed that the Milwaukee strain *could actually live on Clorox.*

In July 1993 some 35,000 residents of New York City had to switch to boiled water when it was discovered that *E. coli* 0157:H7 had made its way into the water supply. The bacteria survived chlorination and a faulty filtration system. And residents of the nation's capital and outlying Virginia suburbs were forced to boil their water for a week in December 1993 because *Cryptosporidium* had made its way into the Washington, D.C., water supply, due to the same set of factors. Similarly, in 1993 in Cabool, Missouri, the water supply, despite chlorination, was contaminated with *E. coli* 0157:H7; three elderly residents of the town died as a result.[91]

In a review of the U.S. water systems the Natural Resources Defense Council, a citizens' action group, concluded that nearly one million Americans were falling ill annually due to water contamination and 900 were dying as a result. The organization named 250,000 violations of federal drinking water laws nationwide.[92] Some 83 percent of the nation's water systems—those that serviced small towns and rural areas—went virtually unmonitored by state and federal agencies, the organization charged.

The precise mechanisms that *Cryptosporidium, Legionella, E. coli,* and other organisms used to resist chlorine weren't fully understood, but indications were that some mirrored membrane pump systems used by microbes to resist other would-be antimicrobials. Special proteins that spanned the protective membrane of a microbe grabbed undesirable chemicals that had managed to get inside, dragged them through the membrane, and pumped the chlorine, antibiotic, detergent, or other compounds back outside before the chemicals could do any harm. It was an expensive way for a microbe to rid itself of a poison because it took molecular energy to operate a pump. But it worked, and when survival was at stake, some energy expenditure was a small price to pay. Bacteria, fungi, and parasites used such pumps to rid themselves of everything from antibiotics to arsenic, from zinc to chloroquine.[93]

As the 1990s dawned, physicians all over the world were recognizing the limitations of their old armamentarium, and again switched to new classes of antimicrobial drugs. Government agencies from Johannesburg to Oslo were at pains to spot newly emerging resistant organisms before they produced epidemics. Pharmaceutical companies were searching for radically new ways of attacking the microbes.

"We're running out of bullets for dealing with a number of these infections," Nobel laureate Joshua Lederberg warned.[94] "Patients are dying because we no longer in many cases have antibiotics that work."

Though he considered emerging viruses a far more significant threat to humanity, Lederberg worried about the sorry state of development of new antibiotics and disinfectants. It was a problem, he said, "much more of an organizational, political, and cultural nature than a technical one. It's a race against the microbes."

With the advent of PCR technology a great deal of scientific attention was devoted to trying to understand how bacteria acquired such resistance and virulence capabilities. The molecular detective work allowed scientists to trace the mobile DNA units from microbe to microbe.

"Bacteria are cleverer than men," concluded Columbia University's Dr. Harold Neu.

"Bugs are always figuring out ways to get around the antibiotics we throw at them," said Harvard Medical School's Dr. George Jacoby. "They adapt, they come roaring back."[95]

The tricks commonly used by bacteria to spread or absorb helpful genes included the plasmids, sexual conjugation, transposons within their own genomes, and mutations at single sites along their DNA. The world, it turned out, was awash with highly mobile segments of DNA. And bacteria were terrific scavengers. Keeping track of all the newly discovered plasmids and mobile DNA pieces seemed an impossible task, though in 1993 the World Health Organization issued contracts to research groups bent on trying.

Thomas O'Brien, whose Harvard Medical School laboratory was among those toiling for WHO to catalogue the world's plasmids, declared in 1992 that what the world faced was not so much an antibiotic resistance crisis as an "epidemic of plasmids."

At the molecular level the microbes possessed multitudinous ways to outwit any given antibiotic.[96] Inside an animal's intestinal tract was a veritable soup of plasmids and resistance factors. Some offered the microbes blueprints for the production of chemical pumps—like microscopic bouncers protecting clientele from undesirable riffraff—that bailed antibiotics out of the cell. Terrific evaders, the bacteria rarely generated overt counterattacks, making enzymes that actually destroyed an antibiotic. Instead, they adapted to the new chemical environment, rich in whatever antibiotic was in use, by building a tougher membrane wall or changing whatever biochemical process the drug was supposed to affect. If, for example, a drug such as tetracycline was designed to inhibit bacterial protein synthesis activity on the microbe's smaller ribosome, the bacteria simply changed the vulnerable protein factors so that the point sensitive to the drug no longer existed.

Every time *Homo sapiens* made a molecular socket wrench to undo some vital bacterial function, the wily microbes simply changed the vulnerable assembly to a Phillips-headed screw.

Most of these powerful defense weapons had probably existed in microbes, and in individual animal or plant cells, for aeons. They performed services for the organisms that extended well beyond resisting antibiotics or attaining greater powers of infectiousness and lethality. For example, species as diverse as yeast, human cancer cells, and malaria parasites all could process a similar set of genes—designated *mdr* or *pgp*—that provided the blueprints for membrane pumps. The yeast used the acquired ability to pump out pheromones (or external hormones) that attracted one yeast to another. Human cancer cells used the genes to expel chemotherapy drugs. *Plasmodium falciparum* used the genetic traits to get rid of chloroquine.[97]

Plasmids played a role in the evolution not only of bacteria but possibly of all species on the planet. Their movement among microbes, or from microbes to plants and animals, was thought by many scientists to have long been crucial to adaptation and change.[98]

Acquisition of one set of genetic characteristics might well have a cascade effect, resulting in an entirely new range of capabilities for the microbe. For example, a plasmid carrying genes for neomycin-kanamycin resistance also had part of a gene for bleomycin resistance. *E. coli* that absorbed this pRAB2 plasmid, as it was called, not only acquired the ability to withstand those antibiotics but also became more fit. Along with the bleomycin resistance came a genetic ability to rapidly repair genetic damage to DNA. With that capability, *E. coli* could live longer and suffer fewer deleterious mutations.[99]

Plasmids and transposons also had some influence over their own expression, once they gained entry into a cell. Many contained genes called integrons that integrated the mobile DNA into the organisms' genomes. Some had regulatory genes that could switch on and off both their own plasmid or transposon genes and key genes inside the microbial chromosome.[100]

In this way, DNA moved not only between various bacterial species but between entire families of organisms: between bacteria and yeasts, between plants and bacteria, between complex parasites and their hosts' cells.[101]

It required a very small leap of logic to conclude that retroviruses such as HIV, HTLV, and feline leukemia virus were originally transposons. Over time, these bits of mobile genes (in the form of RNA, rather than DNA) acquired various regulatory genes from the microbes they inhabited. With passing generations, they gained sufficient genetic sophistication to be able to manufacture hard protective shells or envelopes, inside of which would safely reside their RNA. This gave them, in Bernard Fields's lexicon, both payloads and delivery systems, allowing them to become viruses.[102]

Though dozens of different types of antibiotics had been in use against bacteria and some parasites since the 1940s, humanity had very few antivirals at its disposal. As was the case with drugs aimed at bacteria, resistance was a critical problem shortly after introduction of the key antivirals: acyclovir, ribavirin, amantadine, foscarnet, ganciclovir, and the HIV drugs.

By 1981 the U.S. genital herpes epidemic had reached crisis proportions in much of the world, so word of a drug that might cure the disease raised considerable excitement. Acyclovir, developed by the Burroughs-Wellcome pharmaceutical company, was a terrific success.[103] The drug could prevent latent herpes viruses from resurfacing to produce genital disease, cold sores, shingles, and a variety of other disorders. Both in pill form and as a topical cream, acyclovir brought relief to herpes sufferers and was hailed as a revolution, much as penicillin had been four decades earlier.[104] But even in the original promising studies, physicians noticed that cessation of acyclovir use could immediately—within less than twenty-four hours—result in a herpes surge, typically producing more severe disease in the patients than was seen in people who used placebos. That implied two things: the drug was unable to accomplish much more than forcing the virus to remain in hiding inside nerve cells, and it might be exerting selection pressure on the viral population that resulted in even more virulent pathogens.

The most serious life-threatening herpetic ailment was encephalitis due to infection of the brain. By the late 1980s physicians all over the world were reporting horrendous herpes encephalitis relapses in their patients following cessation of acyclovir use.[105]

Well before acyclovir got the FDA's green light for commercial distribution in the United States, physicians close to its research efforts were worrying publicly about resistance. Some were even concerned about the chemical similarity between acyclovir and other available antivirals, saying that "the possibility of cross-resistance is at least worrisome."[106]

Despite concerns about emergence of highly resistant herpes viruses, surgeons almost immediately began using the drug as a postoperative prophylactic much as they had long done with antibiotics,[107] particularly for patients undergoing transplants and other procedures that required deliberate immunosuppression.[108]

Scientists had long known that some herpes viruses—perhaps fewer than one in ten million—were naturally genetically resistant to acyclovir, meaning that even before acyclovir was invented, some of the viruses had an innate ability to outwit therapy.[109] And researchers soon showed that resistant strains, once emerged and established within the human body, persisted for years, with or without continued use of acyclovir.[110]

At the molecular level there were several different ways herpes viruses could become resistant to acyclovir, and it was clear that some mutations observed in clinical settings were new—that is, the virus mutated during exposure to acyclovir. In many cases the viruses could become invulnerable with a simple point mutation—one tiny change in their DNA. The most successful genetic changes were those that affected one of two key viral enzymes: DNA-polymerase, which the virus used to make copies of itself; and thymidine kinase, a chemical also crucial to viral replication. Mutations in these enzyme genes cost the viruses a great deal: they became resistant, but at the cost of some powers of infectiousness and virulence. The trade-off was overcome, in part, by the intermingling of the mutant viruses with normal ones that still possessed the powerful genes. So, for example, virus populations hidden inside human nerve cells would be protected from the drugs and the immune system until they were activated and exited their protective neural seclusion. The normal viruses would possess the genes that allowed that exiting process and warded off the immune system, while the less virulent mutants were prepared to survive acyclovir.

In 1992 British scientists warned, "There is a possibility that [acyclovir]-resistant strains of herpes simplex virus with epidemic potential eventually will emerge and reduce the efficacy of this drug. The time-scale for the emergence of such resistance is unclear."[111]

That epidemic acyclovir resistance would eventually occur seemed an inevitability to most observers: the only issues were when and where. The answer, it turned out, was AIDS. Because of the overlapping epidemiological risks of AIDS and many herpes viruses, people battling the first disease often suffered horrible bouts of the second. AIDS physicians began in the late 1980s putting patients with histories of prior herpetic illnesses on prophylactic acyclovir, or treating occasional flare-ups of herpes with longer durations of the drug. Not surprisingly, by 1989 virulent acyclovir-resistant mutants appeared in AIDS patients.[112]

In 1990 an otherwise healthy twenty-seven-year-old American came down with genital herpes. Because his case wouldn't respond to normal acyclovir treatment, the young man came under the care of National Institutes of Allergy and Infectious Diseases physician Stephen Straus, a longtime acyclovir expert. Straus treated the ailing man with ever-higher doses of acyclovir, eventually stopping when he reached toxic levels that were six times that normally used to treat the infection.

It was the first time Straus had seen such a case in an otherwise immunologically healthy adult. The patient wasn't infected with the AIDS virus, and hadn't suffered any prior ailments or required surgery. The patient, who was gay, had three sexual partners during the 1990 period in which he got infected. One of the partners was dually infected with HIV and the mutant herpes virus. Straus believed that the herpes mutant had first developed in the immunocompromised HIV-positive man and then had been passed sexually to the unfortunate twenty-seven-year-old.[113] The new mutant was distinctly dangerous because it had *not* traded off virulence for persistence. Furthermore, the mutant possessed the ability to resist another antiviral drug: ganciclovir. It was a multiply resistant, fully pathogenic virus that could be sexually transmitted.

"And now my suspicion is that this [mutant strain] will become more frequent, but the pace of that increase in frequency is unclear," Straus said.[114]

HIV-positive Americans and Europeans were treated to a pharmaceutical cornucopia that helped them survive one microbial onslaught after another. And many individuals were simultaneously taking more than a dozen different drugs. Furthermore, physicians who failed to control herpes infections in their AIDS patients using acyclovir often switched to drugs that were designed to treat other viral diseases. Ganciclovir and foscarnet, for example, were primarily used to treat cytomegalovirus (CMV), which commonly struck HIV-positive people. As acyclovir failures were increasingly reported, many physicians switched to the two anti-CMV drugs.[115]

Unfortunately, resistance to both ganciclovir and foscarnet quickly emerged among cytomegaloviruses that were infecting HIV-positive individuals. It was as if one plague of contagious immunosuppression was fostering a subset of mini-epidemics of viral resistance.

CMV resistance to ganciclovir or foscarnet seemed by 1992 to be an inevitable consequence of prolonged use of either drug in HIV-positive individuals. Physicians had to balance the need to control herpes and mild CMV infections during early stages of HIV disease against the necessity of having something in reserve that would still work—to which the microbes wouldn't be resistant—during the final throes of AIDS. In particular, a significant percentage of long-term AIDS survivors would go blind, victims of CMV retinitis, if the viral strains in their bodies developed resistance to both drugs.[116]

As was the case with acyclovir resistance in herpes viruses, CMV invulnerability to ganciclovir and foscarnet was achieved by single point mutations in the viruses' DNA coding for either DNA polymerase or a key kinase.[117] It was tempting to conclude that a commonality of mutation sites existed for highly divergent types of viruses. And, indeed, in 1993, physicians from around the United States reported treating patients who had foscarnet resistance in *both* their herpes simplex and cytomegalovirus populations. The resistance was of clinical significance, and in some cases the patients also suffered acyclovir-resistance, leaving doctors with no good treatment options.[118]

Further complicating decisions about the uses of these drugs were indications that the herpes-type viruses could directly stimulate the activation signals within the HIV genome, promoting production of more AIDS viruses. Seen first in test-tube experiments, the microscopic partnership between the viral species was confirmed in 1993 in studies of six gay men in Los Angeles who were simultaneously infected with herpes simplex-1 and HIV. Not only did the two viral species stimulate one another, but they shared cellular homes and intermingled so completely that a sort of hybrid virus —part HIV, part HSV-1—appeared.[119]

The most notorious drug resister was the human immunodeficiency virus itself. From the moment azidothymine, or AZT (trade name: Zidovudine) was introduced into use on AIDS patients it was clear that the window of opportunity for its utility was limited by the virus's ability to mutate into a resistant form. Initially thought to be an event that emerged in AIDS patients after two or three years of AZT use, it became clear that some strains developed resistance almost immediately following exposure to the drug. And there were indications that AZT-resistant strains of HIV could be transmitted from one person to another.[120]

Physicians followed the models already set for treating bacteria in the face of resistant antibiotics: they added on other drugs, either in sequence or in combination with AZT. So ddI (dideoxyinosine), ddC (dideoxycytidine), nevrapine, FLT (deoxyfluorothymidine), zalcitabine, 3TC (3-thiacytidine), and carbovir (didehydrodideoxyguanosine) were tried as alternatives or adjuncts to AZT.

Resistance emerged to all of them.

One thing the drugs shared was their target: the key enzyme used by HIV to make a DNA copy of its RNA genome, reverse transcriptase. As Howard Temin had shown during the late 1980s, HIV was one of the most mutable microbes on the planet. And the key to that mutability was the reverse transcriptase enzyme. Helpful mutations could persist for years, even be passed from one human host to another.[121]

Not surprisingly, there were soon HIV strains that were multiply resistant to AZT plus other antiviral drugs, or to combinations of ddI, ddC, and others.[122] So grim was the situation that the head of San Francisco General Hospital, Dr. Merle Sande, threw his hands in the air in a deliberately dramatic gesture at a National Institutes of Health meeting and bellowed, "We need better drugs!"[123]

Such sentiments were echoed by doctors who were trying to treat influenza infections in acutely ill elderly individuals infected with amantadine- or rimantadine-resistant strains of the virus.[124]

In short, it seemed that viruses, because of their rapid reproduction rates and high degrees of inherent mutability, were even more likely than bacteria to find ways around the drugs humans threw at them. Piling on more drugs, or using drugs in higher doses for longer periods of time, hadn't prevented the emergence of untreatable bacterial strains. Why did pharmaceutical companies and physicians believe such tired old antimicrobial tactics could defeat viruses?

"You can't expect physicians to be concerned about public health," Mark Lappé had opined one sunny spring afternoon in his Berkeley office at the University of California. It was 1981 and Lappé's book *Germs That Won't Die* had just been released. No one had yet heard of AIDS or drug-resistant clinical viruses or chlorine-resistant *Legionella*.

"It's hard to put the large view into day-to-day medicine. And it's a real tragedy. And you can't sue a doctor for violating an ecosphere, but you can sue for failure to give an antibiotic that you think would have enhanced

the possibility of patient survival. It's a real dilemma," Lappé had said.
A decade before the resistance crisis was acknowledged by mainstream science, he said that medicine and public health were locked in a conflict over drug-induced emergence of new microbes—a conflict that couldn't easily be resolved. It was the physicians' job, Lappé said, to individuate decisions on a patient-by-patient basis. The mission of the doctor was to cure *individual cases* of disease.[125] In contrast, public health's mission required an ecological perspective on disease: individuals got lost in the tally of microbial versus human populations.

When Lappé looked at American hospitals in 1980 he didn't see the miracles of modern medicine—heart transplants, artificial knees, CT scans. Lappé saw disease, and microbes, and mutations.

"It's incredible," Lappé said. "You can go into a hospital and you will have a four in a hundred chance of getting an infection you've never had before, while in that hospital. In some hospitals the odds are one in ten. What you will get in that hospital will be much worse than what you would have been contaminated with at home. They are the most tenacious organisms you can imagine. They can survive in the detergent. They can actually live on a bar of soap. These are organisms that are part of our endgame."

Decrying improper use of antibiotics as "experiments going on all the time in people, creating genuinely pathogenically new organisms," Lappé occasionally lapsed into a grim global ecological description of the crisis —a perspective that critics charged in 1981 grossly exaggerated the scope of the problem:

Unfortunately, we played a trick on the natural world by seizing control of these [natural] chemicals, making them more perfect in a way that has changed the whole microbial constitution of the developing countries. We have organisms now proliferating that never existed before in nature. We have selected them. We have organisms that probably caused a tenth of a percent of human disease in the past that now cause twenty, thirty percent of the disease that we're seeing. We have changed the whole face of the earth by the use of antibiotics.

By the 1990s, when public health authorities and physicians were nervously watching their antimicrobial tools become obsolete, Lappé's book was out of print. But everything he had predicted in 1981 had, by 1991, transpired.

III

For developing countries, access to still-reliable antibiotics for treatment of everything from routine staph infections to tuberculosis and cholera had reached crisis proportions by the 1990s. In 1993 the World Bank estimated that the barest minimum health care package for poor countries required an annual per capita expenditure of $8.00. Yet most of the least developed countries couldn't afford to spend more than $2.00 to $3.00 per person each year on total health care.[126] With over 100,000 medicinal drugs marketed in the world (5,000 active ingredients), it was possible for government planners to lose sight of their highest-priority needs, purchasing nonessential agents rather than those necessary for their populations' survival. And the scale of global disparity in drug access was staggering: the average Japanese citizen spent $412 in 1990 on pharmaceutical drugs; the typical American spent $191; in Mexico just $28 per year was spent; Kenyans spent less than $4.00 per year; and Bangladeshis and Mozambicans just $2.00 per year, on average.

It was in the wealthy and medium-income countries where billions of dollars' worth of antibiotics and antivirals were used and misused. And it was in the wealthy nations that resistant strains most commonly emerged. But it was the poor nations, unable to afford alternative drugs, that paid the highest price.

"The development of new antibiotics is very costly," wrote Burroughs-Wellcome researcher A. J. Slater, "and their provision to Third World countries alone can never be financially rewarding; furthermore, only about 20% of world-wide pharmaceutical sales are to Third World countries. The industry's interest in developing drugs for exclusive or major use in such countries is declining."[127]

Some poor countries sought to offset rising drug costs and microbial resistance by developing their own pharmaceutical manufacturing and distribution capabilities. In the best-planned situations, the respective governments drew up a list of the hundred or so most essential drugs, decided which could (by virtue of unpatented status and ease of manufacture) be made in their countries, and then set out to make the products. Local manufacture might be carried out by a government-owned parastatal company, a private firm, or—most commonly—a local establishment that was in partnership with or a subsidiary of a major pharmaceutical multinational.

Though such drug policies were strongly supported by all the relevant UN organizations and, eventually, the World Bank, they were considered direct threats to the stranglehold a relative handful of corporations had on the world's drug market. The U.S.-based Pharmaceutical Manufacturers Association, which represented some sixty-five U.S.-headquartered drug and biotechnology companies and about thirty foreign-based multinationals, strongly opposed such policies. In general, the companies—all of which were North American, European, or Japanese—felt that local regulation, manufacturing, marketing restrictions, or advertising limitations infringed on their free market rights.[128]

Given that these companies controlled the bulk of the raw materials required for drug manufacture, and purchase of such materials required hard currency (foreign exchange), most of the world's poor nations were unable to actuate policies of local antibiotic production.[129] At a time when all forms of bacteremia were on the rise in the poorest nations on earth—notably in sub-Saharan Africa[130]—the governments were least equipped to purchase already manufactured drugs or make their own.

Not all the blame for the lack of effective, affordable antibiotics could be justifiably leveled at the multinational drug manufacturers: domestic problems in many poor nations were also at fault. Distribution of drugs inside many countries was nothing short of abominable. In developing countries, most of the essential pharmaceuticals never made their way out of the capital and the largest urban centers to the communities in need. On average, 60 to 70 percent of a poor country's population made do with less than a third of the nation's medicinal drug supply, according to the World Bank.

Perhaps the classic case of the distribution crisis involved not an antibiotic but an antiparasite drug. During the early 1980s the U.S.-based multinational Merck & Company invented a drug called ivermectin that could cure the river blindness disease caused by a waterborne parasite, *Onchocerca volvulus*. About 120 million people lived in onchocerciasis-plagued areas, most of them in West Africa. And WHO estimated that at least 350,000 people were blind in 1988 as a result of the parasite's damage to their eyes.

It was, therefore, an extraordinary boon to the governments of the afflicted region and WHO when Merck issued its unprecedented announcement in 1987 that it would donate—free—ivermectin to WHO for distribution in the needy countries. No drug company had ever exhibited such generosity, and WHO immediately hailed Merck's actions as a model for the entire pharmaceutical industry.

But five years after the free ivermectin program began, fewer than 3 million of the estimated 120 million at risk for the disease had received the drug. Cost was not the issue. Infrastructural problems in transportation and distribution, military coups, local corruption,[131] lack of primary health care infrastructures in rural areas, and other organizational obstacles forced WHO and Merck to privately admit in 1992 that the program to cure the world of river blindness might fail.[132]

The World Bank and many independent economists argued that such problems would persist until developing countries instituted national health care financing policies[133]—a daunting vision given that the wealthiest nation in the world, the United States, only embarked on a course toward implementation of such a policy in 1994. The pharmaceutical industry argued that developing countries had proven woefully unable to produce quality medicinal drugs on an affordable, high-volume basis. Lack of skilled local personnel, overregulation and bureaucratization, corruption, and lack of hard currency for bulk purchase of supplies and raw materials were all given as reasons for developing country inadequacies. Restrictions on multinational access to local markets were doomed, the industry asserted, to exacerbate the situation by denying the populace needed drugs.[134]

From the perspective of developing countries, the pharmaceutical industry and Western governments that acted in its support were solely concerned with the pursuit of profits, and would conduct any practice they saw fit to maintain their monopoly on the global medicinal drug market. Such practices, it was charged, included bribing local doctors and health officials, manipulating pricing structures to undermine local competitors, advertising nonessential drugs aggressively in urban areas, dumping poor-quality or banned drugs onto Third World markets, withholding raw materials and drugs during local epidemics, and declining foreign aid to countries whose drug policies were considered overly restrictive.[135]

While charges and countercharges flew, the crisis in many parts of the world deepened. According to the World Bank, the world spent $330 billion in 1990 on pharmaceuticals, $44 billion of which went to developing countries. The majority of the world's population in 1990 lacked access to effective, affordable antibiotics.

In 1991, with the world facing a tuberculosis crisis, it was suddenly noted that the global supply of streptomycin was tapped out. The second-oldest antibiotic in commercial use was no longer manufactured by any company. Unpatented, cheap, and needed solely in developing countries, it offered no significant profit margin to potential manufacturers. When drug-resistant TB surfaced in major U.S. cities that year, the Food and Drug Administration would find itself in a mad scramble to entice drug companies back into the streptomycin-manufacturing business.

IV

It wasn't just the bacteria and viruses that gained newfound powers of resistance during the last decades of the twentieth century.

"It seems we have a much greater enemy in malaria now than we did just a few years ago," Dr. Wen Kilama said. The director-general of Tanzania's National Institute for Medical Research was frustrated and angry in 1986. He, and his predecessors, had meticulously followed all the malaria control advice meted out by experts who lived in wealthy, cold countries. But after decades of spending upward of 70 percent of its entire health budget annually on malaria control, Kilama had a worse problem on his hands in 1986 than had his predecessors in 1956.

"More than ten percent of all hospital admissions are malaria," Kilama said. "As are ten percent of all our outpatient visits. In terms of death, it is quite high, and it is apparent that malaria is much more severe now than before."

Ten years earlier the first cases of chloroquine-resistant *Plasmodium falciparum* parasites had emerged in Tanzania; by 1986 most of the nation's malaria was resistant to the world's most effective treatment. Like nearly every other adult in the nation, Kilama had suffered a childhood bout with malaria, fortunately in the days before chloroquine resistance surfaced. Natural immunity to malaria among survivors like Kilama was weak, and whenever he was under stress he would be laid up with malarial fevers.

"It is a very unusual individual in this country who doesn't have chronic malaria," Kilama said.

Though he was speaking of Tanzania, Kilama might as well have said the same of most of the nations of Africa, Indochina, the Indian subcontinent, the Amazon region of Latin America, much of Oceania, and southern China. Most of the world's population in 1986 lived in or near areas of endemic malaria.

Since the days when optimists had set out to defeat malaria, hoping to drive the parasites off the face of the earth, the global situation had worsened significantly. Indeed, far more people would die of malaria-associated ailments in 1990 than did in 1960.

For example, the Pan American Health Organization and the Brazilian government had succeeded in bringing malaria cases in that country down to near-zero levels by 1960. In 1983 the country suffered 297,000 malaria hospitalizations; that figure had doubled by 1988. Despite widespread use of DDT and other pesticides, the *Anopheles darlingi* mosquitoes thrived in the Amazon, feeding on the hundreds of thousands of nonimmune city dwellers who were flooding the region in search of gold and precious gems. The situation was completely out of control.[136] By 1989 Brazil accounted for 11 percent of the world's non-African malaria cases.[137]

A 1987 survey of malaria parasites extracted from the blood of nearly 200 Brazilian patients revealed that 84 percent of the Amazon isolates were chloroquine-resistant; 73 percent were resistant to amodiaquine; nearly all the isolates showed some level of resistance to Fansidar (sulfadoxine/pyrimethamine). Only one then-available drug remained effective against malaria in Brazil: mefloquine.[138]

By 1990 more than 80 percent of the world's malaria cases were African; 95 percent of all malarial deaths occurred on the African continent. Up to half a billion Africans suffered at least one serious malarial episode each year, and typically an individual received some 200–300 infective mosquito bites annually. Up to one million African children died each year of the disease.[139] And all over the continent the key drugs were failing.

The first reported cases were among Caucasian tourists on safari in Tanzania and Kenya during 1978–79.[140] As early as 1981 chloroquine's efficacy was waning among Kenyan children living in highly malaria-endemic areas, and higher doses of the drug were necessary to reverse disease symptoms.[141] Within two years, truly resistant parasites had emerged in Kenya, and laboratory tests showed that 65 percent of the *P. falciparum* parasites had some degree of chloroquine resistance.[142]

By 1984 reports of people dying of malaria while on chloroquine, or failing to improve when taking the drug, were cropping up all over the African continent: from Malawi,[143] Namibia,[144] Zambia,[145] Angola,[146]

South Africa,[147] Mozambique,[148] and locations scattered in between. Public health planners watched nervously, wondering how long chloroquine—the best and most affordable of the antimalarials—would remain a useful drug.

Kilama and his counterparts in other African nations tried mosquito control measures, but the insects quickly acquired their own resistance to the pesticides. They tried eliminating watery breeding sites for the mosquitoes, but, as Kilama put it, "what can you do when these creatures can breed thousands of offspring in a puddle the size of a hippo's foot? During the rainy season there is absolutely nothing."

Kilama's staff regularly tested children living in northern equatorial districts of Tanzania for chloroquine resistance, and watched in horror as the parasites' sensitivity to the drug declined *logarithmically* between 1980 and 1986. Furthermore, isolated cases of mefloquine and pyrimethamine resistance were reported in the country.[149]

The CDC developed a simple field test kit for drug resistance that was widely distributed in Africa in 1985. Immediately a picture of the resistance emergence patterns developed. The problem began along coastal areas of East Africa, particularly Zanzibar, Mombasa, and Dar es Salaam. In these areas two factors may have played a role: a highly mobile Asian population that traveled frequently to India and other regions of resistant malaria, and relatively high availability of chloroquine through both legal and black-market venues. From there, resistance spread along the equatorial travel routes connecting traders from Kenya, Tanzania, Malawi, Zambia, Zaire, Burundi, Rwanda, and Uganda—the same trade routes implicated in the spread of the region's AIDS epidemic. The problem eventually spread outward, from Addis Ababa to Cape Town, from Senegal to Madagascar.[150]

Studies of the newly emerging *P. falciparum* strains showed that the mutations involved in resistance, once present, were permanent features in the parasitic line. The resistance mechanisms involved several different genes: partial insensitivity could result from a single mutation, total resistance from two or more. Wherever the single-mutation somewhat insensitive strains emerged, fully resistant mutants soon followed.

The mutants seemed to grow faster in laboratory cultures than did normal *P. falciparum*, indicating that they might have acquired some type of virulence advantage.

And finally, resistance was cropping up not only in regions where chloroquine was heavily used but also among people who rarely took the drug. That implied that the mutation and emergence didn't require heavy selection pressure. And it also posed serious questions about what policies governments should pursue to preserve the utility of the precious drug.[151]

By 1990 chloroquine resistance was the rule rather than the exception in most malarial regions of Africa. In addition, physicians noticed that chloroquine-resistant strains of *P. falciparum* seemed somewhat insensitive to treatment with quinine or quinidine, probably because of the chemical similarities of the three drugs.[152]

Between 1988 and 1990 a seemingly new disease emerged—cases of lethal adult cerebral malaria in East and Central Africa. Individuals who had acquired some degree of immunity during childhood would suddenly as young adults be overtaken with fever and the demented behavior produced by parasitic infection of the brain. The suddenness of both the onset and death in such cases was startling.

In most cases the cerebral malaria victims had lived their adult lives in urban areas, far from their childhood villages and daily exposure to bloodsucking mosquitoes. Once a year, perhaps, they would return to their old village home to visit relatives and would be reexposed to the parasites. As far as the parasites were concerned, these city dwellers, though African, were no less vulnerable to malaria than a Caucasian tourist. The immunity to *P. falciparum* disappeared within twelve months—or less—in the absence of regular reexposure to the parasites. There was no absolute protective immunity in anyone exposed to malaria—nothing akin to the lifelong immunity that resulted from a smallpox vaccine.

As the death toll among young adults increased, so did the economic costs. In 1993 the World Bank estimated that the loss of productive adult workers to malaria could within two years cost African economies $1.8 billion—a staggering figure for such impoverished societies. [153]

Mortality due to malaria in 1993 was at a historic all-time high in Africa.

"Cerebral malaria is now estimated to be responsible for a fatality rate of more than twenty percent of malaria cases, even in urban areas. . . . Mortality and morbidity rates due to malaria, as monitored in specific countries, appear to be increasing. For example, reported deaths due to malaria increased from 2.1 percent in 1984, to 4.8 percent in 1986, to 5.8 percent in 1988 in Zaire. Malaria deaths as a percent of mortality in Zaire increased from 29.5 percent in 1983, to 45.6 percent in 1985, and to 56.4 percent of all mortality in 1986," reported the American Association for the Advancement of Science. [154]

Resistance to chloroquine, mefloquine, Fansidar, quinine, trimethoprim, and quinidine were all rising rapidly, with some areas (particularly Zaire) reporting that virtually all malaria cases were caused by chloroquine-resistant strains by 1990.

Uwe Brinkmann, then at the Harvard School of Public Health, was watching the steady rise in malaria cases, resistance, and deaths, and in 1991 set out to calculate the toll the newly emergent *P. falciparum* were taking in Africa, in both direct medical and indirect societal costs. He predicted that by 1995 malaria would be costing most sub-Saharan African countries 1 percent of their annual GDPs. [155]

By 1995 in Rwanda, Brinkmann's group predicted, "the direct cost of malaria per capita will exceed [Ministry of Health] expenditure per capita." This posed an obvious question: What will societies do when their malaria burden exceeds all available hospital beds, drugs, health providers, and finances?

At the CDC, where Kent Campbell, Joel Breman, and Joe McCormick were devoting their attention by the close of the 1980s entirely to the malaria problem, a new issue cropped up.

"What *is* malaria?" Campbell asked. "If a population is universally infected, and periodically ill, what exactly is the disease we call malaria?"[156]

It wasn't an academic question. By the late 1980s the CDC scientists and their African counterparts were witnessing a dangerous new malaria disease paradigm on the continent. Small children who suffered fevers were immediately given chloroquine by their parents—easily obtained either from government clinics or on the black market. The kids would recover from their fevers, but the partially resistant parasites would remain in their bodies. Unable to mount a strong immune response, the children would suffer more bouts of severe malaria, receive additional doses of chloroquine or quinine, and continue to harbor the parasites. Over time, the parasite load would build to critical levels in their blood, causing significant damage to their red blood cells.

Just a decade earlier all African children had either died or survived severe malaria in their first weeks of life, and from then on been at no greater malarial risk than adults. However, thanks to chloroquine treatments, by the late 1980s tens of thousands—perhaps millions—of African children were surviving those early malarial fever bouts, but lapsing into fatal anemia episodes at later ages ranging from six months to nine years. The only way to save such a child's life would be to transfuse huge amounts of nonmalarial blood as quickly as possible into his or her body.

In 1985, Kinshasa's Mama Yemo Hospital performed about 1,500 such pediatric transfusions; a year later, that figure leapt to 6,000.

By the end of 1986, one out of every three children admitted to Mama Yemo suffered from a chloroquine-resistant strain.

That was also when pediatric AIDS started to soar in Kinshasa.

In an anemia crisis, seconds matter in the race to save a child's life. Even if an impoverished African clinic had the tools to test donated blood, they didn't have the time. Typically doctors would simply grab a relative whose blood type matched appropriately, and pump blood straight from the donor to the child. A review of 200 children who were transfused in this way at Mama Yemo in 1986 showed that 13 percent got infected with HIV as a result.

"The doctors knew they were transmitting AIDS," Campbell explained. "But they were trying to ensure the survival of these children. It's a crapshoot, it really is. They saw them die day in and day out then, so they were making a clinical decision that was the best that they could do."[157]

In field trials in Malawi, David Heymann looked to see if it was possible to use drugs either to eliminate the parasites from pregnant women, thus at least decreasing the chances that babies would be born infected, or to boost maternal immunity so much that mothers would pass powerful antibodies on to their breast-feeding babies. Clearing all parasites from a pregnant woman's body required over $10.00 worth of antimalarials—too expensive in countries that spent less than $3.00 per capita annually for all health needs.[158] And giving chloroquine or quinine to newborns was fruitless, for Joe McCormick showed in Kenya that the average infant or toddler living in East African villages got 50 to 80 infectious mosquito bites *per month*. Other studies showed that half of all infectious bites ended up as malaria cases.

At the same time, the territory of the *Anopheles* mosquitoes was expanding. Usually the mosquitoes had very limited temperature and altitude flexibility; ideally, they preferred tropical sea-level conditions. But as the sheer density of the human population increased, and massive numbers of immune-naïve people moved back and forth between rural and urban areas, the mosquitoes braved their way into previously uncharted territory. Heymann witnessed the expansion of malaria in Rwanda, for example, where for centuries the disease and *Anopheles* mosquitoes were limited to lowland, densely populated areas. Similarly, in Swaziland he saw the expansion of the fruit-canning industry, located in the lowlands, draw people from the remote highland areas of the country, where *Anopheles* mosquitoes weren't found. They were recruited to work in the canneries in the malarial lowlands. Lacking any immunity and confronted with drug-resistant parasites, the highlanders died off in huge numbers. Those few lucky enough to survive carried the parasites in their blood back up to the mountainous communities where local mosquitoes picked up the microbes while feeding on the migrant workers.

Next door in Malawi, Heymann watched throughout the 1980s as the incidence of chloroquine-resistant malaria and malarial deaths rose steadily. In 1980 fewer than 5 percent of all child hospitalizations were due to acute anemia; by 1986 that percentage had tripled.[159]

Back at CDC headquarters in Atlanta, Campbell was still trying to answer his fundamental question: What is malaria? For much of their childhood virtually all Africans were infected with the parasites—did that constitute malaria? Clearly not, he decided. Was it malaria when an infected child developed fevers? Again, Campbell decided, the answer was no, as many ailments could produce fevers in young children. The presence of the malarial parasites did not necessarily mean that those particular bugs were responsible for a child's fever.

"Furthermore," Campbell said, "we cannot continue to treat every fever as if it's malaria, because the roster of drugs is getting shorter." He knew, however, that within twenty-four to forty-eight hours what appeared to be a vague fever could, after a single red blood cell reproduction cycle for the parasites, spell death.

Answering these questions became an obsession for Campbell, as well as the other old Africa hands at CDC: Joel Breman, Joe McCormick, and David Heymann. Stumping from one cluster of malariologists to another, in Geneva and London, at Oxford and Harvard, before U.S. congressional subcommittees and in meetings with the Côte d'Ivoire Ministry of Health, Campbell would ask the same basic question: What is malaria?

And everywhere he went the question was greeted first by disbelief— how could America's supposed leading malaria expert ask such a stupid question? Campbell didn't mind if he sounded naïve. The tall, lanky marathon runner would stretch his long legs out from his seat, and in a style reminiscent of Jimmy Stewart's sly, falsely modest characters, repeat his question in a slow Tennessee drawl. Invariably the experts left these meetings shaking their heads and asking the same question: What is malaria?

Many other disease states were defined by the presence of antibodies against a particular microbe. A vague case of malaise, fever, and nausea could be ascribed to influenza if anti-flu antibodies were found in the ailing human's bloodstream. But here, too, Campbell saw difficulties with defining malaria, because of the strange and transient nature of immunity to the parasites.

When malarial parasites were injected through a female *Anopheles* proboscis into a human's bloodstream, they were in the form of small sporozoites. The sporozoites had a unique coating, bearing a set of specific protein antigens. At each stage in the parasites' life cycles thereafter— schizonts and merozoites—the organisms bore still another set of antigens on their cell membranes. Because the various stages of the parasite's growth presented different sets of antibodies, the human immune system was in a bind. It might make antibodies against one stage and not the others. Or it might mount a T-cell response against one, or none, of the stages.

The sporozoite stage of malaria elicited a particularly strange immune response: people could have millions of antibodies drifting through their bloodstreams, even attaching to the sporozoites, and still die of malaria. One key study involved Kenyan adults who were presumably "immune" to malaria because they were surviving constant exposure to the contaminating mosquitoes. The volunteers were given high doses of chloroquine and Fansidar (pyrimethamine/sulfadoxine) to rid their bodies of parasites. Their antisporozoite antibody levels were measured. And then the individuals were followed for ninety-eight days to see what transpired.

By the end of the study, 72 percent of the supposedly immune adults had become reinfected with *P. falciparum* parasites. In a similar study of small children, 100 percent were reinfected within ninety-eight days.[160] And there was *no* correlation between their likelihood of being reinfected and the levels of anti-malaria antibodies in their blood.[161]

Laboratory and animal studies showed that T-cell responses were critical to controlling sporozoites, particularly the classes of T cells called CD4.[162] The problem with such T-cell responses was that they were usually very specific, capable of recognizing a malaria enemy only if a very particular kind of antigen was present on the sporozoite surface. So narrow was the range of T-cell response that if a single amino acid building block in the proteins protruding from a sporozoite's surface differed, the T cells wouldn't recognize the malarial invasion and no effective immune response would be mounted. Such specificities were called epitopes, and each different malaria epitope could be coded for by a single gene—or even a tiny part of one gene—in the parasite's DNA. That meant that one tiny mutation might be enough to allow the malarial parasite to evade the human T-cell immune responses.[163]

As a result, a person whose T-cell system was able to control malarial parasites found in, say, Bujumbura in 1989 wouldn't be able to handle malaria in Kinshasa or Brazil or Thailand.[164] Indeed, if the individual left Bujumbura and returned twelve months later, his or her T-cell system might not recognize the strains then present in Burundi's capital city.[165]

These findings simply reinforced Kent Campbell's sense of perplexity. The parasites, despite the existence of proteins on their surfaces that clearly signaled alarms to the immune systems,[166] managed to elude all the host's defenses, and nothing short of extraordinary doses of antimalarial drugs could dislodge the microbes.

In the face of rising microbial resistance to the antimalarials, even the drugs were increasingly proving incapable of ridding human bodies of the parasites.

Oddly enough, as time wore on, Campbell reached the conclusion that it was a good thing, if one lived in an area with endemic malaria, to have some parasites in one's body at all times. In the absence of parasites there was soon no immunity whatsoever.[167] If some malarial sporozoites inhabited one's liver, or merozoites one's red blood cells, the body made IgG antibodies. With luck, enough antibodies and activated T cells were present to create a condition of tolerance: each creature tolerated the other. The parasites put up with constant onslaughts from the immune system, and the human managed, most of the time, to accept the infection of some number of red blood cells and hepatocytes.

Once again the dogged Campbell queried: "What is malaria?" If not this state of chronic infection, which occasionally got out of balance to produce fevers and chills, what exactly was the disease that WHO claimed appeared in over 300 million people a year by 1990, killing 3.5 million?

Campbell liked to remind fellow malariologists of the old British colonial scourge, blackwater fever. It was the British who figured out how to mix quinine with water and cover the bitterness with Bombay gin. The result —the gin and tonic—went a long way to prophylaxing the British Army against malaria. But overuse of quinine, both medicinally and in heavy gin-and-tonic doses, created the new disease of blackwater fever. Victims urinated dark fluid, ran high fevers, felt miserable, and often—25 to 50 percent of the time—died.

It was decades before British physicians figured out that blackwater fever was an iatrogenic disease. After repeated episodes of malaria, each of which was treated with escalating doses of quinine, the nonimmune Europeans fell ill with what they thought was another infectious disease. Scientists eventually determined that quinine caused blackwater fever. Overuse of the powerful drug to counteract malaria led to quinine's attachment directly onto the membranes of red blood cells. The protruding quinine molecule attracted the attention of the immune system, which misinterpreted its presence on the red blood cells as an indication of an alien invasion. Antibodies and T-cell killers attacked the quinine-labeled red blood cells, killing them. The dying blackwater fever patient was, thus, the victim of his own medical attempts to cure malaria.

In the early 1980s researchers tentatively demonstrated that chloroquine could also have a deleterious effect on the human immune system. Test-tube studies indicated that the drug could hamper the immune system's ability to recognize some threats and properly stimulate antibody response.[168]

In 1983 a U.S. Peace Corps volunteer died of rabies while working in Kenya, though the twenty-three-year-old had previously been vaccinated against the disease. Because she had been taking chloroquine prophylaxis for malaria for several months prior to her death, CDC researchers decided to see if the drug's dampening effect seen on immune system cells in test-tube studies was hampering similar activities in human beings.

In 1984 CDC scientists tested the immune systems of Peace Corps volunteers working in eight countries, comparing responses to rabies vaccines among chloroquine users with the responses of those not taking the drug. Chloroquine, taken in recommended prophylactic doses, clearly diminished volunteers' ability to make antibodies against rabies following vaccination. Three months after vaccination, the chloroquine users had about half the antibody response to rabies seen in Peace Corps volunteers who didn't take the drug.[169]

Though nobody knew exactly how chloroquine worked, in terms of either limiting malaria or dampening the immune response, Campbell was certain by 1990 that the drug had little or no effect directly on the parasites.[170]

"It treats the fever without directly attacking the parasites," Campbell said.[171] He concluded that the pediatric malarial anemia syndrome sweeping across Africa in the 1990s was iatrogenic, as was the AIDS epidemic in that age group that stemmed from the emergency blood transfusions given to the gravely ill children. The first was a disease that resulted from overuse of chloroquine, the second from use of HIV-infected blood.

The parasites developed partial resistance to the chloroquine, and the drug dampened the children's immune responses. Though chloroquine kept down the kids' fevers, it did nothing to slow red blood cell damage. The children's hemoglobin counts steadily descended despite all treatments, until so few viable hemoglobin molecules remained that the children's bodies were starving for oxygen.[172]

The anemia syndrome, Campbell concluded, was not malaria. It, just like blackwater fever, was a man-made disease that resulted from improper use of an antimalarial drug.

Improper, by the way, was defined as precisely the use patterns suggested by World Health Organization policy. For over two decades WHO told African health planners to train mothers living in endemic areas to assume that all fevers in small children were due to malaria and immediately treat the youngsters with chloroquine.

From all of this Campbell reached two conclusions:

"Life cannot be made into policy guidelines from Geneva," and "Malaria is a disease that responds to antimalarial drugs."

The definition of malaria, then, was that it was a disease whose existence was proven by reversing it with drug treatments.[173] At least in the African context, when the disease no longer responded to chloroquine treatment because of resistance, it was transformed into a different syndrome.

In 1992, having concluded that WHO policies were inadequate and the CDC's analysis better reflected their situation, scientists within Malawi's Ministry of Health recommended a change in policy. And Malawi became the first African nation to abandon chloroquine.

Fortunately, Africa still had options. Though resistances to other antimalarials had emerged, they were not widespread, and alternative, albeit considerably more expensive, medications were available.[174]

In southern Asia, however, where chloroquine resistance first appeared in the 1950s,[175] the malarial parasites were often multiply resistant to *all* the readily available drugs.

Though most Asian malaria was due to the less dangerous *P. vivax* parasite, the sheer density of parasite-carrying mosquito populations in tropical southern Asian areas was far greater than was seen in Africa with *Anopheles gambiae* and *P. falciparum.* In addition, many Asian nations had lively pharmaceutical black markets and/or over-the-counter sales of antimalarials as early as the 1950s. Finally, in some Asian regions both *P. falciparum* and *P. vivax* were present, creating a mixed malarial population.

It wasn't long after the first laboratory reports of chloroquine-resistant strains in Asia that treatment failures were correlated with the emerging mutants. In 1962, for example, three members of an American medical research team working in western Cambodia came down with severe malaria, despite taking prophylactic chloroquine. Analysis of the Cambodian *P. falciparum* strain, as well as a Malaysian strain, showed that they were resistant not only to chloroquine but also to pyrimethamine and proguanil.[176]

A team of National Institutes of Health researchers tested the drug-resistance capabilities of several Cambodian and Malaysian *P. falciparum* strains directly by injecting samples into prisoners confined in the federal penitentiary in Atlanta, Georgia. In this ethically questionable manner (due to the strong potential of coercion among alleged research volunteers in a research environment), the NIH scientists demonstrated that only one of the six Asian strains remained susceptible in 1963 to chloroquine and three strains were resistant to all four of the leading antimalarials of the day (chloroquine, proguanil, mepacrine, and pyrimethamine).[177]

A decade later—still five years before the first resistant parasites emerged in East Africa—chloroquine resistance was widespread in Thailand, Burma, Bangladesh, India, Indonesia, the Philippines, Cambodia, Sri Lanka, Malaysia, Vietnam, Australia, Laos, Japan, Singapore, Papua New Guinea, the Solomon Islands, Vanuatu, and China.[178]

To counter the trend toward nearly universal *P. falciparum* chloroquine resistance in Asia, WHO recommended the use of multiple-drug treatments. The thinking, which paralleled contemporary approaches to antibiotic resistance in bacteria, was that emerging resistance could be snuffed with simultaneous use of other drugs, one of which was sure to kill off the mutant strains.[179]

But it wasn't long before multiple resistance expanded in the *P. falciparum* parasites. In the wake of widespread social disruption, a mass refugee exodus, and a genocidal campaign conducted by the Khmer Rouge, a new malaria strain emerged in Cambodia that was strongly resistant to both chloroquine and Fansidar. The mutant strain struck a refugee camp located along the Thai-Cambodian border, causing widespread disease. The CDC responded by recommending a switch from standard chloroquine plus Fansidar treatments for malaria in the region to a combination of quinine and tetracycline.[180]

Resistance spread and grew stronger all over Asia throughout the 1980s at a pace that was staggering.[181]

During the same time period many of the *Anopheles* mosquito species that carried malarial parasites in Asia developed resistance to DDT, making insect control both more difficult and costly. Some of the insects expanded their territories, appearing in ecologies not previously thought to be suitable for their breeding and feeding.

Even more troubling, the ratio of *P. falciparum* to its less dangerous cousin *P. vivax* changed in many places between the mid-1970s and the late 1980s. In India, where over 90 percent of all malaria was the milder *P. vivax* form in 1976, by 1989 only 65 percent were *vivax*, the remainder *falciparum*. In Sri Lanka, where *falciparum* had been virtually nonexistent, by 1990 close to half of all disease was due to the more dangerous parasite. Burma saw the percentage of *falciparum* jump from 60 percent to more than 90 percent.[182]

One of the great tragedies was Nepal, which had been the success story of America's earlier efforts to eradicate malaria. Between 1950 and 1970, Nepal's malaria rate was reduced by an extraordinary 99 percent, from 2 million cases and 300,000 deaths per year to a mere 25,000 cases and fewer than 200 deaths. But by 1985 the country's malaria incidence had doubled, and mortality had increased due to parasite resistance. Similar patterns were seen throughout southern and western Asia.[183]

During the Vietnam War the U.S. Army invented mefloquine and in the early 1980s it was tested on civilians as an alternative to chloroquine. By 1987 mefloquine had supplanted most other antimalarials, becoming the drug of choice in much of Asia. It was a highly effective drug, offering minimal toxicity, at a time when no other agent appeared to guarantee protection against *P. falciparum* in Asia.[184]

But in 1986, along the Thai-Cambodia border, malarial strains resistant to mefloquine alone, and in combination with chloroquine and Fansidar, emerged. In addition, strains appeared in the crucial area that were less responsive to quinine, rendering some malaria cases untreatable. By 1990 the mefloquine cure rate in Thailand had plummeted from 98 percent to 71 percent. And the following year halofantrine, the only remaining marketed drug, *which had never even been used in the country*, was rendered close to useless in Thailand by virtue of *P. falciparum* resistance.[185]

When the new mefloquine-resistant *P. falciparum* strain emerged, Cambodia was in a state of civil war, a 16,000-strong United Nations peacekeeping force was poised to enter the area, and 360,000 refugees bivouacked across the border in Thailand were scheduled for imminent repatriation. It was an opportune moment for the microbe.

The new strain was resistant to chloroquine, Fansidar, mefloquine, and their chemical cousins. That left only two available drugs, quinine and tetracycline. Neither was ideal. Quinine was a poor prophylactic drug, tetracycline a weak treatment.

By March 1992, WHO estimated that more than half of all malaria cases in Cambodia involved the new strain, and control seemed impossible because years of civil war had left the country's public health system in a state of ruin. The National Cambodian Malaria Control Office, such as it was, estimated that some 10,000 people died of the new malaria in that country in 1991. But officials conceded it was just a conservative guess.

At her laboratory in the Harvard School of Public Health, Dyann Wirth worked at a feverish pace in 1992–93, trying to understand the nature of the mutation in the new strain and find a way to defeat it. She concluded that the parasite had produced a large, unique protein that nestled in its membrane. When the drugs entered the parasite's environment, the protein acted as a pump, shunting the chemicals out of the *P. falciparum*. Though such a mechanism had previously been seen for chloroquine resistance, Wirth and several other scientists were convinced that the new strain had a pump that evicted nearly all the antimalarial drugs.[186]

Evidence that such a pump existed was strong. A heart disease drug called verapamil, which blocked calcium pumping across cellular membranes, could reverse drug resistance. Some scientists urged WHO to release drugs that combined verapamil and chloroquine: one drug would shut down the pumps, the other would stop the parasite. Researchers saw evidence of such pumps in test-tube studies: when they compared resistant malaria to nonresistant, the resistant strains had forty to fifty times more chloroquine on the *outside* of the parasite.

The pump mechanism was genetically controlled by at least two so-called *mdr* (multidrug-resistance) genes. No one knew how the malaria parasites got *mdr* genes—such genes had previously been seen operating in mammalian cancer cells, pumping out chemotherapy drugs. Nevertheless, Wirth was convinced that *mdr* genes not only were present in the super-resistant bugs but were amplified so that the parasite made huge numbers of the pumps.

"Once this kind of mechanism occurs, it means resistance will emerge even before the drug can be invented," Wirth said.

It was tempting to conclude that the pump mechanism explained why the pace of resistance had so accelerated. Though chloroquine resistance first emerged in the 1950s, it and most of the other early antimalarials remained effective worldwide for decades. But in the 1980s resistance emerged at an extraordinary pace, seemingly from the moment drugs were introduced in some Asian areas, particularly Thailand and Cambodia. Most early mutant *P. falciparum* strains resulted from apparently random point mutations, and those strains were resistant to one drug at a time. But by the end of the 1980s, Indochina seemed to be awash in multidrug-resistant parasites.

Perhaps, malariologists whispered nervously, the presence of a pump mechanism provided the parasites with a way to quickly outwit new drugs, by fine-tuning their pumps to adapt to each new agent. If that were so, the resistance trend would only worsen, and accelerate, wherever in the world the malaria parasites possessed such pumps.[187]

In 1989, after forty years of effective trouble-free use, physicians treating people in Papua New Guinea who were infected with *P. vivax* parasites noticed that chloroquine no longer cured that type of malaria.[188] Though resistance problems had been apparent with *P. falciparum* almost from the beginning of the chloroquine revolution, *P. vivax* had always remained vulnerable to the drug.

Malariologists had hoped that the Papua New Guinea *P. vivax* cases were nothing more than an odd fluke—perhaps even a failure due to improper treatment—but in 1993 chloroquine-resistant *P. vivax* appeared in Indonesia (Irian Jaya).[189] Because *P. vivax* had a more complicated life cycle inside humans, and spent far longer in the liver, the parasite's vulnerability to drugs differed from that of *P. falciparum*.

"There is no obvious replacement for chloroquine," researchers said.[190]

"We are in a crisis," declared WHO's parasitic disease expert Tore Godal. "The situation is truly alarming."

Only one alternative drug remained. For over two thousand years Chinese herbalists had treated malaria with extracts from the sweet wormwood, or

Artemisia annua, plant—called qinghao in Chinese. The plant's chemicals had long been used to bring down fevers. In 1972 Chinese scientists succeeded in isolating the responsible chemical in qinghao, giving it the name qinghaosu (or, alternatively, arteether and artemether). In the 1980s, as they witnessed the rapid rise of drug resistance throughout the malarial world, WHO and the Walter Reed Army Institute of Research teamed up to conduct studies of the drug. And in 1994, with much fanfare, WHO announced completion of successful field trials of the drug in Vietnam.

WHO possessed the patent. And no drug company was interested. Critics wondered why WHO was embarking on such a mammoth project, given that synthesized slightly altered versions of the chemical were in development elsewhere.[191]

Meanwhile, WHO officials spoke candidly about the need to "protect" the Chinese drug from the social conditions in Southeast Asia that led to the downfalls of chloroquine, halofantrine, mefloquine, quinine, Fansidar, proguanil, and every other antimalarial. But in late 1993 a French traveler picked up a strain of *P. falciparum* in Mali, West Africa, that was resistant to *everything*, including the new Chinese drug. Researchers discovered four strains of *P. falciparum* in Mali that were resistant to qinghaosu, though the drug wasn't widely available in the country.[192]

Insiders like Brinkmann and Campbell were skeptical. They had seen the conditions in Asian malarial regions up close, understood how the mosquitoes and parasites spread, and doubted that without strong political will in key countries any drug could be protected.

Political will was in short supply. And biology was working against public health.

Ten percent of Southeast Asia was rain forest, involving sixteen ecologically distinct types of forests. Malaria in the region was *forest* malaria, which meant that standard mosquito control measures were ineffective. How could one spray DDT in a wet, humid, dense tropical jungle?

At least thirty major species of mosquitoes carried malaria in Asia, many of which fed on a range of other animals as well as humans. Among them, these mosquitoes had adapted to breed and feed in every possible Asian forest context: a bamboo stump filled with rainwater, an irrigation canal, jungle pools, puddles of muddy water left by the feet of marching soldiers, elephant footprints, wheel ruts, rice paddies, lagoons.

These sturdy insects spanned most elevations of southern Asia. And they fed on people at all sorts of times of day and night. Many were resistant to the key pesticides, and most were "wild mosquitoes," meaning that they stayed away from open spaces and human habitations, preferring the safety of dense tropical foliage.

People who lived or worked in the forested areas were constantly bitten by mosquitoes. For centuries an ecological balance existed between the humans and the parasites, via the mosquitoes. A large percentage of the humans would die of the disease during infancy, but survivors, who were "vaccinated" every day by mosquitoes that injected parasites into their blood, were immune, or, as Kent Campbell would put it, tolerant.

Efforts to eradicate malaria severely disrupted that balance. Temporarily successful mosquito control programs eliminated the daily "vaccinations," and immunity immediately disappeared. Prophylactic use of antimalarials fended off disease, but also lowered immunity. In periodic times of drug scarcity, surges of malaria cases could be seen.

The female mosquitoes, which fed voraciously on a range of creatures (from reptiles to humans), absorbed all kinds of microbes from animal blood. Different strains of malarial parasites co-inhabited the insect's gut, and there was evidence of genetic exchange and shuffling occurring between the various microbes inside the mosquitoes.

Entomologists felt certain that the roughly thirty identified species of malaria-carrying mosquitoes represented only a small percentage of all the Asian forest insects that were capable of serving as vectors for the parasites. It was, however, extremely difficult to study and taxonomically identify insects in such densely forested ecologies.

"Malaria is an ecological disease," wrote Indian scientist V. P. Sharma.[193]

Human beings in these regions were highly mobile in the 1970s, 1980s, and 1990s. Warfare and civil strife, religious persecution, economic necessity, and natural disasters prompted tens of millions of families and individual laborers to migrate within countries and between nations. More than half a million Cambodians alone were refugees during the 1980s. The Indonesian government transplanted nearly seven million people in 1990 to colonize forested outer islands.

With this mass movement came great risk for malaria. Most of the migrating humans either came from nonmalarious regions and had no immunity or were moving between areas inhabited by distinctly different strains or species of parasites. When a concentration of such immune-naïve *Homo sapiens* settled alongside a forest area, the mosquito population swelled and malaria was soon rampant.

Warfare and civil strife, such as the Vietnam War or the long Khmer Rouge insurgency in Cambodia, not only produced mass human migrations but directly disrupted the ecology in ways that were advantageous to the mosquitoes. Rain-filled bomb craters, abandoned water-soaked military vehicles, and such leavings of war created ideal breeding sites for insects. As Asia's human population exploded in the 1980s, desperate people pushed into forest lands, chopped and burned their way to the creation of farmlands. Public health and medical systems were nonexistent in much of Asia's forested area because the human inhabitants were usually poor, often migratory, and increasingly resided in areas not previously inhabited by people.

In many regions flagrant overuse of antimalarial drugs resulted, as adults and children alike swallowed whatever they could afford in an attempt to protect or cure themselves. Poorly trained paramedics widely dispersed drugs to anyone who was suspected of having malaria.

"I have never seen such a low level of health infrastructure, even in Africa," Ethiopian malariologist Awash Teklehaimont, a scientific consultant to Indochina for WHO, said. "Chloroquine injections are done openly, in the marketplaces, by quacks, under full view of the police," he added, referring to Cambodia in 1992. With local physicians paid only five dollars a month, it was perhaps unremarkable that the shelves of government clinics seemed always to be empty, while the black market had no shortage of supplies.

If there was a single Asian focus of all this social/ecological/medical interaction, a place where resistant strains most often appeared, it was the gem-mining area straddling the Thai-Cambodian border. On the Thai side were squadrons of underpaid police and soldiers, anxious to look the other way when fortune seekers illegally entered the area, and equally eager to grab them as they exited, taking a percentage of the ruby and emerald harvest. On the Cambodian side was the Khmer Rouge army of Pol Pot, which exacted their percentage from the gem miners to support continued insurgency.

As word of the lawless access to fortunes spread during the 1980s, men poured into the area from all over Asia: Indians, Burmese, Thais, Cambodians, Lao, Vietnamese, Chinese. They moved surreptitiously, avoiding border patrols, police, soldiers, and, of course, health authorities. The area they moved into was one of dense rain forest inhabited by more than a dozen different species of wild, *falciparum*-carrying mosquitoes.

"It's a remarkable situation," Uwe Brinkmann said, having spent time observing the gem miners. "All day long they sit or squat in the streams and rivers zigzagging through the rain forest. It's steaming hot and humid—you can't imagine the heat. They wear no protective clothing, and they stand in mosquito breeding areas all day long sifting the water and mud for gems. At night they sleep in open sheds."

It was these fortune seekers who proved to be the best customers for the antimalarial black market. They purchased anything, and used anything, to keep the disease at bay. Most of the time they used drugs improperly, encouraging development of resistant strains.[194]

Not surprisingly, it was there that multidrug resistance emerged. In 1983 the combined use of mefloquine and Fansidar cured 96.7 percent of all malaria in the gem-mining area. By 1990 the same drug combination cured less than 21 percent of all cases. In practical terms, malaria acquired in the region was incurable.

Men who contracted malaria in the area but weren't too sick to travel did their best to sneak past layers of police, armies, and border guards to get home, clutching whatever riches they had sifted from the Khmer streams. But it was tragedy that they carried with them, for in their bloodstreams lurked resistant parasites that were soon sucked up into the probosci of feeding mosquitoes from Bangladesh to Nepal. And so the epidemic spread.

Despite expenditures of billions of dollars by governments,[195] the UN, and numerous Western agencies, malaria was completely out of control in Asia in 1994.

In 1977, WHO finally abandoned all hopes of eradicating malaria. In 1978 it outlined a global strategy that linked malaria *control* to primary health care. But in the absence of adequate primary health systems in most of the affected area, that policy, too, failed.

By 1992, WHO was forced, reluctantly, to admit that there was no *global* strategy for malaria control. Rather, every individual ecology in each endemic nation needed to develop its own environmentally and socially tailored plan of action. What might work in an African savanna certainly would not be effective in a swampland or an Asian mahogany forest.

WHO had discovered ecology.

Weary of failure and angry about corruption in malaria vaccine development efforts,[196] the U.S. government moved in 1993 to slash its financial commitment to malaria control efforts. Federal expenditures declined steadily between 1987 and 1990, and in the winter of 1993 two agencies of the government were at loggerheads over whether or not to completely cease funding overseas malaria programs.[197]

When Teklehaimont viewed the crisis in southern Asia he couldn't help but worry about his home, Ethiopia. In 1992 Ethiopia experienced its worst malaria epidemic in *Homo sapiens* history, with more than 20,000 people killed by the parasites in less than six months. At least 10 percent of the cases were chloroquine-resistant, and the victims were of all ages. When Teklehaimont personally surveyed households in an area of 13,000 people, he found 759 dead.

And that terrible epidemic was in the face of *only* chloroquine resistance.

What will happen if the Cambodian multiresistant parasite gets to Africa? Teklehaimont wondered.

14

Thirdworldization

THE INTERACTIONS OF POVERTY, POOR HOUSING, AND SOCIAL DESPAIR WITH DISEASE

The States Parties to this Constitution declare, in conformity with the charter of the United Nations, that the following principles are basic to the happiness, harmonious relations and security of all peoples:
—Health is a state of complete physical, mental, and social well-being and not merely the absence of disease or infirmity. . . .
—Unequal development in different countries in the promotion of health and control of disease, especially communicable disease, is a common danger.

— *Constitution of the World Health Organization*, July 22, 1946

■ Heavy, purple-tinged clouds filled the equatorial sky, blotting out the harsh noon sun. It was stifling hot, and the air was so moist that beads of condensation mixed with sweat on the skin. Three men struggled to push a bicycle uphill along a road made of thick clay mud, rutted deeply from the two or three vehicles that had passed from Bukoba, bound for Uganda, since yesterday.

Wrapped in white cloth and elephant grass, a five-foot bundle lay stretched across the handlebars. The somber trio maneuvered their way past a steady stream of pedestrians, most of whom bore enormous bundles upon their heads or carried a huge Nile perch dangling from their shoulders, the fish so massive that its mud-covered tail trailed along the road.

Each time the men hit a large rut, one of them carefully steadied the bundle, while the other two gave the bicycle a strong shove. Passersby, recognizing the nature of the bundle, carefully avoided staring and ceased their laughter or chatter. Even the wild young boys who dodged school and helped smugglers get their goods across the Ugandan border grew silent when they spotted the bicycle's burden.

The journey eased when the men reached a plateau and turned off the road onto a well-beaten footpath. Winding their way through dense, verdant banana groves, they occasionally passed a mud-and-thatch home. Residents greeted them with nods or a quiet *"Jambo,"* children scurried to their mothers' sides, staring wide-eyed at the bicycle and its load. As the trio moved on, clusters of people gathered up specially wrapped bundles, called children to their sides, and fell in line. Soon a procession of a few dozen residents of Kanyigo had formed.

In the distance could be heard the high-pitched ululating of female voices. The procession drew near to the mournful sound and a child stationed along the path spotted the bicycle and ran ahead to alert others of its approach. The keening suddenly stopped, and for a moment the only sounds were those of squawking Lake Victoria birds and human feet tromping over mud.

The villagers of Kanyigo reached a small clearing, surrounded by banana trees. To one side was a round thatched-roof house. On the opposite end of the clearing a group of men took turns shoveling out a large hole in the clay soil. In the center of the clearing stood a thirty-five-year-old man wearing a button-down shirt, dark cotton pants, and a brightly colored print sash. As the bicycle trio approached, the man drew close to him five small children, aged two to seven years, each of whom wore sashes identical to his.

Without exchanging words, the trio greeted the man and his children, silently untied their elephant grass-wrapped bundle, and carried it into the hut. As they entered the home, the women's wailing resumed inside. Its volume and pitch were at first painful to hear, and some of the gathered children, unfamiliar with social propriety, cupped their hands over their ears. Mothers quietly clucked disapproval; the children obediently dropped their hands and stared with apparent fear at their sash-adorned counterparts.

The families took turns approaching the father and children. Whispered greetings, bowed heads, proffered gifts, some tears, an occasional hug.

Some of the adults stepped into the hut, stopping for a moment until their eyes adjusted to the darkness, and then groped their way through the crowded one-room home to a seat upon the clean floor of packed dirt. They sat in concentric circles surrounding the five-foot bundle that had quietly been set in place by the bicycle trio. An older woman occasionally lost control, wailing loudly and flailing about so wildly that her friends were forced to restrain her.

AIDS had claimed another life in Kanyigo. The thirty-two-year-old woman, who now lay upon her floor, swaddled in cloth and elephant grass, left behind a husband and five children.

"She was suddenly attacked by stomach pain four months ago," the widower said. "So she went to her birth village, the next village over, to stay with her family. She had no appetite. She wasn't eating anything. We tried to force her to drink tea, eat bread. We really tried to force it on her. But it was no use. At eleven o'clock yesterday morning she collapsed and died."

The man spoke in a monotone, too overwhelmed to express emotion. He looked down at his children, who stoically stood by his side, stifling their tears. His eyes swept over them and then settled on the visiting *Mzungu*. He studied the American for an instant before speaking.

"It is a great deal of work for me to feed them, care for them, and do my work. Why don't you take the children? I give them to you."

I

Jonathan Mann was tremendously excited. True, there were any number of things that could still go awry; diplomatic noses might start bleeding, political shenanigans could well break out. But he and his highly energetic—and sly—staff of the World Health Organization's Global Programme on AIDS had for months carefully and strategically planned for this day.

"We are entering a new era," Mann had assured an international press corps. "We will make 1988 the year we turn the tide against the AIDS virus."

And here he sat, his bow tie straight, hair brushed, as usual, straight back off his forehead, wearing a natty tailored European suit, giving him the air not of a CDC epidemiologist but of a French diplomat. He looked out over the largest gathering of Ministers of Health ever assembled. Of the representatives of 148 nations who now sat before him in the vast Queen Elizabeth II Conference Center in London, 117 were Ministers of Health or their country's equivalent. Every key nation, save one, was represented by the most politically powerful health official in their land: Mann was ashamed to say that the exception was his own country. Still not wishing to give AIDS a priority status, the Reagan administration sent Dr. Robert Windom, who ranked two notches down the power ladder from the Secretary of Health and Human Services, Otis Bowen.

Never in history had the majority of the world's top health officials gathered to discuss an epidemic. No scourge—not malaria, smallpox, yellow fever, or the plague—had ever commanded such diplomatic attention. Some 700 delegates and 400 journalists were also present in the London hall on this ice-cold January morning in 1988 to witness the World Summit of Ministers of Health on Programmes for AIDS Prevention. Mann felt that it was a coup for his program, for WHO, and for millions of powerless people with AIDS.

Mann urgently hoped to drive home a message to the world's health leadership: AIDS is spreading; if it hasn't yet emerged in your country, it will, unless you plan now, follow our recommendations, educate your populations, and embrace condom-based programs as a prevention strategy.

As of January 26, 1988, some 75,392 cases of AIDS had officially been reported to the World Health Organization. But that figure was a gross understatement of the true dimensions of the pandemic: most nations lacked genuine systems for amassing and recording such health statistics. Mann tactfully didn't mention from the podium what everyone in the audience knew to be true; namely, that many nations were deliberately covering up their epidemics for political and economic reasons. Such delicate issues would be dealt with later, in private arm-twistings and minister-to-minister preplanned strategic confrontations.

Mann differentiated the ways in which the AIDS virus was spreading from person to person. In what he called Pattern I countries, such as those of North America and Western Europe, AIDS was spreading primarily via the sharing of needles between intravenous drug users and sexually among gay men. In Pattern II countries, such as those of Africa, AIDS was a heterosexual disease.

Though he was cautious in his public choice of words, it was Pattern III nations that most concerned Mann as he spoke in London. Asia, the communist bloc, the largely Muslim Middle East, and much of the Pacific region had only very tiny outbreaks of AIDS. Some of these countries were truthfully reporting no cases of the disease, and several more were accurately stating that the handful of cases in their countries all involved foreigners or citizens who had acquired HIV while living overseas. In those Pattern III countries, the relative handfuls of cases were equally likely to have resulted from heterosexual, homosexual, needle, or blood exposure.

Pattern III, in other words, represented the potential future of the worldwide AIDS epidemic. There was still a window of opportunity for public health action that might successfully prevent HIV from emerging in the majority of the world's populations.

Many of the Pattern III political leaders had already recognized the threat of HIV importation, of course, and taken their own steps to curb such events. However, Mann and his staff, which included smallpox hero Daniel Tarantola, were appalled by many of the anti-emergence measures some countries had taken. Privately, Tarantola had already spent months flying all over the world in attempts to convince many of the same ministers who now sat in the London conference hall that AIDS wasn't anything like smallpox. There was no vaccine that one could require that immigrants and visitors receive. The virus didn't manifest itself symptomatically for years—perhaps over a decade—in ways that indicated its presence even to the infected individual. And the AIDS blood test wasn't foolproof.

"What are you going to do, test every immigrant five or six times a year, every visiting student once a week? If you think you can keep the virus out of your country with legislation and testing, you are wrong," Tarantola told public health officials.

Mann was worried that the world would become a patchwork of repressive public health regimes with laws aimed at keeping a virus, as well as its potential carriers—gays, Africans, prostitutes, drug users, poor immigrants—out. He feared that it would push populations that already existed at the margins of global society further away from the mainstream, medicine, and all hopes of disease control. Indeed, restrictions intended to control populations at greatest danger for HIV infection might actually have the reverse effect, exacerbating the social and economic conditions in their lives that drove them to adopt risky behaviors. Simply put, he felt certain that this moment in London was pivotal to deciding whether HIV's emergence in most countries would be prevented through education of local populaces or temporarily stalled by repressive laws.

"Our opportunity—brought so clearly into focus by this Summit—is truly historic," Mann told his distinguished audience. "We live in a world threatened by unlimited destructive force, yet we share a vision of creative potential—personal, national, and international. The dream is not new—but the circumstances and the opportunity are of our time alone. The global AIDS problem speaks eloquently of the need for communication, for sharing of information and experience, and for mutual support; AIDS shows us once again that silence, exclusion, and isolation—of individuals, groups, or nations—creates a danger for us all."

Though his words were received with thunderous applause and a standing ovation, Mann knew that many of those before him who were loudly slapping their hands together and politely nodding approval were, back home, promoting policies of mandatory quarantine of HIV-positive individuals, escalated repression against homosexuals, even public execution of AIDS sufferers.

As a scientist, Mann knew that the men and women now looking up at him on the dais, studying his smile and careful public modesty, were People of Politics. They might wear the titles of health officials, but their modi operandi were less those of the laboratory or hospital than those of the maneuvering, backstabbing, and power plays seen in parliaments and presidential inner circles. What the ministers said publicly here in London would be at least as much for domestic consumption as for the sake of any global effort to stop the pandemic.

Anticipating such limitations, Mann and his Global Programme on AIDS (GPA) staff had toiled for months in preparation for this moment. Lifelong WHO veterans, and occasional renegades, Tarantola and Manuel Carballo showed Mann how to maneuver around the labyrinthine and often byzantine United Nations bureaucracy. Swiss-American Tom Netter, having spent years covering the rise of Solidarity and the fall of communism in Poland for the Associated Press, plotted every step of the GPA's interactions with the international media. Spanish-born Carballo, who knew every nook and cranny of the World Health Organization even better than WHO Director-General Halfdan Mahler, helped spot the few potentially influential individuals within the bureaucracy who understood the urgency of the AIDS epidemic.

"This is a place where people put URGENT! on requests for pencil supplies," Mann said in wonder. "The concept of genuinely dire emergency has almost no meaning here."

Carballo couldn't have agreed more. One of the happiest days of his life was when he joined the GPA staff. He felt charged up, at the top of his performance and truly impassioned about his work, possibly for the first time in his life.

They all did: American epidemiologists Jim Chin and David Heymann, Venezuelan biologist José Esparza, British public health expert Roy Widdus, Tarantola, Mann, and the dozens of scientists and public education experts who came to Geneva under special contracts to advise the GPA. They shared a mission: stopping the further spread of AIDS. And as Heymann and Tarantola had done before in their efforts to stop smallpox, these men were willing to bend every UN and WHO rule as far as possible to stop the pandemic. They were believers. Between them they shared the ability to write and converse in at least fifteen languages. And they had a camaraderie that was quite uncharacteristic of the usually opportunistic careerist atmosphere pervading most United Nations programs.

When Mann had originally left Kinshasa to take the reins of power in Geneva in November 1986, he had a total working budget of $5 million, a part-time secretary, and three epidemiologists who were borrowed from other programs. Mann's own salary was still paid by the U.S. Centers for Disease Control.

By the time he reached London for the January 1988 Summit, less than two years later, forty-year-old Mann commanded a far-flung AIDS program, a considerable staff, and a budget of over $50 million, with $92 million promised for 1989. It was, by WHO bureaucratic standards, a meteoric rise.

And none of it went unremarked by Mann's WHO peers, who headed other disease programs. With the envy of Cinderella's stepsisters, they watched as the cinder maid grabbed all the attention and the Prince's love at the ball.

That Mann had unique access to Director-General Mahler and could enter the chief's office without first passing through the usual rungs of intermediary power was noticed. That Mahler increasingly mentioned AIDS in his speeches, placing it with each oration higher on the WHO totem pole of priorities, was noticed. That U.S., Canadian, and Western European currency poured into Geneva specifically earmarked for the GPA was noticed. That Dr. Jonathan Mann, this Johnny-come-lately international bureaucrat, was almost daily gracing the front pages of leading newspapers and magazines from Tokyo to Casablanca was inescapable.

While Mann, Tarantola, Heymann, Carballo, and the rest of the AIDS staff did their best to create a highly publicized sense of worldwide emergency and mobilization, jealousy simmered in the hallways of the vast Geneva complex. In the enormous vaulted lobby of WHO headquarters, experts on cholera, malaria, diarrheal diseases, schistosomiasis, health economics, polio, and vaccine development gathered in discreet clusters by the three-story-high glass wall that afforded a view of Lake Geneva and Mont Blanc. And they whispered. They cited the Programme's own statistics on AIDS—those modest numbers of underreported cases—and asked why the new disease should command such resources and attention when other microbes were killing tens of millions of people. They noted that Mann and some other Programme staffers were Americans, and assured one another that all the concern was *only* in place because AIDS was killing homosexuals in New York and San Francisco.

And even that increased their envy: they admired the skills and energy of the American and European gay activists who relentlessly lobbied WHO, knowing that cholera victims in Bangladesh or Cambodian malaria patients would never be able to mount similar campaigns on their behalf.

Mann and his team either were oblivious to the talk behind their backs or chose to ignore it. In either case, when questioned directly about comparisons between WHO commitments to, for example, malaria versus AIDS, the Programme group would say that all global health programs were underfunded and not one dollar or yen of AIDS monies should be gathered at the expense of other health efforts.

And they would politely remind critics that AIDS was a newly emerging epidemic which, by definition, would swell to claim tens of millions of lives if not stopped immediately. On that point Mann enjoyed the full support of the director-general.[1]

The staff of the Global Programme on AIDS discussed quite consciously among themselves the inherent contradictions in the need for a state of emergency to halt a newly emerging disease versus the essential nature of WHO and the United Nations system. Though Ebola, Marburg, Lassa, and other emergencies had received the quick attention of WHO, they couldn't serve as models for action against AIDS. First of all, each had surfaced as seemingly confined local emergencies. Second, they, at least in part, burned out on their own. Third, the microbes caused almost immediate disease in those who were infected, with an alarming level of mortality; there could be no doubt to the populaces or their governments that a state of emergency was warranted. Fourth, fairly simple measures, such as provision of sterile syringes, could stop the primary spread of the diseases.

In contrast, HIV surfaced almost simultaneously on three continents and was quickly a feature on the health horizons of at least twenty different nations. Not only was there no sign that AIDS might burn out on its own; scientists could see no evidence of the famous bell-shaped curve of infection and disease.[2] Far from causing immediate disease and death, HIV was a slow burner that hid deep inside people's lymph nodes, often for over a decade, before producing detectable infections. As a result, a society could already have thousands of infected citizens before any sound of alarm was rung, and even when the first AIDS cases appeared, their numbers were small enough to allow governments to feel comfortable about ignoring the seemingly trifling problem. Denial was all too easy a response to AIDS.

Furthermore, no facile measures could be taken by a government to bring AIDS to a halt. Unlike Ebola, Marburg, drug-resistant cerebral malaria, or Lassa, HIV hit specific social targets. It was a sexual disease. It was associated with homosexuality, promiscuity, and drug abuse. It pitted public health against organized religion and the moral pillars of society.

It was, in short, easy to ignore and uncomfortable to confront.

The World Health Organization, acutely aware of the unsettling aspects of AIDS, initially chose the first route—ignoring the emerging disease.[3] From 1981 to late 1986 barely a whisper about AIDS emanated from Geneva. By the time Mann and his crew started sounding every alarm they could get their hands on, HIV had successfully emerged and reached full-fledged epidemic status in all the major cities of North America and Western Europe, as well as most of the urban centers of sub-Saharan Africa.

The GPA group felt justified in, figuratively speaking, yelling about AIDS at the top of their lungs.

But yelling, figuratively or literally, simply wasn't done inside the World Health Organization. Mahler might approve, but his underlings, and officials elsewhere in the UN system, did not. Indeed, in the entire United Nations system, "yelling" was the exclusive right of the General Assembly and the Security Council. The peripheral UN agencies were *intended* to plod.[4]

Rubbing against the bureaucratic grain, the Programme staff group moved with both haste and deliberation. They decided on a strategy for control of AIDS in which vaccine and drug research efforts, already underway in key wealthy nations, received the Programme's encouragement but not significant emphasis. With no cure in sight, the Programme's best focus, they felt, was on prevention of further spread of HIV. Though details would come much later, during 1987 the GPA outlined the need for national AIDS programs in every country—programs that would coordinate mass education campaigns about the disease. Societal awareness was the first step—that was Tarantola's job. To prevent further spread of HIV it was crucial that every nation's blood banks be free of HIV, sterile syringes had to be available to health providers, and people who were infected with HIV had to be counseled carefully so that they wouldn't pass their virus on to others. Counseling was Carballo's job. Anti-AIDS programs had to be coordinated not only within countries but worldwide.

And perhaps most important in Mann's mind was the need to eliminate the atmosphere of discrimination and prejudice that surrounded every aspect of AIDS.

"Discrimination simply drives AIDS underground," Mann repeatedly asserted. "The epidemic doesn't go away, it simply becomes harder to see, more alienated from public health. If you drive it underground, you guarantee its spread."

With those vague principles in mind, the Programme targeted sequentially each of the international bodies whose support the GPA staff felt was crucial. On May 15, 1987, the Fortieth World Health Assembly, the legislative body of WHO, passed the Global Strategy for the Prevention and Control of AIDS,[5] endorsing the strategy of the Global Programme on AIDS.[6] That gave the Programme its mandate, power, and seal of approval. In subsequent months Mann and his team sewed up further political support by gaining formal endorsements at the Venice Summit of the European Economic Community and the Economic Council of the UN. And on October 26, 1987, Jonathan Mann did something no WHO functionary at his level had ever done: he addressed the General Assembly of the United Nations. For the first time in its history the UN passed a resolution on a specific disease, formally endorsing WHO's leadership in the war against AIDS.[7]

Over the next three months the Programme staff carefully prepared for the London Summit, further detailing its strategy for control of the emerging pandemic, collecting data on the epidemic's scope, and carefully monitoring the AIDS-related activities of governments around the world. Though they loudly decried all attempts to keep AIDS at bay through legislation against HIV-positive individuals or members of social groups considered at greatest risk for exposure to the virus, the GPA members watched helplessly as eighty-one nations passed such laws.[8] As the New Year and the London Summit approached, at least ten more governments were debating passage of similar legislation and the international mood was growing ugly.[9]

In the Middle East tough laws in some Islamic countries passed in 1986–87 made failure to submit to HIV tests and "promiscuous" behavior punishable by imprisonment.

In Western Europe the EEC repeatedly condemned all efforts to legislatively restrict the travels, employment, or reasonable freedoms of people infected with HIV. Nevertheless, laws and condemnations were forthcoming.

Just four days before the opening of the London Summit of Ministers of Health, on January 22, 1988, American Gene Meyer was forcibly detained by British authorities when he attempted to enter the U.K. It was the second time Meyer had faced such problems with British authorities: in September 1987 immigration officers at London's Gatwick Airport, surmising that Meyer was a homosexual, read his diaries, saw references to medical tests, assumed he had AIDS, and designated the man "medically undesirable." Meyer was eventually permitted entry on January 22, 1988, when a very embarrassed Ministry of Health intervened, countermanding immigration officials.

There were contradictions to EEC cries of openness in other Western European government efforts, particularly directed against Africans. Belgium, West Germany, Greece, Finland, and Spain all passed new legislation, or interpreted preexisting public health law, to permit expulsion or visa denial to HIV-positive foreigners who were seeking work permits or student credentials. In practice, these regulations were primarily directed against Africans and, in the case of Germany, Turks.

Germany offered a unique set of challenges to the Global Programme on AIDS and the rest of the European Community. On the one hand, Germany was one of the first countries to respond to AIDS with a national education campaign, distributing 27 million leaflets on the disease during 1985 and promoting condom use.

But a gay, retired U.S. Army sergeant living in Nuremberg was arrested by German authorities in February 1987 and charged with knowingly spreading the disease to his sexual partners. He was ultimately convicted and sentenced to four years' imprisonment. A German homosexual was shortly thereafter brought up on similar charges. As AIDS panic in Bavaria increased, and public support for the arrests was loud and clear, Bavaria's Interior Minister, August Lang, announced that all prostitutes, civil service job applicants, drug addicts, immigrants, prisoners, and foreigners applying for extended residency permits would be required to undergo HIV blood tests. Days later the Bavarian city of Munich announced plans to dismiss all HIV-positive civil servants.

The Bavarian actions received a surprising amount of support from German residents of other, typically more liberal states. Federal Interior Minister Friedrich Zimmermann was prompted to order the nation's border patrol to deny entry to all foreigners who carried the AIDS virus.[10] The European Community was outraged. Officials denounced the German actions as clear violations of Community principles of freedom of movement on the continent.

In November 1987 the president of the German Federal Court of Justice, Gerd Pfeiffer, announced that in the absence of an HIV vaccine it might soon prove necessary to tattoo and quarantine people who were infected with the virus. The last time Germany had carried out tattoo-and-quarantine measures on its residents was during World War II, when "misfits," Jews, and other "undesirables" were placed in concentration camps and exterminated. Not surprisingly, Judge Pfeiffer's pronouncement sent shock waves throughout the world.[11]

The Eastern bloc and the Soviet Union posed special difficulties for Mann and his colleagues, because, in general, the communist states claimed not to have much—or any—AIDS, and they wanted to keep it that way.

The Soviet Union, after long denying that it had any indigenous AIDS cases, issued fiats in late 1987 requiring testing of most foreigners and giving the KGB and the police powers to order HIV tests—refusal punishable by imprisonment—on its citizens.[12]

Elsewhere in the communist world two nations clearly stood out: Cuba and China.

No nation on earth had ordered as broad a sweep of AIDS regulations as had Cuba. Between March 1986 and January 1988 the government conducted 1,534,993 HIV tests, according to the Ministry of Health, and the intention was to test every citizen and nontourist visitor to the country, or 10.4 million people.[13]

By January 1988 the Cuban government had identified 174 HIV-positive individuals and placed them under lifetime quarantine. Several of the infected people were recently returned veterans of the Angolan civil war, in which Cuban military advisers played a pivotal role in defense of the Luanda government. More than 300,000 Cubans returned from Angola between 1975 and 1987; HIV-1 was clearly present and causing AIDS in Angola at least as early as 1983.

In the People's Republic of China there were also practices underway that troubled the GPA. Beginning in December 1986, the Chinese government instituted mandatory testing for all foreign students: in reality, the edict was carried out with greatest vigor on Americans and Africans.[14] Students who failed to comply with the tests were barred entry or deported. The mandatory testing list had expanded by 1987 to include all foreigners who wished to stay in China for more than a year and all Chinese citizens returning from overseas.

By the time the world's Health Ministers gathered in London for their AIDS Summit, China had already tested more than 10,000 foreign students, 20,000 returning Chinese students, and thousands of foreign businessmen: all in a period of less than four months. In addition, the Chinese government issued strict laws against "illicit sexual contacts with foreigners," which included all forms of nonmarital sex. All foreigners caught having such relations with Chinese citizens could be deported, and the government stipulated that entertaining local citizens in one's hotel room—regardless of what actually transpired in the room—would be considered in violation of the law.

Asia was a very special concern for the World Health Organization; though AIDS hadn't yet emerged in most of the area, those familiar with social and medical practices in much of the region felt sure that the virus could easily overwhelm the continent. WHO Director-General Mahler was so worried that in mid-1987 he broke with the usual diplomatic UN niceties that precluded mentioning countries by name when sounding an alarm. He predicted that a "major catastrophe" loomed for Asia if the continent failed to come to grips with AIDS, and specifically named India, Bangladesh, Thailand, Indonesia, and the Philippines as countries at greatest risk. Mahler did not break with UN decorum far enough to enunciate the reasons for naming those particular countries, but WHO officials privately voiced deep concerns about rapidly rising heroin and prostitution markets in urban centers of those countries.

Some of the countries in question seemed to recognize the veracity of Mahler's comments and responded aggressively. The nature of their responses, however, troubled WHO.

Thailand, for example, had a thriving sex and prostitution trade. Long a major source of foreign exchange for the nation, the prostitution and "entertainment services" industry swelled radically during the Vietnam War, as Thailand was designated an official R&R (rest and recreation) site for U.S. military personnel. By the end of the war Thailand's revenues from the sex trade equaled a quarter of all rice trade income.[15] Not wishing to call attention to potential problems in so lucrative an industry, the Thai government ignored all WHO pleas to institute nationwide AIDS education campaigns and promote condom use. Instead, Thailand alternately tried to repress or ignore the virus, imprisoning some HIV-positive foreigners while issuing so-called AIDS-free certificates to male and female prostitutes who serviced tourists.

India also perceived AIDS as a foreign problem and declined to conduct any form of domestic AIDS education. By the end of 1986 India had in place laws requiring HIV tests of all foreign students. As was the case in so many other countries, these laws were almost exclusively—and often brutally—enforced against African students.

Despite such measures, by mid-1987 scattered surveys of female prostitutes in India were already revealing that AIDS was emerging in the country. As the numbers of documented AIDS cases in India rose during 1987, the Ministry of Health declared that foreign students and tourists were chiefly responsible, as were "foreign priests attending Christian conventions."

By the end of 1987 fear of foreigners with AIDS had reached such heights that villagers in Goa fell upon a group of German tourists, smearing them with dung because of their allegedly filthy foreign ways.

Other Asian countries responded with similar anti-foreigner laws and actions, notably Japan, South Korea, Indonesia, Malaysia, and Singapore.

The Global Programme on AIDS staff scrambled to convince Asian leaders that such policies would only hinder efforts to prevent the emergence of an enormous AIDS epidemic on the continent. But the Asian nations correctly pointed out that their policies were modeled after those of the most powerful nation on earth, the home of Jonathan Mann, the country leading the world's AIDS research effort, the nation with the greatest number of officially reported AIDS cases: namely, the United States of America.

The Reagan administration's decision to follow an overall policy of trying to control AIDS through the use of legal instruments was a huge thorn in Jonathan Mann's side. At a time when the GPA was stressing public education as the primary tool for preventing the spread of HIV, the U.S. government was torn asunder by sentiments that *no* form of tax-funded AIDS education should be permitted. And tensions at the White House mirrored a severely dichotomous response toward the AIDS epidemic at the level of grass-roots America. All across the country by 1986 the populace was deeply divided between those who favored a nonjudgmental education-driven approach to the epidemic and those who wanted HIV-positive people and members of identified high-risk groups segregated by some means from the rest of society.

During 1987 more than 350 items of AIDS-related legislation were debated by politicians in U.S. states, most of them aimed at restricting the activities of HIV-positive individuals or at mandating testing of various population groups.

In June 1987, Howard Phillips, who had influence at the White House because he chaired a powerful right-wing group dubbed the Conservative Caucus, called for passage of a federal law giving "every hospital, every private business, every property owner, every school . . . the right to [HIV] test people who seek to use its facilities." And he said that "quarantining is something that we have to consider."

The foci of attack were homosexuals, "immoral lifestyles," drug users, and sinners—the purported purveyors of viral ruin. Like their Islamic counterparts in the Middle East, many Christian political leaders in the United States were convinced that there was a religious message to be derived from AIDS, an epidemic that would best be stopped through moral virtue.

The year 1987 was unique in recent American history in that Christian moralists ran against one another in national elections, and a disease rose to the dubious status of a pivotal issue in state, federal, even presidential elections.

Ronald Reagan's second term in the White House wasn't scheduled to end until January 1989, but campaigning for the November 1988 election began extraordinarily early. His Vice President wanted to be next in line for the job, but George Bush was no shoo-in. Sensing that the national mood was volatile, and no single issue or candidate had yet captured widespread support, more than a dozen men were already stumping for office in the spring of 1987, a full year before the first round of scheduled primary elections. And right up until election day in November 1988, AIDS would figure prominently in their campaigns.

Pat Robertson, a Baptist minister and founder of television's Christian Broadcasting Network, ran against Vice President George Bush in the Republican Party primaries. Robertson maintained that scientists were "frankly lying" when they claimed that HIV could be transmitted heterosexually, and asserted that condoms were useless to prevent infection. He supported the right of employers to fire and landlords to evict people who were infected with the virus. And he told his Christian followers that they were engaged in "a holy war" against the debauchery and decadence that he said were at the root of AIDS.[16]

The Moral Majority, a Christian fundamentalist political body led by the Reverend Jerry Falwell, had long proclaimed that AIDS was the wrath of God against homosexuals. By 1987 the organization, which had backed the previous presidential election of Ronald Reagan, was nervous about supporting Reagan's heir apparent. AIDS was one reason for that nervousness, as the organization felt that George Bush might cave in to the "AIDS Lobby," as patient advocates were called, and allow sexually explicit education about the disease. Even federally funded basic research on AIDS was opposed by the group.

"What I see is a commitment to spend our tax dollars on research to allow these diseased homosexuals to go back to their perverted practices without any standards of accountability," declared Moral Majority director Ronald S. Godwin.

In his first major speech addressing the AIDS epidemic, delivered before the College of Physicians in Philadelphia on April 2, 1987, President Reagan assured the nation—for the first time—that he was concerned about AIDS and considered it "Public Enemy Number One."

"The federal role must be to give educators accurate information about the disease. How that information is used must be up to schools and parents, not government," Reagan said. "But let's be honest with ourselves. AIDS information cannot be what some call 'value neutral.' After all, when it comes to preventing AIDS, don't medicine and morality teach the same lessons? . . . I think that abstinence has been lacking in much of the education."

The President's comments reflected an ongoing dispute within his administration over the proper tactics for control of AIDS and prevention of the emergence of HIV in geographic and demographic parts of the country not yet touched by the virus. Reagan's Surgeon General, Dr. C. Everett Koop, wanted frank discussion of abstention, the AIDS epidemic, and safe sex to be conducted in the nation's schools. But Reagan's Secretary of Education, William Bennett, adamantly opposed such plans, favoring instead efforts to identify and control HIV carriers through compulsory HIV testing of all hospital patients, marriage license applicants, prison inmates, and foreigners applying for immigration visas.[17]

Vice President Bush was straddling his roles as adviser to Reagan and candidate for the presidency. He played to voters on the right, calling for mandatory marriage license HIV tests and public identification of people who were infected.

It all came to a head in Washington, D.C., June 1–5, 1987. More than 10,000 scientists, physicians, and reporters descended upon the nation's capital for the Third International Conference on AIDS.

In the vast expanse of the Hilton Hotel's conference room, scientists searched for seats, shielding their squinting eyes from the glare of television lights that created, alternately, bright areas and deep, eerie shadows. Around the periphery milled clusters of activists, dressed in black jeans and black-and-white T-shirts emblazoned with Act Up slogans. The hall was filled with nervous energy that confused most of the foreign scientists: for four days they would witness a uniquely American exercise in democracy and confrontation that some would find distasteful, others inspiring.

The keynote speaker, U.S. Surgeon General C. Everett Koop, dressed in his starched white Public Health Service uniform, looked at the sea of enthusiastic scientists and activists with genuine surprise. What began as a polite reception swelled into nearly hysterical cheering, chanting, shouting, and foot-stomping as thousands of activists and American scientists signaled their support for Koop's dissident position within the Reagan administration. Koop was stunned. Just two years earlier most of the people in the room would have booed him off the stage because of his staunch, often radical opposition to abortion. But now they gave him a hero's welcome unlike any the seventy-one-year-old Brooklyn-born physician had experienced.

"Stop it! You're embarrassing me!" he shouted, and like obedient schoolchildren, the crowd fell silent, took their seats, and behaved themselves.

In contrast, when presidential candidate George Bush took to the podium, activists stood silently, their backs turned to the Vice President, many holding placards aloft condemning Reagan administration policies. Cameras rolled, photographers' bulbs flashed, and hundreds of scientists stood one by one to join the activists in turning their backs on the Vice President of the United States.

On the final day of the gathering, American and French scientists took the podium together to denounce not only the Reagan administration's policies but those of governments all over the world that, they said, were "based on irrational fears rather than science." They urged scientists to sign a petition calling for an end to discriminatory HIV-testing policies, an end to immigration and travel restrictions for people with HIV, and all other forms of what they considered repressive approaches to AIDS control.

"AIDS is a touchstone of politics, of racism, of bigotry," Mann told the conferees. "We see a rising wave of stigmatism around the world. AIDS has become a threat to free travel and global movement. People all over the world are seeking answers—simple answers—as the pandemic spreads. People are promoting sex cards, tattoos, quarantines, police lists, deportations, home burnings, incarcerations of select population groups.

"How our societies treat HIV-positive individuals will test our collective moral strength. This test will present itself with increasing challenge in the coming years."

Though Mann's remarks received thunderous applause that day in Washington, and were carried by the media worldwide, the message many powerful politicians derived from the Third International Conference on AIDS was quite the opposite. They saw shouting homosexuals showing disrespect for national leaders, and upstart scientists daring to tell them how to govern. And they didn't like it.

Two weeks after the close of the AIDS conference the U.S. Congress voted unanimously—96 to 0—to mandate HIV tests for all applicants for legal immigration to the United States.[18] The same week governors of three states—Minnesota, Texas, and Colorado—signed laws permitting local authorities to quarantine indefinitely HIV-positive individuals who seemed by virtue of their sexual activities to pose a threat to society.

Throughout the summer of 1987 debates raged in state and federal legislative assemblies over restrictive versus educative approaches to controlling the spread of AIDS. And with the autumn came both more action inside the U.S. Congress and an escalation of presidential electioneering. In October the U.S. Senate voted nearly unanimously—94 to 2—to cut off all federal funds for AIDS education efforts targeted at homosexuals. At issue was a comic book designed by a New York men's group that depicted graphically how men could safely have sex with one another in the midst of the AIDS epidemic.

"Christian ethics cry out for me to do something," Republican Senator Jesse Helms said, claiming the comic books would promote sodomy in America. "I call a spade a spade, a perverted human being a perverted human being. This subject matter is so obscene, so revolting, it is difficult for me to stand here and talk about it. I may throw up."

By the time the world's Ministers of Health gathered two months later in London, the United States had federal laws requiring HIV tests of foreign students, immigrants, long-time visitors, all military personnel and applicants for military service, U.S. overseas foreign service personnel, and applicants for the domestic youth employment service called the Job Corps. Entry to the United States could be barred to any noncitizen known to be HIV-positive, and though Bush had in oratory opposed discrimination against people with AIDS, HIV-positive applicants for foreign service, military, or Job Corps positions were, by law, denied employment.

Before the London meeting the GPA staff had reviewed all the legal and political activities surrounding AIDS and concluded that they were witnessing, in slow motion, many of the same social responses that had followed the arrival of the plague in fourteenth-century Europe. In both cases there were actually three different social epidemics within the larger biological epidemic.

First, with the initial emergence of the microbe—plague bacteria or HIV—came denial in all tiers of society. The tendency was to ignore the microbial threat, or assume only "they"—some distinct subpopulation of society—were at risk. The microbes exploited such denial, spreading rapidly while humans made no attempts, through their personal or collective behaviors, to block any of the avenues of transmission of the organisms.

The second social epidemic was fear. Some event in the biological epidemic would suddenly shock a society out of its state of denial, propelling people into a state of group terror. In fourteenth-century Europe it was often the plague death of a popular cleric or a local lord or the sudden public expiration of a child that prompted panic. The timescale was quick: plague-infested rats might arrive in a town on Tuesday, local human deaths might begin in the harbor area by Thursday or Friday, and a riveting event could spark widespread panic by the middle of the following week.

But AIDS was a slow killer, and the biological epidemic unfolded in each country over a span of many months or years. So the first social denial stage might persist for over a decade. Fear might also linger for years, giving rise to all sorts of panic responses and inappropriate actions, such as setting fire to the home of two HIV-positive children with hemophilia in Florida.

Eventually, the Programme staff knew, the social epidemic of fear usually yielded to a wake of repression. Fear-driven government response was usually irrational, prompting attacks on the victims of disease, rather than the microbes. During the plague such fear-driven repression led to the wholesale slaughter of Jews and of women accused of witchcraft. Though outright genocide certainly hadn't surfaced in response to AIDS, Mann's staff felt certain that in the absence of strong political leadership guiding populaces toward rationality, the epidemic could have violent consequences in some societies.

As HIV emerged in new areas of the world, Mann hoped to find a way to break this chain of social epidemics; to push governments out of denial before they had an epidemic on their hands; or failing that, to move a society out of fear to effective action, rather than panic-driven repression. The GPA group knew that they were breaking new ground, that few societies had ever in history responded wisely or rationally to major epidemics, and that lessons learned with AIDS could be applied to combating future emergences of all sorts of microbes. They searched for answers.

In Nigeria, Dr. F. Soyinka studied his society's response to AIDS in 1987. Nigeria had very few cases of the disease, as it was located far from Africa's AIDS epicenter. Nevertheless, Soyinka and other physicians knew it was only a matter of time before HIV took its toll in Nigeria, so they waged a massive monthlong television, radio, and newspaper campaign to warn the public. At the campaign's end, Soyinka surveyed residents of Lagos.

He was sadly surprised to discover that "85 percent believe AIDS is a disease of the white man. They believe it can only be gotten if you have sex with a white man."

A 1987 Gallup poll conducted in thirty-five nations showed that 96.5 percent of the people questioned had heard of AIDS, but most respondents were deeply confused about how dangerous the virus might be, how one got infected, and which activities put a person at risk. Similarly, U.S. CDC surveys year after year revealed that nearly every adult American had heard of AIDS and knew that it was caused by a virus. But about half thought one could become infected by donating blood, by being bitten by an insect, and/or by sitting on a public toilet.

Throughout the world there was an alarming confusion between the myths and the realities of AIDS, producing either continued denial or highly exaggerated fear.

A complicating factor unique to AIDS and other sexually transmitted diseases was the nearly universal dislike of condoms. All over the world, men felt that condoms diminished their pleasure and women had little or no control over their use. Nobody enjoyed talking about condoms during lovemaking, and it could be dangerous for a woman to request that her lover or husband use one: there were widespread reports of men beating their wives or partners in response to such requests.

Studies of gay male behavior in San Francisco showed that crucial to individual protective action, such as consistent use of condoms, was a high level of fear, brought about by witnessing the deterioration and AIDS death of a close friend, relative, or lover. Similarly, on a societal scale it was apparent that few cultures were able to confront AIDS until the death toll had become sufficiently high to have given more than 10 percent of all adults a firsthand view of the horrendous disease.

But that was unacceptable. How could Jonathan Mann, the GPA staff, the World Health Organization, or the planet's citizenry sit back and wait for a massive death toll before taking effective action? How could they allow the microbes to emerge in one geographic or cultural place after another, infect tens of thousands of people, slowly—over a period of years—cause visible disease and deaths, and be utterly *endemic* to the societies before action was then taken to stave off an *epidemic*?[19]

Studies all over the world were revealing the scale of the problem. For example, by 1987 more than 5 percent of the adult population of Brazzaville, Congo, were infected with HIV, and the visible AIDS death toll was already obvious to even the casual observer. Yet researcher Marc Lallemont found that pregnant women in the city were in "an almost complete state of denial, perhaps the most complete I've ever seen." Lallemont surveyed hundreds of women who were making prenatal visits to local clinics and discovered that more than half of them insisted AIDS was caused by mosquito bites, despite numerous government educational campaigns stating just the opposite and warning about sexual transmission of the disease.[20]

In 1986 the U.K. government launched one of the highest-profile AIDS education campaigns seen anywhere in the world. It was a case where most of the elements for success appeared to be in place: top-level political will, resources, national television accessibility, and a heightened media interest. Yet the campaign was eventually judged a failure, as it succeeded in raising AIDS awareness and fear but failed to put a dent in public misperceptions about how the virus was transmitted or general disdain for those who carried HIV.[21]

In no country, it seemed, had a government found the secret to preventing further spread of HIV once the epidemic became endemic.[22]

At the Global Programme on AIDS, Manuel Carballo said that the epidemic was forcing researchers all over the world to evaluate—and reevaluate—the effectiveness of a whole battery of standard public health weapons, in hopes that something besides a chilling death toll could motivate individuals and governments to take rational steps to protect themselves from the virus.

"What makes the AIDS effort especially difficult," Carballo said one afternoon shortly before the London Summit, "is that those who are at greatest risk are those who are divorced from traditional values and culture. They have had to innovate new cultures. They find friends in bars and clubs. And nothing in the relationships is stable."

Without social stability, people were hard to reach, whether they were gay men frequenting bars in San Francisco, migrant workers in Mexico, newly urbanized young women in Kinshasa, Burmese prostitutes in Bangkok, or injecting drug users in the Bronx. Such people were deeply separated from the traditional mores of their respective societies, often cut off from their families and mainstream workplaces.

In the 1960s, René Dubos wrote extensively about the special vulnerability to the microbes among people who lived lives of poverty. History demonstrated repeatedly that, with rare exceptions, the microbes exploited the weak points of economically bereft lives: chronic malnutrition, prostitution, alcoholism, dense housing, poor hygiene, and egregious working conditions.

Carballo and his colleagues recognized that there was more to microbial vulnerability than the social-class arguments put forward by Dubos. When information was the key to self-protection, there were gradations of *Homo sapiens* vulnerability that, yes, could be rooted in economic class, but could also stem from social alienation. People who were treated as outcasts from the dominant culture in which they lived could be denied vital life-protecting information or public health tools. If the larger society reviled a particular subgroup, its marginalization could be a risk factor, Carballo argued, every bit as crucial as a contaminated syringe.

Carballo saw a confluence of social factors at play in the emergence of HIV in societies: marginalization, social alienation, poverty, and discrimination. In his mind, they united to form a social bridge across which HIV traveled into one society after another.

As Panos Institute AIDS researcher Renée Sabatier put it: "I think there is a very real danger that we're going to end up as a [world] society divided between those who were able to inform themselves first and those who were informed late. Those who have access to information and health care, and those who don't. Those who are able to change, and those who aren't. I think there is a real danger of half of us turning into AIDS voyeurs, standing around watching others die."

On January 28, 1988, the London Summit endorsed the GPA's fifteen-point declaration that called for openness and candor between governments and scientists, opposed AIDS-related discrimination, gave primacy to national education programs as means to limit the spread of AIDS, and reaffirmed the GPA's role in international leadership. Mann and Mahler viewed it as a triumph.[23]

But even as they smiled for the cameras and signed the declaration, seeds of failure were being sown. The declaration said nothing directly about quarantines, immigration policies, or forced deportations, delegates to the Summit having concluded that no agreement on those pivotal issues could be reached between the 149 nations. Worse yet, representatives of critically important countries—like China and the U.S.S.R.—openly scoffed at the GPA's attempts to promote educative efforts over restrictive measures. China's delegate denied the existence of homosexuals, drug users, and prostitutes in his country, thus insisting AIDS couldn't threaten the People's Republic. And Soviet Minister of Health Yevgeny Chazov insisted that Slavic genetic superiority had rendered the populace immune to the virus.

Despite the efforts of the GPA, the pandemic spread relentlessly, always emerging first in communities that were on the outer periphery of societies' margins. Mann, Tarantola, Carballo, and the rest of the GPA staff zigzagged madly about the planet, living in a perpetual state of jet lag, as they frantically tried to squelch the tandem fears of HIV emergence and social denial, fear, or repression.

With each passing day in 1988, Mann became more strongly convinced that disease emergence was a human rights issue, in the strictest legal sense of the phrase. Though the physician/scientist had never before been exposed to international human rights law, some of those working around him had—particularly Katarina Tomasevski, an attorney and public health expert who served as a consultant to the GPA. Tomasevski introduced Mann to the body of international human rights law. And Mann, in turn, increasingly framed GPA policy pronouncements on such issues as international freedom to travel, HIV screening of refugees, access to health care for prostitutes, and discrimination against homosexuals in the context of the major instruments of human rights law.[24] Tomasevski demonstrated that most of the government actions the GPA found repugnant, such as deportation of HIV-positive Africans from Asian countries following enforced testing and detention, were violations of international legal pacts to which the offending nations had previously agreed.

In the United States, attorney Larry Gostin, of the Boston-based American Society of Law and Medicine, was carefully documenting the astonishing growth in AIDS-related legislation and precedent-setting legal decisions. He, too, felt that basic tenets of international human rights and national civil rights law were being violated or eroded.[25]

While the staff of the Global Programme on AIDS became more outspoken about the connection they perceived between human rights and the spread of HIV, anger and jealousy were building all around them. Some critics began dropping hints to the international press corps about "left-wingers in Geneva." Among Mahler's top aides were men who made no bones about their feelings that the GPA was reflecting "homosexual politics." Human rights, though a topic of serious discussion within most other UN agencies, had never received much attention at the World Health Organization.

"WHO human rights policies were characterized as incoherent, fragmented, inconsistent. We really didn't get moving on human rights until it was thrust upon us," WHO rights expert Sev Fluss said. What thrust human rights up to WHO's front burner was AIDS, and specifically references in the London declaration to abolishing discrimination and inequity.

"Medical people think of human rights as torture and so on. They don't think of it as what they do. And they certainly don't think of a constitutional right to health care," Fluss explained.

"When AIDS first emerged, our response was disastrous," Fluss conceded. "People thought it was like Ebola and Marburg, which went away without creating a global epidemic. A flash in the pan, that's what they thought."

But as early as 1983 ten countries passed legislation specifically targeting AIDS, and Fluss thought it rather intriguing that a new disease was prompting so many laws. By the time the GPA was established, twenty-one more countries had passed major AIDS legislation, and Fluss had an office designated as the WHO Health Legislation Unit. But the HIV pandemic kept spreading, right past all those laws, national border patrols, HIV-testing centers, and alleged human genetic superiority. Within nations it spread to new population groups, made its way from urban centers to rural areas, crossed class boundaries. Between nations it surmounted virtually every obstacle, save condoms, that humans placed in its way—and certainly each legislative barrier.

II

By 1988 Western economists and African leaders were asking out loud, "Will this epidemic slow, or even destroy, African development? Is it possible that AIDS will destroy all the development programs we have spent the last three decades building?"

The disease, which so recently had been added to the agenda of international human rights, was also becoming a bona fide macroeconomics issue, threatening both fiscal and social development in the world's poorest nations. It seemed too horrible to contemplate, yet inescapably apparent, that the global AIDS pandemic might well make the world's poorest nations much, much poorer. After years of struggling to rise above Third World status, these nations might be slipping backward on a wave of Thirdworldization.

The World Bank's Mead Over pioneered much of the research on the economic impact of AIDS in Africa, which between 1988 and 1993 was supplemented greatly by the research of economists, mathematical modelers, and epidemiologists in the United States, U.K., France, and at WHO.

They began their calculations with several key assumptions: first, that African nations entered the AIDS era already severely impoverished. For example, the 1987 GNP per capita in the United States was $16,690. In Tanzania it was $290, in Zaire a mere $170.

Second, no African nation faced a single epidemic crisis. Since the 1970s a host of new microbes had successfully emerged and swept across the continent: drug-resistant malaria, drug-resistant tuberculosis, urbanized yellow fever, Rift Valley fever, and waves of measles epidemics, to name a few. That meant that the health care systems of African nations were already stretched to their limits. Given scarce resources for health care— averaging $1.00 to $10 per capita annually—any additional burden seriously endangered the viability of entire national medical systems.

Compounding the problem was the seeming synergy between microbial epidemics. Wherever AIDS became endemic, tuberculosis followed closely. One epidemic sparked another: malaria and HIV fed upon one another, as did cytomegalovirus, Epstein-Barr virus, syphilis, gonorrhea, chancroid, and a host of others. Though no one had a detailed empirical grasp of the relationship, it was clear throughout Africa that wave upon wave of infectious diseases influenced one another, and further taxed the health care systems and economies of afflicted nations.

A third assumption was that AIDS would have a uniquely harsh impact because of who in Africa were the microbe's primary targets. Studies all over the continent showed that among the hardest-hit social groups was the well-educated urban elite. These were the young adults who had attended universities in Boston, Oxford, Moscow, and Paris, acquiring skills that could be used to navigate their countries out of postcolonial stagnation into prosperity and infrastructural order. But they were also among the few Africans who possessed disposable incomes and could afford to indulge in the carefree nightlife of cities like Kinshasa, Nairobi, Harare, and Yaoundé. Long before anyone had heard of AIDS, the continent's educated elite was unknowingly becoming infected in the discos, brothels, and nightclubs of Africa's glittering nocturnal ambience. To economists, who placed productivity values on human lives, that meant that AIDS was taking a particularly sharp toll on Africa's future.

A fourth consideration was the familial nature of the epidemic. In Africa, whole families seemed to die off, each survivor's burden increased by the need to care for the sick and compensate for the decline in family income brought about by the deaths of adult providers. In some devastated areas, such as the Lake Victoria region, familial destruction led to the economic collapse of whole villages. And, with time, that could have a ripple effect through all tiers of the regional economy.

All economic forecasts had to begin with estimates of the size and forecasted scope of a country's current epidemic. Nobody, however, including those who reported countries' AIDS statistics, believed that the officially reported numbers came close to reflecting the true scope of the HIV/AIDS epidemics in developing countries. But what was the reality?[26]

Some African countries were still holding back accurate information about the scope of their epidemics as late as 1990, particularly when sensitive groups—such as the military—evidenced high infection rates. Still other countries were overwhelmed by famines, civil wars, and political instabilities that rendered the business of disease record keeping all but impossible. And all African countries were hampered by severe infrastructural problems that hindered diagnosis, treatment, and reporting of AIDS.[27]

HIV infection rates in some groups were already staggering by 1988, and would reach positively horrendous proportions by 1993, when some studies would find, for example, that upward of 40 percent of women of reproductive age in key African cities carried the virus.[28]

Even without solid epidemic estimates economists who were paying attention to Africa's pandemic were, as early as January 1988, predicting financial hard times for the continent: patchworks of small-scale famines;[29] "an economic disaster" based on the direct costs of AIDS care, HIV-testing costs,[30] a year's supply of condoms,[31] AZT and other drugs for opportunistic infections (where such pharmaceuticals were at all available); and loss of net industrial and agricultural productivity due to deceased workforce. They warned that AIDS was creating "a global underclass," over and above the previously existent world community of impoverished individuals.[32]

Direct AIDS costs—drugs, hospitalization, health care personnel—were very low in the African countries when compared with the United States, simply because of the differences in availability of such resources and lower labor costs, according to studies by the World Bank's Over and collaborators in Tanzania and Zaire. They estimated that direct HIV-positive lifetime costs for the United Kingdom topped $20,000; under the U.S. health care system it averaged more than $50,000.[33] In contrast, Zaire spent less than $600 in direct AIDS costs per average patient, Tanzania about $800.[34]

But when the researchers compared various African diseases in terms of years of productive life lost—economically significant life for society as a whole—HIV infection ranked roughly equal to the other top scourges, sickle-cell anemia, birth injury, and neonatal tetanus. And what were the monetary values of those lost productive lives? In 1985 dollars, the group estimated the average Zairian life lost to AIDS had a top value of $3,230; the equivalent Tanzanian loss was valued at $5,316.

When those values were compared with national GNPs per capita, that meant that a typical Zairian AIDS death equaled about 19 years of per capita GNP, a Tanzanian about 18.3 years. If such numbers were multiplied by thousands or tens of thousands of losses in the two nations' epidemic futures, it was clear that the result could be financial ruin for the already desperately poor countries.[35]

But such an analysis had its limits because it assumed that costs and values would be stagnant over time. In an expanding epidemic, however, costs were compounded over time as family and workplace burdens increased due to multiple deaths: their combined impact was more than additive. For example, a farming family might be able to compensate for the loss of productivity due to the death of one adult, but after two or three deaths it would no longer be possible to till the soil or harvest a crop, particularly in areas lacking all forms of agricultural machinery.

From Mead Over's point of view, the real compounding crisis was loss of skilled and professional labor. A national bank in a country like Zaire would typically be operated by a handful of well-educated men, with no surplus labor pool upon which to draw for replacements. For many professions Africa's generation of twenty-five- to forty-year-olds was the first in the continent's history to achieve expertise. With colonialism so recently defeated, this was not surprising, but it did place most sub-Saharan economies in extraordinarily vulnerable positions in the face of an expanding epidemic.

"Indirect costs are twenty times as important as direct costs, because AIDS is striking people in their productive years. That is the real problem. I think the impact of the indirect costs on a typical East African country over the next twenty years could be to reduce the growth rates of the national economies from two or three percent, where they are now, to close to zero percent," Over said. "That means a zero GNP growth. That's a worst-case scenario. So what we've got is a menace on the horizon."

The real question was whether the AIDS epidemic might destroy the Third World's arduous efforts to pull itself out of perpetual poverty and disease into political stability and economic growth. After the expenditures of billions of dollars of foreign aid and loans from wealthy nations—and after accruing massive debts—some of the world's poorest nations were just beginning to turn the tide.

Jonathan Mann felt it essential to get a handle on the development issue, not only because it was intrinsically important but also because solid empirical answers to the economic question would most likely affect investments in AIDS prevention programs at the international, national, and local levels.

The task fell to the GPA's Jim Chin. A year earlier Chin had been running infectious disease programs for the state of California, living a comfortable, albeit generally routine, life in Berkeley. There, he had commanded a staff of about 400 people and oversaw an annual $65 million budget. In 1989, however, the cautious American found himself facing the formidable task of forecasting the fate of a continent. With a total staff of five people and a tiny piece of the GPA's $90 million budget, Chin toiled in a cramped Geneva office. Though by nature an affable social animal, Chin approached his new job with introspection and conservatism, consciously lowballing his estimates lest he later be accused of playing Chicken Little.

Chin collaborated with Tanzanian scientist S. K. Lwangwa to develop models that, first, could determine how many unreported AIDS cases were currently occurring in Africa; second, how large the current pre-AIDS HIV epidemic might be; and, finally, what might be the epidemic's growth rate and future toll.

In 1989 the pair published a study that predicted that a typical East or Central African country already in the grips of a severe AIDS epidemic could expect by 1991 to have HIV infection in one out every five of its citizens.[36]

"That's lowball," Chin said. "It's the high-end estimate based on an overall conservative set of assumptions. It could be a lot worse. Our most conservative estimate is that there will be 575,000 new AIDS cases in Africa in 1991, for a cumulative total of more than 800,000."

Sitting at Chin's side, Mann listened attentively, then said with a heavy voice that the 1990s would be far worse.

"I would like to be optimistic," Mann said, "but I think we must be realistic. Not until 1985 did the message really come home that AIDS was a global problem. In retrospect, probably historians will say it took too long. We are consistently faced with situations where the reality far exceeds our grasp. It's legitimate to ask, 'Are we able to see clearly enough? Or, when we look into the future, is the horror of it all simply too much even for us to confront?' "[37]

But by 1990 Chin's estimates were even grimmer. He was saying that 8 to 10 million were infected, perhaps 5 million of them in Africa. It would prove the first of many upward revisions.[38]

By the time WHO's July 1990 revised forecast was released, Jonathan Mann and much of his GPA staff were gone. They had lost a power struggle within the Geneva-based organization, and Mann had developed a contentious relationship with the new WHO director-general, Hiroshi Nakajima. Mann's enemies within WHO were legion: all those months of greening with envy over the upstart American's meteoric rise finally paid off.

Japanese physician Nakajima, who had headed WHO's Asian regional office during the period when multidrug-resistant malaria spread across the southern region, was clearly uncomfortable with Mann's very public persona and high-profile AIDS program. He shared the views of those who had long whispered derisive comments about the GPA in the WHO hallways. Nakajima felt that disease programs should be managed in accordance with established WHO protocol. It was a reasonable expectation, except for one key point: established protocol did not provide for the contingencies presented by a rapidly expanding worldwide epidemic.

In Mann's stead Nakajima placed another American physician, Michael Merson. For most of his professional life Merson had worked for WHO in Geneva, managing programs for respiratory and diarrheal diseases. Merson understood WHO protocol.

An unfortunate political battle ensued, with leaders in the world's AIDS control effort taking sides for or against Mann, Merson, Nakajima, and the professional positions each took on approaches to the pandemic.

In Merson's first six months heading GPA, the program upwardly revised its estimates of the size of the global pandemic three times. By September 1990 the official WHO estimate of the cumulative number of AIDS cases was 1.2 million, 400,000 of which were infants and small children—90 percent of whom were in sub-Saharan Africa. And the new WHO year 2000 projection was for 25 to 30 million HIV infections worldwide.

With concern mounting about the Thirdworldization that AIDS might bring upon Africa, Peter Piot teamed up with Mead Over to do a systematic analysis of the relative cost of HIV compared to other, better-understood diseases. After carefully computing the per capita burden in terms of productive healthy years of life lost, Piot and Over concluded that the direct costs of treating HIV disease, even in the absence of AZT and other expensive drugs widely available in North America and Western Europe, far outstripped those of any other common ailment in Africa.[39]

The impact was already being felt keenly in some sub-Saharan countries. Malawi's entire health care system, for example, was in genuine peril of collapse under the burden of HIV, and the nation's public health leadership in 1990 issued desperate pleas to WHO, the World Bank, U.S. AID, and other Western organizations for funds.[40]

Even as Africa's leaders began to absorb the dire economic implications of the WHO and World Bank AIDS studies, critics were emerging who charged that the well-intended analyses grossly underestimated the epidemic's impact. For example, nurse Eunice Muringo Kiereini, a Kenyan woman who chaired the WHO Regional Nursing/Midwifery Task Force, claimed that the studies failed to consider the special economic roles women and children played in African economies. Ever since the beginning of Africa's mass urbanization it was the continent's young men who left the farms and villages in favor of jobs in the cities. Few village women had the option of abandoning their traditional lifestyles. As a result, in many parts of Africa villages were populated by females of all ages, male children, and elderly men, many of whom were too feeble to work. Young men would return to their wives and children periodically, but their lives were elsewhere.

So it was the women and children of Africa who maintained the continent's agricultural economies, Kiereini said.

"Women are hit the hardest by the international structural injustices prevailing in the Third World," Kiereini explained.[41] "The majority of countries in Africa are dependent for foreign earnings on the export of one or two agricultural products such as cocoa, tea, coffee, etc. Trade in these products is grossly imbalanced in the favour of the rich countries. Prices are so low and unstable and the market is controlled by foreign interests. A country in this situation sinks even deeper into poverty.

"It is true to say that women and children who provide 80%–90% of the labour force earn extremely low income at the end of the day. The little money they are paid is controlled fully by men. Consequently, women and children are trapped in the vicious cycle of structural poverty. In this kind of situation there is little or no money available to meet the basic needs of the family."

Though in some parts of Africa women were less valuable than local livestock—as evidenced by prevailing bride-prices and dowries—it was they who raised the continent's futures: its crops and children. When husbands contracted HIV in the cities and passed the virus on to their wives during periodic return visits to the villages, AIDS appeared in rural areas that were completely lacking in health care and social support systems. The affected women continued to plow the soil with their hand hoes, lugging babies on their backs, until their AIDS-devastated bodies collapsed. And with each female death Africa's agricultural productivity declined another, barely perceptible notch. The cumulative burden of these declines, Kiereini warned, could, by the year 2000, be more desperate for some countries than their losses of professional elites. The demise of Africa's female agricultural workers could, she warned, lead to acute food shortages.

Even uninfected, healthy African women were being forced out of productive roles in agricultural sectors by the AIDS epidemic. In most African societies, both traditionally and under modern codified law, women had virtually no basic rights. They were, legally, their husband's property, as surely as were his cows or goats.

If the husband died, his property reverted not to the wife but to his relatives. The crops that the widow had tilled became a new source of prosperity for the in-laws. The widow and children, now landless, often lacking even changes of clothing, had to find a means of survival.

In the short run the village and overall societal economies experienced little if any impact from this process because the in-laws continued to harvest crops. But as the epidemic expanded, and even those in-laws were infected, Africa faced the creation of an unprecedented rural underclass of desperately impoverished, often starving women and children. Further, it was obvious that eventually the cumulative load of deprived widows would exceed the available labor force of in-law inheritors, causing declines in crop production.

Worse yet, one of the few survival options left to widows—perhaps the only way they could feed their children—was prostitution. So, impoverished by AIDS, the woman would be forced into a life that virtually guaranteed that she, too, would die of the disease.[42]

By 1991 the gender ratio of AIDS in Africa was shifting, reflecting higher infection rates among women. For example, researchers from the University of California at San Francisco studied nineteen- to thirty-seven-year-old women in Kigali, Rwanda. A third of the randomly sampled women were HIV-positive. Even among women previously thought to be at very low risk for HIV because they were monogamous throughout their lives, the infection rate was 20 percent. The same group also showed that many women in Rwanda were dying of AIDS but not being counted in national statistics because their symptoms didn't fit with the male-based WHO definition of the disease. The researchers suggested that the true extent of AIDS in African women might be two to three times the diagnosed numbers.[43]

Josef Decosas, of Canada's International Development Agency, argued that the continent's women were caught in "an epidemiologic and demographic trap" from which they could not be freed without greater social equity. Decosas contended that any hope of slowing Africa's devastating epidemic before it brought financial ruin to much of the continent was inextricably tied to improvement in the status of Africa's women.[44]

Researchers at the UN's Food and Agriculture Organization, based in Rome, tried to calculate the impact AIDS would have on African agricultural production. Their best estimate was that Africa's overall labor force—predominantly women—would be reduced by 25 percent by 2010.[45]

Another factor compounding estimates of the socioeconomic costs of AIDS was the epidemic's continuing geographic expansion. Though every political leader on the continent knew by 1989–90 what caused the disease and which social measures might prevent its spread, agonizingly few took steps to warn their populaces prior to HIV's full-scale emergence in their midst. South Africa, for example, was spared significant HIV emergence until the late 1980s. There is no evidence that the virus existed in the country prior to 1986, and for the first three years it was almost exclusively a disease of gay white men who had traveled in Europe and North America.

By 1989, however, HIV was emerging in South Africa's black population. The microbe found advantages in apartheid policies regarding migrant labor: men from throughout the country, as well as nearby Swaziland, Mozambique, and Malawi, were granted work permits for jobs in the gold and diamond mines, but were not allowed to bring their wives and children. Living in squalid barracks, the men frequented brothels in the mining towns whenever possible. Each prostitute became an AIDS amplifier.

By 1991 local experts were estimating that as many as 400,000 black South Africans, mostly men, might be HIV-positive. Given that black infection rates were thought to be near zero two years earlier, this represented explosive growth.[46]

Similarly, Ethiopia, which had long been spared the AIDS scourge, witnessed a phenomenal explosion of cases in 1991–93. As late as 1986 the country had no evidence that HIV had emerged within its borders. In February of that year the first AIDS case was diagnosed in Addis Ababa. By 1992 local experts estimated that more than 800,000 Ethiopians were infected, with the highest rates of infection—nearly 15 percent—seen among military personnel.[47]

Roy Anderson and his team at Imperial College in London predicted that AIDS in fifty-three African nations—including several north of the Sahara—"would reverse the size of population growth rates over timescales of a few to many decades."[48]

In other words, even though African countries had some of the highest population growth rates in the world, the epidemic was likely to outstrip that explosion and some countries might eventually experience negative population growth.

The World Bank predicted two immediate consequences of AIDS in hard-hit African areas: a radical slowdown of national GDPs and tremendous competition for scarce health care resources.

The latter was already occurring, prompting fears that secondary and tertiary disease epidemics would emerge in the wake of AIDS because countries would no longer have the resources to control other bacterial, viral, and parasitic microbes. Given that HIV/AIDS monopolized less than 1 percent of the continent's meager health budgets prior to 1984, the trend toward resource absorption by the AIDS epidemic was disturbing.

The World Bank estimated in 1991 that HIV/AIDS commanded more than 4 percent of Tanzania's health budget, even with fewer than 10 percent of all AIDS cases receiving hospital attention. But the situation was worse elsewhere: AIDS was eating up 7 percent of Malawi's health budget, 9 percent of Rwanda's, 10 percent of Burundi's, and an astonishing 55 percent of the Ugandan health budget.[49]

In early 1991 physicians in Zambia predicted that HIV/AIDS would soon overtake malaria, becoming the number one illness requiring hospitalization in the country.[50] A month later physicians in Lusaka reported that HIV/AIDS had, indeed, overtaken malaria and accounted for the use of 80 percent of the city's hospital beds. Barclays Bank complained of radically increased rates of absenteeism due to employee attendance at AIDS funerals and personal sick days. By 1992 in Lusaka, bus companies were requiring several days' advance notice for booking transport to funerals, due to backlogs. And copper production, Zambia's and Zaire's primary sources of foreign exchange, was slowly declining due to lost labor and AIDS illnesses. In 1990 the countries produced 800,000 tons of copper for export; by 1993 that was down to 600,000. Life insurance companies in Zimbabwe and South Africa faced bankruptcy as AIDS-related claims mounted. Coffee production declined by 5 percent in Uganda's hard-hit Rakai District between 1991 and 1993. Large-scale tobacco farmers in Zimbabwe took to distributing condoms to their workers in hopes of preventing a labor crisis.

The U.S. Census Bureau assisted counterpart African agencies in trying to count the population impact of AIDS. In early 1994 the agency announced that infant and child mortality was 15 percent higher in Zambia than had been the case in 1984. Since infant mortality rates were used worldwide as a gauge of development, the finding meant that Zambia's overall status represented a reversion to pre-1980 levels of development. Similar findings were made for Malawi and Uganda.

Following his dispute with WHO Director-General Nakajima, Mann joined the faculty of the Harvard School of Public Health in Boston and founded the Global AIDS Policy Coalition. Together with former GPA colleagues Tom Netter and Daniel Tarantola, Mann organized yet another analysis of the scope and future of the AIDS pandemic. The Coalition's estimates, which were based on the Delphi Survey technique,[51] were far grimmer than those promulgated by WHO.

By 1992, they estimated, 12.9 million people worldwide had been infected with HIV, 2.7 million of whom had already developed AIDS.

By the year 2000 a minimum of 38 million people would have been infected with HIV, according to the Global AIDS Policy Coalition, but "a more realistic projection is that this figure will be higher, perhaps up to 110 million."

In that scenario, 25 million people would have died of AIDS between 1980 and 2000.[52]

Anderson's group in London predicted a terrifying future for those African societies in which HIV had already, by 1990, become endemic to the general population. Barring development of a vaccine or effective means of controlling further spread of the microbe, HIV infection rates would exceed 80 percent of the sexually active population within forty-five years of the emergence of the virus in a given African society. Following that model, if HIV emerged around the Lake Victoria region, for example, during the period of hostilities between Uganda and Tanzania (1977–79), eight out of every ten adults living in the area in 2020 would either have AIDS, have already succumbed to the disease, or be HIV-positive.

In 1993 the United Nations Development Program estimated that Africa's HIV/AIDS epidemic had already cost the continent (in combined direct and indirect effects) some $30 billion.[53] And the World Bank said that many African countries faced "catastrophically costly consequences."[54]

In the spring of 1994 the U.S. Census Bureau delivered the most horrible prognosis to date. Based on up-to-the-minute seroprevalence data, the Bureau predicted that by 2010 there would be *121 million fewer human beings* in sixteen countries than would have been the case in the absence of AIDS. The countries—Brazil, Burkina Faso, Burundi, Central African Republic, Congo, Côte d'Ivoire, Haiti, Kenya, Malawi, Rwanda, Tanzania, Thailand, Uganda, Zaire, Zambia, and Zimbabwe—were expected to also experience radical reductions in overall population growth rates and increases in infant mortality, child mortality, and premature death rates.[55]

Life expectancies for several countries were expected to plummet: Uganda, without AIDS, would have had an average life expectancy of 59 years by 2010. With AIDS, that was forecast to drop to a mere 32 years. Haiti's life expectancy would decline from a projected 59 years to 44. Meanwhile, premature death rates, already climbing in the early 1990s because of AIDS, were expected to have doubled, compared with 1985 levels.

Hope had to rest with the children of Africa, the continent's next generation of potential bankers, lawyers, economists, farmers, business financiers, and planners. But studies in Zambia, Zaire, and Malawi revealed that many AIDS orphans died shortly after their mothers' demises, even though the children were not themselves infected. The causes of death were numerous, usually falling under the pediatric catchall "failure to thrive." Many of the children hadn't been fully vaccinated against measles, polio, and other common diseases. Most were malnourished. And many languished without their mothers, lacking the love and attention of any alternative adult.

"There is no doubt, AIDS threatens to alter the economic and social fabric of many societies, which may affect the development process," Uganda's United Nations representative James Baba said in December 1991. "The major problem AIDS presents today is the factor of creating an increasing number of orphans which traditional societies are beginning to fail to cope with. The traditional extended family systems that once absorbed and catered for such orphans are being stretched to the limit by the sheer enormity of the problem. As a result, the extended traditional family system is breaking down. The social and economic cost it imposes is simply too demanding."[56]

The U.S. Agency for International Development used mathematical models to estimate the impact of orphans in East Africa. In the year 2015 alone, the agency scientists predicted, 2.4 million mothers would die of AIDS, each leaving an average of three orphans. Thus, it was possible that in a *single year* in East Africa 7.2 million AIDS orphans would be generated.[57] Other studies forecast 355 million AIDS orphans in Central and East Africa by the year 2000, or up to 11 percent of the region's entire population of children up to the age of fifteen.[58]

Meanwhile, the U.S. Census Bureau predicted dire upturns in infant and child mortality in several African nations, due both to direct AIDS deaths and to neglect of children orphaned by the deaths of parents who succumbed to the disease. Hard-won improvements in those two key measures of national development were expected to evaporate. By 1994, the Bureau said, Zambia had already experienced a staggering 15 percent increase in infant mortality, compared with 1984, and Malawi, Uganda, and Zaire had suffered nearly comparable increases.[59]

"The concept of a war on AIDS with its goal to stop HIV is seriously flawed and should be discarded," Decosas wrote.[60] "Most regions in the world have a well-established epidemic of HIV. This epidemic requires a social response ranging from a review of legislation to a rethinking of the national industrial development plans. It also urgently requires new programmes, new approaches, and some radical reforms in health care and public health."

For the exhausted few adults of Kanyigo all the forecast and debated numbers for Africa's future AIDS toll, loss of productivity, and abandoned orphans were just so much hand-waving by abstract people living in even more ephemerally imagined places, like Washington, London, and Geneva. But there was nothing surreal about AIDS or the tragedy it had created.

What was harder to imagine every day, Kanyigo elder Cosmos Bilasho said, was the future: How could there be a future if no one was here today to raise the children?

III

As the train pulled out of the Rome station Subhash Hira made another quick scan about the floor, making certain that he and his Indian colleagues were still in possession of all the suitcases, valises, shopping bags, and carryalls they had already toted over so many thousands of miles. It was the natural reflex of an experienced Third World traveler.

Physically, Hira had changed little over the years. He sported the same—or identical—round wire-rimmed glasses that had been perched upon his nose thirteen years earlier in 1978 when he had first arrived in Lusaka to head up Zambia's sexually transmitted disease program. Aside from some gray hairs, Hira hadn't aged much; he still possessed boundless energy.

But inwardly Subhash Hira was a very different man. Keeping track of Zambia's horrific AIDS epidemic had taken away a bit of his soul, left scars on his spirit. He sighed a lot and didn't seem to notice it until someone asked, "Hira, what's wrong?"

"People born in the post-plague era never could imagine what it had been like then," Hira said, speaking above the chugging train's din. "People said to me when AIDS started in Zambia, 'You are looking at the bubonic plague in the Middle Ages, and ten years down the line you will see the same kinds of mass deaths.' And I thought it was exaggeration. How could we even be thinking of thirty to forty percent HIV seropositivity? Six years ago, in 1985, it was only three percent in pregnant women in Lusaka."

Hira looked out the window at the spectacular countryside of Tuscany, but seemed not to see anything. His mind's eye was on the wards of Lusaka's University Teaching Hospital. As he spoke, Hira's Indian colleagues eavesdropped, worried expressions filling their faces. They were all on their way to the Seventh International Conference on AIDS in Florence, where they hoped to spark discussion of HIV's emergence in Asia.

"I am moving home soon, to Bombay," Hira said with a hedged smile. There was no escaping the homesickness he had felt all those years in Lusaka. The circumstances of his return were less than ideal. But when he glanced at his four colleagues, two sari-adorned women and two men wearing Western-style suits, Hira's face lightened up. Obviously, he was content with the notion of working with his own people.

But his smile soon evaporated and his voice was muted when he explained, "AIDS has come to India. I must do everything in my power to ensure that what I have witnessed this last decade in Lusaka does not occur in Bombay or Calcutta or Delhi or Madras. HIV is emerging all over India. It may even be too late already. It may even be too late."

It was. By 1991 HIV had already appeared in several Asian populations. Dr. I. S. Gilada, secretary-general of the Bombay-based Indian Health Organization, estimated that 100,000 female prostitutes in his city were infected, 2 million nationwide, with the highest rates—up to 70 percent —seen among India's Tamil women who worked as prostitutes in Bombay. Dr. Jacob John, of Christian Medical College in Vellore, reckoned that a third of the female prostitutes in that Indian city were HIV-positive, as were 6 percent of the men tested in sexually transmitted disease clinics.

WHO's Jim Chin estimated in 1991 that about 250,000 Indians were infected in toto, but characteristically added, "That's a lowball guess-timate."

In Asia's most prosperous countries AIDS remained a stranger. A nationwide 1991 survey of blood donors in Japan, for example, found that the infection rate was less than 0.002 percent; Japan seemed virtually free of HIV. Similarly, Singapore had by mid-1991 seen only eighty HIV infections, according to Dr. Roy Chan of the Singapore AIDS Commission.

But wherever poverty was high, HIV seemed to have made its entry into Asia well before 1991.

Shortly before the opening of the Seventh International Conference on AIDS in Florence, during June 1991, Representative Jim McDermott, a physician and Democrat from the state of Washington, released the results of an AIDS investigation he conducted for the House of Representatives. The report drew appalling conclusions about the Asian pandemic, prompting many fellow politicians to discreetly voice concerns that McDermott was deliberately exaggerating the situation to skew foreign aid commitments. As time would show, however, McDermott's report underestimated the scope of the Asian pandemic.[61]

After touring India, Thailand, and the Philippines at the request of Speaker Tom Foley, McDermott reached the conclusion that "Asia is the sleeping giant of a worldwide AIDS epidemic." He estimated that as of June 1991 some 1 million Indians were already infected with HIV and in the year 2000 India and Thailand combined would have 12 million infected citizens. McDermott predicted that Asia's epidemic would, within perhaps just five years' time, outstrip that of Africa.

With all the prior warnings, prognostications, and clear evidence of the devastation AIDS was inflicting upon Africa, how could the microbe have so overwhelmed Asia? Why hadn't humanity succeeded in preventing HIV's emergence on the continent? As late as the fall of 1989 valid surveys of Thai drug users and prostitutes revealed infection rates below 0.04 percent—seemingly negligible. Yet within a mere twenty months that 0.04 percent infection rate among Chiang Mai prostitutes had soared to more than 70 percent. In just twenty months the virus emerged, spread in an epidemic fashion, and became endemic among key population groups in

Thailand. That constituted the most rapid HIV emergence in the history of the global epidemic.

How could this have happened? In retracing the virus's pathway across Asia, scientists and public health experts gained greater evidence supporting the GPA's earlier theories that human rights violations, poverty, and the behavior of *Homo sapiens* played crucial roles in the emergence of disease. Indeed, the *only* way to comprehend Thailand's astonishingly rapid HIV emergence was to recognize the intimate coupling of social, political, biological, and economic factors.

African history, tragically, repeated itself in Asia. Lessons went unlearned. As had many African societies, most Asian countries initially tried to legislate away the virus by restricting the activities and movement of potential carriers. When that appeared to fail, governments simply refused to acknowledge the virus in their midst, penalizing physicians and experts who raised public alarm about AIDS. Official AIDS figures reported to the World Health Organization reflected attempts by most governments to downplay the impact of AIDS.[62]

During the last weeks of 1987 a meeting on AIDS in Asia was convened in Manila, under the partial sponsorship of WHO. Few cases of the disease had, at that point, surfaced in any Asian nation except one, and that country was populated predominantly by Caucasians: Australia. Though Australia was geographically in the Pacific Rim, most Asians considered the country, and its epidemic, European. But Dr. John Dwyer, the avuncular director of AIDS research at the University of South Wales in Sydney, did his best to convince those in attendance at the Manila conference that the pandemic was coming, and it would hit Asia not in the manner of its attack upon Europe but as it had Africa.

Dwyer pointedly reminded his colleagues that incidences of syphilis, gonorrhea, and other sexually transmitted diseases were rapidly rising throughout Asia; that female prostitution was rampant in almost all of the continent's sprawling centers, and male prostitution in several cities; that opium smokers were abandoning that drug and their pipes in favor of heroin and syringes; and that many parts of Asia were suffering levels of poverty and malnutrition comparable to those seen in Africa.

India's first AIDS cases included recipients of contaminated U.S. blood products manufactured by Cutter Biological, a California-based company, and of anti-RhD vaccines made by Bharat Serums and Vaccine, Ltd., an Indian firm.[63]

During 1985–89 the Indian Council of Medical Research tested more than 2 million people, finding 764 who carried the virus; half of them were female prostitutes. By the end of 1989 the infection rate was soaring. A Bombay survey revealed that 4.9 percent of the city's female prostitutes were infected.[64] As evidence of HIV's presence in India mounted, proposed legislation outlawing sexual intercourse with foreigners was introduced into the Maharashtra state legislature. Though it was defeated, the proposed law reflected a strong mood at that time in Indian society.

So poor were educational efforts that a 1989 survey of a sampling of India's AIDS patients revealed that even they hadn't heard of the disease. Only 4 percent professed to have heard of AIDS before contracting it; most, long after their diagnosis, still had no idea what the disease was. An important factor contributing to ignorance was illiteracy—94 percent of those who were interviewed were unable to read the few AIDS brochures or news articles that were published in India.[65]

By mid-1990 the infection rate among Bombay's prostitutes had risen to 10 percent and 5.6 of every 1,000 blood donors in the city carried the virus. The director of the Indian Medical Research Council, Dr. A. S. Paintal, estimated that Bombay's infection rates had reached such proportions that every day 100,000 sexual acts were performed with HIV-positive female prostitutes.[66] One Bombay STD clinic was finding infection rates among prostitutes of 40 percent.

At the same time, blood donor infection rates rose to 1 percent, and India saw its first cases of HIV involving injecting drug users. Sixty-two heroin users in Manipur were cited in government notices in April 1990. Concern about the blood supply grew when a government survey uncovered 510 HIV-positive blood donors in the state of Gujerat. Among them, 430 were "professional donors," individuals so poor that they subsisted off the meager funds earned by regularly selling their blood. Despite such clear evidence of the microbe's presence in the national blood supply, by the Indian government's admission less than 5 percent of all commercial blood was screened for HIV in 1991. That figure wouldn't budge much in 1992.

Data on HIV infection rates grossly underestimated India's crisis because most high-risk individuals were by 1991 actively avoiding testing. Their reluctance stemmed from widespread knowledge that in Manipur some 100 HIV-positive people were placed in permanent seclusion, chained to their beds, and barred from further social interaction.[67] That drove other potentially infected people underground, away from the public health system.

One group that was able to penetrate the mistrust was the government's cholera program, which enjoyed widespread respect among India's poor. Their 1991 survey in Manipur revealed that an astonishing 80 percent of heroin injectors were HIV-positive.

The microbe had been handed another bit of good fortune. Beginning around 1987, when Burma, once the richest nation in Southeast Asia, was given the World Bank's least developed country status, the traditional opium trade was transformed into a heroin market. It was no longer necessary to ship raw opium paste to Marseilles or other European locales for processing into heroin, thus reducing Burmese profits. But with the shift in opiate processing, heroin was suddenly available for local consumption. Within the so-called Golden Triangle—Burma, southern China, and Laos—opium, and now heroin, production outstripped the 1960s market share held by Turkey and Afghanistan.

In Manipur, which bordered Burma, the sudden availability of the far more powerful heroin drew opium users like bees to honey. Needles, however, were in short supply.

HIV appeared in Manipur riding the crest of the heroin wave. Former opium smokers clumsily experimented with tourniquets, cookers, and syringes, clustering in groups to share not only the knowledge of how to get high but the equipment with which to do so. In less than sixteen months opiate users went from less than 10 percent heroin injectors with under 1 percent HIV seroprevalence to more than 95 percent heroin addicts, mainlining the purest and most powerful smack in the world. And 80 percent of them had within that time also mainlined HIV.[68]

Stunned by the rapidity of HIV's march across India, the World Health Organization mustered $20 million and the World Bank $100 million for the most aggressively funded AIDS education campaign ever planned. But from the start the effort seemed doomed, as political leaders throughout India failed to lend their support, some states refused to participate, and allegations of impropriety, even embezzlement, buzzed about the health system. For example, reluctant to face the political flak that would shower down from all over India's business community if the foreign aid millions were spent outside the country, the government purchased more than a billion defective condoms from a local manufacturer and raised prices on quality imported products.

"We're sitting on a volcano. We won't be able to cope," Maharashtra AIDS researcher Geeta Bhave declared. When all the hundreds of thousands of HIV cases progress to AIDS, she predicted, India's health care system would collapse.

Even HIV-2, previously found almost exclusively in West Africa, emerged in India. By June 1993, STD clinics in Tamil Nadu, Bombay, and Goa reported that 2 to 3 percent of their clients carried the second species of AIDS virus.

German researchers studied the genetics of HIV-1 and HIV-2 viral strains found in various parts of India, finding further evidence for quite recent emergence of the viruses in the country and extraordinarily rapid spread. No matter where they looked, they found infected Indians, and there was no sign that the viruses' spreads were concentrated geographically, as they were in North America and Europe.

The HIV-1 strains were all quite similar and matched closely to a strain of the virus found in South Africa. Given the large number of Indian-descended people living in South Africa and their frequent travels back to India, this was not surprising. But it was astonishing, the researchers said, to discover so little genetic difference between HIV-1 strains in Bombay, Goa, Manipur, and other locations separated by thousands of miles.

"We conclude that these [HIV-1] strains must have been introduced into

India very recently and are spreading *very* rapidly," the German research team said.[69]

HIV-2 also showed little genetic diversity in India, again indicating that the virus had arrived in the country very recently. Further, all HIV-2 strains appeared to be descended from a common ancestor, indicating that a single infected individual brought the virus from West Africa; its emergence and spread within Indian society occurred with extraordinary speed.

By comparing HIV-1 and HIV-2 incidence rates throughout India in early 1993 with the amount of genetic diversity seen in the various viral strains, the Frankfurt research team estimated that India's epidemic was growing by 1 million new infections a year. If Congressman McDermott's estimate of 1 million infected Indians was correct for 1991, and the Frankfurt growth rate held true, the world's oldest continuous civilization would be confronting about 10 million HIV cases in 2000.

But, of course, epidemics couldn't be expected to grow at a stagnant rate over time because the more people infected in a society, the greater the potential for additional infections. Thus, growth rates themselves grew with time. When officials at WHO plotted India's AIDS forecast they were reluctant to put precise figures on the nation's future epidemic. But they were able to compare its growth rate with Africa's: while the slope of Africa's pandemic arched upward at a gentle angle for the 1990s, India's forecast was a sharp line soaring up at a 60-degree angle.

"This is threatening to clear the world," Kenyan AIDS physician Mboya Okeyo said. "Africa first. Then India, then Southeast Asia. Then, who knows?"

In 1993 Subhash Hira moved back to Bombay. Having witnessed the emergence of AIDS in Zambia he was now determined to do all in his power to slow the deadly virus's race across his homeland.

If India's epidemic was racing, Thailand's was moving at supersonic speed. Thai Ministry of Health studies showed that HIV-1 infection rates in nearly every sector of society were well below 2.5 percent in 1989. Eighteen months later double-digit infection rates were the norm all over the country.[70]

Something particularly strange and troublesome happened in Thailand: two separate lineages of HIV-1 emerged, each exploiting entirely different population groups. Among Bangkok's heroin injectors there appeared a B-class virus that looked genetically like a typical American HIV. But a very different HIV emerged in Thailand's prostitute and heterosexual populations, one that closely resembled a virulent virus seen in Uganda. The two strains moved on separate paths in Thailand, and as of 1993 there was no evidence of cross-mixing of their genetic material.[71]

So Thailand, biologically speaking, had two separate epidemics, both of which grew at unprecedented rates.

The Thai situation demonstrated the folly of dismissing the threat of an emerging microbe merely on the basis of initially small numbers of cases.

And it showed, once again, the links between human rights and the emergence of microbes new to a particular society. In the beginning of its epidemic the Thai government took many of the toughest steps advocated by hard-liners elsewhere in the world. A special HIV quarantine unit was established in Lard Yao Prison in Bangkok. When, by June 1989, tests indicated that up to 44 percent of the female prostitutes in Chiang Mai were HIV-positive, the government issued decrees in an attempt to crack down on the brothels. As rates of infection soared among heroin addicts, the government ordered Thai police to come down hard on the drug trade and narcotics injectors. Infected foreigners were deported.

Thailand also took positive steps that drew praise from WHO, including establishing the first national HIV sentinel surveillance program in the developing world. By carefully and continuously monitoring levels of HIV infection in key subpopulations of Thai society, the Ministry of Health kept close tabs on the nation's burgeoning epidemic.[72] It may well have been the best-documented HIV emergence in any society in the world.

Despite these efforts the virus spread at record speed throughout the Southeast Asian nation, primarily via its enormous sex industry. As word of the new plague spread, few Thais took steps to protect themselves. Denial, Thai health official Dr. Chai Podhista said in 1992, was the number one problem.

"We have an expression in Thailand," Podhista explained. "It goes, 'If you don't see the body in the coffin, you don't shed a tear.' Rapid spread of the virus is possible—is ignored—because there hasn't yet been mass death. And there won't be for a few years. Hundreds of thousands of people are all getting infected at once, in a clandestine epidemic. Years from now when they all get AIDS the entire Thai society will go into a state of shock."

In early 1990 a variety of nongovernmental organizations waged impressive AIDS education campaigns, particularly among female prostitutes, and by late 1990 more than 90 percent of the prostitutes working in Chiang Mai were using condoms. But for the majority of the women it was too late: they were already infected.[73]

At the most crucial moment in its emergence into Thai society, HIV was handed a social gift: human chaos. In February 1991 there was a coup in Thailand, bringing a military junta to power.[74] AIDS programs came to a grinding halt; the flow of nearly all foreign aid, including monies earmarked for HIV control, stopped abruptly. AIDS programs generally fell apart, and the military regime responded to the HIV threat with the sorts of repressive actions that typify juntas: conducting raids on brothels, shutting down those that failed to provide adequate bribes, and rounding up children, alleged slaves, and foreign men and women working in the houses of prostitution.[75]

During this time there was little apparent change in the sexual appetite of male customers. Foreign sex tourists continued to flock into Thailand from all over the world, particularly Japan[76] and Germany. And local Thai men showed no signs of slackening their demand. A 1989–90 survey showed that more than a quarter of randomly queried Thai men had sex outside their marriage that year, most of them with male or female prostitutes.[77] A year later no apparent change was observed, and upon compulsory entry to the Thai military more than 97 percent of the twenty-one-year-old recruits admitted to having frequented brothels.

As more and more of Thailand's prostitutes became infected, and concern about AIDS rose, brothel owners began actively recruiting virgins and young girls. This allowed them to market safety for their male clientele, though, of course, it remained extremely risky for the women/girls. Various studies indicated that between 1991 and 1993 the demographics of Thailand's female prostitute population shifted dramatically, particularly in the northern Chiang Mai area, which bordered on Burma. The average ages of the prostitutes plummeted (to include nine- to twelve-year-olds), and the number of Burmese women working in the brothels soared, topping 40 percent by 1993.

According to Amnesty International and Human Rights Watch, nearly all the Burmese female prostitutes were slaves, either sold outright by their parents to brothel brokers or signed on to indentured servant contracts from which they couldn't extricate themselves once they reached Thailand. Few of the girls, most of whom were under eighteen years of age at the time of their sales/recruitments, understood that they were to be prostitutes. The vast majority were illiterate, spoke no Thai, and were virgins when they reached their new brothel homes.

Periodically, Thai police would raid the brothels, round up Burmese nationals, and march them off to the border. Some women, fearing what lay in store for them on the other side, gave sexual favors to the police in exchange for allowing them to return to lives of prostitution.

But what could possibly be more horrible than the lives of sex slavery to which they had been subjected in Thailand?

In September 1988 the Burmese government was overthrown in a coup that brought the most corrupt elements of the country's business and military communities to power. Ne Win took the reins of control, running an authoritarian state that cracked down mercilessly on its citizenry while assiduously protecting the nation's opium/heroin producers. The country, which was renamed Myanmar,[78] sank into chaos. Amid reports of torture and mass executions, as well as economic despair, demonstrations broke out all over Burma in 1990, led by supporters of Aung San Suu Kyi. Though she won the national presidential elections in May 1990 and subsequently received the Nobel Peace Prize, Aung was placed under house arrest. As of mid-1994 she remained a homebound prisoner of the military state.

The government's actions after the 1990 elections only worsened, and the nation became dangerous for all vocal advocates of human rights. Small wonder, then, that Burmese poured illegally across the Thai border by the hundreds every day, and some 300,000 were estimated to have immigrated by 1993. It was perhaps less than surprising also that impoverished parents were willing to sell their daughters to brothel brokers.

In April 1992, Commander Bancha Jarujareet of the Thai Crime Suppression Division announced that twenty-five HIV-positive Burmese brothel girls that had been rounded up by his officers and deported back to Burma were dead. According to the Thai policeman, Burmese officials injected cyanide into the women and set their bodies afloat in a border stream as a warning that Burma would take whatever steps necessary to keep AIDS out of the country.

In Burma, heroin was locally produced and could therefore be purchased cheaply with the internationally worthless Burmese currency. But syringes required foreign exchange, and the abusive Burmese state had become an international pariah, cut off economically from the rest of the world.[79] By 1992, WHO estimates put HIV infection rates among Rangoon's heroin injectors at over 76 percent, but that was a conservative guess. Even if these people knew about HIV, understood how the virus was spread, and were motivated to protect themselves, they couldn't do so.[80]

Fortunately for the Thai people, their nation had, in contrast, a courageous local hero who was willing to take politically dangerous steps to slow the country's skyrocketing epidemic. Mechai Viravaidhya worked within the Ministry of Health and outside the government (depending on who was in power) tirelessly promoting condom use. Equally comfortable arguing with a brothel owner in a Bangkok red-light slum or twisting the arm of a member of the Thai cabinet during a celebrity golf match, Mechai forcefully pushed a "100 percent condom use" policy.[81] But even Mechai knew that the real battle had been lost. AIDS was endemic in Thailand, and in 1993 the government predicted that 3 million adults (out of a population of 25 million over the age of fourteen) would be HIV-positive by the year 2000.

As was the case in Burma and India, the Golden Triangle heroin connection was having an effect on promoting HIV emergence in southern China. Though the government denied it, China had serious heroin, prostitution, and sexually transmitted disease incidences that were readily apparent to even casual observers as early as 1987.[82] The most severe problem was in China's southern Yunnan province, which shared borders with Laos and Burma and had long been an opium center. Yunnan narcotics traffickers, like their counterparts in Burma, had learned how to process opium into heroin. By 1991 heroin was in ready supply in Yunnan; syringes were not. The pattern there mirrored what had occurred with HIV among heroin injectors in Manipur and Rangoon, and by 1993 the World Health Organization was estimating that up to a third of Yunnan heroin users were infected.[83]

Less than a year later WHO announced that heroin was driving a terrible HIV epidemic in Ho Chi Minh City, Vietnam. Among heroin users the

HIV rate climbed from less than 2 percent to more than 30 percent in about nine months' time.[84]

As had been the case with Africa's AIDS epidemic, Asia watchers wondered aloud whether the pandemic might reverse the region's famed "Economic Miracle," causing a Thirdworldization effect. If local epidemics continued to expand at their breathtaking 1989–93 rates, Asia could be expected to overtake Africa in HIV numbers before the turn of the century. And ironically, the fiscal cost to Asia would be greater precisely because the continent's economy had boomed so impressively during the 1980s. With greater prosperity came higher costs. The dollar value of productive capacity lost due to worker illness and death was greater in Asia (compared with Africa) simply because there was a larger highly skilled labor force and incomes across the board were higher. Direct medical costs were higher as well, because of the availability of more sophisticated—and costly—health care systems.

Still, it seemed at first glance unimaginable that AIDS could make a dent in Asia's economic boom. Only a handful of countries (the Philippines, Papua New Guinea, Burma, and Cambodia) experienced negative GNP growth during the 1980s, and many Asian countries had growth rates that were five to seven times greater than those of the United States and Switzerland.[85]

Like Africa, however, much of Asia was simultaneously undergoing other disease emergences that could be expected to compound or synergize with HIV/AIDS. These included dengue, hepatitis (A, B, C, D, and E), multiply drug-resistant malaria, tuberculosis, drug-resistant cholera, and virtually every known sexually transmissible microbe. Though no one knew how to calculate the additive or multiplicative economic impacts the interlocking epidemics might have, it was clear, biologically and epidemiologically speaking, that interconnections existed.

In mid-1993 the GPA estimated that 1.5 million residents of South Asia were HIV-positive, most of them Indian or Thai citizens. For Thailand, specifically, WHO estimated that 450,000 people were infected by late 1992.[86]

But WHO's numbers were surely overly conservative. Newer data demonstrated that the rate of expansion of the country's epidemic, far from slowing as many hoped, was accelerating alarmingly.[87] The only hopeful slowdown in Thailand's plague was seen among injecting drug users in Bangkok, who readily snapped up sterile syringes when they were made available.[88]

Mechai Viravaidhya estimated that Thailand's cumulative AIDS death burden by the year 2000 would be 470,000 to 560,000 adults. Based on an average productivity loss of $22,000 per dead worker, that could inflict an indirect loss of $7.3–$8.5 billion on the Thai economy. Direct treatment costs for those people would be between $61 and $167 million out of a total annual Ministry of Health budget of just over $40 million.[89] In 1992 a single day of AIDS hospitalization in Bangkok cost an average of $298.73;

Thailand was in the unfortunate position of having reached Western standards of curative medicine and hospitalization while its populace still earned Third World wages. By 1992 Thai officials were predicting that the epidemic would push the nation's health and medical advances backward, as an overwhelmed system collapsed under the economic costs and the sheer load of AIDS cases.[90]

Drawing on slightly different estimates of both the forecast epidemic size and indirect costs, WHO predicted a total economic burden from AIDS of $9 billion for Thailand by the year 2000.[91] And the GPA told World Bank officials as early as October 1991 to expect a possible economic downturn in South Asia during the latter half of the decade: it was a view shared by the Thailand Development Research Institute in Bangkok.[92]

The U.S. Census Bureau issued dire forecasts for Thailand, based on HIV-prevalence rates as of early 1994. The Bureau predicted that by 2010 Thailand would have experienced such severe devastation due to AIDS that the country's population growth rate would have plummeted to −0.8 percent (from a pre-AIDS predicted +0.9 percent); there would be 25 million fewer people in the country than would have been the case in the absence of AIDS; life expectancy would take a nosedive from what would have been 75 years to a mere 45 years; child mortality rates would more than triple (reaching some 110 per 1,000 children born); and the nation's crude death rate would soar from about 6 deaths per 1,000 to more than 22 per 1,000.[93]

Though few analogous economic analyses had been done for India, Burma, the Philippines, or other Asian countries in the grips of HIV, there was a clear consensus in international public health circles by 1993 that the pandemic would, at the very least, exert a Thirdworldization effect upon the health care systems, tourist industries, and government-funded social service sectors of hard-hit countries. Worst-case scenarios forecast sharp declines in both agricultural and industrial productivity with resultant declines in GDPs.[94] The United Nations Development Program and the Asian Development Bank predicted in late 1993 that the HIV epidemic would increase general levels of poverty and, by the year 2000, cause local famines in key areas.

Such dire economic forecasts, whether they concerned the projected impact of AIDS on Africa, Asia, or Latin America,[95] were always intended to draw the attention of wealthy donor states. The nations of North America, Western Europe, and, to a lesser degree, Japan and the Soviet Union had always been forthcoming with cash when a crisis struck, even if the quantities were more symbolic than substantial. If the cash was offered at interest, Africa and Latin America might cringe, but Asia had an excellent debt-repayment record.

Throughout the world hope for a global AIDS bailout rose as Berliners clawed away at the wall that had physically divided their city for three decades. What began with dockworkers in Gdansk in the early 1980s built slowly for years in pockets of antiauthoritarian resistance that spanned from Prague to Riga, from Vladivostok to Berlin. Once the Berlin Wall fell there was no turning back: the ideal and reality of communism were dead. And with the end of communism came capitalist dominance and Western victory in the Cold War. Threat of global thermonuclear annihilation suddenly seemed quite remote. Politicians all over the world spoke of a Peace Dividend. And suddenly the world had surplus cash, they claimed, and long-neglected social programs could now be subsidized. For a few moments in history, it seemed, people around the world were remarkably optimistic.

But no Peace Dividend appeared. People craned their necks looking for it, soon spotting a shadow emerging on the horizon. Excitement yielded to despair and frustration as they recognized the shadowy Dividend for what it was: international recession.

After all the celebrations and dancing in the streets of Prague and Berlin settled down, the West got a good, hard look at what lay behind the Wall, inside the long-sequestered world of communism. And they discovered that Stalinists from Uzbekistan to the Baltics had been juggling the books for decades. The East was broke.

Worse yet, its populations, which had long had nearly every aspect of their lives controlled by the state, were ill prepared to build strong civic societies. With their economies in a shambles, cynicism quickly overcame the brief sensation of elation for most Europeans.

Reunification of the two Germanys was concretized on October 3, 1990, amid fetes and fireworks, but by the end of that year official unemployment in the former GDR had soared from zero to more than 350,000 adults.

Overnight the former Cold War multitrillion-dollar spending became a latter-day Marshall Plan for reconstruction of the ex-communist world. Ten billion dollars shifted from coffers in Bonn to national bank vaults in Moscow in a single day. And that was just one of many West-to-East transfers.

Not only was there no Peace Dividend, there was newfound, long-term structural agony. Even the booming Asian economies felt the pain as demand for autos, electronics, and consumer goods dropped in Europe and North America.

While much of the world watched the demolition of the Berlin Wall during the fall of 1989 with astonishment and elation, Hans Seyfarth-Hermann dashed about Checkpoint Charlie tossing condoms at the crowds of East Germans as they poured through. The bewildered East Germans snapped the packets out of the air and examined what looked like matchbook covers. They read this brightly colored inscription: "You will see many tantalizing things during your visit to West Berlin. Enjoy yourself, but remember, we have AIDS."

Inside each putative matchbook was a latex condom. Political openness, it appeared, could carry a price tag. If it was true, as the old Stalinist leaders claimed,[96] that AIDS hadn't made its way yet into most of Eastern Europe, the fall of the Wall would surely put an end to the political barriers that had allegedly kept the microbe at bay.

Seyfarth-Hermann and fellow AIDS activist Julian Eaves were in a Berlin gay bar called the Dark Cellar the night of November 9, 1989, when they overheard Germans speaking in a startling accent. The two of them realized that the men were from Saxony, part of East Germany, and that they were filled with excitement that night, Eaves later recalled, "trying to enjoy the wild life in the big city." Eaves and Seyfarth-Hermann recognized that their Saxon gay counterparts knew nothing about AIDS and safe sex. They also were sadly aware that some West Berlin gays, sick and tired of "latex sex," might take advantage of the Easterners' ignorance.

The two activists spent the following day making hundreds of the special packets which they later tossed at the hordes of Saxons and other Easterners crossing through Checkpoint Charlie.

"Everything that is new is welcome now in East Germany," Eaves explained. "The old stigmas have been thrown away, and everything is possible. We hope East Germany will achieve a world level in everything, except AIDS deaths."

No one could imagine that just four months later prostitutes in West Berlin would be on the verge of staging a protest strike over the thousands of competitors that flooded in from the East every Friday night to earn valuable deutsche marks over the weekend. Hungry for hard currency, young women, most of whom didn't really consider themselves prostitutes, would pour into Berlin to turn a few quick tricks, often for as little as five deutsche marks. The regular hookers would be outraged because the newcomers would charge far less than the former going rate, and they wouldn't require that their customers wear condoms.[97]

Within three years the Eastern prostitutes would be a regular feature of red-light districts all over the wealthier West.

HIV would also ride Europe's new heroin trail. Opening up the formerly secluded states rang bells of opportunity for organized criminal elements on both sides of the former Wall. Poland, in particular, would become both a center for a locally produced opiate called *kompot* and a transfer point for pure heroin imported from other parts of the world and destined for distribution in Central Europe.

The first serious emergences of HIV in Eastern Europe were not via either prostitution or heroin injection, however. Rather, they came by means that reflected the tragic state of medicine in much of the communist bloc.

Though there had been isolated AIDS cases in Russia for at least four years, HIV really emerged during the early spring of 1988 in Elista, capital of the Kalmyk Republic, located on the Caspian Sea. A baby languished on the pediatric ward of the town's hospital, suffering every imaginable ailment. Doctors were stumped, unable to reach a diagnosis, until one suggested sending blood samples from the infant to Valentin Pokrovsky, a virologist doing AIDS research in Moscow. Pokrovsky confirmed that the child was infected with HIV.

The child's father, it turned out, had visited the Congo in 1981, where he apparently was exposed to HIV. He passed the virus sexually to his wife, who, in turn, transmitted HIV to the child.

It was tantamount to treason to publicly acknowledge shortages of vital goods during the regime of Joseph Stalin, and forty years after the dictator's death many Soviet citizens remained reluctant to step outside normal bureaucratic channels in order to draw attention to production deficiencies. In 1988, however, prior to news from Elista, U.S.S.R. Minister of Health Alexander Kondrusev publicly decried the country's sorry state of medical supplies. In particular, he warned that the nation needed to use 3 billion syringes per year, but was only manufacturing 30 million annually, and importing none. Simple mathematics indicated, then, that the average syringe was being used 100 times. Kondrusev warned that this syringe shortage could spell disaster.

He would soon prove remarkably prescient.

The AIDS baby at the Elista hospital was treated by staff who used the same syringes to withdraw blood samples from and administer drugs to all the babies on the neonatal ward. For more than three months the nurses unknowingly injected HIV into all of the babies and, in a few cases, their mothers.

As the numbers of AIDS babies mounted, the overwhelmed Elista doctors ordered some of the infants shipped to a hospital in Volgograd. And again, the medical staff reused syringes over and over, soon having infected nearly every child on the Volgograd baby ward.

The incidents were kept quiet until early 1989 when a Russian trade union newspaper, *Trud*, broke the story. According to *Trud*, Health Minister Kondrusev had grossly underestimated the enormity of the gap in the Soviet Union between the number of injection procedures of one kind or another that were performed by health providers and the annual production rate of sterile syringes.[98] While leaders in Moscow received single-use sterile injections, the masses living in outlying areas relied on hospitals that suffered permanent supply shortages. So in Elista and Volgograd, for example, health care workers had little choice but to reuse syringes 400 or 500 times, occasionally honing the needles on a whetstone so that they would still pierce skin.

It was horribly reminiscent of the events in Yambuku Hospital in 1976, where Belgian nuns used a handful of syringes hundreds of times per week, unwittingly spreading the deadly Ebola virus. That, however, occurred in a remote, impoverished region of Central Africa; the Soviet Union was, allegedly, part of the advanced industrialized world.

For three years Soviet health leaders counted the numbers as similar hospital outbreaks of HIV surfaced in Rostov, Astrakhan, and Stavropol.

By June 1990, Vadim Pokrovsky was telling the world that 260 children had become infected with HIV as a result of unsterile needles.[99]

Moscow's Second City Hospital for Infectious Diseases was designated the nation's AIDS treatment center and half the patients on its wards were children under five years of age. As fear of AIDS mounted in the Soviet medical community, widespread shortages were reported not only of syringes but of latex gloves, sterile catheters, surgical gowns, transfusion equipment, dental drills and probes, and other essential supplies. In the new atmosphere of *perestroika*, young physicians for the first time spoke frankly about the inadequacies of the Soviet medical system.

The result was widespread public panic and a sharp decline in willingness to undergo invasive medical procedures. Dentists, vaccinators, physicians—all health providers noted a drop in attendance, particularly in large cities where the media gave serious attention to the young physicians' disclosures.

Dr. Mikhail Narkevich, newly appointed head of AIDS education in the Ministry of Health, was forced to concede that the nation's economic difficulties were so grave that adequate medical supplies could not possibly be available until 1992–93. By 1994 Russian physicians would be crying out even more loudly for supplies that still hadn't materialized.

In the absence of supplies sufficient to limit the spread of HIV within medical facilities, panic further increased. There were anecdotal reports of people beating AIDS patients and of health care workers refusing to go near people who carried the virus. The Ministry of Health was forced in 1991 to offer higher salaries to doctors and nurses who worked with HIV/AIDS patients as compensation for the perceived risks involved.

But Soviet leaders were preoccupied with far more pressing issues than supplies of syringes. The country was literally falling apart. Food shortages, riots, separatist uprisings, political instability, and a face-off between the hero of *glasnost*, Mikhail Gorbachev, and upstart leader Boris Yeltsin monopolized national attention. By 1991 the Soviet Union no longer existed. By 1993 two major coup attempts had threatened the stability of the Russian Republic, and insurrections had occurred inside most of the former Soviet socialist states.

AIDS was overshadowed by history. And the microbe spread, unfettered by any serious efforts on the part of human beings to limit its modes of transmission. Prostitution and drug abuse stepped into the economic vacuum of social restructuring. Criminal elements gained control of many foreign trade sectors, and syringes remained in short supply.

By late 1993 the microbial situation was clearly out of control. Before the Berlin Wall fell, Russia's syphilis rate was 4.3 cases per 100,000 people annually. Amid the national chaos, health officials said they were witnessing a syphilis epidemic. In St. Petersburg, for example, the incidence of syphilis increased eightfold between 1989 and 1993, with most of the newly infected individuals young, destitute female prostitutes. In the same city the incidence of gonorrhea among teenagers had soared 150 percent by 1993, as compared with 1976 levels. And in the same subpopulation syphilis incidence was up 400 percent.

Dr. Nikolai Chaika, of the St. Petersburg Pasteur Institute, announced that all Russian disease data, including numbers of HIV/AIDS cases, were unreliable due to the "complete collapse of Russian medicine." The social fabric of Russian society was unraveling, he said, and people were turning to behaviors that virtually guaranteed the spread of disease.

Thirdworldization had set in. Russia, as well as nearly all of the other former Soviet states, was rolling backward on the development scale. Epidemics of all sorts of diseases were reported anecdotally, though most were impossible to verify given the collapse of epidemiological systems. In the summer of 1992 cholera outbreaks were reported in Makhachkala, Nizhny Novgorod, Krasnodar, Naberezhnye Nizhny, and Moscow. The Tass news agency reported an outbreak of anthrax among peasants in the Altai region and typhoid fever in Volgodonsk. Even a case of bubonic plague was reported from Kazakhstan.[100]

In March 1993 special counsel to President Boris Yeltsin, Dr. A. V. Yablokov, addressed the grave state of the Russian people's health in a speech before the nation's Security Council.[101] He revealed that in 1991 Russia's "total losses due to premature mortality amounted [to] 2.23 million person-years of labour activity. . . . It is obvious *that prevention of population health losses due to premature mortality from socio-economically conditioned causes is the most important strategic direction in improving safety and security of life of peoples of Russia* [his emphasis]."

The primary cause of Russia's massive excess death burden was suicide, which rose by 20 percent between 1991 and 1992. Alcoholic self-destruction, drunk-driving accidents, and homicides ranked as the remaining top causes of the excess death rates.

Meanwhile, he said, the nation's medical and public health system had deteriorated to the point where in 1991, 70 percent of all pregnancies involved serious complications, "only half of deliveries were considered normal," anemia rates among pregnant women had increased by 61 percent in just three years, and maternal mortality rates were five times those in Western Europe. And preventable deaths—those ascribed directly to drug shortages or medical and public health failures—had risen sharply since 1990.

"Among these are all forms of tuberculosis, some infectious diseases (measles, whooping cough, tetanus, typhoid fever) . . . respiratory diseases, pregnancy complications, diseases of the perinatal period," Yablokov said.

Life expectancy in Russia was lower in 1990 than in 1964 (70.1 years versus 70.4) and real life-span measurements for some areas of the country were as low as 44 years.

Separate EC studies of Russian health revealed that tuberculosis rates were climbing sharply. In Siberia in 1990 there was a TB incidence of 43 cases per 100,000 people (as measured by positive sputum). By 1993 that ratio had more than doubled, to 94:100,000. Over the same period Moscow's TB rate jumped from 27:100,000 to 50:100,000. The principal cause of the escalation was said to be the lack of foreign exchange with which to purchase European- and American-made antituberculosis drugs; without treatment an ever-expanding pool of contagious individuals was spreading the disease to others.[102]

Perhaps the most striking example of Russian Thirdworldization was the 1993 outbreaks of diphtheria in St. Petersburg and Moscow.

A hallmark of the old Soviet Union had been its tremendous success in universal vaccination and resultant declines in the incidence of former scourges such as measles, whooping cough, polio, and diphtheria. By 1976 the numbers of diphtheria cases diagnosed in the U.S.S.R. approached zero.

But in 1990 diphtheria reemerged in Russia, with 1,211 cases reported from St. Petersburg, Kaliningrad, Orlovskaya, and Moscow. The epidemic took off, with reported cases and geographic spread increasing steadily well into 1994. In 1991 nearly 1,900 diphtheria cases and 80 deaths were reported in Russia. Though the bacterial disease could be treated with antibiotics, deaths occurred due to the sorry state of the nation's health care systems.

During the summer of 1993, when nearly 1,000 cases were reported in a single month in Moscow and St. Petersburg, the British government issued travel advisories recommending that its citizens be revaccinated prior to traveling in the former U.S.S.R. And the numbers kept rising: between January and August 1993, nearly 6,000 Russians came down with diphtheria, 106 died.[103]

There had been massive waves of migration from outlying rural and rust-belt areas of Russia into Moscow, St. Petersburg, and, to a lesser degree, Kaliningrad and Orlovskaya. Most of the migrants were economic refugees, hoping to find work in the country's largest cities. But they soon discovered quite the opposite, according to Russian authorities, and many thousands ended up living inside public transport stations—train depots, airports—in squalid conditions. Over 40 percent of the diphtheria cases occurred among these homeless.

Diphtheria had been virtually eradicated from the United States because of strict rules about preschool vaccination of children with the so-called DTP shots. But DTP shots had also been meticulously administered in Russia since the early 1960s. Nearly every new diphtheria case in the country had involved individuals who were previously vaccinated.

Officials concluded that the vaccine didn't, as previously thought, work for a lifetime. It might offer less than five years' protection against the disease. The reason, they said, was not a failure of the vaccine, but its success.

It seemed that thirty years of worldwide vaccination had drastically reduced the numbers of diphtheria microbes in the world, and most people lived their lives never being naturally exposed to the bacteria. Natural exposure in the 1960s, however, acted like booster shots, constantly rejuvenating lagging immunity: that explained why health officials had then mistakenly concluded that the vaccine provided lifetime protection. But by the 1980s most people's immune systems never saw diphtheria, and the natural booster effect didn't take place.

In response to global concern that the Russian epidemic might spread to other parts of the former Soviet Union, the Baltic States, or Scandinavia, the Russian Ministry of Health announced in 1993 a five-year plan to revaccinate up to 90 percent of all the nation's citizens. Some UN officials privately questioned whether the Russians were responding with the proper amount of urgency and haste: a handful of diphtheria cases were reported during the summer of 1993 in Finland and the Baltic States.[104] Still other skeptics questioned the wisdom of a mass adult vaccination campaign in Russia, given the country's acute shortage of syringes. Considering the lesson of Elista, they asked, might such an effort only hasten emergence of blood-borne microbes, such as hepatitis B and HIV?

The Elista tragedy was closely mirrored by events in Romania, where the government of communist dictator Nicolae Ceausescu covered up the existence of thousands of institutionalized orphans who were the legacy of decades of strict bans on all forms of contraception. Further, the Ceausescu regime hid evidence that many of these children were infected with HIV,[105] the tragic outcome of common use of contaminated syringes[106] and the primitive belief that injecting adult blood into children gave them strength.[107]

When the Iron Curtain was lifted, it revealed the Third World status of the old communist regimes, and conditions which only worsened amid the infrastructural chaos. And with that revelation came recognition of countless opportunities for the further emergence of not only HIV but all manner of microbes.

But there was no need to search behind the Iron Curtain, the Bamboo Curtain, or below the Sahara to witness microbial exploitation of Thirdworldization. The process was occurring during the 1980s and the early 1990s inside the wealthy nations of North America and Western Europe.

Despite the AIDS epidemic, most of the public health community, which was not involved in infectious diseases work, remained optimistic during the 1980s. So much so that health became a matter of personal responsibility. Health economists tallied up the costs of diseases that were preventable through diet, exercise, cessation of tobacco or illicit drug use, elimination of alcoholism, and the like, reaching the conclusion that personal health decisions were no longer the exclusive purview of individual choice. Smokers, they concluded, cost the rest of society billions of dollars. So did alcoholics. And fat people.

"The cost of sloth, gluttony, alcoholic intemperance, reckless driving, sexual frenzy, and smoking is now a national and not an individual responsibility," wrote Dr. John Knowles, president of the Rockefeller Foundation. "This is justified as individual freedom—but one man's freedom in health is another man's shackle in taxes and insurance premiums. I believe that a right to health should be replaced by the idea of an individual moral obligation to preserve one's own health—a public duty if you will."[108]

Public health advocates warned, however, that it was exceedingly unfair, and unrealistic, to hold poor Americans responsible for their health—to condemn them, as it seemed Knowles did, for their inability to afford ideal foods, membership in exercise clubs, and temperance in all sexual and intoxicant affairs. Further, they warned that the medical triumphs that had sparked such rosy calls for personal responsibility were fleeting. In the face of rising poverty, they said, the old scourges would return.[109]

It wasn't necessary to go to Africa to see AIDS orphans or whole families buried side by side. New York City alone would have more than 30,000 AIDS orphans by the end of 1994, Newark over 10,000. The U.S. Department of Health and Human Services predicted that there would be 60,000 AIDS orphans in the country by the year 2000.[110] Just as AIDS was exhausting the extended-family networks in much of Africa, so it was taxing the social support systems in America's poorest communities.

With every passing year in America's AIDS epidemic the impact upon the nation's poorest urban areas grew more severe. It compounded the effects of other plights—homelessness, drug abuse, alcoholism, high infant mortality, syphilis, gonorrhea, violence—all of which conspired to increase levels of desperation where dreams of urban renewal had once existed.

As the virus found its way into communities of poverty, the burden on urban public hospitals was critical. Unlike Canada and most of Western Europe, the United States had no system of national health care. By 1990 an estimated 37 million Americans were without any form of either public or private health insurance. Too rich to qualify for government-supported health care, which was intended only for the elderly and the indigent, but too poor to purchase private insurance, millions of Americans simply prayed that they wouldn't fall ill. Another 43 million Americans were either chronically uninsured or underinsured, possessing such minimal coverage that the family could be bankrupted by the required deductible and co-payments in the event of serious illness.[111]

Any disease that hit poor urban Americans disproportionately would tax the public hospital system. But AIDS, which was particularly costly and labor-intensive to treat, threatened to be the straw that broke the already weakened back of the system.[112]

"We are fighting a war here," declared Dr. Emilio Carrillo, president of the New York City Health and Hospitals Corporation, which ran the city's network of public medical facilities. "People are sick and dying from AIDS, tuberculosis is rampant, malnutrition, drug addiction, and other diseases resulting from poverty are also at epidemic levels, while at every level of government, city, state, and federal, the health care system is facing cutbacks. Only the number of sick people and people in need of basic health care is not being cut back. Among them there have been no reductions, no downsizing. They are still coming in to us for treatment."

A 1990 survey of 100 of the nation's largest public hospitals (conducted by the National Association of Public Hospitals) revealed worsening situations in all American cities and predicted collapse of the "public safety net" offered by the system. A microbe that had emerged in America only a decade earlier was threatening to topple the system.

By 1987, 3 percent of the women giving birth in hospitals in New York City were HIV-positive, as were some 25 percent of their babies, according to the U.S. Public Health Service. Nearly two-thirds of those mothers and babies were born in public hospitals located in largely African-American or Hispanic neighborhoods of Brooklyn and the Bronx. The following year the state of New York concluded that one out of every 61 babies born in the state was infected with the virus. But that rate varied radically by neighborhood: in posh, semi-rural communities located far from New York City fewer than one out of every 749 babies was born HIV-positive in 1988. But in desperately poor neighborhoods of the South Bronx one out of every 43 newborns, or 2.34 percent, was infected—and every one of them was born in a public hospital.[113] Those numbers could only be expected to worsen as the epidemic's demographics shifted into younger, predominantly heterosexual population groups.[114]

A significant percentage of the nation's HIV-positive population was also homeless, living on the streets of American cities. A 1991 study, led by Andrew Moss, of homeless men and women in San Francisco found that 3 percent of those who had no identifiable risk factors for HIV exposure were infected. Another 8 percent of the homeless were HIV-positive due to injecting drug use, prostitution, or sex with an infected individual. Overall, more than one out of every ten homeless adults in San Francisco carried the virus.[115]

HIV wasn't the only microbe that was exploiting opportunities in America's urban poor population: hepatitis B (which by 1992 was responsible for 30 percent of all sexually transmitted disease in America), syphilis, gonorrhea, and chancroid were all appearing less commonly in Caucasian gay men and with alarming, escalating frequency in the heterosexual urban poor, particularly those who used crack cocaine or heroin. By 1990 two-thirds of New York State's syphilis cases, for example, were African-Americans residing in key areas of poverty, and within that population male and female infection rates were equal.

In 1993 the New York City Health Department announced that life expectancy for men in the city had *declined*, for the first time since World War II, from a 1981 level of 68.9 years to a 1991 level of 68.6 years. This occurred even though outside New York City life expectancies for men in the state *had risen* during that time from 71.5 years to 73.4 years. Though rising homicide rates played a role, city officials credited AIDS with the bulk of that downward shift. By 1987 AIDS was already the leading cause of premature death for New York City men of all races and classes; by 1988 it was the number one cause for African-American women as well.

Well before AIDS was claiming significant numbers of Americans, Harlem Hospital chief of surgery Dr. Harold Freeman calculated that men growing up in Bangladesh had a better chance of surviving to their sixty-fifth birthday than did African-American men in Harlem, the Bronx, or Brooklyn. Again, violence played a significant role in the equation, but it was not critical to why a population of hundreds of thousands of men living in the wealthiest nation on earth were living shorter lives than their counterparts in one of the planet's poorest Third World nations. Average life expectancy for Harlem's African-American men born between 1950 and 1970 was just 49 years. Freeman indicted disease, poverty, and inequitable access to medical care as the primary factors responsible for the alarming death rate among African-American men.[116]

Well before a new tuberculosis epidemic struck several U.S. cities, the warning signs were there for all to see: rising homelessness, fiscal reductions in social services, complacency in the public health sector, rampant drug abuse, and increases in a number of other infectious diseases. The emergence of novel strains of multiply drug-resistant TB came amid a host of clangs, whistles, and bells that should have served as ample warning to humanity. But the warning fell on unhearing ears.

During the Ronald Reagan presidency American fiscal policies favored expansion of the investment and monetary sectors of society and simultaneous contraction of social service sectors. Economist Paul Krugman of the Massachusetts Institute of Technology estimated that 44 percent of all income growth in America between 1979 and 1989 went to the wealthiest 1 percent of the nation's families, or about 800,000 men, women, and children. On the basis of Federal Reserve Board data, Krugman calculated that total wealth (which included far more than the cash income measured above) was more concentrated in the hands of the nation's super-rich than at any time since the 1920s. By 1989, the top 1 percent richest Americans controlled 39 percent of the nation's wealth.

Several studies showed that by the end of 1993 more than 25 million Americans were hungry, consuming inadequate amounts of food. In 1993 one in ten Americans was compelled to stand at least once a week on a breadline, eat in a soup kitchen, or find food through a charitable agency. And the numbers of people living below the federally defined poverty line increased three times faster between 1982 and 1992 than the overall population size. In 1992 some 14.5 percent of all American citizens lived in conditions of legally defined poverty. Most were single mothers and their children.[117]

Though difficult to measure precisely, the numbers of homeless people in America rose steadily between 1975 and 1993,[118] and the demographics of the population shifted from the traditional hard-core group of older male vagrants and alcoholics to a younger, more heterogeneous contingent that included large numbers of military service veterans, chronically institutionalized mental patients, individuals with severe cocaine or heroin habits, and newly unemployed families and individuals. Estimates of the size of the nation's homeless population ranged from about 200,000 to 2,200,000, based on head counts in emergency shelters and a variety of statistical approaches to the problem.[119]

Even more difficult to calculate was the rise in housing density in urban areas. As individuals and whole families faced hardships that could lead to homelessness, they moved in with friends and relatives. One estimate for New York City during the 1980s suggested that 35,000 households were doubled up in public housing, along with 73,000 double-density private households. Assuming each family averaged four members, that could mean that more than 400,000 men, women, and children were packed into double-density housing.[120]

Finally, a large percentage of the urban poor population cycled annually in and out of the criminal justice system. Young men, in particular, were frequently incarcerated in overcrowded jails and prisons. In 1982 President Ronald Reagan called for a war on drugs: by 1990 more men were in federal prisons on drug charges alone than had comprised the entire 1980 federal prison population for all crimes combined. The pace of federal, state, and county jail construction never came close to matching the needs created by the high arrest rates. As a result, jail cells were overcrowded, and judges often released prisoners after shortened terms, allowing them to return to the community. This, too, would prove advantageous to the microbes.

Some of the microbial impact of this urban Thirdworldization might have been controllable had the U.S. public health system been vigilant. But at all tiers, from the grass roots to the federal level, the system was by the mid-1980s in a very sorry state. Complacent after decades of perceived victories over the microbes, positioned as the runt sibling to curative medicine and fiscally pared to the bone by successive rounds of budget cuts in all layers of government, public health in 1990 was a mere shadow of its former self.

An Institute of Medicine investigation determined that public health and disease control efforts in the United States were in a shambles. Key problems included "a lack of agreement about the public health mission" between various sectors of government and research; a clear failure of public health advocates to participate in "the dynamics of American politics"; lack of cooperation between medicine and public health; inadequate training and leadership; and severe funding deficiencies at all levels.

"In the committee's view," they wrote, "we have let down our public health guard as a nation and the health of the public is unnecessarily threatened as a result."[121]

An example of public health's disarray that proved painfully embarrassing to officials during the 1980s was provided by measles. In 1963 a safe, effective measles vaccine became widely available in the United States and childhood cases of the sometimes lethal disease plummeted steadily thereafter. In 1962 half a million children in the United States contracted measles; by 1977 fewer than 35,000 cases were reported annually and many experts forecast that virtual eradication of the disease would soon be achieved.

But problems were already apparent in 1977: many children who were vaccinated before the age of fourteen or fifteen months later developed measles, and researchers soon understood that timing was crucial to achievement of effective immunization. Vaccination schedules were adjusted accordingly, executed nationwide with vigor, and the number of measles cases in the country continued to decline. The only serious emergences of the microbe took place in communities where a significant number of parents refused, for religious reasons, to have their children vaccinated.[122]

By the early 1980s the United States had achieved 99 percent primary measles vaccination coverage for young children and fewer than 1,497 measles cases occurred in the country in 1983.

In 1985, however, a fifteen-year-old girl returned from a trip to England to her Corpus Christi, Texas, home and promptly developed the roseola rash that was characteristic of measles. The virus quickly spread through her high school and the local junior high school. Ninety-nine percent of the students had, during infancy, received their primary live-measles immunizations; 88 percent had also had their recommended boosters. Nevertheless, fourteen students developed measles.[123]

Blood tests performed during the outbreak on more than 1,800 students revealed that 4.1 percent of the children, despite vaccination, weren't making antibodies against the virus, and the lowest levels of antibody production were among those who hadn't had boosters. All the ailing teens fit that category. The clear message was: (1) primary immunization, in the absence of a booster, was inadequate to guarantee protection against measles; and (2) having even a handful of vulnerable individuals in a group setting was enough to produce a serious outbreak.[124]

The crucial importance of proper timing of vaccination and booster follow-up was further supported by other measles outbreaks among groups of youngsters whose primary vaccination rates exceeded 97 percent.[125] In 1989 the measles rate in the United States climbed considerably. More than 18,000 cases of measles occurred, producing 41 deaths: a tenfold increase since 1983. Forty percent of the cases involved young people who had received their primary, but not booster, vaccinations; the remainder had had no shots, or their vaccinations were administered at improper times.

Though some pediatricians and policy makers found the 1989 numbers worrisome, nobody forecast an epidemic. Measles epidemics were considered Third World problems by 1989.

But an epidemic did occur. The incidence of measles in the United States leapt by 50 percent between 1989 and 1990. More than 27,000 U.S. children, half of them under four years of age, contracted measles during 1990; 100 died of the disease.

Hardest hit was New York City, with 2,479 reported measles cases.

CDC investigators were baffled by the severity of illnesses in the 1990–91 epidemic.

"These kids are much sicker, and death rates are definitely higher," the CDC's Bill Atkinson said. "We don't know whether it's because the strain of measles out there is more virulent, or the kids are more susceptible."

Many of the ailing children, particularly in New York City, had never been vaccinated. They hadn't even received their primary shots, much less boosters.

"Now the majority of cases are in unvaccinated children," Dr. Georges Peter, chair of the American Academy of Pediatrics, said. "Measles is the most contagious of all the vaccine-preventable diseases. The nature of the problem has clearly changed—it is undoubtedly a failure to vaccinate. And what this really is, is indication of a collapse in the public health system, of lack of access to health care."

What was going on? Were parents deliberately keeping their children away from doctors? Were Americans suddenly phobic about immunizations?

The answers, it turned out, could be found in the demographics of the population of children with measles. The vast majority lived in large cities—New York, Chicago, Houston, Los Angeles—and were nine times more likely to be African-American or Hispanic.

As the epidemic persisted in 1991, worsening in New York City's African-American and Hispanic populations, it was evident that the microbe had successfully emerged in populations of poor urban people with little or no access to health care. This underlying social weakness also facilitated surges in whooping cough and rubella cases during 1990–93.[126]

In 1978 the U.S. Surgeon General had declared that measles would be eradicated from the country by 1982, and an ambitious immunization campaign was mounted. By 1988, however, conditions of poverty, health care collapse, and public health disarray had grown so acute that the United States had a poorer track record on *all* childhood vaccination efforts than did war-torn El Salvador and many other Third World countries.[127]

In some inner-city areas—notably in New York City—only half of all school-age children had been vaccinated. For much of the urban poor in America the only point of access to the health care system was the public hospital emergency room. Families spent anxious, tedious hours queued up in urban ERs because they felt that they had no choice: there were no clinics or private physicians practicing in the ghettos, few alternative sources of basic care. But few poor families were willing to put up with a daylong line in the ER simply to get their children immunized, particularly if it meant loss of a day's pay.[128]

Further study of the measles crisis revealed that some deaths and many cases—indeed, most at the key hospitals—went unreported. The city of New York uncovered up to 50 percent underreporting in the region's largest inner-city hospitals during the 1991 epidemic. It was possible that up to 5,000 cases of the disease occurred in New York City, though only half that number were officially reported.[129]

In 1993, World Health Organization adviser Dr. Barry Bloom, of the Albert Einstein School of Medicine in the Bronx, announced that the United States had fallen behind Albania, Mexico, and China in childhood vaccination rates.[130]

At the World Summit on Children convened by the United Nations in September 1990, the Bush administration was in the dubious position of having, on the one hand, to pledge sweeping concern for the health and survival of the world's children while hoping no one would publicly note that the health status of America's impoverished kids rivaled that of children in much of Africa and South Asia.

"This society is so wealthy, obviously this country is better off than the Third World. But this country should be ashamed of the child mortality rates and health," decried Jim Weill, of the Children's Defense Fund, at the Summit. "The U.S. ranks 19th in the world on infant mortality, 29th in low birthweight babies, 22nd on child mortality for children under five, and, perhaps most amazing, 49th in the world on child immunization, for our non-white children. We kill our children.

"Let's face it, when it comes to America's children we live in the Third World."

Not only had America's cities sunk to Third World levels of childhood vaccination and access to health care, but its surveillance and public health systems had reached states of inaccuracy and chaos that rivaled those in some of the world's poorest countries.[131]

Weill's words had barely been uttered when officials at the CDC acknowledged that America's public health system was also doing a worse job of handling tuberculosis than did many African nations.

Multiply drug-resistant TB had arrived. Microbes had emerged that were so broadly resistant to antibiotics that, in practical terms, they were invulnerable.

Tuberculosis didn't reemerge overnight in the United States. On the contrary, the new mutant microbes made numerous tentative incursions into the *Homo sapiens* population over a period of years. It wasn't a surprise attack.

It almost seemed as if human beings were deliberately ignoring the plentiful warning signs.

Though tuberculosis had never disappeared, its incidence had declined steadily in the United States since the 1880s, and hit record lows following the introduction of antibiotic treatment. The robust *Mycobacterium tuber-*

culosis was impossible to eradicate, as half the world's population at any given time was infected with the bacteria. For most people *M. tuberculosis* infection was a benign event: the microbe was kept in check by the immune system and the individual never, throughout his or her life, fell ill.

On average, infected people had a 10 percent chance of developing active disease sometime during their lives, and a 1 percent chance of coming down with a lethal TB illness. Thus, statistics would indicate that about 2 billion human beings were infected with the microbe in 1988; 200 million would during their lives suffer tuberculosis and 2 million would die of the disease.

But those neatly averaged numbers belied the true nature of the risks of TB and the disease's extremely unequal distribution worldwide.

From the earliest days of Western tuberculosis research, scientists and physicians had recognized that the microbe moved hand in hand with poverty. Though there were famous cases of TB among more affluent individuals, most of the world's tuberculosis victims had always been the poorest citizens.

The nature of the association between TB and poverty was hotly debated throughout the nineteenth and twentieth centuries,[132] but the salient points were clear. The *M. tuberculosis* bacterium was, like its close cousin the *M. leprae*, which caused Hansen's disease, an extremely slow-growing microbe that under most circumstances spent its life either under attack from the human immune system or lying low, causing no disease. Its best hopes of vigorously reproducing, developing a large microbial colony within a human being, and causing disease lay with either a diminished host immune capacity or continuous reinfection of the human being.

Diminished immune systems were plentiful wherever *Homo sapiens* lived in squalor and poverty. Malnutrition played an important role, though chronic infections with other microbes, such as tropical parasites, influenza, and amoebas, were also factors. Any ailment that taxed the immune system could create opportunities for *M. tuberculosis*.

M. tuberculosis exploited vulnerabilities. It was an opportunist. For decades it might silently lurk inside a *Homo sapiens* awaiting a moment when defenses were down, and then, when the victim's immune system was preoccupied with malaria or cancer, famine or pneumonia, it would strike.

It was also possible for people living in densely crowded situations to be continuously reexposed to the *M. tuberculosis* exhaled by others, which greatly increased their risk for developing an active case of the disease. That was why TB had historically been so strongly linked with urbanization and, in particular, slum housing and institutionalization.

Certainly it could have been predicted that the arrival of a new disease that produced severe immune deficiency and struck particularly hard in communities of poverty would spawn a reemergence of tuberculosis. If such communities had already been witnessing a slow, steady rise in TB cases, well before the new wave of immunodeficiency arrived, a resurgence of tuberculosis seemed a virtual certainty, unless public health mitigating actions were taken.

In 1947, when antibiotic therapy for TB was still considered a novel treatment and disease prevention technique, 134,946 cases of tuberculosis were reported in the United States. By 1985 the uses of streptomycin, rifampin, isoniazid, and other antibiotics, coupled with an aggressive public health effort to identify and treat TB cases, had brought the U.S. caseload down to 22,201. Fewer than 30,000 Americans had actually contracted tuberculosis each year since 1977, and the majority were elderly individuals of European descent who had carried the *M. tuberculosis* microbes in their bodies for decades, only falling ill as their aging immune systems failed to keep the bacteria in check.

Well before the actual numbers of TB cases began to swell, the demographics of the disease shifted. Between 1961 and 1969 more than 80 percent of all active TB cases in the United States were among people over sixty-two years of age, most of them readily treated without hospitalization through basic long-term antibiotic therapy. During that time the U.S. federal government spent $69,287,996 on TB control.[133]

Between 1975 and 1984, however, the numbers of active TB cases reported among elderly Americans and Caucasians of all ages declined sharply. White male cases dropped 41 percent, white female cases 39 percent. In contrast, though TB was declining across the board, its downturn among non-whites was far slower: only 25 percent for males and 26 percent for females. And the age distribution of cases had shifted: by 1984 only 29 percent were over sixty-two years of age. In the non-white population less than one out of every five active TB cases that year involved a person over sixty-two and fully 20 percent were between the ages of twenty-five and thirty-four.[134]

As early as the mid-1970s, Lee Reichman, then head of tuberculosis control for New York City, was seeing a marked increase in active cases among injecting drug users and vagrants living in Harlem, most of them young men. Reichman's attempts to sound alarms about the new trend were muffled by a medical establishment that had already written TB off as a historical artifact.[135]

There were other clear warning signs. Between 1980 and 1986 five different surveys documented a relationship between the rise of homelessness in America and surges of TB in young adult populations. The spread of tuberculosis within emergency homeless shelters was demonstrated, and it was even clear to the CDC by 1984 that new mutant strains of drug-resistant TB were spreading among the urban indigent.[136] A striking 1980 study of young adult men living in subsidized single-room occupancy housing for the otherwise homeless in New York City found that 98 of 101 came up positive in skin tests for TB infection, and 13, or 6 percent, had active disease as measured by laboratory analysis of their sputum. The 13 were carrying contagious pulmonary disease, meaning they could exhale the microbes onto others.[137]

By 1986 nearly half of all active TB cases reported in the United States were among nonwhites, most of them African-American. There could be no doubt that dramatic changes were underway by the mid-1980s. Tuberculosis had clearly shifted to younger, predominantly African-American and urban populations. Geographically, it had shifted from areas such as Virginia to New York City, Miami, and scattered urban sites. The CDC itself noted the shift in 1986, which coincided with the first upward trend in TB cases reported in the United States since 1953. The agency also believed that "HIV infection may be largely responsible for the increase in tuberculosis in New York City and Florida."[138]

From the beginning of the AIDS epidemic, researchers in both the United States and Haiti had noted that HIV-positive Haitians had a high rate of tuberculosis. Indeed, published reports stated as early as 1982 that Haitians suffering from AIDS in Port-au-Prince were more likely to die of tuberculosis than of any other opportunistic infection. But American officials took little notice of this observation. Like their counterparts throughout the Western world, U.S. physicians tended to view the TB risk for people with HIV as a Third World problem.

They were partly right; tuberculosis was an enormous, and escalating, problem in the developing world.

In 1990 Africa's most famous contemporary hero, Nelson Mandela, developed acute tuberculosis during his twenty-sixth year of imprisonment. Spitting up blood during the bitter Cape Town winter, Mandela was gravely ill. At the age of seventy at the time, Mandela fit three classic risk groups for active tuberculosis: elderly, living in cramped, densely populated quarters, and black. In South Africa, 15 percent of infected blacks went on to develop active TB, compared with only 3 percent of whites, largely because of inequities in housing and health care.

As early as 1984, Project SIDA researchers in Zaire had seen a direct link between rising TB rates in that country and the HIV epidemic. Five years later, the World Health Organization's TB and AIDS programs issued a joint statement calling attention to the linkage and warning of growing parallel pandemics. In particular, the WHO report noted that 60 percent of all AIDS patients in Haiti had active TB, as did 20 to 60 percent of all African AIDS patients (rates varying geographically across the continent).[139]

Though many developing countries quickly took steps to follow the WHO recommendations, the United States and most of Western Europe were unmoved.

There were several disturbing facets to Africa's new TB epidemic—again, offering clues that should have served as warnings to officials in the wealthy nations. Some HIV-positive patients seemed to suffer not only activated disease from long-dormant *M. tuberculosis* infection but also new infection. That meant the disease was spreading and could be posing an increased risk for general populations, not just those who were infected with HIV.[140] Where endemic tuberculosis rates were high, TB was "the single most important opportunistic disease related to HIV infection in the developing world," according to researchers based in Côte d'Ivoire.[141] HIV-positive patients did not respond well to the two cheapest antituberculosis drugs, thiacetazone and streptomycin; the drugs were four times more toxic in people with HIV, even lethal. This posed enormous problems in terms of the cost of tuberculosis treatment.[142] And the relative severity of tuberculosis in HIV-positive people did not vary appreciably with the stage of HIV disease. Indeed, for many Africans tuberculosis was the first ailment that tipped off physicians that they might have AIDS. Thus, hundreds of thousands—perhaps millions—of people in developing countries, who didn't yet realize that they were infected with HIV, were at tremendous risk for tuberculosis.[143]

By 1990 public health experts in some African countries were predicting not only utter defeat in their decades-old tuberculosis control efforts, but also potentially dire economic impacts that would further compound the grim damage the AIDS epidemic was expected to cause.[144] On the wall of the Geneva headquarters of the Global Programme on AIDS hung a graph tracking the AIDS and TB epidemics of Burundi, Malawi, Zambia, and Tanzania. The two epidemics tracked in clear tandem, each growing at exactly the same rates.

Despite all these observations the CDC concluded in early 1989 that the goal of eliminating tuberculosis from the United States by the year 2010 remained attainable and the nation's TB control efforts were essentially on track.[145]

The following year, however, the CDC's tone changed to one of alarm as fuller assessment of American TB reports revealed that the decade of the 1980s had witnessed a 28,000-person excess caseload of tuberculosis. Indeed, the downward slope TB had been following since 1953 plateaued in 1984–85 and climbed steadily, so that by the end of the decade the United States had almost as many cases of the disease as had been seen in 1980. The biggest increase was among inner-city African-Americans— TB cases in that group skyrocketed by 1,596 percent between 1985 and 1990.[146] Between 1985 and 1991 there was an overall 18.4 percent increase in tuberculosis cases in the United States,[147] most of it attributable to the HIV epidemic.[148]

When the crisis hit, Dr. Karen Brudney was one of those who could say, "I told you so." Not that it gave her much satisfaction. She was far too overwhelmed with her huge tuberculosis caseload to spend a lot of time wagging her finger at public health bureaucrats. The street-savvy, tough-talking physician made up in spades with attitude for what respect her thin, wiry female frame might otherwise fail to muster from the kinds of clients she served every day in the city's Lincoln Hospital, located in the Bronx. Equally comfortable conversing in English, Spanish, French, or Haitian

Creole, Brudney barked her commands and castigations just as freely to the drug dealers, alcoholics, thieves, and ex-convicts as she did to New York's model citizens. If any of them took this thirty-something white lady for a pushover, they were in for a big surprise.

On an icy late-winter day in 1992, Brudney paced the hospital's out-patient TB clinic, clearly agitated. The waiting room was packed with people of all ages who chattered loudly, mostly in Spanish, or watched the Puerto Rican soap opera flickering from the television that was secured to the wall by two separate sets of locks and chains. Unfortunately, none of the men, women, and children crammed into the Health Stat 10 waiting room were Brudney's patients.

As she angrily moved up and down the clinic hallway, avoiding the crowds and gurneys with the skill of an experienced rush-hour driver, Brudney grumbled.

"Clinic's been open an hour and not one single client is here. We'll be lucky if two out of the twelve clients that are supposed to be here actually show up for their TB checkups. We're only open once a week, they can't get their meds without coming to clinic, but we never get a better than fifty percent turnout," Brudney said, taking yet another look at her client list. "If they don't show up, it means they're not taking their meds. And if they're not taking their medication, they're contagious."

Her eye caught sight of a particular name—"Joanne"—and Brudney's aquiline face screwed up into an expression of disgust.

"This one! Ugh!" Brudney exclaimed. "This one is somebody they should lock up. She's out there infecting everybody. She's already been responsible for one outbreak, one where people died. And the strain she's carrying is multiply drug-resistant. If she showed up right now I wouldn't even want her in clinic, exposing everyone.

"What the hell would I do with Joanne if she did show up—which, of course, she won't. If I ordered a mandatory detention on her I'd need a bed here in the hospital. That's a whole day's work, a mountain of paperwork, a real nightmare. Then suppose I succeed in getting a bed, who's going to pay for the twenty-four-hour guard on her? And she's not going to stay, guard or no guard. What's security going to do, shoot her? Chain her in shackles in her bed?

"That woman is carrying a mutant TB strain that is virtually untreatable, 50 percent fatal. She's spreading it all over New York City. And there's nothing—*nothing*—I can do about it," Brudney exclaimed as she snapped Joanne's chart shut.

Minutes later Vernon, a thirty-three-year-old African-American male, strolled in unannounced. He didn't have an appointment, but so what—nobody else had shown up. Even an amateur could tell that Vernon had tuberculosis: his six-foot-one frame was down to 149 pounds, his movements were slow, from deep in his lungs came periodic painful coughing fits, and his eyes had that ghostly look that comes with acute illness. Characteristically, Vernon compensated for his illness with a forced kinetic energy that could be mistaken for an amphetamine high.

"You've lost more weight, Vernon. You taking your pills?" Brudney asked.

Vernon launched into an earnest, lengthy description of his daily medication routine, insisting that, despite all their side effects and the painful injections involved with one of his four medications, he was taking all fifteen pills and one shot a day, just as instructed. Brudney rolled her eyes, grunted a smirking sound, and let it be known that she'd heard all this before from Vernon.

"I'm not ashamed," Vernon insisted. "I'm dealing with it. I really am. This time."

"Yeah, *this time*," Brudney responded. The physician called in a social worker and, in front of Vernon, told the patient's story. Vernon enthusiastically added details along the way, seemingly proud of his dubious battle with tuberculosis. In early 1989 Vernon had been hospitalized with what appeared to be pneumonia. Three weeks later the hospital lab returned a different verdict: tuberculosis. There was nothing special at the time about Vernon's strain of *M. tuberculosis*; it was garden-variety TB.

So Vernon was released from the hospital and ordered to take two relatively inexpensive, extremely effective drugs every day for six months: isoniazid and rifampin.

"But you screwed up, didn't you, Vernon?" Brudney said.

Shrugging his shoulders, Vernon said, "I figured anytime I felt bad, I'd just go to the emergency room and get more pills."

After a year of sporadic, improper use of the drugs, Vernon's tuberculosis bacteria mutated, becoming resistant to both drugs. Since he had long disappeared off the City Health Department's radar screen, investigators were sent out in search of Vernon.

But he had disappeared.

"I move around a lot," Vernon said, vaguely referring to several emergency homeless shelters and the apartments of friends and relatives.

Then he had suffered a major tuberculosis relapse and in November 1991 ended up back in Lincoln Hospital, spitting up blood. For ninety-four days Vernon struggled at death's doorstep in Lincoln, his lung mucus coming up clear.

"That's bad," Vernon said, though he deferred to Brudney for an explanation. The TB colonies in his lungs had formed a hard, calcified cavity inside of which they thrived, protected from his immune system and from the four powerful drugs that dripped via an intravenous line into his bloodstream all day, every day, for three months.

Since his discharge from Lincoln Hospital in January 1992, Vernon had been having night sweats and felt fatigued. "But I'm alive, and I'm gonna stay that way."

"You are, if you take all of your medication," Brudney scolded.

Vernon swore that every morning he was swallowing eleven pills, comprising three different antibiotics. And he insisted that he was always home after breakfast when the public health nurse came to inject amikacin into his shoulder.

"Man, that hurts," he said. "Stings, man. Burns going in, and takes its time getting there."

Brudney, for the first time since he arrived, fully agreed with something Vernon said.

"It's a four-cc injection, and it's excruciating. And you wouldn't have to be putting up with it if you had taken your pills in the first place," she said.

Vernon was now living at home in the South Bronx with his mother and older siblings. He had a girlfriend and a fifteen-month-old daughter, both of whom, so far, were free of TB. Until he got well, Vernon would live on welfare and social security funds, but, he said, "I'm gettin' a job working on a movie that's shooting in Harlem, just as soon as I lick this TB."

Brudney made a few notes on Vernon's chart, handed the patient his prescriptions, and shook her head as he exited.

"Everything that man says is a lie. It's amazing. Every single word," Brudney insisted. "For months he's been checking in and out of homeless shelters, using false names so the Health Department couldn't find him. And why? So he could deal drugs. I don't know, he may even be selling his TB meds on the street. Some of the patients do."

Brudney noted that since 1989 Vernon had missed more than 75 percent of his appointments, was hospitalized four times, and was found hiding under an alias on two occasions.

"That's what we're up against."

Two years earlier, Brudney and Columbia College of Physicians and Surgeons colleague Dr. Jay Dobkin had warned government officials that men like Vernon were breeding drug-resistant tuberculosis. The pair studied TB treatment records for Harlem Hospital, a public facility located in the middle of one of New York's poorest neighborhoods, which, more than a decade earlier, Lee Reichman had identified as one of the communities with the highest incidence of TB in the United States.[149] By 1985 it was also a neighborhood ranked in the top ten nationally for homelessness and narcotics use.

Brudney and Dobkin examined the records of all patients hospitalized for tuberculosis between January 1, 1988, and September 30, 1988. Eight out of ten of the patients were men twenty-five to forty-five years of age, half of them were homeless, the remainder were listed as "unsteadily housed." More than 80 percent of the patients were unemployed, 79 percent were alcoholic, and 40 percent were HIV-positive.

More than a quarter of the patients—26 percent—were hospitalized for tuberculosis relapses, meaning that they had failed to properly take their medications. And a startling 89 percent of the patients disappeared some-time after hospital discharge, never returning for their mandated checkups and drug prescriptions. A subgroup of the patients—women who were addicted to crack cocaine—were 97 percent noncompliant with tuberculosis medication.

"Within 12 months of discharge, 48 of 178 (27%) patients were read-mitted with confirmed active tuberculosis at least once," Brudney and Dobkin wrote.[150] "Almost all of those discharged were again lost to follow-up, with 20 percent admitted a third time as of April 1989."

The two physicians noted that New York City spending for tuberculosis control stood at $40 million in 1968, more than 80 percent of which was spent on outpatient services, tracking patients, and ensuring their com-pliance with medication orders. In addition, the federal government added $1.4 million annually to New York's TB effort during the 1970s.

By 1988 that federal commitment had fallen below the $200,000 mark and New York City officials had dropped their fiscal expenditures for tuberculosis control to less than $2 million a year. In addition, at a time when the patient population was largely homeless and extremely difficult to follow, nearly all resources were directed to hospitalization costs rather than outpatient services and patient compliance issues.

Meanwhile, the CDC had been monitoring laboratory tests on tuberculosis antibiotic resistance, finding a clear correlation between the number of times an individual had been treated for TB and the levels of resistance in the patient's tuberculosis bacterial population. For example, based on lab data amassed between 1982 and 1986 on patients with resistant TB strains, the individuals were four times more likely to have isoniazid re-sistance if they had been previously treated for TB, more than three times more likely to be resistant to streptomycin if previously treated, and so on for all available drugs.

In 1986, just as tuberculosis was making its reemergence in America, the federal government pulled the plug on the CDC's drug-resistance track-ing program. That explained, in part, why the new TB epidemic blindsided the watchdog agency.

If significant numbers of TB patients in New York City were, as Brudney and Dobkin demonstrated, failing to adhere to proper medication schedules, the CDC's findings indicated that widespread drug resistance was a virtually guaranteed outcome.

When multiply drug-resistant strains of tuberculosis spread from the largely impoverished homeless population of New York City to their doctors, jail guards, and fellow patients inside hospitals, panic broke out. Though the first incidents occurred as early as 1989, word of the full extent of the prob-lem and the number of health providers and patients so afflicted didn't get out until early 1992.[151] When the statistics were released by the CDC and the New York City Department of Health, nurses, physicians, people infected with HIV, and the general population were briefly shaken out of their complacency.

During the first quarter of 1991, it turned out, 42.5 percent of all new tuberculosis cases diagnosed in New York City were caused by mutant strains that were resistant to the primary treatment drugs, isoniazid and rifampin. Worse yet, 60 percent of the relapse cases seen during the first twelve weeks of 1991 were multiply drug-resistant. Nowhere else in the nation were *M. tuberculosis* resistance levels that extreme. New Jersey and Florida ranked second and third nationally with 6.3 and 5.3 percent MDR (multiply drug-resistant) TB rates, respectively. Averaged nationally, 21.5 percent of all relapse TB cases were MDR, as were 8.2 percent of new cases.

By 1989, New York had become the nation's epicenter of four epidemics, each of which fed upon the other: HIV/AIDS, MDR tuberculosis, heroin addiction, and crack cocaine use.

Three dreadful hospital tuberculosis outbreaks in New York and a fourth in Miami drew sharp attention to the interconnection between MDR-TB and HIV. In each instance a patient with active drug-resistant tuberculosis was in the same clinic or ward with HIV patients, and the immunodeficient individuals were terribly susceptible to both infection and death. Death rates among the newly infected HIV-positive patients ranged from 91 to 100 percent, most dying less than sixteen weeks after infection.[152]

So grim were the prospects for the newly infected HIV-positive patients that officials referred to them as individuals who posed no direct public health threat: they didn't survive to leave the hospital. They could, however, pose a risk for those who cared for them in the hospital.[153]

When scientists with the CDC, various New York-based institutions, and research centers around the United States worked their way backward to understand why and how drug-resistant tuberculosis had emerged in the United States more than forty years after the invention of curative drugs, they were forced to conclude that the nation's public health system had failed on every front.

Twenty-six people caught TB in three Boston homeless shelters between February 1984 and February 1985; two died. Laboratory analysis revealed that fourteen of the individuals were newly infected with a strain of TB that was resistant to isoniazid and streptomycin. Searching for the source of the outbreak, researchers found two candidates, both of whom had MDR-TB. The first was a thirty-three-year-old alcoholic who had been in and out of TB treatment for ten years. The other was a fifty-seven-year-old diagnosed schizophrenic who had suffered two bouts of TB since 1980.[154] The outbreak demonstrated both that tuberculosis readily spread inside homeless shelters and that individuals who failed therapy could become carriers of chronically active MDR-TB.

The most important points of vulnerability in the public health system were made apparent when a thirty-two-year-old man died of multidrug-resistant tuberculosis in Davidson County, North Carolina, on April 20, 1984. The cause of death was not confirmed as TB until over three months after his funeral; it took the North Carolina State Laboratory more than five months to determine the drug-resistance characteristics of the man's TB strain. By then the individual had been six feet under for four months, his doctors having treated him with drugs rendered useless by a strain that proved resistant to isoniazid, rifampin, ethambutol, and streptomycin.[155]

The system failures proved even more embarrassing when investigators from the CDC tested the North Carolina victim's close friends, discovering that the dead man's next-door neighbor had suffered chronic tuberculosis since 1978, passed it on to his live-in girlfriend, her brother living in Washington, D.C., and a drinking partner. All the cases had escaped the public health safety net, though they had been seen by physicians. And all were infected with powerfully drug-resistant mutant bacteria. All male members of the cluster died of the disease—only the female survived. The man who appeared to have been the first TB case was an alcoholic, and the group spent hours drinking together in a local bar. Because the anti-tuberculosis drugs could not be tolerated with alcohol, the individuals failed to follow medication instructions.

And nobody from the city, county, or state public health systems took steps at any time between 1978 and 1985 to track the recalcitrant patients or force medication compliance.

The 1990s witnessed dangerous epidemics of MDR-TB first in Miami, San Juan (Puerto Rico), and New York City, later scattered all over the nation. Retrospective analysis of the New York City outbreak showed it began in September 1989 and continued well into 1994. In every case laboratory analysis of patient sputum and tissue samples was so slow that many victims were long dead by the time physicians knew which drugs might kill the particular TB strains in the victims' bodies. Median time for laboratory diagnosis of tuberculosis was nine weeks, and median additional lab time for determining the bacteria's drug-resistance patterns was six weeks. In other words, half of all New York City medical laboratories took nearly four months to reach a definitive diagnosis, and many required five to six months' lab time. New York's lab times were considered typical for the nation as a whole.[156] Though HIV-positive people were the most vulnerable victims in the epidemic, health care workers, prison guards, homeless shelter employees, fellow HIV-negative patients, and relatives were also infected as the airborne mycobacteria spread.[157]

To save money in the mid-1980s federal and state politicians had slashed TB control and surveillance budgets. By the time officials realized what had hit them, TB was draining financial resources at an astonishing rate. In 1991 direct tuberculosis treatment costs in the United States topped $700 million,[158] and the costly cases kept coming well into 1994. In the state of New York, in 1991 direct hospital expenditures for TB ran to more than $50 million.[159] In response to the MDR-TB epidemic the city of New York had to build a special 140-bed tuberculosis unit in the Rikers Island jail, at a total cost over three years of $115 million. The city's public hospitals spent $4 million to construct air-flow-controlled isolation rooms for TB patients that, for the first time, guaranteed that no other hospital employees or patients would be compelled to breathe air that was contaminated by an individual with tuberculosis.

In addition, the federal government had to increase TB spending from $17 million in 1991 to $54.9 million in 1992, much of which went to New York City.[160] When all the costs of the 1989–94 MDR-TB epidemic were totaled it was clear that more than $1 billion was spent to rein in the mutant mycobacteria. Saving perhaps $200 million in budget cuts during the 1980s eventually cost America an enormous sum, not only in direct funds but also in lost productivity and, of course, human lives.

Amazingly, even as federal concern escalated, and TB reports from all over the country demonstrated a national upward trend in tuberculosis, cities and states, other than New York, continued to slash their TB budgets. A survey of 25 large-city health departments revealed that between 1988 and 1992 sixteen of them slashed their TB budgets.[161] Though TB caseloads rose during that period in twenty-three of the cities, MDR-TB appeared in virtually all urban centers, expensive hospitalization was required in nearly twice as many cases, and the length of average treatment time increased by two months, cuts were the order of the day in most municipalities.

By 1993 the MDR-TB epidemic had made its way to the suburbs, such as New York's Long Island and Westchester County.[162] Jails and prisons all over the country reported MDR-TB outbreaks similar to that seen in Rikers in 1990–91. And Los Angeles, Chicago, Dallas, Detroit, and Miami all reported surges in the incidence of tuberculosis generally and MDR-TB in particular. Though New York City succeeded in bringing its TB incidence down that year, it remained fifty times greater than the national average—which itself was pretty bad. The CDC determined that 14.2 percent of the nation's tuberculosis cases in 1993 involved MDR-TB.

Further, studies showed that any diminution in the number of reported tuberculosis cases could only be considered a brief respite so long as the underlying conditions responsible for the emergence of MDR-TB remained unchanged. For example, Dr. Fred Gordin led a seventeen-center federal study in 1991–92 for the National Institute of Allergy and Infectious Diseases, looking at 4,314 indigent individuals around the country who were infected with the human immunodeficiency virus. About a quarter of the individuals came from poor communities of New York City, notably Harlem, the South Bronx, and eastern Brooklyn.

Skin tests of New York individuals were 28 percent positive for TB infection, compared with a national infection rate among HIV-positive poor people of less than 8 percent. Because it had long been known that HIV-positive people failed to respond to the TB skin test due to the beleaguered state of their immune systems, Gordin went a step further. He conducted anergy tests on the individuals aimed at determining whether they could give skin-test responses to *anything*, and then used a mathematical model to estimate what percentage of the anergic patients were infected with tuberculosis.

The result, Gordin said, was alarming: 51 percent of the New York area individuals were TB-infected.

"It is really very scary in New York," Gordin said. "We have found 10.2 percent of the New York cohort have actually had TB already, which is mind-boggling. It's what you'd expect in a Third World country."

Another disturbing finding came out of the National Jewish Center for Immunology and Respiratory Medicine in Denver, Colorado. Established at the turn of the century when physicians believed that fresh mountain air held curative powers for people with tuberculosis, National Jewish was, by 1990, the last fully operational TB sanitarium and research center left in the United States. Its chief TB physician, Dr. Michael Iseman, was widely considered the preeminent expert in the United States on diagnosis and treatment of the disease, and doctors all over the country typically sent their most desperately ill tuberculosis patients to National Jewish.

It came as grim news, indeed, when Iseman announced that even in his hands, in the best TB treatment center in the entire world, MDR-TB was extremely lethal. Of 171 patients (all HIV-negative) suffering from *M. tuberculosis* strains that were resistant to isoniazid and rifampin—as well as other drugs, in most cases—35 percent showed no response whatsoever to treatment with remaining, theoretically effective, drugs. And among those who did initially improve under Iseman's care, many suffered relapses. Despite radical treatments, including surgical removal of TB-filled lungs, more than half the patients never recovered from the disease; either they fell into the sort of lifelong tubercular state that Edgar Allan Poe and Charles Dickens had described eloquently more than a century earlier, or they died. Most of Iseman's patients were *not* HIV-positive, and the Denver physician blamed the poor efficacy of the second- and third-string anti-tuberculosis drugs for the dismal treatment outcome.[163]

When MDR-TB struck the United States in 1991, the CDC was swamped with requests for assistance from state agencies that were searching for second- and third-string drugs. The agency identified twenty-nine regions of the United States (of fifty-nine questioned) that were experiencing extreme shortages in anti-TB drug supplies. The U.S. government scrambled to persuade multinational drug companies to rapidly increase their production capacities.[164]

A confluence of events had played key roles in the emergence of New York's MDR-TB epidemic. First, President Ronald Reagan's declaration of a war on drugs and call for mandatory imprisonment for a range of drug-associated crimes coincided with an enormous surge in heroin and crack cocaine use in New York. Studies showed that some 80 percent of all MDR-TB index cases in 1989–90 (not including the secondary HIV-positive cases) were injecting drug and crack users, many of whom, as a result of federal and local crackdowns, drifted in and out of the jail and prison system. In 1991 some 295,000 arrests were made in New York City, 120,000 of which resulted in some period (days to years) of incarceration.

Most city inmates were incarcerated for only short periods while they awaited arraignment or trial, so the situation from a microbial point of view, between the densely crowded jail ecology and the general community, was quite fluid. On any given day, 26 percent of the female inmates and 16 percent of the males were HIV-positive, providing the microbes with an enormous pool of unusually vulnerable *Homo sapiens*.

Thus, what may have begun as isolated cases of MDR-TB among handfuls of scattered recalcitrant tuberculosis patients—men and women like Vernon—was amplified inside the city's jails into a full-scale epidemic.[165]

The social revolutions that would be necessary to reverse years of heroin and cocaine infiltration into the very fabric of the lives of hundreds of thousands of Americans staggered the imagination, as did the scale of what would be required to properly house all the homeless, employ the jobless, end the cycle of mass incarcerations, and stem all the other social tides that doomed most of America's urban poor to lives of tremendous microbial vulnerability. The public health community, overwhelmed by the social dimensions of the crisis, turned to Science and beseeched researchers to find simpler solutions in their laboratories.

Perhaps Thirdworldization of American cities couldn't be stopped; TB's reemergence might, however, be aborted with the proper magic bullets.

But the scientific community was woefully ill prepared to meet the challenge. Having long since switched most medical research priorities to chronic diseases, and only recently having developed an infrastructure for AIDS research, the NIH was caught with its pants down.

Impressed by the urgency of pleas for assistance emanating from both the public health community and a terrified HIV-positive population, National Institute for Allergy and Infectious Diseases (NIAID) director Dr. Anthony Fauci convened an emergency meeting on tuberculosis in Bethesda on February 10, 1992. All of America's leading tuberculosis experts were invited—all forty or fifty of them.

Looking around the sparsely attended room, Barry Bloom, a TB expert for WHO and researcher at the Albert Einstein School of Medicine in the Bronx, addressed Fauci directly, saying, "If I were you, I'd ask myself how there could possibly be scientific expertise in this country on tuberculosis if you're only handing out twenty-three research grants a year."

Acknowledging that total NIH expenditures on TB research had amounted to just $3.5 million a year, Fauci asked, "Yes, but if we throw $50 million at it next year would there be expertise, would we be able to seduce new investigators into this area of research on an urgent basis?"

Bloom sighed.

"It's true, we can't get rolling fast. There's a generation gap of people who know something about this disease," Bloom, himself in his fifties, said. "Essentially everything that is known about tuberculosis was figured out before 1948, when antibiotics came into use. And virtually all research stopped after that. Dead stop."[166]

The situation was no better overseas. Though TB claimed 3 million lives a year, newly infected 8 million people annually, and was the single largest cause of infectious disease deaths during the 1980s, it drew little scientific attention in the wealthy world. Until the U.S. MDR-TB epidemic began there was virtually no scientific interest in pursuing the developing world's big killer. The cries of years of neglect voiced at the NIAID meeting in 1992 were echoed in the halls of science in London, Tokyo, Paris, Geneva, Amsterdam, Stockholm, indeed worldwide.

Once money was thrown their way, scientists did succeed in 1992–94 in discovering the genetic basis of at least one type of *M. tuberculosis* antibiotic resistance,[167] identifying 500 genetically distinct tuberculosis strains in twenty-nine U.S. outbreaks occurring in 1991–92,[168] developing an ingenious way to "see" drug-resistant strains in the laboratory using the luciferase chemical found in fireflies to light up resistance genes,[169] and figuring out how the bacteria managed to hide inside CD4 cells of the immune system.[170]

But these were just first shots out of a scientific cannon that was in for a long siege. Everybody knew that. If the emerging MDR-TB epidemic was to be stopped, public health would have to use methods immediately at hand.

When U.S. and European experts cast their eyes about in search of successful tuberculosis control programs that had managed to prevent significant emergence of drug resistance, they were a bit embarrassed to see that the best efforts were carried out in the poorest nations.[171] In Tanzania, war-torn Nicaragua, the Zululand province of South Africa, China, even Mozambique in the midst of a civil war, tuberculosis was better managed than it was in the wealthy world.

Brudney and Dobkin compared the dismal 11 percent patient compliance rate they saw in Harlem Hospital with treatment successes in Nicaragua during the same period (late 1980s) and reached the startling conclusion that the tiny Central American country, with per capita incomes of less than $585 per year, had achieved a far better level of TB control than had New York. Using a basic strategy of finding active tuberculosis cases and putting the individuals on two months of carefully monitored daily medication (isoniazid, streptomycin, or thiacetazone), followed by ten months of lower-dose continued daily treatment, the Nicaraguan Ministry of Health achieved an almost 75 percent cure rate during a civil war. In contrast, New York's cure rate was below 50 percent.[172]

In Zululand, South Africa, between mid-1991 and the close of 1992, health care workers managed to successfully treat 83 percent of all tuberculosis patients, lost track of only 13 percent, and had a mere 7 percent mortality rate. This level of success was achieved despite a large local HIV epidemic and major tribal conflicts that often disrupted local social services. As had been the case in Nicaragua, the key to success was careful monitoring of patient medication.[173]

Tanzania and Mozambique employed similar methods of community-monitored medication to keep their incidences of tuberculosis down, and at very low cost. Before the East African nations were overwhelmed by HIV, their TB rates were extremely well controlled and treatment compliance exceeded 80 percent. As the HIV epidemics exploded, however, the incidence of TB also climbed. Still, both countries managed to prevent significant TB spread in the HIV-negative community.[174]

Harvard medical economist Christopher Murray did a cost-effectiveness analysis of the East African TB control efforts, and concluded that they made far more fiscal sense than any programs in the United States. He then teamed up with Karel Styblo, who had designed the East African programs, and Annik Rouillon of the International Union Against Tuberculosis and Lung Diseases to assess the success rates and costs of TB control programs all over the world. The team's conclusions, submitted to the World Bank in mid-1991, were striking.[175]

"There is no country in the developing world that has a treatment compliance rate as bad as New York City," Murray said. "New York has around 10 percent compliance. While India, which is very bad, has 25 percent compliance. China has 80 to 90 percent. Mozambique in a civil war attained 80 percent."

Treatment success rates of 80 percent or better were the norm in many of the world's poorest nations.[176] No nation's TB control system did a poorer job than did the United States in identifying tuberculosis cases,[177] successfully treating those cases, and keeping track of their outcome and possible contacts for spread of the disease.

In 1992, the CDC and the New York City Department of Health adopted what amounted to a Third World tuberculosis control strategy. Millions of dollars were spent to train nonprofessionals to work as Directly Observed Therapy (DOT)[178] officers, monitoring patient compliance with medication. When patients continued to refuse their treatments, incarceration in designated medical facilities was used as a last resort.[179]

The plan went into action too late to spare Dr. Frantz Meedard from acquiring MDR-TB from one of his patients in Metropolitan Hospital in Harlem. Too late to prevent his suffering a year of undiagnosed illness followed by twenty-seven months of multidrug therapy that included injections of amikacin—"so painful that I used to cry," Meedard said. But once he was cured, in late 1992 Meedard eagerly jumped at the chance to run the Harlem Hospital DOT program. Within ten months he had cut the hospital's dismal noncompliance record, losing track of only 8 percent of his TB patients and getting 18 percent successfully through their entire medication program.

"We're still worse than most of the Third World," Meedard said in late 1993, "but I'm determined. I tell the patients, 'Look, I went through it. So can you.'"

15

All in Good Haste

HANTAVIRUSES IN AMERICA

*Neither rat nor man has achieved social, commercial, or economic
stability. This has been, either perfectly or to some extent, achieved
by ants and by bees, by some birds, and by some of the fishes in the
sea. Man and the rats are merely, so far, the most successful animals
of prey. They are utterly destructive of other forms of life. Neither
of them is of the slightest use to any other species of living things.*
—Hans Zinsser, *Rats, Lice, and History*, 1934

■ Long-distance runner Merrill Bahe was on his way to his girlfriend's
funeral on May 14, 1993, when he found himself gasping for air. Suddenly,
and quite dramatically, Bahe was overcome with fever, headache, and
respiratory distress. In the presence of his grief-stricken relatives, Bahe
gulped desperately for air in their car, en route south to Gallup, New
Mexico.

Minutes later the nineteen-year-old Navajo athlete was dead.

His twenty-four-year-old girlfriend had died in a small Indian Health
Service clinic located sixty miles away from Gallup a few days earlier after
an identical bout of sudden respiratory illness. And within the week her
brother and his girlfriend, also young, athletic Navajos, who lived in trailers
near Bahe's, fell mysteriously ill; the young woman died.

Word spread across the Navajo Nation of 175,000 people, living in an
area of seventeen million acres spanning four states—Arizona, New Mexico, Colorado, and Utah. Because the borders of the four states met in the
area, the region was called Four Corners. The locale for many John Wayne
Westerns, Four Corners was surrounded by massive tracts of sparsely populated sandstone landscape that plunged into majestic canyons and arched
upward forming dramatic ridges and peaks. It was a place where people
spoke of "big sky" as they gazed across the psychedelically colored desert
to the wide expanse that reached to the horizon.

The entire Navajo Nation was soon buzzing with the news of three strong,
young members of the community who suddenly found themselves gasping
in vain for air.[1]

Before Merrill Bahe was carried into the emergency room of the Indian
Health Center in Gallup, resuscitation attempts in the ambulance had

failed. Bahe was declared DOA, dead on arrival. His death shocked the already bereaved families and sent a chill through the medical staff.

Attending IHS internist Bruce Tempest was struck by Bahe's youth and athleticism, and he recalled discussing a similar case over the phone with an IHS colleague at another Navajo clinic. When he realized that the other case was Bahe's fiancée, Tempest took three decisive steps that eventually cast Bahe's death in the light of a national epidemic investigation, rather than a mere routine case of unexplained illness.

First, he called the state medical examiner, alerting the forensic pathologist to the possibility of a communicable disease problem. The New Mexico examiner, Richard Malore, immediately declared jurisdiction, placing Bahe's body under an autopsy order. Then the investigator walked across the street and similarly took custody of Bahe's fiancée's corpse.

While pathologists prepared the bodies for full autopsies that would keep them working around the clock over the May 14 weekend, Tempest took his second decisive step, reaching again for the telephone. The Navajo IHS was unique in that its clinics were spread out over an area so vast that some physicians never had an opportunity to meet one another. But they were in constant telecommunication, and IHS physicians known for their particular expertise received dozens of calls a day from other doctors working in American Indian clinics from Colorado all the way down to Window Rock, Arizona, hundreds of miles to the south.

Tempest, who had worked in the area for the IHS since 1967, was known for his unique problem-solving facility in confusing situations. As a result, he already had on his desk in Gallup the medical files on a Navajo woman who had died mysteriously in a distant clinic around Christmastime of an apparently similar acute respiratory distress, and he had served as a telephone adviser on a couple of other puzzling pulmonary cases during the spring.

Now he got on the horn and called all those attending physicians, asking for details on the earlier respiratory death cases.

"So by the end of the day, Friday [May 14], I was able to compile a list of five healthy young people who had died of acute respiratory distress syndrome," Tempest later said.

He called in the New Mexico Department of Health and IHS epidemiologist Dr. Jim Cheek. The state set its laboratories in motion, testing the autopsy samples and reviewing medical charts, looking for evidence of respiratory diseases that had haunted the Navajos for decades: bubonic plague, *Hemophilus* influenza, viral pneumonia, and influenza.

The obvious and immediate autopsy finding was that the lungs of Bahe and his girlfriend were so severely fluid-filled that they weighed twice as much as would normally be expected for young adults of their sizes.

If Tempest or someone else in Four Corners hadn't spotted the cases and sent alerts immediately in the proper directions, the mini-epidemic would have gone unnoticed, according to sources at all levels. Tempest never hesitated, however. Nor did his counterparts in the New Mexico Health Department.

By May 16, the state medical examiner and the labs were unable to find evidence of flu, or any other common viruses or bacteria, in the autopsied materials. On Wednesday, May 19, when Tempest alerted Jim Cheek, the IHS chief epidemiologist had already heard rumors of "weird deaths" in the northeastern part of the Navajo Nation from a Navajo co-worker in his office. Cheek hadn't paid them much attention at the time. New Mexico state epidemiologist C. Mack Sewell told Cheek the initial conclusion was that the first couple had died of pneumonic plague.

Isolated cases of the bubonic form of the plague had occurred sporadically among the Navajos for decades, carried by wild prairie dogs. Since the early 1970s, when Jonathan Mann ran the New Mexico epidemiology program, the state had maintained a strong and vigilant plague surveillance program, quickly spotting the occasional case. Far less common was pneumonic plague, in which the bacteria grew in the victims' lungs and could be spread through the air from one person to another.

New Mexico had an extraordinary plague laboratory—possibly the best in the world—and the state had seen enough cases over the years to be able to rapidly diagnose and stop an outbreak. On the basis of their symptoms, state health officials therefore hypothesized that Bahe and his girlfriend had died of pneumonic plague.

But that wasn't what the laboratory concluded. No plague bacteria could be found in the victims' blood or tissue samples.

Cheek set his small IHS team into action, immediately exploring three avenues: hospital records on other recent unexplained respiratory deaths in the area, a computer search on chemicals known to cause such symptoms, and an investigation of the Bahe home and community.

ARDS, or acute respiratory distress syndrome, was typically the final cause of death of millions of people worldwide every year. Most cases, 50 to 90 percent, occurred in elderly people, burn patients, victims of traumatic injuries, or other individuals for whom a clear cause of the rapid lung fluid buildup was evident. But in a minority of cases, no obvious basis for the respiratory distress could be found, and doctors usually listed cause of death as "ARDS of unknown etiology."

Cheek's team scoured IHS medical records for the spring of 1993 looking for unknown etiology ARDS cases. Five popped up—overlapping with Tempest's list—and Cheek had them investigated.

Meanwhile, Cheek suspected a toxic chemical as the culprit for the first two cases. A computer search turned up several possibilities, but "the one that fit the bill perfectly," Cheek said, was phosgene. Used during World War I by the Germans, phosgene could cause symptoms of ARDS over twenty-four hours after exposure. A sister compound, phosphene, produced more rapid symptoms, but was also known to cause ARDS. After snooping around a bit, disease detective Cheek learned that phosgene had long been banned in the United States and it would be exceedingly difficult to produce toxic levels of the chemical through such practices as arc welding, which could create trace amounts of the compound.

But phosphene, he discovered, was legally used to kill prairie dogs. Over the winter CDC scientists working out of the agency's laboratory in Fort Collins, Colorado, had predicted that record snowfalls in the 1992–93 season in the Four Corners region would result in an increase in the springtime prairie dog population. And with that, the scientists forecast, would come increased plague. As it turned out, the region was inundated with record levels of snowfall. Putting the phosphene pieces of the puzzle together, Cheek thought, "Aha! We have something here. Somebody has been doing some prairie dog eradication."

But his enthusiasm was soon dampened by an investigative visit to the Bahe dwelling, a trailer. He found no sign of phosphene containers, chemical spray apparatuses, or residual chemicals. In fact, he found nothing out of the ordinary in the empty trailer, except, perhaps, an unusual amount of mouse feces scattered here and there. Cheek assumed that the rodents had invaded the trailer after it was abandoned by the ailing humans.

While he poked about Bahe's trailer, gathering rodent feces, dishes, clothing, and other articles for laboratory scrutiny, Cheek took no special precautions for his own safety. It hadn't occurred to him that whatever killed the three Navajos might still be present, in a transmissible form. So he wore no respirator mask, no special latex gloves, no protective unit.

Cheek would later marvel about his foolishness and luck.

By May 20, Cheek had a list of ten suspected cases, all from the Four Corners area, and he was stumped. The thirty-five-year-old physician had been in the New Mexico area for only seven months, and he was running out of ideas. So, having spent the previous two years working as a CDC Epidemic Intelligence Service officer, he called the agency's top epidemiologist and his old friend Rob Breimen.

"I wondered if it might be some kind of mycoplasma [bacteria], because they're so hard to culture in the laboratory. I thought maybe that's why we weren't finding anything," Cheek later said.

Breimen, who had been involved in previous investigations of equally puzzling outbreaks, including Legionnaires' Disease, was intrigued, until he tossed ideas around for a while, sighed, and tried to dismiss the issue as "a small problem."

But try as he might to focus on the more pressing issues on his Saturday working agenda, Breimen couldn't shake Cheek's intriguing puzzle from his mind, and on Monday, May 24, he phoned Albuquerque for a faxed rundown of the cases.

That same day the New Mexico Department of Health sent letters to all the state's physicians, describing the mysterious disease and requesting immediate notification should other cases be seen.

The next day Breimen shared the curious list of ten suspect ARDS cases with a few CDC colleagues, all of whom agreed that there was something awfully odd about such sudden deaths among healthy young people.

The following night Breimen was on a four-hour conference call with the New Mexico and Arizona state epidemiologists. The trio went over the details of some nineteen suspected cases found in the two states—not all of which were among Navajos. Twelve had died. The victims ranged in age from nineteen-year-old Bahe to a fifty-eight-year-old woman. Most had taken ill during the month of May. With the Memorial Day weekend approaching, the three epidemiologists were anxious to figure out whether or not they had a genuine epidemic on their hands—one that might suddenly explode over the federal holiday when most government scientists and physicians were on vacation.

That same Wednesday, the CDC's physician-scientist Louisa Chapman was going over some old chronic fatigue syndrome data when an anonymous caller from the Navajo Nation rang up, requesting urgent advice. Nervous, the man identified himself only as a dentist, and wanted to know if he should close down his office.

"Why would you want to do that?" asked Chapman, a tough scientist whose baby face belied her nearly ten years of experience in infectious disease investigations.

"Well, we've got young people dropping dead all over the reservation," the panicked dentist told her, adding that he was worried that whatever was killing the Navajo Nation's youth might be transmissible during dental procedures.

"I don't know what you're talking about," Chapman said, wondering whether she was the victim of a crank caller.

"Okay, turn on CNN right now," the dentist said, just before he hung up.

Chapman didn't have a television in her office, so she didn't see the report that described an outbreak of "Navajo disease." Instead, she stepped next door and mulled the situation over with a colleague. They decided the responsible thing to do was to send a heads-up notice on the dentist's queries by computer to the agency's viral disease group. A few minutes later Chapman found herself in the middle of the Four Corners puzzle, as computer messages poured in from other CDC staffers and her old friend IHS epidemiologist Cheek.

The next day, Thursday, May 27, Chapman went to her regularly sched-uled grand rounds at the local Veterans Administration Hospital, where she saw Dr. Jim Hughes, the director of the CDC's Center for Infectious Diseases. Chapman filled Hughes in on the mysterious outbreak, adding, "I'd like in on this one." Hughes, an affable leader who always tried to "keep the troops happy," smiled and acknowledged Chapman's eager interest.

When Hughes returned late Thursday morning to the CDC, his deputy, Dr. Ruth Berkelman, told him that the Indian Health Service and the state of New Mexico had formally requested CDC assistance in solving the Four Corners mystery. Now the vaguely interesting puzzle which Chapman had expressed a desire to work on just an hour earlier was a matter of official urgency. Still, what little he knew at that point led Hughes to believe that it was a fairly small problem.

Berkelman, whose specialty was newly emerging diseases, begged to differ.

"There's something about certain calls," Berkelman told her boss. "You get a feeling for these things."

Respecting Berkelman's expert opinion, Hughes told her to put together an emergency meeting. As they spoke, residents of New Mexico were opening their morning paper, the *Albuquerque Journal*, which carried the headline: "Mystery Flu Kills 6 in Tribal Area."

When Louisa Chapman walked into the infectious disease conference room that Thursday afternoon, she was stunned. Never in her six years at the CDC had she seen so many top officials and scientists crammed into one room for a disease investigation. She felt a familiar thrill as her adrenaline started pumping.

All the people in that room, and the staff they represented, were already stretched thin with a dizzying array of epidemics—more than the agency had ever handled over a six-month period in the CDC's history.[2]

This came at a time when the CDC's 1994 budget reauthorization, though large at $2.2 billion, required the agency for the first time to conduct several ambitious immunization and disease prevention programs. (The agency's name was changed to the Centers for Disease Control and Prevention, underscoring the shift.) As a result, Hughes and his top-level counterparts throughout the agency were in the process of eliminating 518 jobs in nonprevention divisions of the CDC—about 7 percent of the agency payroll.

C. J. Peters, who had left the U.S. Army's laboratory at Fort Detrick a year earlier due to Department of Defense budget cutbacks, told meeting participants that his CDC Special Pathogens Branch was already so overextended that he was going to have to borrow scientists to process blood and tissue samples as they came in from the Four Corners area. All suspect materials sent to the CDC were first studied in the high-security P3 and P4 laboratories, until proven safe enough for use in standard research facilities.

Inside the P4 component of the CDC's security lab, scientists wore fullbody respirator-fed suits when they worked with animals or conducted experiments on bench tops. Most of the cellular work was done in airtight glass-and-steel boxes that scientists accessed by inserting their hands into heavy rubber gloves that were permanently attached to the boxes. All research animals were kept in similarly air-sealed housings and scientists took special precautions to avoid being bitten or scratched by primates or rodents.

It took a special kind of person to work under such restrictive and tense conditions. Some P4 workers likened it to spending a lifetime in outer space because even the tiniest of invisible holes in a hose, glove, or respirator suit could let in a lethal atmosphere.

Peters knew it was going to be hard to find additional personnel capable of working in his P3/P4 facilities.

The meeting resolved to send epidemiologist Dr. Jay Butler, along with two EIS officers, out to Four Corners immediately. A shy, blue-eyed marathon runner, Butler first learned of the mysterious Four Corners deaths on Friday morning, and was in Albuquerque that afternoon. In the few hours between the staff meeting and his flight to New Mexico, Butler hastily prepared for the unknown.

The next day, May 29, he pored over X rays and medical records in Albuquerque. Together with colleagues from the New Mexico Department of Health, the University of New Mexico medical staff, and the IHS, Butler and his two CDC assistants made up a list of twenty-five possible explanations for the deaths and posted it on a large board. Then, drawing from their collective experience and knowledge of strange diseases, they eliminated most of the options.

By five o'clock the forty experts in the room had a short list of hypothetical causes that included some unknown chemical toxin, a new virulent flu strain, a new coxiella (sheep) bacterium, anthrax, Crimean-Congo hemorrhagic fever virus, Hantaan virus, or "something completely new." Though the researchers had no evidence as yet that any of these microbes or chemicals were rampaging through Four Corners, all fit the disease patterns that had been observed.

That pattern typically started with flu-like symptoms: fever, muscle aches, headaches. After a period of a few hours to two days, those symptoms escalated to include coughing and irritation in the lungs. These were caused by leaks in the capillary network feeding the lungs, through which poured fluids. Within a matter of hours patients would become highly hypoxic, unable to absorb oxygen that they hungrily inhaled. Starving for oxygen, the heart would slow down and death could soon follow, caused by either cardiac failure or pulmonary edema.

Butler noticed that those doctors in the meeting who had personally handled such cases were clearly emotionally affected by the drama of patients' deaths and the futility of their medical efforts.

While the New Mexico meeting was getting underway, a similar gathering of experts was winding down in Hughes's office at the CDC. Breimen and other physicians in the room were going over the latest medical reports faxed from Cheek's Albuquerque office.

After an hour of discussion, the CDC list of hypothetical causes of the outbreak was almost identical to the one then being compiled in Albuquerque. And Hughes could see no way to narrow the scope. It troubled him that two completely different categories of agents—toxic chemicals and infectious microbes—were on the list. He remembered that the 1976 Legionnaires' Disease outbreak, which claimed fifty-nine lives in Philadelphia among those attending a summer convention, was bogged down for months with a similarly broad range of causes under consideration.

Hughes grabbed the phone and called the man who solved the Legionnaires' Disease puzzle in 1977, discovering a new bacterium, dubbed *Legionella*, in the hotel air conditioning system. Thirty minutes later, Joe McDade strolled into the meeting. After they gave him a brief status report, McDade quietly narrowed the options.

"It's unlikely to be a toxic chemical because few chemicals cause fevers," McDade said, suggesting that the effort focus on the microbes.

"You've got to develop an algorithm of the disease," McDade said, using terms most of those in the room had heard from him before. "You start by ruling out what is known. And then you get to work on isolating the virus. My guess is it's a virus."

He suggested that C. J. Peters's group test the patient samples they'd received from Four Corners against antibodies for every virus they had in stock.

"Throw out a big net and see what comes in," McDade urged.

Peters agreed, but reiterated his urgent plea for additional personnel. Over the next few days, "loaners" would join the fifteen staff scientists in the P3/P4 lab, some coming from state agencies as well as other sectors of the CDC.

While the Special Pathogens staff toiled in maximum security over petri dishes full of patients' blood and some twenty different types of viral reagents, Cheek, Butler, and the state epidemiologists in the Four Corners area were having a hard time getting useful information from the friends and relatives of the deceased. The local press, having gotten wind of the story several days earlier, was crawling all over the Navajo Nation asking questions many residents found offensive. Furthermore, they were publishing information that the privacy-minded Navajo considered distasteful.

In Navajo culture it was taboo to speak of the dead or utter their names for several days after their demise, yet reporters were doing their jobs, knocking immediately on relatives' doors to ask for details about the lives and deaths of Merrill Bahe and the eighteen other known ARDS victims. Matters worsened when an Arizona newspaper published details about one patient, drawing extensively from the deceased's medical chart.

"The obvious conclusion people drew was that [the paper] got it from us," IHS physician Tempest said. "Here we were trying desperately to protect patient confidentiality, and the public trust was eroding. We were getting it from both sides, being accused of giving the press confidential information on the one hand, and charged with some conspiratorial cover-up on the other. There was so little trust that some people called for an independent investigation."

The situation reached a boiling point when some officials and media referred to the mysterious ailment as "Navajo flu" or "Navajo disease," ignoring the fact that non-Navajos were also falling ill, and marking the American Indians with what the Navajos considered a grossly unfair stigma.

By the first week of June the situation was out of control, as anti-Indian racism mixed with fears of disease. Non-Navajos stayed away from Indian-owned businesses, schoolchildren from the Navajo Nation were denied a field trip to California that had long since been planned, waitresses reportedly wore rubber gloves when serving Navajo customers, and there were rumors of tourists driving across the Navajo mesas wearing surgical masks.[3]

Shortly after Memorial Day there were reports of health investigators and journalists being run off the Navajo Nation at gunpoint by angry residents, and Cheek feared that the entire disease investigation might collapse. Cheek, a Cherokee Indian, sympathized with the Navajos and worked with the IHS area director, Dr. John Hubbard, to relieve tensions in the Four Corners area. Hubbard, a Navajo physician, took Cheek with him to a meeting of the Tribal Council, where they made their case. Tribal president Peterson Zah promised full cooperation, and Hubbard vowed there would be no further violations of tribal sensitivities. Zah also issued an unusual plea to the press, asking that they stay off reservation land until the investigation was completed.

"We decided to have Navajo people involved in every step of the investigation. I insisted on it," Cheek said. "Because I could sense this feeling of betrayal, that we [the Indian Health Service] had betrayed them. We were seen as conduits to the media, even though it wasn't true."

What followed was an investigation unprecedented in its integration of community members into every aspect of the inquiry. Tribal medicine men and elders were respectfully consulted, and they provided the investigators with two vital clues: the piñon nut harvest was unusually large that spring, as was the mouse population. Not since the great epidemics of 1918 and 1936, the elders said, had piñon, mouse, and disease conditions all been so high. The elders' insights steered Cheek, Butler, and other investigators toward searching for a link between the ailments and mice.

By fortuitous coincidence, the University of New Mexico's Robert Parmenter was heading up the massive Sevilleta Long Term Ecological Research survey of the region's flora and fauna, and his team of forty scientists had recently focused on the local rodent population. They had been startled to note a sudden population explosion among the deer mice—a tenfold increase that began in May 1992 and was peaking as the CDC's disease investigation began.

On June 2, with the death toll up to twelve and suspected cases numbering twenty-one, U.S. Health and Human Services Secretary Donna Shalala turned to her staff in a morning meeting in Washington, D.C., and asked, "Are we on top of it? Do you need more resources?"

Assured that the CDC was mobilized, Shalala requested regular briefings. And she expressed concern that Navajos were being improperly labeled as the source of the disease. Recalling the early, incorrect assumptions that AIDS was a "gay disease," she warned her staff to shun the use of terms that linked Navajos to the ailment and asked that special steps be taken at the highest levels to demonstrate sensitivity to American Indian concerns.

The New Mexico State Health Commissioner soon requested federal assistance in handling public response to the outbreak, asking that the CDC send out an expert in media relations to handle local panic and the press. Again, an unprecedented step was taken, and CDC information officer Bob Howard flew out to New Mexico on June 5 to coordinate all press operations.

By then the field team included more than a hundred scientists, physicians, animal trappers, and paraprofessionals who were scouring the Four Corners area. Among them were the CDC's Chapman, Breimen, Childs, Butler, McDade, and dozens more.

And by then they knew what was causing the illnesses and deaths.

C. J. Peters's lab group struck pay dirt during the predawn hours of Thursday, June 3, when antibodies against a family of viruses called hantaviruses cross-reacted in test tubes with blood from the patients. Furthermore, patient blood carefully injected into mice housed in the P4 laboratory showed even stronger antibody reaction to hantavirus reagents. That proved that the agent was infectious and that the virus could reproduce and multiply inside mice.

"That raised some serious eyebrows," Peters recalled. "As soon as we knew where to focus, we got the molecular biologists into the act."

Peters's old Fort Detrick comrade Tom Ksiazek, also a DOD budget cuts refugee recently arrived at the CDC, helped coordinate what would turn out to be the fastest new viral identification ever carried out during an epidemic. In just seven days Ksiazek's laboratory team had serologically narrowed their clues down to hantaviruses; now they needed to determine which specific virus strain was causing the Four Corners epidemic.

Identifying the exact strain of hantavirus responsible for the outbreak required infected wild animals—the microbes' reservoirs. Only in the animal reservoirs would virus levels be sufficient to make isolation and identification possible, Peters knew. So at a CDC staff meeting that day Peters told McDade, Breimen, Chapman, and other scientists who were about to head out to Four Corners that the culprit was probably a hantavirus, and he wanted them to send back plenty of wild rodent samples. Applause and disbelief followed. While the staff praised the lab's speedy solution of the cause of the epidemic, some were dubious. Breimen noted that all known hantaviruses caused kidney problems, none produced respiratory distress.

"It doesn't match," he said, garnering support from most of the physicians in the room. Those in the meeting who were familiar with hantavirus history were skeptical that the virus—first noted in Korea—could have found its way to a landlocked, remote region of North America.

Peters told the group that his molecular biology team was working on a detailed genetic analysis of the Four Corners virus, and he suspected it might turn out to be something new.

"Look, we know there are different types of hantas out there [in the world]," he said. "So let's not rule things out, or in. Finding antibodies isn't enough. We know that."

Hantaviruses first came to world attention during the Korean War.[4] Between 1951 and 1954, more than 2,500 GIs and an unknown number of Korean soldiers fell ill to a mysterious disease that caused fevers, weakness, fatigue, and kidney failure: 121 GIs died of the ailment.[5] U.S. Army researchers fairly quickly figured out that the disease was caused by a virus that was normally carried by field mice.

But it took over twenty years for scientists to successfully isolate the virus—dubbed Korean Hantaan—from the lungs of infected *Apodemus agrarius* mice. Dr. Karl Johnson, then with the U.S. Army Medical Research Institute of Infectious Disease (USAMRIID) at Fort Detrick, collaborated with Dr. Ho Wang Lee of Korea University Medical School in Seoul to discover the virus, using electron microscopes to spot the round microbes that were neatly stacked in rows along the epithelial lining of *Apodemus* lungs.[6]

The natural territory of *A. agrarius* included large parts of Japan, Korea, northeastern China, and southeastern and central Russia. In South Korea between 1955 and 1977, over 9,000 cases of Hantaan were documented; 6.5 percent were fatal. Far more cases were suspected, but were thought to have escaped diagnosis because of their similarity to milder, common ailments, such as influenza.

During the 1970s eleven other forms of hantaviruses were discovered in Eastern Europe and Asia, all linked to usually mild kidney diseases with fatality rates ranging from 10 down to 0.1 percent of all infected people. The viruses were always carried by some type of wild rodent, and people came in contact with the microbes through skin exposure or inhalation of infected animal feces or urine.

In 1977 Belgian researcher Guido van der Gröen, having maintained his interest in hemorrhagic viruses since the Ebola investigation in Zaire, discovered in his laboratory at the Institute of Tropical Diseases in Antwerp hantavirus-induced mild cases of muscle pain, hypertension, and kidney dysfunction among residents of his city. The apparently urban viral strain was very similar to one previously discovered in Sweden,[7] called Puumala virus, carried by voles (*Clethrionomys glareolus*) that inhabited riverbanks. Between 1977 and 1986 van der Gröen identified seventy-six cases of Antwerp-type hantaviral disease in Belgium and France, and tried to warn local physicians that many hanta cases were undoubtedly going undiagnosed because they were confused with occupational back problems, flu, and other minor ailments.[8]

Johnson and Lee, intrigued by van der Gröen's urban hanta findings, tested rats in downtown Seoul, finding that the two most common species in the world—*Rattus rattus* and *Rattus norvegicus*—carried a form of virus only slightly different from the Korean Hantaan strain. They were convinced that the rat infections were relatively recent, having occurred sometime around the Korean War when aerial bombing campaigns drove the *A. agrarius* field mice out of their natural habitats into urban areas, where they got into turf battles with the rats and probably passed the virus on to the larger rodents in biting and clawing fights.

Since Korea was rapidly becoming one of America's biggest trading partners, Johnson wondered whether infected Seoul rats might have found their way into the cargo holds of Korean ships and then escaped into U.S. harbor cities. In 1982, at Johnson's urging, Fort Detrick and CDC scientists combed the harbor areas of Baltimore, Houston, Philadelphia, San Pedro, and New Orleans looking for rats. Everywhere the Army scientists looked, Seoul virus turned up in both black *R. rattus* and their brown cousins, *R. norvegicus*.

That year, 1982, Lee teamed up with NIH Nobel laureate Carleton Gajdusek to test common North American voles for hantaviruses. They scoured Gajdusek's property in Frederick, Maryland, capturing local rodents. And the scientists found a new strain of hantavirus—dubbed Prospect Hill virus after the site of its discovery—and showed that it was carried by two vole species, *Microtus pennsylvanicus* and *M. californicus*. Between them, these voles spanned territory encompassing most of North America.

One of the U.S. Army scientists most intrigued by evidence of hantaviruses in North American rodents was a tall, lean microbiologist named James LeDuc. During the early 1980s LeDuc reasoned that somebody in the United States must suffer Hantaan illness if the virus was infecting domestic rats, so he teamed up with other Army scientists and Jamie Childs, then at Johns Hopkins Medical School, to search for evidence of hanta disease in Baltimore. First, they carefully tested the local rats, discovering to their amazement that virtually every rat over two years of age was infected with a hanta virus. The team then tested 1,788 adults admitted either to Johns Hopkins or a Baltimore sexually transmitted disease clinic in 1986; four were infected with the Korean Hantaan virus. Because the individuals hadn't traveled outside the United States, LeDuc and Childs concluded that they had acquired their infections from local rats.[9]

That led LeDuc and Childs to consider focusing their research on people who suffered symptoms that could be produced by hantaviruses. They knew that the viruses could cause chronic kidney disease in Korea and parts of Europe, so they tested the blood of 1,766 people who were undergoing proteinuria blood chemistry analysis at Johns Hopkins, as well as 254 kidney dialysis patients. They discovered that 6.5 percent of the dialysis patients who were suffering hypertensive kidney disease had serum that reacted with Seoul hantavirus antibodies, indicating that they had been infected.[10]

LeDuc and Childs also found that common Baltimore house mice—*Mus musculus*—carried the Seoul virus.

In August 1986 a Mexican immigrant working in the town of Leakey, Texas, died of internal hemorrhaging and kidney failure. Scientists from the NIH suspected a hantavirus was responsible, and trapped rodents found in areas known to have been frequented by the deceased individual. They discovered that local house mice—again *M. musculus*—were infected with another type of hantavirus, which they designated Leakey virus.[11]

By 1992, LeDuc was convinced that hantaviruses of various types were prevalent in rodents throughout North America, and strongly believed that they were responsible for the higher rates of hypertension and kidney disease seen among America's inner-city poor, particularly African-Americans.[12] Furthermore, he feared that the problems might be worsening, as rat infestation of American inner-city areas increased. Between 1989 and 1991, for example, citizen complaints about rat infestation increased 33 percent in Baltimore and rodent control staff over the same period declined by 50 percent. All the lost staff had been funded under a federal program which was severely slashed by the Bush administration. Similarly, New York City took significant cuts in rodent control funds between 1989 and 1994, during which federal and local funding plummeted from $10.3 million to $5.2 million. Federal funding completely evaporated by 1992, prompting the New York City Health Commissioner, Dr. Margaret Hamburg, to formally express grave concern to the CDC that hantaviruses and other rodent-borne disease agents might get out of control in America's largest metropolis. Despite the commissioner's warnings, the newly elected mayor of New York City, Rudolph Giuliani, slashed the city's rodent control budget by a further 50 percent in early 1994.

Budget cuts in 1991–92 at the U.S. Department of Defense forced closure of most Army medical research programs; Childs, Peters, and Ksiazek went to the CDC and LeDuc ended up working at the World Health Organization headquarters in Geneva. Army hantavirus research slowed radically, leaving only the fortunately prolific molecular biology labs of Dr. Connie Schmaljohn and Peter Jahrling at Fort Detrick to carry the load.

The moment Peters and his CDC laboratory staff had hints of hantaviruses in the Four Corners outbreak, Childs and LeDuc were excited and intrigued. LeDuc was in daily telephone communication with his former Army colleagues, providing insight and gathering information to pass on to interested WHO scientists.

Childs and LeDuc were the first scientists to apply polymerase chain reactions, or PCR, to the diagnosis and study of hantaviruses. LeDuc and Childs developed PCR techniques for hantaviral searches in 1991,[13] and Peters's Special Pathogens Laboratory would benefit enormously from that legacy in 1993. As the second wave of CDC field investigators, led by Breimen and McDade, headed out to Four Corners over the first June

weekend, Peters, Ksiazek, and PCR expert Stuart Nichol eagerly anticipated receipt of rodent samples, upon which they intended to perform PCR
analysis to identify which species of hantavirus was causing the Southwest
outbreak.

Breimen was skeptical of the hanta connection, though he did note that
all earlier animal studies of various hantaviruses found high concentrations
of the viruses in rodent lungs. One LeDuc/Childs study even indicated that
lung tissue was the only site from which viruses could easily be extracted
over the full course of a mouse infection.

During their first forty-eight hours in Albuquerque, Breimen and Chapman hammered out a standard description of the disease, setting out the
criteria for designating a suspected ARDS case as a possible hantavirus
infection. And they created the questionnaires to be used by Navajo and
CDC field investigators making door-to-door surveys on the outbreak.

No sooner had Chapman completed those tasks, sleeping no more than
four hours in two days, than she was asked to assess the utility of ribavirin
as a treatment for the mystery disease. LeDuc and his colleagues had tested
the antiviral drug on Hantaan patients in China in 1987, finding that it
decreased the likelihood of dying if ribavirin was taken within the first
three days of the disease. After that time, it wasn't clear whether or not
the drug was beneficial.[14]

For physicians in the Four Corners area it was a desperate matter to find
something—even a drug of dubious value—to give their ailing patients.
Short of ribavirin, all the doctors could offer was good hospital management
and TLC. And the mortality rate from the virus was very high, appearing
to exceed 70 percent.

By June 7 the CDC had confirmed twenty-four cases of the strange
sickness, twelve lethal, all occurring in the Four Corners area.

The most important clue the investigators had was the CDC lab's hints
of hantaviruses, which, based on the history of such viruses, pointed to
rodent disease carriers. Childs, Chapman, and CDC ecologist John Krebs
were among the three dozen investigators who, teamed with Navajo trappers
and health workers, went out to every site where people had become ill
and set hundreds of rodent traps. They had two types of spring-action traps;
one, a heavy steel case, could handle animals as large as raccoons and
skunks, while the smaller aluminum traps were designed for mice, prairie
dogs, and the like.

The crews had to be careful, avoiding exposure not only to the mystery
virus but also to bubonic plague. Despite a temperature of over 100 degrees,
they wore respirator masks, goggles, paper body suits, double-layered latex
gloves, and disposable booties over their shoes. These precautions might
protect them against airborne bits of hanta-contaminated dust, but Chapman
was much more worried about getting bitten by a plague-carrying flea that
might crawl under her protective garments to reach her vulnerable skin.
She slathered generous layers of insect repellent on her skin as well,

knowing that she wouldn't be able to swat pests when she was wearing the gloves and body suit.

When an inhabited trap was opened, scientists always stood upwind of the animal and carefully placed a plastic bag containing anesthetic over the door of the trap. As the animal fell into the bag, it quickly went unconscious. Later the scientists would withdraw blood and tissue samples, place them on dry ice, and ship the materials back to Peters's P3/P4 laboratory for analysis.

It seemed that the Navajo elders and Robert Parmenter's scientific team were both right about the huge piñon harvest and a very large rodent population. After five years of severe drought, the Four Corners area had record snowfalls during the 1992–93 winter season, followed by an extraordinarily moist spring. Even in June the scientists could still see greenery in the desert and piñon trees standing flush with nutted cones.

With the abundant vegetation came apparently unprecedented rodent populations.

Nearly half the traps contained all sorts of creatures, from mice and prairie dogs to fat rats and smelly skunks, but by far the most common were *Peromyscus maniculatus*, a brown, big-eared mouse with white belly and tail and huge, black eyes sunk into the skull. Because of its brown-and-white coloring, *P. maniculatus* was called a deer mouse.

In initial blood antibody tests the CDC investigators found evidence of hanta infection in the deer mice, as well as two other *Peromyscus* species, two types of chipmunks, common house mice, and *Neotoma albigula* pack rats.

In the lab, however, where CDC scientists had just perfected the use of PCR techniques for diagnosing these infections, only the deer mice were found to be commonly carrying the virus. With time the CDC lab would confirm by PCR the presence of the virus in a third of the first 770 *P. maniculatus* caught in the Four Corners area, as well as in 19.7 percent of *P. truei* (of 314 tested) and 6.9 percent of *P. boylii* mice (of 59 tested).

The *P. maniculatus* were not mice restricted to habitats in Four Corners; on the contrary, these deer mice could be found all over Canada, as far north as the Arctic Circle, and, to the south, throughout the United States and northern Mexico. Only the Deep South states of the United States seemed to be excluded from the deer mouse's natural territory. *P. maniculatus* were, in other words, ubiquitous North American field rodents.

Knowing that, Breimen and McDade wondered how long it might be before mysterious ARDS cases turned up in other states.

By the end of the first week in June, Peters's lab already had suspicions, based on PCR analysis, that the virus infecting both people and *Peromyscus* in Four Corners was yet another newly discovered hanta strain. By mid-June they were sure: the Four Corners virus was, as they wrote in the CDC weekly publication, "a previously unrecognized hantavirus."[15]

If the virus was genetically different from previously identified hantas, its ability to produce acute respiratory symptoms, as opposed to the kidney problems caused by all other hantaviruses, became less questionable. The CDC lab spent the rest of June, July, and August comparing bits of the genetic sequences of the Four Corners virus with the eleven other known hantaviruses, discovering to their amazement that the new virus cross-reacted most strongly with Prospect Hill, the strain found in Maryland voles that had never been associated with human disease.

"That surprised us," Ksiazek said, "because here you had human mortality rates of seventy or eighty percent in Four Corners, and no disease of any kind with Prospect Hill."

The PCR analysis eventually revealed some significant differences in the genetic sequences of the two viral strains, yet Prospect Hill remained Four Corners' closest known relative. [16]

By mid-July the CDC lab had received over 10,000 animal and human samples for analysis; by late August that number was approaching 20,000. In addition to the tremendous burden of simply storing and labeling all those samples, Peters and Ksiazek felt they should pull out of their Atlanta freezers archived rodent samples collected during previous years around Four Corners for plague studies. They wanted to use the archive material to answer two key questions: (1) Is there an epidemic of the virus this year among the rodents or are the animals always carrying the virus at about the same frequency of infection? (2) Has this sudden apparent set of infections been the result of the virus's mutating recently from a benign Prospect Hill-like form to a new type capable of causing ARDS?

The latter question had some purely circumstantial evidence behind it. Hantaviruses were of a class of viruses whose genetic material was stored in the form of three discrete pieces of RNA. Other segmented RNA viruses, such as influenza, were known to mutate frequently because during viral replication the copying of these big pieces of RNA was often a sloppy business. One RNA piece might cross over another, mixing up their genes. Extraneous bits of RNA in the cell the virus infected might get picked up and incorporated into the virus's genetic blueprint. The result for many segmented RNA viruses was a sort of natural crapshoot, with each viral replication event carrying some odds of mutation.

But by August the CDC lab was far too overwhelmed to consider an immediate foray through the archives. The questions would have to remain unanswered for the time being.

Meanwhile, a woman seemed to have died of the mysterious ARDS a thousand miles away from Four Corners, in East Texas. The epidemic had expanded. In a matter of hours, Dr. Ali Khan, having just completed two years of postmedical training as a CDC Epidemic Intelligence Officer, was wandering around the East Texas town of Lufkin, working with state investigators to ensure that all clues at the death site were properly collected and none was overlooked.

The victim, a youthful grandmother, had lived in a fairly elegant, neat house located in a rural area not far from the Louisiana border. No rodents were in or around her house, but Khan discovered telltale mouse feces in a back shed the woman used for putting up fruits and vegetables. Her husband said that she spent hours out there, canning all sorts of produce.

The autopsy revealed that she had suffered a classic case of ARDS of unknown etiology, and Khan had her tissue and blood samples sent to C. J. Peters's lab, where Four Corners virus infection was confirmed by PCR.

Over the next four weeks Khan was constantly on airplanes, flying into suspected ARDS-hanta outbreaks in California, Nevada, Oregon, Louisiana, Arizona, Utah, and Idaho. His second case involved a fifty-one-year-old woman who "lived in smack-of-the-middle-of-nowhere Nevada," as Khan later described it. She survived acute ARDS, thanks to what Khan insisted was "brilliant medical care by her rural doctors." According to the woman, all spring and summer her pet cat kept dragging rodents into the house. Her area of Nevada had unusual rainfall, and most of the locals felt the rodent population was just about out of control.

Khan's third hantavirus victim wasn't as lucky; a resident of a remote area straddling the California-Nevada border, she died of hanta-ARDS. Khan's team found infected mice around her property.

The fourth Khan case involved a twenty-nine-year-old ranch hand who worked the range along the northern California coast. He, too, died of acute ARDS.

An Oregon physician alerted the CDC to the possibility that one of his patients, a sixteen-year-old boy, had died mysteriously of ARDS a year earlier and might have been a hantavirus victim. By the end of August the CDC's lab had confirmed that hantaviruses were in the dead boy's body.

A particularly emotional case for Khan involved a fellow scientist: a promising twenty-seven-year-old female graduate student, Jeanne Messier, who was conducting ecology research in an isolated part of the California Sierras. Ailing, Messier made her way down to a small medical clinic in Mammoth Lake, California, on July 31, and was immediately airlifted to a hospital in Reno. She died shortly after arrival. Scientists found her Sierras cabin overrun by mice.

The U.S. Congress took notice of these events, and the eight senators representing the Four Corners states pushed through legislation during the dog days of July that allocated $6 million in emergency funds to assist the state and federal investigation; $2.6 million went to the CDC. Some members of Congress remarked that such an allocation might not have been necessary if DOD budget cutters hadn't gutted the Army's hanta program two years earlier.[17]

In August, New Mexico authorities requested additional laboratory assistance from USAMRIID scientists, and Peter Jahrling and Connie Schmaljohn began separate efforts to isolate the strange new hantaviruses that were seemingly cropping up all over the United States. Soon the Army researchers were in a competition—later a rivalry—with their CDC colleagues. They discovered the virus in the body of an eight-year-old Mississippi girl who died of an ailment whose symptoms didn't match the CDC definition of hanta-ARDS that Breimen and Chapman had written three months earlier. The CDC refused to add the Mississippi case to its growing list of Four Corners virus victims.

"The CDC is claiming it did not meet their case criteria," Jahrling would later say with some bitterness. "They cannot refute, however, our evidence that we have a replicating hantavirus from that case, in cell culture."

In early June, another alert rural physician spotted what would prove to be one of the most puzzling hanta cases, this one occurring in Louisiana. A fifty-eight-year-old bridge inspector had died of sudden ARDS, and rural physicians called Khan in late July. While the Louisiana doctors frankly doubted the case could be another example of Four Corners disease, they had read the CDC's bulletins and thought Khan should know that the symptoms matched.

Khan was dubious. The *Peromyscus* mouse didn't inhabit Louisiana, and the area was over 200 miles away from the East Texas site of Khan's first investigation. All doubt vanished, however, when Khan saw the patient's medical records and Peters's lab group confirmed that the blood samples were antibody-positive for hanta infection. The victim's farmhouse was neat as a pin—no signs of rodent infestation. But the victim's co-workers told Khan that they ran into rats and mice every day as they crawled around culverts and ditches to inspect western Louisiana's bridges.

When the lab did PCR analysis on the Louisiana man's virus, the mystery deepened: it didn't match with the Four Corners strain or any other known hantavirus. Fellow bridge workers hadn't seen the distinctive white-bellied deer mice around their work site, but they had encountered plenty of the big brown *Rattus norvegicus*—the same rat species LeDuc discovered years earlier carrying Seoul-strain hanta in Baltimore.

Ksiazek was one of the few CDC scientists who weren't surprised by the discovery.

"These viruses are all pretty close to one another, as viruses go. And all these rodents have common ancestors. The genetically closest viruses are carried by close-relative rodents. Personally, I think this is an indication of co-evolution of rodents and their passenger viruses," Ksiazek said.

By summer's end Ksiazek suspected that more strains of hantaviruses remained to be uncovered in North America, and many hundreds of cases of ARDS, kidney disease, and hypertension in the United States every year would turn out to have been caused by these rodent viruses. While only forty-two cases of hanta-ARDS were confirmed by the CDC as of October 29, Ksiazek was convinced that they represented the tiny tip of a vast disease iceberg. Retrospective analysis showed that the earliest identifiable case occurred in July 1991. Since then, patients had ranged in age from twelve to sixty-nine years, and 62 percent had died. Half the victims were American Indians: presumably due to lifestyle, not genetic, factors.[18]

Throughout history, rats and mice have taken advantage of human movements to gain access to new ecologies around the world. The arrival of *R. rattus* and its cousins in the Americas is undoubtedly recent, probably having occurred less than 500 years ago when rodent stowaways made their way to American soil from the boats of European explorers and slave traders. And they may have been on the European continent for only some 1,200 to 1,500 years prior to that, having stowed away on traders' ships and caravans from the Middle East and the Far East.[19]

Given how recently, on a scale of evolutionary time, these rodents have spread around the world, it should come as no surprise that they carry viruses which, whether found on the Volga steppes or in the deserts of Arizona, bear remarkable resemblances. And if the origins of these hantaviruses can be traced back to the earliest periods of mouse and rat evolution, it would seem logical to assume that careful inspection of *Rattus*, *Peromyscus*, and *Mus* cousins all over the world would reveal still more hanta strains.

The Four Corners outbreak prompted scientists to rethink diseases once labeled as "unknown etiology" and consider the possibility that millions of people worldwide may needlessly suffer ailments and death caused by the rodent-carried viruses.

Kidney expert Dr. Guy Neild of Middlesex Hospital in London was moved to ask during the Four Corners investigation whether the long-mysterious "trench nephritis" that claimed the lives of hundreds of soldiers hunkered down in trenches during the American Civil War and World War I may not have been due to hantaviruses carried by co-entrenched rats or mice.[20]

German physicians from the University of Würzburg reported that a surge in European hantavirus cases occurred during the spring of 1993, leading to hemorrhagic fever with kidney complications. They noted that rodent control efforts had slowed a bit since German unification, and wondered whether the surge in the German rat population could have been prevented.[21] German records indicated that a first, fairly small hantavirus epidemic surfaced in that country during mid-1990, causing no fatalities, but 88 confirmed renal illnesses. A second German hantavirus epidemic ran its course from September 1992 to October 1993: some 183 illnesses, no deaths. Researchers were certain that the outbreaks were the result of local surges in the rodent and vole populations, and that illnesses in the 1993 outbreak were more severe. They were also certain that hanta cases were underreported by German physicians.[22]

Similar surges in Puumala infection were noted in France, Belgium, and the Netherlands in 1993.[23]

Though the Four Corners type of respiratory hantavirus had never been seen in Russia, classic kidney disease-producing forms were seen between 1985 and 1992 in twenty-three regions of the country, afflicting 68,796 people. The ailments were caused by more than seventy different hantavirus strains that were carried by sixty-three different species of birds and small mammals, including all common rodents.

"These rodents and their viruses have been here for millennia," McDade concluded. "There may have originally been a common ancestor virus infecting a single rodent species, which may have mutated and spread to other rodents over time. But these viruses have almost certainly been among us for centuries."

McDade paused, examined his hands for a moment, and added, "I often wonder with Legionnaires' Disease, if there had not been an association with a particular hotel, a drama, if you will. If there had just continued to be sporadic, scattered, inexplicable pneumonia deaths, would we have ever recognized Legionnaires' as a distinct disease?

"And I now wonder the same thing about Four Corners. If there hadn't been that one cluster of four cases among healthy Navajos, would we have ever recognized the virus among us?" McDade asked, noting that there were many other diseases for which no clear cause is known.

"We should continue to allow for the possibility that they are all due to infectious agents," he said.[24]

Inside the CDC's P4 laboratory Dr. Luanne Elliott was toiling around the clock during the summer and fall of 1993 trying to find a way to grow the Louisiana and Four Corners viruses in test tubes. For four months the slow-growing viruses stubbornly resisted her expert efforts. In the last week of September she finally succeeded in experimentally infecting laboratory mice, and a month later was able to cultivate live virus.

At USAMRIID similar efforts to grow the Four Corners virus in the laboratory, isolate it, and further elaborate its genetic makeup were underway. The two federal laboratories raced to complete the jobs first, eventually reaching a dead heat with both groups declaring victory at the annual meeting of the American Society of Tropical Medicine and Hygiene in Atlanta on November 3, 1993, and at a University of New Mexico hantavirus conference on November 20. Jahrling succeeded in growing virus that came from a New Mexico patient, and University of New Mexico collaborator Kurt Nolte made electron microscope photographs of the Four Corners virus budding from the membranes of monkey cells.[25]

At the same time, Schmaljohn succeeded in growing Four Corners virus that was extracted from a deer mouse trapped near the California mountain cabin of hantavirus victim Jeanne Messier. As was the case with Jahrling's isolate, the Schmaljohn virus grew on monkey Vero cells.

And the CDC team isolated the virus from a New Mexico deer mouse, also successfully growing the microbe on monkey cells.

Unfortunately, each group jostled for credit as "the first," and considerable tensions existed between the CDC and USAMRIID. Eventually they agreed to simultaneous publication of their work, sharing credit for the isolation and identification of the Four Corners virus.

Successful isolation of the virus opened up the next obvious phases of the effort: development of a vaccine and an easy screening test that could be used in rural medical clinics. Schmaljohn's team had already developed an experimental vaccine for the Seoul virus,[26] so there were reasonable grounds for optimism that a similar Four Corners vaccine could be created quickly.

Perhaps more important than an eventual vaccine or diagnostic test was the actual process whereby the collective scientific enterprise identified the cause of the mysterious disease and swiftly brought the epidemic to a halt. Though there were unfortunate tensions between the CDC and USAMRIID, and Navajos felt the sting of discrimination, the overall effort was noteworthy as a demonstration of two old principles of epidemic investigation and as an illustration of an exciting new principle.

The old, but often overlooked principles were simple. All "new" diseases must first be noticed by someone who has the insight and courage to sound an alarm and set in motion a thorough investigation. And once in place, investigations are best conducted in an atmosphere of candor and collectivity, rather than the secrecy, backbiting, rivalry, and mutual contempt that had unfortunately characterized many other scientific pursuits of emerging microbes.

The novel discovery—one that is sure to permanently change the course of emerging microbe and epidemic research—was the utility of molecular biology and, in particular, PCR. Just as police work was forever changed by the discovery that all human beings have unique fingerprints that can be "lifted" from weapons and other objects found at the scene of a crime, so PCR provided a revolutionary tool that, for the first time, put the laboratory scientists in the driver's seat in an epidemic investigation. Before the CDC animal catchers even set foot in the Four Corners area, Stuart Nichol had been able to use USAMRIID genetic primers for various hantaviruses to rapidly screen human samples shipped to Atlanta by the New Mexico authorities. That would have been impossible twelve years earlier, when the AIDS virus made its appearance.

The hantavirus investigation of 1993 proved that things could be done right, that humanity could comprehend and control the microbes, if there was the political and scientific will.

At the beginning of 1994, the CDC reported that a total of fifty-five hantavirus ARDS cases had been confirmed in sixteen of the United States. Thirty-two of the infected individuals had died. And one of the newly confirmed cases occurred in Florida—an area that was definitely bereft of *P. maniculatus* deer mice.[27] Days later the CDC and the states of Rhode Island and New York announced the death of David Rosenberg, a twenty-year-old student at the Rhode Island School of Design in Providence.

Though Rosenberg died of ARDS in a Providence hospital, he had spent the weeks prior to taking ill with his parents on Long Island, and making a student film in his father's abandoned warehouse in Queens. Investigations revealed that one worker in the Rosenberg family electrical supplies factory, located near the warehouse, had developed antibodies to hantaviruses, and Rosenberg was infected with a virus that closely resembled the Four Corners strain. No infected animals were retrieved, but investigations were hampered by an unusually harsh winter that brought seventeen major snowstorms in New York, driving rodents into hiding.[28]

In January 1994 the strange new microbe was officially named Muerto Canyon, after the valley inside the Navajo Nation in which the Four Corners virus first appeared.

Muerto Canyon—Valley of Death.

16

Nature and Homo sapiens

SEAL PLAGUE, CHOLERA, GLOBAL WARMING, BIODIVERSITY, AND THE MICROBIAL SOUP

It is hard to gain historical perspective on an event that is completely unlike any other we have seen before.
—Al Gore, *Earth in the Balance*, 1992

■ That humanity had grossly underestimated the microbes was no longer, as the world approached the twenty-first century, a matter of doubt. The microbes were winning. The debate centered not on whether *Homo sapiens* was increasingly challenged by microscopic competitors for domination of the planet; rather, arguments among scientists focused on the whys, hows, and whens of an acknowledged threat.

It was the virologists, and one exceptional bacteriologist, who started the debate in 1989, but they were quickly joined by scientists and physicians representing fields as diverse as entomology, pediatric infectious disease, marine mammal biology, atmospheric chemistry, and nucleic genetics. Separated by enormous linguistic and perceptual gulfs, the researchers sought a common language and lens through which they could collectively analyze and interpret microbial events.

There had never really been a discipline of medical microbial ecology, though some exceptional scientists had, over the years, tried to frame disease and environmental issues in a manner that embraced the full range of events at the microscopic level. It was far less difficult to study ecology at the level of human interaction—the plainly visible.

There were certainly lessons to be drawn from the study of classical ecology and environmental science. Experts in those fields had, by the 1980s, declared that a crisis was afoot spanning virtually all tiers of earth's macroenvironment, from the naked mole rats that foraged beneath the earth to the planet's protective ozone layer. The extraordinary, rapid growth of the *Homo sapiens* population, coupled with its voracious appetite for planetary dominance and resource consumption, had put every measurable biological and chemical system on earth in a state of imbalance.

Extinctions, toxic chemicals, greater background levels of nuclear and ionizing radiation, ultraviolet-light penetration of the atmosphere, global warming, wholesale devastations of ecospheres—these were the changes of which ecologists spoke as the world approached the twenty-first century. With nearly 6 billion human beings already crowded onto a planet in 1994 that had been occupied by fewer than 1.5 billion a century earlier, something had to give. That "something" was Nature—all observable biological systems other than *Homo sapiens* and their domesticated fellow animals. So rapid and seemingly unchallenged was human population growth, the World Bank predicted that nearly three times more *Homo sapiens*, on the order of 11 to 14.5 billion, would be crowded onto the planet's surface by 2050. Some high-end United Nations estimates forecast that more than 9 billion human beings would be crammed together on earth as early as 2025.

The United Nations Population Fund spoke of an "optimistic" forecast in which the planet's *Homo sapiens* population "stabilized" at 9 billion by the middle of the twenty-first century.[1] But it was hard to imagine what kind of stability—or, more likely, *instability*—the world would then face, particularly given that the bulk of that human population growth would be in the poorest nations on earth. By the 1990s it was already obvious that the countries that were experiencing the most radical population growths were also those confronting the most rapid environmental degradations and worst scales of human suffering.[2]

Biologists were appalled. Like archivists frantic to salvage documents for the sake of history, ecologists scrambled madly through the planet's most obscure ecospheres to discover, name, and catalogue as much flora and fauna as possible—before it ceased to exist. All over the world humans, driven by needs that ranged from the search for wood with which to heat their stoves to the desire for exotic locales for golf courses, were encroaching into ecological niches that hadn't previously been significant parts of the *Homo sapiens* habitat. No place, by 1994, was too remote, exotic, or severe for intrepid adventurers, tourists, and developers.

At Harvard University, Dr. E. O. Wilson was one of the leaders of a worldwide effort to catalogue the world's species and protect as much of the planet's biodiversity as possible. He estimated that there were 1.4 million known species of terrestrial flora, fauna, and microorganisms on earth in 1992, and perhaps as many as 98.6 million yet to be identified. The vast majority of those unknown plants and creatures, he argued, were living in the world's rain forests.[3] There the plentiful supply of rain, tropical sunlight, and nutrient-rich soil bred such striking diversity that Wilson found 43 different *species* of ants living on a single tree in the Amazon.[4] Devoted biologists were literally risking their lives in a mad rush to identify the missing 10 to 98.6 million species, some 50 percent of which were thought to reside in the rain forests of Amazonia, Central Africa, and South Asia.

The pace of the loss was staggering—on the order, by UN estimates, of 4.75 million acres annually.[5]

Whether supplying the highly profitable heroin and cocaine markets, which in the Andes was responsible for devastation of upward of 90 percent of the Colombian forest and only slightly less alarming percentages of the forests of Ecuador, Peru, and Bolivia, the fast-food beef consumption habits, or the coffee needs of the wealthy world, entrepreneurs of the developing nations were responding to all too present economic incentives when they destroyed their natural ecologies.[6] Without competing economic incentives for protecting the ecospheres it seemed unrealistic to expect that local human beings would take meaningful steps to reverse or slow the pell-mell pace of deforestation.[7]

Using Landsat satellite imagery that was enhanced to reflect geographic features that might be hidden in flat photographs, David Skole and Compton Tucker, of the University of New Hampshire and NASA's Goddard Space Flight Center, made computer estimates of destruction in the Amazon between 1978 and 1988. Six percent of the Amazon's upper canopy and 15 percent of its total forest mass had, they concluded, effectively been destroyed.

Though it was well known to biologists that tiny isolated pockets of dense vegetation surrounded by devastation couldn't support a diverse range of species, none of the prior ecosphere calculations had factored for such islets of forestry. When Skole and Tucker studied the Amazon, however, they realized that many areas looked like a checkerboard, with slashes and zigs and zags of devastation slicing the rain forest into ever-thinner islets bordered by constantly thickening swaths of desertification or development. Humanity didn't nibble into the forest from its edges; it built huge super-highways that plunged into the pristine center and side roads that bisected one subsection after another.

So, the two scientists concluded, about 15,000 square kilometers of Amazonia were being directly destroyed by human beings every year, but another 38,000 square kilometers were indirectly destroyed annually by the isolation and fragmentation process.[8] That combined effect represented an annual forest loss of an area larger than the United Kingdom of Great Britain and Northern Ireland. It also implied that between 1978 and 1988 Amazonia effectively lost 15 percent of its productive forest.

When ecospheres were so severely stressed, certain species of flora and fauna that were best suited to adapt to the changed conditions would quickly dominate, often at the expense of less flexible competitors. The net result would be a marked decline in diversity. This could clearly be visualized when, for example, a tropical area was cut to make way for a golf course. Though the golf course was comprised of flora and fauna, its range of diversity was strictly controlled by human beings. At the course's periphery Nature would constantly try to push its way back in, but the aggressive species were usually limited to the healthiest plants and animals. If humans ceased trying to control the golf course, those sturdy aggressor species would swiftly move in, but it would be years before the original scope of diversity would be restored—if ever.

Both deforestation and reforestation could, therefore, give rise to microbial emergence. If an ecology had been entirely devastated, and its eventual replacement species were of inadequate diversity to ensure a proper balance among the flora, fauna, and microbes, new disease phenomena might emerge.

Such was the case in 1975–76 in the Atlantic seaside town of Lyme, Connecticut. Like many New England coastal communities that dated back to the colonial era, Lyme was a quaint town of two-hundred-year-old buildings, birch trees, and homes interspersed with pockets of picturesque pastoral scenery.

During the mid-1970s fifty-one residents of the town came down with what looked like rheumatoid arthritis. The ailment, dubbed Lyme disease, would by 1990 have surfaced in all 50 states and parts of Western Europe. Though scattered reports of Lyme would emanate from states with ecologies as disparate as those of Alaska and Hawaii, more than 90 percent of all cases were reported out of coastal and rural areas between Long Island, New York, and Maine. New York would, by 1988, lead the world in Lyme diagnoses with 6.09 cases per 100,000 adults, and reported cases from the northeastern states would double every year between 1982 and 1990.[9]

The typical Lyme disease patient suffered localized skin reddenings that were indicative of insect bites, followed days to months later by skin lesions, meningitis, progressive muscular and joint pain, and arthritic symptoms. Untreated, the ailment could be lifelong, leading to a range of neurological disorders, amnesia, behavioral changes, serious pain syndromes in the bones and muscles, even fatal heart disease or respiratory failure.[10] Once physicians learned of Lyme, the disease was undoubtedly overdiagnosed in endemic areas of the Northeast,[11] but there remained a clear upward trend in the United States in bona fide cases, and by 1992 Lyme was the most reported vector-borne disease in the country.

Most Lyme sufferers lived in wooded areas that were inhabited by common North American feral animals: deer, squirrels, chipmunks, and the like. Notably absent in these untroubled, quiet woods were the ancient predators, such as wolves, cougars, and coyotes. Keeping deer and small mammal populations in check had, in fact, become a major headache for affluent wooded communities all over North America.

In 1982, Dr. Allen Steere of Tufts University in Boston discovered that Lyme patients were infected with a previously little-studied spirochete bacterium, *Borrelia burgdorferi.*[12] Subsequently he and other physicians showed that many of the dreadful symptoms of the disease were the result of the immune system's protracted battles with the microbe.[13]

Scientists soon determined that the *Borrelia* bacteria were transmitted to people by a tick, *Ixodes dammini.* While the tick was happy to feed on

Homo sapiens, its preferred lunch was deer blood, specifically that of the white-tailed deer then common to the North American woods. About 80 percent of all North American cases were linked to either residing in a deer habitat or hiking through such an area.[14]

Harvard's Andy Spielman showed, however, that getting rid of the deer in a region didn't eliminate Lyme disease. While the incidence of the disease among human beings might decline, it didn't go away. Further, there was a seasonal periodicity to Lyme outbreaks that coincided with the life cycle of the *I. dammini* tick, but not necessarily with that of the deer.[15]

Spielman and his lab staff figured out that the ubiquitous northeastern mouse *Peromyscus leucopus* was the natural reservoir for the *B. burgdorferi* bacterium that caused Lyme disease. The immature ticks lived on the mice and fed on the rodents' blood. The mice, which were harmlessly infected with the bacteria, passed their *B. burgdorferi* on to the ticks. As spring approached, the winter thaw each year witnessed surges in the populations of both the *P. leucopus* mice and their tick passengers. The two species, rodent and insect, shared the ecology of low scrub brush that grew along the sand-duned shores and woodlands of the American Northeast. The deer grazed through these areas, picking up *I. dammini* ticks, which, while feeding on deer blood, passed on the bacteria.

The deer carried the ticks with them as they made long foraging journeys through woodlands and into suburban yards. Because there were no predators around to keep the deer population in check, their sheer numbers were great enough to force the animals to scour boldly for food, often stepping right into suburban front yards and patios to nibble at carefully cultivated azaleas and lawns. That, in turn, guaranteed that three more species—*Homo sapiens*, felines, and canines—would come in contact with *I. dammini* ticks and the *B. burgdorferi* bacteria they carried.[16]

Studies in New York showed that the territory inhabited by the *I. dammini* tick was expanding at a steady and rapid rate, as deer, pet dogs, humans, rodents, and even some birds carried the insects further and further from the initial outbreak sites. By 1991, Lyme, the disease, and *I. dammini*, its vector, had spread widely throughout wooded and scrub-brush ecospheres all over the Northeast. Their invasion, and the epidemic they spawned, was new.[17]

To understand how, and why, Lyme disease had suddenly emerged in North America, Spielman and his colleagues tried to recapitulate the history of the expansion of *I. dammini*'s territory.[18]

The work took Spielman's group back in time to the arrival of British colonists in North America. When the Pilgrims landed in Massachusetts they set to work with Puritanical fervor clearing local forests and building settlements, Spielman said. By the late eighteenth century Massachusetts was the center of North America's iron industry, and remaining forests of the region were denuded to supply fuel for iron smelting. By the nineteenth century most of the woodlands of the entire Northeast had been so thoroughly devastated that housing construction required importation of wood from what were then the western territories.

"The result was an ecology just as artificial as a concrete parking lot," Spielman said, speaking of the later return of flora and fauna to the denuded areas. The grand tall trees, oaks and larches, never returned, nor did the large carnivorous animals. What did replace the old forests was an ecology similar to what probably had comprised the edges of the woods in the sixteenth century: scrub brush, small birches and other nonshade trees, meadows, deer, chipmunks, voles, squirrels, and birds.

"It's an artificial landscape that we have created, largely by neglect, here in the East," Spielman said, adding that the new ecology was filled with insect and rodent vectors, "lurking out there in this system of change."

Into the denuded forests came aggressor flora and, unchallenged by predators, the deer, rodents, and *I. dammini* ticks. As their numbers soared, bringing the deer, in particular, back from the edge of extinction in the Northeast, a new disease paradigm emerged.[19]

As the invasion of *I. dammini* ticks and deer into artificially reforested areas demonstrated, no matter how hard *Homo sapiens* struggled to pave the world, Nature never ceased trying to force its way back. No area could escape the steady global spread of plant, animal, and insect species. In the absence of natural predators or competitors, alien species introduced into artificial ecologies—including mega-cities—could quickly overwhelm all suitable niches. And with the immigrant species could—and had—come microbes that were new to the local environment.

The Lyme case demonstrated the fallacy of viewing flora and fauna per se as "natural." From the point of view of microbial opportunity, loss of original biodiversity couldn't be corrected merely by introducing a handful of aggressor species.

During the early 1980s ecologists Paul and Anne Ehrlich of Stanford University developed the "Rivet Hypothesis" of diversity. They thought of the ecosphere as a huge airplane held together by steel rivets, or species. As each species died out, the total mass of the "airplane" might remain the same, but rivets were lost, weakening the overall structure. Eventually, a critical number of rivets having been lost, the plane would come apart, crash, and perish.

The epochal "Rivet Hypothesis" was given credibility by several experiments conducted in laboratories around the world. Scientists grew plants in environmentally sealed greenhouses filled with devices that measured carbon dioxide, oxygen, and total biomass. And it turned out that the more diverse the species assortment in a greenhouse—even when total biomass, sunlight, and all other factors were equal—the greater the oxygen production and general vitality of the little ecosphere.

In a survey of nineteen tropical forest ecospheres, researchers from the Missouri Botanical Garden found striking evidence that the changing ratio of oxygen to carbon dioxide was already having dire effects: forest turnover rates were increasing dramatically. Whole sections of forest biota "rivets" were dying and regenerating with radically escalating haste. In several major forests—particularly in Central Africa and Amazonia—turnover rates over the 1970–94 period had increased 150 percent every five years. The result, wrote Al Gentry (who died in a plane crash over Ecuador while making these surveys), was a net decrease in biodiversity as the older, massive hardwood trees, and the multitude of flora and fauna that existed in the ecospheres they created, died off and were replaced by a limited range of aggressive smaller trees and tropical vines. These gas-dependent species had less dense wood, and could transform forests into carbon sinks which would emit chemicals that further exacerbated the CO_2 imbalance and ozone crisis. Gentry predicted an accelerated rate of species extinctions and a radical change in the density and diversity of the world's rain forests, all occurring at astonishing speed.[20]

From an atmospheric scientist's point of view the most crucial issue was the decline in oxygen production from the earth's flora due both to its overall declining mass and to the lowered range of diversity among surviving vegetation. Coupled with increased production of carbon dioxide owing primarily to human fossil fuel consumption and forest burning, and the expected increase in oxygen-dependent *Homo sapiens*, a clear chemical crisis loomed.

The most immediate impact was chemical destruction of the earth's ozone layer. The invisible layer of gas composed primarily of uncoupled oxygen atoms, or ozone, had unique physical properties. The atoms responded to specific wavelengths of light, repelling those in the ultraviolet and infrared bands. Little light in those wavelengths emanated upward from the earth's surface, but the planet was bombarded with such radiation from the sun. If not for the ozone layer, the planet would be a humanly uninhabitable hothouse bathed in mutation-causing ultraviolet light.

Throughout the 1980s researchers, particularly at NASA's Goddard Space Flight Center, amassed evidence that the ozone layer was weakening, especially over the South Pole. Over Antarctica an actual seasonal hole had developed in the ozone layer, through which poured levels of ultraviolet light unprecedented in known human history.

By 1990 a fierce debate raged in scientific circles over the size and significance of that ozone hole. But something was undoubtedly happening to the global ecology. Glaciers were retreating in some parts of the world, skin cancer rates were up in Australia and southern Chile, surface temperatures of oceans in some areas had risen, and mean surface air temperatures were up. Some researchers found, in fossils and deep glacial ore samples, evidence of such periods of warming in the earth's past, indicating that such events could all be part of a historic cycle on the planet. Further, it was possible that the bulk of the warming was induced not by human pollution and rain forest destruction, but by natural catastrophic events such as the 1991 eruption of the Mount Pinatubo volcano in the Philippines.[21]

There was strong evidence, however, that halogen ions, particularly chlorides and bromides, were making their way via human pollution into the atmosphere. These were the breakdown products of thousands of plastics, pesticides, fuels, detergents, and other modern materials. Once inside the ozone layer, the halogens acted as chemical scavengers, attaching themselves to free oxygen atoms to form heavy molecules that then fell out of the protective layer into lower tiers of the earth's atmosphere. In this way, ozone was actively depleted.

Most Western scientists insisted that the pollution- and deforestation-driven ozone depletion and global warming hypothesis was correct, though among believers there were significant differences of opinion about its current and forecast severities. The strongest evidence supported an estimate of a net global temperature increase of half a degree centigrade during the twentieth century, with five degrees centigrade marking the difference between, on the one hand, the Ice Age and, on the other, a severely deleterious greenhouse warming effect.

The first outcome of this warming was a higher surface water evaporation rate, which, in turn, led to greater levels of rainfall and monsoon in key areas of the planet. In places that normally had low levels of rainfall, such as the Sahara Desert, there would be even less precipitation. The net result would be greater extremes in water distribution, with severe droughts afflicting some parts of the planet, flooding and hurricanes hitting others. That, in turn, was expected to alter everything from the migrations of birds to the feeding patterns of blue whales; from habitat ranges of malarial mosquitoes to the amount of the planet's arable land suitable for profitable agricultural growth.[22]

The lesson of macroecology was that no species, stream, air space, or bit of soil was insignificant; all life forms and chemical systems on earth were intertwined in complex, often invisible ways. The loss of any "rivet"—even a seemingly obscure one—might imperil the physical integrity of the entire "plane."

The "plane," in the Ehrlichs' scenario, was destined to crash. What hadn't been anticipated was that the plane would first get sick, heavily encumbered by emerging pathogenic microbes.

In 1987, Siberian fishermen and hunters working around Lake Baikal noticed large numbers of dead seals (*Phoca sibirica*) washing up along the shores of the huge Central Asian lake. By year's end, the seal death toll would top 20,000, or nearly 70 percent of the entire population. The world's deepest lake—a mile deep—was a unique 12,000-square-mile ecosphere inhabited by a number of species of flora and fauna found nowhere else in the world, including the dark gray freshwater seals. Because the Soviet government had long used the country's lakes as waste dumps, it was first assumed that the seals were victims of some toxic chemicals.

But with the spring thaw of 1988 came an apparent epidemic of miscarriages among female harbor seals (*P. vitulina*) in the North Sea along the coasts of Sweden and Denmark. Some 100 spontaneously aborted seal pups were recovered, a few of which survived long enough for scientists to study their symptoms: lethargy, breathing difficulties, nasal congestion.[23] A quick-and-dirty analysis of the pups' blood revealed that the dying and dead seals had antibodies that reacted weakly in laboratory tests against canine distemper viruses.

With the arrival of summer 1988 came hundreds of dead adult seals in the North Sea area. They washed up upon shores separated by huge expanses of land and sea, from the western Baltic Sea area of Sweden and Denmark to the far west coast of Scotland. By August dead seals were even found scattered along the beaches of northern Ireland.

In laboratories in the Netherlands, Ireland, Russia, and the United States, scientists swiftly determined that the seals were dying from a morbillivirus—the same class that included human measles, cattle rinderpest virus, and canine distemper. The die-offs continued well into 1990, eventually claiming more than 17,000 North Sea harbor seals, or more than 60 percent of the entire population.

Scientists working in laboratories from Atlanta to Irkutsk swiftly determined that two different viruses were responsible for what seemed to be separate seal epidemics in Lake Baikal and the North Sea. The virus isolated from the massively infected lungs of Lake Baikal's seals was dubbed phocine distemper virus-2 (PDV-2), and it proved virtually identical to canine distemper virus.

The North Sea seals, however, were suffering from something never before seen. Their microbial assailant, named phocine distemper virus-1 (PDV-1), was distinct from any other known morbillivirus. It appeared to be something new, and the extraordinary death rates among harbor seals indicated that their immune systems had never previously encountered such a virus.

While the seal experts worked on that puzzle, veterinarians in Spain were examining dolphins that were beaching themselves along the Mediterranean coast of Catalonia, Spain. By July 1990 more than 400 Mediterranean dolphins had washed onto the shores of North Africa, Spain, and France, clearly suffering respiratory distress. Autopsies of the animals revealed startling brain damage and acute immunodeficiencies. Similar signs of immune deficiency had already been documented in the North Sea seals, and accounts in the popular Spanish press were soon calling the mysterious marine mammal ailment "dolphin AIDS."

But it wasn't AIDS—it was more like measles. Dolphins were also coming down with a deadly morbillivirus. By 1991 common dolphins (*Delphinus delphis*), striped dolphins (*Stenella coeruleoalba*), white-beaked dolphins (*Lagenorhymchus albirostris*), and porpoises (*Phocoena phocoena*) all over the Mediterranean and Ionian seas were dying.

Dutch scientists determined that at least four newly discovered viruses were attacking Europe's and Central Asia's marine mammals: PDV-1, which was similar to human measles; PDV-2, which appeared to be identical to the virus that caused distemper in dogs; and PMV or porpoise morbillivirus, and DMV, or dolphin morbillivirus.

Russian scientists discovered a possible explanation for the Lake Baikal epidemic of PDV-2. It seemed that an epidemic of distemper swept through the Siberian sled dog population in 1986; local people threw their dead dogs into the lake. Curious seals that investigated the large corpses became infected. In the Siberian case, then, the virus was not new to the world, though it was new to the freshwater seal species. The extraordinary death toll was the result of a microbe jumping from an ancient host species into a new, immunologically vulnerable species.

While the PDV-2 puzzle appeared to have been solved, mystery continued to shroud the origins of PDV-1, DMV, and PMV. Where did the viruses come from? How did they spread so rapidly over such a broad geographic area? Were they old viruses with newfound mutant virulence? Or were the animals particularly vulnerable because of other factors?

Spanish researchers were convinced that PCB pollution of European seas played a key role. All the dolphins they autopsied had high levels of PCBs stored in their body fat; some showed signs of PCB-induced tumors. So one hypothesis was that seals and dolphins that inhabited particularly contaminated waters were already immune-deficient when the virus appeared, making them uniquely vulnerable. Such an explanation might resolve questions about why Canadian seals, though infected with PDV-1, apparently hadn't fallen ill, while their cousins living in polluted North Sea and Baltic waters, had. But still unanswered was the origin of the viruses.

By 1993 the dolphin and porpoise die-offs had slowed considerably, and many scientists felt that the epidemic might be over. But why?

A key difference between 1988–90 and 1993 was the severity of Europe's winter. In the Mediterranean, in particular, the earlier winters were remarkably mild, which meant that small fish populations in the region were unusually low. Spanish researchers examined some 500 dolphin corpses, and concluded that all the animals were unusually skinny and their livers were severely damaged. The scientists decided that PCBs, which are normally stored in human and animal body fat, had flooded the dolphins' livers as the starving animals burned up stored body fat. That, in turn, led to high blood levels of the toxic chemicals and mild immune deficiency. The virus subsequently exploited the dolphins' vulnerability.

By the close of 1990 at least 1,000 Mediterranean and Ionian dolphins had succumbed.

The morbillivirus mystery deepened still further when bottlenose dolphins, beluga whales, Atlantic harbor seals, and porpoises were washed ashore on beaches stretching from the Gulf of Mexico to Quebec's St.

Lawrence Seaway. Those that were discovered alive often appeared dazed and distressed, as if suffering brain damage or high fevers.

Again, scientists looked for viral and pollution explanations, finding a confluence of factors at play. As had been the case with European sea mammals, the North American die-offs came during unusual weather. An El Niño weather pattern gave rise to extraordinary rainfall in the Midwest, which led to high levels of pesticide, pollutant, and human and livestock fecal waste runoff into the Mississippi and other major arteries. That waste made its way to the Gulf of Mexico, where some of the first bottlenose die-offs occurred. The polluted waters moved through the Gulf, up the Florida coastline to the Carolinas, where more marine mammal die-offs ensued. From there, the water mass wended its way along the coasts of New Jersey, New York's Long Island, Massachusetts, Maine, the St. Lawrence Seaway, and Nova Scotia, everywhere claiming a toll.

Debates about what factors in that water led to the die-offs raged well into 1993. PCBs and other chlorinated hydrocarbon toxic chemicals were in the river runoff water, but few scientists believed the chemicals were directly responsible; many accepted the notion that chemically induced immunodeficiency served to aggravate some other underlying cause of disease in the dolphins, seals, whales, and porpoises.[24]

Off the Carolinas scientists discovered massive colonies of algae of the species *Ptychodiscus brevis* that secreted a powerful neurotoxin called brevetoxin. They hypothesized that the unusual weather, coupled with high levels of nitrogen-rich human and livestock fecal matter, had led to the formation of huge "red tides," or algal blooms, that contained such toxic algae.

It was a tempting explanation, not only for the situation in the Americas but also for the European epidemic. Whether the animals were killed directly by some species of algae or indirectly by morbilliviruses that were spread around the world by hiding inside such algae, it would no longer appear terribly mysterious that seals in the Ionian Sea, Mediterranean, Baltic, North Sea, Gulf of Mexico, and off the shores of Long Island should all experience lethal epidemics at roughly the same time and under similar climatic conditions.[25]

For more than two decades biologist Rita Colwell of the University of Maryland had been amassing evidence that bacteria and viruses lurked inside algae, and by the late 1980s other scientists were not only acknowledging the tremendous body of evidence but also correlating her algal findings with disease outbreaks in marine life and humans. Colwell knew that hundreds—perhaps thousands—of species of predatory algae were capable of secreting toxins designed to kill or paralyze their larger marine prey, allowing groups of the microscopic beings time to consume their conquest at leisure.

Algae were the oldest living species on earth, thought to have developed out of the planet's primordial soup more than three billion years ago. As creatures they resembled protozoa, but their use of chlorophyll to convert the sun's energy into useful chemicals made them, technically, plants. Algae clustered in both fresh and salt water, sometimes forming visible discolorations and "tides" on the surface. Three broad categories of algae were designated on the basis of their colors, which, in turn, reflected the nature of their internal chemistry: blue-green, red, and brown.

Algae could, during times of environmental stress or food shortages, encyst themselves in a protective coating, go dormant, and drop into hiding for extended periods. Once activated, however, algae needed sunlight and plenty of nitrogen. Most species preferred warmer waters, and under ideal conditions could multiply rapidly, drifting about in massive colonies that, in the case of oceanic red or brown tides, could span surface distances larger than Greater Los Angeles.

Just as E. O. Wilson speculated about the tremendous numbers of terrestrial species of all sizes that had yet to be discovered in the earth's rain forests, Colwell was concerned about the mysteries of the planet's marine world.

"Of the some 5,000 species of viruses known to exist in the world, we've characterized less than four percent of them," Colwell said. "We've only characterized 2,000 bacterial species, most of them terrestrial. That's about 2,000 of an estimated 300,000 to one million thought to exist. Less than one percent of all ocean bacteria have been characterized."[26]

Colwell had devoted years to the study of microorganisms living in Maryland's Chesapeake Bay, where she discovered that viruses were seasonal: they reached their nadir in population during the icy months of January, when there were about ten thousand viruses per milliliter of water, and increased steadily in numbers as the bay warmed. By October, after three hot summer months, there were as many as a billion per milliliter, and the viruses outnumbered algae and bacteria. Even more profound variation in viral populations was seen in the waters of Norway's fjords, and Norwegian researchers were convinced that viruses passed genetic material on to algae to assist in their adaptation to change.[27]

In Colwell's Chesapeake some of the swollen summer viral population was indigenous to the bay, having simply multiplied in number as the water warmed. But increasingly over her more than thirty years of studying the Chesapeake, Colwell saw viral intrusion occurring, as human and animal waste washed into the bay, carrying with it a variety of pathogens. The greatest density of intrusion was around dump and sewage sites, where Colwell found veritable stews of viruses, plasmids, transposons, and bacteria intermingling.

"The probability of genetic exchange is very great," Colwell said. Indeed, lab studies had shown that some ocean bacteria possessed antibiotic-resistance genes, presumably acquired under just such conditions. Those newly antibiotic-resistant bacteria were, in turn, ingested by various mollusks and then eaten by sea mammals and humans. The mollusks and crustaceans—from scallops to lobsters—readily ingested all manner of microorganisms found along the world's polluted coastlines, including a host of enteric human pathogens.[28]

"We have very few places left on earth where we can get pathogen-free mollusks," Colwell said.

Hepatitis, Norwalk virus, polio, and a host of other microbes were turning up in shellfish caught in the world's coastal waters, particularly around waste dump sites. And strange microbes appeared that burned through the shells of mollusks, killed off salmon, and made lobsters lose their sense of direction.

By one calculation a single gram of typical human feces contained one billion viruses. And in a liter of raw human waste there were more than 100,000 infectious viruses—none of which were vulnerable to mere chlorine treatment. Chlorine might eliminate the bacteria—though increasing chlorine resistance in bacterial populations was rendering such chemical sanitation insufficient—but viral elimination required more extensive filtration and tertiary treatment.[29]

Ocean pollution due to raw sewage, fertilizers, pesticides, and other chemical waste was increasing steadily, producing tremendous changes in coastal marine ecospheres. Though the World Bank and the United Nations had designated sewage and sanitation systems a top priority for development during the 1980s, it was estimated that at least two billion *Homo sapiens* had no access to a sanitary fecal waste disposal system, most of them residents of Africa and southern Asia.[30] Their fecal waste, as well as that of their domestic animals, ended up in nearby rivers, streams, and seas.

Algal blooms, as a result, increased in frequency and size worldwide throughout the four post-World War II decades. The nutrient supply provided by steady flows of fecal matter, garbage, fertilizers, silt, and agricultural runoff gave the algae plenty of food. Many scientists thought that the thinning ozone layer warmed the sea surfaces to temperatures suitable for microbial growth and reproduction. Algal blooms grew so rapidly on the surface of lakes, ponds, and the open sea that they actually blocked all oxygen and sunlight for the creatures swimming below, literally suffocating fish, marine plants, and mollusks. And some scientists believed there was evidence that the additional load of ultraviolet light making its way through the ozone layer was driving a higher mutation rate in sea surface organisms, possibly allowing for more rapid rates of adaptive evolution. If such a mechanism were in effect, it would favor microorganisms, which, on a population basis, were well positioned to make use of helpful mutations and tolerate individual die-offs due to disastrous mutations. The reverse would be the case for more complex marine creatures, such as fish, whales, and dolphins.

"The oceans have become nothing but giant cesspools," declared oceanographer Patricia Tester, "and you know what happens when you heat up a cesspool."[31]

Jan Post, a marine biologist at the World Bank, used a similar metaphor when announcing the release of the Bank's 1993 report on the condition of the seas: "The ocean today has become an overexploited resource and mankind's ultimate cesspool, the last destination for all pollution."[32]

Tester, who worked for the U.S. National Marine Fisheries Service in Beaufort, North Carolina, had been monitoring weather patterns and algal blooms. She was one of many oceanographers who noted that the die-offs of dolphins, seals, porpoises, and whales during 1987–92 coincided with massive algal-induced bleaching of coral reefs worldwide and enormous red tides. She felt that there was compelling evidence for not only increased frequency and size of algal blooms but also their territorial expansion into latitudes of the seas formerly considered too cold for such algal growth. Using satellites to track the algal blooms, scientists documented increases—in some cases doublings—in size and scope of algal blooms during the 1980s and early 1990s.[33]

Meanwhile, the overall diversity of the marine ecosphere was declining at a dramatic rate. With more than 95 percent of all marine life adapted to coastal regions, their susceptibility was high: human interference in the form of coastal development, sewage, and fishing was claiming a huge toll. U.S. Fish and Wildlife Service biologist Kenneth Sherman calculated that biomass production off the shores of New England, for example, had declined by more than 50 percent between 1940 and 1990 due, primarily, to overfishing.[34]

A feedback loop of oceanic imbalance was thus in place. As the populations of plankton/algae eaters—whales, for example—declined, only the viruses remained to keep blooms in check. But raising the sizes of viral populations in the world's saline soup held out other dangers to marine and, ultimately, human health.

Rita Colwell was convinced that the entire oceanic crisis was already directly imperiling human health by permitting the emergences of cholera epidemics. During the 1970s she showed that the tiny resilient cholera vibrio could live inside of algae, resting encysted in a dormant state for weeks, months, perhaps even years. Colwell, a gritty, energetic woman, fought hard for years to convince the world's public health establishment that the key to forecasting emergence of cholera lay in tracking algal blooms that drifted from the shores of Bangladesh and India, key endemic sites for the microbe.

"But the bloody stupid physicians have this idée fixe that cholera is only directly transmitted, from person to person," Colwell said. "They just couldn't wrap their minds around the concept of microbial ecology. They fight me tooth and nail at every turn."

It was the emergence of cholera in Peru in January 1991 that compelled the World Health Organization and the global medical community to take notice of Colwell's message.

The global Seventh Pandemic of cholera[35] began in the Celebes Islands in 1961, with the new strain, dubbed *Vibrio cholerae* 01, biotype El Tor.

By the late 1970s the El Tor microbe had made its way into all the developing coastal countries of southern Asia and eastern Africa, and it was impossible to control it. It would not be until the 1991 Peruvian outbreak, however, that WHO and health experts would publicly acknowledge what Colwell had been saying for years: namely, that the El Tor strain was particularly well equipped, genetically, for long-term survival inside algae.

Since the early 1980s Colwell had been collaborating with the International Centre for Diarrhoeal Disease Research in Dhaka, Bangladesh, eventually becoming its research chair. There, in the heart of cholera endemicity—perhaps the cradle of all cholera epidemics[36]—Colwell and her colleagues discovered that the El Tor strain was capable of shrinking itself 300-fold when plunged suddenly into cold salt water. In that form it was the size of a large virus, very difficult to detect. But the presence of hibernating cholera vibrio in a water source, or inside algae, could be verified by simply taking a sample and, in the laboratory, changing the conditions: add nitrogen, raise the temperature, decrease the salinity, and bingo! instant cholera.

They further discovered that the vibrios could feed on the egg sacs of algae: up to a million vibrios were counted on the surface of a single egg sac.[37] That explained why health authorities couldn't manage to eliminate El Tor once it had entered their communities. The organism simply hid in algal scum floating atop local ponds, streams, or bays, lurking until an opportune moment arrived for emergence from its dormant state.[38]

When El Tor hit the coastal parts of Peru in early 1991 the country was caught completely unprepared for such an occurrence. In Peru's hot summer January—made hotter still by an El Niño event—a Chinese freighter arrived at Callao, Lima's port city. Bilge water drawn from Asian seas was discharged into the Callao harbor, releasing with it billions of algae that were infected with El Tor cholera.[39]

The first human cases of the disease offered the microbes terrific opportunities for spread in Peru. A national summertime delicacy was ceviche, or mixed raw fish and shellfish in lime juice. The bilge-dumped vibrio had quickly infected Peru's shellfish, so uncooked ceviche represented an ideal vehicle of transmission.

The second ideal opportunity for transmission of the microbes was provided by Lima's largely unchlorinated water supply. Because of both cost constraints and U.S. Environmental Protection Agency documents that indicated there was a weak connection between ingesting chlorine and developing cancer, Peru had abandoned the long-standing disease control practice of using the chemical to disinfect public drinking water. Later, CDC studies would show that the majority of Peru's cholera microbes were transmitted straight into people's homes, dripping from their water faucets.

The first cholera cases hit Lima hospitals on January 23; days later cholera broke out some 200 miles to the north in the port town of Chimbote.

As the El Niño water spread out along the Pacific coast of the continent, carrying with it bilged algae, cholera appeared in one Latin American port after another.[40] Eleven months into the Western Hemisphere's pandemic, cholera had sickened at least 336,554 people, killing 3,538. Throughout those months the microbe's emergence was aided by obsolete or nonexistent public water purification systems, inadequate sewage, and airplane travel. Cases reported in the United States involved individuals who boarded flights from Latin America unaware that they were infected, and fell gravely ill either in flight or shortly after landing.[41]

Colwell and her colleagues demonstrated that the vibrio in the algae and those recovered from ailing patients were genetically identical. Further, they showed that the El Tor substrain, Inaba, which was raging across Latin America, possessed genes for resistance to the antibiotics ampicillin, trimethoprim, and sulfamethoxazole. The same substrain was highly antibiotic-resistant in Thailand, where it was invulnerable to eight drugs.[42]

In Latin America the epidemic raged on well into 1994, with, according to WHO officials, "no end in sight." More than $200 billion would be spent by Latin American governments by 1995, according to the Pan American Health Organization, for emergency repairs of water, sanitation, and sewage systems. Only about 2 percent of all cholera cases were actually reported to authorities, WHO said, and across the continent 900,000 cases were officially reported as of October 1993, involving more than 8,000 deaths. Officially reported numbers of cholera cases were so grossly understated that, by 1994, the only accurate statement one could make was this: between January 1991 and January 1994 millions of Latin Americans fell ill with cholera, thousands died, and the epidemic continues.

Once chlorine was vigorously introduced into Peruvian water supplies, the 01 strain proved fairly resistant to the chemical.[43]

Though it was obvious to scientists all over the Americas by 1992 that the El Tor epidemic had succeeded in becoming *endemic* cholera in much of Latin America largely because the microbe was carried in algae, the real challenge to rigid old analyses of the spread of the vibrio came in December 1992 when an entirely new strain of cholera emerged in Madras, India. Dubbed Bengal cholera, or *V. cholerae* 0139, the newly emergent microbe competed with El Tor for control of the Bay of Bengal ecology. By June 1993 Bengal cholera had claimed over 2,000 lives and caused severe illness in an estimated 200,000 people. It had spread across much of the coastal region of the Bay of Bengal, encompassing the Indian metropolises of Calcutta, Madras, Vellore, and Madurai, as well as most of southern Bangladesh.[44]

This new Bengal cholera appeared to be spreading far faster than the Seventh Pandemic. It took three years for that cholera strain to spread from India to Thailand, but the Bengal cholera had already turned up in Thailand's capital, Bangkok, by mid-1993, and threatened to spread nationwide, according to researchers from Mahidol University in Bangkok.

In March the leading hospital in Dhaka was treating 600 Bengal cholera cases a day—three times their normal daily cholera rate. In rural parts of Bangladesh cholera victims were reportedly falling ill at rates up to ten times those seen with the previous year's classic cholera outbreak.

Prior to the Bengal cholera outbreak there were two types of cholera in the world: classic and El Tor. Classic cholera, which was endemic in parts of India and Bangladesh, was extremely virulent and easily passed from one person to another via contact with microscopic amounts of feces. The El Tor type, in contrast, was less virulent but could survive in the open environment far longer. A hallmark of the El Tor strain was its ability to move in the open oceans, as a silent passenger inside algae.

The Bengal cholera appeared to represent a combination of characteristics found in both the El Tor and the classic vibrio. Researchers from the International Centre for Diarrhoeal Disease Research in Bangladesh reported that the new mutant "may be hardier than and probably has survival advantage over" the classic strain of the bacteria. They found thriving colonies of the Bengal organism in 12 percent of water samples they tested, and the bacterial toxin was in 100 percent of all waters examined in Bangladesh.

One genetic trait was clearly missing in the new Bengal strain: that which coded for antigens that were usually recognized by the human immune system. As a result, people did not seem to have antibodies to the new mutant, and even adults who had survived previous cholera outbreaks appeared to be susceptible to the Bengal strain.

Genetic analysis of the new mutant vibrio suggested a terrible scenario: that it was essentially the El Tor strain possessing the virulence genes of classic cholera. As such, it would represent an entirely new class of cholera microbe, the like of which had never been seen. The emergence of 0139 "hit epidemiologists and physicians like a two-by-four between the eyes, because there is no explanation for its emergence and spread but ecology," Colwell said in the fall of 1993.

In 1993 Colwell teamed up with two Cambridge, Massachusetts, physicians to try to pull together the Big Picture, an explanation of how global warming, loss of oceanic biodiversity, ultraviolet radiation increases, human waste and pollution, algal blooms, and other ecological events joined forces. Together, they theorized that the cholera microbe defecated by a man in Dhaka, for example, got into algae in the Bay of Bengal, lay dormant for months on end, made its way via warm water blooms or ship bilge across thousands of miles of ocean, and killed a person who ate ceviche at a food stand in Lima. Drs. Paul Epstein and Timothy Ford, both members of a group of physicians and scientists at the Harvard School of Public Health calling themselves the Harvard Working Group on New and Resurgent Diseases, were convinced that essential to protecting their Boston patients in the twenty-first century was a better understanding of what was transpiring in the oceans. They saw a complex interplay at work, involving global climate changes, pollution, and the microorganisms.

In Epstein's view, algal blooms were giant floating gene pools in which antibiotic-resistance factors, virulence genes, and plasmids moved about between viruses, bacteria, and algae. He thought that ultraviolet radiation might be hastening the mutational pace. And terrestrial microbes were constantly being added to the gene pool, he said, in the form of human waste and runoff.[45]

Epstein lobbied scientists working in fields as varied as oceanography, atmospherics, satellite imagery, plankton biology, and epidemiology to find ways to collaborate, and answer questions about the links between the marine environment and human health. Epstein discovered that many other scientists had already reached the conclusion that changes in global ecology—particularly those caused by warming—were too often working to the advantage of the microbes.

For example, at Yale, where he still ran the Arbovirus Laboratory, Robert Shope was considering the impact of global warming on disease-carrying insects. On the basis of his nearly forty years of arbovirus research, Shope was convinced that even a minor rise in global temperature could expand the territory of two key mosquito species: *Aedes aegypti* and *A. albopictus*. Both species were limited geographically in the 1990s by climate. *A. aegypti* couldn't withstand prolonged exposure to temperatures below 48°F and died after less than an hour of 32°F weather. *A. albopictus* was only slightly heartier in cold climes. As a result, in the Northern Hemisphere *A. aegypti* couldn't live above 35°N latitude, or roughly the levels of Memphis, Tennessee, Tangier, Morocco, and Osaka. *A. albopictus* couldn't survive above 42°N latitude during the 1990s, roughly equivalent to Madrid, Istanbul, Beijing, and Philadelphia.

Shope expected that warming would allow both mosquito species to comfortably move northward, invading population centers such as Tokyo, Rome, and New York. *A. albopictus*, the Asian tiger mosquito, could carry the dengue virus. *A. aegypti* was more worrisome because it carried both dengue and yellow fever; the latter was typically fatal 50 percent of the time.[46] Historical analysis seemed to confirm such a hypothesis, as malaria had shifted geographically over the millennia in accordance with major climate changes.[47]

British experts on insect-borne diseases felt certain that global warming would greatly expand the territory and infectivity ratio of the East African tsetse fly, which carried the trypanosomes responsible for sleeping sickness. The researchers concluded that even a moderate increase—on the order of 1° to 2°C—could result in a higher rate of disease spread because the tsetse flies were known to be more active, to feed at a higher pace, and to process trypanosomes more rapidly at higher ambient temperatures. Thus, each tsetse fly could infect more people daily.[48]

The same principles held true for *Anopheles* mosquitoes and the spread of malaria. In 1993, Uwe Brinkmann, who headed the Harvard Working Group on New and Resurgent Diseases, was trying to figure out ways to predict not only latitude movements of mosquitoes in response to global warming but also their altitude changes. He felt there was an urgent need for research to determine which factors played a greater role in limiting *Anopheles* activities at altitudes above 500 feet: air pressure or cooler temperatures. If the latter was more important, he predicted, malaria could quickly overtake mountainous areas of Zimbabwe, Botswana, Swaziland, Rwanda, Tanzania, Kenya, and other geographically diverse parts of Africa. Further, the disease might with global warming climb its way further up the foothills of the Himalayas, the Sulaiman Range, the Pir Panjal, and other mountainous regions of Asia.

A detailed WHO Task Group report in 1990 offered a broader range of expected disease impacts from global warming. Even a moderate net temperature increase—on the order of 1°C—would alter wind patterns, change levels of relative humidity and rainfall, produce a rise in sea levels, and widen the global extremes between desert regions and areas afflicted with periodic flooding. These conditions would, in turn, radically alter the ecologies of microbes that were carried by insects. Furthermore, expected changes in vegetation patterns could, the WHO Task Group said, radically alter the ecologies of microbe-carrying animals, such as monkeys, rats, mice, and bats, bringing those vectors into closer proximity to *Homo sapiens*.[49]

There was also a strong consensus among immunologists that heightened exposure to ultraviolet light—particularly UV-B radiation—suppressed the human immune response, thus increasing *Homo sapiens*' susceptibility to all microbes.[50] Just as PCBs and other hydrocarbon pollutants were thought to have played a role in increasing microbial susceptibility in marine mammals, so many physicians felt there was ample evidence that air, water, and food pollutants affected the human immune system.

Another feature of global warming would be an increased dependence in wealthier nations on air conditioning. In order to conserve energy, buildings in the industrialized world had specifically been designed to minimize outward and inward air flow. It was much cheaper to heat or cool the same air repeatedly in a sealed room than to pump in fresh air from the outside, alter its temperature, circulate it throughout a structure, and at the same time expel old air. As the numbers of hot days per year increased, necessitating longer periods of reliance upon air conditioning, the economic pressures to recirculate old air repeatedly, to the limits of reasonable oxygen depletion, could be expected. Such practices for winter heat conservation in large office buildings had already been linked to workplace transmission of influenza and common cold viruses. Spread of Legionnaires' Disease and other airborne microbes was expected to increase with global warming.

Even in the absence of serious global warming, energy conservation practices were, for purely economic reasons, spurring architects and developers toward construction of buildings that lacked any openable windows and were sealed so tightly that residents were apt to suffer "sick building syndrome": the result of inhaling formaldehyde, radon, and other chemicals present in the building foundation or structure. Such chemicals posed little threat to human health if diluted in fresh air, but were significant contributors to health problems in residents and employees who inhaled levels that were concentrated in recirculated or thin air. Obviously, a building that was capable of concentrating such trace chemicals in the air breathed by its inhabitants would also serve as an ideal setting for rapid dissemination of *Mycobacterium tuberculosis*, if an individual who suffered from active pulmonary disease was residing or working within the structure.

The human lung, as an ecosphere, was designed to take in 20,000 liters of air each day, or roughly 60 pounds. Its surface was highly variegated, comprised of hundreds of millions of tiny branches, at the ends of which were the minute bronchioles that actively absorbed oxygen molecules. The actual surface area of the human lung was, therefore, about 150 square meters, or "about the size of an Olympic tennis court," as Harvard Medical School pulmonary expert Joseph Brain put it.

Less than 0.64 micron, or just under one one-hundred-thousandth of an inch, was all the distance that separated the air environment in the lungs from the human bloodstream.

All a microbe had to do to gain entry to the human bloodstream was get past that 0.64 micron of protection. Viruses accomplished the task by accumulating inside epithelial cells in the airways and creating enough local damage to open up a hole of less than a millionth of an inch in diameter. Some viruses, such as those that caused common colds, were so well adapted to the human lung that they had special proteins on their surfaces which locked on to the epithelial cells. Larger microbes, such as the tuberculosis bacteria, gained entry via the immune system's macrophages. They were specially adapted to recognize and lock on to the large macrophages that were distributed throughout pulmonary tissue. Though it was the job of macrophages to seek out and destroy such invaders, many microbes had adapted ways to fool the cells into ingesting them. Once inside the macrophages, the microbes got a free ride into the blood or the lymphatic system, enabling them to reach destinations all over the human body.

The best way to protect the lungs was to provide them with 20,000 liters per day of fresh, clean, oxygen-rich air. The air flushed out the system.

Dirty air—that which contained pollutant particles, dust, or microbes —assaulted the delicate alveoli and bronchioles, and there was a synergism of action. People who, for example, smoked cigarettes or worked in coal mines were more susceptible to all respiratory infectious diseases: colds, flu, tuberculosis, pneumonia, and bronchitis.

Because of its confined internal atmosphere, the vehicle responsible for the great globalization of humanity—the jet airplane—could be a source of microbial transmission. Everybody on board an airplane shared the same air. It was, therefore, easy for one ailing passenger or crew member to pass a respiratory microbe on to many, if not all, on board. The longer the flight, and the fewer the number of air exchanges in which outside air was flushed through the cabin, the greater the risk.

In 1977, for example, fifty-four passengers were grounded together for three hours while their plane underwent repairs in Alaska. None of the passengers left the aircraft, and to save fuel the air conditioning was switched off. For three hours the fifty-four passengers breathed the same air over and over again. One woman had influenza: over the following week 72 percent of her fellow passengers came down with the flu; genetically identical strains were found in everyone.[51]

Following the worldwide oil crisis of the 1970s, the airlines industry looked for ways to reduce fuel use. An obvious place to start was with air circulation, since it cost a great deal of fuel to draw icy air in from outside the aircraft, adjust its temperature to a comfortable 65°–70°F, and maintain cabin pressure. Prior to 1985 commercial aircraft performed that function every three minutes, which meant most passengers and crew breathed fresh air throughout their flight. But virtually all aircraft built after 1985 were specifically designed to circulate air less frequently; a mix of old and fresh air circulated once every seven minutes, and total flushing of the aircraft could take up to thirty minutes.[52] Flight crews increasingly complained of dizziness, flu, colds, headaches, and nausea.

Studies of aircraft cabins revealed excessive levels of carbon dioxide—up to 50 percent above U.S. legal standards. Air quality for fully booked airliners failed to meet any basic standards for U.S. workplaces.[53]

In 1992 and 1993 the CDC investigated four instances of apparent transmission of tuberculosis aboard aircraft. In one case, a flight attendant passed TB on to twenty-three crew members over the course of several flights.

Similar concerns regarding confined spaces were raised about institutional settings, such as prisons and dormitories, where often excessive numbers of people were co-housed in energy-efficient settings.

In preparation for the June 1992 United Nations Earth Summit in Rio de Janeiro, the World Health Organization reviewed available data on expected health effects of global warming and pollution.[54] WHO concluded that evidence of increased human susceptibility to infectious diseases, due to UV-B immune system damage and pollutant impacts on the lungs and immune system, was compelling. The agency was similarly impressed with estimates of current and projected changes in the ecology of disease vectors, particularly insects.

It wasn't necessary, of course, for the earth to undergo a 1°–5°C temperature shift in order for diseases to emerge. As events since 1960 had demonstrated, other, quite contemporary factors were at play. The ecological relationship between *Homo sapiens* and microbes had been out of balance for a long time.

The "disease cowboys"—scientists like Karl Johnson, Pierre Sureau,

Joe McCormick, Peter Piot, and Pat Webb—had long ago witnessed the results of human incursion into new niches or alteration of old niches.[55] Perhaps entomologist E. O. Wilson, when asked, "How many disease-carrying reservoir and vector species await discovery in the earth's rain forests?" best summed up the predicament: "That is unknown and unknowable. The scale of the unknown is simply too vast to even permit speculation."

Thanks to changes in *Homo sapiens* activities, in the ways in which the human species lived and worked on the planet at the end of the twentieth century, microbes no longer remained confined to remote ecospheres or rare reservoir species: for them, the earth had truly become a Global Village. Between 1950 and 1990 the number of passengers aboard international commercial air flights soared from 2 million to 280 million. Domestic passengers flying within the United States reached 424 million in 1990.[56] Infected human beings were moving rapidly about the planet, and the number of air passengers was expected to double by the year 2000, approaching 600 million on international flights.[57]

Once microbes reached new locales, increasing human population and urbanization ensured that even relatively poorly transmissible microbes faced ever-improving statistical odds of being spread from person to person. The overall density of average numbers of human beings residing on a square mile of land on the earth rose steadily every year. In the United States, even adjusting for the increased land mass of the country over time, density (according to U.S. census figures) rose as follows:

Year	Total Population	Persons per Square Mile
1790	3,929,214	4.5
1820	9,638,453	5.5
1850	23,191,876	7.9
1870	39,818,449	13.4
1890	62,947,714	21.2
1910	91,972,266	31.0
1930	122,775,046	41.2
1950	151,325,798	42.6
1970	203,211,926	57.5
1990	250,410,000	70.3
1992	256,561,239	70.4

In most of the world the observed increases were even more dramatic. In a comparison of 1990 and 1992 census information as collected by the United Nations, the two-year upward trend in population density was unmistakable:

Country	1990 Population	1990 Persons per Square Mile
China	1,130,065,000	288
India	850,067,000	658
Indonesia	191,266,000	255
Mexico	88,335,000	115
Rwanda	7,603,000	715

Country	1992 Population	1992 Persons per Square Mile	% Density Difference 1990–92
China	1,169,619,000	315	8.5
India	886,362,000	700	6.0
Indonesia	195,000,000	262	2.6
Mexico	92,380,000	121	4.9
Rwanda	8,206,000	806	11.3

Though the population was spread unevenly over a country, density trends remained favorable to the microbes. If worst-case projections for human population size came to pass, some regions would have densities in excess of 3,000 people per square mile. At that rate the distinctions between cities, suburbs, and outlying towns would blur and few barriers for person-to-person spread of microbes would remain.

With the passage of time and the increase in travel it was becoming more and more difficult to pinpoint where, exactly, a microbe first emerged. The human immunodeficiency virus was a classic case in point, as it surfaced simultaneously on three continents and spread swiftly around the globe.

Those scientists in the 1990s whose primary focus was viruses believed that the worst scales of disease and death arose from epizootic events: the movement of viruses between species. In such instances, the hosts were usually highly susceptible, as they lacked immunity to the new microbe. Ebola, PDV-2, Marburg, Machupo, Lassa, and Swine Flu were all examples of such apparently sudden emergences into the *Homo sapiens* population.

Rockefeller University's Stephen Morse, who by 1988 was devoting nearly all his professional energies to emerging disease problems, labeled these movements of viruses between host species "viral trafficking." He considered the world's fauna a vast "zoonotic pool," each species carrying within itself an assortment of microbes that might jump across species barriers under the proper circumstances to infect an entirely different type of host.[58]

At Harvard, Max Essex was similarly impressed with the ferocity of new cross-species viral infections. A case he found chilling was the *Herpesvirus saimiri*, which was carried without apparent harm by *Saimiri sciureus* squirrel monkeys living in Amazonia. When the virus was first discovered in captive *Saimiri* monkeys it was thought to be a harmless microbe.

But within months after its discovery in 1968 by scientists at the New England Primate Research Center, located outside Boston, other monkey species at the center took ill. The strange herpes virus turned out to be an extraordinary cancer-causing agent: less than two months after infection Old World monkeys would develop extensive, lethal cancers of their lymphatic systems.

In the early 1970s the same researchers discovered a similar herpes virus, *H. ateles*, in spider monkeys. Like *H. saimiri*, it was harmless in its normal host and infected virtually 100 percent of the host species in the wild. And it was also a potent cancer-causing virus in other monkey species. Both viruses specifically infected cells of primate immune systems, causing lymphomas and leukemias. And both approached the 100 percent lethality mark when they infected primates other than their host species. Experimental infections of rabbits also proved extraordinarily lethal.

What chilled Essex wasn't the viruses' ability to cause cancer, though the appalling certainty and speed of their carcinogenic action were certainly unprecedented and frightening. Essex's concern was the mode of transmission: both herpes viruses were airborne.

In the wild such horrendously dangerous viruses might have, over the millennia, served the squirrel and spider monkeys well, residing harmlessly inside their species but killing off all other species of competitive primates. In captive animal colonies a spider monkey could simply breathe on a howler monkey and five weeks later the victim would die of leukemia.[59]

The viruses appeared to be able to elude monkey immune systems by manufacturing proteins that specifically switched off or dampened cellular immune responses. And the *saimiri* virus contained fifteen genes that were remarkably similar to genes found inside the monkey's DNA.[60]

Lab analyses of *H. saimiri* strains grown on monkey cells revealed an astonishing rate of mutation and gene swapping. The virus's DNA, in the absence of the rest of the microbe, was capable of infecting and destroying a cell. Once whole viruses were inside cells they immediately began a mutation process so pronounced that it was impossible to recover the original strain. So, though no human being was known to have been infected with either *H. saimiri* or *H. ateles*, the viruses' ability to transform themselves at such staggering speed left open the disturbing possibility that, given ample opportunity—such as exposure to an immunodeficient person or implantation into a *Homo sapiens* in the form of a monkey-to-human tissue transplant—the organisms might quickly adapt to human cells, becoming a lethal airborne cancer-causing virus.

Since the establishment of research animal colonies, scientists had unwittingly uncovered many other monkey and ape viruses that proved capable of producing infection and disease in the humans who handled the simians. A herpes virus, designated B virus, infected rhesus macaques and some other Old World monkeys, attacking nerve cells to produce everything from localized pain to encephalitis and death. About 10 percent of all imported rhesus monkeys were typically infected with the B virus; infection rates inside some captive colonies reached 100 percent. Once infected, the animals carried the virus for life, whether or not they developed disease.

From the time of its discovery in 1975 to 1989, twenty-eight animal handlers had contracted B virus infection, twenty-five of whom went on to develop encephalitis. Only five human beings had ever survived known B virus infection.[61]

Other monkey viruses that held out the potential for human infection, either in their natural form or in a mutant form, included type D simian retrovirus (SRV), the simian AIDS virus (SIV), simian sarcoma-associated viruses (SSAVs), paramyxovirus simian virus 5, gibbon ape leukemia virus, and Mason-Pfizer virus (M-PMV).[62]

During the early 1970s, 126 American primate research facility employees were accidentally infected with monkey microbes. The precise etiology of most of their ailments was never determined. Among the microbes known to have been transmitted were tuberculosis, *Shigella*, *Streptococcus*, *Staphylococcus*, and influenza.[63] Every year thereafter animal colony workers all over the world were exposed to, and became infected with, a variety of monkey and ape viruses, bacteria, and parasites.[64]

Despite the clear presence of pathogens dangerous to humans in the simian population, there was much interest in the U.S. medical community in using the animals as sources for organ transplants. Ever since the first successful human heart transplants were performed in 1953 the use of organ transplantation had increased steadily in the United States and Europe. Development of effective drugs to suppress a recipient's immune response greatly improved the success of human-to-human organ transplant procedures, and by 1988 the five-year-survival rate exceeded 50 percent for patients undergoing all common transplants, save those of the lung. Kidney transplants, the most common of all such procedures, enjoyed a 91 percent success rate.[65]

As success rates mounted, so did the demand for organs. By the mid-1980s there was a very real crisis of organ availability and American television and newspapers regularly carried heart-wrenching stories about desperate children who faced imminent death unless a suitable liver, or heart, or other organ was found posthaste. A federal waiting list system was created in order to put some order into an organ procurement system that was spinning dangerously out of control. Order and fairness didn't ensure adequate availability, however. In 1990, for example, 2,206 people on the organ waiting list died before a suitable transplant donor could be found.

In 1963 the first tentative baboon-to-human transplants were performed, with little success.[66] Such experiments continued over the years in the United States and South Africa.[67]

In 1992–93 researchers at the University of Pittsburgh transplanted baboon livers into two men who suffered hepatitis B virus-induced destruction of their own organs. Though both patients succumbed, the transplants were not the causes of their deaths, and physicians hailed the breakthrough.

But infectious disease experts cried foul. The donor baboons came from the Southwest Foundation, the largest research monkey facility in the United States. Officials at the San Antonio-based primate center were shocked to learn that the baboon organ had been transplanted into a human being. The baboon used in the first Pittsburgh transplant experiment was infected with SIV (the simian AIDS virus), CMV (the simian cytomegalovirus), EBV (the simian type of Epstein-Barr virus), and Simian Agent 8 (the baboon form of B virus). If the thirty-five-year-old man had survived for months after receiving the baboon liver, critics asked, what might have happened with those viruses?

"We assume as a given that these primates carry pathogens that are infectious to humans," Southwest Foundation Biomedical Research Center scientist Jon Allan said. "You assume it's something that can kill you. But then in the next breath we turn around and ship a baboon up to Pittsburgh, they open it up, probably every human in the OR is exposed to whatever is in there, and they stick its liver into a human.

"Does that seem rational?"

Another Southwest Foundation virologist, Julia Hilliard, expressed concern that monkey viruses that seemed initially harmless to people might exchange genetic material with human DNA following a transplant, resulting in highly lethal new super-bugs.[68]

Transplant surgeons had long known, of course, that infection was every recipient's greatest enemy. Old, latent infections were often activated by the procedure because, to avoid transplant rejection, doctors used powerful drugs to suppress the patients' immune systems. It was also possible for the transplants themselves to be infected: thus, the recipient got not only the donor's organ but also microbes such as cytomegalovirus,[69] hepatitis B,[70] adenoviruses,[71] Epstein-Barr virus,[72] and HIV.

For most of the world's human population, however, such exotic things as liver transplants were hardly of concern. More likely modes of epizootic disease transmission involved insects.

Yellow fever, for example, could for decades on end afflict virtually no *Homo sapiens* in a given area because the *Aedes aegypti* mosquitoes were busy feeding on monkeys and marmosets in the jungle. But with changes in either the forest environment or the social behaviors of local *Homo sapiens* the mosquito could almost overnight change its feeding patterns and a human epidemic would commence. Such was the case with yellow fever epidemics in Nigeria and Kenya in 1987, 1988, 1990, and 1993.

Tom Monath had seen it happen several times in West Africa, where such simple actions as chopping down a stand of trees and leaving the stumps in place could spawn a yellow fever outbreak. The mosquitoes left their larvae in rainwater that collected in the tree stumps.

Microbes and insect vectors could suddenly appear in areas thousands of miles from their usual habitats. For example, for reasons never understood, the American screwworm fly, which could transmit deadly maggots to livestock, turned up quite suddenly in the deserts of Libya in 1988 and quickly spread throughout North Africa. The insects' normal habitat was the dry Southwest of the United States and northern Mexico.[73]

In temperate ecologies, keeping wild insects at bay was quite easy, provided abatement and control systems remained intact and vigilant. Even a year of slackening in such an effort could, however, permit a sudden surge in insect vector populations, with resultant disease.[74]

In the United States there were several outbreaks of mosquito-borne diseases between 1985 and 1992, each of which could be traced to a breakdown in mosquito control efforts or public health vigilance.[75] In 1990, for example, an epidemic of St. Louis encephalitis caused widespread panic in parts of Florida and southeastern Texas, forcing cancellation of baseball games and other nighttime outdoor activities. Though public health authorities had a year earlier witnessed rises in viral infection rates in chickens and other birds used for monitoring the microbes, few steps were taken to stem the increases in local mosquito populations prior to the summer 1990 epidemics.[76]

For most insect experts it came as no surprise that even a one-year slackening in mosquito control efforts could result in a surge in the bugs and the microbes that they carried. Both insects and microbes had evolved mechanisms over the millennia that ensured their mutual survival. Studies of genetic relationships between particular microbes and their most common insect vectors suggested that the species had co-evolved, developing capabilities that were primarily advantageous to the microbes.

Blood-feeding insects had over millions of years developed traits that served to aid the transmission and the evolution of microbes. When the female insects bit into human flesh they spit into the site a fluid that contained vasodilators that opened up local capillaries, anticoagulation enzymes that would prevent clotting of the wounded capillaries, and a variety of factors that destroyed immune system cells and chemicals. This ensured the insect a steady flow of food, without toxic human immune system chemicals or cells. As these chemicals were secreted out of the insect's salivary glands, the proboscis drew blood into a separate set of lobes, and eventually into the insect's midgut.

The process represented a passage made in heaven for microbes: all blood flow was unobstructed, the local human immune system was shut down, and the destination—the insect midgut—was a very comfortable microbial ecology.[77]

Once in the insect's midgut, microbes could swiftly multiply, make their way back up to the salivary gland, and be injected into an unwitting host. Or they might remain in the insect, exerting unusual pressures on the creature. For example, only female insects fed on blood: to ensure a plentiful supply of females, some microbes made their way to the insect's ovaries, where they genetically manipulated the male chromosomes, ensuring that offspring would be female. The organisms would then be passed on to the adult insect's female offspring, which would be born already infected.

Evolution was a very dynamic and active process for microbes housed in an insect's midgut. Some viruses changed very slowly over time, probably because they possessed extremely accurate mechanisms for replication and repair of their genetic material. But there were insect-borne viruses that were capable of sorting and resorting their chromosomes, shuffling RNA about seemingly at random. And under conditions of co-infection of an insect by more than one species of microbe, exchanges might occur. The end result could be new mutant organisms.

Barry Beaty of Colorado State University in Fort Collins pointed out that it was a simple matter to get a single mosquito infected with two different strains of bluetongue virus. The virus, which produced disease in ruminant livestock, was comprised of ten RNA segments, or chromosomes, each of which had to be properly duplicated and assorted each time the virus reproduced itself. Beaty's group showed that insects that were co-infected with two strains of bluetongue virus rarely injected back into animals a viral strain that was identical to either of the original strains: rather, it was a mélange of the two.[78]

Beaty noted that the usually mild snowshoe hare virus was in the 1980s producing serious human disease in northern Russia. It was due, Beaty said, to such a recombination event, only the exchange resulted, not in a wider vector range, but in greater virulence. On the basis of genetic analysis of the pertinent viruses, Beaty believed that the new Russian epidemic— which by 1992 was causing encephalitis in more than 100,000 people a year—was the result of a gene swap between the Inkoo and Tahyna viruses. The two parent viruses produced little more than mild flu-like symptoms in human beings, but their recombination proved potentially deadly.

How commonly other viruses, or bacteria, exchanged genetic material while inside vectors wasn't known. Nor was it clear how significant a role such mutations might play in the emergence of new diseases.[79]

Many insect-borne viruses were thought to have originally been plant microbes that, thousands of millions of years ago, infected insects as they fed on plant nectar. In the 1990s, amid evidence of rising rates of genetic change in many plant microbes, concern was expressed about the possible emergence of new species that might be absorbed by insects. In such a scenario, a microbe that was genuinely new, to which humans had no natural immunity, might quite suddenly emerge. Genetic change in plant microbes was accelerating due to agricultural practices that exerted strong selection pressures on the microbes; to changing geography of plant growth due to international trading of plant seeds and breeding practices; and to the deliberate release of laboratory genetically altered plant viruses that were intended to offer agricultural crops protection against pests.[80]

To minimize use of toxic pesticides, and to prevent incurable viral diseases in plants, scientists in the 1990s were developing ingenious genetic means to protect plants. Using crippled viruses to carry genes that would help vital food crops fend off dangerous pathogens, researchers were breeding plants that could withstand a range of types of infections. There was a catch, however. Studies showed that, in nature, plants such as corn, wheat, and tomatoes were commonly co-infected with up to five different viruses, and those viruses could exchange genetic material.[81] A review of 125 plant strains produced through such laboratory manipulation showed that 3 percent of the time the crippled virus that was used to carry such genes into plant cells could swap genes with other viruses in the plant, producing active, pathogenic—*new*—viral species.[82]

"Microbes are masters at genetic engineering," wrote Canadian microbiologist Julian Davies.[83] He was referring to mechanisms bacteria use to become resistant to antibiotics, but Davies's comment could just as well apply to viruses in an insect's midgut, malarial parasites responding to chloroquine, or influenza cyclically reinventing itself. That recognition prompted many virologists in the late 1980s to ask, "What is the likelihood that a truly new virus capable of causing human disease will emerge?"

One approach to answering that question was to use molecular techniques to sequence the DNA or RNA of a group of viruses and try to trace their family trees, searching for evidence of such recombination events: if they had occurred in the past, it seemed logical that dangerous gene swappings might in the future occur again. Researchers in San Diego concluded that several human and animal retroviruses shared sequences of RNA, which could have been the result of crossover RNA recombination events. However, those events had to have occurred hundreds or thousands of years ago, because most of the retrovirus species were as different from one another as were humans from fungus.[84]

Nobel laureate Howard Temin took a different approach to answering the question, which, at the outset, he said was "inherently unpredictable." Temin tried to calculate rates of mutations or incremental RNA changes in the human immunodeficiency virus type 1. Overall, he estimated that the HIV viruses made a significant mutational change in seven out of every 100,000 viral replications. Considering that an ailing person might have millions of HIVs in his or her body at any given moment, all of which were undergoing constant replication, Temin's figure was far from comforting. But it was also not a worst-case scenario. Scientists discovered so-called hypermutation sites on HIV that were particularly prone to change: there, the RNA would mutate significantly in one out of every 1,000 viral replications.[85]

Temin doubted that such mutations would result in a more dangerous form of HIV—he felt the virus was already perfectly adapted by virtue of having combined certain lethality with a decade-long period of invisible infection during which the microbe would be passed on to other human beings. Still, he felt the essentially labile nature of the virus made its future incalculable.[86]

At the 1989 "Emerging Viruses" conference convened in Washington, D.C., by Rockefeller University and the National Institutes for Allergy and Infectious Diseases, another Nobel laureate, Joshua Lederberg, questioned Temin's confidence that HIV's rapid mutation rate probably wouldn't result in greater viral virulence.

"My concern is not what we know, but what we don't know," Lederberg said. HIV was capable of infecting macrophage cells, he noted, asking, "Could the virus evolve the ability to infect macrophages in the lungs and, thus, become a respiratory disease?"

At Caltech in southern California Jim and Ellen Strauss busied themselves with the task of mapping the evolution of *all* viruses whose genetic material was in the form of RNA. They concluded that all RNA viruses were descended from a single ancestor virus and that over the millennia the viruses had mutated a million times more rapidly than had their DNA-based hosts. Though rates of change varied from RNA virus to RNA virus, the Strausses were convinced that each and every one of the microbes had at some point come into existence through such a process.

"We now recognize that RNA viruses will continue to evolve rapidly as they have over the millennia," the Strausses wrote. "As the recent epidemic of AIDS makes clear, new pathogens can and will arise."[87]

The Strausses felt that scientists, when referring to RNA viruses, shouldn't really speak of species; rather, they should refer to "consensus sequences." The rate of mutation was so high that RNA viral populations were actually pools of genetic concoctions, some particular form of which might dominate at any given time. It was a widely shared view. Many researchers spoke of "quasispecies" of viruses that moved about in "swarms." John Holland, of the University of California at San Diego, felt that the high error rate of RNA polymerase, the enzyme responsible for making copies of RNA viruses, was the key to the extraordinary mutation rate. The polymerase was constantly "jumping" and "stuttering," to use the official vernacular, to make different viruses.

"Natural selection among viruses isn't about ending up with a specific genome that you call a species," Holland said. "It's about statistics." More specifically, it was about the statistical odds that any specific genotype would dominate a "quasispecies swarm" at any given time.

RNA was nothing more than a long sequence of four different chemicals—base pairs, or nucleotides—the order of which comprised the genetic code. Microbiologist Peter Palese, of Mount Sinai School of Medicine, discovered in laboratory tests that if he examined a pool of 100 clones of flu viruses—clones being supposedly identical organisms—there were on average seven mutations for every 91.6 nucleotides. Similarly, in polio virus clones he found about one mutation in every 95.3 nucleotides.

The rate of mutation, of course, depended on the number of times the viruses reproduced themselves, as all the changes occurred during replication. That meant that mutations were most likely to occur among infectious microbes inside an extremely sick individual or when the rate of spread among people was very high.

"The greater the number of people infected," Palese said, "the greater the rate of mutation."

As the *Homo sapiens* population swelled, greater opportunities would present themselves for both viral spread and mutation. It seemed perfectly reasonable therefore to assume that the evolution of microbes capable of infecting *Homo sapiens* would accelerate, perhaps dramatically.

Most mutations were, of course, deleterious to the individual microbes. But given such a high rate of change, it was inevitable that the microbes would occasionally hit on a mutation that increased their edge against the human immune system, gave them a wider range of cellular targets, allowed them to pass more efficiently from person to person, or rendered them in other ways more dangerous to *Homo sapiens*.

In laboratory studies these processes appeared to function by more stringent rules than pure happenstance. The polymerase and replicase enzymes that controlled replication of viral RNA and DNA seemed to jump about. The polymerase could be seen snaking its way down an RNA nucleotide chain like some molecular zipper slithering along the zipper track. But then the enzyme would jump tracks, taking the portion of new RNA it had already manufactured and joining it to another nucleotide stretch. The result would be a genetic hybrid. This had been seen with viruses as varied as polio and a microbe that infested tobacco plants. It sometimes occurred because there were "bumps" or "kinks" in the original RNA/DNA strand, which prompted the busy polymerase to jump tracks. These bumps themselves weren't entirely random events, as they seemed to exist at significant points in the microbes' genes.[88]

There were parts of microbial genomes that required constant mutation, particularly the genes that coded for proteins on their outer surfaces which were recognized by the human immune system. HIV, influenza, polio, schistosomes, *Plasmodium falciparum*, and staphylococci all had hypervariable mutation sites in the genetic regions that coded for such proteins. Change, at one pace or another, was essential to survival in the midst of human antibodies, T cells, and macrophages.[89]

Researchers at Louisiana State University felt they had evidence of "virus gumbos," or mixtures of species of viruses which, combined, produced diseases that they seemed incapable of causing alone. In particular, they saw that some quasispecies of the feline leukemia virus were harmless to cats unless the cats were co-infected with other strains of the leukemia virus or the feline immunodeficiency virus (cat AIDS).[90]

In human beings similar "virus gumbos" were known to affect the AIDS disease process. Two herpes viruses, for example, were capable of directly activating replication genes inside HIV: herpes simplex Type 1 and HHV-6 (human herpes virus 6). There was even evidence that the polymerase zipper-jumping effect could occur inside patients, with strange hybrid viruses appearing: apparent mixtures of herpes and AIDS viruses.[91]

Cancer viruses were known to bring about their deleterious effects either by inserting themselves into specific oncogene sites in human or animal DNA or by manufacturing special proteins that switched on those cancer-causing genes. If the proper oncogenic signals were turned on (or switched off, in the case of cancer-suppression genes), a cascade effect might result in which other oncogenes and cellular signals were altered, ultimately transforming the cell into a cancerous entity. Most biologists believed that it was far from coincidental that such an intimate relationship between viruses and oncogenes had evolved over the millennia. And some went so far as to question whether genetic experiments with oncogenes and cancer cells might not result in infectious release of oncogene-carrying viruses or bacteria: perhaps altogether new species of viruses might result.[92]

DNA viruses also possessed special stretches of nucleotides that seemed to command polymerases to act with greater care, making accurate, often multiple copies of the stretch of genes located next along the "zipper." These DNA sections were called "enhancers." Studies showed that a key viral characteristic coded for by enhancers was the infectivity of the virus: the range of cell types it could invade and the ways in which the microbe could spread.[93]

Given such a vast range of mutational options for change, Harvard's Dr. Bernard Fields asked his colleagues, "Why haven't viruses wiped out all life on Earth?" It was, he felt, the crucial question.

And the answer, Fields said, lay in the difference between studying viruses in test tubes and studying them in the animals or humans that they infected. The venerable scientist, who had written *the* book on virology,[94] chastised his colleagues for being "overly reductionist," deriving too much from the fact there was one mutation in every 10,000 viral replications—in a test tube. In the real world those mutants still had to deliver their genetic payloads to the proper types of cells inside an animal or person in order to cause disease. And that necessary leap proved too great an obstacle for most mutant microbes, he said.

On the macrolevel, as Fields called it, little was well understood. There was no discipline of microbial ecology dedicated to studies of the behavior of microbes inside the human body.

"That's the big black box," Fields said, "and it's the secret to evaluating all the conjectured risks of emergence of a new pathogenic virus."

Some of the uncertainties in that black box included knowing how, exactly, viruses gained entry into the human body via alveoli in the lungs, M cells in the intestines, or lymphatic cells in the bloodstream; the roles various immune system chemicals played in either stifling or promoting viral activity; how viruses got past the thick membrane of cellular nuclei and past the chromatin mélange of proteins and carbohydrates to gain access to the host's DNA; which host chemical systems viruses exploited to their advantage; and how viruses got back out of hosts in order to be spread to other animals or humans.

Also in the black box were factors that seemed to make hosts more susceptible to viruses, Fields said, such as starvation, stress, and additional disease burdens. Though the catchall phrase "lowered immune response" was traditionally used to sidestep the mystery, little was known at the microbial level about how such factors influenced events. A starving child might make less protective mucus for his intestinal and stomach linings, for example, exposing more M-cell receptors to passing viruses. Was that a genuine phenomenon in nature, linking starvation and disease? Or was the mucosa depleted as a result of the infection?

"We know what many of the instruments are," Fields said, "but we haven't a clue about the orchestration. The problem isn't a flute problem; it's an orchestra problem."[95]

Virtually all the questions raised about viruses could also be directed toward bacteria and parasites, although the black box was somewhat smaller. There was plenty of evidence that bacterial and parasitic mutations were occurring in nature and that they often conferred new advantages on the microbes. Resistance to antibiotics and antimalarial drugs spoke volumes on the matter.

There, debate centered on the question originally raised in 1988 by John Cairns: was all bacterial mutation random, or were there directed changes that occurred in response to specific environmental pressures in the microbe's ecosphere?

The general dogma had it that such evolutionary events were random. New types of organisms emerged by chance mutations and haphazard genetic exchanges. If chance favored a certain type of organism when it surfaced, the microbe would thrive. In the meantime, the endless DNA dance of transposons, mutations, plasmids, and sexual conjugation went on, its pace essentially unaltered by environmental events. Genes shuffled and recombined, swapped and moved, whether or not the microbes were threatened.

"All DNA is recombinant DNA," said the bible of biology, *Molecular Biology of the Gene*.[96] "Genetic exchange works constantly to blend and rearrange chromosomes." The emergence of, say, a penicillin-resistant *Streptococcus* was a "rare event, typically occurring in less than one per million cell divisions." Of course, *Streptococcus* underwent more than a million individual cell divisions in twenty-four hours, starting with a single bacterium and expanding exponentially.

Multiply resistant organisms that carried plasmids with dozens of advantageous genes were also considered the result of chance mixtures of DNA pieces that combined and recombined over microbial generations.[97] If randomness was at the root of microbial evolution, humanity needn't fear unexpected changes in the rates of emergence of new mutants.

Well before scientists appreciated the extreme mobility of discrete pieces of DNA, seminal laboratory experiments were done with *Escherichia coli* proving that mutant abilities to withstand attacks from either viruses or antibiotics *preceded* the appearances of those threats in the bacteria's environment.[98] Roughly one out of every 10 million *E. coli* in a petri dish might randomly mutate to be resistant to, say, penicillin. Then, if the drug were poured into the petri dish, 9,999,999 bacteria would die, but that one resistant *E. coli* would survive, and divide and multiply, passing its genes for resistance on to its progeny.

In 1988, however, John Cairns of the Harvard School of Public Health challenged that central dogma of biology.[99] Using recombinant DNA techniques, his laboratory made a set of specific *E. coli* mutants that had unusual nutritional needs. They then altered the bacteria's environments, making them deficient in chemicals the mutants couldn't manufacture on their own. And they showed that the *E. coli* would specifically change two separate sets of genes to adapt to the situation and survive, doing so in far less time than random mutation would permit.

"That such events ever occur seems almost unbelievable," Cairns wrote, "but we have also to realize that what we are seeing probably gives us only a minimum estimate of the efficiency of the process, since in these cases the stimulus for change must fairly quickly disappear once a few mutant clones have been formed. . . . It is difficult to imagine how bacteria are able to solve complex problems like these—and do so without, at the same time, accumulating a large number of neutral and deleterious mutations—unless they have access to some reversible process of trial and error."

Cairns used computer metaphors to describe what he believed was going on in the microbial world. The essential genetic material that made an *E. coli* an *E. coli* was the organism's hard disk. The bacteria had an almost endless number of ways to scan that basic disk, turning off and on various genetic programs and data bases. Plasmids and transposons were "drifting floppy disks," carrying additional bits of genetic data and programming.

There was a limit, Cairns argued, to how large any given organism's hard disk could be. Furthermore, energy needs placed restrictions on how many genes could be expressed, or turned on, at any given time. Some genetic programs would remain silent most of the time, stored against emergencies in the bacteria's data bank. Among those, he felt, were programs that actually ordered mutations, or changes, in elements of the basic hard disk. Since the bacteria couldn't afford to contain enough DNA to carry programs in anticipation of every possible crisis, a direct mutation command was, Cairns argued, the next-best alternative.[100]

The British-born biologist was convinced that such mechanisms were at play in some cases of drug resistance, as well as microbial evasion of the human immune system. In laboratory experiments it was possible to induce lysis—or rupture—of bacterial cells and see the microbes' DNA "hard disk" flood into the fluid petri dish. There, other healthy bacteria would absorb the roaming DNA. And if antibodies were added to the mixture the scavenger bacteria would use the newly absorbed DNA to make new proteins to coat their membranes. In this way, the bacteria would disguise themselves from the antibodies, successfully evading immune system attack.

Even the scavenging activity was less random than it seemed. Studies by Rockefeller University's Alexander Tomasz of *Neisseria gonorrhoeae* and *Hemophilus influenzae* showed that these organisms had special proteins on the outer surface of their cell walls. The proteins scanned passing DNA, looking for useful genetic sequences. When something good drifted past, the protein grabbed it and pulled the DNA into the bacterium. And the pneumococci, which absorbed any "promiscuous DNA," as Tomasz called it, had a special internal enzyme system that scanned the scavenged genetic material and rejected useless chunks of DNA.

There were, by 1992, several identified "mutator alleles" along the *E. coli* genome—sites in the hard disk that ordered neighboring programs to alter themselves. And under experimental conditions it was possible to see a sort of "trial and error" mechanism in play, in which the microbe rejected useless or harmful mutations, but placed beneficial mutations in its permanent bacterial hard disk.[101]

A number of stress proteins were discovered in microbes—proteins (or the genes that coded for them) that were activated when the cell was challenged by a range of threats: heat, fevers, some human hormones, arachidonic acid (an immune system activator), and a variety of human disease states. When activated, these proteins acted rapidly to protect vital biochemical functions inside the microbe. Termed "molecular chaperones," the proteins guided fragile compounds through their duties. The stress proteins could be turned on and off experimentally by inflicting definable changes upon their environments. There was no clearer example of a microbe's adaptation to its environment—adaptation that required genetic as well as chemical change.[102]

Studies of vancomycin resistance in *Staphylococcus aureus* strains found in a handful of European clinical settings revealed that seven separate genes were required to render the bacteria invulnerable to the drug. The seven genes prompted one simple alteration in the chemistry of the microbe's cell wall, replacing an ester bond in a structural protein with an amide one. The ester bond was the target for vancomycin.

Here was the amazing thing: those seven resistance genes were switched on *only* when vancomycin was in the bacteria's environment. How the bacteria knew of the threat's presence was an utter mystery.[103]

Researchers noted that many extremely divergent microbial species shared genetic signaling sites, called operons, that with very minor mutation conferred multiple antibiotic resistances on the organisms. For example, seven very different microbes (*E. coli, Salmonella, Shigella, Klebsiella, Citrobactr, Hofnia,* and *Enterobactr*) naturally shared an operon which, with a single point mutation, made the organisms resistant to tetracycline, chloramphenicol, norfloxacin, ampicillin, and quinolones.[104] In Cairns's terms, this implied that all seven bacterial species shared a few bytes of hard disk space that was specifically designed to undergo a single data-bit alteration when necessary to respond to an antibiotic threat.

Studies of various pathogenic *E. coli* strains showed that there was often a trade-off between genes for extreme virulence and those for antibiotic resistance. Rarely could the organisms carry enough genetic baggage to render them both highly lethal and resistant. Highly virulent strains didn't usually need resistance genes, however, because they could produce disease—and reproduce themselves—so rapidly that *Homo sapiens* didn't have the opportunity to make antibodies before the bacteria had accomplished their essential tasks of invasion, reproduction, and spread.[105]

While infectious disease biologists debated questions of random versus directed mutations among the microbes, the overall evolutionary role of jumping genes was the subject of great debate among biologists of all stripes. Some scientists had, by the 1990s, come to believe that transposons and plasmids were a driving force—perhaps *the* driving force—of evolution, even in plants and animals. The grand biological soup of shifting genes, it was suggested, was constantly giving one creature the capabilities normally carried by another. Human beings, in fact, might be nothing more than four billion years of gene jumping.[106]

But pure random chaos in such a mutation soup seemed terrifying. How could any species survive if its cells absorbed any chunk of DNA that came their way, no matter how dangerous it might be? Most random mutations were lethal, or at least deleterious, to the altered organism.

A series of startling experiments performed in a variety of laboratories during the early 1990s significantly raised the stakes of that debate. Amber Beaudry and Gerald Joyce, of the Scripps Research Institute located in southern California, succeeded in forcing a particular protein, called a ribozyme, to evolve in a test tube. Normally the ribozyme's job was to make specific cuts and slices in the organism's RNA. But Beaudry and Joyce showed that after ten generations of reproduction the ribozyme could mutate, becoming capable of chopping DNA as well.[107]

Critics of Cairns's experiments on bacteria and yeast grown under starvation conditions, which gave rise to directed mutations, charged that the British scientist's conclusions were unjustified: even in the Cairns model, they said, the mutations could have been due to random events.[108] The arguments heated up as researchers found evidence of seemingly strange behaviors in microbes. For example, some transposons seemed to be able to sense when it was a good time to pop out of bacterial DNA and go their separate ways in search of a safer genome. How did they "know" that the bacterium was under fatal attack? Or was it possible the transposons didn't "know" anything and scientists were simply witnessing the results of successful, though utterly random, gene jumping?[109] In fungi, it was noted, environmental stress could induce a process called "ripping," in which a massive number of single point mutations were suddenly made. Again, was the fungus responding to a stress by mutating in a specific, directed manner, or was it simply randomly mutating at a feverish pace?[110]

On an even more basic level, many scientists argued that utterly random mutation and absorption and use of mobile DNA would be prohibitively expensive for microbes. It cost chemical energy to scavenge plasmids and transposons, to sexually conjugate, or to move pieces of DNA around inside cells. It seemed inconceivable that stressed organisms, in particular, would waste energy soaking up all sorts of DNA completely at random. Several genes had to be switched on and membrane changes had to be made in order, for example, for *E. coli* to absorb useful antibiotic-resistance factors from another species, *Bacteroides fragilis*.[111] And though such horizontal transfers of genes between entirely different species of organisms were costly, they clearly occurred, spreading advantageous traits for resistance and virulence among microbes.[112] In some cases the plasmids themselves seemed to improve as they moved about between species, recombining and adding new pieces of DNA as they went.[113] Chemicals such as anesthetics, detergents, and environmental carcinogens seemed, for example, to influence bacterial sexual conjugation.[114]

In 1994 the Cairnsian view of directed mutation got a boost from experiments performed at Rockefeller University and the University of Alberta, Canada. Researchers first confirmed Cairns's initial experiments, showing that there was a specialized pathway of mutations that was switched on during *E. coli* starvation. Further, they showed that genetic recombination and resultant adaptive mutation occurred *in the absence of bacterial reproduction*. In other words, bacteria altered themselves not just through a process of random, error-prone reproduction that eventually yielded a surviving strain—the classic Darwinian view. In addition, they changed themselves, in some concerted manner, without reproducing.[115]

The differences in the Darwinian and Cairnsian views were not trivial. If, for example, an *E. coli* bacterium residing in the human gut were suddenly exposed to a flood of tetracycline, would it occasionally mutate and perhaps become resistant after generations of bacterial reproduction? Or could it acquire instant resistance via some directed recombination or transposon mechanism?

As issues of emerging diseases drew greater attention within the scientific community, theoretical debates centered on key questions: How likely was it that a previously unknown microbe would suddenly appear out of some stressed ecosphere? What were the odds that a fundamentally new pathogenic organism would emerge, the result either of recombination among other microbes or of large-scale mutation? Was it likely that old, well-understood microbes might successfully mutate into more dangerous forms? The first two questions were the subjects of mathematical models and extensive theoretical discussion, though the numbers of unknowns involved in such computations were enormous and significantly impeded conclusive analysis. Most scientists involved in such exercises felt that further basic research on microbial ecology and human behavior was needed in order to obtain enough data points to solve these quandaries.[116]

As to the question of virulence, it was considered axiomatic that all pathogenic microbes would seek a state of moderate virulence in which they didn't kill off their unwitting hosts too rapidly, giving themselves plenty of time to reproduce many times over and spread to other would-be hosts.[117] Over time, even a rapid killer such as the 1918–19 Swine Flu would evolve toward lower virulence. Or so it was thought.

But in the 1990s the world saw two viral cousins take off on very different virulence pathways. HIV-2 in West Africa became markedly less virulent between 1981 and 1993, infecting fewer people (despite the lack of safe sex practices) and possibly causing less severe disease in those it did infect.[118] In contrast, over the same period there emerged strains of HIV-1 that seemed to be *more* transmissible and to cause more rapid disease. Thus, tendencies toward both less and greater virulence seemed to be occurring simultaneously in the global AIDS epidemic.

Max Essex, Phyllis Kanki, and Souleymane MBoup studied HIV-2 closely and felt that there were inherent differences in the two species of AIDS viruses that could explain their opposite tendencies in virulence. Kevin DeCock felt, on the basis of his studies in Côte d'Ivoire, that HIV-2 was less transmissible than HIV-1, and probably always had been.

Biology theorist Paul Ewald of Amherst College in Massachusetts believed HIV-1 was also becoming less virulent. He argued that Kaposi's sarcoma, which was primarily seen among gay men with AIDS, was caused by a more virulent form of the virus that existed during early years of the epidemic. In Australia, Kaposi's sarcoma and AIDS deaths had declined markedly over the course of the epidemic, due, Ewald thought, to a shift toward less virulent HIV-1 strains.[119] But Australia's situation was not mirrored in the rest of the world in 1994: globally HIV-1 was spreading at an extraordinary pace, and strains of the virus had recently emerged that seemed to be especially adapted to rapid heterosexual or intravenous transmission. A Ugandan strain surfaced sometime in 1992 that appeared to cause full disease within less than twelve months after the time of infection.[120]

On the basis of mathematical models, British researchers predicted that HIV-1 would continue its trend toward greater virulence so long as the rates of multiple partner sexual activity remained high in a given area. As sexual activity declined, or as it became more monogamous, the rates of successful mutation, the number of quasispecies, and the virulence of HIV-1 would decrease.[121] And on that one point Ewald agreed: namely, that multiple partner sex was the key to virulence for sexually transmissible microbes.[122]

At the root of much of the new thinking about virulence lay a key assumption: that microbes would be extremely virulent if long-term survival of the host wasn't important for the spread and survival of the microbial species.[123] If host population density increased, the microbes could afford to become more virulent, as they were guaranteed greater exposure to secondary and tertiary victims.

That theoretical view received some experimental support in 1993 when Allen Herre, of the Smithsonian Tropical Research Institute in Panama, made a startling observation on the relationship between fig tree wasps and the minute roundworms that parasitized the insects. After ten years of observation and manipulation, Herre concluded that the worms became more virulent when the size of the wasp population, and the number of broods occupying any given fig tree niche, grew. When population size was low, the parasites were of low virulence and were passed from female wasps to their offspring via infected eggs laid in the figs. When the wasp population size swelled, and various broods intermingled, the parasites spread horizontally, from wasp to wasp. This allowed the parasites to become more virulent and, among other things, to destroy the insects' eggs. The difference could be seen in the paradoxical observation that the figs might be healthier, and suffer less wasp larvae infestation, at times when the adult wasp population was at its peak.[124]

At Harvard Medical School, John Mekalanos studied a host of known virulence factors and developed a technique for teasing out unknown bacterial virulence genes. He concluded that many microbes stored virulence factors, just as they did resistance genes, on plasmids and transposons, snapping them up when conditions were ripe for all-out activity, and discarding them as excess baggage when the time was right. Such virulence factors could be shared across microbial species.

Things that seemed to turn on known virulence factors included calcium fluxes, warmer temperatures (98.6°F inside a human body versus an external 60°F), the presence of iron, and a number of key chemicals.[125] But Mekalanos also showed that for every known virulence factor in a given microbe there were dozens awaiting discovery. What mechanisms might switch those genes on, or cause them to mutate, weren't known.

Mekalanos disagreed with Ewald's theory that virulence was tightly linked to transmissibility. There were exceptions. For example, a huge dose of cholera vibrio was needed to cause a human infection—on the order of one million. In contrast, *Shigella* could cause infection and disease with fewer than a hundred bacteria. Nevertheless, cholera was far more lethal than shigellosis.

"It's more complicated than mere transmissibility," Mekalanos said. "Microorganisms respond to a more complex array of pressures that decide levels of virulence."

The most blatant source of pressure was the host's immune system. In most cases the microbial advantage might look like virulence because the host's disease progressed badly, but from the microbe's point of view what was transpiring could better be described as escape. Microbes had discovered a long list of ways to escape the immune system, including disguise, Trojan Horse-like use of immune system cells as modes of entry and avoidance, constant mutation of genes coding for their outer surfaces so that the immune system failed to recognize them, and manipulation of immune system chemicals to set off false alarms that would occupy the system while the microbes slipped into safe hiding places.[126]

Theorists were busy trying to determine whether the balances between human immunity and microbial virulence were tipped by any particular identifiable contemporary factors. Nobel laureate Dr. Thomas Weller expressed concern that the ever-increasing numbers of severely immunosuppressed people on the planet posed a real threat for emergence of new disease problems. Cancer patients treated with high doses of chemotherapy or radiation, people infected with HIV, and individuals undergoing transplant operations all represented potential breeding sites for new or mutated microbes. Weller worried about a possible "piggyback" effect, with one microbial population taking advantage of severe immunodeficiencies produced by another microbe or medical treatment.[127]

Another population of immunosuppressed individuals consisted of those suffering from chronic malnutrition. Wherever a significant percentage of the *Homo sapiens* population was starving was likely to be a spawning ground for disease.[128]

Vaccines, where available, protected people against disease, but not against infection. Microbes could enter the body, but even highly virulent organisms found themselves facing an immune system that was primed and ready to mass-produce antibodies. Battles ensued; the invader was vanquished.[129] If a sufficient number of *Homo sapiens* in a given area possessed such immunity it would be possible to essentially eliminate the microbe. Unable to find a *Homo sapiens* host in which it could replicate, the microbial population would nearly disappear. Nearly. In this state, known as herd immunity, humans (or livestock animals) never suffered disease, though they might be infected, unless the necessary level of immunity in the overall population slacked off. For that reason, schoolchildren vaccine campaigns had to reach a critical threshold of successful completion or the unvaccinated children would be a great risk for disease.[130]

Herd immunity faced tough challenges in the age of air travel because individuals who carried microbes to which they were personally immune could fly into geographic areas where herd immunity was extremely low. Under such circumstances, even organisms not generally thought to be particularly virulent could produce devastating epidemics.

The best example of the phenomenon was the estimated 56 million American Indians who succumbed to disease following the arrival of Europeans—and their microbes. That die-off continued 500 years later, into the 1990s, as Old World microbes reached the Xikrin, Surui, and other Amazon Indians.

Yale University epidemiologist Francis Black argued forcefully in the 1990s that the terrifying death toll among New World natives was not a straightforward question of their having naïve immune systems that hadn't previously been exposed to the European microbes. Such an explanation was, he said, overly facile and flew in the face of evidence that new diseases commonly afflicted other populations of peoples without exacting such horrendous tolls. For example, new diseases were also introduced into sub-

Saharan Africa by European explorers, and though they claimed many lives, nowhere were there wholesale microbial genocides, as were witnessed in South America.

Black's theory was that what did in the American Indians was their own lack of biodiversity. Since all Amerindians were descended from two fairly small waves of migration from Asia, their gene pool was small. For the microbes this meant that the range of genetic diversity, not only in immune response but also in a host of other factors that affected the appearance and behavior of target cells in the *Homo sapiens*, was very limited. The microbes were, therefore, able to adapt swiftly to the very narrow set of obstacles before them, attacking the American Indians with unusual ferocity.

Black calculated that as the microbe was passed from Amerindian to Amerindian, it had a 32 percent chance of encountering a human with the same immune system genetics (major histocompatibility complex) as its prior host had possessed.[131]

A contemporary example of such a biodiversity mechanism at work was discovered by Michel Garenne and Peter Aaby during measles studies in Senegal. Aaby and Garenne noticed that measles became steadily more lethal as the microbe spread from one child to another, biologically related child. This was true for cousins as well as siblings. Mortality rates rose so markedly from child to child that they couldn't possibly be ascribed to chance or random immunity. It wasn't the children's immune systems that varied, *it was the measles virus*, which adopted ever more acute virulence capabilities as it passed from one genetically similar person to another.

The virus, in short, evolved to become a tailor-made killer for particular extended *Homo sapiens* families.[132]

Overall, that seemed to argue that increased mixing of *Homo sapiens*, both through intermarriages across racial lines and through travel and immigration, would eventually bolster the collective *Homo sapiens* immune response. That was the good news. But microbiologist Avrion Mitchison, director of Berlin's Deutsches Rheuma Forschungszentrum, was less than convinced that greater biodiversity in the human race could guarantee success over the microbes, particularly if the overall *Homo sapiens* population size did one day exceed ten billion.

"Even old pathogens invent new tricks," Mitchison wrote in a *Scientific American* article entitled, grimly, "Will We Survive?"[133] He continued: "Recently evolved drug-resistant strains of the tuberculosis bacillus have been plaguing industrial urban centers. Will such developments change the comfortable deadlock? Will *Homo sapiens* and the microbes continue to coexist, or will one side win?"

The answer, Mitchison concluded, was not at all clear.

Human activities that didn't seem amenable to positive change were, by the 1990s, playing significant roles in the spread and possible creation of emerging diseases. Between 1980 and 1989, for example, the number of refugees fleeing natural disasters, wars, famine, or oppression increased by 75 percent every year. By the end of 1992, according to the United Nations, 17.5 million *Homo sapiens* were refugees, most of them living in squalor in the world's poorest countries.

Thirdworldization had set in all over the globe. Millions of abandoned children roamed the streets of the world's largest cities, injecting drugs, practicing prostitution, and living on the most dangerous margins of society. Western European unemployment soared, from less than 3 percent in 1970 to more than 11 percent in 1993, and a sense of hopelessness cast a pall over much of the continent.[134] Civil war in the horribly overcrowded nation of Rwanda broke into inconceivable carnage during the spring of 1994. Serb invasions of Bosnia devolved into little more than slaughter of civilians.

Conservative Harvard University political analyst Samuel P. Huntington opined that the world had entered a stage of conflict that superseded nation-states, economic competition, and ideologies, becoming something far more insidious: cultural conflict. Wars and battles were fought over religion, over historic enmities that in some cases traced back to slights that had transpired between opponents more than 2,000 years ago.

In such a context, it seemed difficult to discuss *E. coli* virulence mutation probabilities. If men in the former Yugoslavia considered multiple acts of gang rape of civilian women justified acts of war, how could there be rational discussion of probabilities of sexual transmission of disease?

Still, the scientists pushed on, determined to remain cool in the face of global disarray, perhaps *because* of the chaos which threatened to abet the microbes. Studies demonstrated the rapid spread of disease among refugees and the emergence of antibiotic-resistant bacteria and drug-resistant parasites in such clusters of humanity.[135] The health risks of famine were carefully tallied.[136]

A concern shared by all public health observers was the shift globally from low-intensity geopolitical nuclear confrontation to high-intensity local conflicts. While the former had posed the threat of thermonuclear war, little actual conflict occurred. With the fall of the Berlin Wall dawned an era of extremely high-intensity conventional and guerrilla conflict which took a tremendous toll on civilian populations: direct losses of life, homelessness, refugee migrations, demolition of basic infrastructures, destruction of hospitals, and, in some cases, pointed deliberate assassinations of health providers.

When such conflicts occurred in developing countries, they created new possibilities for reemergence of old scourges such as typhus, cholera, tuberculosis, and measles—the classic wartime opportunists. Where sex became an economic component of strife, microbes that could exploit sexual transmission emerged. And along the peripheries of human battle and despair lurked the unexpected. In the flora and fauna of remote ecospheres they resided, human events affording them ever-greater opportunities for jumping from their ancient hosts to the warring *Homo sapiens*.

17

Searching for Solutions

PREPAREDNESS, SURVEILLANCE, AND THE NEW UNDERSTANDING

I don't even recognize the CDC anymore. It's a bunch of politicized pencil-pushers who make all the decisions without ever hitting the ground, never going into the field, never seeing things up close. I'm sick of it. I quit.
        —Joe McCormick, March 1993

Joe McCormick? I'm not familiar with that name, and I've asked around—nobody around here has ever heard of him. You're the first reporter I know of who's ever asked for him. Are you sure he works at CDC?
        —a public relations spokesperson for CDC, January 1993

The lesson I learned in Cairo still applies. The only way to deal with bureaucrats is with stealth and sudden violence.
        —United Nations Secretary-General Boutros Boutros-Ghali, 1993

■ On April 6, 1994, an airplane was shot down over Rwanda during the final leg of its flight from Tanzania to the Rwandan capital, Kigali. Aboard the plane were Rwandan President Juvénal Habyarimana and President Cyprien Ntaryamira of Burundi.

Three weeks later the carnage following the deaths of the two heads of state was staggering. Long-standing ethnic, economic, political, and cultural hatreds between the two nationalities living in the region, the better-educated Tutsis and the far more numerous and historically less advantaged Hutus, erupted in Rwanda and threatened the stability of neighboring Burundi. Tutsi rebel forces, reportedly backed by the Museveni government of Uganda, surged toward Kigali. The Hutu-dominated government forces and gangs of Hutu thugs responded by slaughtering Tutsi civilians living in the capital in a manner so wanton and barbaric that the global community was flabbergasted. Images of young Rwandan men filled television newscasts: men who grabbed innocent children, slashed off their heads with machetes, and then turned unashamedly to international camera crews, grinning and shouting in triumph.

The Tutsi rebel forces retaliated with equally brutal massacres of Hutu civilians living in the Rwandan countryside.

By April's end, with the carnage still continuing, the United Nations estimated that anywhere from 100,000 to 500,000 civilians had been slaughtered, and more than a million had fled their homes in search of safe havens. On April 29 more than a quarter of a million Rwandan refugees poured across the corpse-laden Kagera River into Tanzania during a twenty-five-hour period, making it the largest short-term refugee migration in world history. Tens of thousands more made their way to Zaire, Uganda, and Burundi.

In international relief tents they awaited their uncertain futures, huddled against the rain upon muddy hillsides located less than two degrees below the equator.

In 1989 the HIV infection rate among young adults living in Kigali exceeded 30 percent, and WHO observers were certain that it had continued to escalate radically over the four subsequent years. In rural Rwanda, however, HIV rates were below 10 percent. The lines between urban and rural Rwandans blurred with the refugee exodus, however, and the people poured into areas of Tanzania and Uganda that ranked as the most hard-hit rural AIDS centers in the entire world. International health officials made an anxious prediction: if the populations remained uprooted for weeks or months on end, and refugee poverty promoted prostitution, another explosive surge in the Lake Victoria region's already horrendous AIDS epidemic would ensue. Before that could transpire, however, cholera would come, spread as people drew their water from rivers clogged with rotting corpses.

What else lurked in the refugees' new environs? If a novel epidemic appeared, was the international public health community prepared to handle the crisis?

Five years earlier, just before Christmas 1989, some 800 tropical disease experts gathered in Honolulu for the annual meeting of the American Society of Tropical Medicine and Hygiene. They staged an extraordinary war games scenario, envisioning a horrendous epidemic in a mythical African region. The hope was that such a role-playing scenario would reveal weaknesses in the public health emergency system that could later be corrected.

What transpired was an event eerily prescient of the Rwandan crisis. And an event that proved disheartening.

In the war games scenario three mythical equatorial African countries, designated Changa, Lubawe, and Basangani, were interlocked in a crisis that threatened *Homo sapiens* worldwide. Civil war inside Changa had devolved into brutal, high-intensity struggle, with both sides in the ethnically divided dispute venting their hatred upon innocent civilians. Over six months' time an estimated 125,000 civilians had been slaughtered, virtually the entire national infrastructure destroyed, and about a quarter of a million people had fled into neighboring Lubawe and Basangani.

Most of the refugees were in a squalid encampment in Basangani, less than a mile from the Changa border. Conditions were atrocious, with drug-resistant malaria, malnutrition, and tuberculosis rampant. Some 25 percent of the adult refugees were HIV-positive. An international relief effort was underway, with physicians, nurses, and advisers from all over the world treating the ailing refugees. In addition, a United Nations peacekeeping force, comprised of military personnel from the U.S., France, Italy, Finland, the U.K., and Malaysia, was guarding the Basangani and Lubawe borders, protecting the refugees from possible Changan attacks.

As key scientists played their roles in Honolulu, a terrible epidemic unfolded among the refugees, multinational health providers, and UN forces. Before it was even noticed, ailing individuals infected with a mysterious microbe had traveled to the U.S., the Philippines, Thailand, Germany, and neighboring African countries.

And although every imaginable effort was made to swiftly identify and control the mysterious microbe, within a month a global pandemic of what appeared to be an airborne, nearly-100-percent-lethal virus was underway.

Antibody tests were positive for Ebola, and Karl Johnson, who took part in the war games scenario, declared, "You say this might be a mutant strain of Ebola that is respiratorily transmitted. Well, if that is the case, it would be very close to Andromeda" (named for the Michael Crichton medical thriller *The Andromeda Strain*). "You may say 'ridiculous,' but I don't think we can disregard that possibility," Johnson said. "It was, and still is, a potential."

Audience members in Honolulu began murmuring to one another. Though all knew it was only a scenario, tension was high because it bore such a close resemblance to past disease emergences.

Ebola was a particular sticking point for infectious disease experts in December 1989 because just a month prior to their Honolulu gathering the virus broke out in a primate colony located in Reston, Virginia. Ebola, the scourge of Yambuku and N'zara, had surfaced in the United States.

Fortunately the Reston Ebola outbreak involved a strain of the virus that, though highly lethal to monkeys, was harmless for *Homo sapiens*. Nevertheless, there had been a few tense days in Virginia when scientists weren't sure what they had on their hands, and fear ran high.

Ebola, therefore, was very much on the minds of the 800 experts gathered in Honolulu. Though the tropical sun and Waikiki beaches beckoned, nobody left the cavernous hotel conference room. The Reston outbreak had shocked these experts into taking the question of readiness very seriously.

Unfortunately, what the war games revealed was an appalling state of *nonreadiness*. Overall, the mood in Honolulu after five hours was grim, even nervous.[1] The failings, weaknesses, and gaps in preparedness were enormous.

There were no prepackaged infectious disease hospitals anywhere in the United States or at WHO in Geneva that were ready at a moment's notice to be airlifted into an epidemic. Virtually no civilian hospitals in the United States were equipped to handle a highly contagious, lethal microbe, either in patients or inside petri dishes in their laboratories.

Only one permanent maximum-containment facility existed inside the U.S. Public Health Service system, and the vast network of overseas high-security laboratories that had been run by the Rockefeller Foundation and the CDC no longer existed. The Public Health Service and WHO would, therefore, be forced in such an epidemic crisis to choose between two unsavory options: deploying all security research capabilities and personnel to the epidemic site, thus putting large numbers of personnel at risk; or shipping all the patients, blood samples, and tissue biopsies to the CDC's P4 laboratory, the Institut Pasteur, and Fort Detrick, risking the chance of civilian exposure should samples break open during transport.

In the 1960s when biological warfare research was underway in the United States and the Soviet Union, both the U.S. military and the civilian Public Health Service maintained supplies of special respirators that used ultraviolet light to decontaminate air before it was inhaled. "Where are those masks now?" Johnson asked. "Does anybody know?"

None of the experts had the slightest idea.

Johnson noted that no one in either the military or the civilian sector in the scenario wore protective "space suits" or respirators. "I hope that's not a mistake," he said, using a tone of voice that indicated he felt that it was, indeed, a grave error.

Malaria expert Dr. Ruth Nussenzweig of the New York University Medical Center complained about the "paucity of expertise in this situation." And she asked out loud the question many had murmured throughout the scenario: "Who should be in charge in such a situation? Who knows enough to make these kinds of decisions?"

General Philip Russell, then commanding general of the U.S. Army Medical Research and Development Command, said that in the Army the expertise "just isn't there, and the military is now strained far beyond the breaking point. The armed forces of the U.S. are organized for the defense of the country and are not organized for civilian medical emergencies." Military supplies of tropical disease vaccines, medicines, and diagnostic equipment were limited, Russell said, and probably could not be shared with the civilian sector in such an emergency.

Even in 1989, before the pace of military budget cutting would become frenetic, the Pentagon had just two portable biological containment facilities, which during the war games scenario were deployed to Fort Bragg and Fort Campbell to handle ailing U.S. soldiers just returned from Basangani. With only one such facility available in the civilian sector, the United States was ill prepared to deal with an epidemic if contagion was spreading in more than three locations.

"The bottom line is that we have insufficient expert manpower to sustain appropriate levels of health care, and inadequate supplies," Russell concluded in Honolulu. "In this situation we would have to put very junior people in and hope they learn very, very fast."

International agencies would not be in a position to significantly augment U.S. expertise and supplies in a crisis such as was depicted in the scenario, Dr. Adetukunbo Lucas said in Honolulu. As the former head of WHO's infectious disease control efforts (and now with the Harvard School of Public Health), Lucas was familiar with the capabilities of the various UN agencies and nongovernmental relief organizations. "At all times the infectious disease unit at WHO is running on a shoestring," he told the Honolulu assemblage. WHO's role would be limited to controlling international dissemination of information about the epidemic and smoothing out any political difficulties between authorities from the nations involved.

That situation, too, would worsen after 1989. By the time former U.S. Army scientist Jim LeDuc would in 1993 take command of WHO emergency responses to viral outbreaks, his entire annual budget would be a mere $25,000. "In a real crisis," LeDuc would say in 1993, "I'd spend that in the first fifteen minutes."

Canada had always assumed that the United States would handle such an emergency: "We have always counted on you," Dr. Robert Wydess of Health and Welfare Canada said to his U.S. colleagues gathered in Honolulu, "and I am shocked to discover that you aren't prepared."

Expertise in tropical disease was dwindling in the United States. Of the roughly 1,000 members of the American Society of Tropical Medicine and Hygiene the majority were retired or approaching retirement age. "There is no question that human resources are dwindling," Dr. Stephanie Sagabiel of the National Academy of Sciences said. "There is a real dearth of shoe-leather scientists with actual experience of working in the tropics."

The CDC's Duane Gubler and Joe McCormick had separately reached the conclusion that America and Europe no longer offered the kinds of training that would leave the world another generation of "disease cowboys." Everybody was overly specialized, they argued, unable to handle themselves in a crisis that required a broad range of skills.

"Twenty years ago field epidemiologists were the real article," Gubler explained. "They could do it all: field study on the ground, laboratory work, organism isolation, vector analysis. There's a paucity of that now. And I can't understand it, really. To me what's sexy is to go out in the field—that's where the excitement is. Maybe I'm a romantic, but to me what's hot is going out, kicking around in the field, and seeing a disease in its natural ecology."

The problem was money, or the lack thereof. Any bright twenty-five-year-old junior scientist could see that there was no financial future in preparing to be a "disease cowboy." At Harvard University, Dr. John David ran the Department of Tropical Public Health and spent increasing amounts of his time trying to find words of encouragement for his students and young faculty.

"I tell them, 'First you have to decide, do you want to do biomedical research, practice medicine, or do research in developing countries, field-work,' " David said, noting that it would be almost impossible to have it all in the 1990s. "If you want to do it badly, you'd better realize there will be a lot of sacrifice. A bright young person could be trained very quickly here in epidemiology or health policy for developing countries, but where would he get a job?

"It's a big problem now, and people are leaving the field. It's a very hard time to recruit people to the field. And it's *very* discouraging. I have faculty people here say to me, 'Look, if I don't get a grant this time I'm going to have to go to industry.' And they have gone, many of them.

"There's no reason to be particularly optimistic about these problems," David concluded.

Had the Honolulu scenario been reality, had an airborne Ebola virus broken out in a genuine refugee crisis as depicted, less than ten days after its apparent emergence the virus would have been carried by infected relief workers and soldiers from its Basangani epicenter to the following locations: Bangkok, Manila, Frankfurt, Geneva, Fayetteville, Washington, D.C., New York, Honolulu, and Fort Detrick, Maryland. Quarantine and isolation procedures would have to be strictly adhered to in each location where people interacted with the infected travelers. In addition, since most of the civilian travelers from Basangani flew aboard standard commercial airlines, WHO would need to mount an immediate international effort to trace all passengers who had flown on more than ten intercontinental flights. As the experience nearly two decades earlier with the U.S. Peace Corps volunteer infected with Lassa had shown,[2] such passenger tracings, health examinations, and quarantines were both monumentally difficult to perform and extremely expensive. Most participants in the Honolulu exercise were frankly skeptical that such an effort could be mounted before the virus became pandemic in scope.

Two years after the Honolulu war games Russell would be out of a job, as would most of the experienced scientists who had served under his command, victims of DOD cutbacks. The military's preparedness for such medical emergencies would only worsen after the 1989 war games in Honolulu.

U.S. military preparedness was put to the test in the Persian Gulf war. In that case several months of diplomatic saber rattling would precede actual combat, providing the Department of Defense with ample time to construct portable operating theaters, quarantine units, "space suits," respirators, and other gear to withstand Saddam Hussein's alleged biological weaponry.

But biowarfare would never break out, so the Defense Department would never know whether or not the equipment would have stood up to anthrax, plague, Ebola, or whatever microbe was hurled at U.S. troops. Physicians working in the facilities would complain, however, of the damage wrought by the Saudi desert sands that would find their way inside all the portable hospitals and operating theaters, into the high-tech equipment, and onto complaining patients. Grains of sand were considerably larger than bacterial and viral microbes.

If recent history hadn't offered ample evidence, a startling reminder of the difficulties inherent in trying to limit the spread of emerging microbes in the era of jet travel came from the Reston Ebola outbreak among research monkeys. The incident showed that in a week's time hundreds of people on four continents could be exposed to an apparently new microbe, well before authorities were aware of its existence. Though the particular strain of virus involved in the outbreak at Hazelton Research Products, Inc., in Reston, Virginia, later proved harmless to human beings, public health experts were stunned, and a little bit frightened. What if it *had* been a human pathogen?

Jim Meegan, who in 1989 filled the emergency response slot at WHO that would later be Jim LeDuc's posting, had been responsible for tracing the international aspects of the Reston monkey Ebola outbreak. The Reston incident bared all the public health system's weaknesses, and the naked truth was deeply disturbing to the scientific community.

The chain of events began on October 21, 1989, at the Ferlite Company in the Philippines, an animal distribution center. One hundred cynomolgus monkeys were shipped from Manila in the cargo hold of a KLM commercial airliner; their ultimate destination was New York City.

On October 24 the monkeys arrived at the Hazelton Primate Center in Reston, Virginia, and within two weeks the caretakers noticed an unusually high die-off rate among the group, according to General Philip Russell. "By November 10 we began to suspect some sort of hemorrhagic fever was killing these animals off," and they took steps to quarantine the infected animals.

By December over 50 monkeys had died of the disease and 300 had been euthanized to control the epidemic. Blood tests showed that the monkeys carried two viruses: the simian hemorrhagic fever virus, which was not infectious to humans, and Ebola, which usually was.[3] Inside the Hazelton facility and nearby Fort Detrick, where Army scientists were trying to figure out what was causing the monkey die-off, panic reigned. Every sniffle, headache, or fever that struck the personnel was taken as a sign of a possible spread from the monkeys to humans. And it was the flu season, so many researchers had some symptoms of illness.

Urgent detective work began, headed by Meegan and Joe McCormick, then head of the CDC's Special Pathogens lab. Where possible, every human being and animal that might have come in contact with the monkeys was examined for infection. Unfortunately, testing in the Philippines was delayed because a rebel uprising made it impossible for CDC investigators to reach the Ferlite Company, located well outside Manila in an area under guerrilla siege.

When Meegan tried to guess how many *Homo sapiens* had been exposed to the cynomolgus monkeys before they reached Reston, he came up with sobering numbers. He warned his colleagues at the Honolulu gathering to pay close attention to the message hidden behind the numbers: namely, that epizootic events could instantly lead to global pandemics if the microbes possessed the ability to infect human beings. As he described the chain of events, General Russell tried to keep a mental tally of how many human beings could have been exposed to the monkey virus and might—had it been a microbe that was dangerous to humans—have fallen ill and/or spread it to others.

The human exposure trail began at the Ferlite Company, where more than a dozen employees handled the animals, and proceeded to the Manila Airport. There, the boxed animals rested in a hangar prior to enplaning, and were fed and watered by a small staff of caretakers. The animals were loaded onto the airplane mechanically, but their plywood boxes were positioned by two cargo men prior to takeoff. It did not appear that additional personnel were exposed to the monkeys during flight stopovers in Bangkok and Dubai.

But in Amsterdam the animals were off-loaded manually by a team of cargo handlers who, without wearing gloves or protection of any kind, carried the monkeys in their boxes to automated delivery conveyer belts. The animals were then housed inside one of the world's largest animal hostels, located in the Amsterdam airport and operated by KLM. More than 23,000 monkeys per year were housed for a day or longer in the KLM facility, species of all types, coming from all over the world.

Stacked near the Philippine monkeys were two monkeys from Ghana that were destined for delivery in Mexico, and another group of African animals bound for Moscow. During their stay in the hostel the Philippine animals shared a common water source, and caretakers fed and monitored the animals, moving around the monkey area without wearing masks or changing gloves.

After their Amsterdam stay was completed, the animals were hand-carried aboard another KLM aircraft, potentially exposing an additional crew of cargo loaders. Hours later the plane landed at JFK International Airport in New York City, where yet another crew of cargo handlers was potentially exposed to the animal viruses. The monkeys were stored in the massive JFK animal holding center, through which more than 50,000 animals were processed every month. They were tended to by animal caretakers who, as had been the case with their KLM counterparts in Amsterdam, took few special precautions to protect themselves from potential primate microbes. Outside the animal center was the huge airport, which typically handled close to 28 million human passengers a year, coming from or bound for nearly every other international airport in the world.

Even as CDC investigators began screening employees at Hazelton Research Products in Reston and at JFK, word came down that an unusual die-off of cynomolgus monkeys was occurring in another research colony, in Philadelphia. Those animals also came from the Philippines, via JFK, arriving in Philadelphia on November 28, 1989.[4] And three other shipments of Philippine monkeys that arrived in the United States between November 1989 and March 1990 were found to contain animals that were infected with the Reston virus. At least 173 people working in the various monkey centers were potentially exposed.[5]

The CDC found that five animal handlers working in the primate centers and one employee of the animal center at JFK had developed antibodies to the Reston virus, indicating that they had probably been infected.[6]

U.S. Army researchers maintained a fairly high level of concern about the Reston virus well past the New Year, but Joe McCormick and his CDC crew at the Special Pathogens Branch had long since decided that the microbe was harmless to human beings. And they had good reasons to be dismissive: McCormick was one of fewer than ten scientists on earth who had ever witnessed an Ebola epidemic and studied patients who were afflicted by the virus. Working with McCormick was British physician Susan Fisher-Hoch, who had studied Ebola extensively, first at England's Porton Down Laboratory and later at the CDC. And McCormick's group had just completed genetic sequencing of key portions of the dangerous human Ebola virus.[7]

"This is not a dangerous pathogen," McCormick and Fisher-Hoch repeatedly said, despite continued panic at Fort Detrick. They were thoroughly convinced that the Reston virus posed no threat to human beings.

But that wasn't satisfactory for the New York State Department of Health. The agency was appalled to learn of the apparent infection of an airport employee, particularly given that 80 percent of all research primates legally imported into the United States passed through JFK. There was also grave concern that the Ebola-like virus might not be limited to monkeys from the Philippines. CDC tests showed that 10 percent of all African and Asian monkeys had antibodies to filoviruses, the class of viruses that included Ebola and Marburg.

Accordingly, the New York Commissioner of Health, David Axelrod, decreed that effective at 12:01 a.m., March 23, 1990, no more monkeys could come through facilities located in the state of New York without documented evidence of sixty prior days of quarantine outside the United States and the understanding that an additional sixty days of quarantine would be required before commercial sale or research use of the animals would be permitted.[8]

The state's action forced the CDC's hand, and a flurry of edicts and national meetings followed. The public health community was reminded of the 1967 Marburg disaster, of a 1989 outbreak of simian hemorrhagic fever virus in New Mexico State University's primate center, and of the 1989 death of a laboratory worker at the International Research and Development Corporation in Mattawan, Michigan, who caught the simian herpes B virus from a Chinese monkey. There were loud calls for a full ban on the importing of wild monkeys.[9]

CDC director William Roper revoked the import licenses of the three biggest primate businesses—Hazelton, Worldwide Primates of Miami, and Charles River Primates Corporation of Port Washington, New York—setting off a loud outcry from the research community.[10] With AIDS researchers charging that the actions were bringing their efforts to a halt, and drug companies claiming that the CDC's steps amounted to a ban on pharmaceutical research and development, the federal agency was caught in the middle. The standoff dragged on until June, when CDC surveys of randomly obtained human blood samples revealed that many people who had never been near monkeys—including lifelong residents of the state of Alaska— had antibodies in their blood that would neutralize the Reston virus in a test tube.[11] The finding greatly reduced the significance of the infections seen in people exposed to the Philippine monkeys.

Later studies by McCormick, Fisher-Hoch, and their colleagues in the Special Pathogens Branch of the CDC showed that most monkeys survived even high experimental doses of infection with the Philippine virus, and that all survivors (27 of 42 animals) completely cleared virus from their bodies, as measured by PCR.[12] They also showed that the relative dangers of filoviruses varied dramatically depending upon whether they originated in Asia or Africa: African Ebola-like viruses were far more lethal.[13]

Despite reassuring findings about the low pathogenicity of the Reston virus, the scientific community had experienced a rude awakening. Monkey importers were back in full operation by July 1990, but under far more stringent testing and quarantine guidelines. The CDC and WHO were forced to reexamine the primate handling guidelines that had been issued twenty years earlier, in response to the Marburg outbreak. The airline industry, which had briefly refused to participate in further transport of primates, resumed animal shipping, but with a new sense of the risks involved.

The Honolulu war games exercises and the Reston virus incident were pieces in a larger picture of sharply heightened concerns in some scientific circles about preparedness for confronting the emergences of new disease. Five major U.S. government studies addressed the issue between 1988 and 1994.[14]

In addition, several international agencies and organizations addressed various aspects of the emerging disease preparedness issue.[15]

These reports, though produced by different groups of scientists and physicians, shared a sense of urgency and despair over the status of public health infrastructures and infectious diseases research in the United States and Europe. The solutions varied strikingly, however, reflecting the agendas of the various institutions involved.

American scientists, particularly virologists and those who were practitioners of the fledgling field of microbial ecology,[16] tended to support large-scale monitoring and surveillance schemes. Satellites, biological containment laboratories, computers, and PCR devices were the tools they hoped to use to spot changes in ecologies that might promote microbial emergences. Failing that, they hoped to be equipped to swoop in with a scientific rapid strike force that would identify and destroy emerging microbes before an outbreak progressed to an epidemic.

The most ambitious of these proposals, ProMED,[17] sponsored by the Federation of American Scientists, was the brainchild of Stephen Morse. In his claustrophobic, cluttered office at Rockefeller University, the bearded, bespectacled driven Morse burned midnight oil for years searching for answers to how best to help humanity stay one step ahead of the microbes.[18] When he and Nobel laureate Joshua Lederberg discussed that matter for hours on end in 1988 while planning the historic 1989 "Emerging Viruses" conference, Morse thought a fairly modest approach would suffice. Resurrecting the old Rockefeller Foundation international network of tropical laboratories would, he then thought, provide adequate protection.

But as the enormity of the scope of the emerging disease problem became apparent, the scale of Morse's envisioned surveillance net grew. The ProMED scheme involved a vast international network of monitoring systems that would keep an eye on diseases emerging not just in hospitals and clinics but also in agricultural crops, livestock, wild-caught animals, and sampled water supplies. The system Morse imagined would serve as a watchdog not only for natural emergences but also for uses of biological weapons.

Such a far-flung network could only work if supported politically by the United Nations. Accordingly, Morse and his ProMED colleagues, drawn from the ranks of biologists from all over the world, convened at WHO headquarters in Geneva during September 1993 in hopes of mustering more formal support for the initiative.

"The perception is growing that more needs to be done to prevent the emergence of new epidemics," the Federation of American Scientists' Dr. Barbara Rosenberg told the gathering. "This perception comes from *both* the bioweapons and public health communities. . . . There is a deep worldwide undercurrent of concern about emerging diseases, and an obvious need to develop a comprehensive, global plan."

D. A. Henderson, who had once led efforts to eradicate smallpox, told the Geneva gathering that "there is a growing belief that mankind's well-being, and perhaps even our survival as a species, will depend on our ability to detect emerging diseases. . . . Where would we be today if HIV were to become an airborne pathogen? And what is there to say that a comparable infection might not do so in the future?"

Years earlier, Karl Johnson had voiced darker concerns. After conversing at length with colleagues at a tropical diseases meeting in Seattle, he pulled Joe McCormick and a reporter aside, drawing the pair into a cranny away from crowds.

"I worry about all this research on virulence," Johnson had said, his tone deadly serious. "It's only a matter of months—years, at most—before people nail down the genes for virulence and airborne transmission in influenza, Ebola, Lassa, you name it. And then any crackpot with a few thousand dollars' worth of equipment and a college biology education under his belt could manufacture bugs that would make Ebola look like a walk around the park."

With genetic engineering it was a simple enough matter to insert genes coding for just about anything into the DNA or RNA of a virus.[19] Johnson believed that discovery of the Ebola genes for hemorrhagic disease could lead to their insertion into a virus, such as influenza or measles, that was adapted for respiratory transmission. And he wasn't alone among biologists in expressing that concern.[20]

By 1993 some 125 nations had signed the Bioweapons Convention,[21] yet the agreement had no teeth.

As a result, scientists living in countries with historic border and regional tensions worried that even a poor, backward nation could develop bugs that would produce famine by wiping out crops, cause widespread veterinary or human disease, or target economically crucial commodity crops to cripple a rival's economy.

"It can easily be done," Dr. A. N. Mukhopadyay said in Geneva. As dean of agriculture for G. S. Pant University in Pantnagar (Nainital), India, Mukhopadyay was particularly concerned that tensions between India and its neighbors could lead some country in the Indian subcontinent to carry out agricultural sabotage against its enemies. "This is not science fiction," he said.

Barbara Rosenberg asserted that biological weapons posed special diplomatic problems not encountered with their nuclear or chemical counterparts. "None of the equipment is so high-tech that it could not be homemade by any nation intent on developing BW capacity," she warned, adding that "no nation is immune to the dangers."

Microbiologist Mark Wheelis, of the University of California at Davis, was among those who believed that PCR technology could be used to finger bioweapons culprits.

"It's the molecular equivalent of finding the murderer's fingerprints on the gun," Wheelis said, noting that even as technology was creating new opportunities for bioweaponry, it was also opening up novel options for detection and deterrence.

The ProMED leaders ardently believed that the same international mechanisms that would permit monitoring and verification of bioweapons violations would also be ideal for watchdogging natural emergences of dangerous microbes.

But that made many scientists from developing countries nervous.

"I think a critical aspect of emerging disease questions is global partnership. It is crucial, essential, for people living in developing countries," Dr. Natth Bhamarapravati, president emeritus of Mahidol University in Bangkok, said. "We must do nothing to undermine that sense of partnership."

Japan's Isao Arita, a former leader of smallpox eradication efforts, felt that it was already extremely difficult to get past nationalist and cultural suspicions in order to carry out entirely beneficial programs, such as vaccination campaigns; if public health efforts were linked with punitive arms enforcement issues, many countries would deny access to both enterprises.

"The efforts must be separated," Arita concluded.

If public health disease emergence were to be separately executed on a global scale, what might a system look like, and who—what agency— would be at its helm? Arita wasn't sure.

Neither Arita nor D. A. Henderson were terribly enthusiastic about the obvious solution—namely, handing over control to the World Health Organization. After their experiences leading the smallpox eradication efforts, both men were fed up with WHO.

"We conquered smallpox *in spite of* WHO," Henderson said.

"By the time WHO realized there was an AIDS epidemic it already existed on four continents," Henderson added. "That's WHO preparedness and emergency response for you."

But if WHO wasn't adequate to the task, who, or what, was?

Henderson felt that the U.S. Centers for Disease Control was best suited for the job.

"WHO has pathetically few resources of its own," Henderson said.[22] In addition, the Geneva headquarters was often at odds with its scattered regional offices, which, he asserted, were "staffed by one or two [virologists] only. Inevitably, those who staff such units are prized more for their administrative skills in bringing experts together rather than for their own professional expertise. . . . I therefore see no option but to acknowledge CDC as an international resource, to fund it appropriately, and to acknowledge its mandate in legislation."

In Henderson's view, worldwide preparedness could be coupled structurally with such programs as the South American polio eradication effort and UNICEF's global campaign to vaccinate the world's children against the leading preventable pediatric diseases. And active surveillance would best be conducted through a series of fifteen tightly networked tropical outpost laboratories, staffed by CDC scientists, colleagues from local public health institutions in the host country, and academic researchers drawn from some fifty U.S. universities.

Henderson estimated that the entire system would cost $150 million per year to operate, adding, "Can we afford to invest in such a program? A better question is whether we can afford *not* to invest in a program that could be a determinant in our own survival as a species."

The Henderson proposal was similar to one that had been outlined fifteen years earlier by Jordi Casals,[23] and had over the years received support from Tom Monath, Robert Shope, Frederick Murphy,[24] and most of the scientists who had played roles in outbreaks of hemorrhagic or arboviral diseases.[25] It was formally endorsed by the U.S. Institute of Medicine.[26]

In response to the Institute of Medicine's report on emerging diseases, the CDC gave Dr. Ruth Berkelman the task of formulating plans for surveillance and rapid response to emerging diseases. For a year and a half Berkelman coordinated an exhaustive effort, identifying weaknesses in CDC systems and outlining a new, improved system of disease surveillance and response.

Berkelman and her collaborators discovered a long list of serious weaknesses and flaws in the CDC's domestic surveillance system and determined that international monitoring was so haphazard as to be nonexistent. For example, the CDC for the first time in 1990 attempted to keep track of domestic disease outbreaks using a computerized reporting system linking the federal agency to four state health departments. Over a six-month period 233 communicable disease outbreaks were reported. The project revealed two disturbing findings: no federal or state agency routinely kept track of disease outbreaks of any kind, and once the pilot project was underway the ability of the target states to survey such events varied radically. Vermont, for example, reported outbreaks at a rate of 14.1 per one million residents versus Mississippi's rate of 0.8 per million.[27]

Minnesota state epidemiologist Dr. Michael Osterholm assisted the CDC's efforts by surveying the policies and scientific capabilities of all fifty state health departments. He discovered that the tremendous variations in outbreak and disease reports reflected not differences in the actual incidence of such occurrences in the respective states, but enormous discrepancies in the policies and capabilities of the health departments.[28] In the United States all disease surveillance began at the local level, working its way upward through state capitals and, eventually, to CDC headquarters in Atlanta. If any link in the municipal-to-federal chain was weak, the entire system was compromised. At the least, local weaknesses could lead to a skewed misperception of where problems lay: states with strong reporting networks would appear to be more disease-ridden than those that simply didn't monitor or report any outbreaks. At the extreme, however, the situation could be dangerous, as genuine outbreaks, even deaths, were overlooked.

What Osterholm and Berkelman discovered was that nearly two decades of government belt tightening, coupled with decreased local and state revenues due to both taxation reductions and severe recessions, had rendered most local and regional disease reporting systems horribly deficient, often completely unreliable. Deaths were going unnoticed. Contagious outbreaks were ignored. Few states really knew what was transpiring in their respective microbial worlds.

"A survey of public health agencies conducted in all states in 1993 documented that only skeletal staff exists in many state and local health departments to conduct surveillance for most infectious diseases," the research team concluded. The situation was so bad that even diseases which physicians and hospitals were required by law to report to their state agencies, and the states were, in turn, legally obligated to report to CDC, were going unrecorded. AIDS surveillance, which by 1990 was the best-funded and most assiduously followed of all CDC disease programs, was at any given time underreported by a minimum of 20 percent. That being the case, officials could only guess about the real incidences in the fifty states of such ailments as penicillin-resistant gonorrhea, vancomycin-resistant enterococcus, *E. coli* 0157 food poisoning, multiply drug-resistant tuberculosis, or Lyme disease. As more disease crises cropped up, such as various antibiotic-resistant bacterial diseases, or new types of epidemic hepatitis, the beleaguered state and local health agencies loudly protested CDC proposals to expand the mandatory disease reporting list—they just couldn't keep up.

Osterholm closely surveyed twenty-three state health department laboratories and found that *all but one had had a hiring freeze in place since 1992* or earlier. Nearly half of the state labs had begun contracting their work out to private companies, and lacked government personnel to monitor the quality of the work.[29] In a dozen states there was no qualified scientist on staff to monitor food safety, despite the enormous surge in *E. coli* and *Salmonella* outbreaks that occurred nationwide during the 1980s and early 1990s.

At the international level the situation was even worse. The CDC's Jim LeDuc, working out of WHO headquarters in Geneva, in 1993 surveyed the thirty-four disease detection laboratories worldwide that were supposed to alert the global medical community to outbreaks of dangerous viral diseases. (There was no similar laboratory network set up to follow bacterial outbreaks or parasitic disease trends.) He discovered shocking insufficiencies in the laboratories' skills, equipment, and general capabilities. Only half the labs could reliably diagnose yellow fever; the 1993 Kenya epidemic undoubtedly got out of control because of that regional laboratory's failure to diagnose the cause of the outbreak. For other microbes the labs were even less prepared: 53 percent were unable to diagnose Japanese encephalitis; 56 percent couldn't properly identify hantaviruses; 59 percent failed to diagnose Rift Valley fever virus; 82 percent missed California encephalitis. For the less common hemorrhagic disease-producing microbes, such as Ebola, Marburg, Lassa, and Machupo, virtually no labs had the necessary biological reagents to even try to conduct diagnostic tests.

As a first line of defense against emerging diseases—at least the viruses—LeDuc advocated a modest $1.8 million one-shot program to upgrade all the laboratories and tighten the WHONet voluntary reporting system that linked key hospitals and medical systems worldwide.[30] LeDuc's proposal was formally endorsed by WHO and a panel of disease experts chaired by Joshua Lederberg on April 26, 1994. Months after the proposal went out to the wealthy nations of the world LeDuc was still waiting for some dollars, marks, yen, or other solid currency.

Berkelman's plan for bolstering CDC capabilities rested on the successful funding of LeDuc's global program, major improvements in domestic surveillance programs in all tiers of government, and vast advances in federal research, infrastructure, laboratory efforts, training, and general commitment to the problem.

That cost money: perhaps $125 million a year.[31] And any requests for funds immediately threw the fate of disease surveillance and preparedness in the hands of politicians. Thus, what began as a scientific concern ended up as fodder for congressional debate at a time when legislators were under public pressure to reduce the huge U.S. national debt.

Any vision of global health monitoring that ultimately rested in the hands of a U.S. agency was bound to be controversial in the court of international public opinion. The CDC had a track record of playing that role reasonably well for four decades with everything from Ebola to yellow fever. And when a crisis occurred the first call WHO generally made was to Atlanta.

But Francophile nations were likely to call the Institut Pasteur, which also had an established track record, particularly in West Africa. Members of the Commonwealth were, similarly, likely to contact the London Institute of Hygiene and Tropical Medicine. And nongovernmental organizations, such as Médecins Sans Frontières, Médecins du Monde, the International Red Cross/Red Crescent, and Oxfam were increasingly playing the role of disease early-warning systems. It was Médecins Sans Frontières, for example, that spotted the 1992–93 epidemic of extremely lethal visceral leishmaniasis in southern Sudan. With the country in a state of civil war and virtually all public health systems having collapsed, there was no Sudanese agency that was even monitoring the health of people in the rebel-held south, much less reporting disease outbreaks to Khartoum or Geneva. If not for the outsiders—Médecins Sans Frontières, in this case—the epidemic, though it afflicted tens of thousands of people, might well have remained invisible to the global public health community.

Indeed, as the 1990s witnessed an overwhelming number of high-intensity local conflicts between political, ethnic, and religious rivals, it became apparent to organizations most involved in relief work that *no* government-based disease surveillance systems had a prayer of success in regions of conflict. In 1993 alone, a massive measles epidemic swept over war-torn Angola; the Luanda government officially denied its existence. Médecins Sans Frontières identified ten populations at high risk for starvation and disease in 1993: non-Muslim Sudanese (700,000 people at risk), Afghani civilians (more than 10 million at risk), Tajikistani Muslims (more than 300,000 of whom were refugees in a bloody, ongoing civil war), Caucasus minorities (numbers not stated), Liberian civilians (some 820,000 at high risk), Angolan civilians (some 8 million imperiled by ongoing civil war), Cambodian noncombatants (millions subject to drug-resistant malaria and TB, as well as famine, in Khmer Rouge-held western parts of the country), Bosnian civilians (more than a million Muslims and Serbs endangered by ongoing civil war), Nagorno-Karabakh (more than 700,000 refugees fleeing war between Armenia and Azerbaijan), and Somalis.

All told, it seemed in 1993 that more than 21 million people on earth were living under conditions ideal for microbial emergence: denied governmental representation that might improve their lot; starving; without safe, permanent housing; lacking nearly all forms of basic health care and sanitation.

The situation only worsened in 1994, as more than two million Rwandans fled their country, most of them ending up in perilous refugee encampments lacking even the most rudimentary sanitation or safe water supplies.

On June 17, 1993, Médecins Sans Frontières filed an official protest with the United Nations Security Council, documenting numerous examples in war-torn areas of relief workers being endangered by local military forces, outlaw gangs, or United Nations troops. Further, the group charged that civilians were routinely denied access to hospitals and medical care—in some cases hospitals were deliberately targeted by warring forces.

By charter the United Nations was proscribed from doing anything that might be viewed as disrespecting national sovereignty. In times of crisis the UN interpreted that to mean that its agencies—including WHO—could not intervene in a nation without the official invitation of its recognized government. Without such permission, WHO could no more deploy a team of physicians to investigate an unusual disease outbreak in Kigali than it could in Los Angeles or Paris.

For Jonathan Mann, Daniel Tarantola, and most of the other former members of the Global Programme on AIDS, these concerns only heightened their conviction that disease emergence was inextricably bound to human rights. Mann wanted the world community to examine ways to use already existing human rights laws as leverage for UN and WHO access to health-imperiled populations.

During the Cold War there were far fewer such civil wars and nationalistic clashes because the superpowers imposed an overriding layer of control over most global conflict. In the absence of that global supervision, some governments felt free to slaughter their own people, exterminate rival minorities, eliminate all social (including medical) facilities for key population groups, and deny the existence of disease.

While some emerging disease specialists spoke of setting up NASA satellite networks to monitor rainfall and mosquito populations, red tides, or rain forest destruction in order to keep tabs on the microbes, physicians working in the midst of crises argued that what was needed was far more fundamental.

"There will always be a need for an emergency response effort, and that will probably always primarily mean the CDC," Joe McCormick argued. "But you need people on the ground to spot these things first. You need a health care system. And you need a place to call."

If the government is your enemy—if you and your people are victims of oppression—whom do you call?

"In candor, there is discomfort," Henderson said. "I'm all too conscious of the constraints that are upon us."

Former CDC director Dr. William Foege felt that new disease emergence was tightly linked to Thirdworldization: the overall status of health care, immunizations, sanitation, education, and total burden of disease in a society. Working with the Atlanta-based Carter Center for International Peace, Foege argued that structural adjustments ordered by the World Bank and the International Monetary Fund, coupled with a genuine capital crisis following the fall of the Berlin Wall, had severely worsened the human condition and improved odds for the microbes. More than $178 billion a year was flowing from the world's poorest nations to the richest in the form of debt payments, while a third of that amount (just under $60 billion) flowed in the other direction in the forms of loans and foreign aid.[32]

"This is a public health crisis," Foege argued. "One trillion dollars is spent on weapons annually. Of the fourteen million kids who died in 1989, nine million [deaths] could have been prevented for two and a half billion dollars. And that's what's spent in the United States annually for cigarette advertising."

Foege felt that international and domestic American health were so thoroughly integrated by the 1990s due to globalization of the microbes that it was impossible to ensure a disease-free existence for people in North America and Western Europe without providing similar assurances for residents of Azerbaijan, Côte d'Ivoire, and Bangladesh.

As the world, and disease threats, became increasingly complex, McCormick and Fisher-Hoch got fed up with both the CDC and WHO. McCormick decried all the "desk jockeys" and "pencil pushers" in Atlanta, Washington, Paris, and Geneva. After years of battling Lassa, McCormick and Fisher-Hoch saw civil war in Liberia and government instability in Nigeria wash away all their efforts and outbreaks of the rat-borne disease become commonplace. Having logged a lifetime of fighting the microbes both in the laboratory and dead center amid epidemics, McCormick had lost all patience. He thought that the links between poverty, lack of basic health care, ecological disturbances, and the emergence of dangerous microbes were so obvious as to be basic tenets of public health. Yet his kind of global thinking—and that of Henderson, Johnson, Monath, and Foege

—was no longer in vogue at CDC and WHO or inside the federal health bureaucracies in Washington, Paris, and London.

In the spring of 1993, McCormick and Fisher-Hoch left the CDC, moving to Pakistan to do what they believed was the last hope in the war against the microbes: train people in poor countries to conduct their own microbial search-and-destroy missions.

It might never be possible to routinely intercede in the "trafficking" events Stephen Morse spoke of, spotting epizootic or other emergences as they occurred and taking steps to bring them to an immediate halt. For the near future it seems that slow microbes such as HIV will continue to successfully emerge globally because *Homo sapiens* have no means for detecting organisms that enjoy years-long latency periods: detection comes only after disease has appeared. Most of the world is simply too bereft of infrastructure or too remote for even rapidly appearing microbial events to be recognized before full-scale outbreaks, or epidemics, have occurred.

Once emerged, however, a microbe could recirculate in a small segment of the human population, producing only occasional and isolated incidents of disease. As such, it could exist for decades, even centuries, avoiding detection and posing little immediate danger to society as a whole. Such, undoubtedly, was the case with HIV-1, HTLV-I, HTLV-II, Lassa, Muerto Canyon virus, Ebola, and many other microbes whose existence came to be acknowledged following striking disease outbreaks.

It might be possible to prevent full-scale epidemics, however, by concentrating efforts on sites of amplification: behaviors or conditions that assist microbes in making the leap from emerging into handfuls of *Homo sapiens* to widespread infection of a given human population. Concretely, amplifiers might make the difference between an infection level of less than 0.1 percent in a group of human beings and a 2 to 10 percent incidence of infection.

Unfortunately, the state of the art in research in the two fields that could best identify amplifiers is primitive. Behavioral science has long been looked down upon by those who worked in the "hard sciences," such as molecular biology and physics. And medical microbial ecology is an all but nonexistent discipline.

Nevertheless, from the information already amassed from disease emergences—from Machupo to Muerto Canyon—it is possible to identify several amplifiers.

At the top of the list in the 1990s has to be sex: specifically, multiple partner sex. The terrifying pace of emergences and reemergences of sexually transmitted diseases all over the world since World War II is testimony to the role that highly sexually active individuals, or places of sexual activity, play in amplifying microbial emergences such as HIV-1, HIV-2, and penicillin-resistant gonorrhea. Sex clubs, crack houses, bathhouses, and brothels act as disease amplifiers.

Epidemiologist King Holmes of the University of Washington in Seattle developed a mathematical equation that described the role multiple partner sex played in amplifying disease:

$$R_0 = B \times C \times D$$

In Holmes's equation, R_0 was the rate at which infection reproduced. A rate of 1 would represent a stagnant situation; if R_0 was greater than 1 the infection was spreading and an epidemic was underway. The B in Holmes's equation stood for the mean efficiency of transmission of the microbe per sexual contact. If B was a low number it signified that the microbe wasn't terribly contagious and the odds of becoming infected through a single act of intercourse were low. A high B number indicated that the organism was highly contagious.

The D in Holmes's equation stood for duration of infectiousness. Some microbes, such as herpes simplex, were highly contagious only during brief episodes when an individual might have sores around the genitalia that were shedding herpes viruses. If such episodes lasted only a few days, D would be a low number. The reverse would be the case with HIV, which could be carried in an infectious state, and transmitted sexually, for a period of more than a decade of an individual's life.

Finally, C reflected the mean number of sexual partners per day. A monogamous, married individual might have a C of less than 1, while a prostitute might have a C of 6.

What was immediately obvious in Holmes's equation was that human beings couldn't do much to alter D or B—those factors were under the microbes' control. C, however, was an entirely *Homo sapiens* issue.

If a sexually transmitted microbe was treatable, such as gonorrhea or syphilis, C would serve to direct physicians and public health authorities to logical sites for intervention. For example, the city of Amsterdam has legal prostitution that, though widespread, is not associated with a high level of disease. The reason is that the city requires licensing of all prostitutes and brothels, and regular medical checkups for periodic relicensing. Any prostitute who comes up positive for gonorrhea, for example, must cease working until cured.

For incurable diseases, such as HIV/AIDS, sexual amplifiers—high C sites or individuals—can serve as key targets for education and condom distribution. If, however, such sites or people are marginalized by the larger society, it may prove difficult, if not impossible, to effectively target their activities. The history of social responses to HIV/AIDS illustrates that banning or imprisoning people who, to use Holmes's terms, have high C factors only drives them away from public health authorities; their C levels remain at the same high numbers, and the epidemic continues spreading.

In addition, there is a gender difference in the ability of individuals to control C, with men having an advantage due both to their higher status in society and to their control of condom use. In countries where female children can be sold into prostitution or women are trapped in abusive marriages, the denigrated the role of women is an important factor in control of disease emergence.

UNICEF discovered during the 1970s that women were the key to most successful public health interventions. The UN agency found that the ability to get compliance with, for example, child vaccination programs was directly correlated with the educational status of the mother. Literate mothers were more likely to comprehend the need for vaccines; women who had completed secondary school were far more likely to embrace a larger perspective of family health that included planned births, hygiene, nutrition, regular doctor visits, and prenatal care. Child survival rates and disease rates in families directly paralleled maternal education levels.[33] Conversely, a woman who lacked education was less likely to appreciate the roles improper immunizations, malnutrition, hygiene, drinking water quality, and reused needles could play in the emergence of disease.

Probably the most efficient amplifier was the reused syringe. The Yambuku 1976 Ebola epidemic was grossly amplified by the mission hospital's use of five syringes on an average of 300–600 patients per day.[34] Though Ebola probably emerged in or around the N'zara cotton factory in 1976 and 1979,[35] it was amplified through reuse of nonsterile syringes in local medical facilities.[36] Tom Monath showed that yellow fever was spread in Nigeria by injectionists—untrained men who sold allegedly curative injections of all sorts of concoctions. Operating openly in shopping areas, the injectionists used the same unsterilized needle and syringe on dozens of clients per day; if one of those clients had yellow fever, subsequent clients were likely to be infected. Similar injectionist practices were credited for amplifying outbreaks of Lassa,[37] hepatitis B, malaria, and HIV.

Reuse of syringes by members of the medical profession was responsible for amplifying pediatric HIV outbreaks in Romania and Russia. Improper syringe use has also played a role in nosocomial amplification of antibiotic-resistant bacterial emergences.[38] Indeed, virtually any potentially blood-borne microbe could be amplified through improper medical use of syringes.[39]

Syringes used for the injection of heroin, cocaine, amphetamines, morphine, or other illicit drugs have also proven major disease amplifiers, spreading microbes that include hepatitis B, delta virus, HIV, HTLV-I, HTLV-II, *Clostridium botulinum*, *Clostridium tetani*, MRSA, *Pseudomonas aeruginosa*, *Pseudomonas cepacia*, *Serratia marcescens*, *Candida albicans*, and malaria.[40]

Countless studies carried out all over the world have demonstrated that most injecting drug users would cease sharing needles and syringes if alternative, sterile equipment was available, particularly if paraphernalia laws were rescinded or ignored by local police. When sterile equipment was offered either through legal sale in pharmacies, via free distribution programs, or through needle-for-needle exchanges, users generally flocked to the supplies. And in communities where needle sharing was discouraged either for cultural reasons or by virtue of ready availability of sterile supplies, the incidence of disease amplification was lower—often markedly so.[41] Microbes that emerged in the injecting drug user population, and were amplified through needle sharing, often spread to the general population via the blood supply or hospital settings. Therefore, the lack of sterile syringes for drug addicts appeared to represent a health threat to society as a whole.

In much of the world, however, injecting drug users were viewed as criminals, and provision of syringes and needles was illegal.

"If needle exchange is creeping legalization of heroin, which I think it is, then I'm opposed to it," Federal Bureau of Investigation's officer Richard Held told a 1992 gathering of public health experts.[42] "The problem is crime and violence. The problem is people living in fear. I think that's a public health issue."

Former San Jose, California, chief of police Joseph McNamara surveyed 488 judges, defense attorneys, prosecutors, police chiefs, and police officers in 1992 on their attitudes concerning criminalization of needles, syringes, and narcotic drugs. Though McNamara's survey was not randomized, the results offered a perspective too striking to dismiss. Nearly all those surveyed said the United States was losing its war on drugs (96 percent of judges, 95 percent of police officers, 85 percent of police chiefs): a war fought almost entirely through the criminal justice system.

Meanwhile, heroin use increased in the United States during the 1990s, having declined slightly during the 1980s. Public health priorities remained a subject of heated debate in 1994, pitting advocates of controversial programs such as needle exchange against those who felt that needle availability would lead to further increases in drug use.

Other devices designed to be inserted or implanted in the human body had proven capable of serving as disease amplifiers, usually in hospital or clinical settings. HIV, hepatitis, malaria, cytomegalovirus, antibiotic-resistant bacteria, Chagas' disease, yellow fever, and numerous other microbes had successfully exploited blood banks, transfusions, and plasma markets to amplify their numbers markedly. The dramatic rates of HIV and hepatitis B infection among the world's population of people with hemophilia offered striking evidence of the rapidity with which the risk for infection by a microbe that infects less than a tenth of a percent of the general society can be amplified many times over through multiple transfusions, decimating a whole generation of blood recipients.

No one has ever attempted to calculate the global burden of disease produced by nonsterile blood and syringes, but it is surely in the tens of millions, on the basis of WHO reports of rising blood-borne disease rates. It seemed reasonable to conclude that an international campaign to provide sterile syringes as needed, and clean up the blood and plasma systems, would go a long way toward eliminating amplification of emerging microbes. Both efforts are feasible: there are no technical roadblocks, nor are the efforts terribly expensive. What is lacking is political will.

In hospital and clinical settings there have been outbreaks of a host of diseases in which an invasive device or medical equipment served as an amplifier, creating foci for fulminant, contagious infections or, through reuse, directly spreading disease. For example, a shared medical inhalation device used for prophylactic treatment of HIV-positive men in a Miami outpatient clinic amplified a single MDR-TB infection many times over, leading to a large, lethal outbreak.[43]

Outbreaks of multidrug-resistant bacteria and mycobacteria have also resulted from amplification directly by medical personnel,[44] syringe or devise reuse,[45] device packaging,[46] septic catheters or IV lines,[47] septic surgical procedures, often involving contaminated implants of heart pacemakers, valves, limbs, joints, or other devices,[48] and respiratory assistance equipment.[49] In some cases an entire room—its walls, tables, beds—could be so thoroughly saturated with microbes that the physical setting itself served as an amplifier.[50]

When hospital-based amplification events occurred with enough frequency, they could lead to endemic nosocomial infection. In such cases the hospitals simply came to accept that newly emergent microbes, for example, drug-resistant bacteria, were permanent features of their environment, and the microbes frequently found their way into the general community. Such was the case for hepatitis B,[51] vancomycin-resistant enterococci,[52] respiratory syncytial virus,[53] MRSA (methicillin-resistant staphylococci),[54] MR *S. pneumoniae* and *S. epidermis*,[55] fluoroquinolone-resistant *Serratia* and *P. aeruginosa*, and a host of aminoglycosides-resistant bacterial species.[56]

In addition, the use of air conditioning or recirculation devices in otherwise airtight facilities has served to amplify airborne infections. Whether the setting was an airplane, nursing home, prison, or office building, the scenario was always the same: constant recirculation of the same air afforded small numbers of microbes enhanced opportunities to infect human beings. Examples include influenza,[57] tuberculosis,[58] Legionnaires' Disease,[59] measles, and influenza.

Research into microbial ecology would help identify more amplifiers, as well as measures that could be taken to mitigate their effects. At Yale, Robert Shope was in 1993 trying to develop ways to spot zoonotic events along the frontiers of *Homo sapiens* encroachment upon rain forest areas. Harking back to his earlier years studying Junín and Oropouche in South America, Shope headed a team of scientists that was studying human residents of small islands in the Amazon River near Belém. Busily carving out farmland from the rain forest, they were being monitored by Shope's group for evidence of novel viruses in their bloodstreams.

Carol Jenkins of the Papua New Guinea Institute of Medical Research headed up a similar effort in that country, following the health of people living in four comparable villages, two of which were employed in large-scale logging and deforestation efforts. Jenkins hoped to witness the emergence of newly recognized microbes or disease trends among the villagers most actively engaged in logging the previously pristine tropical forests.[60]

Stanford University's Gary Schoolnick came up with another novel approach to monitoring disease emergence in Mexico. He set up a modest molecular epidemiology laboratory along the Mexico-Guatemala border and trained local residents how to spot unusual diarrhea cases and collect stool samples. The community straddled the Pan American Highway, the primary artery connecting North and South America. Every year millions of people journey from the remote regions of the Southern Hemisphere up the highway to jobs in the north. They carry their microbes with them—microbes Schoolnick hopes to spot as they make their way to the California and Texas borders.

In the microbe magnets—the world's urban centers—research is needed to determine which aspects of city life most amplify microbial spread. For example, how strong is the correlation between rising rat populations and disease? When, in 1992–93, New York City had a 70 percent jump in reported incidents of rats biting humans, was that a harbinger of coming disease? When city budget cuts led to radical reductions in rodent control programs, and rollbacks in garbage collection left steady supplies of would-be food lying about sidewalks protected only by plastic bags, could a surge in the rat population have been anticipated? Would a surge in disease follow? If New York City already had more than seven million rats in 1992, what was the threshold for a public health crisis? Ten million? Twenty million?[61] Just how remote is the possibility that *Yersinia pestis* could acquire broad powers of antibiotic resistance, making the plague untreatable?

The same questions could be asked about unclean sewage, nonchlorinated drinking water supplies, the role of air pollution in enhancing human lung susceptibility to disease, the role of open water containers in promoting mosquito population growth, failures in urban vaccination programs, overcrowded housing, homelessness, and a plethora of other factors that affect—inversely—the quality of life for *Homo sapiens* and microbes.

In 1991 Zambia became the first country to undergo a hopeful revolution. With the flick of a switch the University of Zambia Medical Library was on-line via satellite ground station to data bases in medical libraries in the United States and Canada. By the middle of 1993 eleven developing countries were connected to medical data bases in the wealthy world and to each other, via SatelLife. It was expected that six more developing countries would be on-line by the close of 1994.[62]

The SatelLife movement was the brainchild of Nobel Peace Prize winner Dr. Bernard Lown. Having led the International Physicians for the Prevention of Nuclear War, Lown had developed a vast network of medical associates all over the world. And he recognized how desperately isolated physicians and scientists were in developing countries. A firm believer that "information is power," Lown worked closely with Russian, Japanese, and Canadian colleagues to develop SatelLife. The Russian government launched a satellite for the program, the NEC Corporation of Japan provided the necessary equipment, and the International Development Research Centre of Canada came up with the funds.

For the first time, physicians in developing countries could consult colleagues in neighboring nations or medical libraries and data banks to help solve puzzling cases and alert one another to disease outbreaks.

So when multidrug-resistant gonorrhea surfaced in Mozambique in April 1993, Dr. Prassad Modcoicar sat before a computer and typed this message: "I would like to know if there are any published studies or current ongoing research about the efficacy of the antibiotic kanamycin in the treatment of acute gonorrhea in men."

Modcoicar's message was carried by a ground-based satellite uplink to the SatelLife's orbiting satellite, which bounced his query back to satellite dishes in fifteen nations. In Lusaka, Zambia, Dr. M. R. Shunkutu saw the message and immediately transmitted to Modcoicar the results of a kanamycin study Subhash Hira had done in Lusaka in 1985. Before SatelLife went into operation, Modcoicar would have been obliged to either write letters to physicians in Europe and North America and wait interminable amounts of time for the results or simply experiment on his patients.

With SatelLife, Modcoicar instantly accomplished two tasks: he notified colleagues that he was seeing an outbreak of drug-resistant gonorrhea, and he found a solution for treatment of his patients. With plans to expand SatelLife into Asia and South America, the cheap, nongovernmental doctor-to-doctor service offered the real possibility of revolutionizing disease treatment and surveillance in the developing world. In the future, the kinds of satellite connections offered by SatelLife might also be used to relay information to public health planners all over the world that would allow them to be proactive—to anticipate potential disease outbreaks. For example, El Niño-type climate shifts often begin in one part of the planet and then spread in a predictable pattern. The events are known to be responsible for ecological changes that heighten risks for emergences of cholera, malaria, nearly all arboviral diseases, and most diarrheal diseases, particularly in poorer countries. Advance climate warnings could go out by satellite, delivered in real time and designed specifically to alert physicians of coming changes that might be relevant to vector or microbe activity.[63]

Genetic data bases, such as the vast GenBank at Los Alamos National Laboratory in New Mexico or GenInfo at the U.S. National Library of Medicine, might also prove invaluable to scientists in developing countries. As PCR technology becomes more widely available, researchers working in poor countries could be able to screen viruses and bacteria found in their patients and compare the strains to ones already archived in genetic data banks. If such technology were widely available, the result could be an avalanche of information on emerging strains of drug-resistant microbes, more virulent HIVs, even apparently new microbes.

At Harvard Medical School, Thomas O'Brien is trying to compile an international computerized data bank of genetic sequences for plasmids, transposons, phages, and resistance factors. Enhanced satellite communications systems that function independently from the frequently unreliable telephone systems of developing countries—as is the case with SatelLife —could allow for far more rapid monitoring, or at least reporting, of drug-resistance microbial activities worldwide.

Of course, such sophisticated systems are useless if no one acts on relayed information, or if there is no real primary medical system in a country. Without trained personnel and a functioning public health system it is almost inconceivable that anyone in a poor nation, or for that matter a poorly funded provincial department inside a rich nation, would be able to make use of GenBank or any such data base.

But no one in the belt-tightening world of the 1990s seemed much interested in contributing dollars, marks, or yen to the development of primary health infrastructures in countries like Armenia, Romania, Albania, Burma, or the Dominican Republic. The scale of the problem seemed too great, the payoff for donors too modest.

For every infectious disease, the preferred, easier option was vaccination. And by 1990 an estimated 70 percent of the world's children had been vaccinated against diphtheria, measles, pertussis, polio, tetanus, and tuberculosis. An estimated 130 million children were vaccinated every year in the developing world at a cost of about $1.5 billion.[64]

But experts were extremely dubious about the likelihood of expanding the vaccine market, developing new products, or using a newly developed vaccine in the midst of an emerging disease crisis. The phenomenal vaccination achievements of Merieux and the Brazilian government during the 1974 meningococcal bacteria epidemic may never be repeated. The reasons are numerous; they all boil down to money.

Pharmaceutical companies saw no profits in making vaccines intended for use by poor people—who would pay for the products? AIDS was revealing the tremendous hurdles involved in tackling new microbes through immunization. The Swine Flu fiasco left the future of government-sponsored mass vaccination in doubt for reasons of litigation. Further, U.S. courts paid huge awards to alleged victims of contaminated vaccines.

By 1990 more than half of all vaccine manufacturers had pulled out of the business, and though biotechnology was pointing the way to exciting new possibilities for vaccine design, enthusiasm was less than lukewarm in corporate circles.[65]

"Serious deficiencies are extant," D. A. Henderson said. "Vaccine pro-

duction for many vaccines is nowhere what is needed and resources for
vaccine purchases are diminishing; vaccine quality control for locally produced vaccines is negligible to nonexistent; and surveillance, the foundation
of disease control, remains seriously deficient and for many diseases, totally
desert."[66]

Ultimately, humanity will have to change its perspective on its place in
Earth's ecology if the species hopes to stave off or survive the next plague.
Rapid globalization of human niches requires that human beings everywhere
on the planet go beyond viewing their neighborhoods, provinces, countries,
or hemispheres as the sum total of their personal ecospheres. Microbes,
and their vectors, recognize none of the artificial boundaries erected by
human beings. Theirs is the world of natural limitations: temperature, pH,
ultraviolet light, the presence of vulnerable hosts, and mobile vectors.

In the microbial world warfare is a constant. The survival of most organisms necessitates the demise of others. Yeasts secrete antibiotics to
ward off attacking bacteria. Viruses invade the bacteria and commandeer
their genetic machinery to viral advantage.

A glimpse into the microbial world, aided by powers of exponential
magnification, reveals a frantic, angry place, a colorless, high-speed pushing and shoving match that makes the lunch-hour sidewalk traffic of Tokyo
seem positively poky. If microbes had elbows, one imagines they would
forever be jabbing neighbors in an endless battle for biological turf.

Yet there are times of extraordinary collectivity in the microbial world,
when the elbowing yields to combating a shared enemy. Swapping genes
to counter an antibiotic threat or secreting a beneficial chemical inside a
useful host to allow continued parasitic comfort is illustrative of this microscopic coincidence.

An individual microbe's world—its ecological milieu—is limited only
by the organism's mobility and its ability to tolerate various ranges of
temperature, sunlight, oxygen, acidity or alkalinity, and other factors in
its soupy existence. Wherever there may be an ideal soup for a microbe,
it will eagerly take hold, immediately joining in the local microbial pushing-
and-shoving. Whether transported to fresh soup by its own micro motor
and flagellae or with the external assistance of wind, human intercourse,
flea, or an iota of dust makes little difference provided the soup in which
the organism lands is minimally hostile and maximally comfortable.

The planet is nothing but a crazy quilt of micro soups scattered all over
its 196,938,800-square-mile surface.

We, as individuals, can't see them, or sense their presence in any useful
manner. The most sophisticated of their species have the ability to outwit
or manipulate the one microbial sensing system *Homo sapiens* possess: our
immune systems. By sheer force of numbers they overwhelm us. And they
are evolving far more rapidly than *Homo sapiens*, adapting to changes in
their environments by mutating, undergoing high-speed natural selection,

or drawing plasmids and transposons from the vast mobile genetic lending library in their environments.

Further, every microscopic pathogen is a parasite that survives by feeding off a higher organism. The parasites are themselves victims of parasitism. Like a Russian wooden doll-within-a-doll, the intestinal worm is infected with bacteria, which are infected with tiny phage viruses. The whale has a gut full of algae, which are infected with *Vibrio cholerae*. Each micro-parasite is another rivet in the Global Village airplane. Interlocked in sublimely complicated networks of webbed systems, they constantly adapt and change. Every individual alteration can change an entire system; each systemic shift can propel an interlaced network in a radical new direction.

In this fluid complexity human beings stomp about with swagger, elbowing their way without concern into one ecosphere after another. The human race seems equally complacent about blazing a path into a rain forest with bulldozers and arson or using an antibiotic "scorched earth" policy to chase unwanted microbes across the duodenum. In both macro- and microecology, human beings appear, as Harvard's Dick Levins put it, "utterly incapable of embracing complexity."

Only by appreciating the fine nuances in their ecologies can human beings hope to understand how their actions, on the macro level, affect their micro competitors and predators.

Time is short.

As the *Homo sapiens* population swells, surging past the 6 billion mark at the millennium, the opportunities for pathogenic microbes multiply. If, as some have predicted, 100 million of those people might then be infected with HIV, the microbes will have an enormous pool of walking immune-deficient petri dishes in which to thrive, swap genes, and undergo endless evolutionary experiments.

"We are in an eternal competition. We have beaten out virtually every other species to the point where we may now talk about protecting our former predators," Joshua Lederberg told a 1994 Manhattan gathering of investment bankers.[67] "But we're not alone at the top of the food chain."

Our microbe predators are adapting, changing, evolving, he warned. "And any more rapid change would be at the cost of human devastation."

The human world was a very optimistic place on September 12, 1978, when the nations' representatives signed the Declaration of Alma Ata. By the year 2000 all of humanity was supposed to be immunized against most infectious diseases, basic health care was to be available to every man, woman, and child regardless of their economic class, race, religion, or place of birth.

But as the world approaches the millennium, it seems, from the microbes' point of view, as if the entire planet, occupied by nearly 6 billion mostly impoverished *Homo sapiens*, is like the city of Rome in 5 B.C.

"The world really is just one village. Our tolerance of disease in any place in the world is at our own peril," Lederberg said. "Are we better off today than we were a century ago? In most respects, we're worse off. We have been neglectful of the microbes, and that is a recurring theme that is coming back to haunt us."

In the end, it seems that American journalist I. F. Stone was right when he said, "Either we learn to live together or we die together."

While the human race battles itself, fighting over ever more crowded turf and scarcer resources, the advantage moves to the microbes' court. They are our predators and they will be victorious if we, *Homo sapiens*, do not learn how to live in a rational global village that affords the microbes few opportunities.

It's either that or we brace ourselves for the coming plague.

Afterword

■ In the summer of 1993, Ron MacKenzie, now comfortably retired in a southern California desert community, was watching the evening news on television. A brief story caught his attention. It concerned shifts from agricultural production to growing coca plants for cocaine in various regions of Latin America, and MacKenzie recognized the place depicted. It was his beloved San Joaquín, Bolivia.

Viewing footage of the old cow town transported the retired physician back to a time when he was a strapping, though naïve, physician in Sausalito, California. And to that day in 1962 in La Paz when the Bolivian Minister of Health asked if he would mind taking a look at a mysterious *typho negro* outbreak deep in the Bolivian interior.

MacKenzie sat in his living room for a few moments, recalling the terror that struck Bolivia's Machupo River region when the strange hemorrhagic fever swept through. And he wondered how, after the passage of thirty years, the people had fared.

He reached for his telephone and called Montana. Karl Johnson, also long retired and living the way he preferred, in rugged cowboy country surrounded by prime trout-fishing streams, answered the phone. The old colleagues resolved to revisit San Joaquín.

MacKenzie, recalling the difficulties involved in reaching the remote area during the 1960s, called a colleague in La Paz for transportation advice, and made arrangements for a September trip. MacKenzie told the colleague that he and Johnson just wanted to have a little look-see after all these years.

For several months a new outbreak of Bolivian hemorrhagic fever had been raging in areas near San Joaquín, and CDC investigators had assisted the government in proving that the Machupo virus had made a comeback. Once again, for the first time in thirty years, Bolivia was waging an aggressive mouse control campaign in the region. That, too, piqued Johnson and MacKenzie's interest.

When the Americans arrived in La Paz, they were surprised by their warm, high-level reception. Thirty years was a long time—neither of them expected anyone to remember their efforts so long ago in the remote savanna region. But ceremonies, praise, and medals were lavished on the stunned scientists during their days in the capital—so much so that their journey to San Joaquín was delayed by nearly forty-eight hours.

Finally, the government arranged for a plane to fly the pair of scientists to the remote region, and MacKenzie reexperienced the dizzying ascent from the mile-high city, the passage through the Andes, and the descent into the steamy headlands of the Amazon.

As the plane approached San Joaquín, Johnson and MacKenzie wondered what was going on: there seemed to be quite a crowd gathered around the airstrip. When they landed, MacKenzie asked if their trip might be coinciding with the arrival of some dignitary or politician.

The moment they stepped out of the aircraft, a local marching band struck up a stirring tune and more than 300 peasants, cowboys, children, and farmers cheered the Americans. MacKenzie and Johnson were overwhelmed by the exuberant, joyous crowd, the hugs and handshakes, the flowers and gifts. Neither man could believe that the people of San Joaquín recognized them, or remembered their search, so long ago, for the source of Bolivia's plague.

"Most of these people weren't even alive back then," MacKenzie said to Johnson, who shared his sense of amazement.

But for the people of San Joaquín the names of MacKenzie, Kuns, Johnson, and Webb were emblazoned permanently on their cultural memory. Streets, some of which were now paved, were named after the near-mythic heroes from North America who had spared San Joaquín from doom. Even the schoolchildren knew who the white-haired "disease cowboys" were.

These were the men who stopped their plague.

And that was why for more than forty-eight hours the people of San Joaquín had waited patiently through the intermittent rain, standing on the landing strip and staring hopefully into the western sky.

Notes

Introduction

1. R. M. Krause, Washington, D.C.: National Foundation for Infectious Diseases, 1981.
2. R. M. Krause, Foreword to S. S. Morse, ed., *Emerging Viruses* (Oxford, Eng.: Oxford University Press, 1993).
3. J. Lederberg, R. E. Shope, and S. C. Oaks, Jr., Washington, D.C.: National Academy Press, 1992.
4. E. O. Wilson, *The Diversity of Life* (Cambridge, MA: Harvard University Press, 1992). 31.
5. A. Gore, *Earth in the Balance* (New York: Houghton Mifflin, 1992).
6. J. Lederberg, "Medical Science, Infectious Disease, and the Unity of Mankind," *Journal of the American Medical Association* 260 (1988): 684–85; and Institute of Medicine, *Emerging Infections: Microbial Threats to Health in the United States* (Washington, D.C.: National Academy Press, 1992).

1. Machupo

1. K. M. Johnson, et al., "Virus Isolations from Human Cases of Hemorrhagic Fever in Bolivia," *Proceedings of the Society of Experimental Biology and Medicine* 118 (1965): 113–18.
2. H. W. Lee, "Korean Hemorrhagic Fever," in S. R. Pattyn, ed., *Ebola Virus Hemorrhagic Fever* (Amsterdam: Elsevier Press, 1978), 331–43.
3. R. G. Gordon, et al., "Bolivian Hemorrhagic Fever Probably Transmitted by Personal Contact," *American Journal of Epidemiology* 82 (1965): 85–91.
4. Several years later Kuns would toil over insect samples he collected in San Joaquín and discover that a species of soft tick, *Ornithodoros boliviensis*, carried a virus. It was not, however, the virus responsible for the human epidemic. Nor did it match any other type of South American virus. So Kuns dubbed his new microbe Matucare virus, after a stream near San Joaquín. He showed the virus was lethal for baby mice and guinea pigs but harmless to human beings. Matucare virus would never have been discovered had there not been a human disease outbreak near the habitat of the *O. boliviensis* tick.
5. R. B. MacKenzie, P. A. Webb, and K. M. Johnson, "Detection of Complement-Fixing Antibody After Bolivian Hemorrhagic Fever, Employing Machupo, Junín and Tacaribe Virus Antigen," *American Journal of Tropical Medicine and Hygiene* 14 (1965): 1079–84; K. M. Johnson et al., "Isolation of Machupo Virus from Wild Rodent *Calomys callosus*," *American Journal of Tropical Medicine and Hygiene* 15 (1966): 103–6; P. A. Webb, "Properties of Machupo Virus," *American Journal of Tropical Medicine and Hygiene* 14 (1965): 799–801; P. A. Webb et al., "Some Characteristics of Machupo Virus, Causative Agent of Bolivian Hemorrhagic Fever," *American Journal of Tropical Medicine and Hygiene* 16 (1967): 531–38; and K. M. Johnson, "Arenaviruses," in B. N. Fields et al., eds., *Virology* (New York: Raven Press, 1985), 1033–53.
The U.S. Army provided logistical support to the ongoing Machupo investigation. In 1964 a Sergeant Lowery got infected and very nearly died of the disease. As had Webb months earlier, Lowery's wife comforted the ailing sergeant and also developed the disease. Lowery's condition was so severe that one Sunday morning the hospital director ordered the Gorgas pathologist put on standby to perform an autopsy. Meanwhile, MacKenzie gave Lowery a pint of his blood, hoping it contained helpful antibodies. Whether or not the antiserum made the difference will never be known, but later that Sunday Lowery started to recover. The sergeant and his wife survived Machupo.

6. A. S. Parodi, et al., "Sobre la Etiología del Brote Epidémico de Junín," *Día Médica* 30 (1958): 2300–1.

Jordi Casals, of the Rockefeller Foundation, isolated the actual virus for Junín disease. He discovered it closely resembled another mouse virus called LCM (lymphocytic choriomeningitis virus), which he had accidentally stumbled upon in 1958 while studying rabies-infected mouse cells. Casals dubbed LCM, and later Junín, Tacaribe, and Machupo, "arenaviruses" because under high-powered microscopes tissues infected with these agents "looked as though they had been sprinkled with sand." The word "arena" is Latin for sand. In 1964 Casals isolated another new arenavirus from Brazilian mice, dubbed Ampari virus. Fortunately, Ampari proved harmless to human beings. N. Mettler, S. M. Buckley, and J. Casals, "Propagation of Junín Virus, the Etiological Agent of Argentine Hemorrhagic Fever, in HeLa Cell Cultures," *Proceedings of the Society of Experimental Biology and Medicine* 107 (1961): 684–88; and F. P. Pinheiro et al., "Amapari, a New Virus of the Tacaribe Group from Rodents and Mites of Amapa Territory, Brazil," *Proceedings of the Society of Experimental Biology and Medicine* 122 (1966): 531–35.

7. Knowing the cause of a disease, even a terrible hemorrhagic one, is only the first step. Despite such knowledge, over 20,000 people have suffered acute Junín since its 1958 discovery, and the territory of the virus-carrying mouse has steadily expanded from the Buenos Aires province to, by 1985, three neighboring provinces of Argentina. Case fatality rates in recent outbreaks have been as high as 30 percent of all infected people. After Johnson and his colleagues completed their Machupo work in San Joaquín, another outbreak of that disease occurred in Beni, Bolivia, some years later. See World Health Organization, "Viral Haemorrhagic Fevers: Report of a WHO Expert Committee," *Technical Report Series* 721 (1985).

8. In 1972 Machupo broke out again in an area about forty miles from San Joaquín. Though details of exactly how the epidemic began remained elusive, Kuns's investigations pointed to the daughter of a wealthy ranchero owner. She apparently got Machupo after visiting her boyfriend, a young man from the village of Magdalena. Magdalena was the first Machupo-plagued town MacKenzie visited nine years earlier, and Kuns has always been convinced the ranchero owner's daughter got infected through sexual contact with a person carrying the virus. Sadly, none of the published accounts of Machupo directly stated the possibility that the virus could be transmitted through sexual intercourse, although official accounts of Webb's illness spelled out the likelihood Machupo could be passed from person to person by a kiss.

9. "The Search for the Invisible Killer," *Saturday Evening Post*, December 3, 1966: 92–96.

2. Health Transition

1. A. R. Hinman et al., "Live or Inactivated Poliomyelitis Vaccine: An Analysis of Benefits and Risks," *American Journal of Public Health* 78 (1988): 291–95.

2. J. A. Najera, "Malaria and the Work of WHO," *Bulletin of the World Health Organization* 67 (1989): 229–43.

3. Ibid.

4. J. S. Horn, *"Away with All Pests": An English Surgeon in People's China: 1954–1969* (London: Modern Reader, 1969).

5. Schistosomiasis is a deadly intestinal and liver disease caused by parasitic worms. The disease process begins when adult snails insert worm eggs into the bloodstream of a human being who works or plays along the edges of canals, streams, irrigation ditches, or lakes. The eggs grow, becoming worms, or flukes, that reside in the liver.

6. Horn (1969), op. cit.

7. U.S. Department of Health, Education, and Welfare, *Selected Disease Control Programs* (Washington, D.C., 1966).

8. U.S. Department of Health, Education, and Welfare, *Study Group on the Mission and Organization of the Public Health Service: Final Report* (Washington, D.C., 1960).

9. W. H. Stewart, "A Mandate for State Action," presented at the Association of State and Territorial Health Officers, Washington, D.C., December 4, 1967.

10. J. Lederberg et al., *Emerging Infections: Microbial Threats to Health in the United States* (Washington, D.C.: National Academy Press, 1992).

11. Further information can be found in: J. D. Watson et al., *Molecular Biology of the Gene* (4th ed.; Menlo Park, CA: The Benjamin/Cummings Publishing Company, 1987); P. Berg and M. Singer, *Dealing with Genes: The Language of Heredity* (Mill Valley, CA: University Science Books, 1992); and J. D. Watson, *The Double Helix* (New York: New American Library, 1969).

12. F. J. Fenner et al., *The Biology of Animal Viruses* (New York: Academic Press, 1968).

13. For excellent renditions of the history of antibiotics and controversies concerning the rise of bacterial resistance to the chemicals, the reader is referred to two highly readable books: M. Lappé, *Germs That Won't Die* (Garden City, NY: Anchor Press, 1982); and S. B. Levy, *The Antibiotic Paradox* (New York: Plenum Press, 1992).

14. J. Lederberg and E. M. Lederberg, "Replica Plating and Indirect Selection," *Journal of Bacteriology* 63 (1952): 399–406.

15. *Webster's New Twentieth Century Dictionary Unabridged* (New York: Collins World, 1975).

16. J. Farley, "Parasites and the Germ Theory of Disease," in *Framing Disease*, eds. C. E. Rosenberg and J. Golden (New Brunswick, NJ: Rutgers University Press, 1992), 33–49.

17. Among the monkey species known to serve as malaria reservoirs in Africa, Asia, and South America are *P. knowlesi, P. cynomolgi, P. brasilianum, P. invi, P. scwetzi, and P. simium.*

18. Two well-organized, excellent texts provide a quick thumbnail description of all infectious diseases prominent on the planet at this time. Both provide basic information on the hosts, reservoirs, and life cycles of parasites. The reader is referred to A. S. Benenson, ed., *Control of Communicable Diseases in Man* (15th ed.; Washington, D.C.: American Public Health Association, 1990); and M. E. Wilson, *A World Guide to Infections: Diseases, Distribution, Diagnosis* (Oxford, Eng.: Oxford University Press, 1991).

19. T. McKeown, *The Origins of Human Disease* (Oxford, Eng.: Basil Blackwell, 1988), 52.

20. W. H. McNeill, *Plagues and Peoples* (New York: Doubleday, 1976), 103.

21. F. Fenner, D. A. Henderson, I. Arita, et al., *Smallpox and Its Eradication* (Geneva: World Health Organization, 1988).

22. Smallpox is a relatively recent human disease, seeming to have arisen in India less than 2,000 years ago. In ancient times medical observers could not clearly discriminate between smallpox and other human-to-human epidemic diseases such as measles, bubonic plague, and typhus. As a result, controversy reigns over modern interpretations of ancient medical records.

Nevertheless, several major epidemics that claimed a quarter to a third of the affected populations were, according to historians familiar with medical records, likely to have been smallpox. These include:

Epidemic Site		Year
China	A.D.	49
Rome		165
Cyprus		251–66
Greece		312
Japan		552
Mecca		569–71
Arabia		683
Europe, various sites		700–800

Compiled from A. Patrick, "Diseases in Antiquity: Ancient Greece and Rome," in D. Brothwell and A. T. Sandison, eds., *Diseases in Antiquity* (Springfield, IL: Charles C Thomas, 1967), 238–46; R. Hare, "The Antiquity of Diseases Caused by Bacteria and Viruses: A Review of the Problem from a Bacteriologist's Point of View," ibid., 115–31; and McNeill (1977), op. cit., 2, 103, 118, 124.

23. W. M. Denevan, *The Native Population of the Americas in 1492* (Madison: University of Wisconsin Press, 1992).

Smallpox may have been the most useful weapon of biological warfare in world history. European colonialists repeatedly took advantage of the special susceptibility of the Amerindian population, deliberately spreading the deadly virus among Indians who were successfully defending their rights to the lands and resources of the Americas. For example, in 1763 Sir Jeffrey Amherst, commander in chief of all British forces in North America, was having great difficulty controlling the Pontiac Indians in the western territories. At Amherst's insistence, blankets inoculated with live smallpox viruses were distributed to the Pontiac, obliterating the tribe. The deliberately induced epidemic quickly spread to the northwest, claiming large numbers of Sioux and Plains Indians, crossed the Rockies and inflicted huge death tolls among Native Americans from southern California all the way north to the Arctic Circle tribes of Alaska. This devastation was cited in the official WHO history of smallpox: Fenner, Henderson, Arita, et al., (1988), op. cit.

24. Fenner, Henderson, Arita, et al. (1988), op. cit.

25. English physician Edward Jenner discovered in 1796 that cowpox, which was harmless to people, could be used as a vaccine against smallpox. The riskier idea of inoculating people with small amounts of human smallpox to raise immunity goes back in some cultures to ancient times, although some people developed the disease after injection. The British royal family was so immunized in 1722, as was the entire American Revolutionary Army under the command of General George Washington in 1776.

26. Spectacular accounts of the smallpox campaign can be found in June Goodfield's *Quest for Killers* (Boston: Birkhauser, 1985), 191–244; and Horace G. Ogden's *CDC and the Smallpox Crusade*, HHS Publication No. (CDC)87–8400 (Washington, D.C.: U.S. Government Printing Office, 1987).

27. In Ghana, for example, there were at least six major smallpox epidemics between 1901 and 1960, and in some the mortality rate among those infected with the virus was a staggering 50 percent. Whole villages were decimated, and the disease disrupted the national economy. Yet in no year were more than 1,950 cases registered in federal records. See D. Scott, *Epidemic Disease in Ghana 1901–1960* (London: Oxford University Press, 1965), 65–84.

28. In 1978 two people in Birmingham, England, contracted smallpox as a result of a laboratory accident. The Birmingham laboratory illegally possessed smallpox samples, and lab photographer Janet Parker died as a result of a containment failure. Her mother also was infected, but survived. Today samples of the two types of smallpox exist in just two high-security laboratories: the CDC's Atlanta lab and a Moscow facility. The CDC is currently sequencing the entire genetic code of the more dangerous variola major virus. And Moscow scientists are sequencing the less lethal variola minor form. By international agreement, both the Atlanta and Moscow smallpox samples are scheduled to be destroyed when gene sequencing is complete, although the elimination of all surviving viruses is a matter of considerable controversy.

29. The real 1991 dollar values of historic public health programs were derived using the latest United States CPI data, found in U.S. Department of Labor, *Consumer Price Index, Detailed Report* (Washington, D.C.: U.S. Government Printing Office, February 1992).

In 1991 dollars, the smallpox eradication effort cost $759.5 million, still a remarkably low cost when compared with the hundreds of billions now expended worldwide annually to combat just three diseases: AIDS, cancer, and heart disease.

30. Thirty-third World Health Assembly, *Declaration of Global Eradication of Smallpox*, Geneva, May 8, 1980.

31. U.S. Department of Labor. (1992), op. cit.

32. Ibid.

33. Institute of Medicine, *Malaria: Obstacles and Opportunities*, eds. S. C. Oates et al. (Washington, D.C.: National Academy Press, 1991).

34. DDT is dichloro-diphenyl-trichloroethane. Its insecticidal properties were discovered in 1939 by Swiss researcher Paul Müller, who received the 1948 Nobel Prize in medicine for his efforts. In the following years chemists developed several sister compounds that were also potent organochlorines, including dieldrin, chlordane, heptachlor, aldrin, and endrin. None was as effective in killing *Anopheles* mosquitoes as was DDT.

A second class of insecticides, organophosphates, was developed by the German Third Reich as nerve gases. It was discovered after World War II that these compounds could block crucial enzymes in insects, and parathion, malathion, and related chemicals came into use. Because of their acute human toxicity, the organophosphates were not widely used for malaria control in the 1950s and 1960s.

35. G. R. Coatney, "Simian Malarias in Man: Facts, Implications, and Predictions," *American Journal of Tropical Medicine and Hygiene* 17 (1968): 147–55.

36. General Douglas MacArthur, who led Allied operations in the World War II Pacific theater, said, "This will be a long war, if for every division I have facing the enemy, I must count on a second division in the hospital with malaria, and a third division convalescing from this debilitating disease."

See P. F. Russell, C. S. West, and R. D. Manwell, *Practical Malariology* (Philadelphia: W. B. Saunders, 1946); and W. Hockmeyer, personal interview, Walter Reed Army Institute of Research, Washington, D.C., September 1986.

37. Institute of Medicine (1991), op. cit.

38. Several species of mosquitoes are capable of carrying malarial parasites, but *A. gambiae* is the best suited to spreading the disease.

39. IDAB later became the U.S. Agency for International Development, or US AID.

40. Anonymous, *Malaria Eradication: Report and Recommendations of the International Development Advisory Board, April 13, 1956* (Washington, D.C.: ICA).

41. About $103.7 million in 1991 dollars. See U.S. Department of Labor, op. cit.

42. A. Spielman, U. Kitron, and R. J. Pollack, "Time Limitation and the Role of Research in the Worldwide Attempt to Eradicate Malaria," *Journal of Medical Entomology* 30 (1993): 6–19; and J. A. Najera, "Malaria and the Work of WHO," *Bulletin of the World Health Organization* 67 (1989): 229–43.

43. Written in deceptively simple prose, *Silent Spring* raised havoc when it was released in 1962, spawning the contemporary environmental movement and massive public outcry about the ecological destruction caused by improper pesticide use. The importance of Carson's book cannot be overstated: it was to the budding environmental movement what Charles Darwin's *Origin of the Species* was to evolution. Many have credited her book and the movement it started for the institution of environmental regulatory systems and laws in nations throughout the Western world.

44. R. Carson, *Silent Spring* (Boston: Houghton Mifflin Company, 1962).

45. Institute of Medicine (1991), op. cit.

46. The time lags between the commencement of DDT use and the apparent swelling of resistant mosquito populations varied dramatically around the world, depending on how the chemical was used. In countries where DDT use was enormous, resistant mosquitoes were obvious within a couple of growing seasons.

47. D. V. Moore and J. E. Lanier, "Observations on Two *Plasmodium falciparum* Infections with an Abnormal Response to Chloroquine," *American Journal of Tropical Medicine and Hygiene* 10 (1961): 5–9.

48. M. D. Young and D. V. Moore, "Chloroquine Resistance in *Plasmodium falciparum*," *American Journal of Tropical Medicine and Hygiene* 10 (1961): 317–20.

49. T. Harinasuta et al., "Chloroquine Resistance in *Plasmodium falciparum* in Thailand," in *UNESCO First Regional Symposium on Scientific Knowledge of Tropical Parasites* (University of Singapore, 1962), 148–53.

50. S. Maberti, "Desarrollo de Resistencia a la Primetamino: Presentación de 15 Casos Estudiados en Trujilo, Venezuela," *Archivos Venezolanos de Medicina Tropical y Parasitología Médica* 3 (1960): 239–59.

51. D. C. Rodrigues, "Cases of Malaria Caused by *Plasmodium falciparum* Resistance to Treatment with Chloroquine," *Arquivos de Higiénica (S. Paulo)* 26 (1961): 231–35.

52. G. R. Coatney, "Pitfalls in a Discovery: The Chronicle of Chloroquine," *American Journal of Tropical Medicine and Hygiene* 12 (1963): 121–28.

53. D. J. Wyler, "The Ascent and Decline of Chloroquine," *Journal of the American Medical Association* 251 (1984): 2420–22.

54. Between 1943 and 1984 the U.S. Army screened over 250,000 potential antimalarial drugs.

55. J. H. Meeuwissen, "Campaign Against Malaria with Medicated Salt in New Guinea," *Nederland T. Geneesk* 108 (1964): 1241–43; and J. H. Meeuwissen, "Malaria Prevention in New Guinea with Drugs Added to Kitchen Salt," *Annals de la Société Belgique de Médecine Tropique* 43 (1963): 209–12.

56. Julius Richmond served as U.S. Surgeon General during the administration of Jimmy Carter. Though he happily celebrated the smallpox victory, serving as U.S. representative to ceremonies held at WHO in Geneva, he warned American political leaders in 1978 that most infectious diseases would not so readily succumb to human attack. J. Richmond, personal interview, September 18, 1992.

57. Najera (1989), op. cit.

3. Monkey Kidneys and the Ebbing Tides

1. G. A. Martini and R. Siegert, *Marburg Virus Disease* (Frankfurt: Springer Verlag, 1971).

2. Ibid.

3. Ibid.

4. World Health Organization, "Viral Haemorrhagic Fevers: Report of a WHO Expert Committee," *Technical Report Series* (1985): 721; and Anonymous, "Méthodes de Surveillance et de Prise en Charge du Monkeypox et des Fièvres Hémorrhagiques d'Origine Virale," CDS/80.1 (Geneva: World Health Organization, 1980).

5. Martini and Siegert (1971), op. cit.

6. Since 1980 Rhodesia has been called Zimbabwe, in recognition of the Shora culture that preceded the British colonization of the area, which was spearheaded by Cecil Rhodes.

7. J. S. S. Gear et al., "Outbreak of Marburg Virus Disease in Johannesburg," *British Medical Journal* 4 (1975): 489–93.

8. J. S. S. Gear, "Clinical Aspects of African Viral Hemorrhagic Fevers," *Reviews of Infectious Diseases* 11, Supplement 4 (1989): s777–s782.

9. L. Lapeyssonie, *An Outbreak of Meningococcal Meningitis in Brazil*, EM/BD/8 (Geneva: World Health Organization, September 1974).

10. The federal medical training program provided most of the physicians that staffed Public Health Service clinics in rural areas, on Indian Nation lands, and in inner-city areas. The Reagan administration would terminate the program in 1986, citing budgetary concerns and a physician surplus as reasons.

11. A. S. Benenson, *Control of Communicable Diseases in Man* (15th ed.; Washington, D.C.: American Public Health Association, 1990).

12. C. O. Bastos et al., "Meningitis in São Paulo," *Revue Associacion Medicale Brasiliera* 19 (1973): 451–56.

13. The Type B meningococcus was more common in developed countries. Years later, as Brazil developed, it would experience Type B epidemics too. See J. C. DeMoraes, B. A. Perkins, M. C. C. Camarago, et al., "Protective Efficacy of a Serogroup B Meningococcal Vaccine in São Paulo, Brazil," *Lancet* 340 (1992): 1074–78.

14. Data from the International Air Transport Association, London. By 1980, 163 million people were flying annually between countries, and that figure reached 280 million by 1990. Within the United States, 42 million domestic tickets were sold in 1990.

15. World Health Organization, *Epidemic Report*, EM/EPID/39 (Geneva: World Health Organization, 1974).

16. L. Lapeyssonie, *Report on a Second Visit to Brazil, 28 April to 28 May 1975*, EM/BD/11 (Geneva: World Health Organization, September 1975).

17. The methods used to make such vaccines were fairly standard by 1974. See E. C. Gotschlich, M. Rey, W. R. Sanborn, et al., "The Immunological Responses Observed in Field Studies in Africa with Group A Meningococcal Vaccines," *Progress in Immunobiological Standards* 5 (1972): 485–91.

18. In its 1976 summary of the Brazilian pestilence, PAHO concluded: "It has been shown that populations belonging to lower economic classes which are exposed to overcrowding, poor housing, and poor environmental sanitation, and which have low levels of personal hygiene, are the populations most affected. Improvement of living conditions, housing and personal hygiene—which should be coupled with general education and health education—would certainly be beneficial, although the effectiveness of such improvement has not been definitely assessed in quantitative terms." See Pan American Health Organization, *Report on the Meeting on Meningococcal Disease, São Paulo–Brasília, Brazil, 23–28 February 1976* (Washington, D.C.: Pan American Health Organization, 1976).

19. Since the Brazilian epidemic, other outbreaks of equally mysterious origins have occurred in Mali, Niger, Papua New Guinea, Senegal, Nigeria, Finland, Norway, Cuba, Mongolia, Vietnam, South Africa, Sudan, Gambia, and the United States. In January 1993 panic developed in the largely Dominican Washington Heights section of New York City when a six-year-old boy died of Type A meningococcal meningitis. The boy left school on a Thursday feeling fine, and was dead a day later. See Pan American Health Organization (1976), op cit; B. M. Greenwood, "Selective Primary Health Care: Strategies for Control of Disease in the Developing World. XIII: Acute Bacterial Meningitis," *Reviews of Infectious Diseases* (Chicago: University of Chicago Press, 1984), 374–89; and M. Howe, "After a Meningitis Death, Striving to Calm the Fears of Other Pupils' Parents, *New York Times*, January 29, 1993.

20. Centers for Disease Control, "Yellow Fever Vaccine: Recommendations of the Immunization Practices Advisory Committee," *Morbidity and Mortality Weekly Report* 39 (1990): 1–6.

21. M. S. Pernick, "Politics, Parties and Pestilence: Epidemic Yellow Fever in Philadelphia and the Rise of the First Party System," in J. W. Leavitt and R. L. Numbers, eds., *Sickness and Health in America* (Madison: University of Wisconsin Press, 1985), 356–71.

22. W. H. McNeill, *Plagues and Peoples* (New York: Doubleday, 1977).

23. Roughly translated, one such tune repeats these lines:

> *Only mosquito can save Nigeria.*
> *Only mosquito can save South Africa.*
> *Only mosquito can save Zimbabwe*
> *Only mosquito can save Namibia.*
> *Only mosquito can save Africa.*
> *Only malaria can save Africa.*
> *Only yellow fever can save Africa.*

Translation courtesy of Bunmi Makinwa, John F. Kennedy School of Politics and Government, Harvard University.

24. R. M. Taylor, "Epidemiology," in G. K. Strode, ed., *Yellow Fever* (New York: McGraw-Hill, 1971), 442.

25. G. Strode, *The Conquest of Yellow Fever* (New York: McGraw-Hill, 1951).

26. F. L. Soper et al., "Yellow Fever Without *Aedes aegypti*: Study of Rural Epidemic in Valle do Chanaan, Espírito Santo, Brazil, 1932," *American Journal of Hygiene* 18 (1933): 555.

27. J. Boshell, "Marche de la Fièvre Jaune Sylvatique vers les Régions du Nordouest de l'Amérique Centrale," *Bulletin of the World Health Organization* 16 (1957): 431.

28. M. E. Wilson, *A World Guide to Infections* (Oxford, Eng.: Oxford University Press, 1991), 697–700.

29. T. P. Monath, "Yellow Fever: Victor, Victoria? Conqueror, Conquest? Epidemics and Research in the Last Forty Years and Prospects for the Future," *American Journal of Tropical Medicine and Hygiene* 45 (1991): 1, 43.

30. Ibid.

31. Scientists in these outposts investigated eight more yellow fever outbreaks during the 1960s, ranging in size from those involving fewer than twenty people (Zaire) to massive epidemics involving 20,000 people (Senegal, 1965), and to a West African pandemic which struck well over 100,000 people (Nigeria, Mali, Burkina Faso, Togo, and Ghana, 1969). In these epidemics 25 to 35 percent of those infected with the virus perished. See ibid.

32. V. H. Lee et al., "Arbovirus Studies in Nupeko Forest, a Possible Natural Focus of Yellow Fever Virus in Nigeria. II: Entomologic Investigations and Viruses Isolated," *Transactions of the Royal Society of Tropical Medicine and Hygiene* 68 (1974): 39–47.

33. T. P. Monath et al., "The 1970 Yellow Fever Epidemic in Okwoga District, Benue Plateau State, Nigeria. 2: Immunity Survey to Determine Geographic Limits and Origins of the Epidemic," *Bulletin of the World Health Organization* 49 (1973): 123–28.

34. T. P. Monath et al., "The 1970 Yellow Fever Epidemic in Okwoga District, Benue Plateau State, Nigeria. 1: Epidemiological Observations," *Bulletin of the World Health Organization* 49 (1973): 113–21.

35. Monath (1991), op. cit.

4. Into the Woods

1. According to Brinkmann, at least one member of the group attempted suicide. Brinkmann chose to protect the individual's identity.

2. J. D. Frame et al., "Lassa Fever, a New Virus Disease of Man from West Africa," *American Journal of Tropical Medicine and Hygiene* 19 (1970): 670–76.

3. Several sources of excellent information on the events of 1969 in Nigeria were consulted in preparation of this chapter. The reader is particularly referred to the following: J. G. Fuller, *Fever: The Hunt for a New Killer Virus* (New York: Reader's Digest Press, 1974); and J. D. Frame, "The Story of Lassa Fever," *New York State Journal of Medicine:*

"Part I: Discovering the Disease," Vol. 92 (1992): 199–202;
"Part II: Learning More About the Disease," Vol. 92 (1992): 264–67;
"Part III: The Disease in the Community," Vol. 92 (1992): 440–44;
"Part IV: The Politics of Research," Vol. 93 (1993): 35–40;
"Part V: The Fruits of Research" (in press).

4. The bubonic plague that claimed hundreds of millions of lives worldwide in two major sweeps across Europe and Asia was spread originally by fleas. But there was a form of the disease that spread pneumonically, or in the air from one person to another.

5. Frame (1992), Part I, op. cit.

6. S. M. Buckley, J. Casals, and W. Downs, "Isolation and Antigenic Characterization of Lassa Virus," *Nature* 227 (1970): 174.

7. R. W. Speir et al., "Electron Microscopy of Vero Cell Cultures Infected with Lassa Virus," *American Journal of Tropical Medicine and Hygiene* 19 (1970): 692–94.

8. E. Leifer, D. J. Locke, and H. Bourne, "Report of a Laboratory-Acquired Infection Treated with Plasma from a Person Recently Recovered from the Disease," *American Journal of Tropical Medicine and Hygiene* 19 (1970): 677–79.

9. Frame (1992), Part I, op. cit.

10. F. P. Pinheiro et al., "Amapari, a New Virus of the Tacaribe Group from Rodents and Mites of Amapa Territory, Brazil," *Proceedings of the Society of Experimental and Biological Medicine* 122 (1966): 531–35.

11. J. Casals et al., "A Current Appraisal of Hemorrhagic Fevers in the U.S.S.R.," *American Journal of Tropical Medicine and Hygiene* 15 (1966): 751–64.

12. Buckley et al. (1970), op. cit.

13. S. M. Buckley and J. Casals, "Isolation and Characterization of the Virus," *American Journal of Tropical Medicine and Hygiene* 19 (1970): 680–91; and Speir et al. (1970) op. cit.

14. The CDC has, over the years, upgraded its high-security laboratories repeatedly as the technology of biological containment improved. In the late 1960s the CDC's top security facility housed labs-within-labs that formed a sort of concrete onion, layer upon layer of which had to be penetrated to reach the central core. At each level double air-lock doors sealed the air space tightly, and the chambers were kept under pressure that directed all air—and microbes—toward special ventilators where they were destroyed by ultraviolet light and filtered through several layers of sheets that strained out anything bigger than a large molecule. All personnel showered before and after entry and wore protective head-to-toe suits.

15. Very detailed renditions of these events can be found in John Fuller's *Fever* and John Frame's "The Story of Lassa Fever. Part II: Learning More About the Disease." Both are cited above.

16. G. M. Edington, and H. A. White, "The Pathology of Lassa Fever," *Transactions of the Royal Society of Tropical Medicine and Hygiene* 66 (1972): 381–401.

17. J. M. Troup et al., "An Outbreak of Lassa Fever on the Jos Plateau, Nigeria, in January–February, 1970," *American Journal of Tropical Medicine and Hygiene* 19 (1970): 695–96.

18. D. E. Carey et al., "Lassa Fever—Epidemiological Aspects of the 1970 Epidemic, Jos, Nigeria," *Transactions of the Royal Society of Tropical Medicine and Hygiene* 66 (1972): 402–8.

19. B. E. Henderson et al., "Lassa Fever: Virological and Serological Studies," *Transactions of the Royal Society of Tropical Medicine and Hygiene* 66 (1972): 409–16.

20. D. Cummins et al., "Acute Sensorineural Deafness in Lassa Fever," *Journal of the American Medical Association* 264 (1990): 2093–96; L. P. Ryback, "Deafness Associated with Lassa Fever," *Journal of the American Medical Association* 264 (1990): 2119; J. D. Frame, "Clinical Features of Lassa Fever in Liberia," *Review of Infectious Diseases* 11, Suppl. 4 (1989): s783–s789; and J. B. McCormick et al., "A Case-Control Study of the Clinical Diagnosis and Course of Lassa Fever," *Journal of Infectious Diseases* 155 (1987): 445–55.

21. J. B. Dibble, *Outlaw for God* (Hanover, MA: Christopher Publishing House, 1992).

22. Frame (1992), Part III, op. cit.

23. It is unfortunate that in historic outbreaks, such as this one, the names of non-Western disease victims and medical staff often go unrecorded in published literature. Whenever possible, the author has endeavored to include such names.

24. P. E. Mertens et al., "Clinical Presentation of Lassa Fever Cases During the Hospital Epidemic at Zorzor, Liberia, March–April 1972," *American Journal of Tropical Medicine and Hygiene* 22 (1973): 780–84.

25. Ibid.

26. The team's predictions proved quite accurate. Lassa is now known to be endemic to many regions of Liberia, and treatment of the disease had, prior to its most recent civil war, become a nearly routine feature of the country's medicine.

27. In 1867 English physician Lister revolutionized medicine by publishing and promoting details on sterile hospital practices. Prior to Lister, patients were extremely likely to die of bacterial infections following surgery.

28. T. P. Monath et al., "Lassa Virus Isolation from *Mastomys natalensis* Rodents During an Epidemic in Sierra Leone," *Science* 185 (1974): 263–65.

29. D. W. Fraser et al., "Lassa Fever in the Eastern Province of Sierra Leone, 1970–1972. I: Epidemiological Studies," *American Journal of Tropical Medicine and Hygiene* 23 (1974): 1131–39; and T. P. Monath et al., "Lassa Fever in the Eastern Province of Sierra Leone, 1970–1972. II: Clinical Observations and Virological Studies on Selected Hospital Cases," *American Journal of Tropical Medicine and Hygiene* 23 (1974): 1140–49.

30. Monath et al. (1974), "Lassa Fever in the Eastern Province."

31. Nevertheless, Casals and Pinneo both donated blood countless times over subsequent years, and CDC antisera made from their plasma undoubtedly saved many lives, particularly in Nigeria. Nigerian physicians and nurses who survived Lassa fever also generously donated their blood over the years, providing a steady supply of antiserum specific to the Jos strain.

32. Upper Volta is now called Burkina Faso.

33. J. D. Frame, "Surveillance of Lassa Fever in Missionaries Stationed in West Africa," *Bulletin of the World Health Organization* 52 (1975): 593–98.

34. The April 9, 1974, edition of the *New Nigerian*, published out of Lagos, carried a headline: "How Alien Tried to Spread Killer Disease—Lassa Epidemic Narrowly Avoided."

35. *Diepahtzer Nachrichter*, March 18, 1974: 1.

36. "This Doctor Risks His Life for a Deadly Ill Patient," *Bild Zeitung*, March 18, 1974.

37. This is the same Dr. Lehmann-Gruber described in Chapter 3 as playing a key role in the battle against the Marburg virus.

38. "Is the Fever Physician Dangerous for Us?" *Bild-Hamburg*, March 22, 1974.

39. "U.S. Foreign Minister Kissinger Will Save the Courageous Fever Doctor from Hamburg," *Bild Zeitung*, March 19, 1974.

40. Had Monath's final conclusions on the Sierra Leone outbreak been published before these events took place, the group's fears might have eased. Monath and Casals found that the terrifying fatality rates seen in the hospital outbreaks of Jos and Zorzor were atypical. About half of all infected people in those cases died. But surveys of the Sierra Leone villages showed that most people survived the disease: only 5 percent of those who were infected by inhaling the virus from another person or rat feces/urine developed fatal disease. Monath was convinced direct blood-to-blood exposure to the virus was more dangerous, and might account for the far higher death rates in hospital settings.

41. K. Muller, "Flug in einer Plastikfolie," *Welt*, March 22, 1974.

42. "Ghostly Arrival," *Hamburger Abendblatt*, March 22, 1974.

43. *Hamburger Morgenpost*, March 19, 1974: 1.

44. H. Nannen, "One Can See How One May Err," *Stern*, March 28, 1974: 3.

45. *New Nigerian* (1974), op. cit. Years later the Nigerian government and press positions concerning Lassa would change, as outbreaks of the disease continued. Though there would continue to be regret that the virus had been named after a Nigerian town, the press and government would no longer claim the disease was carried by foreigners. For good examples of later Nigerian press accounts, see A. Adenmosum, "The Lassa-Fever Scourge," *Pharmacy World Journal* 6 (1989): 95; and U. Fabian, "Combating Lassa Fever," *Daily Sketch*, February 23, 1993: 5.

5. Yambuku

1. In addition to interviews, the author drew generally from the following written materials: P. Sureau, *Yambuku* (1992), unpublished; and S. R. Pattyn, *Ebola Virus Haemorrhagic Fever* (Amsterdam: Elsevier Press, 1978).

2. International Commission, "Ebola Haemorrhagic Fever in Zaire, 1976," *Bulletin of the World Health Organization* 56 (1978): 271–93.

3. R. D. Estes, *The Behavior Guide to African Mammals* (Berkeley: University of California Press, 1991).

4. The Belgian Catholic missionaries who worked in the Yambuku Mission Hospital were:

Sister Béata	(Jeanne Vertommen, age 42)
Sister Myriam	(Louise Ecran, age 42)
Sister Edmonda	(Jeanne DeRoover, age 56)
Sister Romana	(Angeline Geerts, age 54)
Father Germain Lootens	(age 67)

In addition to the Belgians employed in hospital work, there were in Yambuku:

Sister Genoveva	(Annie Ghysebrechts)
Sister Mariette	(Mariette Witrouwen)
Mother Superior Aloydis	(Bumba Zone)
Sister Marcella	(Julienne Ronsmans)
Father Carlos Rommel	(district head for Bumba Zone)
Father Germain Moke	(Bumba Zone)
Brother Adolphe Litepu	
Father Augustin Siegers	(superior for Yambuku)
Father Léon Claes	(aka "Père Dubois")

The Zairian staff included seven male nurse assistants:

Amane Ehumba	(age 26)
Anangi Mabelesama	(age 25)
Mgwanga Magbuka	(age 30)
Ambaya Mandu	(age 30)
Sukato Manzomba	(age 24)
Magbo Mazebele	(age 35)
Masangaya Alola	(age 40)

5. It was common for people to bury their dead inside the village, even under the clay floor of their houses. Ngoi feared such practices could spread the disease.

6. Untreated rabies ranks as number one: it is 100 percent fatal. Many experts believe HIV/AIDS has a lethality that approaches 100 percent, but, because of the extremely protracted course of HIV disease and the difficulty in knowing how many people are asymptomatic HIV carriers, it is impossible to say with certainty what percentage of HIV cases are fatal.

7. Dr. Close's daughter, the movie actress Glenn Close, reportedly hopes to make a Hollywood film based on the Zairian disease story. William Close has written a novelized version of these events, entitled *Yambuku: A Novel*. It is, as yet, published only in Dutch. The author also drew comments from W. T. Close, "A Medical Perspective from Africa and the Rocky Mountains," presentation to the University of California Medical School at Davis, 1993.

8. World Health Organization, "Viral Haemorrhagic Fever," *Weekly Epidemiological Report* 52 (1977): 177–84; and WHO Expert Committee, "Viral Haemorrhagic Fevers," *Technological Report Series* (1985), No. 721.

9. Isolated by Guido van der Gröen and Peter Piot.

10. Translated from the French by the author.

11. CDC memo, October 14, 1976, signed by Karl Johnson and issued on that date internally; several days later sent to WHO in Geneva and the U.S. Embassy in Kinshasa.

12. G. A. Martini and R. Siegert, *Marburg Virus Disease* (Frankfurt: Springer Verlag, 1971).

13. Despite deliberate attempts to render press accounts banal, some reports swung hyperbolically in the opposite direction. The October 28 issue of *L'Express* claimed: "Huit cents morts en un mois au Soudan. Responsable: le virus de Marburg. Un des trois cents 'arbovirus' de l'Afrique Centrale contre lesquelle la médecine n'a pas d'armes." And the highly exaggerated account concluded by asking:

"Que se passera-t-il le jour ou des voyageurs commenceront à exporter la virus? D'ores et déjà, l'Afrique du Sud a annoncé que des mesures sanitaires impitoyables ont été prises aux frontières. Mais l'extension des transports aériens permet-elle de contrôler efficacement l'état de santé de tous les voyageurs circulant dans les régions infesteés par ces maladies méconnues, venues du fond des temps? La civilisation industrielle a commencé d'unifier la planète. Les philosophes s'en réjouissent autant que les hommes d'affaires. Les médecins, eux, ne cachent pas qu'ils y a aussi des raisons de s'inquiéter."

14. To this day members of the CDC team refuse to name the American individual. As McCormick puts it, "You can't blame a guy for being scared as hell. And I respect that he had the guts to be honest about it. It would have been a lot worse if he tried to be macho about it and came to Zaire, and then was an impediment to our work because he was so frightened that we couldn't count on him."

15. The species gathered for analysis in the Yambuku area were:

Bedbugs, *Cimex hemipterus*	818	samples
Mosquitoes, *Culex cinereus*	8	
Culex pipiens	2	
Mansonia africana	5	
Pigs, domestic	10	
Cows, domestic	1	
Bats, species unidentified	7	
Rodents, *Mastomys*	69	
Rattus rattus	30	
Squirrels	8	
Monkeys, *Cercopithecus*	6	
Duikers, *Cephalophus monticola*	2	

None of the samples was infected with the Ebola virus.

16. That day Sureau summarized his feelings after a month in the eye of the epidemic storm. "There it is, the virus responsible for the Yambuku hemorrhagic fever is baptized: 'Ebola'! I would have preferred 'Yambuku' but Karl prefers to give the names of rivers where the virus is discovered: he called 'Machupo' after the name of a nearby river, the virus that caused hemorrhagic fever in Bolivia. . . . But no matter; we don't ask genius to be right; we ask it to be provocative," Sureau wrote in his diary.

17. A clinical account of Platt's case appears in: R. T. D. Emond, et al., "A Case of Ebola Virus Infection," *British Medical Journal* 2 (1977): 541–44.

18. "The disease was hitherto unknown to the people of the affected region. Intensive search for cases in the area of northeastern Zaire between the Bumba Zone and the Sudan frontier near N'zara and Maridi failed to detect definite evidence of a link between an epidemic of the disease in that country and the outbreak near Bumba. Nevertheless, it was established that people can and do make the trip between N'zara and Bumba in not more than four days: thus it was regarded as quite possible that an infected person had travelled from Sudan to Yambuku and transferred the virus to a needle of the hospital while receiving an infection at the outpatient clinic," stated the official report of an International Commission. See World Health Organization, "Ebola Haemorrhagic Fever in Zaire, 1976" (1978), op. cit.

19. N. J. Cox, J. B. McCormick, K. M. Johnson, and M. P. Kiley, "Evidence for Two Subtypes of Ebola Virus Based on Oligonucleotide Mapping of RNA," *Journal of Infectious Diseases* 147 (1983): 272–75.

20. J. G. Breman et al., "Human Monkeypox Update 1970–79," *Bulletin of the World Health Organization* 58 (1980): 849–68.

21. I. Arita et al., "Human Monkeypox: A Newly Emerged Orthopoxvirus Zoonosis in the Tropical Rain Forests of Africa," *American Journal of Tropical Medicine* 34 (1985): 781–89.

22. Z. Jezek et al., "Human Monkeypox: A Study of 2,510 Contacts of 214 Patients," *Journal of Infectious Diseases* 154 (1986): 551–55.

23. L. Khodakevich, Z. Jezek, and D. Messinger, "Monkeypox Virus: Ecology and Public Health Significance," *Bulletin of the World Health Organization* 66 (1988): 747–52.

24. P. E. M. Fine et al., "The Transmission Potential of Monkeypox Virus in Human Populations," *International Journal of Epidemiology* 17 (1988): 643–50.

25. World Health Organization, *Weekly Epidemiologic Record* (1977), op. cit.

26. WHO Expert Committee, "Viral Haemorrhagic Fevers," *WHO Technical Report Series* 721 (1985).

27. International Commission (1978), op. cit.

28. The brief official CDC summaries of Yambuku and Maridi events appeared in the agency's *Morbidity and Mortality Weekly Report*: October 15, 1976 (p. 324); October 26, 1976 (p. 339); November 5, 1976 (p. 342); December 3, 1976 (pp. 378–79); and July 1, 1977 (pp. 209–10).

6. The American Bicentennial

1. A great deal has been written about the Swine Flu crisis of 1976, and though interpretations of events differ markedly, the basic facts are fairly consistent. Key sources of such information include: P. M. Boffey, "Anatomy of a Decision: How the Nation Declared War on Swine Flu," *Science* 192 (1976): 636–41; P. M. Boffey, "Swine Flu Vaccination Campaign: The Scientific Controversy Mounts," *Science* 193 (1976): 559–63; P. M. Boffey, "Swine Flu Vaccine: A Component Is Missing," *Science* 193 (1976): 1224–25; P. M. Boffey, "Guillain-Barré: Rare Disease Paralyzes Swine Flu Campaign," *Science* 195 (1976): 155–59; W. Dowdle and J. LaPatra, *Informed Consent: Influenza Facts and Myths* (Chicago: Nelson-Hall, 1983); General Accounting Office, *The Swine Flu Program: An Unprecedented Venture in Preventive Medicine*, report to the U.S. Congress, Washington, D.C., 1977; R. M. Henig, "Flu Pandemic, a Once and Future Menace," *The New York Times Magazine*, November 29, 1992: 28–31, 55–67; R. M. Henig, "The Emergence of a New Flu," Chapter 7 in R. M. Henig, *A Dancing Matrix* (New York: Alfred A. Knopf, 1993); R. E. Neustadt and H. V. Fineberg, *The Swine Flu Affair: Decision-Making on a Slippery Slope* (Washington, D.C.: U.S. Department of Health, Education, and Welfare, 1978); R. E. Neustadt and H. V. Fineberg, *The Epidemic That Never Was: Policy-Making and the Swine Flu Affair* (New York: Random House, 1982); J. E. Osborn, ed., *Influenza in America 1918–1976* (New York: Prodist, 1977); A. M. Silverstein, *Pure Politics and Impure Science: The Swine Flu Affair* (Baltimore: Johns Hopkins University Press, 1981); and Swine Flu Supplement, *Journal of Infectious Diseases* 136 (December 1977).

2. According to base officials, windchills at Fort Dix during January 1976 ranged from 0°F to −45°F, and snowfall was heavy and frequent.

3. Centers for Disease Control, "Current Trends: Influenza—United States," *Morbidity and Mortality Weekly Report* 25 (1976): 47–48.

4. W. I. B. Beveridge, "The Chronicle of Influenza Epidemics," *History and Philosophy of Life Sciences* 13 (1991): 223–35.

Influenza is an ancient microbe that has appeared in millions of different forms over the millennia, periodically producing devastating epidemics. Some outbreaks caused debilitating illnesses in well over half the populations of vast areas, such as all of continental Europe. Only a few epidemics are thought to have also produced mass mortality, claiming the lives of over 5 percent of those infected with the influenza virus. A rough estimate of historic pandemics follows:

Date	Probable Origin	Eventual Scope	Estimated Mortality
1173	Unknown	Europe	—
1387	Unknown	Europe	—
1510	Unknown	Europe	Serious
1557	Unknown	Europe	Serious

		Unknown		Europe		Serious
._-30		Sweden?		Europe, global		"True pandemic"
• 1732–33	†	Connecticut or Moscow	†	Global	‡	High morbidity, Low mortality
1742–43	†	Germany, Switzerland		Europe	†	Moderate
1761–62	‡ †	N. America or Breslau in Silesia		N. America, Caribbean, Europe	†	True pandemic
1767	†	New England?		N. America and continental Europe		Moderate
1775–76		Germany, Austria		Europe, Asia	†	Moderate
• 1781–82	‡	China or † Siberia		Asia, Europe, N. America— global	†	True pandemic Extremely high morbidity, moderate mortality
1788–89	†	St. Petersburg, Russia		N. America, Europe		Low
• 1800–02		Moscow, Russia		Europe, China, Brazil	†	Moderate
• 1830–33		Canton, China		Global	†	True pandemic, low mortality
1836–37		Sydney, Australia, or recurrent 1830 epidemic		Europe, Africa, Australia		True pandemic, high mortality
• 1847–48		Moscow, Russia		Europe, Brazil, Caribbean	†	True pandemic, high mortality
1850–51	‡	Recurrent 1847 epidemic		The Americas, Caribbean, Australia, and Germany	†	Mild
• 1857–58		Panama		Europe and the Americas	†	Mild
1873–75		—		N. America and continental Europe		Mild
• 1889–90	‡	Bukhara, Russia		Global	†	True pandemic, severe, at least 250,000 deaths in Europe alone
• 1918–19	‡	?China or U.S.A.		Global		15 to 25 million *21,640,000 **100,000,000
1946		Australia		Global		
• 1957	‡	Kweichow, China		Global		Major
• 1968–70		Hong Kong		Global		Major pandemic
1977		China		Global		

The above was compiled from several sources, including:

** F. M. Burnet, "Portraits of Viruses: Influenza Virus A," *Intervirology* 11 (1979): 203.

† W. I. B. Beveridge, *Influenza: The Last Great Plague* (London: Heinemann, 1977).

* Musée de Condé, "Les Très Riches Heures du Duc de Berry" (Chantilly, France).

‡ D. K. Patterson, *Pandemic Influenza 1700–1900* (Totowa, NJ: Rowman and Littlefield, 1986).

• Pandemics which were, according to the historic record, particularly massive, affecting millions of people.

5. Beveridge (1991), op. cit., note 1.

6. It may well be the case that the AIDS epidemic will by 1999 surpass the horrible toll taken by influenza in 1918–19, making AIDS the worst pandemic of the twentieth century. In early 1993, the World Health Organization forecast a cumulative total of 40 million HIV infections by the end of 1999, and 8 million cumulative AIDS deaths. A more frightening projection from the Harvard-based Global AIDS Policy Coalition predicts up to 120 million cumulative HIV infections by that time and 20.4 million deaths. Regardless of the horrible eventual toll of AIDS, however, it will undoubtedly remain the case that the influenza pandemic merits the gruesome ranking of the number one global killer in this century, as the bulk of the planet's HIV-infected population will not die until after the millennium.

7. Beveridge (1977), op. cit.

8. D. Scott, *Epidemic Disease in Ghana 1901–1960* (London: Oxford University Press, 1965).

9. A. W. Crosby, "The Influenza Pandemic of 1918," in Osborn (1977), op. cit., pp. 5–13.

10. Excellent anecdotal accounts of the 1918 epidemic can be found in A. A. Hoehling, *The Great Epidemic* (Boston: Little, Brown, 1961); and A. W. Crosby, *Epidemic and Peace, 1918* (London: Greenwood Press, 1976).

11. R. E. Shope, "Swine Influenza," *Harvey Lectures* 31 (1936): 183–213.

12. The CDC's *Morbidity and Mortality Weekly Report* chronicles the spread of A/Victoria/75 week after week. The week Private Lewis fell ill at Fort Dix, for example, there were A/Victoria/75 outbreaks in Johannesburg, the United Kingdom, and the American states of Arizona, Washington, Iowa, Minnesota, Massachusetts, New York, and New Jersey. See *MMWR* 25 (January 24, 1976): 23–24.

13. H. M. Rose, "Influenza: The Agent," *Hospital Practice*, August 1971: 49–56.

14. A. D. Langmuir, "Influenza: Its Epidemiology," *Hospital Practice*, September 1971: 103–8.

15. E. D. Kilbourne, "An Explanation of the Interpandemic Antigenic Mutability of Influenza Viruses," *Journal of Infectious Diseases* 128 (1973): 668–70. For an excellent detailing of widely accepted influenza theories of the day, see Edwin Kilbourne's *Influenza* (New York: Plenum, 1987).

16. E. D. Kilbourne, *New York Times*, February 13, 1976: A33.

17. N. Masurel and W. M. Marine, "Recycling of Asian and Hong Kong Influenza A Virus Hemagglutinins in Man," *Scientific American* 97 (1973): 48–49.

18. Rose (1971), op. cit.

19. Beveridge (1977), op. cit.

20. The proceedings of the Rougemont gathering were later published by the Sandoz Institute for Health and Socio-Economic Studies, *Influenza: Virus Vaccines and Strategy*, ed. Philip Selby (New York: Academic Press, 1976).

21. In 1980 over 100 wild seals washed ashore around Boston and Cape Cod, victims of influenza. Webster and his colleagues studied the viral RNA, matching it to influenza strains then common among ducks and sea gulls. Three years later, influenza, apparently spread by migratory bird droppings, broke out in Pennsylvania in a massive commercial chicken house. Every single chicken died. From "Influenza," in S. S. Morse, ed., *Emerging Viruses* (Oxford, Eng.: Oxford University Press, 1993, R. G. Webster, 37–45.

22. During a 1993 interview, Sencer took a deep breath when the subject of Swine Flu was broached, smiled, and discussed an old *Peanuts* cartoon strip in which the pooch is sitting atop his doghouse typing a manuscript entitled "Swine Flu." The first words of the manuscript read, "It was a dark and stormy night . . ."

23. Centers for Disease Control, "Current Trends: Influenza—United States," *Morbidity and Mortality Weekly Report* 25 (1976): 47–48.

24. Ibid., 55–56.

25. Ibid., 124.

26. Dowdle and LaPatra (1983), op. cit.

27. A. S. Beare and J. W. Craig, "Virulence for Man of a Human Influenza-A Virus Antigenically Similar to 'Classical' Swine Viruses," *Lancet*, July 3, 1976, 4–5.

28. C. Stuart-Harris, "Swine Influenza Virus in Man—Zoonosis or Human Pandemic?" *Lancet*, July 3, 1976: 31–32.

29. "Planning for Pandemics," *Lancet*, July 3, 1976: 25–26.

30. Centers for Disease Control, "Influenza Vaccine—Supplemental Statement," *Morbidity and Mortality Weekly Report* 25 (1976): 221–27; and Boffey, "Swine Flu Vaccination Campaign" (1976), op. cit.

31. There are several sources for valuable insights into the 1976–77 Legionnaires' Disease investigation. Particularly useful are *Annals of Internal Medicine* 90, No. 4 (1979) (special issue devoted to Legionnaires' Disease); Centers for Disease Control, *Legionnaires': The Disease, the Bacterium and Methodology*, (Atlanta: U.S. Department of Health, Education, and Welfare, 1979); F. W. Chandler, M. D. Hicklin, and J. A. Blackmon, "Demonstration of the Agent of Legionnaires' Disease in Tissue," *New England Journal of Medicine* 297 (1977): 1218–20; D. W. Fraser et al., "Legionnaires' Disease: Description of an Epidemic of Pneumonia," *New England Journal of Medicine* 297 (1977): 1189–97; G. L. Lattimer and R. A. Ormsbee, *Legionnaires' Disease* (New York: Marcel Dekker, 1981); J. E. McDade et al., "Legionnaires' Disease: Isolation of a Bacterium and Demonstration of Its Role in Other Respiratory Disease," *New England Journal of Medicine* 297 (1977): 1197–1203; and Silverstein (1981), op. cit., Chapter 10.

32. Some of the most important features of Public Law 94-380 are as follows:

Informed Consent. The law required that the National Commission for the Protection of Human Subjects of Biomedical and Behavioral Research draft "a written informed consent form and procedures for assuring that the risks and benefits from the swine flu vaccine are fully explained to each individual to whom such vaccine is to be administered. Such consultation shall be completed within two weeks after enactment of this Act, or by September 1, 1976, whichever is sooner. Such procedures shall include the information necessary to advise individuals with respect to their rights and remedies arising out of the administration of such vaccine."

Pharmaceutical Industry Profiteering. "Any contract for procurement by the United States of swine flu vaccine from a manufacturer of such vaccine shall . . . be subject to renegotiation to eliminate any profit realized from such procurement . . . as determined pursuant to criteria prescribed . . . and the contract shall expressly so provide."

Litigation. ". . . in order to be prepared to meet the potential emergency of a swine flu epidemic, it is necessary that a procedure be instituted for the handling of claims by persons alleging such injury or death until Congress develops a permanent approach for handling claims arising under programs of the Public Health Service Act."

"The Attorney General shall defend any civil action or proceeding brought in any court against any employee of the Government . . . or program participant . . . based upon a claim alleging personal injury or death arising out of the administration of vaccine under the swine flu program."

Timing. The bill did not take effect until September 30, 1976. Prior to that date, pharmaceutical manufacturers, or their insurers, carried full liability.

33. For details, see the CDC's *Morbidity and Mortality Weekly Report* (September 3, 1976): 270–76.

34. Crosby (1976), op. cit.

35. Eighteen months later, a medical journal, when referring to the Philadelphia outbreak, opened with the following florid language: "The explosive outbreak of acute respiratory disease that occurred in Philadelphia during the summer of 1976 was both mysterious and terrifying. Now, after a year and one half of painstaking investigation, much of the mystery and some of the terror are beginning to be dispelled." (From "Legionnaires' Disease: An Unfolding Riddle," *Hospital Practice,* February 29 [sic], 1978: 24–25.) The above article was followed by an editorial comment that noted: "Along with those of Lassa fever and Ebola-virus disease, the organism of Legionnaires' disease takes its place as one of the newly recognized infectious agents capable of causing severe, life-threatening illness in man."

36. Committee on Interstate and Foreign Commerce, House of Representatives Proceedings, hearing before the Subcommittee on Consumer Protection and Finance, 94th Congress, November 23–24 (Serial No. 94-159) (Washington, D.C.: U.S. Government Printing Office).

37. "U.S. Center Assailed on 'Legion' Disease," *New York Times,* October 29, 1976: A1.

38. Silverstein (1981), op. cit.

39. An excellent analysis of press coverage during October 1976 can be found in D. M. Rubin and V. Hendy, "Swine Influenza and the News Media," *Annals of Internal Medicine* 87 (1977): 769–74.

40. H. F. Retailliau et al., "Illness After Influenza Vaccination Reported Through a Nationwide Surveillance System, 1976–1977," *American Journal of Epidemiology* 111 (1980): 270–78.

41. Centers for Disease Control, "Influenza—Worldwide," *Morbidity and Mortality Weekly Report* 25 (1976): 331.

42. Centers for Disease Control, "Influenza Surveillance—United States," *Morbidity and Mortality Weekly Report* 25 (1976): 391–92.

43. Rubin and Hendy (1977), op. cit.

44. Centers for Disease Control, "Guillain-Barré Syndrome—United States," *Morbidity and Mortality Weekly Report* 25 (1976): 401–2.

45. J. L. Reismann and B. Singh, "Conversion Reactions Simulating Guillain-Barré Paralysis Following Suspension of the Swine Flu Vaccination Program in the U.S.A.," *Australian and New Zealand Journal of Psychiatry* 12 (1978): 127–32.

46. L. B. Schonberger et al., "Guillain-Barré Syndrome Following Vaccination in the National Influenza Immunization Program, United States, 1976–1977," *American Journal of Epidemiology* 110 (1979): 105–23.

Fifteen years later, CDC researchers not directly involved in the events of 1976 reexamined all Guillain-Barré-associated medical records for 1976–77 in Minnesota and Michigan, submitting patient data to panels of neurologists for reanalysis. This was done because doubts still lingered—doubts fueled by the fact that 1.7 million military personnel were vaccinated in 1976 without a single resulting case of the syndrome. They concluded that some cases had been misdiagnosed in 1976, but Guillain-Barré was still clearly linked to the vaccine. Vaccine recipients in those two states were more than seven times more likely to suffer the syndrome. The study can be found in T. J. Safranek et al., "Reassessment of the Association Between Guillain-Barré Syndrome and Receipt of Swine Influenza Vaccine in 1976–1977: Results of a Two-State Study," *American Journal of Epidemiology* 133 (1991): 940–51.

47. J. Axelrod, personal communication, 1993. A total of $2.4 billion was sought in lawsuits, but less than $48 million was paid out as a result of litigation; a track record of which Axelrod is quite proud. The precise settlement payout was $92,833,020.

48. For details on the lasting impact of Swine Flu and other vaccine litigation upon research and development, see Institute of Medicine, *The Children's Vaccine Initiative*, V. S. Mitchell, N. M. Philipose, and J. P. Sanford, eds., (Washington, D.C.: National Academy Press, 1993).

49. Communicable Disease Center, "Institutional Outbreak of Pneumonia," *Morbidity and Mortality Weekly Report* 14 (1965): 265–66. S. B. Thacker et al., "An Outbreak in 1965 of Severe Respiratory Illness Caused by the Legionnaires' Disease Bacterium," *Journal of Infectious Diseases* 138 (1978): 512–19.

50. Centers for Disease Control, "Follow-up on Respiratory Illness—Philadelphia," *Morbidity and Mortality Weekly Report* 26 (1977): 9–11.

51. Osborn (1977), op. cit.

52. Neustadt and Fineberg (1982), op. cit.

53. M. Goldfield et al., "Influenza in New Jersey in 1976: Isolations of Influenza A/New Jersey/ 76 Virus at Fort Dix," *Journal of Infectious Diseases* 136 (1977): S347–S355.

54. H. F. Top, Jr., and P. K. Russell, "Swine Influenza A at Fort Dix, New Jersey (January– February 1976). IV: Summary and Speculation," *Journal of Infectious Diseases* 136 (1977): S376– S380.

55. Dowdle and LaPatra (1985), op. cit., 88–89.

56. Centers for Disease Control, *Laboratory Manual: "Legionnaires',*" *the Disease, the Bacterium and Methodology* (Washington, D.C.: Department of Health, Education, and Welfare, 1979).

57. Centers for Disease Control, "Legionnaires' Disease—United States," *Morbidity and Mortality Weekly Report* 26 (1977): 300.

58. Centers for Disease Control, "Follow-up on Legionnaires' Disease—Ohio," *Morbidity and Mortality Weekly Report* 26 (1977): 308; and J. S. Marks et al., "Nosocomial Legionnaires' Disease in Columbus, Ohio," *Annals of Internal Medicine* 90 (1979): 565–69.

59. Centers for Disease Control, "Follow-up on Legionnaires' Disease—Vermont," *Morbidity and Mortality Weekly Report* 26 (1977): 328.

60. Centers for Disease Control, "Legionnaires' Disease—Tennessee, Vermont," *Morbidity and Mortality Weekly Report* 26 (1977): 336; and C. V. Broome et al., "The Vermont Epidemic of Legionnaires' Disease," *Annals of Internal Medicine* 90 (1979): 573–77.

61. Centers for Disease Control, "Sporadic Cases of Legionnaires' Disease—United States," *Morbidity and Mortality Weekly Report* 26 (1977): 388; and Centers for Disease Control, "Follow-up on Legionnaires' Disease—United States," *Morbidity and Mortality Weekly Report* 26 (1977): 443.

62. Centers for Disease Control, "Legionnaires' Disease—England," *Morbidity and Mortality Weekly Report* 26 (1977): 391.

63. "Hospital Haunted by Legionnaires' Disease," *New York Times*, November 13, 1978: A1; and C. E. Haley et al., "Nosocomial Legionnaires' Disease: A Continuing Common-Source Epidemic in Wadsworth Medical Center," *Annals of Internal Medicine* 90 (1979): 583–86.

64. W. L. L. Wang et al., "Growth, Survival, and Resistance of the Legionnaires' Disease Bacterium," *Annals of Internal Medicine* 90 (1978): 614–18.

65. R. P. Hudson, "Lessons from Legionnaires' Disease," *Annals of Internal Medicine* 90 (1978): 704–7.

7. N'zara

1. After the fall of the Soviet communist state, documents pertaining to the former regime's biological weapons practices slowly came to light, revealing that McCormick and the CDC were justified in suspecting the intentions of some Soviet Lassa researchers, as well as the efforts of their counterparts working on other infectious diseases. Well after 1972 treaty agreements between the U.S.S.R. and the U.S.A. allegedly suspended all such research and development, the Soviets continued trying to develop weapons based on Legionnaires' Disease, anthrax, bubonic plague, tularemia, Lassa fever, and a variety of other diseases. The effort continued into the 1980s, despite an accidental leakage of anthrax from the government's Sverdlovsk laboratory in 1979, and involved over 25,000 scientists toiling in eighteen laboratories dispersed throughout the U.S.S.R. A summary of the findings revealed in newly released Soviet documents appeared in *Newsweek*, February 1, 1993: 40–41.

2. The biological weapons treaty between the United States and the U.S.S.R. was signed by the two parties on April 10, 1972, in London, ratified by the U.S. Congress on December 16, 1974, and proclaimed as law by President Gerald Ford on March 26, 1975. Formally titled the "Convention on the Prohibition of the Development, Production, and Stockpiling of Bacteriological (Biological) and Toxin Weapons and on their Destruction," the treaty had been signed by 125 nations as of January

1993. Article X of the Convention states that all parties to the treaty shall facilitate "the fullest possible exchange of equipment, materials, and scientific and technological information for the use of bacteriological (biological) agents and toxins for peaceful purposes."

3. Throughout the late 1970s and early 1980s Guido van der Gröen followed the Biological Weapons Convention quite literally, personally working inside the top security laboratories on both sides of the Iron Curtain and freely distributing samples of dangerous viruses—including Ebola and Lassa—to scientists in Moscow, Leningrad, Paris, London, Porton Down, Fort Detrick, Atlanta, and anywhere else he went. The Belgian reasoned that "science knows no politics," and told curious intelligence officers from the CIA, Sûreté, Interpol, KGB, and MI5 who repeatedly grilled him that, as a scientist, he simply refused to acknowledge the Cold War.

4. J. B. McCormick, *Lassa Fever Update*, SME/80.5 (Geneva: World Health Organization, 1980).

5. "For a three-year study in Sierra Leone, it became evident that in many villages a substantial portion of the population was infected with Lassa during their lifetime, beginning at a young age. It appears that people in many of these villages live surrounded by virus-infected rodents which are excreting large amounts of virus in the houses both day and night, since many of the houses are closed during the day, creating a nearly twenty-four-hour nocturnal environment. This allows rodents to move around depositing urine at will in many places such as beds, floors, food supplies, etc. Transmission to humans did occur throughout the year. There is a general correlation between antibody prevalence in humans and percent of *Mastomys* found in the village and the proportion excreting virus. Infection may be frequent in both susceptible persons as well as persons with antibody titers who appeared to be boosted by these infections. Human-to-human infection also occurs and some evidence suggests that it may constitute an important source of human cases." G. van der Gröen and J. B. McCormick, *Expert Committee on Viral Haemorrhagic Fevers, Agenda Item 2.3*, VIR/VHF/EC/84.13 (Geneva: World Health Organization, 1984).

6. D. Grigg, *The World Food Problem* (Oxford, Eng.: Basil Blackwell, 1985).

7. T. McKeown, *The Origins of Human Disease* (Oxford, Eng: Basil Blackwell, 1988).

8. F. Moore Lappé and J. Collins, *World Hunger: Ten Myths* (4th ed.; San Francisco: Institute for Food and Development Policy, 1979).

9. In the spring of 1993 the Russian government confirmed that nearly all Soviet heath statistics released during the communist era, and possibly during the prior czarist regime, were "artificially generated." No data could be considered reliable if released before December 1992, according to the Yeltsin government.

10. World Bank, *Annual Report* (Washington, D.C., 1978).

11. World Bank, *World Development Report* (Washington, D.C., 1978).

12. "Carter en Route to Africa," *Africa*, December 1977.

13. J. J. Gilligan, "America's Stake in the Developing World," *U.S. Department of State Bulletin* 77 (1977): 687–91.

14. World Bank, *Health Sector Policy Paper* (Washington, D.C., 1980).

15. Ibid.

16. Among the key water-related diseases are:

Disease	Pathogen (vector)
Amoebic dysentery	protozoa
Ascariasis	helminth
Bacillary dysentery	bacteria
Cholera	bacteria
Clonorchiasis	helminth (snail, fish)
Diarrheal disease	miscellaneous
Diphyllobothriasis	helminth (copepod, fish)
Dracunculiasis	helminth
Enteroviruses	virus
Fasciolopsiasis	helminth (snail, plant)
Gastroenteritis	miscellaneous
Infectious hepatitis	virus
Leptospirosis	spirochete
Paragonimiasis	helminth (snail, crab)
Paratyphoid	bacteria
Schistosomiasis	helminth (snail)
Typhoid	bacteria

In addition, several disease insect vectors thrive in conditions of ample fresh water, including:

Dengue (all types)	virus (mosquito)
Filariasis	helminth (mosquito)
Malaria	protozoa (mosquito)
Onchocerciasis	helminth (blackfly)
Rift Valley fever	virus (mosquito)
Trypanosomiasis	protozoa (tsetse fly)
Yellow fever	virus (mosquito)

Adapted from R. Feachem, M. McGarry, and D. Mara, eds., *Water, Wastes, and Health in Hot Climates* (London: John Wiley & Sons, 1977); and A. Dievler and M. R. Reich, "The Aswan High Dam," distributed by the Pew Curriculum for Health Policy and Management, Harvard School of Public Health, Boston, 1984.

17. World Bank, *Health Sector Policy Paper* (1980), op. cit., 13.

18. Dievler and Reich (1984), op. cit.; A. B. Mobarak, "The Schistosomiasis Problem in Egypt," *American Journal of Tropical Medicine and Hygiene* 31: (1982): 87–91; and WHO Expert Committee, *The Control of Schistosomiasis*, Technical Report Series 728 (Geneva: World Health Organization, 1985).

19. Mobarak (1982), op. cit.

20. R. Daubney, J. R. Hudson, and P. C. Garnham, "Enzootic Hepatitis of Rift Valley Fever: An Undescribed Virus Disease of Sheep, Cattle and Man from East Africa," *Journal of Pathology and Bacteriology* 31 (1931): 546–79.

21. F. Fenner, B. R. McAuslan, C. A. Mims, et al., *The Biology of Animal Viruses* (New York: Academic Press, 1974), 636.

22. Mims, C. A. "Rift Valley Fever in Mice. VI. Histological Changes in the Liver in Relation to Virus Multiplication." *Australian Journal of Experimental Biology and Medical Science* 35 (1957): 595.

23. J. M. Meegan, "Rift Valley Fever in Egypt: An Overview of the Epizootic in 1977 and 1978," *Controversies in Epidemiology and Biostatistics* 3 (1978): 100–13.

24. A. Jouan, I. Coulibaly, F. Adam, et al., "Analytical Study of a Rift Valley Fever Epidemic," *Research Virology* 140 (1989): 175–86; J. Morvan, J. F. Saluzzo, D. Fontenille, et al., "Rift Valley Fever on the East Coast of Madagascar," *Research Virology* 142 (1991): 475–82; A. Jouan, F. Adam, D. Riou, et al., "Evaluation of the Indicators of Health in the Area of Trarza During the Epidemic of Rift Valley Fever in 1987," *Bulletin de la Société de Pathologie Exotique et de ses Filiales* 83 (1990): 621–27; and, R. E. Shope and A. S. Evans, "Assessing Geographic and Transport Factors, and Recognition of New Viruses," in S. S. Morse, ed., *Emerging Viruses* (Oxford, Eng.: Oxford University Press, 1993), 114.

25. World Bank, *Health Sector Policy Paper* (1980), op. cit., 14.

26. Ibid., 35.

27. Data obtained from ibid., 68–85.

28. J. Nyerere, *Arusha Declaration Ten Years After* (Dar es Salaam: Oxford University Press, 1977).

29. "In Tanzania, the Road to Medicine Is Paved with Magic," *Hospital Practice*, April 1974: 133–57.

30. The world has experienced eight pandemic waves of cholera since 1837. The seventh pandemic began in Indonesia in 1961 and spread slowly around most of the world's poor nations. The eighth pandemic began in Madras, India, in December 1992.

31. "East Africa," *Africa*, June 1979.

32. M. Honey, "How Amin Ran His Economy," *African Business*, July 1979.

33. F. J. Bennett, "A Comparison of Community Health in Uganda with Its Two East African Neighbors in the Period 1970–1979," in C. P. Dodge and P. D. Wiebe, eds., *Crisis in Uganda* (Oxford, Eng.: Pergamon Press, 1985), 43–52.

34. F. Rodhain, J. P. Gonzalez, E. Mercier, et al., "Arbovirus Infections and Viral Haemorrhagic Fevers in Uganda: A Serological Survey in Karamoja District, 1984," *Transactions of the Royal Society of Tropical Medicine and Hygiene* 83 (1989): 851–54.

35. A. Enns, "The Clocks Have Stopped in Uganda," in Dodge and Wiebe (1985), op. cit., 53–54.

36. "WHO: How It Is Making Public Health a Global Cause," *Hospital Practice*, September 12, 1973: 205–18.

37. In the 1990s, it would become the capital of the nation of Kazakhstan.

38. G.A. res. 2200A (XXI), and 999 U.N.T.S. 171, March 23, 1976.

39. G.A. res. 2200A (XXI), 993 U.N.T.S. 3, entered into force January 3, 1976.

40. W. H. McNeill, *Plagues and Peoples* (New York: Doubleday, 1976).

41. Recently, McNeill described the microbial decimation of the Amerindians in greater detail, citing it as the key factor responsible for stimulating his initial interest in historic epidemics. Cortez, McNeill wrote, had fewer than 400 soldiers at his disposal when he laid siege to Tenochtitlán (Mexico City), yet they strolled right into the Aztec capital, taking power almost effortlessly. McNeill believes the smallpox devastation of the Aztecs, coupled with a variety of other European microbes, was so overwhelming that the Aztec people surrendered, having decided that their gods had sided with the white-skinned invaders. See W. H. McNeill, "Patterns of Disease Emergence in History," Chapter 3 in Morse (1993) op. cit.

42. For a collection of representative views, see D. Brothwell and A. T. Sandison, eds., *Diseases in Antiquity: A Survey of the Diseases, Inquiries and Surgery of Early Populations* (Springfield, IL: Charles C Thomas, 1967).

43. M. Burnet and D. O. White, *Natural History of Infectious Disease* (4th ed.; Cambridge, Eng.: Cambridge University Press, 1972).

44. R. Dubos and J. Dubos, *The White Plague: Tuberculosis, Man and Society* (New Brunswick, NJ: Rutgers University Press, 1992), 207.

45. R. Dubos, *Mirage of Health: Utopias, Progress, and Biological Change* (Garden City, NY: Anchor Books, 1961), 138–39.

46. T. McKeown, R. G. Record, and R. D. Turner, "An Interpretation of the Decline of Mortality in England and Wales During the Twentieth Century," *Population Studies* 29 (1974): 391–422.

47. R. C. Baron, J. B. McCormick, and O. A. Zubeir, "Ebola Virus Disease in Southern Sudan: Hospital Dissemination and Intrafamilial Spread," *Bulletin of the World Health Organization* 61 (1983): 997–1003.

48. Time would prove McCormick sadly correct, as conditions in southern Sudan worsened steadily year by year. Widespread famine, over a decade of civil war, and massive refugee migrations would render the area a highly vulnerable ecology for the microbes. By mid-1993 the region would be suffering massive epidemics of AIDS, visceral leishmaniasis or kala-azar, tuberculosis, bacterial meningitis, a host of diarrheal diseases, leprosy, measles, and malaria. If Ebola and Marburg diseases were also rampant at that time they were undetectable, hidden under an overlay of so much disease that nearly every southern Sudanese seemed stricken by at least one potentially life-threatening ailment.

For further details, see *Leishmaniasis Epidemic in Southern Sudan*, WHO/6 (Geneva: World Health Organization, January 26, 1993); R. W. Ashford and M. C. Thomson, "Visceral Leishmaniasis in Sudan: A Delayed Development Disaster?" *Annals of Tropical Medicine and Parasitology* 85 (1991): 571–72; W. A. Perea, T. Ancelle, A. Moren, et al., "Visceral Leishmaniasis in Southern Sudan," *Transactions of the Royal Society of Tropical Medicine and Hygiene* 85 (1991): 48–53; R. Rosenblatt, "The Last Place on Earth," *Vanity Fair*, July 1993: 89–91, 114–20; and J. Seaman, D. Pryce, H. E. Sandorp, et al., "Epidemic Visceral Leishmaniasis in Sudan: A Randomized Trial of Aminosidine Plus Sodium Stibogluconate Versus Sodium Stiboglutinate Alone," *Journal of Infectious Diseases* 168 (1993): 715–19.

8. Revolution

1. J. D. Watson and F. H. C. Crick, "A Structure for Deoxyribonucleic Acid," *Nature* 171 (1953): 737.

2. There are many excellent resources for further information about recombinant DNA techniques. They include P. Berg and M. Singer, *Dealing with Genes: The Language of Heredity* (Mill Valley, CA: University Science Books, 1992); M. Singer and P. Berg, *Genes to Genomes* (Mill Valley, CA: University Science Books, 1991); and J. D. Watson, N. H. Hopkins, J. W. Roberts, et al., *Molecular Biology of the Gene* (4th ed.; Menlo Park, CA: Benjamin/Cummings Publishing Co., 1987).

3. For an excellent review of McClintock's work and its subsequent impact on molecular biology, see N. V. Federoff, "Maize Transposable Elements." Chapter 14 in D. E. Berg and M. M. Howe, eds., *Mobile DNA* (Washington, D.C.: American Society for Microbiology, 1989). One of McClintock's seminal papers is B. McClintock, "The Origin and Behavior of Mutable Loci in Maize," *Proceedings of the National Academy of Sciences* 36 (1950): 344–55.

4. James Watson has written four editions of his grand guide to molecular biology, each of which, since the first in 1965, has been considerably larger than its predecessor, reflecting the explosion of scientific discovery. The most recent edition of *Molecular Biology of the Gene*, completed in 1987 (op. cit.), has this marvelous description of the mobile DNA phenomenon: "Moveable DNA segments called transposons occasionally jump around chromosomes, thus fundamentally altering chromosomal structure. In addition to neatly moving genes, transposons also scramble DNA, making deletions, inversions, and other rearrangements. It is becoming clear that such changes are a critical feature of chromosome evolution, particularly in eucaryotic cells. We now appreciate that recombination is not accidental, but is instead an essential cellular process catalyzed by enzymes that cells encode and regulate for the purpose."

5. D. Baltimore, "Retroviruses and Cancer," *Hospital Practice*, January 1978: 49–57.

6. S. S. Morse, "Evolution, Viral," *Encyclopedia of Microbiology*, Vol. 2 (New York: Academic Press, 1992), 141–55.

7. For an excellent list of key transposable elements in common use for genetic engineering, see C. M. Berg, D. E. Berg, and E. A. Groisman, "Transposable Elements and Genetic Engineering," Chapter 41 in Berg and Howe, eds. (1989), op. cit.

8. J. M. Bishop, "Viruses, Genes, and Cancer," *Harvey Lecture*, March 17, 1983.

9. D. H. Spector, H. E. Varmus, and J. M. Bishop, "Nucleotide Sequences Related to the Transforming Gene of Avian Sarcoma Virus Are Present in DNA of Uninfected Vertebrates," *Proceedings of the National Academy of Sciences* 75 (1978): 4102–06; and D. Stehelin, H. E. Varmus, J. M. Bishop, and P. K. Vogt, "DNA Related to the Transforming Gene(s) of Avian Sarcoma Viruses Is Present in Normal Avian DNA," *Nature* 260 (1976): 170–73.

10. Bishop (1983), op. cit.

11. J. M. Bishop, "Oncogenes," *Scientific American* 246 (1982): 80–92; and H. E. Varmus, "Form and Function of Retroviral Proviruses," *Science* 216 (1982): 812–20.

12. K. Takatsuki, T. Uchiyama, K. Sagawa, and J. Yodoi, in S. Seno, F. Takaku, and S. Irino, eds., *Topics in Hematology* (Amsterdam: Excerpta Medica, 1977), 73–77.

13. Though the acronym HTLV would remain, Gallo—at the urging of numerous scientific colleagues—later changed the L in HTLV from "leukemia" to "lymphotropic," reflecting increasing evidence that the virus rarely caused leukemia but always attacked lymphoid cells.

14. B. J. Poiesz, F. W. Ruscetti, M. Reitz, et al., "Isolation of a New Type C Retrovirus (HTLV) in Primary Uncultured Cells of a Patient with Sézary T-Cell Leukemia," *Nature* 294 (1981): 268–75.

15. Robert Gallo's version of these events appears in his 1991 book, *Virus Hunting—AIDS, Cancer, and the Human Retrovirus: A Story of Scientific Discovery* (New York: New Republic). The Japanese perspective has not been published. Also see B. J. Poiesz, F. W. Ruscetti, A. F. Gazdar, et al., "Detection and Isolation of Type-C Retrovirus Particles from Fresh and Cultured Lymphocytes of a Patient with Cutaneous T-Cell Lymphoma," *Proceedings of the National Academy of Sciences* 77 (1980): 7415–19.

16. I. Miyoshi, M. Fujishita, H. Taguchi, et al., "Natural Infection in Non-human Primates with Adult T-Cell Leukemia Virus or a Closely Related Agent," *International Journal of Cancer* 32 (1983): 333–36.

17. I. Miyoshi, S. Yoshimoto, M. Fujishita, et al., "Natural Adult T-Cell Leukemia Virus Infection in Japanese Monkeys," *Lancet* II (1982): 658.

18. An interesting side note to this story concerns the ultimate fate of the MO line. It has always been difficult to grow normal human cells under laboratory conditions; nearly all studies are done on human cancer cells cloned from specific tumor lines because cancer cells will grow under all sorts of conditions, including inside glass tubes. These clones are given names and the cell lines are sold to researchers all over the world. Such a line, if proven free of contamination and relatively indicative of general human cell activity, can fetch good prices. Golde thought MO might prove a fantastic experimental cell line, and filed a U.S. patent petition. If granted a patent, he would be able to reap royalties from all MO sales to researchers worldwide and profit from any products derived from the cell line. Golde's plans backfired, however, when the patient, John Moore, a Seattle salesman, questioned Golde's right to patent *his* cells, which had been removed from his spleen. The issue became more heated as the profit potentials grew. Moore's MO cells were the source of the discovery of the immune system protein GM-CSF, as well as the HTLV-II virus, which greatly increased the value of the line. As of 1993 the dispute continued to wend its way through the American legal system, having grown extraordinarily complex. A good review of the MO controversy appeared in J. Stone, "Cells for Sale," *Discover*, August 1988: 33–39.

19. J. S. Y. Chen, J. McLaughlin, J. C. Garson, et al., "Molecular Characterization of Genome of a Novel Human T-Cell Leukemia Virus," *Nature* 305 (1983): 502–5.

20. V. S. Kalyanaraman, M. G. Sarngadharan, M. Robert-Guroff, et al., "A New Subtype of Human T-cell Leukemia Virus (HTLV-II) Associated with a T-cell Variant of Hairy Cell Leukemia," *Science* 218 (1982): 571–73.

21. A good example of the public perspective at the time can be found in Larry Agran's *The Cancer Connection and What We Can Do About It* (Boston: Houghton Mifflin, 1977). Agran asserted that "we can change the man-made environment that causes 90 percent of all human cancer."

22. G. Hunsmann, J. Schneider, J. Schmitt, and N. Yamamoto, "Detection of Serum Antibodies to Adult T-Cell Leukemia Virus in Non-Human Primates and in People from Africa," *International Journal of Cancer* 32 (1983): 329–32.

23. K. Yamaguchi, "Human T-Lymphotropic Virus Type 1 in Japan," *Lancet* 343 (1994): 213–16.

24. By 1994 Brooklyn, New York, would be an HTLV-I epicenter, due to its sizable Caribbean immigrant population. See N. S. Larsen, "Study Confirms High Rates of Adult T-Cell Leukemia in N. Y. C.," *Journal of the National Cancer Institute* 86 (1994): 85–86.

25. F. A. Vyth-Dreese, P. Rumke, M. Robert-Guroff, et al., "Antibodies Against Human T-Cell Leukemia/Lymphoma Virus in Relatives of a T-Cell Leukemia Patient Originating from Surinam," *International Journal of Cancer* 32 (1983): 337–42.

26. V. Manzari, A. Gradilone, G. Barillari, et al., "HTLV-I Is Endemic in Southern Italy: Detection of the First Infectious Cluster in a White Population," *International Journal of Cancer* 36 (1985): 557–59.

27. Gallo (1991), op. cit.; "Call It Ishmael," *Hospital Practice*, September 15, 1985: 29; and Z. Ben-Ishai, "Human T-Cell Lymphotropic Virus Type-1 Antibodies in Falashas and Other Ethnic Groups in Israel," *Nature* 315 (1985): 665.

28. A. F. Fleming, "HTLV from Africa to Japan," *Lancet* I (1984): 279.

29. W. F. H. Jarrett, E. M. Crawford, W. B. Martin, and F. Davie, "Leukemia in the Cat: A Virus-like Particle Associated with Leukaemia (Lymphosarcoma)," *Nature* 202 (1964): 567. W. D. Hardy, Jr., L. J. Old, P. W. Hess, et al., "Horizontal Transmission of Feline Leukaemia Virus," *Nature* 244 (1973): 266–69; and D. P. Francis, M. Essex, and W. D. Hardy, Jr., "Excretion of Feline Leukemia Virus by Naturally Infected Pet Cats," *Nature* 269 (1977): 252–54. M. Essex, G. Klein, S. P. Snyder, and J. B. Harrold, "Correlation Between Humoral Antibody and Regression of Tumours Induced by Feline Sarcoma Virus," *Nature* 233 (1971): 195–96. M. Essex, W. D. Hardy, Jr., S. M. Cotter, et al., "Naturally Occurring Persistent Feline Oncornavirus Infections in the Absence of Disease," *Infection and Immunology* 11 (1975): 470–75; and M. Essex, A. Sliski, S. M. Cotter, et al., "Immuno-surveillance of Naturally Occurring Feline Leukemia," *Science* 190 (1975): 790–92.

30. The lab's interest in hepatitis B reflected Don Francis's stay, which was briefly interrupted in 1976 by his work on Ebola in Sudan. The hepatitis B virus was discovered in 1965, and shown to cause liver cancer in 1978. See W. Szmuness, "Hepatocellular Carcinoma and the Hepatitis B Virus: Evidence for a Causal Association," *Progress in Medical Virology* 24 (1978): 40–69. Francis planned to set up hepatitis B surveillance around the United States when he returned to the CDC in an effort to determine just how many Americans suffered from tumors caused by the virus. His scientific thinking about hepatitis B was described in D. P. Francis, M. Essex, and J. E. Mynard, "Feline Leukemia Virus and Hepatitis B Virus: A Comparison of Late Manifestations," *Progress in Medical Virology* 27 (1981): 127–32.

31. M. Essex, "Adult T-Cell Leukemia-Lymphoma: Role of a Human Retrovirus," *Journal of the National Cancer Institute* 69 (1982): 981–85.

32. J. Summers, J. M. Smolec, and R. Snyder, "A Virus Similar to Human Hepatitis B Virus Associated with Hepatitis and Hepatoma in Woodchucks," *Proceedings of the National Academy of Sciences* 75 (1978): 4533–37.

33. Forty-fifth World Assembly, "Implementation of the Global Strategy for Health for All by the Year 2000, Second Evaluation; and Eighth Report on the World Health Situation" (Geneva: World Health Organization, 1992), 10.

34. D. Burkitt, "A Lymphoma Syndrome in Tropical Africa," *International Review of Experimental Pathology* 2 (1963): 67.

35. M. A. Epstein, B. G. Achong, and Y. M. Barr, "Virus Particles in Cultured Lymphoblasts from Burkitt's Lymphoma," *Lancet* II (1964): 702.

36. G. Orth, F. Breitburd, M. Faure, and O. Croissant, "Papilloma-virus: A Possible Role in Human Cancer," in H. H. Hiatt, J. D. Watson, and J. A. Winsten, eds., *Origins of Human Cancer* (New York: Cold Spring Harbor Laboratory, 1977).

9. Microbe Magnets

1. Hippocrates, "On Airs, Waters, and Places," *The Genuine Works of Hippocrates*, Francis Adams, trans. (London: Leslie P. Adams, Jr., 1849).

2. An ecological description of cities can be found in John Reader, *Man on Earth* (Austin: University of Texas Press, 1988), Chapter 12.

3. J. Cairns, *The History of Mortality*, unpublished, 1993.

4. W. R. MacDonell, "On the Expectation of Life in Ancient Rome, and in the Provinces of Hispania and Lusitania, and Africa," *Biometrika* 9 (1913): 366–80.

5. A. T. Sandison, "Parasitic Diseases," in D. Brothwell and A. T. Sandison, eds., *Diseases in Antiquity* (Springfield, IL: Charles C Thomas, 1967).

6. W. H. McNeill, *Plagues and People* (New York: Doubleday, 1976), Appendix, "Epidemics in China"; and D. Twitchett, "Population and Pestilence in T'ang China," in W. Baver, ed., *Studia Sino-Mongolia* (Wiesbaden: Franz Steiner Verlag, 1979), 35–68.

7. A. Patrick, "Disease in Antiquity: Ancient Greece and Rome," in Brothwell and Sandison, eds. (1967), op. cit.

8. Jonathan Mann and his colleagues observed cases in the American Southwest during the early 1980s, David Scott studied outbreaks in Ghana in 1908 and 1924, and the World Health Organization reported further outbreaks between 1975 and 1985 in Madagascar, Uganda, Tanzania, Bolivia, Brazil, Peru, Burma, and Vietnam. See World Health Organization. "Human Plague in 1986," *Weekly Epidemiological Record* 62 (1987): 299–300.

9. Among the most accessible and fascinating accounts of the 1346 Black Death and the later 1665 plague are D. Defoe, *A Journal of the Plague Year* (1722), available in many published forms; McNeill, op. cit., 134–54; B. W. Tuchman, *A Distant Mirror: The Calamitous 14th Century* (New York: Alfred A. Knopf, 1978); and Philip Ziegler, *The Black Death* (London: Collins, 1969).

10. The trend of blaming the Jews for the Black Death began in what is today called Switzerland and spread quickly throughout Europe. In most cities, the persecutions were started by lower-class tradesmen and peasants who were whipped into frenzies by overzealous monks and priests. In some cities these activities were officially sanctioned by local authorities. In Basel, for example, the town leadership voted to kill all Jews, destroy their homes, and ban Jews from entering the city for another two centuries. On the other hand, there were those in power who opposed such actions and sought to protect Europe's Jewish population. Pope Clement VI gave Jews safe haven inside his papal residence in Avignon. Emperor Charles IV of France tried to stop persecutions in his country, but was overridden by nobles who hoped to avoid paying off debts by letting the mobs kill their Jewish creditors. Duke Albert of Austria was labeled a "Jew-master" because he protected hundreds of Jewish families, allowing them sanctuary inside his fortress. For further details, see J. F. C. Hecker, *The Epidemics of the Middle Ages*, B. G. Babington, trans. (London: Sydenham Society, 1844).

11. Such a level of microbe-induced mass destruction would not be achieved again until the influenza epidemic of 1918–19. It would be surpassed by AIDS, which over a twenty-year period between 1980 and 2000 is projected to claim an estimated 24 million people, 20 million of whom will have died of the disease by the year 2000. See J. Mann, D. J. M. Tarantola, and T. W. Netter, *AIDS in the World* (Cambridge, MA: Harvard University Press, 1992), 127–32.

12. Earlier circumstantial evidence led Joseph Needham to conclude that China had leprosy prior to A.D. 500, but skeletal studies found no clear leprotic remains in Asia until well after the medieval leprosy epidemic of Europe. See V. Moller-Christensen, "Evidence of Leprosy in Earlier Peoples," in Brothwell and Sandison, eds. (1967), op. cit. Indeed, the greatest leprosy epidemics of Asia followed European colonialism of the region during the eighteenth century.

13. For these and many other cogent details on the history of tuberculosis, see F. Ryan, *The Forgotten Plague: How the Battle Against Tuberculosis Was Won—and Lost* (Boston: Little, Brown, 1993; and R. Dubos, "Tuberculosis," *Scientific American* 181 (1949): 31–40.

14. J. B. Bass, Jr., L. S. Farer, P. C. Hopewell, et al., "Diagnostic Standards and Classification of Tuberculosis: Official Statement of the American Thoracic Society," *American Review of Respiratory Diseases* 142 (1990): 725–35.

15. R. Riley, "Airborne Infection," *American Journal of Medicine* 57 (1974): 466–75.
In one particularly ingenious mid-twentieth-century study researchers placed caged guinea pigs in the sealed rooms of human tuberculosis patients. The scientists calculated that, provided the patient's room wasn't aired out, the tiny guinea pig lungs inhaled thirty infectious TB particles a day. The far larger human lung would presumably absorb ten to twenty times as many particles daily under the same circumstances. See R. L. Riley, C. C. Mills, F. O'Grady, et al., "Infectiousness of Air from a Tuberculosis Ward: Ultraviolet Irradiation of Infected Air: Comparative Infectiousness of Different Patients," *American Review of Respiratory Diseases* 84 (1962): 511–25.

16. Riley (1974), op. cit.

17. W. L. Salo, A. C. Aufderheide, J. Buikstra, and T. A. Holcomb, "Identification of *Mycobacterium tuberculosis* DNA in a Pre-Columbian Peruvian Mummy," *Proceedings of the National Academy of Sciences* 91 (1994): 2091–94.

18. Consumption was a particular concern. The numbers of cases rose steadily. In Massachusetts, for example, death certificate reports listed consumption (tuberculosis) as the cause of death for 1,634 individuals in 1844; by 1846 that figure was 2,567. And the consumption death reports jumped to 4,593 in 1853. The state's total population in the 1850 census was 994,665. Between 1849 and 1853, a total of 20,000 people—most of them residents of Boston—died of consumption. That was about 2 percent of the population. Far more people were sick with the disease, which, like AIDS 120 years later, killed so slowly that statistics never actually reflected at a given moment the full toll of the disease on society.

19. There are several excellent sources for the history of the nineteenth-century cholera pandemics, including D. Barua and W. B. Greenough III, *Current Topics in Infectious Disease: Cholera* (New York: Plenum, 1992); J. Duffy, "Social Impact of Disease in the Late 19th Century," in J. W. Leavitt and R. L. Numbers, eds., *Sickness and Health in America* (2nd ed.; Madison: University of Wisconsin, 1985), Chapter 29; and R. J. Evans, *Death in Hamburg: Society and Politics in the Cholera Years 1830–1910* (Oxford, Eng.: Clarendon Press, 1987).

20. T. McKeown, R. G. Record, and R. D. Turner, "An Interpretation of the Decline of Mortality in England and Wales During the Twentieth Century," *Population Studies* 29 (1969): 391–422; and T. McKeown and R. G. Record, "Reasons for the Decline of Mortality in England and Wales During the Nineteenth Century," *Population Studies* 16 (1962): 94–122.

21. Select Committee on Population, "Domestic Consequences of United States Population Change," report prepared for the U.S. House of Representatives, 1978.

22. Noteworthy for the future would be a largely ignored fact in 1970: namely, that tuberculosis skin tests showed that the highest rates of infection in the United States that year were among poor African-American residents of seven Deep South states, and 85 percent of New York City residents who tested positive in TB skin tests that year were blacks who had recently moved to the city from the Deep South.

23. S. O. Freedman, "Tuberculin Testing and Screening: A Critical Evaluation," *Hospital Practice*, May 1972: 63–70.

24. T. McKeown, *The Origins of Human Disease* (Oxford, Eng.: Basil Blackwell, 1988).

25. R. Dubos and J. Dubos, *The White Plague: Tuberculosis, Man and Society* (Boston: Little, Brown, 1952).

26. B. Bates, *Bargaining for Life: A Social History of Tuberculosis, 1876–1938* (Philadelphia: University of Pennsylvania Press, 1992).

27. B. Bates, "Tuberculosis in Pennsylvania," in C. E. Rosenberg and J. Golden, eds., *Framing Disease: Studies in Cultural History* (New Brunswick, NJ: Rutgers University Press, 1992), 229–47.

28. For example, between 1937 and 1947 the numbers of South African families on waiting lists for housing in Johannesburg's black- and colored-designated communities rose from 11 to 16,195. During the subsequent decade the apartheid government decreased its commitment to subsidized housing construction for poor and working-class families from a 1949 high of 7,407 houses to 1957's low of 155. During the same time construction of so-called economic housing, built at government expense for white working- and middle-class families, rose from 348 houses in 1947 to 15,364 in 1957.

29. R. M. Packard, *White Plague, Black Labor: Tuberculosis and the Political Economy of Health and Disease in South Africa* (Berkeley: University of California Press, 1989).

30. E. H. Hudson, "Treponematosis and Anthropology," *Annals of Internal Medicine* 58 (1963): 1037.

31. Ziegler (1991), op. cit.

32. When the spirochete enters sores on the skin's surface it remains in the vicinity of its initial site of infection. Over time the organism may invade local bone, cartilage, and skin, but infection is rarely systemic. In contrast, sexual transmission of the syphilis spirochete provides immediate access to the blood system, allowing for disease in every organ in the body. The localized yaws infections often resolved without treatment in a matter of weeks, but once syphilis gained entry to the bloodstream most people were fated to suffer systemic illness and, in many cases, slow death.

33. "France: Pitchforked," *The Economist*, November 28, 1992: 56–57.

34. "Pollution in Asia," *The Economist*, October 6, 1990: 19–21.

35. United Nations, "The Prospect of World Urbanization," *Population Studies*, No. 101, ST/ESA/SER/101 (New York, 1987).

36. A. Pryer and N. Crook, *Cities of Hunger: Urban Malnutrition in Developing Countries* (London: Oxfam, 1988).

37. I. Tabibzadeh, A. Rossi-Espagnet and R. Maxwell, *Spotlight on the Cities: Improving Urban Health in Developing Countries* (Geneva: World Health Organization, 1989).

38. T. Harpham, T. Lusty, and P. Vaughan, *In the Shadow of the City: Community Health and* *the Urban Poor* (Oxford, Eng.: Oxford University Press, 1988); and Tabibzadeh, Rossi-Espagnet, and Maxwell, op. cit.

The cities projected to have attained megacity status by 2000 are:

	City/Metro Area	Country	Est. Population (in millions)
AFRICA	Cairo	Egypt	12.9
AMERICAS	Buenos Aires	Argentina	12.1
	Rio de Janeiro	Brazil	19.0
	São Paulo	Brazil	25.8
	Mexico City	Mexico	31.0
	Los Angeles	U.S.A.	13.9
	New York	U.S.A.	22.4
EAST ASIA	Beijing	China	20.9
	Shanghai	China	23.7
	Osaka-Kobe	Japan	10.9
	Tokyo-Yokohama	Japan	23.7
SOUTH ASIA	Dhaka	Bangladesh	10.5
	Bombay	India	16.8
	Calcutta	India	16.4
	Delhi	India	11.5
	Madras	India	12.7
	Jakarta	Indonesia	15.7
	Baghdad	Iraq	11.0
	Tehran	Iran	11.1
	Karachi	Pakistan	11.6
	Bangkok	Thailand	10.6
	Manila	Philippines	11.4
	Istanbul	Turkey	10.8
EUROPE	Paris	France	10.6

London is expected to drop off the list, as its population is forecast to decline to 9.2 million due to lowering birth rates and middle-class suburban outward migration.

39. Tabibzadeh, Rossi-Espagnet, and Maxwell (1989), op. cit.

40. "Pollution in Asia," op. cit.

41. Aspen Institute, U.S.A., 1989.

42. United Nations, "The Prospect of World Urbanization," op. cit.

43. D. B. Ottaway, "Cairo Is Plagued with Environmental Disasters," *San Francisco Chronicle,* January 12, 1983: F1.

44. D. J. Stanley, and A. G. Warne, "Nile Delta: Recent Geological Evolution and Human Impact," *Science* 260 (1993): 628–34.

45. A rich litany of such details of human urban existence in developing countries can be found in Harpham, Lusty, and Vaughan (1988), op. cit.

46. N. Coulibaly, "Place et Approches des Problèmes de la Tuberculose à Abijan," *Médecine d'Afrique Noire* 28 (1981): 447–49.

47. A. Rossi-Espagnet, "Health and the Urban Poor," *World Health,* July 1983; and P. Khanja-nasthiti and J. D. Wray, "Early Protein-Calorie Malnutrition in Slum Areas of Bangkok Municipality, 1970–1971," *Journal of the Medical Association of Thailand* 57 (1974): 357–66.

48. A vivid account of those years appears in Agnes Brinkmann's *Unter Afrikanischem Zauber* (Hanover: Landbuch Verlag, 1992).

49. For a description of the disease and its treatment, see T. E. Nash and F. A. Neva, "Recent Advances in the Diagnosis and Treatment of Cerebral Cysticercosis," *New England Journal of Medicine* 311 (1984): 1492–96.

50. F. O. Richards, P. M. Schantz, E. Ruiz-Tuben, and F. J. Sorvillo, "Cysticercosis in Los Angeles County," *Journal of the American Medical Association* 254 (1985): 3444–48.

51. A. Benyoussef, "Santé, Migration et Urbanization: Une Etude Collective au Sénégal," *Bulletin of the World Health Organization* 49 (1973): 517–37.

52. M. E. Wilson, *A World Guide to Infectious Diseases, Distribution, Diagnosis* (Oxford, Eng.: Oxford University Press, 1991).

53. K. E. Mott, P. Desjeux, A. Moncayo, et al., "Parasitic Diseases and Urban Development," *Bulletin of the World Health Organization* 68 (1990): 691–98.

54. R. S. Desowitz, *The Malaria Capers: More Tales of Parasites and People, Research and Reality* (New York: W. W. Norton, 1991).

55. K. M. Rahman, and N. Islam, "Resurgence of Visceral Leishmaniasis in Bangladesh," *Bulletin of the World Health Organization* 61 (1983): 113; and R. D. Pearson, "Leishmaniasis: The Pathologic Spectrum," *Hospital Practice*, May 1984: 100e–100x.

56. W. Hammow, A. Rudnick, and G. E. Sather, "Viruses Associated with Epidemic Hemorrhagic Fevers of the Philippines and Thailand," *Science* 131 (1960): 1102–3.

57. S. B. Halstead and C. Yamarat, "Recent Epidemics of Hemorrhagic Fever in Thailand: Observations Related to Pathogenesis of a 'New' Dengue Disease," *American Journal of Public Health* 55 (1965): 1386–94.

58. D. M. Morens, "Dengue Fever and Dengue Shock Syndrome," *Hospital Practice*, July 1982: 103–13; Halstead and Yamarat (1965), op. cit.; and S. B. Halstead, "Immunological Enhancement of Dengue Virus Infection in the Etiology of Dengue Shock Syndrome," Third Asian Congress of Pediatrics, Bangkok, Thailand, 1974.

59. A. Morales, H. Groat, P. K. Russell, and J. M. McCown, "Recovery of Dengue-2 Virus from *Aedes aegypti* in Colombia," *American Journal of Tropical Medicine and Hygiene* 22 (1973): 785–87.

60. G. P. Kouri, M. G. Guzmán, J. R. Bravo, and C. Triana, "Dengue Haemorrhagic Fever/ Dengue Shock Syndrome: Lessons from the Cuban Epidemic, 1981," *Bulletin of the World Health Organization* 67 (1989): 375–80.

61. Centers for Disease Control, "Suspected Dengue—Laredo, Texas," *Morbidity and Mortality Weekly Report* 29 (1980): 503.

62. D. J. Gubler, "Dengue Haemorrhagic Fever: A Global Update," *Virus Information Exchange Newsletter* 8 (1991): 2–3.

63. According to the Centers for Disease Control, *A. albopictus* had, by mid-1987, taken hold in the states of Texas, Kentucky, Delaware, Maryland, Ohio, California, Alabama, Arkansas, Florida, Georgia, Illinois, Indiana, Louisiana, Mississippi, Missouri, North Carolina, and Tennessee.

10. Distant Thunder

1. Detailed accounts of these and other events in the nascent years of the New York gay liberation movement can be found in two sources: *The Village Voice* chronicled events closely, particularly as followed by reporters Lucian Truscott and Howard Smith. See also D. Teal, *The Gay Militants* (New York: Stein & Day, 1971); and M. Duberman, *Stonewall* (New York: Penguin, 1993).

2. Randy Shilts was the first openly gay reporter hired to work for a major American daily newspaper. *The San Francisco Chronicle* consciously put Shilts on staff to follow these historic developments. Out of that journalistic enterprise came two seminal books, each marking different extremes in recent U.S. gay history: *The Mayor of Castro Street: The Life and Times of Harvey Milk* (New York: St. Martin's Press, 1981); and *And the Band Played On* (New York: St. Martin's Press, 1987).

3. In reviewing the global sexually transmitted disease (STD) situation in developing countries, one group summarized the situation as follows:

Population shifts from rural to urban areas, where STD rates are higher, are proceeding rapidly in most parts of the world. Rural to urban migration generally results in an excess of men in urban areas and of women in rural areas. The sex ratio imbalances created in places of origin and places of destination may further increase STD risks. In addition, increasing educational opportunities in developing countries, particularly for women, may delay marriage, also increasing STD risks. Similar demographic changes have caused dramatic social changes contributing to high STD rates in the United States.

The differences between the values of youth and parents have been referred to as a generation gap. Many less developed countries are undergoing rapid social transition, with corresponding changes in values between generations. . . . Among the many new values adopted, altered sexual behaviors will place this population at greater risk for acquiring an STD. From S. T. Brown, F. R. K. Zacarias, and S. O. Aral, "STD Control in Less Developed Countries: The Time Is Now," *International Journal of Epidemiology* 14 (1985): 505–9.

4. J. E. Sutherland, V. W. Persky, and J. A. Brody, "Proportionate Mortality Trends: 1950 Through 1986," *Journal of the American Medical Association* 264 (1990): 3178–84.

5. Center for Policy Studies, "The Federal Health Dollar, 1969–76," Washington, D.C., 1977.

6. For an excellent history of the success—and failure—of America's attempts to control sexually transmitted diseases, see A. M. Brandt, *No Magic Bullet: A Social History of Venereal Disease in the United States Since 1880* (Oxford, Eng.: Oxford University Press, 1987).

7. H. H. Handfield, "Sexually Transmitted Diseases," *Hospital Practice*, January 1982: 99–116;

and Anonymous, "VD: Gonorrhea Incidence Put at 2 Million Annually in U.S.," *Hospital Practice*, June 1971: 27–40.

According to the CDC, the reported numbers of sexually transmitted diseases between 1940 and 1982 were as follows:

Year	Syphilis (All Stages & Types)	Hepatitis B	Gonorrhea	Chancroid
1940	472,900	NR	175,841	NR
1942	479,601	NR	212,403	5,477
1944	467,755	NR	300,676	7,878
1946	363,647	NR	368,020	7,091
1948	314,313	NR	345,501	7,661
1950	217,558	NR	286,746	4,977
1952	167,762	NR	244,957	3,738
1954	130,697	Not rep. sep. from Hep. A	242,050	3,003
1956	130,201	Not rep. sep. from Hep. A	244,346	2,135
1958	113,884	Not rep. sep. from Hep. A	232,386	1,595
1960	122,538	Not rep. sep. from Hep. A	258,933**	1,680
1962	126,245	Not rep. sep. from Hep. A	263,714	1,344
1964	114,325	37,740*	300,666	1,247
1966	105,159	32,859*	351,738	838
1968	96,271	45,893*	464,543	845
1970	91,382	56,797*	600,072	1,416
1972	91,149	54,074*	767,215	1,414
1974	83,771	10,631*	906,121	945
1976	71,761	14,973*	1,001,994	628
1978	68,875	15,016*	1,013,436	521
1980	68,832	19,015*	1,004,029	788
1982	75,579	22,177*	960,633	1,392

NR = Not reported.   Not rep. sep. from Hep. A = Not reported separately from Hepatitis A.

* Overestimated because a hepatitis test that could specifically diagnose hepatitis B (versus Type C, non-A non-B, D, etc.) was not available.

** Year the downward trend of gonorrhea cases reverses.

8. S. A. Morse, A. A. Moreland, and S. E. Thompson, eds., *Atlas of Sexually Transmitted Diseases* (Philadelphia: J. B. Lippincott, 1990).

9. A. E. Washington, P. S. Arno, and M. A. Brooks, "The Economic Cost of Pelvic Inflammatory Disease," *Journal of the American Medical Association* 255 (1986): 1735–38.

10. H. H. Handsfield, L. L. Jasman, P. L. Roberts, et al., "Criteria for Selective Screening for *Chlamydia trachomatis* Infection in Women Attending Family Planning Clinics," *Journal of the American Medical Association* 255 (1986): 1730–34.

11. Centers for Disease Control, "Penicillinase-Producing *Neisseria gonorrhoeae*," *Morbidity and Mortality Weekly Report* 25 (1976): 261.

12. I. Phillips, "Beta-Lactamase-Producing Penicillin-resistant Gonococcus," *Lancet* II (1976): 656.

13. Centers for Disease Control, "Follow-up on Antibiotic Resistant *Neisseria gonorrhoeae*," *Morbidity and Mortality Weekly Report* 26 (1977): 29–30.

14. P. F. Sparling, K. K. Holmes, P. J. Wiesner, and M. Puziss, "Summary of the Conference on the Problem of Penicillin-resistant Gonococci," *Journal of Infectious Diseases* 135 (1977): 865–67.

15. Centers for Disease Control, "Follow-up on Penicillinase-Producing *Neisseria gonorrhoeae*—Worldwide," *Morbidity and Mortality Weekly Report* 26 (1977): 153–54.

16. Centers for Disease Control, "Tetracycline-resistant *Neisseria gonorrhoeae*—Georgia, Pennsylvania, New Hampshire," *Morbidity and Mortality Weekly Report* 34 (1985): 569–70; and Centers for Disease Control, "Plasmid-mediated Tetracycline-resistant *Neisseria gonorrhoeae*—Georgia, Massachusetts, Oregon," *Morbidity and Mortality Weekly Report* 35 (1986): 304–5.

17. Centers for Disease Control, "Penicillinase-producing *Neisseria gonorrhoeae*—United States, 1986," *Morbidity and Mortality Weekly Report* 36 (1986): 107–8; and H. Faruki, R. N. Kohmeschler, P. McKinney, and P. F. Sparling, "A Community-based Outbreak of Resistant *Neisseria gonorrhoeae* Not Producing Penicillinase (Chromosomally Mediated Resistance)," *New England Journal of Medicine* 313 (1985): 607–11.

18. L. Corey and P. G. Spear, "Infections with Herpes Simplex Viruses," *New England Journal of Medicine* 314 (1986): 685–90.

19. A. J. Nahmias and B. Roizman, "Infection with Herpes-Simplex Virus 1 and 2," *New England Journal of Medicine* 289 (1973): 667–74, 719–25, 781–89.

20. A. J. Nahmias, H. L. Keyserling, and G. M. Kerrick, "Herpes Simplex," in J. S. Remington and J. O. Klein, eds., *Infectious Diseases of the Fetus and Newborn Infant* (Philadelphia: W. B. Saunders, 1983).

21. M. F. Goldsmith, "Possible Herpesvirus Role in Abortion Studied," *Journal of the American Medical Association* 251 (1984): 3067–70.

22. Using sophisticated genetic mapping techniques for the first time, University of Chicago scientists showed in 1979 that most neonatal herpes cases came not from infected mothers but from other babies. Bernard Roizman and Timothy Buchman compared the genetic sequences of herpes viruses found in babies co-housed in intensive-care wards at several major U.S. urban hospitals with the strains seen in their mothers. They discovered that many of the mothers were uninfected, and the babies on any given hospital ward all had exactly the same viral strain in their bodies. The conclusion was that most pediatric herpes simplex in the United States was nosocomial in origin; that is, it was spread from baby to baby by the hospital medical staff. Presumably nurses tending a primary case of maternally derived herpes failed to observe sterile hygiene practices as they moved from patient to patient in the neonatal intensive-care wards. See B. Roizman and T. Buchman, "The Molecular Epidemiology of Herpes Simplex Virus," *Hospital Practice*, January 1979: 95–104.

23. M. F. Goldsmith, "Sexually Transmitted Diseases May Reverse the 'Revolution,' " *Journal of the American Medical Association* 255 (1986): 1665–72.

24. "There are many unanswered questions regarding sexually transmitted CMV infections," noted H. Hunter Handsfield in 1982. "One of the most intriguing revolves around the markedly high prevalence of past or present CMV infection in homosexual men and the fact that these men probably have a higher than average risk of the rare malignancy, Kaposi's sarcoma." See H. H. Handsfield, "Sexually Transmitted Diseases," *Hospital Practice*, January 1982: 99–106.

25. Morse, Moreland, and Thompson (1990), op. cit.; and G. P. Schmid, "The Treatment of Chancroid," *Journal of the American Medical Association* 255 (1986): 1757–62.

26. Handsfield (1982), op. cit.

27. A. DeSchryver and A. Meheus, "Epidemiology of Sexually Transmitted Diseases: The Global Picture," *Bulletin of the World Health Organization* 68 (1990): 639–54.

28. R. K. St. John and S. T. Brown, eds., "International Symposium on Pelvic Inflammatory Disease," *American Journal of Obstetrics and Gynecology* 138 (1980): Supplement.

29. O. Frank, "Infertility in Sub-Saharan Africa: Estimates and Implications," *Population and Development Review* 9 (1983): 137–44.

30. S. K. Hira, "Sexually Transmitted Diseases: A Menace to Mothers and Children," *World Health Forum* 7 (1986): 243–47; S. K. Hira et al., "Congenital Syphilis in Lusaka, II. Incidence at Birth and Potential Risk Among Hospital Deliveries," *East African Medical Journal* 59 (1982): 306–10; and T. E. Watts et al., "A Case-Control Study of Stillbirths at a Teaching Hospital—Zambia 1979–1980: Serological Investigations for Selected Infectious Agents," *Bulletin of the World Health Organization* 62 (1984): 803–8.

31. M. Callen, *Surviving AIDS* (New York: HarperCollins, 1990).

32. For a detailed discussion of this new ecology and the implications for microbial transmission, see "Biological and Social Conditions," Chapter 14 in M. D. Grmek, *History of AIDS: Emergence and Origin of a Modern Pandemic* (Princeton, NJ: Princeton University Press, 1990).

33. Michael Callen summarized the mood beautifully in *Surviving AIDS*, op. cit.:

No one took heed of the warning signs that were all around us. No one asked what the cumulative consequences might be of continually wallowing in what was, to put it bluntly, an increasingly polluted microbiological sewer. Rumors that the [New York City] Health Department had been able to culture cholera and other exotic microbes from the greasy stair rails of the Mineshaft (a notorious Manhattan sex club) were dismissed as apocryphal.

We took each new disease in stride. I can even recall that the "invention" of a disease dubbed as "gay bowel syndrome" [*E. histolytica*] was, in some quarters, almost a matter of pride; now we even had our own *diseases*, just like we had our own plumbers and tax advisers. A whole new breed of physicians, affectionately known as "clap doctors," grew rich treating our STDs. Many of these physicians could themselves be observed in the bathhouses and back rooms leading the same fast-lane life-style as their patients. Even if they had warned us, who would have listened?

34. T. J. John, G. T. Ninan, M. S. Rajagopalan, et al., "Epidemic Hepatitis B Caused by Commercial Human Immunoglobulin." *Lancet* I, 8125 (1979): 1074.

35. S. C. Hadler, D. L. Sorley, K. H. Acree, et al., "An Outbreak of Hepatitis B in a Dental Practice," *Annals of Internal Medicine* 95 (1981): 133–38.

36. A. L. Reingold, M. A. Kane, B. L. Murphy, et al., "Transmission of Hepatitis B by an Oral Surgeon," *Journal of Infectious Diseases* 145 (1982): 262–68.

37. M. Carl, D. L. Blakey, D. P. Francis, and J. E. Maynard, "Interruption of Hepatitis B Transmission by Modification of a Gynaecologist's Surgical Technique," *Lancet* I, 8274 (1982): 731–33.

38. W. L. Heyward, T. R. Bender, A. P. Lanier, et al., "Serological Markers of Hepatitis B Virus and Alpha-fetoprotein Levels Preceding Primary Hepatocellular Carcinoma in Alaskan Eskimos," *Lancet* II, 8304 (1982): 889–91.

39. N. E. Reiner, F. N. Judson, W. W. Bond, et al., "Asymptomatic Rectal Mucosal Lesions and Hepatitis B Surface Antigen at Sites of Sexual Contact in Homosexual Men with Persistent Hepatitis B Virus Infection," *Annals of Internal Medicine* 96 (1982): 170–73.

40. According to CDC statistics amassed by Dr. Miriam Alter, the more easily transmitted hepatitis A dominated all reported cases of the liver disease until 1983, when it was surpassed by hepatitis B. The increase was largely in gay Caucasian males. Between 1966 and 1983, hepatitis A rates declined, as surely as B rates escalated.

	HEPATITIS A		HEPATITIS B	
	Number of	Rate per 100,000	Number of	Rate per 100,000
Year	Cases	Americans	Cases	Americans
1966	32,859	16.77	1,497	0.79
1968	45,893	22.96	4,829	2.49
1970	56,797	27.87	8,310	4.08
1972	54,074	25.97	9,402	4.52
1974	40,358	19.54	10,631	5.15
1976	33,288	15.51	14,973	7.14
1978	29,500	13.53	15,016	6.89
1980	29,087	12.84	19,015	8.39
1982	23,403	10.11	22,177	9.58
1983	21,532	9.20	24,318	10.39

41. R. H. Purcell, "The Viral Hepatitides," *Hospital Practice*, July 1978: 51–63.

42. M. Robertson, "Joining the War Against Hepatitis B," *San Francisco Chronicle*, February 5, 1982: A24.

43. Laws governing drug paraphernalia and possession vary widely around the world and from state to state within a country. In Amsterdam, for example, it is legal to purchase both the narcotic drugs and sterile equipment for injection. In the United States narcotics use is governed at both the state and federal levels. In general, the commerce or business end of narcotics—import, transport, and distribution—falls under federal law, while the consumer end—use and possession—is under state law. The exceptions relate to quantities, with federal jurisdiction coming into play where the amount of drugs in one's possession is large. A cogent review of the complexities of U.S. narcotics laws can be found in L. Gostin, "The Interconnected Epidemics of Drug Dependency and AIDS," *Harvard Civil Rights–Civil Liberties Law Review* 26 (1991): 114–84.

44. In 1975, *Newsday* and reporters Les Payne, Knut Royce, and Bob Greene won the Pulitzer Prize for their series of reports tracing the deaths of suburban Long Island, New York, residents back through international crime networks all the way to the poppy fields of Turkey. The series was republished as a book: *The Heroin Trail* (New York: Holt, Rinehart and Winston, 1974). The author has drawn liberally from their work in describing heroin distribution during the 1970s.

45. Heroin was first marketed internationally by the Bayer Chemical Company of Germany, which sold the compound in 1898 as a cough suppressant. By 1906 the product had become so popular that the American Medical Association officially recommended its use as a painkiller.

But by 1924 some 200,000 Americans and an unknown number of Europeans were cough syrup addicts. That year the U.S. Congress outlawed further importation of the Bayer cough syrup and all other heroin-containing products.

The second international heroin marketing effort came on the heels of World War II, reaching the lucrative urban centers of Europe and North America during the 1950s. From its outset, the second heroin marketing effort, which was entirely illegal, was run by well-established, traditional organized crime elements in each country.

In 1961 the Mafia distributors who controlled the U.S. heroin trade created increased consumer demand and profits by manufacturing a false drug panic. Heroin was deliberately held off the market, causing tens of thousands of addicts, most of them then living in New York City and nearby cities such as Newark, to experience the physical pangs of withdrawal. Once the desired pressure had been exerted, the wholesalers released drugs diluted still further, setting the standard of 90 to 95 percent dilution that would remain in place for years. To maintain the high to which they had become accustomed, addicts had to purchase more heroin. On the other hand, the diluted formulation seemed less threatening to potential new consumers.

46. Meanwhile, there was one city on the planet in which nearly pure heroin crystals could be purchased relatively cheaply at the retail level: Saigon. The heroin was of such strength that the thousands of U.S. soldiers stationed in the city who indulged in the drug refrained from injecting the substance because they obtained a powerful high from merely smoking the Saigon product.

The congressional Subcommittee on Public Health heard testimony that at least 12,000 returning GIs were severely addicted and in need of detoxification and treatment. A U.S. medical officers' survey put that figure higher, estimating that 25,000 to 37,000 GIs had become heroin addicts during the Vietnam War. See "The Drug Abuse Problem in Vietnam," *Report to the Office of the Provost Marshal*, U.S. Military Assistance Command Vietnam, 1971; and *New York Times*, May 16, 1971: A16.

In 1975 the U.S. government spent $360 million on drug treatment and prevention programs, and $320 million on law enforcement aimed at narcotics control. By 1980 those figures had shifted to $410 million for treatment and prevention, half a billion dollars for law enforcement. By 1989 those figures would have soared to $1.2 billion for treatment and prevention efforts and $2.66 billion for law enforcement. Institute of Medicine, *Treating Drug Problems*, Vol. 1 (Washington, D.C.: National Academy of Sciences, 1990).

47. New Jersey Department of Health, "Statistical Perspectives on Drug Abuse Treatment in New Jersey, 1985," CN 362, Office of Data Analysis and Epidemiology, Trenton, NJ, 1986.

48. C. E. Cherubin, "Infectious Disease Problems of Narcotic Addicts," *Archives of Internal Medicine* 128 (1971): 309–13.

49. D. P. Levine and J. D. Sobel, *Infections in Intravenous Drug Abusers* (Oxford, Eng.: Oxford University Press, 1991).

50. Ibid.

51. E. Drouhet, B. Dupont, C. Lapresle, and P. Ravisse, "Nouvelle Pathologie: Candidose Folliculaire et Nodulaire avec les Localisations Ostéo-Articulaires et Oculaires au Cours des Septicémies à *Candida albicans* chez les Héroïn Omanes, Mono et Polythérapic Antifongue," *Bulletin de la Société Française Mycologique Médecin* 10 (1981): 179–83.

52. J. Mills and D. Drew, "*Serratia marcescens* Endocarditis: A Regional Illness Associated with Intravenous Drug Abuse," *Annals of Internal Medicine* 84 (1976): 29–35.

53. Bacterial subtypes 29/52/80/95 in Detroit; 29/77/83A/85 in Boston. See L. R. Crane, D. P. Levine, M. J. Zervos, and G. Cummings, "Bacteremia in Narcotic Addicts at the Detroit Medical Center: I. Microbiology, Epidemiology, Risk Factors, and Empiric Therapy," *Review of Infectious Diseases* 8 (1986): 364–73; and D. E. Craven, A. I. Rixinger, T. A. Goularte, and W. R. McCabe, "Methicillin-Resistant *Staphylococcus aureus* Bacteremia Linked to Intravenous Drug Abusers Using a 'Shooting Gallery,' " *American Journal of Medicine* 80 (1986): 770–76.

54. L. B. Reichman, C. P. Felton, and J. R. Edsall, "Drug Dependence: A Possible New Risk Factor for Tuberculosis Disease," *Annals of Internal Medicine* 139 (1979): 337–39.

55. The exceptions, of course, were the middle- and upper-class cocaine addicts. By 1985 a multimillion-dollar industry for treatment of such individuals would flourish worldwide. One would even bear the imprimatur of the wife of a former U.S. President, Betty Ford.

11. Hatari

1. Centers for Disease Control, "*Pneumocystis* Pneumonia—Los Angeles," *Morbidity and Mortality Weekly Report* 30 (1981): 250–52.

2. Centers for Disease Control, "Kaposi's Sarcoma and *Pneumocystis* Pneumonia Among Homosexual Men—New York City and California," *Morbidity and Mortality Weekly Report* 30 (1981): 305–8.

3. The often-cited first popularly published account of these cases was written by Lawrence Altman: "Cancer Outbreak in Homosexuals," *New York Times*, July 3, 1981: 20. In point of fact, the first published account of a mysterious new ailment in the gay community was authored by Dr. Lawrence Mass and appeared on May 18, 1981, in the *New York Native*.

4. Many details of the first days of the AIDS epidemic, particularly in the United States, can be found in D. Black, *The Plague Years* (New York: Simon & Schuster, 1986); S. Connor and S. Kingman, *The Search for the Virus* (London: Penguin, 1988); M. Daly, "AIDS Anxiety," *New York*, June 20,

1983: 23–29; L. Kramer, *The Normal Heart* (New York: New American Library, 1985); R. Shilts, *And the Band Played On* (New York: St. Martin's Press, 1987); and F. P. Siegal and M. Siegal, *AIDS: The Medical Mystery* (New York: Grove Press, 1983).

5. By 1990 the CDC's budget was $1 billion. The nearly fourfold increase was largely related to AIDS surveillance, state health department support for HIV work, and AIDS education campaigns.

6. S. E. Follansbee, D. F. Busch, C. B. Wofsy, et al., "An Outbreak of *Pneumocystis carinii* Pneumonia in Homosexual Men," *Annals of Internal Medicine* 96 (1982): 705–13.

7. S. K. Dritz, "Medical Aspects of Homosexuality," *New England Journal of Medicine* 302 (1980): 463–64.

8. Centers for Disease Control, "Acquired Immune Deficiency Syndrome (AIDS): Precautions for Clinical and Laboratory Staffs," *Morbidity and Mortality Weekly Report* 31 (1982): 577–80.

9. Larry Kramer, Joseph Sonnabend, Michael Callen, and Richard Berkowitz penned dire warnings to the gay community that appeared in the *New York Native*, then the city's leading gay newspaper. Sadly, the most outspoken leaders of the gay community at the time, and the *Native* itself, roundly denounced Kramer and Callen. The pair were reviled as homosexuals filled with self-loathing; anti-sex gays who wanted their fellow travelers to return to the quiet, closeted days before Stonewall. Kramer, in particular, was called everything from a homophobe to a hateful fearmonger. The campaign against Callen would start slowly, but would snowball until, in 1983, false rumors would spread that the gay activist was secretly a Christian fundamentalist, working for the Reverend Sun Myung Moon.

10. K. B. Hymes, J. B. Greene, A. Marcus, et al., "Kaposi's Sarcoma in Homosexual Men: A Report of Eight Cases," *Lancet* II (1981): 598–600.

11. M. S. Gottlieb, R. Schroff, H. M. Schanker, et al., "*Pneumocystis carinii* Pneumonia and Mucosal Candidiasis in Previously Healthy Homosexual Men: Evidence of a New Acquired Cellular Immunodeficiency," *New England Journal of Medicine* 305 (1981): 1425–31; H. Masur, M. A. Michelis, J. B. Greene, et al., "An Outbreak of Community Acquired *Pneumocystis carinii* Pneumonia: Initial Manifestation of Cellular Immune Dysfunction," *New England Journal of Medicine* 305 (1981): 1431–38; and F. P. Siegal, C. Lopez, G. S. Hammer, et al., "Severe Acquired Immunodeficiency in Male Homosexuals, Manifested by Chronic Perianal Ulcerative Herpes Simplex Lesions," *New England Journal of Medicine* 305 (1981): 1439–44.

12. W. L. Drew, L. Mintz, R. C. Miner, et al., "Prevalence of Cytomegalovirus Infection in Homosexual Men," *Journal of Infectious Diseases* 143 (1981): 188–92.

13. H. Masur, M. A. Michelis, J. B. Greene, et al., "An Outbreak of Community-Acquired *Pneumocystis carinii* Pneumonia," *New England Journal of Medicine* 305 (1981): 1431–38.

14. W. W. Darrow, D. Barrett, K. Jay, et al., "The Gay Report on Sexually Transmitted Diseases," *American Journal of Public Health* 71 (1981): 1004–11.

15. Gottlieb, Schroff, Schanker, et al. (1981), op. cit.

16. Siegal and Siegal (1983), op. cit.

17. W. Rozenbaum, J. P. Coulaid, A. G. Saimot, et al., "Multiple Opportunistic Infection in a Male Homosexual in France," *Lancet* I (1982): 572–73.

18. Most budgetary information outlined in this chapter was drawn, unless noted otherwise, from U.S. Congress, Committee on Government Operations, "Federal Response to AIDS" (Washington, D.C.: Government Printing Office, November 30, 1983); and U.S. House of Representatives, Hearings Before a Subcommittee of the Committee on Government Operations, "Federal Response to AIDS," August 2, 1983.

19. No laboratory funds were specifically earmarked for the group, though lab research is generally the most expensive component of any scientific investigation.

20. Original members included Drs. Willy Rozenbaum, Jacques Leibowitch, Serge Kernbaum, Jean-Claude Gluckman, David Klatzmann, Odile Picard, and Charles Mayaud, as well as Claude Villalonga and Jean-Baptiste Brunet. For a detailed description of French anti-AIDS efforts, see M. D. Grmek, *History of AIDS: Emergence and Origin of a Modern Pandemic* (Princeton, NJ: Princeton University Press, 1990).

21. A. R. Moss, P. Bacchetti, M. Gorman, et al., "AIDS in the 'Gay' Areas of San Francisco," *Lancet* I (1983): 923–24.

22. A good synopsis of the early San Francisco Cohort findings can be found in the CDC's *Morbidity and Mortality Weekly Report* 34 (1985): 573–75, entitled "Update: Acquired Immunodeficiency Syndrome in the San Francisco Cohort Study, 1978–1985."

23. In his revealing book, *Koop: The Memories of a Family Physician* (New York: Random House, 1991), former Surgeon General C. Everett Koop described in detail his struggles over AIDS. He asserted, "Within the politics of AIDS lay one enduring, central conflict: AIDS pitted the politics of the gay revolution of the seventies against the politics of the Reagan revolution of the eighties."

24. E. N. Brandt, "Implications of the Acquired Immunodeficiency Syndrome for Health Policy," *Annals of Internal Medicine* 103 (1985): 771–73.

25. According to the U.S. Public Health Service, AIDS-related actual spending in FY 1982 and 1983 broke down as follows:

Agency	(In thousands)	
	FY 1982	FY 1983
ADAMHA (Alcohol, Drug Abuse, and Mental Health Administration)	$ 0	$ 516
CDC (Centers for Disease Control)	$2,050	6,202
FDA (Food and Drug Administration)	150	350
NIH (National Institutes of Health):		
NCI (Nat'l Cancer Institute)	2,400	9,790
NHLBI (Nat'l Heart, Lung and Blood Institute)	5	1,202
NIDR (Nat'l Institute of Dental Research)	25	25
NINCDS (Nat'l Institute of Neurological, Coronary Disease and Stroke)	31	684
NIAID (Nat'l Institute of Allergy and Infectious Diseases)	297	9,223
NEI (Nat'l Eye Institute)	33	45
DRR (Department of Research Resources)	564	699
National Institutes of Health Totals	$3,355	$21,668
Grand Total Public Health Service Spending	$5,555	$28,736

26. Memorandum, Dr. James Wyngaarden, Director, National Institutes of Health, to Board of Institute Directors, July 13, 1982.

27. G. Bosker, "Gays and Cancer—Blaming the Victims?" *In These Times*, August 25–September 7, 1982: 2.

28. The events are described in D. M. Auerbach, W. W. Darrow, H. W. Jaffe, and J. W. Curran, "Cluster of Cases of the Acquired Immune Deficiency Syndrome," *American Journal of Medicine* 76 (1984): 487–92; W. W. Darrow, "AIDS: Socioepidemiologic Responses to an Epidemic," in R. Ulak and W. F. Skinner, eds., *AIDS and the Social Science: Common Threads* (Lexington: University Press of Kentucky, 1991), 82–99; and Centers for Disease Control, "A Cluster of Kaposi's Sarcoma and *Pneumocystis carinii* Pneumonia Among Homosexual Male Residents of Los Angeles and Orange Counties, California," *Morbidity and Mortality Weekly Report* 31 (1982): 305–7.

29. R. Shilts, "Patient Zero: The Man Who Brought the AIDS Epidemic to California," *California*, October 1987: 96–99, 149–60; and R. M. Henig, *A Dancing Matrix* (New York: Alfred A. Knopf, 1993).

30. The early AIDS gay clusters looked like this:

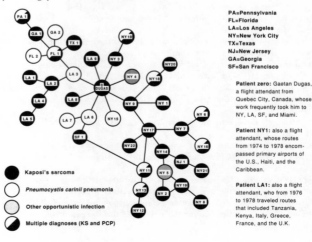

PA=Pennsylvania
FL=Florida
LA=Los Angeles
NY=New York City
TX=Texas
NJ=New Jersey
GA=Georgia
SF=San Francisco

Patient zero: Gaetan Dugas, a flight attendant from Quebec City, Canada, whose work frequently took him to NY, LA, SF, and Miami.

Patient NY1: also a flight attendant, whose routes from 1974 to 1978 encompassed primary airports of the U.S., Haiti, and the Caribbean.

Patient LA1: also a flight attendant, who from 1976 to 1978 traveled routes that included Tanzania, Kenya, Italy, Greece, France, and the U.K.

Kaposi's sarcoma

Pneumocystis carinii pneumonia

Other opportunistic infection

Multiple diagnoses (KS and PCP)

Sources: D. M. Auerbach, W. W. Darrow, H. W. Jaffe, and J. W. Curran, "Clusters of Cases of the Acquired Immune Deficiency Syndrome," *American Journal of Medicine* 76 (1984): 487–92; W. W. Darrow, "AIDS: Socioepidemiologic Responses to an Epidemic," in R. Ulack and W. F. Skinner, eds., *AIDS and the Social Sciences* (Lexington: University Press of Kentucky, 1991), 82–99; and W. W. Darrow, M. E. Gorman, and B. P. Glick, "The Social Origins of AIDS: Social Change, Sexual Behavior, and Disease Trends," in D. A. Feldman and T. M. Johnson, eds., *The Social Dimensions of AIDS: Method and Theory* (New York: Praeger, 1986), 95–107.

31. In Auerbach, Darrow, Jaffe, and Curran (1984), op. cit., the question of latency was depicted as follows:

Possible Source Patient	Linked Patient	Date of Exposure	Symptom Onset	Latency (months)
LA 4	LA 5	12/80	7/81	7
FL 1	GA 2	10/80	6/81	8
Patient 0	LA 9	11/80	8/81	9
NY 18	NY 20	7/80	7/81	12
Patient 0	LA 8	2/80	3/81	13
Patient 0	NY 15	4/80	6/81	14

Latency Average = 10.5 months

32. L. Corey, "The Diagnosis and Treatment of Genital Herpes," *Journal of the American Medical Association* 248 (1982): 1041–49.

33. Centers for Disease Control, "Update on Kaposi's Sarcoma and Opportunistic Infections in Previously Healthy People in the United States," *Morbidity and Mortality Weekly Report* 3 (1982): 294–301.

34. Centers for Disease Control, "Opportunistic Infections and Kaposi's Sarcoma Among Haitians in the United States," *Morbidity and Mortality Weekly Report* 31 (1982): 353–61.

35. For a detailed account of Haiti's AIDS epidemic and its political dimensions in the United States, see P. Farmer, *AIDS and Accusation: Haiti and the Geography of Blame* (Berkeley: University of California Press, 1992).

36. R. Altema and L. Bright, "Only Homosexual Haitians, Not All Haitians," *Annals of Internal Medicine* 99 (1983): 877.

37. Jaffe's rough pictograph of the epidemic appeared as follows:

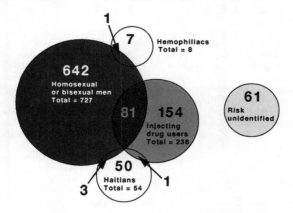

Source: H. W. Jaffe, D. J. Bregman, and R. M. Selik, "Acquired Immune Deficiency Syndrome in the United States: The First 1,000 Cases," *Journal of Infectious Diseases* 148 (1983): 339–45.

38. Centers for Disease Control, "*Pneumocystis carinii* Pneumonia Among Persons with Hemophilia A," *Morbidity and Mortality Weekly Report* 31 (1982): 365–67.

39. Centers for Disease Control, "Possible Transfusion-Associated Acquired Immune Deficiency Syndrome (AIDS)—California," *Morbidity and Mortality Weekly Report* 31 (1982): 652–54.

40. Centers for Disease Control, "Unexplained Immunodeficiency and Opportunistic Infections in

Infants—New York, New Jersey, California," *Morbidity and Mortality Weekly Report* 31 (1982): 665–67.

41. B. D. Colen, "Epidemic Baffles U.S. Experts," *Newsday*, September 12, 1982: 14, 27.

42. H. W. Jaffe, D. J. Bregman, and R. M. Selik, "Acquired Immune Deficiency Syndrome in the United States: The First 1,000 Cases," *Journal of Infectious Diseases* 148 (1983): 339–45.

43. In defense of the lassitude of government responses vis-à-vis the blood supplies of the United States, Canada, and Western European countries, it would later be argued that such surveys were impossible until the etiologic agent of AIDS was discovered. However, in the absence of a viral screening test, much could have been done to study those individuals listed as "hemophiliacs" or "unknown risk factor" cases on government AIDS rosters to identify contaminated units of blood. Requests for funding of such research in the United States were consistently denied until mid-1984.

44. T. Beardsley, "British AIDS: Whose Blood Can Now Be Safe?" *Nature* 303 (1983): 102.

45. In 1983 blood was collected in the United States by 180 regional centers and 1,800 hospitals, all members of the American Association of Blood Banks. In addition, some 3,800 smaller hospitals not members of the AABB collected and transfused blood. See D. M. Surgenor, E. L. Wallace, S. H. S. Hao, and R. H. Chapman, "Collection and Transfusion of Blood in the United States, 1982–1988," *New England Journal of Medicine* 322 (1990): 1646–51.

46. The series for which Gilbert Gaul won the Pulitzer Prize ran over several days in the *Philadelphia Inquirer* in September 1989.

47. In 1980, about 11,600 of the people with hemophilia had a genetic deficiency related to Factor VIII; about 3,000 had a Factor IX deficiency. Both products were available for treatment use.

48. It would later be shown that at least 10 percent of the injecting drug users of New Haven, Connecticut, were already infected with the AIDS virus by 1982. Since other areas have proven to have even greater numbers per capita of AIDS cases among drug users, it is assumed that far more than 10 percent of the users of East Brooklyn, Harlem, the South Bronx, and Newark were infected in 1982. See R. D'Aquila, A. B. Williams, H. D. Kleber, and A. E. Williams, "Prevalence of HTLV-III Infection Among New Haven, Connecticut, Parenteral Drug Abusers in 1982–1983," *New England Journal of Medicine* 314 (1986): 117.

49. There were some laboratories that pooled plasma from over 30,000 donors to make a batch of Factor VIII. See P. H. Levine, "HIV Infection in Hemophilia," *Journal of Clinical Apheresis* 8 (1993): 120–25.

50. D. L. Aronson, "Infection of Hemophiliacs with HIV," *Journal of Clinical Apheresis* 8 (1993): 117–19.

51. These figures were calculated by the author based on 1980 STD data, plus hemophilia blood-use data found in L. M. Aledort, "Current Concepts in Diagnosis and Management of Hemophilia," *Hospital Practice*, October 1982: 77–92; and National Institutes of Health, "Pilot Study of Hemophilia Treatment in the U.S.," report to the Department of Health, Education, and Welfare, June 30, 1972.

52. In lawsuits against the blood bank industry in 1992–94, Francis so testified.

53. With the advantage of hindsight, there were other courses of action open to the FDA and the blood industry in 1983. Without knowing the cause of AIDS, they could have lowered the danger of the blood supply through:

- Closing all plasma and blood-for-money sites. This would only be done for blood years later in the United States, following rising pressure from the voluntary donor segment of the industry. It would never be done for plasma. By 1994, however, most European, Asian, and Latin American countries would still allow commercial blood banks to pay donors. In some countries (such as Brazil and India), most blood would be purchased, and rates of HIV and hepatitis contamination would, as a result, be very high.
- Shutting down mobile and permanent blood collection operations in neighborhoods known to have higher numbers of injecting drug users and gay men. Some blood banks—notably those in San Francisco—took such steps as early as mid-1983; most worldwide did not. See H. A. Perkins, "Safety of the Blood Supply," *Journal of Clinical Apheresis* 8 (1993): 110–16.
- Use only female-donated blood in pooled samples destined to be used for Factors VIII or IX. In 1983 over 90 percent of all AIDS cases in Europe and North America were male, and though that gender disparity would narrow over the years, the odds of infection among women were dramatically less than among men during the early 1980s.
- Verbally counsel all donors about the risks of contaminating the blood supply. Request that those who may be in a "risk group" for AIDS exclude themselves from donating blood.
- Actively lobby surgeons, with the aim of decreasing both unnecessary surgery and the amounts of blood used during necessary procedures.
- Heat the donated blood. Heat treatment of Factor products, already routine for albumin, would

be proven effective for Factor VIII by Jay Levy in mid-1984. Only Cutter Laboratories, which sponsored Levy's study, would then begin sterilization. Most products would not be so treated until mid-1985.

54. At that time HTLV-I was simply called HTLV, because the discovery of HTLV-II hadn't yet been announced. To avoid confusion, however, the author will refer to the various HTLVs in their numbered forms.

55. M. Essex, "Adult T-Cell Leukemia-Lymphoma: Role of a Human Retrovirus," *Journal of the National Cancer Institute* 69 (1982): 981–85.

56. J. J. Goedert, W. C. Wallen, D. L. Mann, et al., "Amyl Nitrite May Alter T Lymphocytes in Homosexual Men," *Lancet* I (1982): 412–15. See also, for speculation on "poppers," I. Gorin, O. Picard, L. LaRoche, et al., "Kaposi's Sarcoma Without the U.S. or 'Popper' Connection," *Lancet* I (1982): 908; G. R. Seage, K. H. Mayer, C. R. Horsburgh, et al., "The Relation Between Nitrite Inhalants, Unprotected Receptive Anal Intercourse, and the Risk of Human Immunodeficiency Virus Infection," *American Journal of Epidemiology* 135 (1992): 1–11; Letters to the Editor (several authors), "Re: An Autopsy of Epidemiologic Methods: The Case of 'Poppers' in the Early Epidemic of the Acquired Immunodeficiency Syndrome (AIDS)," *American Journal of Epidemiology* 131 (1990): 195–200; K. A. Jørgensen, "Amyl Nitrite and Kaposi's Sarcoma in Homosexual Men," *Lancet* 307 (1982): 893–94; and R. O. Brennan and D. T. Durack, "Gay Compromise Syndrome," *Lancet* II (1981): 1338–39.

57. " 'Highly Abnormal' B-Cell Function Found in AIDS," *Hospital Practice*, October 1983: 32–40; and H. C. Lane, H. Masur, A. H. Rook, et al., "Abnormalities of B Lymphocyte Activation and Immunoregulation in Patients with the Acquired Immunodeficiency Syndrome," *New England Journal of Medicine* 309 (1983): 453.

58. Letter to the Honorable Margaret Heckler, May 19, 1983, signed by Donald Abrams, Jay Levy, W. J. W. Morrow, Conrad Casavant, Andrew Moss, Marcus Conant, William Drew, Daniel Stites, Paul Volberding, John Ziegler, and John Greenspan, all of the University of California at San Francisco.

59. Robert S. Walker (Pennsylvania), Frank Horton (New York), John N. Ehlenborn (Illinois), Lyle Williams (Ohio), William F. Clinger, Jr. (Pennsylvania), Judd Gregg (New Hampshire), Dan Burton (Indiana), Alfred A. McCandless (California), Larry Craig (Idaho), and Dan Schaefer (Colorado).

60. Groupe de Travail Français sur le SIDA, "Le Syndrome d'Immuno-déficit Acquis," *La Presse Médicale* 12 (1983): 2453–56.

61. Centers for Disease Control, "Experimental Infection of Chimpanzees with Lymphadenopathy-Associated Virus," *Morbidity and Mortality Weekly Report* 33 (1984): 442–44. By the time the CDC finally mustered the resources for animal research, the Pasteur Institute's research team had already identified the presence of a retrovirus in AIDS patients, which they dubbed lymphadenopathy-associated virus, or LAV. Therefore, the CDC actually injected viral inoculum into the test animals. The virus did enter the animals' cells, and reproduced, but no symptomatic disease had been produced by August 1984.

62. J. Gerstoft, A. Malchow-Møller, I. Bygbjerg, et al., "Severe Acquired Immunodeficiency in European Homosexual Men," *British Medical Journal* 235 (1982): 17–19.

63. Gorin, Picard, LaRoche, et al. (1982), op. cit.

64. Groupe de Travail Français sur le SIDA, "Sarcome de Kaposi et Infections Opportunistes chez des Subjets Jeunes sans Antécédent Susceptible d'Entraîner une Immuno-dépression," *La Presse Médicale* 12 (1983): 2431–34. Other cases with which Liebowitch was familiar are described in J. B. Brunet, E. Bouvet, J. Liebowitch, et al., "Acquired Immunodeficiency Syndrome in France," *Lancet* I (1983): 700–01.

65. N. Clumeck, F. Mascaret-Lemone, J. deMaubeuge, et al., "Acquired Immune Deficiency Syndrome in Black Africans," *Lancet* I (1983): 642.

66. R. M. DuBois, J. R. Mikhail, and J. C. Batten, "Primary *Pneumocystis carinii* and Cytomegalovirus Infections," *Lancet* II (1981): 1339; J. L'Age-Stehr, R. Kunze, and M. A. Koch, "AIDS in West Germany," *Lancet* II (1983): 1370–71; G. Rezza, G. Ippolito, G. Marasca, and D. Greco, "AIDS in Italy," *Lancet* II (1984): 642; W. Rozenbaum, D. Klatzmann, C. Mayaud, et al., "Syndrome d'Immunodépression Acquire chez 4 Homosexuals," *La Presse Médicale* 12 (1983): 1149–54; O. Tello, "AIDS in Spain," *Lancet* II (1984): 1472; and H. K. Thomsen, M. Jacobsen, and A. Malchow-Møller, "Kaposi Sarcoma Among Homosexual Men in Europe," *Lancet* II (1981): 688.

67. Working in the lab with Montagnier were Françoise Barré-Sinoussi, Jean-Claude Chermann, David Klatzmann, Jean-Claude Gluckman, Marc Alizon, Simon Wain-Hobson, Pierre Sonige, and Christine Rouzioux.

68. The history of the discovery of the human immunodeficiency virus, or HIV-1, is mired in extraordinary controversy that still, in the 1990s, defies absolutely objective reporting. Indeed, the longer the disputed history has been debated in the popular press, scientific journals, halls of Congress and Assemblé, U.S. patent courts, inside the NIH and Pasteur Institute, the more difficult it has become to separate myth from fact, personality from genuine achievement, nationalism from legitimate scientific competition, and politics from science.

For the purpose of this book it is not necessary to reinterpret the history of the discovery of HIV-1 and the Franco-American dispute in detail; nor is it germane to the overarching issues of disease emergence.

Nevertheless, curious readers anxious to reach their own conclusions about who discovered HIV-1, and how damaging the scientific duel may have been to the overall AIDS research efforts, are directed to the following:

- M. Chase, "French Scientists Sue U.S. on AIDS Research Royalties," *Wall Street Journal*, December 16, 1985: A1.
- Connor and Kingman (1988), op. cit., Chapters 3 and 4.
- J. Crewdson, "The Great AIDS Quest," *Chicago Tribune*, November 19, 1989: A1.
- F. J. Dyson, "Science in Trouble," *American Scholar* 62 (1993): 513–25.
- A. G. Fettner, *Viruses: Agents of Change*, Part III (New York: McGraw-Hill, 1990).
- Gallo, *Virus Hunting* (1991), op. cit., Part III.
- R. C. Gallo, G. M. Shaw, and P. D. Markham, "The Etiology of AIDS," Chapter 2 in V. T. DeVita, S. Hellman, and S. A. Rosenberg, eds., *AIDS: Etiology, Diagnosis, Treatment and Prevention* (New York: J. B. Lippincott, 1985).
- M. D. Grmek (1990), op. cit., Chapters 6 and 7.
- Groupe de Travail Français sur le SIDA, "Le Syndrome d'Immuno-déficit Acquis: Une Nouvelle Maladie d'Origine Infectieuse?" *La Presse Médicale* 12 (1983): 2453–56.
- C. Marwick, "French, U.S. Viral Isolates Compared in Search for Cause of AIDS," *Journal of the American Medical Association* 251 (1984): 2901–9.
- J. Palca, "Hints Emerge from the Gallo Probe," *Science* 253 (1991): 728–31.
- D. Remnick, "Robert Gallo Goes to War," *Washington Post*, August 9, 1987: 11, 43.
- Research Integrity Adjudications Panel, Docket No. A-93-100, Decision No. 1446, U.S. Department of Health and Human Services, 1993.
- "Settling the AIDS Virus Dispute" and "The Chronology of AIDS Research," both in *Nature* 326 (1987): 425–26 and 435–36.
- B. Seytre, "British Say Pasteur Institute Slighted Their Help on AIDS Test," *Nature* 358 (1992): 358.
- R. Shilts (1987), op. cit.
- "The One True Virus," *The Economist*, June 8, 1991: 83–84.
- B. Werth, "By AIDS Obsessed," *GQ Magazine*, August 1991: 144–208.

69. Working on the effort in 1982–85 in the Gallo lab at the National Cancer Institute were Flossie Wong-Stahl, M. G. Sarngadharan, S. Zaki Salahuddin, Mikulas Popovic, Beatrice Hahn, George Shaw, Howard Streicher, and Genoveffa Franchini.

70. J. Maurice, "Human 'T' Leukemia Virus Still Suspected in AIDS," *Journal of the American Medical Association* 250 (1983): 1015–21.

71. Jay Levy's laboratory group included Anthony Hoffman, Susan Kramer, Jill Landis, Joni Shimabukuro, and Lyndon Oshiro.

72. J. A. Levy and J. S. Ziegler, "Acquired Immunodeficiency and Kaposi's Sarcoma Results from Secondary Immune Stimulant," *Lancet* II (1983): 78–81.

73. R. T. Ravenholt, "Role of Hepatitis B Virus in Acquired Immunodeficiency Syndrome," *Lancet* II (1983): 885–86.

74. M. H. Poleski, "Kaposi's Sarcoma and Hepatitis B Vaccine," *Annals of Internal Medicine* 97 (1982): 5; M. I. McDonald, J. D. Hamilton, and D. T. Duract, "Hepatitis B Surface Antigen Could Harbour the Infective Agent of AIDS," *Lancet* II (1983): 882–84; letters from various authors, *New England Journal of Medicine* 312 (1985): 375–76; H. S. Sacks, D. N. Rose, and T. C. Chalmers, "Should the Risk of Acquired Immunodeficiency Syndrome Deter Hepatitis B Vaccination?" *Journal of the American Medical Association* 252 (1984): 3375–77; and Centers for Disease Control, "Hepatitis B Vaccine: Evidence Confirming Lack of AIDS Transmission," *Morbidity and Mortality Weekly Report* 33 (1984): 685–87.

75. H. L. Coulter, *AIDS and Syphilis: The Hidden Link* (Berkeley, CA: North Atlantic Books, 1987).

76. H. H. Neumann, "Use of Steroid Creams as a Possible Cause of Immunosuppression in Homosexuals," *New England Journal of Medicine* 306 (1982): 935.

77. P. N. Goldwater, B. J. L. Synek, T. D. Koelmeyer, and P. J. Scott, "Scrapie-Associated Fibrils and AIDS Encephalopathy," *Lancet* II (1985): 1300.

78. J. Teas, "Could AIDS Agent Be a New Variant of African Swine Fever Virus?" *Lancet* I (1983): 922–23; and J. Beldekas, J. Teas, and J. R. Hebert, "African Swine Fever and AIDS," *Lancet* I (1986): 564–65.

79. For an excellent analysis of these and other theories on the origin and cause of AIDS, see R. Sabatier, *Blaming Others: Prejudice, Race and Worldwide AIDS* (London: Panos Institute, 1988).

80. R. J. Ablin and M. J. Gonder, "Possible Immunosuppressive Factors in Blood Products," *Annals of Internal Medicine* 100 (1984): 155–56.

81. T. J. Greenwalt, "Blood-Products Transfusion and the Acquired Immunodeficiency Syndrome," *Annals of Internal Medicine* 100 (1984): 155.

82. A. J. Amman, D. W. Wara, S. Dritz, et al., "Acquired Immunodeficiency in an Infant: Possible Transmission by Means of Blood Products," *Lancet* I (1983): 956–58; G. Angarano, G. Pastore, L. Monno, et al., "Rapid Spread of HTLV-II Infection Among Drug Addicts in Italy," *Lancet* II (1985): 1302; J. R. Bove, "Transfusion-Associated AIDS: A Cause for Concern," *New England Journal of Medicine* 310 (1984): 115–16; J. R. Jett, J. N. Kuritsky, J. A. Katzmann, and H. A. Homburger, "Acquired Immunodeficiency Syndrome Associated with Blood-Product Transfusion," *Annals of Internal Medicine* 99 (1983): 621–24; E. Lissen, I. Wichmann, J. M. Jimenez, and F. Andrew-Kern, "AIDS in Haemophilia Patients in Spain," *Lancet* I (1983): 992; M. Malbye, R. J. Biggar, J. C. Chermann, et al., "High Prevalence of Lymphadenopathy Virus (LAV) in European Haemophiliacs," *Lancet* II (1984): 40–41; and J. Wood, "AIDS Mystery: Why It Misses Many Blood Recipients," *San Francisco Chronicle*, July 10, 1983: A4.

83. R. D. deShazo, A. Andes, J. Nordberg, et al., "An Immunologic Evaluation of Hemophiliac Patients and Their Wives," *Annals of Internal Medicine* 99 (1983): 159–64.

84. H. W. Jaffe, K. Choi, P. A. Thomas, et al., "National Case-Control Study of Kaposi's Sarcoma and *Pneumocystis carinii* Pneumonia in Homosexual Men: Part 1, Epidemiologic Results," *Annals of Internal Medicine* 99 (1983): 145–51.

85. M. F. Rogers, D. M. Morens, J. A. Stewart, et al., "National Case-Control Study of Kaposi's Sarcoma and *Pneumocystis carinii* Pneumonia in Homosexual Men: Part 2, Laboratory Results," *Annals of Internal Medicine* 99 (1983): 151–58.

86. *Nature* 302 (1983): 749–50.

87. D. P. Francis, J. W. Curran, and M. Essex, "Epidemic Acquired Immune Deficiency Syndrome: Epidemiologic Evidence for a Transmissible Agent," *Journal of the National Cancer Institute* 71 (1983): 1–4.

88. Press release of the Office of Cancer Communications, National Cancer Institute, May 12, 1983. The press release refers to the *Science* studies, which were: R. C. Gallo, P. S. Sarin, E. P. Gelmann, et al., "Isolation of Human T-Cell Leukemia Virus in Acquired Immune-Deficiency Syndrome (AIDS)," *Science* 220 (1983): 865–67; E. P. Gelmann, M. Popovic, D. Blayney, et al., "Proviral DNA of a Retrovirus, Human T-Cell Leukemia Virus, in Two Patients with AIDS," *Science* 220 (1983): 862–65; M. Essex, M. F. McLane, T. H. Lee, et al., "Antibodies to Cell Membrane Antigens Associated with Human T-Cell Leukemia Virus in Patients with AIDS," *Science* 220 (1983): 859–62; and F. Barré-Sinoussi, J. C. Chermann, F. Rey, et al., "Isolation of a T-Lymphotropic Retrovirus from a Patient at Risk for Acquired Immunodeficiency Syndrome (AIDS)," *Science* 220 (1983): 868–71.

89. J. Oleske, A. Minnefor, R. Cooper, et al., "Immune Deficiency Syndrome in Children," *Journal of the American Medical Association* 249 (1983): 2345–49; and A. Rubinstein, M. Sicklick, A. Gupta, et al., "Acquired Immunodeficiency with Reversed T4/T8 Ratios in Infants Born to Promiscuous and Drug-Addicted Mothers," *Journal of the American Medical Association* 249 (1983): 2350–56.

90. Centers for Disease Control, "Immunodeficiency Among Female Sexual Partners of Males with Acquired Immune Deficiency Syndrome (AIDS)—New York," *Morbidity and Mortality Weekly Report* 31 (1983): 697–98.

91. There are several sources of reference for motivated readers. See A. Goldstein, *Addiction: From Biology to Drug Policy* (New York: W. H. Freeman, 1994); and E. M. Brecher, *Licit and Illicit Drugs* (Boston: Little, Brown, 1972). The Goldstein book contains an extremely useful reading list.

92. R. V. Henrickson, D. H. Maul, K. G. Osborn, et al., "Epidemic of Acquired Immunodeficiency in a Colony of Macaque Monkeys," *Lancet* I (1983): 388–90.

93. M. D. Daniel, N. W. King, N. L. Letvin, et al., "A New Type D Retrovirus Isolated from Macaques with an Immunodeficiency Syndrome," *Science* 223 (1984): 602–5.

94. G. Weissman, "AIDS and Heat," *Hospital Practice*, October 1983: 136–49.

95. I. Braveny, "AIDS—A New Plague?" *European Journal of Clinical Microbiology* 2 (1983): 183–85.

96. See D. Grady, "AIDS: A Plague of Fear," *Discover*, July 1983: 73–77; J. E. Groopman and M. S. Gottlieb, "AIDS: The Widening Gyre," *Nature* 303 (1983): 575–76; and Shilts (1987), op. cit.
97. Maurice (1983), op. cit.
98. Groupe de Travail Français sur le SIDA, "Le Syndrome d'Immunodéficit Acquis," *La Presse Médicale* 12 (1983): 2453–56.
99. H. W. Jaffe, D. P. Francis, M. F. McLane, et al., "Transfusion-Associated AIDS: Serologic Evidence of Human T-Cell Leukemia Virus Infection of Donors," *Science* 223 (1984): 1309–12.
100. E. Vilmer, F. Barré-Sinoussi, C. Rouzioux, et al., "Isolation of New Lymphotropic Retrovirus from Two Siblings with Haemophilia B, One with AIDS," *Lancet* I (1984): 753–57.
101. M. Chase, "Cancer Virus Tied to AIDS May Be Disclosed Soon," *Wall Street Journal*, April 16, 1984: A1.
102. M. Popovic, M. G. Sarngadharan, E. Read, and R. Gallo, "Detection, Isolation, and Continuous Production of Cytopathic Retroviruses (HTLV-III) from Patients with AIDS and Pre-AIDS," *Science* 224 (1984): 497–500; R. C. Gallo, S. Z. Salahuddin, M. Popovic, et al., "Frequent Detection and Isolation of Cytopathic Retroviruses (HTLV-III) from Patients with AIDS and at Risk for AIDS," *Science* 224 (1984): 500–2; J. Schüpbach, M. Popovic, R. V. Gilden, et al., "Serological Analysis of a Subgroup of Human T-Lymphotropic Retroviruses (HTLV-III) Associated with AIDS," *Science* 224 (1984): 503–5; and M. G. Sarngadharan, M. Popovic, L. Bruch, et al., "Antibodies Reactive with Human T-Lymphotropic Retroviruses (HTLV-III) in the Serum of Patients with AIDS," *Science* 224 (1984): 506–8.
103. C. Marwick, "French, U.S. Viral Isolates Compared in Search for Cause of AIDS," *Journal of the American Medical Association* 251 (1984): 2901–09.
104. L. G. Gürtler, D. Wernicke, J. Eberle, et al., "Increase in Prevalence of Anti-HTLV-III in Haemophiliacs," *Lancet* II (1984): 1275–76; and J. E. Groopman, S. Z. Salahuddin, M. G. Sarngadharan, et al., "Virologic Studies in a Case of Transfusion-Associated AIDS," *New England Journal of Medicine* 311 (1984): 1418–22.
105. G. M. Shaw, B. H. Hahn, S. K. Arya, et al., "Molecular Characterization of Human T-Cell Leukemia (Lymphotropic) Virus Type III in the Acquired Immune Deficiency Syndrome," *Science* 226 (1984): 1165–71; and M. Alizon, P. Sonigo, F. Barré-Sinoussi, et al., "Molecular Cloning of Lymphadenopathy-Associated Virus," *Nature* 312 (1984): 757–60.
106. J. A. Levy, A. D. Hoffmann, S. M. Kramer, et al., "Isolation of Lymphocytopathic Retroviruses from San Francisco Patients with AIDS," *Science* 225 (1984): 840–42.
107. P. A. Luciw, S. J. Potter, K. Steimer, et al., "Molecular Cloning of AIDS-Associated Retrovirus," *Nature* 312 (1984): 760–63.
108. A. G. Dagliesh, P. C. L. Beverley, P. R. Clapham, et al., "The CD4 (T4) Antigen Is an Essential Component of the Receptor for the AIDS Retrovirus," *Nature* 312 (1984): 763–66.
109. S. Wain-Hobson, P. Sonigo, O. Danos, et al., "Nucleotide Sequence of the AIDS Virus, LAV," *Cell* 40 (1985): 9–17; L. Ratner, W. Haseltine, R. Patarca, et al., "Complete Nucleotide Sequence of the AIDS Virus, HTLV-III," *Nature* 313 (1985): 277–84; and I. M. Chiu, A. Yaniv, J. E. Dahlberg, et al., "Nucleotide Sequence Evidence for Relationship of AIDS Retrovirus to Lentivirus," *Nature* 317 (1985): 366–68.
110. S. Broder and R. C. Gallo, "A Pathogenic Retrovirus (HTLV-III) Linked to AIDS," *New England Journal of Medicine* 311 (1984): 1292–1303.
111. In 1993 scientists working with the World Health Organization would conclude that at least one local subtype (dubbed D and found in parts of Tanzania, Uganda, and Rwanda) differed from all other HIVs on the planet. It possessed the genetic ability to immediately infect macrophages, bypassing intermediary cells, and to cause devastation of the human T-cell system in a matter of months, rather than years. As many as 12 percent of those infected with subtype D progressed from asymptomatic infection to full AIDS in less than twelve months.
112. "AIDS in Europe, Status Quo 1983." *European Journal of Cancer and Clinical Oncology* 20 (1984): 155–73; Brunet, Bouvet, Liebowitch, et al. (1983), op. cit.; I. C. Bygbjerg, "AIDS in a Danish Surgeon (Zaire 1976)," *Lancet* I (1983): 925; Clumeck, Mascart-Lemone, deMaubeuge, et al. (1983), op. cit.; N. Clumeck, J. Sonnet, H. Taelman, et al., "Acquired Immunodeficiency Syndrome in African Patients," *New England Journal of Medicine* 310 (1984): 492–97; D. Edwards, P. G. Harper, A. K. Pain, et al., "Kaposi's Sarcoma Associated with AIDS in a Woman from Uganda," *Lancet* I (1984): 631–32; G. Offenstadt, P. Pinta, P. Hericord, et al., "Multiple Opportunistic Infection Due to AIDS in a Previously Healthy Black Woman from Zaire," *New England Journal of Medicine* 308 (1983): 775; and J. Vandepitte, R. Verwilghen, and P. Zachee, "AIDS and Cryptococcosis (Zaire 1977)," *Lancet* I (1983): 925–26.
113. A. Ellrodt, F. Barré-Sinoussi, Ph. LeBras, et al., "Isolation of Human T-Lymphotropic

Retrovirus (LAV) from Zairian Married Couple, One with AIDS, One with Prodromes," *Lancet* I (1984): 1383–85.

114. Clumeck, Sonnet, Taelman, et al. (1984), op. cit.

115. P. Van de Perre, D. Rouvroy, P. Lepage, et al., "Acquired Immunodeficiency Syndrome in Rwanda," *Lancet* II (1984): 62–65.

116. B. Ivanoff, P. Duquesnoy, G. Languillat, et al., "Haemorrhagic Fever in Gabon: I. Incidence of Lassa, Ebola and Marburg Viruses in Haut-Ogooué," *Transactions of the Royal Society of Tropical Medicine and Hygiene* 76 (1982): 719–20.

117. N. J. Cox, J. B. McCormick, K. M. Johnson, and M. P. Kiley, "Evidence for Two Subtypes of Ebola Virus Based on Oligonucleotide Mapping of RNA," *Journal of Infectious Diseases* 147 (1983): 272–75.

118. P. Piot, T. C. Quinn, H. Taelman, et al., "Acquired Immunodeficiency Syndrome in a Heterosexual Population in Zaire," *Lancet* I (1984): 65–69.

119. R. Shilts, "The Heterosexual Connection: AIDS Researchers Look to Africa," *San Francisco Chronicle*, November 7, 1984: 5.

120. SIDA is the French acronym for AIDS.

121. F. Brun-Vézinet, C. Rouzioux, L. Montagnier, et al., "Prevalence of Antibodies to Lymphadenopathy-Associated Retrovirus in African Patients with AIDS," *Science* 236 (1984): 453–56.

122. J. M. Mann, H. Francis, T. Quinn, et al., "Surveillance for AIDS in a Central African City," *Journal of the American Medical Association* 255 (1986): 3255–59.

123. The finding was also published. See P. Van de Perre, N. Clumeck, M. Carael, et al., "Female Prostitutes: A Risk Group for Infection with Human T-Cell Lymphotropic Virus Type III," *Lancet* II (1985): 524–26; and N. Clumeck, M. Robert-Guroff, and P. Van de Perre, "Seroepidemiological Studies of HTLV-III Antibody Prevalence Among Selected Groups of Heterosexual Africans," *Journal of the American Medical Association* 254 (1985): 2599–2602c.

124. The proceedings of the First International Conference on AIDS were published in the *Annals of Internal Medicine*, Vol. 103, No. 5 (1985). The Kenyan data also appeared in R. J. Biggar, B. K. Johnson, C. Oster, et al., "Regional Variation in Prevalence of Antibody Against Human T-Lymphotropic Virus Types I and III in Kenya, East Africa," *International Journal of Cancer* 35 (1985): 763–67.

125. R. J. Biggar, M. Melbye, L. Kestens, et al., "Seroepidemiology of HTLV-III Antibodies in a Remote Population of Eastern Zaire," *British Medical Journal* 290 (1985): 808–10.

126. W. C. Saxinger, P. H. Levine, A. G. Dean, et al., "Evidence for Exposure to HTLV-III in Uganda Before 1973," *Science* 227 (1985): 1036–38.

127. These findings were published. See P. J. Kanki, M. F. McLane, and N. W. King, "Serologic Identification and Characterization of a Macaque T-Lymphotropic Retrovirus Closely Related to HTLV-III," *Science* 228 (1985): 1199–1201; M. D. Daniel, N. L. Letvin, N. W. King, et al., "Isolation of T-Cell Tropic HTLV-III-like Retrovirus from Macaques," *Science* 228 (1985): 1201–4; and P. J. Kanki, R. Kurth, W. Becker, et al., "Antibodies to Simian T-Lymphotropic Virus Type III in African Green Monkeys and Recognition of STLV-III Viral Proteins by AIDS and Related Sera," *Lancet* I (1985): 1330–32.

128. For excellent descriptions of the blame-counterblame atmosphere that surrounded discussions of AIDS in Africa from late 1983 to 1990, see Sabatier (1988), op. cit.; and T. Barnett and P. Blaikie, *AIDS in Africa: Its Present and Future Impact* (New York: Guilford Press, 1992).

During elections in South Africa in the mid-1980s, opposing white politicians made use of inflated African AIDS data. One right-wing presidential candidate even claimed that ANC guerrillas in Zambia were deliberately getting infected with HIV, sneaking back into South Africa, and then having sex with white women in order to spread AIDS among Boers.

Inflated AIDS reports were a common feature of the apartheid regime. See, for example, S. Simmie, "One in Ten Africans Has AIDS," *Weekly Mail*, Johannesburg, May 20–June 5, 1986: 9.

129. United Nations Development Programme, *Human Development Report 1993* (Oxford, Eng.: Oxford University Press, 1993). According to the United Nations, military expenditures in non-Arab Africa were as follows:

Rank/Country†	Military Expenditures as % of GDP 1960	1990	Military Expenditures as % of Combined Health and Education Expenditure 1977	1990	Annual Average Imports of Conventional Arms, 1987–91, in U.S. Millions
63 Seychelles	N/A	N/A	N/A	N/A	N/A
104 Botswana	N/A	2.5	N/A	16	18
107 Algeria	2.1	1.5	26	18	220

109 Gabon	N/A	4.5	16	63	33
114 Cape Verde	N/A	N/A	N/A	N/A	N/A
117 Swaziland	N/A	N/A	5	20	N/A
120 Lesotho	N/A	N/A	N/A	N/A	1
121 Zimbabwe	N/A	7.3	116	65	58
125 São Tomé & Principe	N/A	N/A	N/A	N/A	N/A
126 Congo	0.3	3.2	51	50	0
127 Kenya	0.5	2.4	52	31	43
128 Madagascar	0.3	1.4	44	34	N/A
130 Zambia	1.1	3.2	140	43	N/A
131 Ghana	1.1	0.6	14	13	11
133 Cameroon	1.7	2.1	36	51	4
135 Namibia*	N/A	N/A	N/A	N/A	N/A
136 Côte d'Ivoire	0.5	1.2	10	14	12
138 Tanzania	0.1	6.9	58	108	N/A
139 Comoros	N/A	N/A	N/A	N/A	N/A
140 Zaire	N/A	1.2	54	67	N/A
142 Nigeria*	0.2	0.9	92	65	75
144 Liberia*	1.1	N/A	18	29	N/A
145 Togo	N/A	3.2	28	46	5
146 Uganda*	N/A	0.8	63	N/A	N/A
149 Rwanda*	N/A	1.7	77	35	1
150 Senegal	0.5	2.0	51	N/A	6
151 Ethiopia*	1.6	13.5	121	239	121
153 Malawi	N/A	1.5	50	31	2
154 Burundi	N/A	2.2	65	65	N/A
155 Equatorial Guinea	N/A	N/A	125	N/A	N/A
156 Central African Rep.	N/A	1.8	36	41	1
157 Mozambique*	N/A	N/A	132	N/A	N/A
158 Sudan*	1.5	2.0	94	N/A	46
160 Angola*	N/A	20.0	N/A	N/A	721
161 Mauritania*	N/A	N/A	154	N/A	N/A
162 Benin	1.1	2.0	22	N/A	1
164 Guinea-Bissau	N/A	N/A	N/A	N/A	N/A
165 Chad*	N/A	N/A	150	N/A	14
166 Somalia*	N/A	3.0	91	500	N/A
167 Gambia	N/A	N/A	N/A	N/A	N/A
168 Mali	1.7	3.2	62	83	6
169 Niger	0.3	0.8	18	21	N/A
170 Burkina Faso*	0.6	2.8	92	85	N/A
172 Sierra Leone	N/A	0.7	18	11	2
173 Guinea*	1.3	N/A	27	N/A	5

† Listed in order of UNDP Human Development World Ranking. Ranking number precedes country name.
* These countries were in a state of war, civil war, or insurrection during some or all of the period 1983–87.
N/A = not available.
Note: South Africa was not listed by the UN.

130. A similar pattern of rain forest encroachment and subsequent desertification was later observed along South America's Amazon Basin. See J. deOnis, *The Green Cathedral: Sustainable Development of Amazonia* (New York: Oxford University Press, 1992).

131. R. J. Biggar, P. L. Gigasse, M. Melbye, et al., "ELISA HTLV Retrovirus Antibody Reactivity Associated with Malaria and Immune Complexes in Healthy Africans," *Lancet* II (1985): 520–23.

132. D. Serwadda, R. D. Mugerwa, N. K. Sewankambo, et al., "Slim Disease: A New Disease in Uganda and Its Association with HTLV-III Infection," *Lancet* II (1985): 849–52.

133. See T. C. Quinn, J. M. Mann, J. W. Curran, and P. Piot, "AIDS in Africa: An Epidemiologic Paradigm," *Science* 234 (1986): 955–56.

134. A. E. Greenberg, C. A. Schable, A. J. Sulzer, et al., "Evaluation of Serological Cross-Reactivity Between Antibodies to Plasmodium and HTLV-III/LAV," *Lancet* II (1986): 247–48; A. Srinivasan and D. York, "Lack of HIV Replication in Arthropod Cells," *Lancet* I (1987): 094–95; D. Connelley, "Bedbugs and HIV," *New Scientist* (June 5, 1986): 69; J. C. Chermann, "Isolation of

AIDS Virus from Insects," *Comptes Rendues d'Académie des Sciences*, Series A, 303: 303–6; Office of Technology Assessment, *AIDS-Related Issues*, Staff Paper 1, September 1987, report to the U.S. Congress (Washington, D.C.: Government Printing Office); and M. J. Blaser, "Acquired Immunodeficiency Syndrome Possibly Arthropod-Borne," *Annals of Internal Medicine* 99 (1983): 877.

135. P. J. Kanki, M. F. McLane, and N. W. King, "Serologic Identification and Characterization of a Macaque T-Lymphotropic Retrovirus Closely Related to HTLV-III," *Science* 228 (1985): 1199–1201.

136. J. M. Mann, H. Francis, F. Davachi, et al., "Risk Factors for Human Immunodeficiency Virus Seropositivity Among Children 1–24 Months Old in Kinshasa, Zaire," *Lancet* II (1986): 654–56.

137. J. M. Mann, K. Bila, R. L. Colebunders, et al., "Natural History of Human Immunodeficiency Virus Infection in Zaire," *Lancet* II (1986): 707–9; Quinn, Mann, Curran, and Piot (1986), op. cit.; and K. Kayembe, J. M. Mann, H. Francis, et al., "Prévalence des Anticorps Anti-HIV chez les Patients Non Atteints de SIDA ou de Syndrome Associé au SIDA à Kinshasa, Zaïre," *Annals de la Société Belge de Médecine Tropique* 66 (1986): 343–48.

138. J. M. Mann, H. Francis, F. Davachi, et al., "Human Immunodeficiency Virus Seroprevalence in Pediatric Patients 2 to 14 Years of Age at Mama Yemo Hospital, Kinshasa, Zaire," *Pediatrics* 78 (1986): 673–78.

139. J. M. Mann, H. Francis, T. C. Quinn, et al., "HIV Seroprevalence Among Hospital Workers in Kinshasa, Zaire," *Journal of the American Medical Association* 256 (1986): 3099–3102.

140. M. Melbye, E. K. Njelesani, A. Bayley, et al., "Evidence for Heterosexual Transmission and Clinical Manifestations of Human Immunodeficiency Virus Infection and Related Conditions in Lusaka, Zambia," *Lancet* II (1986): 113–15.

141. Tanzania, Zaire, Central African Republic, Zambia, Congo, Kenya, Rwanda, Burundi, Uganda.

142. M. Grmek, *Histoire du SIDA: Début et Origine d'une Pandémie Actuelle* (Paris: Editions Payout, 1989). English language version, 1990, op. cit.

143. Sabatier (1988), op. cit.

144. American political scientist Alfred Fortin summarized the atmosphere in a 1986 speech to the "Challenge of AIDS" conference (Miami, Florida, November 12–16): "And so it is that out of a rather ordinary intellectual inquiry into health issues surfaces the politics of East and West, of right and left, of the exotic and the mundane, of violence and peace, of life and death itself. In these issues of health and health care we can see the struggle against colonialism and neocolonialism, against the financial exploitation by trans-national corporations, the armies, against the ravages of nature, and most of all, against the tragedies built into poverty and ignorance. It is within all of this, the great politics of struggle and survival, that the question of AIDS in Africa must be examined."

145. That sort of molecular epidemiology would become possible in 1986 with Kary Mullis's invention of PCR (polymerase chain reaction). The technique allowed scientists to extract from a biological mass a piece of DNA or RNA and make unlimited numbers of copies of the material. Thus, the seemingly impossible became quite simple. In 1991 the CDC would use the technique to determine whether Florida dentist David Acer infected some of his patients with HIV. The strain of HIV found in Acer's blood could be "fingerprinted" using PCR, and compared with the genetic "fingerprints" of the HIV strains infecting his patients. In this way, the agency proved that Acer's virus was, indeed, identical to those found in the patients—a circumstance that could not have been due to chance.

An excellent description of PCR can be found in K. B. Mullis, "The Unusual Origin of the Polymerase Chain Reaction," *Scientific American*, April 1990: 56–65.

146. B. L. Evatt, E. D. Gompert, J. S. McDougal, and R. B. Ramsey, "Coincidental Appearance of LAV/HTLV-III Antibodies in Hemophiliacs and the Onset of the AIDS Epidemic," *New England Journal of Medicine* 312 (1985): 483–86.

147. J. D. Moore, E. J. Cone, and S. S. Alexander, "HTLV-III Seropositivity in 1971–1972 Parenteral Drug Abusers: A Case of False Positives or Evidence of Viral Exposure?" *New England Journal of Medicine* 314 (1986): 1387–88.

148. H. Nelson and R. Steinbrook, "Drug Users—Not Gays—Called First AIDS Victims," *Los Angeles Times*, October 18, 1985: A1; and D. Perlman, "Drug Users Started AIDS Epidemic, Doctor Says," *San Francisco Chronicle*, October 18, 1985: 28.

149. Darrow, Gorman, and Glick (1986), op. cit.

150. M. Elvin-Lewis, M. Witte, C. Witte, et al., "Systemic Chlamydial Infection Associated with Generalized Lymphedema and Lymphangio-sarcoma," *Lymphology* 6 (1973): 113–21; and M. H. Witte,

C. L. Witte, L. L. Minnich, et al., "AIDS in 1968," *Journal of the American Medical Association* 251 (1984): 2657.

151. R. F. Garry, M. H. Witte, A. Gottlieb, et al., "Documentation of an AIDS Virus Infection in the United States in 1968," *Journal of the American Medical Association* 260 (1988): 2085–87.

152. G. R. Hennigar, K. Vinijchaikul, A. L. Roque, and H. A. Lyons, "*Pneumocystis carinii* Pneumonia in an Adult: Report of a Case," *American Journal of Clinical Pathology* 35 (1961): 353–64.

153. I. C. Bygbjerg, "AIDS in a Danish Surgeon (Zaire 1976)," *Lancet* I (1983): 925.

154. C. F. Lindboe, S. S. Froland, K. W. Wefring, et al., "Autopsy Findings in Three Family Members with a Presumably Acquired Immunodeficiency Syndrome of Unknown Origin," *Acta Pathology, Microbiology and Immunology, Scandinavia* 94 (1986): 117–23; and S. S. Froland, P. Jenum, C. F. Lindboe, et al., "HIV-1 Infection in a Norwegian Family Before 1971," *Lancet* I (1988): 1344–45.

155. G. Williams, T. B. Stretton, and J. C. Leonard, "Cytomegalic Inclusions Disease and *Pneumocystis carinii* Infection in an Adult," *Lancet* II (1960): 951–55; and G. Williams, T. B. Stretton, and J. C. Leonard, "AIDS in 1959?" *Lancet* II (1983): 1136.

156. J. R. Leonidas and N. Hyppolite, "Haiti and the Acquired Immunodeficiency Syndrome," *Annals of Internal Medicine* 98 (1983): 1020–21.

157. L. Gazzolo, A. Gessain, A. Carrel, et al., "Antibodies to HTLV-III in Haitian Immigrants to French Guiana," *New England Journal of Medicine* 311 (1984): 1252–53.

158. A. E. Pitchenik, M. A. Fischl, G. M. Dickinson, et al., "Opportunistic Infections and Kaposi's Sarcoma Among Haitians: Evidence of a New Acquired Immunodeficiency State," *Annals of Internal Medicine* 98 (1983): 277–86; and J. W. Pape, B. Liautaud, F. Thomas, et al., "Characteristics of the Acquired Immunodeficiency Syndrome (AIDS) in Haiti," *New England Journal of Medicine* 309 (1983): 945–50.

159. R. Colebunders, H. Taelman, and P. Piot, "AIDS: An Old Disease from Africa?" *British Medical Journal* 289 (1984): 765.

160. J. Seligmann, M. Hager, and D. Seward, "Tracing the Origin of AIDS," *Newsweek*, May 7, 1984: 101–2; and P. Van de Perre, D. Rouvroy, P. LePage, et al., "Acquired Immunodeficiency Syndrome in Rwanda," *Lancet* II (1984): 62–69.

161. J. K. Kreiss, D. Koech, F. A. Plummer, et al., "AIDS Virus Infection in Nairobi Prostitutes," *New England Journal of Medicine* 314 (1986): 414–18.

162. J. Emmanuel, director of the Zimbabwe Blood Transfusion Service, personal communication, 1986. Also see "Health Education a Must in AIDS Fight," *The Sunday Mail* (Harare), June 15, 1986: 9.

163. T. C. Quinn, J. M. Mann, J. W. Curran, and P. Piot, "AIDS in Africa: An Epidemiologic Paradigm," *Science* 234 (1986): 955–63.

164. J. A. Levy, L. Z. Pan, E. Beth-Giraldo, et al., "Absence of Antibodies to the Human Immunodeficiency Virus in Sera from Africa Prior to 1975," *Proceedings of the National Academy of Sciences* 83 (1986): 7935–37. R. Sher, S. Antunes, B. Reid, et al., "Seroepidemiology of Human Immunodeficiency Virus in Africa from 1970 to 1974," *New England Journal of Medicine* 317 (1987): 450–51. E. Tabor, R. Gerety, J. Cairns, and A. C. Bayley, "Did HIV and HTLV Originate in Africa?" *Journal of the American Medical Association* 264 (1990): 691–92.

165. M. Baldo and A. J. Cabral, "Low Intensity Wars and Social Determination of the HIV Transmission: The Search for a New Paradigm to Guide Research and Control the HIV/AIDS Pandemic," in Z. Stein and A. Zwi, eds., *Action on AIDS in Southern Africa: Maputo Conference on Health Transition in Southern Africa, April 1990* (New York: Committee for Health in Southern Africa).

166. The most outstanding compilation of available information on this sad period in Uganda's history and its impact on AIDS can be found in Barnett and Blaikie (1992), op. cit. Other excellent sources include R. Winter, "Uganda: Creating a Refugee Crisis," United States Committee for Refugees Newsletter, 1983; and C. P. Dodge, "The West Nile Emergency," in C. P. Dodge and P. D. Wiebe, eds., *Crisis in Uganda: The Breakdown of Health Services* (Oxford, Eng.: Pergamon Press, 1985).

167. There are numerous sources of more detailed information on military and political activities in the region during the mid-1970s. See, for example, Western Massachusetts Association of Concerned African Scholars, *U.S. Military Involvement in Southern Africa* (Boston: South End Press, 1978).

168. J. P. Getchell, D. R. Hicks, A. Svinivasan, et al., "Human Immunodeficiency Virus Isolated from a Serum Sample Collected in 1976 in Central Africa," *Journal of Infectious Diseases* 156 (1987): 833–37.

169. N. Nzilami, K. M. De Cock, D. N. Forthal, et al., "The Prevalence of Infection with Human

Immunodeficiency Virus over a 10-Year Period in Rural Zaire," *New England Journal of Medicine* 318 (1988): 276–79.

170. K. M. De Cock and J. B. McCormick, "HIV Infection in Zaire," *New England Journal of Medicine* 319 (1988): 309.

171. A. J. Nahmias, J. Weiss, X. Yao, et al., "Evidence for Human Infection with an HTLV-III/LAV-Like Virus in Central Africa, 1959," *Lancet* I (1986): 1279–80.

172. R. V. Henrickson, D. H. Maul, K. G. Osborn, et al., "Epidemic of Acquired Immunodeficiency in Rhesus Monkeys," *Lancet* I (1983): 388–90.

173. W. T. London, J. L. Sever, D. L. Madden, et al., "Experimental Transmission of Simian Acquired Immunodeficiency Syndrome (SAIDS) and Kaposi's-Like Skin Lesions," *Lancet* II (1983): 869–73.

174. See letters to the editor in response to the California studies, in *Lancet* I (1983): 1097–98.

175. "Thus, African green monkeys seem to possess antibodies to STLV-III [SIVagm], unlike African chimpanzees and baboons. These African green monkeys are apparently healthy, suggesting that STLV-III may be non-pathogenic in this species . . . ," the Harvard team wrote. "We suggest that STLV-III [SIVagm] of African green monkeys may have been transmitted to man coincident with the recognition of AIDS in Central Africa. An HTLV-III [HIV] related virus has thus been found in two species of Old World Primates." P. J. Kanki, R. Kurth, W. Becker, et al., "Antibodies to Simian T-Lymphotropic Retrovirus Type III in African Green Monkeys and Recognition of STLV-III Viral Proteins by AIDS and Related Sera," *Lancet* I (1985): 1330–32.

176. See D. Colburn, "Claiming Credit for HIV-2: A 'Sordid Chapter' in the Politics of Research," *Washington Post*, October 27, 1987: Health 19; F. Clavel, K. Mansinho, S. Chamaret, et al., "Human Immunodeficiency Virus Type 2 Infection Associated with AIDS in West Africa," *New England Journal of Medicine* 316 (1987): 1180–85; F. Clavel, F. Brun-Vézinet, D. Guétard, et al., "LAV Type II: A Second Retrovirus Associated with AIDS in West Africa," *Centre Recherche Académie Science Paris* 302 (1986): 485–88; and M. Blanc, "L'Autre Virus du SIDA," *La Recherche* 17 (1986): 974–76.

177. P. J. Kanki, F. Barin, S. M'Boup, et al., "New Human T-Lymphotropic Retrovirus (HTLV-IV) Related to Simian T-Lymphotropic Virus Type III (STLV-IIIagm)," *Science* 232 (1986): 238–43.

178. P. J. Kanki, S. M'Boup, D. Ricard, et al., "Human T-Lymphotropic Virus Type 4 and the Human Immunodeficiency Virus in West Africa," *Science* 236 (1987): 827–31.

179. J. L. Marx, "Probing the AIDS Virus and Its Relatives," *Science* 235 (1987): 1523–25; and M. Essex and P. J. Kanki, Letter, *Nature* 331 (1988): 621–22.

180. P. Kanki, S. M'Boup, R. Marlink, et al., "Prevalence and Risk Determinants of Human Immunodeficiency Virus Type 2 (HIV-2) and Human Immunodeficiency Virus Type 1 (HIV-1) in West African Female Prostitutes," *American Journal of Epidemiology* 136 (1992): 895–907; and P. J. Kanki, "Biologic Features of HIV-2," in P. Volberding and M. A. Jacobson, eds., *AIDS Clinical Review 1991* (New York: Marcel Dekker, 1991); 19–32.

181. See note 145.

182. G. Franchini, R. C. Gallo, H. G. Guo, et al., "Sequence of Simian Immunodeficiency Virus and Its Relationship to the Human Immunodeficiency Viruses," *Nature* 328 (1987): 539–43.

The HIV-1 strain used in the study was HTLV-IIIb, the research strain that was virtually identical to Montagnier's LAV strain, both of which underwent alteration due to multiple cell culture passages in the laboratories.

183. L. Chakrabarti, M. Guyader, M. Alizon, et al., "Sequence of Simian Immunodeficiency Virus from Macaque and Its Relationship to Other Human Simian Retroviruses," *Nature* 328 (1987): 543–47.

184. Marx (1987), op. cit.

Essex, working in collaboration with Gallo's lab and researchers from Sweden's Karolinska Institute and the Université Pierre et Marie Curie in Paris, reached a similar conclusion about the viruses' similarity. See S. K. Arya, B. Beaver, L. Jagodzinski, et al., "New Human and Simian HIV-related Retroviruses Possess Functional Transactivator (tat) Gene," *Nature* 328 (1987): 548–50. But they stressed that it was possible that they were looking at an example of cross-species transmission of the same virus, from monkey to man.

Harking back to disputes a decade earlier between Gallo and Japanese researchers over credit for discovery of HTLV-I, Yorio Hinuma of the Cancer Institute in Tokyo had isolated an SIV virus from wild-caught African green monkeys in 1985. His team, which had been instrumental in earlier HTLV-I efforts, now was credited with discovery of SIVagm, given evidence that Essex's group was studying an SIVmac contaminant.

185. Y. Ohta, T. Masuda, H. Tsujimoto, et al., "Isolation of Simian Immunodeficiency Virus from

African Green Monkeys and Seroepidemiologic Survey of the Virus in Various Nonhuman Primates," *International Journal of Cancer* 41 (1988): 115–22.

186. Fukasawa, M. Miura, T. Hasegawa, A., et al. "Sequence of Simian Immunodeficiency Virus from African Green Monkey, A New Member of the HIV/SIV Group," *Nature* 33 (1988): 457–461.

187. For good reviews of these findings, see R. C. Desrosiers, "A Finger on the Missing Link," *Nature* 345 (1990): 326; M. McClure, "AIDS and the Monkey Puzzle," *New Scientist*, March 25, 1989; and M. McClure, "Where Did the AIDS Virus Come From?" *New Scientist*, June 30, 1990: 54–57.

188. R. F. Khabbaz, W. Heneine, J. R. George, et al., "Brief Report: Infection of a Laboratory Worker with Simian Immunodeficiency Virus," *New England Journal of Medicine* 330 (1994): 172–77.

189. F. Gao, L. Yue, A. T. White, et al., "Human Infection by Genetically Diverse SIVsm-related HIV-2 in West Africa," *Nature* 358 (1992): 495–99.

190. R. Nowak, "HIV-2 and SIV May Be the Same Virus," *Journal of NIH Research* 4 (1992): 38–40.

191. C. Rouzioux, G. Jaeger, F. Brun-Vézinet, et al., "Absence of Antibody to LAV/HTLV-III and STLV-III (mac) in Pygmies," II International Conference on AIDS, Paris, June 23–25, 1986.

192. It would constitute a major breakthrough in AIDS research when University of Washington at Seattle scientists succeeded in infecting *Macaca nemestrina* monkeys with HIV-1, producing immune deficiency and lymphoma in the animals. From the earliest stages of the AIDS epidemic scientists had been infecting chimpanzees with HIV-1, but the animals never developed AIDS.

193. U. Dietrich, M. Adamski, H. Kühnel, et al., "A Highly Divergent HIV-2 Related Strain, HIV-2alt, Defines an Alternative Subtype of the HIV-2/SIV mac/SIVsm Group of Primate Immunodeficiency Viruses," IX International Conference on AIDS, Berlin, June 6–11, 1993.

194. S. Giunta and G. Groppa, "The Primate Trade and the Origin of AIDS Viruses," *Nature* 329 (1987): 22; and A. Karpas, "Origin and Spread of AIDS," *Nature* 348 (1990): 578.

195. T. Huet, R. Cheynier, A. Meyerhans, et al., "Genetic Organization of a Chimpanzee Lentivirus Related to HIV-1," *Nature* 345 (1990): 356–58.

The details were presented as follows:

	Regulatory Genes of the Virus		Viral Outer Envelope Genes
	gag	pol	env
Percentage of Genetic Homology Between SIVepz and:			
HIV-1bru (Paris)	74.8	84.0	62.7
HIV-1mal (Mali)	73.4	84.1	65.8
HIV-2rod (Cape Verde)	56.2	57.2	37.0
SIVmac	54.9	57.4	35.3
SIVagm	55.9	60.4	38.6
SIVmnd	52.1	58.1	32.7

This showed that over two-thirds of the chimpanzee viral genes and HIV-1 were the same, versus far less commonality between SIVcpz and the other simian viruses or HIV-2.

196. J. N. Nkengasong, M. Peeters, B. Willems, et al., "Phenotypic and Antigenic Properties of HIV-1 Isolates from Cameroon," I International Conference on Human Retroviruses and Related Infections, Washington, D.C., December 12–16, 1993.

197. G. Myers, K. MacInnes, and L. Myers, "Phylogenetic Moments in the AIDS Epidemic," Chapter 12 in S. S. Morse, ed., *Emerging Viruses* (Oxford, Eng.: Oxford University Press, 1993).

198. There are about 9,700 nucleotides, or discrete bits of genetic information, inside an HIV virus. Nobel laureate Howard Temin (co-discoverer of reverse transcriptase and retroviruses) estimated that HIV-1 mutates at the rate of 2×10^{-2} substitutions per nucleotide per year. Put another way, if one could follow all the nucleotides in a population of HIV-1—assuming that the viruses were all identical at the outset—a year later there would be about fifty mutated nucleotides in any given virus one examined. And the whole population of viruses, which had been uniform at the outset, would have over the course of a year undergone thousands of generations of replication, becoming a heterogeneous swarm of dozens of quasispecies. An annual mutation rate of 2×10^{-2} was one of the highest seen in any viral species; but the cow virus responsible for foot-and-mouth disease mutated at a rate of 3×10^{-2} and the segment of the influenza virus responsible for its ability to infect human cells was an even more rapid mutator. See H. M. Temin, "Is HIV Unique or Merely Different?" *Journal of Acquired Immune Deficiency Syndromes* 2 (1989): 1–9.

A study of the earliest HIV-1 strains found in people in 1983–85 similarly revealed a mutation rate of between 0.4 and 1.6 percent per year. The highest mutation rates occurred in viruses that were cultured repeatedly in the laboratory, undoubtedly subjected to selection pressures and contamination that wouldn't occur naturally. Nevertheless, a median 1 percent annual metamorphosis rate seemed reasonable for wild HIV-1 viruses. See P. C. Sheng-Yung, B. H. Bowman, J. B. Weiss, et al., "The Origin of HIV-1 Isolate HTLV-IIIB," *Nature* 363 (1993): 466–69.

199. M. A. McClure, M. S. Johnson, D. F. Feng, and R. F. Doolittle, "Sequence Comparisons of Retroviral Proteins: Relative Rates of Change and General Phylogeny," *Proceedings of the National Academy of Sciences* 85 (1988): 2469–73.

200. There are several ways to picture these historical arguments. Paul Ewald, of Amherst College in Massachusetts, sees AIDS viral evolution as follows:

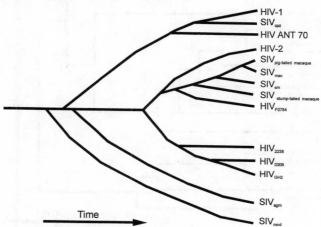

Source: P. W. Ewald, "AIDS: Where Did It Come From and Where Is It Going?" in *Evolution of Infectious Diseases* (Oxford, Eng.: Oxford University Press, 1993), chapter 8.

Gerald Myers pictured the evolution more like this:

and this:

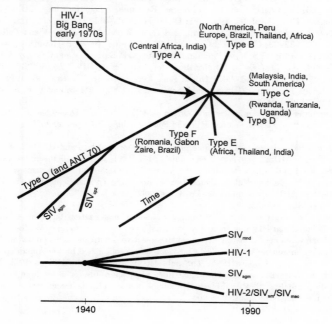

On a larger evolutionary scale, comparing HIV to other similar viruses, Doolittle saw this:

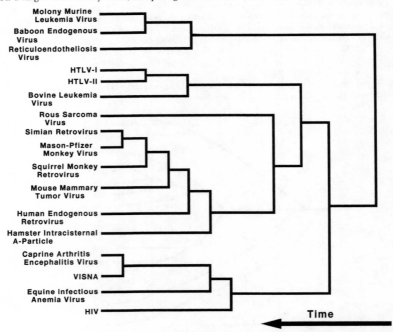

201. The experience of Marburg disease, as described in earlier chapters, left many scientists wondering just how safe monkey-derived vaccines might be.

202. See P. Brown, "US Rethinks Link Between Polio Vaccine and HIV," *New Scientist*, April 4, 1992: 10; T. Curtis, "The Origin of AIDS," *Rolling Stone* 625 (1992): 54–106; T. Curtis and P. Manson, "Do Cold, Hard AIDS Facts Lie in Freezer? Researchers Look for Clues in Old Vials of Polio Vaccine," *Houston Post*, April 16, 1992: A1; C. H. Fox, "Possible Origins of AIDS," *Science* 256 (1992): 1259–60; A. J. Garrett, A. Dunham, and D. J. Wood, "Retroviruses and Poliovaccines," *Lancet* 342 (1993): 932–33; Giunta and Groppa (1987), op. cit.; "In the Beginning," *The Economist*, March 14, 1992: 99–100; W. S. Kyle, "Simian Retroviruses, Poliovaccine and Origin of AIDS," *Lancet* 339 (1992): 600–1; G. Lecatsas and J. J. Alexander, "Origins of HIV," *Lancet* 339 (1992): 1427; A. McGregor, "Poliovaccine and AIDS Origin Link Very Unlikely," *Lancet* 340 (1992): 1090–91; "Panel Nixes Congo Trials and AIDS Source," *Science* 258 (1992): 304–5; and T. F. Schulz, "Origin of AIDS," *Lancet* 339 (1992): 867.

Rolling Stone later printed an apologia: " 'Origin of AIDS' Update," December 9, 1992: 40.

203. N. Touchette, "Fact or Fiction? HIV and Polio Vaccines," *Journal of NIH Research* 4 (1992): 40–41.

204. See J. Rifkin, letter to Dr. James Mason, Director, CDC, undated, 1987; P. M. Boffey, "Cattle Virus Tied to AIDS," *New York Times*, July 7, 1987: A1; J. Rifkin, letter to Dr. Frank Young, Commissioner, Food and Drug Administration, August 3, 1987; and J. B. Wyngaarden and B. W. Hawkins, letter to Mr. Jeremy Rifkin, Foundation on Economic Trends, September 23, 1987.

205. J. Seale, "AIDS Virus Infection: Prognosis and Transmission," *Journal of the Royal Society of Medicine* 78 (1985): 613–15.

206. Using precisely the same reasoning, a right-wing organization based in Orange County, California, concluded that HIV was created by the KGB as part of a plot to take over the United States. The organization favored mandatory quarantine of HIV-positive Americans as the only available measure to stop communist encroachment. See *The Freedom Fighter*, published by the American Information Network of Orange, CA, January 1986. The publication also blamed AIDS on 1959 Nobel Prize recipient Bertrand Russell, saying, "These globalists are responsible for unleashing the AIDS virus upon the world's population, they would be guilty of mass murder at levels heretofore not even conceived of in the most imaginative of monster movies."

207. U.S. Department of State, "The U.S.S.R.'s AIDS Disinformation Campaign," *Foreign Affairs* Note, July 1987.

208. Peter Duesberg's views have been so widely published that it is difficult to narrow a list to key sources. For Duesberg's perspective, see B. Guccione, Jr., *Interview*, September 1993: 95–108; P. H. Duesberg, "Human Immunodeficiency Virus and Acquired Immunodeficiency Syndrome: Correlation But Not Causation," *Proceedings of the National Academy of Sciences* 86 (1989): 755–64; J. Miller, "AIDS Heresy," *Discover*, June 1988: 63–68; P. Duesberg, "A Challenge to the AIDS Establishment," *Biotechnology* 5 (1987): 3; and P. H. Duesberg, "Retroviruses As Carcinogens and Pathogens: Expectations and Reality," *Cancer Research* 47 (1987): 1199–1220.

209. For examples of counterarguments to Duesberg's theories, see J. Cohen, "Keystone's Blunt Message: 'It's the Virus, Stupid,' " *Science* 260 (1993): 292; P. Brown, "MPs Investigate AIDS Maverick," *New Scientist*, June 6, 1992: 9; D. Concar, "Patients Abandon AIDS Drug After TV Shows," *New Scientist*, July 13, 1991: 13; J. E. Groopman, "A Dangerous Delusion About AIDS," *New York Times*, September 10, 1992: A23; J. Weber, "AIDS and the 'Guilty' Virus," *New Scientist*, May 5, 1988: 32–33; and A. G. Fettner, "Dealing with Duesberg," *Village Voice*, February 2, 1988: 25–29.

210. See S. B. Thomas and S. C. Quinn, "Understanding the Attitude of Black Americans," in J. Stryker and M. D. Smith, eds., *Dimensions of HIV Prevention: Needle Exchange* (Menlo Park, CA: Henry J. Kaiser Family Foundation, 1993), 99–128.

211. Estes (1991), op. cit., 489–558.

212. A. J. Pinching, "AIDS and Africa: Lessons for Us All," *Journal of the Royal Society of Medicine* 79 (1986): 501–3.

213. Karpas (1990), op. cit.

214. B. Evatt, D. P. Francis, and M. F. McLane, "Antibodies to Human T Cell Leukemia Virus-Associated Membrane Antigens in Haemophiliacs: Evidence for Infection Before 1980," *Lancet* II (1983): 698–700.

215. Centers for Disease Control, "Recommendations for Counseling Persons Infected with Human T-Lymphotropic Virus, Types I and II," *Morbidity and Mortality Weekly Report* 42 (1993): 1–7.

216. C. Bartholomew, C. Saxinger, J. W. Clark, et al., "Transmission of HTLV-I and HIV Among Homosexual Men in Trinidad," *Journal of the American Medical Association* 257 (1987): 2604–8.

217. H. Lee, P. Swanson, V. S. Shorty, et al., "High Rate of HTLV-II Infection in Seropositive IV Drug Abusers in New Orleans," *Science* 244 (1989): 471–77.

218. P. S. Sarma and J. Gruber, "Human T-Cell Lymphotropic Viruses in Human Diseases," *Journal of the National Cancer Institute* 81 (1990): 1100–6.

U.S. prostitute surveys showed HTLV-II incidence was highest among those women who were injecting drug users. See R. F. Khabbaz, W. W. Darrow, T. M. Hartley, et al., "Seroprevalence and Risk Factors for HTLV-I/II Infection Among Female Prostitutes in the United States," *Journal of the American Medical Association* 263 (1990): 60–64. By 1991 the U.S. blood-bank industry conceded that both HTLV viruses had contaminated the American blood supply. See M. T. Sullivan, A. E. Williams, C. T. Fang, et al., "Transmission of Human T-Lymphotropic Virus Types I and II by Blood Transfusion," *Archives of Internal Medicine* 151 (1991): 2043–48.

219. Serwadda's group noted that in 1985 "the first recognized cases came from a small village on Lake Victoria, just north of the Tanzanian border. This village was one of many from which goods were traded across the border. The notion that the disease may have been transmitted sexually from Tanzania is interesting since it fits historically with the movements of the Tanzanian Army . . . and subsequent regular visits by the Tanzanian traders. Of the 15 traders tested for evidence of HTLV-III [HIV] antibodies, 10 were positive. . . . There have been no studies as yet to show whether the virus is endemic in Tanzania and, if so, whether it has been introduced from Uganda via traders and soldiers." See Serwadda, Mugerwa, and Sewankambo, (1985), op. cit.

12. Feminine Hygiene

1. N. Friedman, "Everything They Didn't Tell You About Tampons," *New West*, October 20, 1980: 33–42; and R. E. Wheatley, M. F. Menkin, E. D. Bardes, and J. Rock, "Tampons in Menstrual Hygiene," *Journal of the American Medical Association* 192 (1965): 113–16.

2. In 1977 a medical team from the University of Colorado Medical Center in Denver reported treating four young women (two of whom were teenagers) for vaginal ulcers, apparently caused by tampons. Two of the cases involved the so-called deodorant products (actually perfumed, rather than deodorized). All four women healed with cessation of tampon use. See K. F. Barrett, S. Bledsoe, B. E. Greer, and W. Droegemueller, "Tampon-Induced Vaginal or Cervical Ulceration," *American Journal of Obstetrics and Gynecology* 127 (1977): 332–33.

3. The industry denied allegations that asbestos was ever used in their products, but also consistently refused to list the stabilizing fibers that were used to prevent the cotton tampons from falling apart.

4. D. E. Marlowe, R. M. Weigle, and R. W. Stauffenberg, "Measurement of Tampon Absorbency: Evaluation of Tampon Brands," Bureau of Medical Devices, U.S. Food and Drug Administration, Rockville, MD, 1981.

5. In her outstanding piece of investigative journalism, Nancy Friedman notes: "The range of blood loss during menstruation is two to six ounces. A single superabsorbent tampon is capable of soaking up an ounce or more of fluid. Since menstruation is a gradual process lasting three to seven days, the question arises: If the tampon has absorbed all the blood leaving the uterus and still hasn't reached saturation, what will it absorb? The answer: the normal secretions of the healthy vaginal walls." Friedman (1980), op. cit.

6. M. Chrapil et al., "Reaction of the Vagina to Cellulose Sponges," *Journal of Biomedical Materials Research* 13 (1979): 1.

7. A. Johnson, "Used Carboxymethyl Cellulose as Chromatographic Purifier for *Staphylococcal Toxins*," *Infection and Immunology* 25 (1979): 1080–85.

8. K. F. Barrett, "Tampon-Induced Ulceration," *American Journal of Obstetrics and Gynecology* 127 (1977): 332; R. K. Collins, "Tampon Induced Vaginal Laceration," *Journal of Family Practice* 9 (1979): 127; and F. K. Beller, "Vaginal Tampon as Drug Carrier," *Medical World* 30 (1979): 709.

9. J. P. Davis, J. Chesney, P. J. Wand, et al., "Toxic-Shock Syndrome," *New England Journal of Medicine* 303 (1980): 1429–35.

10. J. Todd, M. Fishaut, F. Kapral, and T. Welch, "Toxic-Shock Syndrome Associated with Phage-Group-I Staphylococci," *Lancet* II (1978): 1116–18.

11. F. Stevens, "The Occurrence of *Staphylococcus aureus* Infection with a Scarlatiniform Rash," *Journal of the American Medical Association* 88 (1927): 1957.

12. H. Aranow and W. B. Wood, "Staphylococcal Infection Stimulating Scarlet Fever," *Journal of the American Medical Association* 119 (1942): 1491.

13. T. Kawasaki, "Acute Febrile Mucocutaneous Syndrome with Lymphoid Involvement with Specific Desquamation of the Fingers and Toes in Children," *Japanese Journal of Allergology* 16 (1967): 178–222.

14. Between 1975 and 1980 over 10,000 cases of Kawasaki syndrome were diagnosed in Japan.

15. Centers for Disease Control, "Toxic-Shock Syndrome—United States," *Morbidity and Mortality Weekly Report* 29 (1980): 229–30.

16. See Table 1 in K. N. Shands, G. P. Schmid, B. B. Dan, et al., "Toxic-Shock Syndrome in Menstruating Women," *New England Journal of Medicine* 303 (1980): 1436–42.

17. Centers for Disease Control, "Follow-up on Toxic-Shock Syndrome," *Morbidity and Mortality Weekly Report* 29 (1980): 441–45.

18. The CDC's results were:

Tampon Brand	Cases (N = 42)	Controls (N = 114)
Rely	71%	26%
Playtex	19%	25%
Tampax	5%	25%
Kotex	2%	12%
OB	2%	11%

19. Food and Drug Administration, News Release PBD-42, September 25, 1980.

20. Centers for Disease Control, "Toxic-Shock Syndrome—Utah," *Morbidity and Mortality Weekly Report* 29 (1980): 495–96.

21. D. B. Petitti, A. Reingold, and J. Chin, "The Incidence of Toxic Shock in Northern California," *Journal of the American Medical Association* 255 (1986): 368–72.

22. According to the manufacturers, prior to 1977 all tampons were made primarily of rayon and cotton. After 1977, however, 65 percent of all tampons sold in the United States contained polyacrylate fibers, carboxymethyl cellulose, higher-absorbency rayon-cellulose, polyester, or other synthetics.

23. R. W. Tofte, K. B. Crossley, and D. N. Williams, "Clinical Experience with Toxic-Shock Syndrome," *New England Journal of Medicine* 303 (1980): 1417.

24. "Report Rise in Toxic Shock Cases Unrelated to Tampon Use," *Hospital Practice*, July 1982: 197–200; "*S. aureus* Bacteriophage May Be Implicated in Toxic Shock," *Hospital Practice*, May 1983: 36–38; and B. Hanna and P. Tierno, "Staphylococcal Growth on Carboxymethyl Cellulose," presentation to the Annual Meeting of the American Society of Microbiology, 1981.

25. It is likely that true *Staphylococcus* rates were 100 percent. As physicians became more familiar with TSS and recognized the speed with which the ailment could dangerously escalate, it became routine to give symptomatic menstruating females high doses of non-penicillinase antibiotics before taking vaginal samples and awaiting staph culture results.

26. Davis et al. (1980), op. cit.

27. These and many other comments in this chapter were made to the author during interviews conducted over the course of the TSS investigation.

28. J. Langone, "Riddle of the Tampon," *Discover*, December 1989: 26–28.

29. P. M. Schlievert, K. M. Bettin, and D. W. Watson, "Purification and Characterization of Group A Streptococcal Pyrogenic Exotoxin Type C," *Infection and Immunology* 16 (1977): 673–79.

30. P. Schlievert, "Activation of Murine T-Suppressor Lymphocytes by Group A Streptococcal and Staphylococcal Pyrogenic Exotoxins," *Infection and Immunology* 28 (1980): 876–80.

31. Institute of Medicine, "Toxic Shock Syndrome: Assessment of Current Information and Future Research Needs" (Washington, D.C.: National Academy Press, 1982).

32. By that time Schlievert and his collaborators in Minnesota, Wisconsin, Colorado, and California were seeing clear autoimmune disorders in the women who had survived TSS bouts months earlier. Eleven of 123 women surveyed had developed lupus, and 40 percent had early symptoms of arthritis—a striking finding given that most TSS sufferers were under thirty-five years of age.

P. M. Schlievert, K. M. Shands, B. B. Dan, et al., "Identification and Characterization of an Exotoxin from *Staphylococcus aureus* Associated with Toxic-Shock Syndrome," *Journal of Infectious Diseases* 143 (1981): 509–16; and P. M. Schlievert and J. A. Kelly, "Staphylococcal Pyrogenic Exotoxin Type C: Further Characterization," *Annals of Internal Medicine* 96 (1982): 982–86.

33. Centers for Disease Control, "Toxic Shock Syndrome, United States, 1970–1982," *Morbidity and Mortality Weekly Report* 31 (1982): 201–4.

34. Centers for Disease Control, "Update: Toxic-Shock Syndrome—United States," *Journal of the American Medical Association* 250 (1983): 1017.

35. A. L. Reingold, "Epidemiology of Toxic-Shock Syndrome, United States, 1960–1984," *Morbidity and Mortality Weekly Report* 33 (1982): 19ss–22ss.

36. When averaged over the population as a whole for the various states, acute TSS cases occurred, for example, in:

State	Per capita (all ages, both genders)
Utah	1:10,288
Minnesota	1:14,201
Wisconsin	1:17,363
Colorado	1:22,228
Oregon	1:30,978
Ohio	1:73,956
Washington	1:93,917
Indiana	1:99,368
California	1:100,713
Michigan	1:110,262
Texas	1:122,633
New York	1:532,065

Based on cumulative reports, 1975–83, to the CDC.

See Petitti, Reingold, and Chin (1986), op. cit.

37. S. F. Berkley, A. W. Hightower, C. V. Broome, and A. L. Reingold, "The Relationship of Tampon Characteristics to Menstrual Toxic Shock Syndrome," *Journal of the American Medical Association* 258 (1987): 917–20.

38. G. Faich, K. Pearson, D. Fleming, et al., "Toxic Shock Syndrome and the Vaginal Contraceptive Sponge," *Journal of the American Medical Association* 255 (1986): 216–18; and A. L. Reingold, "Toxic Shock Syndrome and the Vaginal Sponge," *Journal of the American Medical Association* 255 (1986): 242–43.

39. S. M. Wolfe, "Dangerous Delays in Tampon Absorbency Warnings," *Journal of the American Medical Association* 258 (1987): 949–51.

40. The chart was as follows:

	Absorbency	Ranges of Absorbency in Grams
Tampons come in the	Junior absorbency	Less than 6
following standardized	Regular absorbency	6 to 9
industry-size	Super absorbency	9 to 12
absorbencies.	Super Plus absorbency	12 to 15

41. L. E. Markowitz, A. W. Hightower, C. V. Broome, and A. L. Reingold, "Toxic Shock Syndrome: Evaluation of National Surveillance Data Using a Hospital Discharge Survey," *Journal of the American Medical Association* 258 (1987): 75–78.

42. K. L. MacDonald, M. T. Osterholm, C. W. Hedberg, et al., "Toxic Shock Syndrome: A Newly Recognized Complication of Influenza and Influenzalike Illness," *Journal of the American Medical Association* 257 (1987): 1053–58; and Centers for Disease Control, "Toxic Shock Syndrome Associated with Influenza," *Morbidity and Mortality Weekly Report* 35 (1986): 143–44.

43. S. J. Sperber and J. B. Francis, "Toxic Shock Syndrome During an Influenza Outbreak," *Journal of the American Medical Association* 257 (1987): 1086–87.
Langmuir and his colleagues hypothesized the existence of such a scourge, combining TSS and influenza, which they dubbed Thucydides syndrome after the great Greek chronicler, and said might have been the cause of the 430–427 B.C. plague of Athens. See A. D. Langmuir, T. D. Worthen, J. Solomon, et al., "The Thucydides Syndrome: A New Hypothesis for the Cause of the Plague in Athens," *New England Journal of Medicine* 313 (1985): 1027–39; and B. B. Dan, "Toxic Shock Syndrome: Back to the Future," *Journal of the American Medical Association* 257 (1987): 1094–95.

44. J. K. Todd, M. Ressman, S. A. Caston, et al., "Corticosteroid Therapy for Patients with Toxic Shock Syndrome," *Journal of the American Medical Association* 252 (1984): 3399–3402.

45. L. K. Altman, "Bacteria Are Linked to Deadly Childhood Disease," *New York Times*, December, 3, 1993: A28.

46. A. L. Bisno, "Staphylococcal Endocarditis and Bacteremia," *Hospital Practice*, April 15, 1986: 139–58.

47. D. Y. M. Leung, H. C. Meissner, D. R. Fulton, et al., "Toxic Shock Syndrome Toxin-Secreting *Staphylococcus aureus* in Kawasaki Syndrome," *Lancet* 342 (1993): 1385–88.

48. J. M. Musser, P. Schlievert, A. W. Chow, et al., "A Single Clone of *Staphylococcus aureus* Causes the Majority of Cases of Toxic Shock Syndrome," *Proceedings of the National Academy of Sciences* 87 (1990): 225–29.

49. TSST-1 can stimulate rapid proliferation of T cells in doses of less than 10^{-9}M, which means a person could have virtually undetectable amounts of the toxin in his or her blood and develop acute Toxic Shock Syndrome.

50. For a succinct description of the immune system effects of TSST-1, see A. K. Abbas, A. H. Lichtman, and J. S. Pober, *Cellular and Molecular Immunology* (Philadelphia: W. B. Saunders, 1991), 304.

51. B. N. Kreiswirth, J. S. Kornblum, and R. P. Novick, "Genotypic Variability of the Toxic Shock Syndrome Exoprotein Determinant," in J. Jeljaszewicz, ed., *The Staphylococci* (Stuttgart: Gustav Fischer Verlag, 1985).

52. M. C. Chu, B. N. Kreiswirth, P. A. Pattee, et al., "Association of Toxic Shock Toxin-1 Determinant with a Heterologous Insertion of Multiple Loci in the *Staphylococcus aureus* Chromosome," *Infection and Immunity* 56 (1988): 2702–8.

53. Plasmids were self-contained units of DNA that could be exchanged from one microbe to another either as random events, or in response to specific survival pressure when a population of bacteria were exposed to antibiotics. For further details of both the TSST-1 and penicillin-resistance transposons, see: Murphy, E. "Transposable Elements in Gram-Positive Bacteria." Chapter 9. Eds. D. E. Berg and M. M. Howe. *Mobile DNA*. Washington, D.C.: American Society for Microbiology, 1989.

13. The Revenge of the Germs

1. There are thousands of natural and synthetic antibiotics, relatively few of which have been tested and marketed. The most commonly used are as follows:

Antibiotic Class	Drugs That fit into the Class	Drugs' Way of Controlling Bacteria	Bacterial Basis of Resistance
Penicillin	amoxicillin, ampicillin carbenicillin, penicillin G, penicillin K, ticarcillin, penicillin V, cyclacillin, bacampicillin, azlocillin, mezlocillin, pipercillin	Act on gram-negative bacterial cell walls, blocking proper formation of membrane polymers. Some, such as penicillin G and V, also act on the cell walls of gram-positive bacteria.	Beta-lactamase enzyme or penicillinase enzyme; due to chromosomal change in bacteria or plasmid.
Penicillinase-Resistant Penicillins	methicillin, naficillin, oxacillin, cloxacillin, dicloxacillin	Same as penicillin.	Altered penicillin-binding proteins, penicillinase enzyme; strains that are resistant to this class are usually *also* resistant to all members of the penicillin, aminoglycoside, tetracycline, cephalosporin, erythromycin, and clindamycin classes.
Cephalosporins	cephalosporin C, cefonicid, cefuroxime, cefaclor, cefoxitin, cefamandole, cephradine, cephalexin, cefazolin, cephapirin	Inhibit bacterial cell wall synthesis.	Beta-lactamase enzyme, blockage of antibiotic binding, enzyme destruction of the drug.
Aminoglycosides	kanamycin, gentamicin, netilmicin, amikacin, tobramycin, streptomycin, neomycin	Primarily target aerobic gram-negative bacteria by blocking bacterial protein synthesis.	Bacterial enzyme phosphorylation or adenylation of the drug—a plasmid-mediated trait. Also via changes in cytoplasmic membrane that render it impermeable to drug.
Tetracylines	chlortetracycline, oxytetracycline, demeclocycline, methacycline, doxycycline, minocycline, tetracycline	Inhabit bacterial protein synthesis at the ribosome level.	Plasmid-mediated decrease in ribosomal susceptibility to the drugs or development of a pump mechanism that shunts drugs out of the organism.
Chloramphenicol	chloramphenicol	Irreversible binding of bacterial ribosome, blocking protein synthesis.	Plasmid-acquired ability to make acetyltransferase enzyme that inactivates drug.
Erythromycin	erythromycin	Irreversible binding of bacterial ribosome, blocking protein synthesis.	Several plasmid-mediated modes, including decrease cell permeability to drug, modified ribosome, and esterase destruction of drug.
Clindamycin	clindamycin, lincomycin	Irreversible binding of bacterial ribosome, blocking protein synthesis.	Plasmid-mediated blockage of ribosome binding or ribosome alteration.
Vancomycin	vancomycin	Inhibit the synthesis of bacterial cell wall.	Production of unique protein that blocks drugs.
Sulfonamides	sulfanilamide, sulfadiazine, sulfamethoxazole, sulfisoxazole, sulfacetamide	Block bacterial synthesis of folic acid.	Plasmid or random-mutation-mediated increase in folic acid synthesizing enzymes or enzymes that destroy the drug.
Trimethoprim	trimethoprim, sulfamethoxazole, usually in combination	Block bacterial folic acid synthesis and folate reduction.	Plasmid or mutation-mediated production of altered folating enzymes.

Quinolones	nalidixic acid, cinoxacin, norfloxacin, ciprofloxacin	Inhibit proper coiling of bacterial DNA.	Unclear; suspected to be chromosomally based change.
Isoniazid	isoniazid	Kill rapidly dividing myco-bacteria, mechanism unknown.	Unclear.
Rifampin	rifampin	Inhibit RNA polymerase, thus blocking protein synthesis.	Alteration of polymerase.
Ethambutol	ethambutol	Unclear.	Unclear.
Streptomycin	streptomycin	Inhibit mycobacterial growth, mechanism unclear.	Unclear.
Pyrazinamide	pyrazinamide	Unknown.	Unknown.
Ethionamide	ethionamide	Unclear.	Unknown.
Aminosalicylic Acid	aminosalicylie acid	Similar to sulfonamides.	Unknown.
Cycloserine	cycloserine	Unclear.	Unclear.
Sulfones	dapsone, sulfoxone sodium	Similar to sulfonamides.	Unclear.
Clofazimine	clofazimine	Blocks mycobacterial DNA.	Unknown.

2. A. L. Bisno, "Staphylococcal Endocarditis and Bacteremia," *Hospital Practice*, April 15, 1986: 139–58.

3. A. L. Panililo, D. H. Culver, R. P. Gaynes, et al., "Methicillin-Resistant *Staphylococcus aureus* in U.S. Hospitals, 1975–1991," *Infection Control and Hospital Epidemiology* 13 (1992): 582–86.

4. M. Truneh, "Phage Types and Drug Susceptibility Patterns of *Staphylococcus aureus* from Two Hospitals in Northwest Ethiopia," *Ethiopian Medical Journal* 29 (1991): 1–6.

5. E. E. Udo and W. B. Grubb, "Transfer of Resistance Determinants from a Multi-Resistant *Staphylococcus aureus* Isolate," *Journal of Medical Microbiology* 35 (1990): 72–79.

6. For overviews of the antibiotic resistance crisis, see M. L. Cohen, "Epidemiology of Drug Resistance: Implications for a Post-Antimicrobial Era," *Science* 257 (1992): 1050–55; S. M. Finegold, "Antimicrobial Therapy of Anaerobic Infections: A Status Report," *Hospital Practice*, October 1979: 71–81; L. O. Gentry, "Bacterial Resistance," *Orthopedic Clinics of North America* 22 (1991): 379–88; A. Gibbons, "Exploring New Strategies to Fight Drug-Resistant Microbes," *Science* 257 (1992): 1036–38; M. Lappé, *Germs That Won't Die* (Garden City, NY: Anchor Press, 1982); S. B. Levy, *The Antibiotic Paradox* (New York: Plenum Press, 1992); H. C. Neu, "The Crisis in Antibiotic Resistance," *Science* 257 (1992): 1064–73; Panlilio et al. (1992), op. cit.; and M. Toner, "When Bugs Fight Back," Pulitzer Prize-winning series of reports in the *Atlanta Constitution*, August 23, 1992–October 16, 1992.

7. D. P. Levine, B. S. Fromm, and B. R. Reddy, "Slow Response to Vancomycin or Vancomycin Plus Rifampin in Methicillin-Resistant *Staphylococcus aureus* Endocarditis," *Annals of Internal Medicine* 115 (1991): 674–80.

8. D. S. Kernodle and A. B. Kaiser, "Comparative Prophylactic Efficacy of Cefazolin and Vancomycin in a Guinea Pig Model of *Staphylococcus aureus* Wound Infection," *Journal of Infectious Diseases* 168 (1993): 152–57; and R. P. Wenzel, "Preoperative Antibiotic Prophylaxis," *New England Journal of Medicine* 326 (1992): 337–39.

In Denmark, where national medical registries are excellent because all the nation's citizens receive their health care from the federal government, a review of the country's meningitis cases from 1986 to 1989 revealed that 104 people had suffered the severe ailment as a result of staphylococcal infection. Sixty-one of the cases were acquired in the hospital following surgery. Fortunately, none of the Danish cases was MRSA, but all involved some degree of penicillin resistance. See A. G. Jensen, F. Esperson, P. Skinhøj, et al., "*Staphylococcus aureus* Meningitis," *Archives of Internal Medicine* 153 (1993): 1902–8.

9. Bacterial subtypes 29/52/80/95 in Detroit; 29/77/83A/85 in Boston. See L. R. Crane, D. P. Levine, M. J. Zervos, and G. Cummings, "Bacteremia in Narcotic Addicts at the Detroit Medical Center: I. Microbiology, Epidemiology, Risk Factors, and Empiric Therapy," *Review of Infectious Diseases* 8 (1986): 364–73; and D. E. Craven, A. I. Rixinger, T. A. Goularte, and W. R. McCabe, "Methicillin-Resistant *Staphylococcus aureus* Bacteremia Linked to Intravenous Drug Abusers Using a 'Shooting Gallery,' " *American Journal of Medicine* 80 (1986): 770–76.

10. Udo and Grubb (1990), op. cit.

11. B. Kreiswirth, J. Kornblum, R. D. Arbeit, et al., "Evidence for a Clonal Origin of Methicillin Resistance in *Staphylococcus aureus*," *Science* 259 (1993): 227–30.

12. For an overview of the global penicillin resistance trends, for example, see C. C. Sanders and W. E. Sanders, Jr., "Beta-Lactam Resistance in Gram-Negative Bacteria: Global Trends and Clinical Impact," *Clinical Infectious Diseases* 15 (1992): 824–39.

13.

Antibiotic(s)	Resistance Present in Pathogens	Resistance Crisis Now	Resistance Future Crisis
Beta-lactams	*S. aureus*	*S. pneumoniae*	*N. meningitidis*
Penicillins	*S. epidermidis*	*S. epidermidis*	*S. marcescens*
Cephalosporins	*S. pneumoniae*	*P. aeruginosa*	*K. pneumoniae*
Monobactams	*S. sanguis*	*E. cloacae*	*Bacterioides*
Carbapenems	*H. influenzae*	*Xanthomonas*	*Enterobacteriaceae*
	N. gonorrhoeae	*Acinetobacter*	*Salmonella*
	N. meningitidis		*Shigella*
	E. coli		*Haemophilus*
	P. aeruginosa		*Enterococcus*
	E. cloacae		
	S. marcescens		
	K. pneumoniae		
	K. oxytoca		
	Enterococcus		
	Enterobacteriaceae		
	Moraxella		
	Bacteroides		
	Acinetobacter		
Fluoroquinolones	*S. aureus*	MRSA	*Pseudomonads*
Norfloxacin	*S. epidermidis*	*Serratia*	*Enterobacteriaceae*
Ofloxacin	*Enterobacteriaceae*	*P. aeruginosa*	*Haemophilus*
Ciprofloxacin	*Pseudomonads*		*N. gonorrhoeae*
Lomefloxacin	*P. aeruginosa*		
Aminoglycosides	*Streptococcus*	*Streptococcus*	*Enterobacteriaceae*
Gentamicin	*Bacterioides*	*Pseudomonas*	*Streptococcus*
Tobramycin	*Pseudomonas*	*Enterococcus*	
Amikacin	*Enterobacteriaceae*	*Enterobacteriaceae*	
	Staphylococcus		
	Enterococcus		
	Pseudomonads		
Macrolideslincosamides	*Streptococcus*	*Enterococcus*	*S. pneumoniae*
Erythromycin	*S. pneumoniae*		*S. pyogenes*
Clindamycin	*Enterococcus*		*S. aglalactiae*
	Staphylococcus		*M. tuberculosis*
Chloramphenicol	*Staphylococcus*		*N. meningitidis*
	Streptococcus		*S. pneumoniae*
	S. pneumoniae		
	Enterobacteriaceae		
	Neisseria		
Tetracyclines	*Staphylococcus*	*Enterobacteriaceae*	*Mycoplasma*
Tetracycline	*Streptococcus*	Cholera	
Minocycline	*Enterococcus*		
Doxycycline	*Enterobacteriaceae*		
	Bacterioides		
	N. gonorrhoeae		
	Mycoplasma		
	Ureaplasma		
Rifampin	*Staphylococcus*	MRSA	*S. aureus*
	Enterococcus		*M. tuberculosis*
	Streptococcus		
	Enterobacteriaceae		
	Pseudomonads		

Folate-inhibitors	Staphylococcus	Enterobacteriaceae	H. influenzae
TMP/SMX	Streptococcus	Shigella	P. cepacia
	S. pneumoniae	Salmonella	MRSA
	Enterobacteriaceae	Neisseria	S. pneumoniae
	Neisseria	Haemophilus	
	Pseudomonads		
	Campylobactr		
Glycopeptides	Enterococcus	E. faecium	MRSA
Vancomycin	Leuconostac		MRSE
Teicoplanin	Lactococcus		Streptococcus
	Pediococcus		
	Lactobacillus		
	S. hemolyticus		
Mupirocin			Staphylococcus
Fusidic acid			Staphylococcus
Fossfomycin	S. aureus		
	Serratia		

Adapted from H. C. Neu, "The Crisis in Antibiotic Resistance," Science 257 (1992): 1066–67.

14. Cohen (1992), op. cit.

15. C. E. Phelps, "Bug/Drug Resistance: Sometimes Less Is More," *Medical Care* 27 (1989): 194–203.

16. D. L. Stevens, M. H. Tanner, J. Winship, et al., "Severe Group A Streptococcal Infections Associated with a Toxic Shock-like Syndrome and Scarlet Fever Toxin," *New England Journal of Medicine* 321 (1989): 1–7.

17. K. Wright, "Bad News Bacteria," *Science* 249 (1990): 22–24.

18. H. C. Dillon, and C. W. Derrick, Jr., "Streptococcal Complications: The Outlook for Prevention," *Hospital Practice*, September 1972: 93–101.

19. S. P. Gotoff, "Emergence of Group B Streptococci as Major Perinatal Pathogens," *Hospital Practice*, September 1977: 85–90.

20. Neu (1992), op. cit.

21. Centers for Disease Control, "Prevalence of Penicillin-Resistant *Streptococcus pneumoniae*—Connecticut, 1992–1993," *Morbidity and Mortality Weekly Report* 43 (1994): 216–23.

22. A survey of 748 cardiac patients treated in Johannesburg, South Africa—all of whom were poor blacks who had suffered childhood bouts of rheumatic fever—illustrated the severity of heart disease produced by the bacteria. Under the apartheid state, black South Africans lived in conditions of acute squalor and health care deprivation. As a result, during the 1980s rheumatic fever was about as common among black South Africans as it had been among white urban Americans in 1920. And like their 1920s counterparts in the United States, black South Africans who survived rheumatic fever during early childhood had a better than 50 percent chance of facing life-threatening heart disease before their twentieth birthday. The primary cause of their cardiac difficulties was mitral valve damage which required open-heart surgery. R. H. Marcus, P. Sareli, W. A. Pocock, and J. B. Barlow, "The Spectrum of Severe Rheumatic Mitral Valve Disease in a Developing Country," *Annals of Internal Medicine* 120 (1994): 177–83.

Dealing surgically with rheumatic fever heart damage has been likened to "attempting to mop up the water on the floor while leaving the faucet open," particularly in the context of poor countries. WHO during the 1990s instituted a trial program of rheumatic fever prevention in sixteen developing countries that involved training local paramedics to recognize streptococcal throat infections in children and treat the kids prophylactically with benzathine penicillin to prevent recurrent infections. M. J. McLaren, M. Markowitz, and M. A. Gerber, "Rheumatic Heart Disease in Developing Countries: The Consequence of Inadequate Prevention," *Annals of Internal Medicine* 120 (1994): 243–44.

23. L. G. Veasy, S. E. Wiedneier, G. S. Orsmond, et al., "Resurgence of Acute Rheumatic Fever in the Intermountain Area of the United States," *New England Journal of Medicine* 316 (1987): 421–27.

24. See "Streptococcal Diseases," in A. S. Benenson, ed., *Control of Communicable Diseases in Man* (Washington, D.C.: American Public Health Association, 1990), 411–18.

25. L. A. Haglund, G. R. Istre, D. A. Pickett, et al., "Invasive Pneumococcal Disease in Central Oklahoma: Emergence of High-Level Penicillin Resistance and Multiple Antibiotic Resistance," *Journal of Infectious Diseases* 168 (1993): 1532–35.

26. F. Shan, S. Germer, and D. Hazell, "Aetiology of Pneumonia in Children in Goroka Hospital, Papua New Guinea," *Lancet* II (1981): 537–41.

27. World Health Organization, "Implementation of the Global Strategy for Health for All by the Year 2000, Second Evaluation; and Eighth Report on the World Health Situation," report to the 45th World Health Assembly, Provisional Agenda Item 17, Geneva, 1992.

28. J. S. Spika, M. H. Munshi, B. Wojtyniak, et al., "Acute Lower Respiratory Infections: A Major Cause of Death in Children in Bangladesh," *Annals of Tropical Pediatrics* 9 (1989): 33–39; and B. J. Selwyn and BOSTID, "The Epidemiology of Acute Respiratory Tract Infection in Young Children: Comparison of Findings from Several Developing Countries," *Review of Infectious Diseases* 12 (1990): S870–S888.

29. World Health Organization, "Acute Respiratory Infections in Children: Case Management in Small Hospitals in Developing Countries," Program for the Control of Acute Respiratory Infections, Geneva, 1990.

30. Ibid.; and M. Harari, F. Shann, V. Spooner, et al., "Clinical Signs of Pneumonia in Children," *Lancet* 338 (1991): 928–30.

31. M. R. Pandey, N. M. P. Dawlaire, E. S. Starbuck, et al., "Reduction in Total Under-Five Mortality in Western Nepal Through Community-Based Antimicrobial Treatment of Pneumonia," *Lancet* 338 (1991): 993–97.

32. B. Sutrisna, R. R. Frerichs, and A. L. Reingold, "Randomised, Controlled Trial of Effectiveness of Ampicillin in Mild Acute Respiratory Infections in Indonesian Children," *Lancet* 338 (1991): 471–74.

33. Ibid.

34. X. Carné, J. M. Arnau, and J. R. Laporte, "Erythromycin Resistance in Streptococci," *Lancet* II (1989): 444–45; Centers for Disease Control, "Penicillin-Resistant *Streptococcus pneumoniae*— Minnesota," *Morbidity and Mortality Weekly Report* 26 (1977): 345; Centers for Disease Control, "Drug-Resistant *Streptococcus pneumoniae*—Kentucky and Tennessee, 1993," *Morbidity and Mortality Weekly Report* 43 (1994): 23–25; A. Marton, M. Gulyas, R. Muñoz, and A. Tomasz, "Extremely High Incidence of Antibiotic Resistance in Clinical Isolates of *Streptococcus pneumoniae* in Hungary," *Journal of Infectious Diseases* 163 (1991): 542–48; L. K. McDougal, R. Facklam, M. Reeves, et al., "Analysis of Multiply Antimicrobial-Resistant Isolates of *Streptococcus pneumoniae* from the United States," *Antimicrobial Agents and Chemotherapy* 36 (1992): 2176–84; and J. Versalovic, V. Kapur, E. O. Mason, Jr., et al., "Penicillin-Resistant *Streptococcus pneumoniae* Strains Recovered in Houston: Identification and Molecular Characterization of Multiple Clones," *Journal of Infectious Diseases* 167 (1993): 850–56.

35. R. Muñoz, A. Marton, and A. Tomasz, Thirtieth International Conference on Antimicrobial Agents and Chemotherapy, 1990, Abstract No. 173; and A. Tomasz, "Multiple-Antibiotic-Resistant Pathogenic Bacteria," *New England Journal of Medicine* 330 (1994): 1247–51.

36. Centers for Disease Control, "Multiple Antibiotic Resistance of Pneumococci—South Africa," *Morbidity and Mortality Weekly Report* 26 (1977): 285–86.

37. H. J. Koornhof, M. Jacobs, M. Isaacson, et al., "Follow-up on Multiple Antibiotic-Resistant Pneumococci," *South African Morbidity and Mortality Weekly Report* 21 (1978): 1–7.

38. A. Tomasz, "Disease Causing Bacteria Resistant to Antibiotics," Annual Meeting of the American Association for the Advancement of Science, San Francisco, February 19, 1994.

39. Dillon and Derrick (1972), op. cit.

40. J. Casal, A. Fenoll, M. D. Vicioso, and R. Muñoz, "Increase in Resistance to Penicillin in Pneumococci in Spain," *Lancet* I (1989): 735.

41. G. M. Caputo, P. C. Appelbaum, and H. H. Liu, "Infections Due to Penicillin-Resistant Pneumococci," *Archives of Internal Medicine* 153 (1993): 1301–10.

42. A. Tomasz, "Multiple-Antibiotic-Resistant Pathogenic Bacteria," *New England Journal of Medicine* 330 (1984): 1247–51; and B. E. Murray, "Can Antibiotic Resistance Be Controlled?" *New England Journal of Medicine* 330 (1994): 1229–30.

43. J. M. H. Pearson, G. S. Haile, and R. J. W. Rees, "Primary Dapsone-resistant Leprosy," *Leprosy Review* 48 (1977): 129–32.

44. C. Jia-Kun, W. Si-Yu, H. Yu-Hong, et al., "Primary Dapsone Resistance in China," *Leprosy Review* 60 (1989): 263–66.

45. N. Honore and S. T. Cole, "Molecular Basis of Rifampin Resistance in *Mycobacterium leprae*," *Antimicrobial Agents and Chemotherapy* 37 (1993): 414–18; and C. Fox, personal communication, 1993.

46. J. W. Boslego, E. C. Tramont, E. T. Takafuji, et al., "Effect of Spectinomycin Use on the

Prevalence of Resistant and of Penicillinase-Producing *Neisseria gonorrhoeae,*" *New England Journal of Medicine* 317 (1987): 272–77.

47. Centers for Disease Control, "Antibiotic Resistant Strains of *Neisseria gonorrhoeae*: Policy Guidelines for Detection, Management and Control," *Morbidity and Mortality Weekly Report* 36 (1987): Supplement 1S–18S.

48. Centers for Disease Control, "Tetracycline-Resistant *Neisseria gonorrhoeae*—Georgia, Pennsylvania, New Hampshire," *Morbidity and Mortality Weekly Report* 34 (1985): 563–65; Centers for Disease Control, "Plasmid-Mediated Tetracycline-Resistant *Neisseria gonorrhoeae*—Georgia, Massachusetts, Oregon," *Morbidity and Mortality Weekly Report* 35 (1986): 304–6; and J. S. Knapp, J. M. Zenilman, J. W. Biddle, et al., "Frequency and Distribution in the United States of Strains of *Neisseria gonorrhoeae* with Plasmid-Mediated, High-Level Resistance to Tetracycline," *Journal of Infectious Diseases* 155 (1987): 819–22.

49. Committee on Public Health, "Statement on Treatment of Gonorrhea: Penicillin Is Passé," *Bulletin of the New York Academy of Medicine* 65 (1989): 243–46.

50. J. W. Tapsall, T. R. Shultz, and E. A. Phillips, "Characteristics of *Neisseria gonorrhoeae* Isolated in Australia Showing Decreased Sensitivity to Quinolone Antibiotics," *Pathology* 24 (1992): 27–31.

51. World Health Organization, "Implementation of the Global Strategy for Health" (1992), op. cit.

52. Wright (1990), op. cit.

53. N. Harnett, "High Level Resistance to Trimethoprim, Cotrimoxazole and Other Antimicrobial Agents Among Clinical Isolates of *Shigella* Species in Ontario, Canada—An Update," *Epidemiology of Infection* 109 (1992): 463–72.

54. A. A. Ries, Abstract, Thirty-first Interscience Conference on Antimicrobial Agents and Chemotherapy, Chicago, September 29–October 2, 1991.

55. One of the clearest examples of the dangers that resistant diarrhea-producing organisms pose to poor countries was Guatemala's 1969 outbreak of antibiotic-resistant *Shigella*. The mutant bacteria were resistant to chloramphenicol, tetracycline, streptomycin, and sulfonamide, then the most commonly used anti-*Shigella* drugs. The epidemic that resulted from the emergence of the new bug sickened 112,000 Guatemalans and killed 12,500, most of them children.

56. According to CDC statistics, acute *Salmonella* food poisoning episodes in the United States skyrocketed. For example, between 1950 and 1980 the number of reported cases jumped from fewer than 1,000 to over 36,000, and the agency felt that physicians reported only one out of every 100 cases. So by 1980 there were probably about 3.5 million *Salmonella* food poisoning incidents in the United States annually. A decade later the CDC would estimate that 5 million Americans came down with *Salmonella* food poisoning annually, and outbreaks that debilitated up to 200,000 at a time were reported throughout the late 1980s.

57. C. Sanchez, E. García-Restoy, J. Garau, et al., "Ciprofloxacin and Trimethoprim-Sulfamethoxazole Versus Placebo in Acute Uncomplicated *Salmonella* Enteritis: A Double-Blind Trial," *Journal of Infectious Diseases* 168 (1993): 1304–7.

58. L. M. Bush, J. Calmon, C. L. Cherney, et al., "High-Level Penicillin Resistance Among Isolates of Enterococci," *Annals of Internal Medicine* 110 (1990): 515–20; J. M. Boyce, S. M. Opal, G. Potter-Bynow, et al., "Emergence and Nosocomial Transmission of Ampicillin-Resistant Enterococci," *Antimicrobial Agents and Chemotherapy* 36 (1992): 1032–39; and F. Caron, J. F. Lemeland, G. Humbert, et al., "Triple Combination Penicillin-Vancomycin-Gentamicin for Experimental Endocarditis Caused by a Highly Penicillin- and Glycopeptide-Resistant Isolate of *Enterococcus faecium,*" *Journal of Infectious Diseases* 168 (1993): 681–86.

Even ampicillin and vancomycin couldn't kill the bacteria. They were bacteriostatic, meaning they controlled the growth of enterococci colonies and prevented their spread throughout the human body. By 1975 *no* drug could actually *kill* the bacteria.

59. T. R. Frieden, S. S. Munsiff, D. E. Low, et al., "Emergence of Vancomycin-Resistant Enterococci in New York City," *Lancet* 342 (1993): 76–79; and R. V. Spera and B. F. Farber, "Multiply-Resistant *Enterococcus faecium*: The Nosocomial Pathogen of the 1990s," *Journal of the American Medical Association* 268 (1992): 2563–64.

60. B. E. Murray, "Can Antibiotic Resistance Be Controlled?" *New England Journal of Medicine* 330 (1994): 1229–30; L. Garrett, "Superbugs," *Newsday*, May 8, 1994: A1, A46, and A47; L. Garrett, "The Ebbing Miracle," *Newsday*, May 8, 1994: A5, A46; and L. Garrett, "Infection Fighters," *Newsday*, May 10, 1994: B25, B28–B29.

61. A. H. C. Uttley, N. Woodford, A. P. Johnson, et al., "Vancomycin-Resistant Enterococci," *Lancet* 342 (1993): 615.

62. M. H. Wilcox, R. C. Spencer, and G. R. Weeks, "Vancomycin-Resistant Enterococci," *Lancet* 342 (1993): 615–16.

63. E. Manso, G. DeSio, F. Biarasco, et al., "Vancomycin-Resistant Enterococci," *Lancet* 342 (1993): 616–17.

64. Centers for Disease Control, "Nosocomial Enterococci Resistant to Vancomycin in United States, 1989–1993," *Morbidity and Mortality Weekly Report* 42 (1993): 597–98.

65. L. L. Livornese, Jr., S. Dias, C. Samel, et al., "Hospital-Acquired Infection with Vancomycin-Resistant *Enterococcus faecium* Transmitted by Electronic Thermometers," *Annals of Internal Medicine* 117 (1992): 112–16.

66. A. H. C. Uttley and R. C. George, "Nosocomial Enterococcal Infection," *Current Opinions in Infectious Diseases* 4 (1991): 525–29.

67. R. Pallares, M. Pujol, C. Peña, et al., "Cephalosporins as Risk Factor for Nosocomial *Enterococcus faecalis* Bacteremia," *Archives of Internal Medicine* 153 (1993): 1581–86.

68. R. Leclerq, E. Derlot, M. Weber, et al., "Transferable Vancomycin and Teicoplanin Resistance in *Enterococcus faecium*," *Antimicrobial Agents and Chemotherapy* 33 (1989): 10–15; and W. C. Noble, Z. Virani, and R. G. A. Cree, "Co-transfer of Vancomycin and Other Resistance Genes from *Enterococcus faecalis* NCTC 12201 to *Staphylococcus aureus*," *FEMS Microbial Letter* 93 (1992): 195–98.

69. L. Garrett, "Superbugs" (1994), op. cit.; and L. Garrett, "Infection Fighters" (1994), op. cit.

70. S. B. Levy, G. B. Fitzgerald, and A. B. Macone, "*Escherichia coli* Transmission from Poultry to Human," *Nature* 260 (1976): 40–42.

71. A very thorough analysis of agricultural and livestock use of antibiotics and the emergence of dangerous bacteria can be found in Levy's book: Levy (1992), op. cit. Levy has dedicated his professional life to the problem, and the reader would be hard pressed to find a more detailed accounting of the multitudinous ways in which overuse and misuse of antibiotics are dooming the drugs to failure and granting victory to the microbes.

72. H. P. Endtz, R. P. Mouton, T. van der Reyden, et al., "Fluoroquinolone Resistance in *Campylobactr spp.* Isolated from Human Stools and Poultry Products," *Lancet* 335 (1990): 787.

73. E. Pérez-Trallero, M. Urbieta, C. L. Lopategui, et al., "Antibiotics in Veterinary Medicine and Public Health," *Lancet* 342 (1993): 1371–72.

74. E. Pérez-Trallero, C. Zigorraga, G. Cilla, et al., "Animal Origin of the Antibiotic Resistance of Human Pathogen *Yersinia enterocolitica*," *Scandinavian Journal of Infectious Diseases* 20 (1988): 573.

75. S. D. Holmberg, M. T. Osterholm, K. A. Senger, and M. L. Cohen, "Drug-Resistant Salmonella from Animals Fed Antimicrobials," *New England Journal of Medicine* 311 (1984): 617–22.

76. The *Salmonella* problem was exacerbated by home microwave oven cooking. It turned out that *Salmonella* could withstand microwaves, and reheated meat dishes often proved dangerously contaminated. See B. D. Gessner and M. Beller, "Protective Effect of Conventional Cooking Versus Use of Microwave Ovens in an Outbreak of Salmonellosis," *American Journal of Epidemiology* 139 (1994): 903–9.

77. S. B. Levy, "Playing Antibiotic Pool: Time to Tally the Score," *New England Journal of Medicine* 311 (1984): 663–64; Brunton, J. "Drug-Resistant Salmonella from Animals Fed Antimicrobials," *New England Journal of Medicine* 311 (1984): 1698–99; and T. H. Jukes, "Drug-Resistant Salmonella from Animals Fed Antimicrobials," *New England Journal of Medicine* 311 (1984): 1698–99.

78. J. S. Spika, S. H. Waterman, G. W. SooHoo, et al., "Chloramphenical-Resistant *Salmonella newport* Traced Through Hamburger to Dairy Farms," *New England Journal of Medicine* 316 (1987): 565–70.

79. L. W. Riley, R. S. Remis, S. D. Helgerson, et al., "Hemorrhagic Colitis Associated with a Rare *Escherichia coli* Serotype," *New England Journal of Medicine* 308 (1983): 681–85.

80. S. Ringertz, B. Bellete, I. Karlsson, et al., "Antibiotic Susceptibility of *Escherichia coli* Isolates from Inpatients with Urinary Tract Infections in Hospitals in Addis Ababa and Stockholm," *Bulletin of the World Health Organization* 68 (1990): 61–68; and S. Harnett, "Transferable High-Level Trimethoprim Resistance Among Isolates of *Escherichia coli* from Urinary Tract Infections in Ontario, Canada," *Epidemiology of Infection* 109 (1992): 473–81.

Some remarkable *E. coli* strains emerged in the early 1990s. For example, outside Cambridge, England, two strains appeared on a hospital transplant ward that were resistant to the antibiotic imipenen, as well as cefotaxime, ceftazidime, ciprofloxacin, gentamicin, ampicillin, azlocillin, coamoxiclav, timentin, cephalexin, cefuroxime, cefamandole, streptomycin, neomycin, kanamycin, tobramycin, trimethoprim, sulfamethoxazole, chloramphenicol, and nitrofurantoin. Only one commonly used antibiotic remained effective: amikacin. If the strains became resistant to that drug, they would be invulnerable to human treatment. See D. F. J. Brown, M. Farrington, and R. E. Warren, "Imipenen-resistant *Escherichia coli*," *Lancet* 342 (1993): 177.

In the Netherlands in 1992 a random survey of fecal samples from 310 healthy people yielded 456 *E. coli* types, nearly all of which bore some level of antibiotic resistance: 89 percent were resistant to ampicillin, 28 percent to trimethoprim, 80 percent to chloramphenicol. Only 19 percent were still susceptible to all eleven frontline antibiotics, and 14 percent were multiply resistant to four or more drugs. See M. Bonten, E. Stobberingh, J. Philips, and A. Houben, "Antibiotic Resistance of *Escherichia coli* in Fecal Samples of Healthy People in Two Different Areas in an Industrialized Country," *Infection* 30 (1992): 258–62.

And it was clear that plasmids and transposons that conferred antibiotic resistance traits in *E. coli* also commonly carried genes for greater virulence. See J. R. Johnson, I. Orskov, F. Orskov, et al., "O, K, and H Antigens Predict Virulence Factors, Carboxylesterase B Pattern, Antimicrobial Resistance, and Host Compromise Among *Escherichia coli* Strains Causing Urosepsis," *Journal of Infectious Diseases* 169 (1994): 119–26.

81. B. Marshall, D. Petrowski, and S. B. Levy, "Inter- and Intraspecies Spread of *Escherichia coli* in a Farm Environment in the Absence of Antibiotic Use," *Proceedings of the National Academy of Sciences* 87 (1990): 6609–13; and M. Singh, M. A. Chaudry, J. N. Yadava, and S. C. Sanyal, "The Spectrum of Antibiotic Resistance in Human and Veterinary Isolates of *Escherichia coli* Collected from 1984–86 in Northern India," *Journal of Antimicrobial Chemotherapy* 29 (1992): 159–68.

82. Several other cases of transmission of *E. coli* from manure to humans have since been described. See P. R. Cleslak, T. J. Barrett, P. M. Griffin, et al., "*Escherichia coli* 0157:H7 Infection from a Manured Garden," *Lancet* 342 (1993): 367; G. M. Morgan, C. Newman, S. R. Palmer, et al., "First Recognized Community Outbreak of Haemorrhagic Colitis Due to Verotoxin-Producing *Escherichia coli* 0157:H7 in the U.K.," *Epidemiology of Infection* 101 (1988): 83–91; and S. A. Renwick, J. B. Wilson, R. C. Clarke, et al., "Evidence of Direct Transmission of *Escherichia coli* 0157:H7 Infection Between Calves and a Human," *Journal of Infectious Diseases* 168 (1993): 792–93.

83. Centers for Disease Control, "Preliminary Report: Foodborne Outbreak of *Escherichia coli* 0157:H7 Infections from Hamburgers—Western United States, 1993," *Morbidity and Mortality Weekly Report* 42 (1993): 85–87; and S. Deresinski, "From Hamburgers to Hemolysis: *Escherichia coli* 0157:H7," *Infectious Disease Alert* 12 (1993): 81–84.

84. A. D. Russell, "Microbial Cell Walls and Resistance of Bacteria and Fungi to Antibiotics and Biocides," *Journal of Infectious Diseases* 168 (1993): 1339–40; and Columbia-Presbyterian Medical Center, *Infection Control: A Training Paradigm for Healthcare Professionals* (New York, 1994).

85. By 1994 the anti-chlorine sentiment would run so high in the United States that the head of the federal Environmental Protection Agency would advocate a virtual ban on all chlorinated products. Despite objections from industries as diverse as plastics manufacturing, cosmetics, electronics, dry cleaning, and petroleum, the EPA would move vigorously for massive reductions. In addition to public concern about cancer, the EPA would cite evidence that free chlorine ions expelled into the atmosphere acted as ozone scavengers, contributing to the depletion of the ozone layer.

86. I. Amato, "The Crusade Against Chlorine," *Science* 261 (1993): 152–54.

87. *Cryptosporidium* are about 4 to 6 microns in size, just a bit smaller than *E. coli*. Most water filters could only screen out organisms and particles of 100 microns or more in size. Disinfection was really the only practical way to sterilize the water.

88. E. B. Hayes, T. D. Matte, T. R. O'Brien, et al., "Large Community Outbreak of Cryptosporidiosis Due to Contamination of a Filtered Public Water Supply," *New England Journal of Medicine* 320 (1989): 1372–76.

89. Centers for Disease Control, "Surveillance for Waterborne Disease Outbreaks—United States, 1991–1992," *Morbidity and Mortality Weekly Report* 42 (1993): SS5–SS22.

90. K. C. Spitalny, R. L. Vogt, L. A. Orciari, et al., "Pontiac Fever Associated with a Whirlpool Spa," *American Journal of Epidemiology* 120 (1984): 809–17; E. J. Mangione, R. S. Remis, K. A. Tait, et al., "An Outbreak of Pontiac Fever Related to Whirlpool Use, Michigan, 1982," *Journal of the American Medical Association* 253 (1985): 535–39; and CDC, "Surveillance for Waterborne Disease Outbreaks" (1993), op. cit.

91. E. Geldreich, "Summary Report: Investigation of the Cabool, Missouri, Outbreak for a Water Supply Connection," U.S. Environmental Protection Agency, Washington, D.C.

92. Natural Resources Defense Council, "Think Before You Drink: The Failure of the Nation's Drinking Water System to Protect Public Health" (New York: NRDC Publication, 1993).

93. S. B. Levy, "Active Efflux Mechanisms for Antimicrobial Resistance," *Antimicrobial Agents and Chemotherapy* 36 (1992): 695–703; and H. Nikaido, "Prevention of Drug Access to Bacterial Targets: Permeability Barriers and Active Efflux," *Science* 264 (1994): 382–88.

94. J. Lederberg, speech before the Irvington Trust, New York City, February 8, 1994.

95. Gibbons (1992), op. cit.

96. J. Davies, "Inactivation of Antibiotics and the Dissemination of Resistance Genes," *Science* 264 (1994): 375–82; and B. G. Spratt, "Resistance to Antibiotics Mediated by Target Alternatives," *Science* 264 (1994): 388–93.

97. M. Raymond, P. Gros, M. Whiteway, and D. Y. Thomas, "Functional Complementation of Yeast *ste6* by a Mammalian Multidrug Resistance *mdr* Gene," *Science* 256 (1992): 232–34.

98. C. F. Amabile-Cuevas and M. E. Chicurel, "Horizontal Gene Transfer," *Scientific American* 81 (1993): 332–41.

99. M. Blot, J. Meyer, and W. Arber, "Bleomycin-Resistance Gene Derived from the Transposon *TnS* Confers Selective Advantage to *Escherichia coli* K-12," *Proceedings of the National Academy of Sciences* 88 (1991): 9112–16.

100. For an excellent overview of the various genes inside normal bacterial chromosomes that control the absorption and use of mobile DNAs, see D. J. Galas and M. Chandler, "Bacterial Insertion Sequences," Chapter 4 in Berg and Howe (1989), op. cit.

101. Amabile-Cuevas and Chicurel [(1992), op. cit.] pictured the movement of transposons and plasmids between various families of organisms as a highly fluid and ongoing process. On the basis of organism interactions and the relatedness of various known plasmids and transposons, they came up with the following representation of likely gene swapping.

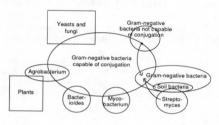

———⟶ = Possible additional avenues of mobile DNA passage, over and above the overlapping regions

102. J. D. Boeke, "Transposable Elements in *Saccharomyces cerevisiae*," Chapter 13 in Berg and Howe (1989), op. cit.; and H. Varmus and P. Brown, "Retroviruses," Chapter 3 in Berg and Howe (1989), op. cit.

103. R. Saral, W. H. Burns, O. L. Laskin, et al., "Acyclovir Prophylaxis of Herpes-Simplex-Virus Infections: A Randomized, Double-Blind, Controlled Trial in Bone-Marrow-Transplant Recipients," *New England Journal of Medicine* 305 (1981): 63–67.

104. G. J. Mertz, C. W. Critchlow, J. Benedetti, et al., "Double-Blind Placebo-Controlled Trial of Oral Acyclovir in First-Episode Genital Herpes Simplex Virus Infection," *Journal of the American Medical Association* 252 (1984): 1147–51; and S. E. Straus, H. E. Takiff, M. Seidlin, et al., "Suppression of Frequently Recurring Genital Herpes," *New England Journal of Medicine* 310 (1984): 1545–50.

105. K. E. VanLandingham, B. Marsteller, G. W. Ross, and F. G. Hayden, "Relapse of Herpes Simplex Encephalitis After Conventional Acyclovir Therapy," *Journal of the American Medical Association* 259 (1988): 1051–53; A. L. Rothman, S. H. Cheeseman, S. N. Lehrman, et al., "Herpes Simplex Encephalitis in a Patient with Lymphoma: Relapse Following Acyclovir Therapy," *Journal of the American Medical Association* 259 (1988): 1056–57; and R. J. Whitley, "The Frustrations of Treating Herpes Simplex Virus Infections of the Central Nervous System," *Journal of the American Medical Association* 259 (1988): 1067.

106. W. I. Whittington and W. J. Cates, Jr., "Acyclovir Therapy for Genital Herpes: Enthusiasm and Caution in Equal Doses," *Journal of the American Medical Association* 251 (1984): 2116–17.

107. Though acyclovir was first used to treat herpes simplex-2, which caused genital herpes, it soon proved effective in controlling the entire family of herpes viruses, including varicella (chicken pox and shingles), herpes zoster, and herpes simplex-1. All of these viruses had the ability to hide latently inside human nerve cells for years, even decades, only surfacing when immunological conditions in the host favored their survival. For example, the same virus that caused childhood chicken pox would hide for five or six decades, resurfacing to produce often excruciating shingles.

108. L. Seale, C. J. Jones, S. Kathpalia, et al., "Prevention of Herpesvirus Infections in Renal

Allograft Recipients by Low-Dose Oral Acyclovir," *Journal of the American Medical Association* 254 (1985): 3435–38.

109. D. Parris and J. E. Harrington, "Herpes Simplex Virus Variants Resistant to High Concentration of Acyclovir Exist in Clinical Isolates," *Antimicrobial Agents and Chemotherapy* 22 (1982): 71–77.

110. E. Katz, O. Rosenblat, and S. Pisanty, "Isolation and Characterization of Herpes Simplex Virus Resistant to Nucleoside Analogs," *Oral Surgery, Oral Medicine and Oral Pathology* 72 (1991): 296–99.

111. H. J. Field and S. E. Goldthorpe, "The Pathogenicity of Drug-Resistant Variants of Herpes Simplex Virus," Fourth Forum in Virology, 1992, 120–24.

112. K. S. Erlich, J. Mills, P. Chatis, et al., "Acyclovir-Resistant Herpes Simplex Virus Infections in Patients with the Acquired Immunodeficiency Syndrome," *New England Journal of Medicine* 320 (1989): 293–96; and R. J. Whitley and J. W. Gnann, Jr., "Acyclovir: A Decade Later," *New England Journal of Medicine* 327 (1992): 782–89.

113. Quite unfortunately, the sexual partner refused to cooperate with the study, so Straus was unable to absolutely confirm this hypothesis by performing PCR analysis of his herpes strain. It is a sorry fact that individuals commonly decline to participate in such studies, which could prove of immense good for the community as a whole. Such lack of participation is evident in all types of people. The failure of cooperation in this case—in a gay man—was actually fairly unusual, as the American gay community had proven remarkably open to scientists and their investigations since the onset of the AIDS epidemic.

114. The study is described in R. G. Kost, E. L. Hill, M. Tigges, and S. Straus, "Brief Report: Recurrent Acyclovir-Resistant Genital Herpes in an Immunocompetent Patient," *New England Journal of Medicine* 329 (1993): 1777–81.

115. S. Safrin, C. Crumpacker, P. Chatis, et al., "A Controlled Trial Comparing Foscarnet with Vidarabine for Acyclovir-Resistant Mucocutaneous Herpes Simplex in the Acquired Immunodeficiency Syndrome," *New England Journal of Medicine* 325 (1991): 551–55; and S. Safrin, "Management of Patients Following Successful Healing of Acyclovir-Resistant Herpes Simplex Infection," Fourth Forum on Virology, 1992, 125–26.

116. J. M. Pépin, F. Simon, M. C. Dazza, and F. Brun-Vézinet, "The Clinical Significance of *in vitro* Cytomegalovirus Susceptibility to Antiviral Drugs," Fourth Forum on Virology, 1992, 126–27; and C. Leport, S. Puget, J. M. Pépin, et al., "Cytomegalovirus Resistant to Foscarnet: Clinicovirologic Correlation in a Patient with Human Immunodeficiency Virus," *Journal of Infectious Diseases* 168 (1993): 1329–30.

117. N. S. Lurain, K. D. Thompson, E. W. Holmes, and G. S. Read, "Point Mutations in the DNA Polymerase Gene of Human Cytomegalovirus That Result in Resistance to Antiviral Agents," *Journal of Virology* 66 (1992): 7146–52.

118. S. Safrin, S. Kemmerly, B. Plotkin, et al., "Foscarnet-Resistant Herpes Simplex Virus Infection in Patients with AIDS," *Journal of Infectious Diseases* 169 (1994): 193–96.

119. M. C. Y. Heng, S. Y. Heng, and S. G. Allen, "Co-infection and Synergy of Human Immunodeficiency Virus-1 and Herpes Simplex-1," *Lancet* 343 (1994): 255–58.

120. The literature on AZT resistance is vast and occasionally contradictory on questions of timing of emergence. Key studies include: A. Erice, D. L. Mayers, D. G. Strike, et al., "Brief Report: Primary Infection with Zidovudine-Resistant Human Immunodeficiency Virus Type 1," *New England Journal of Medicine* 328 (1993): 1163–65; M. S. Hirsch and R. T. D'Aquila, "Therapy for Human Immunodeficiency Virus Infection," *New England Journal of Medicine* 328 (1993): 1686–95; V. A. Johnson, "New Developments in Antiretroviral Drug Therapy for HIV Infection," Chapter 4 in P. Volberding and M. A. Jacobson, *AIDS Clinical Review 1992* (New York: Marcel Dekker, 1992); B. A. Larder, K. E. Coates, and S. D. Kemp, "Zidovudine-Resistant Human Immunodeficiency Virus Selected by Passage in Cell Culture," *Journal of Virology* 6 (1991): 5232–36; and H. Mohri, M. K. Singh, W. T. W. Ching, and D. D. Ho, "Quantitation of Zidovudine-Resistant Human Immunodeficiency Virus Type 1 in the Blood of Treated and Untreated Patients," *Proceedings of the National Academy of Sciences* 90 (1993): 25–29.

121. M. S. Smith, K. L. Korber, and J. S. Pagano, "Long-Term Persistence of Zidovudine Resistance Mutations in Plasma Isolates of Human Immunodeficiency Virus Type 1 Dideoxyinosine-Treated Patients Removed from Zidovudine Therapy," *Journal of Infectious Diseases* 169 (1994): 184–88; and C. P. Conlon, P. Klenerman, A. Edwards, et al., "Heterosexual Transmission of Human Immunodeficiency Virus Type 1 Variants Associated with Zidovudine Resistance," *Journal of Infectious Diseases* 169 (1994): 411–15.

122. Z. Gu, Z. Gao, X. Li, et al., "Novel Mutation in the Human Immunodeficiency Virus Type

1 Reverse Transcriptase Gene That Encodes Cross-Resistance to 2',3'-Dideoxyinosine and 2'3'-Dideoxycytidine," *Journal of Virology* 66 (1992): 7128–35; and Z. Song, G. Yang, S. P. Goff, and V. R. Prasad, "Mutagenesis of the Glu-89 Residue in Human Immunodeficiency Virus Type 1 (HIV-1) and HIV-2 Reverse Transcriptase: Effects on Nucleoside Analog Resistance," *Journal of Virology* 66 (1992): 7568–71.

123. National Institutes of Allergy and Infectious Diseases, State-of-the-Art Conference on Antiretroviral Therapy, Bethesda, MD, June 23–25, 1993.

124. F. G. Hayden, R. B. Belshe, R. D. Clover, et al., "Emergence and Apparent Transmission of Rimantadine-Resistant Influenza A Virus in Families," *New England Journal of Medicine* 321 (1989): 1696–1702; and R. B. Belshe, M. H. Smith, C. B. Hall, et al., "Genetic Basis of Resistance to Rimantadine Emerging During Treatment of Influenza Virus Infection," *Journal of Virology* 62 (1988): 1508–12.

125. Doctors relied almost entirely upon drug companies, directly or indirectly, for advice about use of antibiotics. The companies spent $11 billion a year in the United States alone promoting use of their products. For busy physicians who hadn't the time to sift through medical literature to learn of contrary evidence, it was hard to resist the alluring pull of pharmaceutical promotions. See R. L. Woosley, "A Prescription for Better Prescriptions," *Issues in Science and Technology* (Spring 1994): 59–66.

126. World Bank, *World Development Report 1993: Investing in Health* (New York: Oxford University Press, 1993).

127. A. J. Slater, "Antibiotic Resistance in the Tropics," *Transactions of the Royal Society of Tropical Medicine and Hygiene* 83 (1989): 45–48.

128. A. Chetley, "Bangladesh Drug Policy Hanging in the Balance," *Lancet* 343 (1994): 967.

129. For analyses of drug development policies and politico-economic conflict, see D. E. Bell and M. R. Reich, *Health, Nutrition, and Economic Crises: Approaches to Policy in the Third World* (Dover, MA: Auburn House, 1988); Pan American Health Organization, *Policies for the Production and Marketing of Essential Drugs* (Washington, D.C., 1984); M. R. Reich, "Essential Drugs: Economics and Politics in International Health," *Health Policy* 8 (1987): 39–57; and World Bank, *Financing Health Services in Developing Countries: An Agenda for Reform* (Washington, D.C.: World Bank, 1987).

130. P. Lepage, J. Bogarts, C. Van Goethem, et al., "Community-Acquired Bacteraemia in African Children," *Lancet* I (1987): 1458–61.

131. Corruption and black marketeering were critical problems for antibiotic distribution and misuse in developing countries. In many parts of the world untrained medical "injectionists" earned a living by injecting illegally obtained antibiotics and vitamins into customers, often in the major marketplaces in full view of police. Needles and the medicinal drugs were, therefore, of very high value. Such misuse of the drugs no doubt contributed both to resistance and to lack of supplies for diarrheal epidemics.

The following wire story release from the Associated Press in July 1993 illustrated the problem:

NAIROBI, KENYA (AP)—Patients bring their own medicine, food and syringes to some government hospitals because employees often steal and sell what few supplies are issued.

Inflation is soaring, partly because the government printed billions of new shillings last year to finance the ruling party's campaign and bribe voters.

Much of the social security fund was lost when officials illegally put it into real estate deals and other bad investments. Asked by journalists whether anyone would be held accountable, Treasury Secretary Wilfred Koinange replied: "What do you want, heads to roll?"

132. An excellent summary of the ivermectin saga is in E. Tanouye, "Merck's 'River Blindness' Gift Hits Snag," *Wall Street Journal*, September 23, 1992: A1, B7. Merck was able to reap ample profits off veterinary use of ivermectin: annual sales topped the $500 million mark in 1990, according to the World Bank.

133. See, for example, W. C. Hsiao, "Lessons for Developing Countries from the Experiences of Affluent Nations About a Comprehensive Health Financing Strategy," presented at International Seminar on Comprehensive Financing Strategy in Select Asian Nations, December 10–14, 1990, Bali, Indonesia.

134. International Federation of Pharmaceutical Manufacturers Associations, *The Pharmaceutical Industry: International Issues and Answers* (Washington, D.C.: Pharmaceutical Manufacturers Association, 1979).

135. See, for example, S. Lall and S. Bibile, "The Political Economy of Controlling Transnationals: The Pharmaceutical Industry in Sri Lanka, 1972–1976," *International Journal of Health Services* 8 (1978): 299–328; T. Heller, *Poor Health, Rich Profits: Multinational Drug Companies and the Third*

World (London: Spokesman Books, 1977); UNCTAD Secretariat, "Dominant Positions of Market Power of Transnational Corporations: Use of the Transfer Pricing Mechanism," Geneva, November 30, 1977; J. M. Starrels, "The World Health Organization, Resisting Third World Ideological Pressures" (Washington, D.C.: Heritage Foundation, 1985); R. Deitch, "Commentary from Westminster: More Pressure on the Profits of the Pharmaceutical Industry," *Lancet* I (1984): 521; and International Federation of Pharmaceutical Manufacturers Associations (IFPMA), "Medicines and the Developing World," Geneva, 1984.

136. S. Kingman, "Malaria Runs Riot on Brazil's Wild Frontier," *New Scientist*, August 12, 1989: 24–25.

137. World Health Organization, "Malaria Worsening in Many Areas," PR/WHA/6/May 9, 1991, Geneva.

138. P. G. Kremsner, G. M. Zotter, H. Feldmeier, et al., "*In vitro* Drug Sensitivity of *Plasmodium falciparum* in Acre, Brazil," *Bulletin of the World Health Organization* 67 (1989): 289–93; and S. Reyes, C. H. Osanai, and A. D. Passos, "*In vivo* Resistance of *Plasmodium falciparum* to 4-Aminoquinolones and to Sulfadoxine-Pyrimethamine Combination: II. Study of Manaus, Amazonas, 1983–84," *Revista Brasileira de Malariologia e Doenças Tropicais* 38 (1986): 37–44.

139. American Association for the Advancement of Science, *Malaria and Development in Africa: A Cross-Sectional Approach*, No. AFR-0481-A-00-0037-00, U.S. Agency for International Development, Africa Bureau, Washington, D.C., 1991.

140. Centers for Disease Control, "Chloroquine-Resistant Malaria Acquired in Kenya and Tanzania: Denmark, Georgia, New York," *Morbidity and Mortality Weekly Report* 27 (1978): 463–64; B. H. Kean, "Chloroquine-Resistant *falciparum* Malaria from Africa," *Journal of the American Medical Association* 241 (1979): 395–96; and C. C. Campbell, W. Chin, W. E. Collins, et al., "Chloroquine-Resistant *Plasmodium falciparum* from East Africa: Cultivation and Drug Sensitivity of the Tanzanian I/CDC Strain from an American Tourist," *Lancet* II (1979): 1151–54.

141. H. C. Spencer, S. C. Masaba, and D. Kiaraho, "Sensitivity of *Plasmodium falciparum* Isolates to Chloroquine in Kisumu and Malindi, Kenya," *American Journal of Tropical Medicine and Hygiene* 31 (1982): 902–6.

142. W. M. Watkins, D. G. Sixsmith, H. C. Spencer, et al., "Effectiveness of Amodiaquine as Treatment for Chloroquine-Resistant *Plasmodium falciparum* Infections in Kenya," *Lancet* I (1984): 357–59.

143. D. Overbosch, A. W. vanden Wall Bake, P. C. Stuiver, and H. J. van der Kaay, "Chloroquine-Resistant *falciparum* Malaria from Malawi," *Tropical and Geographic Medicine* 36 (1984): 71–72.

144. A. M. Blumenfeld, W. L. Sieling, A. Davidson, and M. Isaacson, "Probable Chloroquine-Resistant *Plasmodium falciparum* Malaria in South-Western Africa," *South African Medical Journal* 66 (1984): 207–8.

145. K. R. Perry, N. M. Hone, and J. M. Cairns, "Chloroquine Resistant *Plasmodium falciparum* Malaria Confirmed by *in-vitro* Testing in a District Hospital," *Medical Journal of Zambia* 18 (1984): 8–9.

146. J. Linberg, T. Sandberg, B. Bjorkholm, and A. Bjorkman, "Chloroquine and Fansidar Resistant Malaria Acquired in Angola," *Lancet* I (1985): 765.

147. N. J. Visagie and W. L. Sieling, "Chloroquine-Resistant *Plasmodium falciparum* Malaria in South Africa: A Case Report," *South African Medical Journal* 68 (1985): 600–1.

148. G. Charmot, J. Le Bras, P. Sansonetti, et al., "Eight Cases of Drug-Resistant *Plasmodium falciparum* Malaria Contracted in Mozambique," *Bulletin de la Société de Pathologie Exotique et de Ses Filiales* 78 (1985): 500–4.

149. C. C. Draper, G. Brubaker, A. Geser, and V. A. E. B. Kilimali, "Serial Studies on the Evolution of Chloroquine Resistance in an Area of East Africa Receiving Intermittent Malaria Chemosuppression," *Bulletin of the World Health Organization* 63 (1985): 109–18.

150. P. Nguyen-Dinh, "Etudes sur la Chimiorésistance de *Plasmodium falciparum* en Afrique: Données Actuelles," *Annals de la Société Belgique de Médecine Tropique* 65 (1985): Suppl. 2, 105–13.

151. "Chloroquine-Resistant Malaria in Africa," *Lancet* I (1985): 1487–88; and H. C. Spencer, D. C. O. Kaseje, A. D. Brandling-Bennett, et al., "Changing Response to Chloroquine of *Plasmodium falciparum* in Sarididi, Kenya, from 1981 to 1984," *Annals of Tropical Medicine and Parasitology* 81 (1987): Suppl. 98–104.

152. A. Björkman, M. Willcox, N. Marbiah, and D. Payne, "Susceptibility of *Plasmodium falciparum* to Different Doses of Quinine *in vivo* and to Quinine and Quinidine *in vitro* in Relation to Chloroquine in Liberia," *Bulletin of the World Health Organization* 69 (1991): 459–65.

153. "One Bite Is Too Many," *The Economist*, August 21, 1993: 33–34; and American Association for the Advancement of Science (1991), op. cit.

154. Ibid.

Similar observations were made from one end of the continent to the other. In Brazzaville, for example, cerebral malaria in children soared between January 1988 and June 1989, as did chloroquine resistance. Fifteen percent of the cases were immediately lethal. And a third of the survivors went on to suffer debilitating ailments, such as periodic convulsions, as a result of fever damage to their brains. See B. Carme, J. C. Bouquetry, and H. Plassart, "Mortality and Sequelae Due to Cerebral Malaria in African Children in Brazzaville, Congo," *American Journal of Tropical Medicine and Hygiene* 48 (1993): 216–21.

Even more striking were the escalating malaria admissions to hospitals in Kinshasa. Between 1982 and 1986 the percentage of all pediatric admissions that were due to acute malaria rose from 29.5 percent to 56.4 percent. See A. E. Greenberg, M. Ntumbanzondo, N. Ntula, et al., "Hospital-Based Surveillance of Malaria-Related Paediatric Morbidity and Mortality in Kinshasa, Zaire," *Bulletin of the World Health Organization* 67 (1989): 186–9.

155. During that time, malaria incidence in "Zambia, Togo, and Rwanda . . . increased by 7, 10, and 21 percent, respectively, every year over the 1980s," Brinkmann and his colleagues wrote. "Thus, the burden of malaria in 1995 is likely to be two to three times its level in the late 1980s."

Costs included lost workdays due to prolonged bouts with the parasites; funeral expenses; hospital and drug costs; family impoverishment due to the death of a major breadwinner; and a host of other factors.

"Overall, in 1987 the cost of malaria in sub-Saharan Africa was about $791 million per year," the group wrote. "This figure is projected to rise to $1,684 billion by 1995. By comparison, the entire 1990 health assistance to Africa of a major bilateral donor, the U.S. Agency for International Development, was only $52 million for all conditions. In 1987 malaria represented 0.6 percent of GDP; by 1995, if current trends continue, this share will rise to 1.0 percent of GDP." See D. S. Shepard, M. B. Ettling, U. Brinkmann, and R. Sauerborn, "The Economic Cost of Malaria in Africa," *Tropical Medicine and Parasitology* 42 (1991): 199–203. See also in the same volume U. Brinkmann and A. Brinkmann, "Malaria and Health in Africa: The Present Situation and Epidemiological Trends": 205–13; M. B. Ettling and D. S. Shepard, "Economic Cost of Malaria in Rwanda": 214–18; and R. Sauerborn, D. S. Shepard, M. B. Ettling, et al., "Estimating the Direct and Indirect Economic Costs of Malaria in a Rural District of Burkina Faso": 219–23.

156. See S. C. Redd, P. B. Bloland, P. N. Kazembe, et al., "Usefulness of Clinical Case-Definitions in Guiding Therapy for African Children with Malaria or Pneumonia," *Lancet* 340 (1992): 1140–43.

157. See P. B. Bloland, E. M. Lackritz, P. N. Kazembe, et al., "Beyond Chloroquine: Implications of Drug Resistance for Evaluating Malaria Therapy Efficacy and Treatment Policy in Africa," *Journal of Infectious Diseases* 167 (1993): 932–37; C. C. Campbell, P. Nguyen-Dinh, and J. G. Breman, "Epidemiological and Operational Considerations in the Use of Antimalarial Drugs for Chemotherapy and Chemoprophylaxis of Malaria in Africa," *Annals de la Société Belgique de Médecine Tropique* 65 (1985): 165–70; and E. M. Lackritz, C. C. Campbell, T. K. Ruebush, et al., "Effect of Blood Transfusion on Survival Among Children in a Kenyan Hospital," *Lancet* 340 (1992): 524–88;

158. D. L. Heymann, R. W. Skeketee, J. J. Wirima, et al., "Antenatal Chloroquine Chemoprophylaxis in Malawi: Chloroquine Resistance, Compliance, Protective Efficacy and Cost," *Transactions of the Royal Society of Tropical Medicine and Hygiene* 84 (1990): 496–8.

159. Malawi Ministry of Health, "Malaria Control in Malawi, 1984–1988: Policy Modification and Program Development in Response to Chloroquine Resistance," unpublished, 1993.

160. S. L. Hoffman, C. N. Oster, C. V. Plowe, et al., "Naturally Acquired Antibodies to Sporozoites Do Not Prevent Malaria: Vaccine Development Implications," *Science* 237 (1987): 639–42.

161. Similar results were obtained by L. W. Pang, N. Limsomwong, J. Karwacki, and H. K. Webster, "Circumsporozoite Antibodies and *falciparum* Malaria Incidence in Children Living in a Malaria Endemic Area," *Bulletin of the World Health Organization* 66 (1988): 359–63.

162. A. W. Taylor-Robinson, R. S. Phillips, A. Severn, et al., "The Role of T_H1 and T_H2 Cells in a Rodent Malaria Infection," *Science* 260 (1993): 1931–34.

163. Y. Charoenvit, W. E. Collins, T. R. Jones, et al., "Inability of Malaria Vaccine to Induce Antibodies to a Protective Epitope Within Its Sequence," *Science* 251 (1991): 668–72; J. E. Egan, J. L. Weber, W. R. Ballou, et al., "Efficacy of Murine Malaria Sporozoite Vaccines: Implications for Human Vaccine Development," *Science* 236 (1987): 453–56; and R. Rosenberg, R. A. Wirtz, D. E. Lanar, et al., "Circumsporozoite Protein Heterogeneity in the Human Malaria Parasite *Plasmodium vivax*," *Science* 245 (1989): 973–76.

164. J. Lines and J. R. M. Armstrong, "For a Few Parasites More: Inoculum Size, Vector Control and Strain-Specific Immunity to Malaria," *Parasitology Today* 8 (1992): 381–83.

165. S. Kumar, L. H. Miller, and I. A. Quakyi, "Cytotoxic T Cells Specific for the Circumsporozoite Protein of *Plasmodium falciparum*," *Nature* 334 (1988): 258–60; and Institute of Medicine, *Malaria: Obstacles and Opportunities* (Washington, D.C.: National Academy Press, 1991).

166. S. L. Hoffman, V. Nussenzweig, J. C. Sadoff, and R. S. Nussenzweig, "Progress Toward Malaria Preerythrocytic Vaccines," *Science* 252 (1991): 520–21.

167. One theory had it that chloroquine resistance could only get so bad in a society before the humans would develop sufficient baseline immunity to keep the parasites in check. A state of tolerance developed, and the partially immune humans were able to prevent the parasites from developing *complete* resistance. See J. C. Koella, "Epidemiological Evidence for an Association Between Chloroquine Resistance of *Plasmodium falciparum* and Its Immunological Properties," *Parasitology Today* 9 (1993): 105–8.

168. D. Hurvitz and K. Hirschhorn, "Suppression of *in vitro* Lymphocyte Responses by Chloroquine," *New England Journal of Medicine* 273 (1965): 23–26; G. Salmeron and P. E. Lipsky, "Immunosuppression Potential of Antimalarials," *American Journal of Medicine* 75 (1983): 19–24; and D. N. Taylor, C. Wasi, and K. Bernard, "Chloroquine Prophylaxis Associated with a Poor Antibody Response to Human Diploid Cell Rabies Vaccine," *Lancet* I (1984): 1405.

169. M. Pappaionou, D. B. Fishbein, D. W. Dreesen, et al., "Antibody Response to Preexposure Human Diploid-Cell Rabies Vaccine Given Concurrently with Chloroquine," *New England Journal of Medicine* 314 (1986): 280–84.

The immune system dampening effect was not true for all types of immunizations. Nigerian children responded similarly to pneumococcal vaccines regardless of whether or not they took chloroquine. See C. Van Der Straeten and J. H. Klippel, "Antimalarials and Pneumococcal Immunization," *New England Journal of Medicine* 315 (1986): 712.

170. An extensive study of the use of chloroquine prophylaxis on Nigerian children in 1984 yielded a puzzling result: the prophylaxed children suffered fewer cases of malaria, but were just as likely as kids who never took chloroquine to die of malaria. In this study malaria was defined as fever plus laboratory-confirmed parasite infection of the child's blood. Given that definition of malaria, there may be less contradiction than first meets the eye between a lower incidence of "malaria" in the chloroquine-using group and an equal or higher death rate. See A. Bradley-Moore, D. E. Bidwell, A. Voller, et al., "Malaria Chemoprophylaxis with Chloroquine in Young Nigerian Children," *Annals of Tropical Medicine and Parasitology* 79 (1985): 549–62.

171. For further evidence of the antifever effect of chloroquine, see A. D. Brandling-Bennett, A. J. Oloo, W. M. Watkins, et al., "Chloroquine Treatment of *falciparum* Malaria in an Area of Kenya of Intermediate Chloroquine Resistance," *Transactions of the Royal Society of Tropical Medicine and Hygiene* 82 (1988): 833–37.

172. *Plasmodium falciparum* could produce some of this effect. German researchers were convinced that the parasites stimulated an autoimmune response in which red blood cells were destroyed by the victim's own immune system. Their study did not, however, control for chloroquine use, so they couldn't rule out the possibility that the drug induced the observed autoimmunity. See K. Ritter, A. Kuhlencord, R. Thomssen, and W. Bommer, "Prolonged Haemolytic Anaemia in Malaria and Autoantibodies Against Triosephosphate Isomerase," *Lancet* 342 (1993): 1333–34.

There was also a hypothesis that malaria induced the immune system to release large amounts of tumor necrosis factor (TNF), a powerful human chemical that had a broad range of effects on the body. Usually researchers claimed that TNF played a role in cerebral malaria, but some also speculated it was involved in anemia. The evidence was contradictory and controversial. See R. E. Phillips and T. Solomon, "Cerebral Malaria in Children," *Lancet* 336 (1990): 1355–60; I. A. Clarke, G. Chaudri, and W. B. Cowden, "Roles of Tumour Necrosis Factor in the Illness and Pathology of Malaria," *Transactions of the Royal Society of Tropical Medicine and Hygiene* 83 (1989): 436–40; and D. Kwiatkowski, A. V. S. Hill, I. Sambou, et al., "TNF Concentration in Fatal Cerebral, Non-Fatal Cerebral, and Uncomplicated *Plasmodium falciparum* Malaria," *Lancet* 336 (1990): 1201–4.

173. P. B. Bloland, E. M. Lackritz, P. N. Kazembe, et al., "Beyond Chloroquine: Implications of Drug Resistance for Evaluating Malaria Therapy Efficacy and Treatment Policy in Africa," *Journal of Infectious Diseases* 167 (1993): 932–37.

174. Even as early as 1984 there was widespread resistance in East Africa to Fansidar, particularly its component pyrimethamine. But amodiaquine and mefloquine remained useful alternatives in most of Africa well into the late 1980s. See H. C. Spencer, "Drug-Resistant Malaria: Changing Patterns Mean Difficult Decisions," *Transactions of the Royal Society of Tropical Medicine and Hygiene* 79 (1985): 748–58.

175. T. Harinasuta, S. Migasen, and D. Boonag, *UNESCO First Regional Symposium on Scientific Knowledge of Tropical Parasites.* (Singapore: UNESCO, 1962).

176. A. A. Sandosham, "Chloroquine-Resistant *falciparum* Malaria in Malaya," *Singapore Medical Journal* 4 (1963): 3–5.

177. P. G. Contacos, J. S. Lunn, and G. R. Coatney, "Drug-Resistant *falciparum* Malaria from Cambodia and Malaya," *Transactions of the Royal Society of Tropical Medicine and Hygiene* 57 (1963): 417–24.

178. UNDP/World Bank/WHO, Report of a Meeting Held in Kuala Lumpur, Malaysia, August 10–15, 1981 (Geneva: World Health Organization, 1981).

179. Ibid.

180. Centers for Disease Control, "*Plasmodium falciparum* Malaria Contracted in Thailand Resistant to Chloroquine and Sulfonamide-Pyrimethamine—Illinois," *Morbidity and Mortality Weekly Report* 29 (1980): 493–94.

181. G. Watt, G. W. Long, L. P. Padre, et al., "Chloroquine and Quinine: A Randomized, Double-Blind Comparison of Efficacy and Side Effects in the Treatment of *Plasmodium falciparum* Malaria in the Philippines," *Transactions of the Royal Society of Tropical Medicine and Hygiene* 82 (1988): 205–8; and V. P. Sharma, C. Prasittisuk, and A. V. Kondrashin, "Magnitude of Forest Malaria," in V. P. Sharma and A. V. Kondrashin, eds., *Proceedings of an Informal Consultation Meeting WHO/MRC 18–22 February 1991* (New Delhi: World Health Organization, 1991).

182. Sharma, Prasittisuk, and Kondrashin (1991), op. cit.

183. M. K. Banerjee, N. Palikhe, B. L. Shestha, et al., in Sharma and Kondrashin, eds. (1991), op. cit.

In Burma the 1976 malarial death rate was 2.5 per 100,000. By 1989 that had risen to 12.3 per 100,000. And as had been seen in Africa, adult cerebral malaria cases increased in both number and severity. By 1989, one out of every three cerebral cases was fatal, despite drug treatment.

184. T. Chongsuphajaisiddhi, A. Sabchareon, P. Chantavanich, et al., "A Phase III Clinical Trial of Mefloquine in Children with Chloroquine-Resistant *falciparum* Malaria in Thailand," *Bulletin of the World Health Organization* 65 (1987): 223–26; H. O. Lobel, M. Miani, T. Eng, et al., "Long-Term Malaria Prophylaxis with Weekly Mefloquine," *Lancet* 341 (1993): 848–51; and R. Steffen, E. Fuchs, J. Schildknecht, et al., "Mefloquine Compared with Other Malaria Chemoprophylactic Regimens in Tourists Visiting East Africa," *Lancet* 341 (1993): 1299–1303.

185. S. Looareesuwan, C. Viravan, S. Vanijanonta, et al., "Randomised Trial of Artesunate and Mefloquine Alone and in Sequence for Acute Uncomplicated *falciparum* Malaria," *Lancet* 339 (1992): 821–824; and F. O. Ter Kuile, G. Dolan, F. Noster, et al., "Halofantrine Versus Mefloquine in Treatment of Multidrug-Resistant *falciparum* Malaria," *Lancet* 341 (1993): 1044–49.

186. C. M. Wilson, A. E. Serrano, A. Wasley, et al., "Amplification of a Gene Related to Mammalian *mdr* Genes in Drug-Resistant *Plasmodium falciparum*," *Science* 244 (1989): 1184–86; S. K. Martin, A. M. J. Oduola, and W. K. Mihous, "Reversal of Chloroquine Resistance in *Plasmodium falciparum* by Verapamil," *Science* 235 (1987): 899–901; C. J. Newbold, "The Path of Drug Resistance," *Nature* 345 (1990): 202–3; T. E. Wellems, L. J. Panton, I. Y. Gluzman, et al., "Chloroquine Resistance Not Linked to *mdr*-Like Genes in a *Plasmodium falciparum* Cross," *Nature* 345 (1990): 253–55; and S. J. Foote, D. E. Kyle, R. K. Martin, et al., "Several Alleles of the Multidrug-Resistance Gene Are Closely Linked to Chloroquine," *Nature* 345 (1990): 255–58.

187. Similar pumps are thought to exist in other parasites. *Entamoeba histolytica*, the gay bowel disease agent, appears to have acquired an *mdr* gene that is 35 percent identical to the malaria *mdr*. The parasite uses the pump to protect itself from the drug emetine. See J. Samuelson, P. Ayala, E. Orozco, and D. Wirth, "Emetine-Resistant Mutants of *Entamoeba histolytica* Overexpress mRNAs for Multidrug Resistance," *Molecular and Biochemical Parasitology* 38 (1990): 281–90.

188. K. H. Riechmann, D. R. Davis, and D. C. Hutton, "*Plasmodium vivax* Resistance to Chloroquine?" *Lancet* II (1989): 1183–84.

189. G. S. Murphy, H. Basri, Purnomo, et al., "*Vivax* Malaria Resistant to Treatment and Prophylaxis with Chloroquine," *Lancet* 341 (1993): 96–100.

190. Ibid.

191. "Rediscovering Wormwood: Ginghaosu for Malaria," *Lancet* 339 (1992): 649–51; and E. Tanouye, "Chinese Tea Yields Secret of Its Success Against Malaria Bug," *Wall Street Journal*, July 1, 1993: A1; Looareesuwan, Viravan, Vanijanonta, et al. (1992), op. cit.; and J. Karbwang, K. N. Bangchang, A. Thanavibul, et al., "Comparison of Oral Artemether and Mefloquine in Acute Uncomplicated *falciparum* Malaria," *Lancet* 340 (1992): 1245–48.

192. F. Gay, L. Ciceron, M. Litaudon, et al., "In-Vitro Resistance of *Plasmodium falciparum* to Qinghaosu Derivatives in West Africa," *Lancet* 343 (1994): 850–51.

193. Sharma and Kondrashin, eds. (1991), op. cit.

194. See, for example, World Health Organization, "Global Malaria Control Strategy," Ministerial Conference on Malaria, Amsterdam, October 26–27, 1992; and W. Rooney and K. Thimasarn, "Development of Multi-Drug Resistance in Forest Related *falciparum* Malaria," in Sharma and Kondrashin, eds. (1991), op. cit.

195. In addition to the millions of dollars spent on early malaria eradication efforts, substantial amounts of money poured into WHO control efforts and international research. Between 1973 and 1988, for example, WHO received from a variety of sources an average of $73 million annually for malaria control. Most of that was spent on training people at the local level to do such things as apply pesticides or count parasite levels in people's blood.

The U.S. government spent similar amounts of money, primarily for drug and vaccine research through either the Army or the Agency for International Development. USAID spending averaged around $22 million annually during the period; military research spending was about $10 million a year. The CDC and the NIH each spent well under $1 million a year on malaria research. See Institute of Medicine (1991), op. cit.

196. By far the majority of all dollars spent on malaria research from 1960 to 1994 were dedicated to the search for a vaccine. Though a promising product failed to appear, the effort pushed on relentlessly, amid indictments and corruption. The biggest funder was the U.S. Agency for International Development, which spent for vaccine efforts over and above other forms of possible control or drug development. As R. S. Desowitz, formerly of the University of London, then at the University of Hawaii, put it: "AID failed because it was run by amateurs who would not heed the advice of professionals. AID failed because it succumbed to sleaze and corruption. AID failed because it fostered mediocre science and over-inflated the meaning of experimental results. It may also be that AID failed because the human constitution is such that no vaccine can a confer protective immunity." R. S. Desowitz, *The Malaria Capers* (New York: W. W. Norton, 1991).

In the spring of 1993, Colombian scientist Dr. Manuel Elkin Patarroyo announced results of a field trial of a synthesized protein vaccine against *P. falciparum*. In 1,500 Colombians the vaccine proved 38 percent effective in preventing infection, he said. A year later, amid much WHO fanfare, Patarroyo announced similar results from a field trial in Tanzania. Skeptics questioned the timing of the Tanzania announcement, which was coincident with USAID plans to cut the vaccine research budget. And they said that such claims had been made before. See "Malaria Vaccine a 'Good Chance' for a Breakthrough," *World Bank News* XIII (February 17, 1994); World Health Organization, "Malaria Vaccine Could Be Developed Soon: Global Effort Needed," Press Release WHO/13/13 February 1994; M. V. Valero, L. R. Amador, C. Galindo, et al., "Vaccination with SPf66, a Chemically Synthesized Vaccine Against *Plasmodium falciparum* Malaria in Colombia," *Lancet* 341 (1993): 705–10; and P. Brown, "Malaria Vaccine Passes Key Test," *New Scientist*, February 19, 1994: 7.

197. See E. Marshal, "Malaria Parasite Gaining Ground Against Science," *Science* 242 (1991): 190–91; and P. J. Hilts, "U.S. Plans Deep Cuts in Malaria Vaccine Program," *New York Times*, February 13, 1994: A17.

14. Thirdworldization

1. We're running scared," Mahler told *New York Times* reporter Lawrence Altman in 1986, adding that he could "not imagine a worse health problem in this century. . . . We stand nakedly in front of a very serious pandemic as mortal as any pandemic there has ever been. I don't know of any greater killer than AIDS." See L. K. Altman, "Global Program Aims to Combat AIDS 'Disaster,' " *New York Times*, November 21, 1986: A1.

2. The bell-shaped curve typically seen with epidemics indicated that in any population of people infected with a given microbe, some would eventually have a natural immunity and survive, even if enormous numbers of other people died. The only other disease on earth in 1994 that similarly failed to exhibit a bell-shaped curve was rabies, which was 100 percent lethal in all people in the absence of emergency vaccination. The classic curve was pictured as follows:

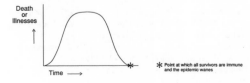

In 1986 the bell was still on its upward curve everywhere in the world. In 1994 it remained so everywhere with the exception of a handful of small population groups that took preventive steps to avoid infection and an even smaller set of human beings scattered around the globe who appeared to have been infected and naturally cleared HIV from their bodies, never contracting AIDS.

3. The full extent of all WHO activities relevant to AIDS prior to establishment of the Global Programme on AIDS is outlined in the following: World Health Organization, Executive Board, Seventy-seventh Session, Provisional Agenda item 20, "WHO Activities for the Prevention and Control of Acquired Immunodeficiency Syndrome (AIDS)," EB 77/42, November 25, 1985.

4. The UN agencies were often viewed by government leaders as graceful dumping grounds for powerful foes, corrupt politicians, burned-out or less than brilliant cronies, or influential political allies who happily received payoffs in the form of cushy jobs in wealthy countries for years of successfully bolstering the power bases of their leaders. Certainly not all the 50,000 professionals employed in the UN system were of that ilk; many were bright visionaries who ardently believed in the need for a global community that sought collective solutions to its problems rather than resorting to wars. But all too often in UN history the bright and idealistic were stifled by the bureaucratic, corrupt, and dim-witted. See "The United Nations Agencies: A Case for Emergency Treatment," *The Economist*, December 2, 1989: 23–26; R. N. Wells, Jr., *Peace by Pieces—United Nations Agencies and Their Roles: A Reader and Selective Bibliography* (Metuchen, NJ: Scarecrow Press, 1991); G. Hancock, *Lords of Poverty* (New York: Atlantic Monthly Press, 1989); and V. Navarro, "A Critique of the Ideological and Political Positions of the Willy Brandt Report and the WHO Alma Ata Declaration," *Social Science and Medicine* 18 (1984): 467–74.

5. Fortieth World Health Assembly, Agenda item 18.2, WHA 40.26, 12th Plenary, 3 pages.

6. The statement read:

1) CONFIRMS that WHO should continue to fulfill its role of directing and coordinating the global, urgent and energetic fight against AIDS;

2) ENDORSES the establishment of a Special Programme on AIDS and stresses its high priority.

3) FURTHER ENDORSES the global strategy and programme prepared by WHO to combat AIDS. . . .

and encouraged the nations of the world to openly share all germane information and cooperate in efforts to combat AIDS.

7. See Annex 2, Resolution 42/8 of the Forty-second General Assembly of the United Nations, "Prevention and Control of Acquired Immune Deficiency Syndrome (AIDS)," WHO/GPA/DIR/89.4, 1987.

8. THE GROWTH OF HIV/AIDS LEGISLATION, 1983–92 (Source: World Health Organization, Health Legislation Unit)

Countries, etc., known to have legislation as of December 1983: Austria, Canada (Alb.; B.C.; Ont.), Denmark, France, Germany, Greece, Israel, Italy, New Zealand, Norway, Sweden, Turkey, U.S.A. (CA; NJ; NY)

Additional jurisdictions introducing legislation between 1984 and 1987: Angola (1987), Australia (1984), Barbados (1985), Belgium (1985), Belize (1987), Benin (1987),* Bermuda (1985), Brazil (1985), Brunei Darussalam (1987), Bulgaria (1985), Burundi (1987), Canada (1985), Chile (1984), China (1987), Costa Rica (1985), Cuba (1986), Cyprus (Sovereign Base Areas) (1987), Czech and Slovak Federal Republic (1984), Denmark (1985), Dominican Republic (1987), Ecuador (1985), Egypt (1986), Finland (1985), [German Democratic Republic (1986)], Grenada (1986), Guatemala (1986), Haiti (1987), Honduras (1987), Hungary (1985), Iceland (1986), India (Goa) (1987), Indonesia (1987), Iraq (1987), Jordan (1987), Kenya (1987), Libyan Arab Jamahiriya (1987), Liechtenstein (1987), Luxembourg (1984), Malaysia (1985), Malta (1986), Mauritius (1987), Mexico (1985), Monaco (1986), Mozambique (1986), Netherlands (1987), Niger (1987), Panama (1985), Paraguay (1985), Peru (1987), Philippines (1986), Poland (1986), Portugal (1986), Republic of Korea (1987), Romania (1985),* Russian Federation (1985), Rwanda (1987), Singapore (1985), South Africa (1987), Spain (1985), Switzerland (1986), Syrian Arab Republic (1987), Thailand (1985), Togo (1987), United Kingdom (1984), Uruguay (1984), Venezuela (1984), Yugoslavia (1986)

Additional jurisdictions introducing legislation between 1988 and 1992: Albania (1992), Algeria (1989), Argentina (1988), Bahrain (1990), Bolivia (1988), China (Province of Taiwan) (1988),* Colombia (1988), Comoros (1988), El Salvador (1988), Equatorial Guinea (1988), Estonia (1992), Gabon (1989), Guinea-Bissau (1989), Hong Kong (1988), Japan (1988), Lebanon (1990), Madagascar (1990), Mongolia (1989), Oman (1990), Saint Lucia (1991), Saudi Arabia (1990), Senegal (1990), Tunisia (1989), Ukraine (1991),* Vietnam (1989)

Date of legislation unknown: Bahamas,* Cyprus,* United Republic of Tanzania*
*Text unavailable to WHO.
See chart on facing page.

9. For a flavor of the period, see Panos Dossier, *The Third Epidemic: Repercussions of the Fear of AIDS* (London: Panos Institute and Norwegian Red Cross, 1990).

10. In addition, Zimmermann revealed that the names of HIV-positive German residents had been forwarded to police authorities, who were closely watching the individuals.

11. These and other details are compiled from a large variety of news, medical, and interview sources. Citing these points would so severely increase the size of this book that I must request the readers' forgiveness and refer, in addition to the previously cited Panos Dossier, to back issues of two invaluable publications: *CDC AIDS Weekly*, P.O. Box 5528, Atlanta, Georgia; and *AIDS Newsletter*, published by the Bureau of Hygiene and Tropical Diseases, London, WC1E7HT.

12. According to a BBC translation, the Soviet decree stated:

The citizens of the U.S.S.R., as well as foreign citizens and stateless persons living or staying in the territory of the U.S.S.R., may be bound to take a medical test for the AIDS virus. If they refuse taking the test voluntarily, the persons, in relation of whom there are grounds for assuming that they are infected with the AIDS virus, may be brought to medical institutions by health authorities with the assistance in the necessary cases of authorities from the Interior Ministry.

The infection of another person with AIDS by a person aware of having AIDS shall be punishable by up to eight years in prison.

13. In fact, according to 1993 Cuban government statistics, more tests were eventually conducted than there were Cubans, meaning some people were repeat-tested. More than 14 million tests were conducted between 1986 and 1993. Some 930 Cubans would test positive during that time, according to Dr. Jorge Pérez, director of Cuba's Los Cocos AIDS sanitarium in Havana.

14. Anti-African sentiments were so high that AIDS rumors sparked mini-riots in Beijing and around universities located in other Chinese cities. Deported African students described raids upon their dormitories by citizens' groups, beatings, and tauntings when they appeared in public places.

15. For details on the evolution of the Thai sex trade and its influence on the AIDS epidemic, see Asia Watch Women's Rights Project, *A Modern Form of Slavery* (New York, Washington, Los Angeles, London: Human Rights Watch, 1994).

16. C. Decker, "Robertson Tailors His Message to Audiences," *Los Angeles Times*, November 23, 1987: A1.

17. This argument generated an enormous amount of press, and no short list of citations can adequately capture the flavor of this often vociferous debate. For a hint of the atmosphere, see T. Monmaney, P. Wingert, G. Raine, and M. Gosnell, "AIDS: Who Should Be Tested?" *Newsweek*, May 11, 1987: 64–65; M. Cimons, "Candidates Forced to Deal with AIDS Issue," *Los Angeles Times*, November 2, 1987: A1; M. Cimons, "Bowen Against Federal AIDS Legal Protection," *Los Angeles Times*, September 22, 1987: A16; R. Shilts, "U.S. Backtracks on AIDS Brochure," *San Francisco Chronicle*, August 28, 1987: A1; C. Thomas, "Fight AIDS with a National Health Card," Editorial, *Los Angeles Times*, May 5, 1987; M. Cimons, "Conservatives Split as Some Attack Koop," *Los Angeles Times*, May 14, 1987: A1; D. Whitman, "A Fall from Grace on the Right," *U.S. News & World Report*, May 25, 1987: 27–28; J. Helms, "Only Morality Will Effectively Prevent AIDS from Spreading," Editorial, *New York Times*, November 23, 1987; and C. E. Koop, *Koop: Memories of America's Family Doctor* (New York: Random House, 1991).

18. For a flavor of the U.S. immigration debate, see R. M. Wachter, *The Fragile Coalition* (New York: St. Martin's Press, 1991).

19. The distinction between "epidemic" and "endemic" is crucial to all discussion of emerging diseases. Ideally, one hopes to spot an emerging microbe and bring it under control when its impact on *Homo sapiens* is limited to small outbreaks. Barring that, there may still be hope for effective action against an epidemic, which, by definition, is a new and potentially short-lived phenomenon.

With microbes like HIV, human papillomavirus, and herpes simplex, it was very difficult to spot emergence at either the outbreak or early epidemic stages because the majority of infected human beings were asymptomatic for months or years. Thus, an epidemic could smolder and spread, unnoticed, for years. Slow-acting viruses offered the greatest challenge to public health advocates, therefore, because it took so long for the public to recognize a threat.

In the absence of alert public health authorities, such simmering epidemics could easily evolve into endemic diseases in a society before the emergence of the microbe was even noticed. Once a disease was endemic to some segment of a society, it was extremely difficult to defeat.

The GPA hoped in 1988 to prevent HIV from becoming endemic to most of the societies in the

(...governmental Health Policy Project, George Washington University)

* Some of these laws were subsequently revoked. In addition, some countries passed analogous legislation or edicts after 1989.

Country	Immigrants	Work Permit	National Military	Visiting Students	Visitors, All (30 days or more)	Visitors (>year)	All "Suspect" Citizens	All Citizens	Mandatory Quarantine for HIV+	Optional Imprisonment of HIV+	Entry into Country May Be Barred if HIV+
Antarctica											X
Australia	X										
Belgium				X							
Bulgaria					X						
Canada											X
China	X	X		X	X	X				X	
Costa Rica				X			X				
Cuba	X	X	X	X	X			X	X		
Cyprus				X	X						
Ecuador					X						
Egypt					X						
Fed. Rep. of Germany				X	X (over 3 mos.)						
German Dem. Rep.											
Finland				X							
Greece		X		X							
India	X					X	X (returning from overseas)				
Iraq					X (over 5 days)	X					
Israel			X								
Japan										X	X
Republic of Korea					X (90 days)						
Kuwait	X	X				X					
Libya	X					X					
Pakistan						X					
Philippines					X (6 mos.)						X (U.S. mil. personnel)
Poland				X							
Qatar	X	X		X		X					
Saudi Arabia	X	X				X					
South Africa		X	X								X
Spain		X		X							
Syria		X									
Thailand		X									X
United Arab Emirates		X									
United Kingdom											X
U.S.A.	X	X	X	X	X (1 day if HIV+ changed to 30 days)				X		X
U.S.S.R.				X	X (3 mos.)						
Yugoslavia				X			X				

Compulsory Testing of: spans the columns Immigrants through All Citizens.

world by shaking governments out of denial and prompting effective action while HIV was still in an outbreak or early epidemic stage of emergence. By 1988 *endemicity* was already the reality in much of sub-Saharan Africa, North America, and Western Europe. But there was hope for the majority of the world's populations, residing in Asia, Eastern Europe, Oceania, and Latin America.

In a worst-case scenario HIV would reach endemic status on one or two continents, spread as a pandemic across the rest of the world, and eventually reach a stage of global endemicity, becoming permanently entrenched in every society on earth.

20. Much has been written about behavioral and societal responses to AIDS, though few serious scientific studies of the matter have been funded. The National Academy of Sciences in the United States identified lack of information about human sexual and drug use behavior and responses to epidemics as the *key* factors responsible for failure to find ways to stop AIDS. Yet most governments remained loath to sponsor such studies, even thirteen years after the pandemic began, because of the religious, political, and cultural sensitivities involved in queries about sex and drugs.

See, for example, Centers for Disease Control, "Assessment of Broadcast Media Airings of AIDS-Related Public Service Announcements—United States, 1987–1990," *Morbidity and Mortality Weekly Report* 40 (1991): 543–46; Centers for Disease Control, "Attitudes of Parents of High School Students About AIDS, Drug, and Sex Education in Schools—Rome, Italy, 1991," *Morbidity and Mortality Weekly Report* 41 (1992): 201–9; Centers for Disease Control, "HIV Prevention in the U.S. Correctional System, 1991," *Morbidity and Mortality Weekly Report* 41 (1992): 389–97; Centers for Disease Control, "Street Outreach for STD/HIV Prevention—Colorado Springs, Colorado, 1987–1991," *Morbidity and Mortality Weekly Report* 41 (1992): 94–102; A. A. Ehrhardt, "Sex Education for Young People," presentation to the Ninth International Conference on AIDS, Berlin, June 7–11, 1993; Family Health International, "Strategies for Behavioral Change—Slowing the Spread of AIDS," *Network* 12, 1 (1991): 1–28; P. Lamptey, T. Coates, G. Slutkin, et al., "HIV Prevention: Is It Working?" presentation to the Ninth International Conference on AIDS, Berlin, June 7–11, 1993; M. V. Nadel, "AIDS Education: Gaps in Coverage Still Exist," testimony before the Senate Committee on Governmental Affairs, GAO/T-HRD-90-26, 1990; National Research Council, *AIDS: Sexual Behavior and Intravenous Drug Use*, C. F. Turner, H. G. Miller, and L. E. Moses, eds. (Washington, D.C.: National Academy Press, 1989); National Research Council, *The Social Impact of AIDS in the United States*, A. R. Jonsen and J. Stryker, eds. (Washington, D.C.: National Academy Press, 1993); Office of Technology Assessment, "How Effective Is AIDS Education?" Report to the Congress of the United States, Staff Paper 3, 1988; Panos Dossier (1990), op. cit.; Panos Dossier, *AIDS and the Third World* (London: Panos Institute and Philadelphia, PA: New Society Publishers, 1989); H. Schietinger, "Good Intentions: A Report on Federal AIDS Prevention Programs" (Washington, D.C.: AIDS Action Council, 1991); J. Sepulveda, H. Fineberg, and J. Mann, *AIDS: Prevention Through Education: A World View* (New York and Oxford, Eng.: Oxford University Press, 1992); U.S. General Accounting Office, *AIDS Education: Public School Programs Require More Student Information and Teacher Training*, report to the Chairman, Committee on Governmental Affairs, U.S. Senate, GAO/HRD-90-103, 1990; and U.S. General Accounting Office, *AIDS-Prevention Programs: High-Risk Groups Still Prove Hard to Reach*, report to the Chairman, Subcommittee on Human Resources and Intergovernmental Relations, Committee on Governmental Operations, House of Representatives, GAO/HRD-91-52, 1991.

21. British Market Research Bureau Limited, "AIDS Advertising Campaign," prepared for the Central Office of Information, London, July 1987.

22. Cuba may be an exception. Through mandatory testing and quarantine, Cuba did succeed in keeping its HIV incidence well below 1 percent of the population for ten years. However, as economic collapse set in following the dissolution of the Soviet state and subsequent cessation of U.S.S.R. subsidization of Cuban industry and agriculture, prostitution and tourism rose in the island nation. A new wave of AIDS, entirely divorced from that spawned years earlier by veterans of the Angolan civil war, emerged in Cuba during the early 1990s. Most observers, including the Cubans and representatives of the Pan American Health Organization, predicted in 1993 that Cuba's AIDS incidence would rise as its starving economy was increasingly dependent upon tourism. According to that analysis, the earlier success of a quarantine approach to AIDS depended upon a high level of isolation of the Cuban society as a whole. As Cuba became more open to outsiders, critics said, such restrictive approaches could no longer be expected to succeed.

See R. Goldstein, "AIDS Arrest: The Cuban Solution," *Village Voice*, February 14, 1989: 18; N. Caistor, "Treatment for Life," *New Scientist*, February 18, 1989: 65; A. M. Gordon and R. Paya, "Controlling AIDS in Cuba," *New England Journal of Medicine* 321 (1989): 829; R. Bayer, "Controlling AIDS in Cuba: The Logic of Quarantine," *New England Journal of Medicine* 320 (1989): 1022–24; N. Scheper-Hughes, "AIDS, Public Health, and Human Rights in Cuba," *Lancet* 342 (1993): 965–67; and J. Glesecke, "AIDS and the Public Health," *Lancet* 342 (1993): 942.

23. World Summit of Ministers of Health on Programmes for AIDS Prevention, "London Declaration on AIDS Prevention," January 28, 1988, in World Health Organization, *AIDS Prevention and Control* (Oxford, Eng.: Pergamon Press, 1988).

24. See F. Newman and D. Weissbrodt, *Selected International Human Rights Instruments* (Cincinnati, OH: Anderson Publishing Co., 1990); K. Tomasevski, S. Gruskin, Z. Lazzarini, and A. Hendriks, "AIDS and Human Rights," in Mann, Tarantola, and Netter, eds. (1992), op. cit.; P. Siegert, *AIDS and Human Rights: A UK Perspective*, British Medical Foundation for AIDS, London; and "Declaration on Respect for Human Rights and Dignity in Addressing the AIDS Pandemic," Global Expert Meeting, The Hague, Netherlands, May 21–24, 1991.

25. See L. O. Gostin, "The Americans with Disabilities Act and the U.S. Health Care System," *Health Affairs* 11, 3 (1992): 248–57; L. O. Gostin, "Public Health Powers: The Imminence of Radical Change," *The Milbank Quarterly* 69 (1991): 268–90; L. O. Gostin, "A Decade of a Maturing Epidemic: An Assessment and Directions for Future Public Policy," *American Journal of Law and Medicine* 15, 1 and 2 (1990); L. O. Gostin, "The AIDS Litigation Project: A National Review of Court and Human Rights Commission Decisions. Part I: The Social Impact of AIDS," *Journal of the American Medical Association* 263 (1990): 1961–70; and L. O. Gostin, "The AIDS Litigation Project: A National Review of Court and Human Rights Commission Decisions. Part II: Discrimination," *Journal of the American Medical Association* 263 (1990): 2086–93.

26. African AIDS Cases Officially Reported to the World Health Organization

(Cumulative, as of dates indicated)

Country	January 12, 1988	February 28, 1989	June 1, 1989	June 1, 1990	June 1, 1991	June 1, 1992
Algeria	5	13	13	45	45	92
Angola	6	85	104	104	104	421
Benin	3	15	36	124	124	185
Botswana	13	34	49	87	87	277
Burkina Faso	26	26	107	978	978	978
Burundi	569	1,408	1,408	3,305	3,305	3,305
Cameroon	25	62	62	78	429	429
Cape Verde	4	18	18	32	32	32
Cen. Afr. Rep.	254	432	662	662	662	1,864
Chad	1	11	11	35	59	130
Comoros	0	1	1	2	2	2
Congo	250	1,250	1,250	1,940	2,406	2,405
Côte d'Ivoire	250	250	250	3,647	6,836	8,297
Equatorial Guinea	0	0	2	3	3	9
Ethiopia	19	81	81	531	636	1,818
Gabon	13	27	27	64	117	117
Gambia	14	62	62	81	123	180
Ghana	145	227	402	1,732	1,732	2,852
Guinea	4	10	33	161	161	338
Guinea-Bissau	16	48	48	123	157	157
Kenya	964	2,732	2,732	9,139	9,139	9,139
Lesotho	2	2	5	11	11	44
Liberia	2	2	2	5	5	24
Madagascar	0	0	0	2	2	2
Malawi	13	2,586	2,586	7,160	7,160	12,074
Mali	0	29	29	338	338	338
Mauritania	0	0	0	16	16	26
Mauritius	1	1	2	5	5	9
Mozambique	4	27	29	151	198	288
Niger	26	43	56	149	149	497
Nigeria	5	13	15	48	48	84
Réunion	1	8	20	49	49	49
Rwanda	705	987	1,302	3,407	3,407	6,578
São Tomé/Principe	0	1	1	1	1	6
Senegal	27	149	181	307	307	648
Seychelles	0	0	0	21	0	21
Sierra Leone	0	5	20	15	35	40

South Africa	93	195	226	590	764	1,019
Swaziland	7	14	14	14	14	71
Togo	0	2	2	100	100	100
Tanzania	1,608	3,055	4,158	7,128	8,163	27,396
Uganda	2,369	5,508	6,772	17,422	21,719	30,190
Zaire	335	335	335	11,732	11,732	14,762
Zambia	536	1,296	1,296	3,494	4,036	5,802
Zimbabwe	0	119	119	5,249	6,716	10,551
*Namibia				311	311	311
Cumulative cases for Africa	8,693	21,169	24,528	80,598	92,422	143,949

* No totals prior to 6/90; not included in regional total until then either.

27. "The Politics of AIDS," *New African*, February 1990: 28.

28.

REPORTED HIV INFECTION RATES

(Based on studies presented at various AIDS in Africa annual conferences, International Conferences on AIDS, and readily available from published medical literature)

As of December 1988 epidemiological surveys, using valid blood tests, of key groups in sub-Saharan Africa revealed incidences as follows:

Country	Group Tested	% HIV-1 Positive	% HIV-2 Positive	% HIV-1 & HIV-2 Positive
Senegal	Female prostitutes, Dakar	0	4	0
Mali	Hospital survey, Bamako	0.25	1.4	0
Guinea-Bissau	General population	0.001	10.1	—
	Blood donors, Bissau	—	25.9	—
	Hospital staff, Bissau	—	40.0	—
	Malaria patients, Bissau	—	75.0	—
Guinea	Hospital survey, Conakry	0.8	0.3	0
Ghana	Female prostitutes, Accra	0.7	—	—
Benin	Female prostitutes	4.5	3.7	—
	Blood donors	0	0	0
Côte d'Ivoire	Female prostitutes, Abidjan	51.3	—	—
	Pregnant women, Abidjan	5.6	—	—
	General population	2.3	4.2	—
Nigeria	Blood donors, Lagos	0.1	—	—
Zaire	TB patients, Kinshasa	33.0	—	—
	Female prostitutes, Kinshasa	40.0	—	—
	Hospital survey, Kinshasa	22.0	—	—
	General population, Kinshasa	12.7	—	—
	General population, rural	2.5	—	—
Angola	STD clinics, Luanda	11.0	10.5	3.5
	Pregnant women, Cabinda	11.0	—	—
	Blood donors, Luanda	9.6	*	—
	TB patients, Luanda	4.0	—	—
	Hospital patients, Luanda	22.0	*	—
Zambia	Blood donors, Lusaka	10.1	—	—
	Blood donors, copper belt	13.0	—	—
Namibia	Blood survey, Windhoek	0.3	—	—
Botswana	STD clinic, Gabarore	2.6	—	—
	Hospital patients, Gabarore	4.3	—	—
South Africa	General population	0.02	—	—
Lesotho	General population	0.0	—	—
Mozambique	General population, Maputo	—	1.89	—
Swaziland	No data	—	—	—
Zimbabwe	STD clinic, Harare	18.5	—	—
Madagascar	No data	—	—	—
Congo	General population, Brazzaville	5.0	—	—
	Female prostitutes, Brazzaville	34.0	—	—

Cameroon	Female prostitutes, Yaoundé	7.1	—	—
Cen. Afr. Rep.	Female prostitutes, Bangui	7.5	—	—
Niger	Blood donors	0.4	—	—
Chad	No data	—	—	—
Ethiopia	Female prostitutes, Addis Ababa	6.7	—	—
Somalia	General population	0.0	—	—
Kenya	Female prostitutes, Nairobi	85.0	0	0
	(lower-price prostitutes)			
	Men, STD clinic, Nairobi	11.2	—	—
Uganda	Hospital patients, Kampala	45.0	NS	19.0
	Pregnant women, Kampala	20.0	—	—
Tanzania	Young adults, Bukoba	50.0	0	0
	Household survey, Bukoba	32.8	0	0
	Children, Bukoba	3.9	0	0
	Female prostitutes, Bukoba	90.0	0	0
	Pregnant women, Mwanza	6.0	—	—
	Female prostitutes, Arusha	75.0	0	0
	Pregnant women, Arusha	7.0	—	—
	Barmaids, Dar es Salaam	22.0	—	—
Burundi	Random survey, Bujumbura	15.0	0	0
	STD clinic, Bujumbura	18.0	—	—
Rwanda	Random survey, Kigali	30.0	0	0

* Some HIV-1 is HIV-2, mixed assay.

Surveys in 1989–91 in various African nations, based on similar sources, revealed rates of HIV infection were rising, as follows:

Country	Group Tested	% HIV-1 Positive	% HIV-2 Positive	% HIV-1 & HIV-2 Positive
Tanzania	Barmaids, Dar es Salaam	42.0	—	—
Uganda	General population, Rakai	12.0	—	—
Côte d'Ivoire	General population, Abidjan	6.4	—	—
	Hospital patients, Abidjan	19.7	—	—
	Barmaids and prostitutes, Abidjan	21.7	—	—
	STD clinics, Abidjan	60.0	—	—
	Pregnant women, Abidjan	12.0	—	—
	Blood donors, Abidjan	15.0	—	—
Burkina Faso	Female prostitutes, Ouagadougou	7.6	18.0	2.3
Senegal	Prostitutes, Dakar	1.0	10.0	—
Guinea-Bissau	Rural survey (mean of varied village-by-village rates)	0.26	—	—
Benin	Rural survey	6.6	0.2	0
	Students applying for overseas study	3.5	—	—
Cameroon	General population, Ghipponi	0.5	—	—
	Female prostitutes, Ghipponi	6.6	—	—
Nigeria	General population	0.02	0.05	—
Uganda	Rakai, trading centers			
	Males (ages 20–29)	43.0	—	—
	Females (ages 20–29)	52.0	—	—
Rwanda	Pregnant women, Kigali	32.0	—	—
	Unmarried pregnant women, Kigali	64.0	—	—
South Africa	General population, blacks	0.76	—	—
	General population, whites	0.01	—	—

HIV Infection Rates Seen in Selected Populations, 1992–93

Country	Group Tested	% HIV-1 Positive	% HIV-2 Positive	% HIV-1 & HIV-2 Positive
Tanzania	Women, aged 20–29, nationwide	8.0	—	—
Zimbabwe	Adult workforce, nationwide	18–20	—	—
	Pregnant women, general	20.0	—	—
	Pregnant women, trucking routes	42.0	—	—
	Hospital patients, national	65.0	—	—

South Africa	STD clinics, Johannesburg	15.0	—	—
	Random worker survey,			
	Johannesburg	0.5	—	—
Uganda	Voluntary HIV testing centers	30.0	—	—
	Pregnant women, Kampala	28.0	—	—
Nigeria	General population	2.0	1.3	0
	Female prostitutes	10.0	6.7	—
	Health care workers	1.6	—	—
	Blood donors	3.69	—	—
Sudan	Female prostitutes, Khartoum	23.0	—	—
	Blood donors, southern region	30.0	—	—
Zambia	Pregnant women, Lusaka	24.0	—	—
Malawi	Pregnant women, Lilongwe	22.5	—	—
Kenya	Pregnant women, Nairobi	15.0	—	—
Côte d'Ivoire	Pregnant women, Abidjan	17.0	—	—

29. R. Shilts, "Economists Predict Hard Times for Africa's 'AIDS Belt,' " *San Francisco Chronicle*, March 10, 1988: A4.

30. About $1.00 for a first-round ELISA blood test and $30 to $50 for Western Blot confirmatory assays, based on technology available in 1988, plus the purchase of an ELISA screening machine: $15,000.

31. Estimated in 1988 to cost $5.00 for 100 condoms if bulk-purchased at discount, or up to twenty times that amount if purchased individually at retail cost.

32. J. Tinker, "AIDS in Developing Countries," *Issues in Science and Technology* IV (Winter 1988): 1–7.

33. There were a variety of estimates for U.S. AIDS treatment costs, some topping $200,000. In general, annual costs decreased over the years as physicians gained skills in handling AIDS cases. But lifetime cumulative AIDS costs increased because patients lived longer and accrued greater expenses. See G. J. Alpauch, "AIDS-Related Claim Survey Results," *Best's Insurance Management Reports*, November 18, 1991: 1–3; L. S. Rosenblum, J. W. Buehler, M. Morgan, and M. Moien, "Increasing Impact of HIV Infection on Hospitalizations in the United States, 1983–1988." *Journal of Acquired Immune Deficiency Syndromes* 5 (1992): 497–504; General Accounting Office, "AIDS Forecasting: Undercount of Cases and Lack of Key Data Weaken Existing Estimates," Report to U.S. Congress, Washington, D.C., 1989; "AIDS: Met Life's Experience, 1986–89," *Met Life's Statistical Bulletin*, October–December 1990: 2–9; A. A. Scitovsky, "The Economic Impact of AIDS in the United States," *Health Affairs*, Fall 1988: 1–14; F. J. Hellinger, "Forecasting the Medical Care Costs of the HIV Epidemic: 1991–1994," *Inquiry* 28 (1991): 213–225; and L. T. Bilheimer, "Modeling the Impact of the AIDS/HIV Epidemic on State Medical Programs," presentation to the Conference on New Perspectives on HIV-Related Illness: Program in Health Services Research, U.S. Public Health Service, Rockville, MD, 1989.

34. M. Over, S. Bertozzi, J. Chin, et al., "The Direct and Indirect Cost of HIV Infection in Developing Countries: The Cases of Zaire and Tanzania," presentation to the Fourth International Conference on AIDS, Stockholm, June 12–16, 1988.

35. Using similar calculations, the average 1988 AIDS patient cost the United States three years of per capita GNP, and as the American epidemic increasingly shifted into communities of extreme poverty that cold value fell further. Using these admittedly crude early attempts at pricing out the AIDS epidemic, 250,000 U.S. cases of AIDS might cost the economy 750,000 years of per capita GNP. But 250,000 Zairian cases would tax that economy to the tune of 4,750,000 years of per capita GNP. See also M. Over, S. Bertozzi, and J. Chin, "A Proposed Approach to Making Preliminary Estimates of the Cost of HIV Infection in a Developing Country," presentation to the Third International Conference on AIDS and Associated Cancers in Africa, Arusha, Tanzania, September 16, 1988.

36. S. K. Lwangwa and J. Chin, "Projections of Non-Paediatric HIV Infection and AIDS in Pattern II Areas," presentation to the Fifth International Conference on AIDS, Montreal, June 4–9, 1989.

37. For a sense of the mood in the Global Programme on AIDS, see S. Kingman, "AIDS Brings Health into Focus," *New Scientist*, May 20, 1989: 37–42; J. M. Mann, "Global AIDS: Into the 1990s," presentation to the Fifth International Conference on AIDS, Montreal, June 4–9, 1989; and World Health Organization, "Global Programme on AIDS," prepared for Delegates at the Fifth International Conference on AIDS, Montreal, June 4–9, 1989.

38. World Health Organization, "The Global AIDS Situation," *In Point of Fact* 68, 1990; and

World Health Organization, "WHO Revises Global Estimates of HIV Infection," WHO Press, WHO/ 38, July 31, 1990.

39. M. Over and P. Piot, "HIV Infection and Other Sexually Transmitted Diseases," Chapter 10 in D. T. Jamison and W. H. Mosley, eds., *World Health Report* (Washington, D.C.: World Bank, 1990).

40. H. M. Ntaba, "Access to Health Care—AIDS in the Developing World," presentation to the Sixth International Conference on AIDS, San Francisco, June 20–24, 1990.

41. E. M. Kiereini, "Women and Children in Africa: AIDS Impact," presentation to the Sixth International Conference on AIDS, San Francisco, June 20–24, 1990.

42. The role of women in African societies and its relationship to AIDS proved to be the greatest stumbling block to efforts to control the expanding epidemic. Women could not in most African societies insist that their partners use condoms. To do so could mean death, for wives were often of such low status compared with their husbands that they had little right to question any of his sexual practices. Even after a man had been diagnosed as having AIDS, in many African countries he might legally insist that his wife yield to unprotected intercourse. Such rights have been challenged of late in the courts of Zambia, Zimbabwe, Kenya, Côte d'Ivoire, Nigeria, and other countries on the continent.

Thankfully, much has been written over the last five years about this subject, and the once taboo issue of women's rights is becoming the subject of discussion for Africa.

For further insight, see L. Garrett, "AIDS in Africa," *Newsday*, December 26 and 27, 1988: A1; L. Garrett, "AIDS: What Women Don't Know," *Elle*, December 1992: 86–96; Global Programme on AIDS, "International Conference on the Implications of AIDS for Mothers and Children: Technical Statements and Selected Presentations," Paris, November 27–30, 1989; B. Grundfest-Schoepf, W. Engundu, R. waNkera, et al., "Research on Women with AIDS," presentation to the First International Conference on AIDS Education and Information, Ixtapa, Mexico, October 16, 1989; C. G. Moreno and L. C. Rodrigues, "Safer Sex and Women in Africa," *Lancet* 340 (1992): 57–58; E. Ojulu, "Uganda Prostitutes Are Now Wiser," *New African*, September 1988: 34; Panos Dossier, *Triple Jeopardy—Women and AIDS* (Washington, D.C.: Panos Institute, 1990); J. Perlez, "Toll of AIDS on Uganda's Women Puts Their Roles and Rights in Question," *New York Times*, October 26, 1990: A14; A. Petras-Barvazian and M. Merson, "Women and AIDS: A Challenge for Humanity," *World Health*, November–December 1990: 1–32; and "Women and Prevention Strategies," *AIDS Newsletter*, 1992: Item 16, Item 515.

43. C. P. Lindan, S. Allen, A. Serufilira, et al., "Predictors of Mortality Among HIV-Infected Women in Kigali, Rwanda," *Annals of Internal Medicine* 116 (1992): 320–28.

44. J. Decosas, "Demographic AIDS Trap for Women in Africa," presentation to the Seventh International Conference on AIDS, Florence, June 16–21, 1991.

45. See Food and Agriculture Organization (FAO) publications, 1991 to 1993, by David Norse. They are varied and available upon request to FAO, Rome, Italy.

46. S. Armstrong, "South Africa Wakes Up to the Threat of AIDS," *New Scientist*, February 16, 1991: 19.

47. C. Hemery, "Spectaculaire Propagation du SIDA en Ethiopie," *Afrique Nouvelle*, July–August 1993: 38–39.

48. R. M. Anderson, R. M. May, M. C. Boily, et al., "The Spread of HIV-1 in Africa: Sexual Contact Patterns and the Predicted Demographic Impact of AIDS," *Nature* 352 (1991): 581–89.

49. World Bank, "The Economic Impact of Fatal Adult Illness from AIDS and Other Causes in Sub-Saharan Africa," Research Project, World Bank, Washington, D.C., 1991.

50. M. King and R. Hall, "AIDS Soon to Overtake Malaria," *Lancet* 337 (1991): 166.

51. The Delphi approach, like Chin's model at GPA, tried to factor for the enormous discrepancy between the numbers of officially reported AIDS cases and HIV infections in the world, on the one hand, and elusive reality, on the other. Chin's approach involved creation of mathematical models of various types of national epidemics and infection spread rates. The Global AIDS Policy Coalition used Delphi techniques of surveying local experts all over the world. Prominent AIDS scientists and physicians were asked to give low- and highball estimates of their country's epidemics, regional pandemics, and the global situation. Statistical methods were used to derive a regional range of estimated pandemic size and future proportions.

Neither technique was perfect. Both lacked crucial data and had to be considered educated guesses.

52. J. Mann, D. J. M. Tarantola, and T. W. Netter, "The Impact of the Pandemic," *AIDS in the World* (Cambridge, MA: Harvard University Press, 1992), 9–132.

53. United Nations Development Program, *Human Development Report 1993* (New York: Oxford University Press, 1993).

54. World Bank, *World Development Report 1993: Investing in Health* (New York: Oxford University Press, 1993).

This represented a marked policy shift for the World Bank. Just five years earlier Bank management, still unable to see how the AIDS epidemic imperiled development, withheld a $15 million loan for AIDS education efforts in Zambia because the Kaunda government had fallen behind on repayment of other, non-AIDS loans.

55. According to *World Population Profile: 1994* (Washington, D.C.: U.S. Census Bureau, Department of Commerce, 1994), population growth rates between 1994 and 2010 for the sixteen hardest-hit countries would be as follows:

Country	Projected Annual Without AIDS	Growth With AIDS
Brazil	0.9%	0.6%
Burkina Faso	3.1	1.6
Burundi	3.0	1.9
Cen. Afr. Rep.	2.4	1.9
Congo	2.3	1.0
Côte d'Ivoire	3.1	2.5
Haiti	2.1	1.3
Kenya	2.5	1.0
Malawi	3.2	1.6
Rwanda	3.5	1.7
Tanzania	3.0	1.5
Thailand	0.9	-0.8
Uganda	3.3	1.5
Zaire	3.3	2.9
Zambia	3.4	1.4
Zimbabwe	2.1	0.5

56. A World Bank study demonstrated that families that absorbed AIDS orphans in Côte d'Ivoire were unlikely to provide the foster children with the same opportunities afforded to their own children. In a survey, foster children performed on average 20 percent more housework and 15 percent more fieldwork than their counterparts who were the natural offspring of the foster parents. And foster children were 30 percent less likely to be sent to school.

57. "Africa Will Suffer 'Millions' of AIDS Orphans," *New Scientist*, February 23, 1991: 23.

58. E. A. Preble, "AIDS and African Children, *Social Science and Medicine* 31 (1990); 671–80. Other estimates of the region's AIDS orphan burden include:

Country	Est. Number of Orphans	For Year Forecast	Source
Zambia	79,300	1991	UNICEF
Zambia	600,000	2000	UNICEF
Uganda	26,000 (Rakai Dist.)	1991	Save the Children Fund
Kenya	300,000	2000	UNICEF
Tanzania	50,000	1990	SOS Children
Tanzania	500,000	1997	SOS Children
Zimbabwe	47,000 (Manicaland only)	1991	Zimbabwe Ministry of Health

59. Center for International Research, *World Population Profile: 1994*, op. cit.

60. J. Decosas, "Fighting AIDS or Responding to the Epidemic: Can Public Health Find Its Way?" *Lancet* 343 (1994): 1145–46.

61. J. McDermott, *Report to the Speaker of the House of Representatives: The AIDS Epidemic in Asia*, International AIDS Task Force, U.S. House of Representatives, June 6, 1991.

62. AIDS in Asia

Cases Officially Reported by Respective Governments to the World Health Organization

Country	1985	1986	1987	1988	1989	1990	1991
Bangladesh	NR	NR	NR	0	0	1	1
Brunei	0	0	0	0	0	0	1
Burma (Myanmar)	0	0	0	0	0	0	0
China (PRC)	1	18	NR	3	NR	2	5

China (Taiwan)	0	1	NR	NR	14	—	—
Hong Kong	3	NR	NR	13	NR	22	27
India	NR	1	9	9	12	44	52
Indonesia	NR	0	0	3	3	7	9
Japan	1	11	26	90	112	285	194
Malaysia	NR	0	1	2	8	13	15
(Kuala Lumpur)							
Nepal	NR	0	0	0	0	2	4
Pakistan	NR	0	3	3	7	13	14
Philippines	1	3	9	12	20	28	37
N. Korea	NR	0	0	0	0	0	0
S. Korea	1	2	—	4	—	100	131
Singapore	0	2	2	6	11	17	21
Sri Lanka	0	0	0	1	1	4	8
Thailand	0	2	6	10	30	67	80
Cambodia (Kampuchea)	NR	0	0	0	0	0	0
Laos	NR	0	0	0	0	0	0
Vietnam	NR	0	0	NR	NR	NR	0

Note: Some reports were adjusted over the years, retrospectively, by the reporting governments.

The key exception was Japan. Though Japanese HIV infection rates had remained quite low, social response to AIDS was striking. Even before AIDS appeared on the public health radar screen, Japan had two cultural traditions in place that protected most of its citizens from emerging sexually transmissible microbes: condoms were the preferred mode of birth control, and very few Japanese ever had sex with a non-Japanese. Fear of AIDS only strengthened both those cultural traditions. See T. Kurima, "AIDS in Japan," presentation to the Conference of Asian Solidarity Against AIDS, Florence, June 18, 1991; K. M. Chysler, "Japan, Alarmed at Arrival of AIDS, Blames Outsiders," *San Francisco Chronicle*, April 26, 1987: A22; D. Rosenheim, "Spread of AIDS Threatens Japan," *San Francisco Chronicle*, December 8, 1986: A10; and L. Garrett, "AIDS in Asia," Morning Edition, National Public Radio, November 23, 1986.

63. B. Mangla, "India: HIV-Positive Blood Donors," *Lancet* 341 (1993): 1527–28. For an uncanny analysis of India's nascent AIDS epidemic and future crisis, see B. Mangla, "AIDS in India: An Alarming Diagnosis," *Express Magazine* in *Sunday Express* (Delhi), March 19, 1989: 1, 7.

64. Rates of infection among prostitutes found in that survey included 2.65 percent in Madras and 2.7 percent in Poona. Just six months earlier, less than 0.4 percent of Poona's prostitutes were HIV-positive.

65. K. S. Jacobs, H. Jayakumari, J. K. John, and T. J. John, "Awareness of AIDS in India: Effect of Public Education Through the Mass Media," *British Medical Journal* 299 (1989): 721.

66. "India: Prostitutes and the Spread of AIDS," *Lancet* 335 (1990): 1332.

67. A. Kumar, "AIDS in India: Fear and Ignorance Are Combining to Produce the Public Health Crisis of the Century," *India Currents*, August 1991: 17–18.

68. O. Sattaur, "Doubts over Testing Hamper India's AIDS Efforts," *New Scientist*, April 20, 1991: 18; and O. Sattaur, "India Wakes Up to AIDS," *New Scientist*, November 2, 1991: 25–29.

69. M. Grez, U. Dietrich, J. Maniar, et al., "High Prevalence of HIV-1 and HIV-2 Mixed Infections in India," presentation to the Ninth International Conference on AIDS, Berlin, June 6–11, 1993.

70. Thailand: HIV Infection Rates in Key Groups (Source: Thai Ministry of Public Health)

Population Group	Percent Positive for HIV Infection	Year Tested
Prostitutes nationwide	3.5	June 1989
(average)	6.8	December 1989
	9.6	June 1990
People attending	0.0	June 1989
STD clinics	2.0	December 1989
	2.5	June 1990
IV drug users	39.0	Early 1989
	46.0	Late 1989
	50.0	1990
Prisoners	12.0	1989

Female prostitutes in	0.4	1989
Chiang Mai province	50.0	Late 1990
	70.0	1991*
Female prostitutes in Bangkok	18.0	Late 1990
Female prostitutes in Phuket	0.0	1989

* Prostitutes employed as sex workers for over 6 months.

71. G. L. Myers, "Global Variation of HIV Sequences," presentation to the First National Conference on Human Retroviruses and Related Infections, Washington, D.C., December 12–16, 1993.

72. The populations that were regularly tested—some on a voluntary basis, some compulsorily—included cohorts of injecting drug users in Bangkok, Cholburi, Pattaya, Chiang Mai, and Rayong; prisoners; all army recruits (which amounted to every twenty-one-year-old male in the nation); and prostitutes and barmaids in several cities.

73. For a sampling of reports on Thailand's early epidemic, see C. Woodard, "Imperiled on Two Fronts," *New York Newsday, Discovery* section, March 6, 1990: 1, 6–7; B. G. Weniger, K. Limpakarnjanarat, K. Ungehusak, et al., "The Epidemiology of HIV Infection and AIDS in Thailand," *AIDS* 5 (1991): S71–S85; W. Sittitrai, S. A. Obremskey, T. Brown, and P. O. Way, "HIV/AIDS Projections for Thailand, 1990–2005," Thai Working Group on HIV/AIDS Projections, Bangkok, 1991; R. Rhodes, "Death in the Candy Store," *Rolling Stone*, November 28, 1991: 62–70, 113–14; W. Sittitrai and T. Brown, "The Asian AIDS Epidemic," presentation to the Congressional Forum on the HIV/AIDS Pandemic, Washington, D.C., June 23–25, 1992; M. Sweat, T. Nopkesorn, T. D. Mastro, et al., "AIDS Knowledge and Risk Perception at Baseline in a Cohort of Young Men in Northern Thailand," presentation to the Eighth International Conference on AIDS, Amsterdam, July 19–24, 1992; T. D. Mastro, D. Kitayaporn, B. Weniger, et al., "Estimate of the Number of HIV-Infected Injecting Drug Users in Bangkok Using Capture-Recapture Method," presentation to the Eighth International Conference on AIDS, Florence, July 19–24, 1992; K. Limpakarnjanarat, T. D. Mastro, W. Yindeeyoungyeon, et al., "STDs in Female Prostitutes in Northern Thailand," presentation to the Ninth International Conference on AIDS, Berlin, June 6–11, 1993.

74. In February 1991 the military installed civilian front man Anand Panyarachun as Prime Minister. His reign was brief: by April the military had decided that Anand was in the way, and coup leader General Suchinda Kraprayoon took over. Mass demonstrations and resistance activities spread over Thailand, building over eleven months' time to a confrontational peak in May 1992. Realizing they could not maintain power without slaughtering thousands of civilians and imposing a costly authoritarian regime, the military leaders stepped aside. The civilian front man, Anand, resumed office and scheduled national elections for September 1992. The military regime was swept out of power in those elections, replaced by a civilian pragmatist, Chuan Leekpai.

In terms of AIDS, the period of military rule and instability, February 1991–September 1992, was characterized by repression, chaotic to nonexistent education efforts, and general disarray.

75. Asia Watch Women's Rights Project (1993), op. cit.

76. A joint Japanese/Thai study in 1992 showed that some of the HIV strains turning up in Japan were genetically identical to those circulating among female prostitutes and their customers in Thailand. Sixty-seven percent of HIV-positive non-Japanese males residing in Japan (immigrant workers) carried the Thai strain, as did 85 percent of their female counterparts.

In addition, five Japanese men were found infected with the same virus. Three had traveled to Thailand, but two had acquired the viruses in Japan, as a result of heterosexual intercourse with immigrant women. See Y. Takebe, C. P. Pau, S. Oka, et al., "Identification of Thailand and HIV-1 Subtypes in Japan," presentation to the International Conference on AIDS, Berlin, June 6–11, 1993.

77. W. Sittitrai, P. Phanuphak, J. Barry, et al., "Survey of Partner Relations and Risk of HIV Infection in Thailand," Seventh International Conference on AIDS, Florence, June 7–11, 1991.

78. In keeping with the practice of human rights advocates inside and outside the country, I have used the name Burma rather than Myanmar throughout this book. It is thought that recognizing the military's change of Burma's ancient name lends international credibility to the outlaw regime.

79. An exception to Burma's otherwise universal pariah status was China. The Chinese government, which was accustomed to ignoring international cries of human rights violations, allowed vigorous trade with Burma. Among the items traded between the nations, openly or on the Burmese-sanctioned black market, were condoms, syringes, heroin, and military arms for the junta's elite forces. See P. Shenon, "Burmese Cry Intrusion (They Lack a Great Wall)," *New York Times*, March 29, 1994): A4.

80. Sittitrai and Brown (1992), op. cit.

81. The *Far Eastern Economic Review* ran a strong summary of the Thai situation in its February 1992 issue, including a profile of Mechai.

82. "VD Cases Soar in China as Prostitution Returns," *San Francisco Chronicle*, May 7, 1987: A10; J. Mann, "China Starts Drive Against Once-Vanquished Scourge—Venereal Disease," *Los Angeles Times*, July 4, 1987: A10; E. A. Gargan, "China Taking Stringent Measures to Prevent Introduction of AIDS," *New York Times*, December 22, 1987: A1; and N. D. Kristof, "Heroin Spreads Among Young in China," *New York Times*, March 21, 1991: A1.

83. Global Programme on AIDS, "The HIV/AIDS Pandemic: 1993 Overview," World Health Organization WHO/GPA/CNP/EVA/93.1, 1993.

84. M. H. Merson, "HIV/AIDS: Epidemic Update and Corporate Response." Presentation to the AETNA/WHO Asia AIDS Seminars, Hong Kong, April 14, 1994.

85. INDICATORS OF ASIAN ECONOMIC GROWTH (Source: World Bank.)

Country	Annual GNP Per Capita (1980–91)	% Annual GNP Per Capita Growth 1980–91	% Annual Inflation Rate (1980–91)	Illiteracy (%)	Life Expectancy Female	Life Expectancy Male
Bhutan	180	—	8.4	48	75	62
Nepal	180	2.1	9.1	53	87	74
Bangladesh	220	1.9	9.3	51	78	65
Laos	220	—	—	50	—	—
India	330	3.2	8.2	60	66	52
China	370	7.8	5.8	69	38	27
Sri Lanka	500	2.5	11.2	71	17	12
Indonesia	610	3.9	8.5	60	32	23
Philippines	730	− 1.2	14.6	65	11	10
Papua New Guinea	830	− 0.6	5.2	56	62	48
Thailand	1,570	5.9	3.7	69	10	7
Malaysia	2,520	2.9	1.7	71	30	22
South Korea	6,330	8.7	5.6	70	7	4
New Zealand	12,350	0.7	10.3	76	<5	<5
Hong Kong	13,430	5.6	7.5	78	—	—
Singapore	14,210	5.3	1.9	74	—	—
Australia	17,050	1.6	7.0	77	<5	<5
Japan	29,930	3.6	1.5	79	<5	<5
By comparison:						
Mozambique	80	− 1.1	37.6	47	79	67
United States	22,240	1.7	4.2	76	<5	<5
Switzerland	33,610	1.6	3.8	78	<5	<5

Note: All figures adjusted to 1991 dollar values. Data for Vietnam, North Korea, Cambodia, Burma, and Taiwan not reported.

86. Global Programme on AIDS (1993), op. cit.

87. T. D. Mastro, G. A. Satten, T. Nopkesorn, et al., "Probability of Female-to-Male Transmission of HIV-1 in Thailand," *Lancet* 343 (1994): 204–7; and "AIDS: The Third Wave," *Lancet* 343 (1993): 186–88.

88. D. C. DesJarlais, K. Choopanya, S. Vanichsenia, et al., "AIDS Risk Reduction and Reduced HIV Seroconversion Among Injection Drug Users in Bangkok," *American Journal of Public Health* 84 (1994): 452–55.

89. P. Handley, "Pumping Up Condoms," *Far Eastern Economic Review*, February 19, 1992: 31; and M. Viravaidhya, S. A. Obremsky, and C. Myers, "The Economic Impact of AIDS on Thailand," Working Paper Series, Department of Population and International Health, Harvard School of Public Health, Number 4, 1992.

90. S. Kongsin, S. Rerks-ngarm, L. Suebsaeng, et al., "Hospital Care Cost Analysis of ARC/AIDS Patients, Thailand," presentation to the Ninth International Conference on AIDS, Berlin, June 6–11, 1993.

91. M. H. Merson, "Slowing the Spread of HIV: Agenda for the 1990s," *Science* 260 (1993): 1266–68.

92. C. N. Myers and T. Ashakul, "AIDS in Thailand: Some Preliminary Findings," *TDRI Quarterly Review* 6 (1991): 8–12.

93. Center for International Research (1994), op. cit.

94. See World Bank. *World Development Report 1993*, op. cit.; P. Shenon, "After Years of Denial, Asia Faces Scourge of AIDS," *New York Times*, November 8, 1992: A1; Asian Development Bank,

Annual Report, 1992; U.S. State Department, "The Global AIDS Disaster: Implications for the 1990s" (Washington, D.C.: Government Printing Office, 1992); D. W. FitzSimons, "Further Asian Spread in 1994?" *AIDS Newsletter* 8 (1993): 14: 1; and P. Piot, "AIDS: The State of the Epidemic," speech delivered at the Opening Ceremony of Biotech 94, Florence, April 10, 1994.

95. Latin America was, of course, also experiencing a rapidly growing HIV pandemic during the late 1980s and the 1990s. The microbe successfully emerged in every island nation of the Caribbean well before 1986, and reached endemicity in that region before the close of the decade. Some Caribbean nations, notably Haiti, the Dominican Republic, Bermuda, and the Bahamas, had per capita HIV/ AIDS rates by the late 1980s that ranked among the highest in the world, exceeding most of Africa.

Mexico's evolving AIDS epidemic was strongly linked with that of the United States, as tens of thousands of Mexicans traveled back and forth between the two countries every year.

Of greatest concern in Latin America was Brazil, the largest nation on the continent. With its economy in a shambles and external debt astronomical, Brazil was in no shape to take on an additional burden. AIDS hit Brazil hard and fast, spread initially through the country's blood supply. As late as 1993 there were still private blood banks that failed to properly screen potentially contaminated blood.

Brazil's long tradition of sensuality and overt sexuality also contributed to the spread in that many young adults had several sexual partners each year. In addition, the society had long-standing ambivalence about homosexuality: men who self-identified as "gay" were vilified and scorned, yet a sizable percentage of Brazil's married "heterosexual" men engaged in gay anal intercourse outside their marriages. This duality and secrecy made the task of AIDS education extremely difficult.

See: R. E. Koenig, J. Pittaluga, and M. Bogart, "Prevalence of Antibodies to the Human Immunodeficiency Virus in Dominicans and Haitians in the Dominican Republic," *Journal of the American Medical Association* 257 (1987): 631–34; S. Siebert, A. Guillermoprieto, and R. Marshall, "An Epidemic Like Africa's," *Newsweek*, July 27, 1987: 38; T. Golden, "AIDS Is Following Mexican Migrant Workers Back Across the U.S. Border," *New York Times*, March 8, 1992: A3; M. Schecter, L. H. Harrison, N. Halsey, et al., "Coinfection with Human T-Cell Lymphotropic Virus Type 1 and HIV in Brazil," *Journal of the American Medical Association* 271 (1994): 353–57; Centers for Disease Control, "Isolation of Human T-Lymphotropic Virus Type III/Lymphadenopathy-Associated Virus from Serum Proteins Given to Cancer Patients—Bahamas," *Morbidity and Mortality Weekly Report* 34 (1985): 489–91; R. Howell, "AIDS in Puerto Rico," *Newsday, Discovery* section, December 11, 1990: 69, 74–75; M. Hernandez, P. Uribe, S. Gortmaker, et al., "Sexual Behavior and Status for Human Immunodeficiency Virus Type 1 Among Homosexual and Bisexual Males in Mexico City," *American Journal of Epidemiology* 135 (1992): 883–94; R. G. Parker, "AIDS Education and Health Promotion in Brazil: Lessons from the Past and Prospects for the Future," in J. Sepulveda, H. Fineberg, and J. Mann, eds., *AIDS Prevention Through Education: A World View* (New York: Oxford University Press, 1992), 109–26; and J. Sepulveda, "Prevention Through Information and Education: Experience from Mexico," ibid., 127–44.

96. As of March 1990 the Eastern bloc had officially reported cumulative AIDS cases and other related findings as follows:

GDR (East Germany)	277 AIDS cases. Government reports a rise in heroin use since the fall of the Wall.
Czechoslovakia	19 AIDS cases
Hungary	32 AIDS cases. Homosexual groups given government approval January 1, 1990, to begin AIDS education campaigns.
Albania	0 AIDS
Soviet Union	18 AIDS cases. 30 million people allegedly HIV-tested since 1987, of whom 722 were found to be positive.
Romania	74 AIDS cases (50 babies). In addition, 799 of 2,850 children tested are HIV-positive.
Poland	348 AIDS cases. Government reports sharp increase in heroin in the country since the fall of the Wall.
Bulgaria	10 AIDS cases. Total of 2.75 million tested, 23 found HIV-positive.

97. L. Garrett, "AIDS in Eastern Europe: Tragedies Out in the Open," *Newsday*, March 6, 1990: 1.

98. *Trud* investigators estimated that 6 billion injection/withdrawal procedures were performed each year, but only 45 million sterile syringes were manufactured. Thus, each syringe was reused an average of 133 times.

99. Vadim Pokrovsky headed up the Specialized Research Laboratory on AIDS, located in Moscow. His father, Valentin, was head of the National Virology Institute in Moscow, President of the Soviet Academy of Sciences, and an early pioneer in AIDS research in that country. Their names were often confused in Western press accounts.

100. Based on data compiled by the Russian State Committee on Sanitary and Epidemiologic Surveillance, the Russian Republic Information and Analytic Center, and *Science* magazine ["Resurging Infectious Diseases in Russia," *Science* 261 (1993): 415], disease rates soared between May 1992 and May 1993:

Disease	Percentage Increase	(Rate per 100,000 people) Russian 1993 Incidence	Comparative U.S. Incidence
Measles	282	30.0	3.8
Diphtheria	163	1.9	less than 0.1
Syphilis	136	9.5	17.3
Typhoid	82	0.2	0.2
Gonorrhea	60	77.4	249.5
Malaria	34	less than 0.1	0.4
Influenza	28	3,651.1	10,000
Tuberculosis	24	13.6	10.4
Whooping cough	16	5.9	1.1

101. A. V. Yablokov, "The Need for a New Approach to Definition of Priority Health Problems of Populations of the Russian Federation," address to the Security Council, March 17, 1993.

102. E. Taylor, C. P. Besse, and T. Healing, "Tuberculosis in Siberia," *Lancet* 343 (1994): 968.

103. Centers for Disease Control, "Diphtheria Outbreak—Russian Federation, 1990–1993," *Morbidity and Mortality Weekly Report* 42 (1993): 840–47; World Health Organization, "Diphtheria in the Former Soviet Union: The Epidemic Continues," Press Release WHO/70 (September 10, 1993); C. Bohlen, "Diphtheria Epidemic Sweeps Russia," *New York Times*, January 29, 1993: A12; and S. Erlanger, "Diphtheria Afflicting Russia, Kills 100 This Year," *New York Times*, August 22, 1993: A8.

104. J. Lumio, M. Jahkola, R. Vuento, et al., "Diphtheria After Visit to Russia," *Lancet* 342 (1993): 53–54; and A. Dezoysa, A. Efstantious, R. C. George, et al., "Diphtheria and Travel," *Lancet* 342 (1993): 446.

105. For details on the Romanian cases, see C. Bohlen, "Romania's AIDS Babies: A Legacy of Neglect," *New York Times*, February 8, 1990: A1; A. Purvis and M. Hornblower, "Rumania's Other Tragedy," *Time*, February 19, 1990: 74; I. V. Patrascu, St. N. Constantinescu, and A. Dublanchet, "HIV-1 Infection in Romanian Children," *Lancet* I (1990): 672; World Health Organization, "World Health Organization Announces Emergency AIDS Action Plan for Romania," Press Release WHO/10/16 February 1990; S. Dickman, "AIDS in Children Adds to Romania's Troubles," *Nature* 343 (1990): 579; L. Garrett, "The Baby Experiments," *Newsday*, October 28, 1990: 7; and J. Pope, "Tulane Probing Doctor's Role in AIDS Test," *Times-Picayune* (New Orleans), October 30, 1990: A1, A6.

106. N. Beldescu, presentation to the Sixth International Conference on AIDS, San Francisco, June 19–23, 1990. (Not formally on the conference agenda.)

107. B. S. Hersh, J. M. Oxtoby, F. Popovici, et al., "Acquired Immunodeficiency Syndrome in Romania," *Lancet* 338 (1991): 645–49.

108. J. Knowles, "The Responsibility of the Individual," in *Doing Better and Feeling Worse: Health in the United States*, Special Issue of *Daedalus*, 1977.

109. J. B. McKinlay and S. M. McKinlay, "The Questionable Contribution of Medical Measures to the Decline of Mortality in the United States in the Twentieth Century," *Milbank Memorial Fund Quarterly*, Summer 1977; D. Fife and C. Mode, "AIDS Incidence and Income," *Journal of Acquired Immune Deficiency Syndromes* 5 (1992): 1105–10; New York State Department of Health, "AIDS in New York State—1992," Public Affairs Group, Albany, NY, 1993; and Committee on AIDS Research and the Behavioral, Social, and Statistical Sciences. National Research Council, "The AIDS Epidemic in the Second Decade," Chapter 1 in H. G. Miller, C. F. Turner, and L. E. Moses, eds., *AIDS: The Second Decade* (Washington, D.C.: National Academy Press, 1990).

110. A private survey forecast 80,000 children under age eighteen would be AIDS orphans by the year 2000. See D. Michaels and C. Levine, "Estimates of the Number of Motherless Youth Orphaned by AIDS in the United States," *Journal of the American Medical Association* 268 (1992): 3456–61.

The researchers made high and low estimates of the New York City orphan problem as follows:

	Cumulative Number	
Year	Low Range	High Range
1991	15,200	22,000
1995	39.200	56,700
2000	72,000	125,000

111. According to Harvard School of Public Health economist Kathy Schwartz, U.S. medical insurance/noninsurance broke down in 1990 as follows:

> *Completely Uninsured Population:* 37 million people
> (29%) 10.7 million—below poverty* incomes
> (20%) 7.4 million—incomes 100–199% above poverty
> (18%) 6.6 million—incomes 200–300% above poverty
> (22%) 8.1 million—incomes 300% above poverty
> * poverty = an income of less than $14,000 per year for a family of four.
>
> 26% were children under 17
> 25% were 18–24 years of age
> 10% were 45–54 years of age
> 7% were 55–64 years of age
>
> 60% of adults were employed
>
> *Underinsured Population:* 40 million people
>
> *Chronically Uninsured Population:* 3–5 million people

Reinhardt estimated that in 1987 Americans spent $500 billion on health care, or 11.1 percent of the GNP.

According to the U.S. government's Health Care Financing Administration (HCFA), health care spending as a percentage of U.S. GNP rose steadily between 1975 and 1987; from 8.3 percent of GNP in 1975 to 11.1 percent in 1987.

Between 1980 and 1987 actual health care expenditures, in all forms, leapt from about $641.9 billion to $1,306,600,000,000, or $1.3 trillion. That was a 181.5 percent increase in seven years, making health care the most rapidly inflating sector of the U.S. economy, and U.S. health care the most inflationary of any industrialized nation.

See U. E. Reinhardt, "The United States: Breakthroughs and Waste," *Journal of Health Politics, Policy and Law* 17 (1992): 637–66.

112. The American Hospital Association estimated that 14 percent of American hospitals—all of them large teaching facilities and public institutions—handled 80 percent of all inpatients in the United States in 1988.

113. B. Lambert, "One in 61 Babies in New York City Has AIDS Antibodies, Study Says," *New York Times*, January 13, 1988: A1; and L. F. Novick, ed., "New York State Seroprevalence Project," *American Journal of Public Health* (Supplement) 81 (1991): 1–61.

114. P. S. Rosenberg, R. J. Biggar, and J. J. Goedert, "Declining Age at HIV Infection in the United States," *New England Journal of Health* 330 (1993): 789; and T. A. Green, J. M. Karom, and O. C. Nwanyanwu, "Changes in AIDS Incidence Trends in the United States," *Journal of Acquired Immune Deficiency Syndromes* 5 (1992): 547–55.

115. R. Gorter, R. Meakin, A. Keffelew, et al., "Homelessness and HIV Infection: A Population Based Study," presentation to the Seventh International Conference on AIDS, Florence, June 16–21, 1991.

116. Again, absent the impact of AIDS, racial disparities in health were glaring throughout the United States. Consider the following (1988) data:

- White babies were 70 percent more likely than black babies to survive to age four.
- Average life expectancies (genders combined) for whites were 76 years, blacks 70.3 years.
- Black men were 50 percent more likely to die of heart attacks than their white counterparts.
- Black children were three times more likely than whites to die of pneumonia or meningitis due to an assortment of microbial infections.
- Thirty-three percent of blacks lived below the official poverty line, compared to 12 percent of whites.
- Infant mortality was 9.2 per 1,000 live births for whites, but 18 per 1,000 for blacks.
- Fifty percent of black women received a first breast cancer diagnosis after the malignancy had become untreatable, compared with 8 percent of white women.

- Between a third to a half of all black men in the United States (rates vary geographically) were unemployed, a rate at least 10 percent higher than the general rate during the Great Depression.
- More black men died as a result of homicide in the United States in 1977 than did fighting for over ten years in the Vietnam War.

See J. Taylor Gibbs, ed., *Young, Black and Male in America: An Endangered Species* (Berkeley, CA: Auburn House, 1988).

117. The 1990 U.S. Census revealed that half of the nation's five million divorced single mothers were not receiving court-ordered child support payments because the fathers were delinquent: the majority of these women and children ended up among the nation's officially counted impoverished citizens.

See 1993 annual reports of Second Harvest, the Urban Institute, and Food Research and Action Center. See also 1990 U.S. Census Report; R. Pear, "Poverty in U.S. Grew Faster Than Population Last Year," *New York Times*, October 5, 1993: A20; D. Wallace, "Poverty and Disease in the USA," *Lancet* 343 (1993): 238–39; J. Freedman, *From Cradle to Grave: The Human Face of Poverty in America* (New York: Atheneum, 1993); and E. L. Bassuk, "Homeless Families," *Scientific American*, December 1991: 66–74.

118. Institute of Medicine, *Homelessness, Health, and Human Needs* (Washington, D.C.: National Academy Press, 1988); Committee on Public Health, "Housing and Health: Interrelationship and Community Impact," *Bulletin of the New York Academy of Medicine* 66 (1990): 379–591; and S. L. Neibacher, ed., "Homeless People and Health Care: An Unrelenting Challenge," United Hospital Fund, Paper Series 14 (December 1990), New York.

119. R. Rosenheck, L. Frisman, and A. M. Chung, "The Proportion of Veterans Among Homeless Men," *American Journal of Public Health* 84 (1994): 466–69; and Institute of Medicine (1988), op. cit.

120. K. Hopper, "New Urban Niche," *Bulletin of the New York Academy of Medicine* 66 (1990): 435–50.

121. Committee for the Study of the Future of Public Health, Institute of Medicine, *The Future of Public Health* (Washington, D.C.: National Academy Press, 1988).

122. Centers for Disease Control, "Measles—New York," *Morbidity and Mortality Weekly Report* 29 (1980): 452–53; and Centers for Disease Control, "Measles—United States, First 39 Weeks of 1980," *Morbidity and Mortality Weekly Report* 29 (1980): 501–2.

123. Averaged across all ages, encephalitis, which can lead to permanent brain damage and/or death, occurs in one out of every 2,000 measles cases in industrialized nations. Death due to either respiratory failure or neurological damage occurs in one out of every 3,000 measles cases. See Centers for Disease Control, "Measles Prevention," *Morbidity and Mortality Weekly Report* 36 (1987): 409–25.

124. T. L. Gustafson, A. W. Lievens, P. A. Brunell, et al., "Measles Outbreak in a Fully Immunized Secondary-School Population," *New England Journal of Medicine* 316 (1987): 771–74.

125. M. B. Edmonson, D. G. Addiss, J. T. McPherson, et al., "Mild Measles and Secondary Vaccine Failure During a Sustained Outbreak in a Highly Vaccinated Population," *Journal of the American Medical Association* 263 (1990): 2467–71.

126. S. H. Lee, D. P. Ewert, P. D. Frederick, and L. Mascola, "Resurgence of Congenital Rubella Syndrome in the 1990s," *Journal of the American Medical Association* 267 (1992): 2616–20; D. E. Shalala, "Giving Pediatric Immunizations the Priority They Deserve," *Journal of the American Medical Association* 269 (1993): 1844–45; and Centers for Disease Control, "Resurgence of Pertussis—United States, 1993," *Morbidity and Mortality Weekly Report* 42 (1993): 952–65.

127. V. S. Mitchell, N. M. Philipose, J. P. Sanford, *The Children's Vaccine Initiative: Achieving the Vision* (Washington, D.C.: Institute of Medicine, National Academy Press, 1993); G. Peter, "On the Measles Epidemic," testimony before the U.S. House of Representatives Subcommittee on Health and the Environment of the Committee on Energy and Commerce, March 11, 1991; and National Vaccine Advisory Committee, "The Measles Epidemic: The Problems, Barriers and Recommendations," Executive Report, January 8, 1991, Washington, D.C. [a summary can be found in the *Journal of the American Medical Association* 266 (1991): 1547–52].

128. A great deal has been written elsewhere about this alarming trend, which worsened markedly after 1985. The lack of primary health care in impoverished urban areas, overutilization of tertiary facilities, and greater cost to society have been well documented elsewhere.

See, for example, D. E. Rogers and E. Ginzberg, *Medical Care and the Health of the Poor* (Boulder, CO: Westview Press, 1993); L. K. Abraham, *Mama Might Be Better Off Dead: The Failure of Health Care in Urban America* (Chicago: University of Chicago Press, 1993); A. Carper, "Ailing Grades: Index Shows Need for More Primary Care," *New York Newsday*, October 10, 1993: 7, 58; J. Hadley, E. P.

Steinberg, and J. Feder, "Comparison of Uninsured and Privately Insured Hospital Patients," *Journal of the American Medical Association* 265 (1991): 374–79; V. R. Fuchs and D. M. Reklis, "America's Children: Economic Perspectives and Policy Options," *Science* 255 (1992): 41–46; J. Mangaliman and K. Freifeld, "City Hell for Kids," *New York Newsday*, September 30, 1993: 4; and C. Woodard, "Report: Many Kids Not Immunized," *New York Newsday*, September 1, 1993: 12.

129. S. F. Davis, P. M. Strebel, W. L. Atkinson, et al., "Reporting Efficiency During a Measles Outbreak in New York City, 1991," *American Journal of Public Health* 83 (1993): 1011–15; and S. Friedman, "Measles in New York City," *Journal of the American Medical Association* 266 (1991): 1220.

130. According to Bloom, in 1992 the following were the rates of successful completion of childhood vaccination recommended by age two in selected areas of the United States:

City/State	% of Children Age 2 Who Have Fulfilled WHO Vaccination Recommendations (excluding BCG)
Houston	10
Bronx, NY	38
Miami	27
New Hampshire	79
Vermont	68
Tennessee	70
Massachusetts	57
California	48
New York State	56
Arkansas	42
Utah	36
Missouri	44
New York City	38
Chicago	27

See also J. Cohen, "Childhood Vaccines: The R and D Factor," *Science* 259 (1993): 1528–29; and "Vaccine Demand and Supply," in Mitchell, Philipose, and Sanford (1993), op. cit.

131. Centers for Disease Control, "Vaccination Coverage of 2-Year-Old Children—United States, 1991–92," *Morbidity and Mortality Weekly Report* 42 (1994): 985–88; E. R. Zell, V. Dietz, J. Stevenson, et al., "Low Vaccination Levels of U.S. Preschool and School-Age Children," *Journal of the American Medical Association* 271 (1994): 833–39; and Centers for Disease Control, "Reported Vaccine-Preventable Diseases—United States, 1993, and the Childhood Immunization Initiative," *Morbidity and Mortality Weekly Report* 43 (1994): 57–61.

132. See R. Dubos and J. Dubos, *The White Plague: Tuberculosis, Man, and Society* (New Brunswick, NJ: Rutgers University Press, 2nd paperback edition, 1992); and F. Ryan, *The Forgotten Plague: How the Battle Against Tuberculosis Was Won—and Lost* (Boston: Little, Brown, 1992).

133. U.S. Department of Health, Education, and Welfare, "The Project Years 1961–69," *Tuberculosis Program Reports* (December 1970).

134. Centers for Disease Control, "Tuberculosis in the United States: 1981–84," U.S. Department of Health and Human Services, Washington, D.C., 1986.

135. See Chapter 9 of this book for further discussion of tuberculosis in urban centers of the Western world prior to 1981.

136. M. A. Barry, C. Wall, L. Shirley, et al., "Tuberculosis Screening in Boston's Homeless Shelters," *Public Health Reports* 101 (1986): 487–94; P. W. Brickner, B. Scanlan, A. Conan, et al., "Homeless Persons and Health Care," *Annals of Internal Medicine* 101 (1986): 405–9; Centers for Disease Control, "Drug Resistant Tuberculosis Among the Homeless," *Morbidity and Mortality Weekly Report* 34 (1985): 429–32; R. Glickman, "Tuberculosis Screening and Treatment of New York City Homeless People," *Annals of the New York Academy of Science* 435 (1984): 19–21; E. B. Narde, B. McInnis, B. Thomas, and S. Weidhass, "Exogenous Reinfection with Tuberculosis in a Shelter for the Homeless," *New England Journal of Medicine* 315 (1986): 1570–75; and A. Pablos-Mendez, M. C. Raviglione, R. Battan, and R. Ramos-Zuñiga, "Drug Resistant Tuberculosis Among the Homeless in New York City," *New York State Journal of Medicine* 90 (1990): 351–55.

137. M. N. Sherman, "Tuberculosis in Single-Room-Occupancy Hotel Residents: A Persisting Focus of Disease," *New York Medical Quarterly* 1 (1980): 39–41.

138. Centers for Disease Control, "Tuberculosis, Final Data—United States, 1986," *Morbidity and Mortality Weekly Report* 36 (1988): 817–19; and Centers for Disease Control, "Tuberculosis—

United States, 1985—and the Possible Impact of Human T-Lymphotropic Virus Type III/Lymphade-nopathy-Associated Virus Infection," *Morbidity and Mortality Weekly Report* 33 (1986): 74–79.

139. Global Programme on AIDS and Tuberculosis Programme, "Statement on AIDS and Tuber-culosis," World Health Organization, WHO/GPA, INF/89.4, March 1989; and A. D. Harries, "Tu-berculosis and Human Immunodeficiency Virus Infection in Developing Countries," *Lancet* 335 (1990): 387–90.

140. C. L. Daley, "Tuberculosis Recurrence in Africa: True Relapse or Re-Infection?" *Lancet* 342 (1993): 756–57; and J. D. Klausner, R. W. Ryder, E. Baende, et al., *"Mycobacterium tuberculosis* in Household Contacts of Human Immunodeficiency Virus Type-1-Seropositive Patients with Active Pulmonary Tuberculosis in Kinshasa, Zaire," *Journal of Infectious Diseases* 168 (1993): 106–11.

141. K. M. DeCock, "Tuberculosis and HIV Infection in Sub-Saharan Africa," *Journal of the American Medical Association* 268 (1992): 1581–87.

142. M. Hawken, P. Nunn, S. Gathua, et al., "Increased Recurrence of Tuberculosis in HIV-1-Infected Patients in Kenya," *Lancet* 342 (1993): 332–37; and A. L. Pozniak, "The Influence of HIV Status on Single and Multiple Drug Reactions to Antituberculosis Therapy in Africa," *AIDS* 6 (1992): 809–14.

143. Y. Mukadi, J. H. Perriens, M. E. St. Louis, et al., "Spectrum of Immunodeficiency in HIV-1-Infected Patients with Pulmonary Tuberculosis in Zaire," *Lancet* 342 (1993): 143–46; and Centers for Disease Control, "Tuberculin Reactions in Apparently Healthy HIV-Seropositive and HIV-Seronegative Women in Uganda," *Morbidity and Mortality Weekly Report* 39 (1990): 638–46.

144. D. S. Shepard, R. N. Bail, and A. Bucyendore, "Costs of AIDS Care in Rwanda," Report to the Bigel Institute for Health Policy, Brandeis University, Waltham, MA, 1992; and K. B. Noble, "AIDS Linked to TB Outbreak in Africa," *New York Times*, April 29, 1990: A14.

145. H. L. Rieder, G. M. Cauthen, G. W. Comstock, and D. E. Snider, "Epidemiology of Tuberculosis in the United States," *Epidemiologic Reviews* 11 (1989): 79–96.

146. Centers for Disease Control, "Tuberculosis Morbidity in the United States: Final Data, 1990," *Morbidity and Mortality Weekly Report* 40 (1991): SS23–SS27.

147. J. J. Ellner, A. R. Hinman, S. W. Dooley, et al., "Tuberculosis Symposium: Emerging Problems and Promise," *Journal of Infectious Diseases* 168 (1993): 537–51.

148. G. Sunderam, R. J. McDonald, T. Maniatis, et al., "Tuberculosis as a Manifestation of the Acquired Immunodeficiency Syndrome (AIDS)," *Journal of the American Medical Association* 256 (1986): 362–66; Centers for Disease Control, "Tuberculosis and Acquired Immunodeficiency Syndrome—Florida," *Morbidity and Mortality Weekly Report* 35 (1986): 587–90; J. Garrison, "AIDS Fuels Sharp Rise in TB Cases," *San Francisco Sunday Examiner and Chronicle*, December 27, 1987: A1; Centers for Disease Control, "Tuberculosis and Acquired Immunodeficiency Syndrome—New York City," *Morbidity and Mortality Weekly Report* 36 (1987): 785–95; and Centers for Disease Control, "Tuberculosis and AIDS—Connecticut," *Morbidity and Mortality Weekly Report* 36 (1987): 133–35.

149. Tuberculosis rates in Harlem had always been exceptionally high.

NEW YORK CITY TUBERCULOSIS (cases per 100,000):

Year	Harlem	New York City	U.S.A.
1970	135.0	32.8	18.22
1975	105.0	27.2	15.95
1980	78.6	19.9	12.2
1985	110.9	26.0	9.3
1986	130.4	31.4	9.4
1987	134.9	31.1	9.4
1988	158.9	32.8	9.1
1989	169.2	36.0	9.5

150. K. Brudney and J. Dobkin, "Resurgent Tuberculosis in New York City," *American Review of Respiratory Disease* 144 (1991): 745–49.

151. C. Woodard, "TB in New York," *New York Newsday*, March 8, 1992: 1; C. Woodard, "Bitter Medicine to Swallow," *New York Newsday*, March 8, 1992: 38; M. Gelman, "A Prison Breeding Ground," *New York Newsday*, March 11, 1992: 23; M. Gelman, "City Races to Finish 'Cutting Edge' TB Jail," *New York Newsday*, March 11, 1992: 87; L. Garrett and C. Woodard, "New Risk in Hospitals," *New York Newsday*, March 10, 1992: 6; L. Garrett, "HIV/TB—Tandem Epidemics Breed a Contra-diction in Control," *New York Newsday*, March 31, 1992: 57; L. Garrett, "Jobs That Carry a High Risk of TB," *New York Newsday*, March 31, 1992: 61; L. Garrett, "Tackling the TB Puzzle," *New York Newsday*, March 12, 1992: 8; and G. Cowley, E. A. Leonard, and M. Hager, "A Deadly Return," *Newsweek*, March 16, 1992: 53–57.

152.

NOSOCOMIAL OUTBREAKS	HIV-RELATED Hospital A	MULTIDRUG-RESISTANT Hospital B	Hospital C	TUBERCULOSIS Hospital D
Location	Miami	NYC	NYC	NYC
Outbreak period	1988–91	1989–91	1989–91	1990–91
Early cases	29	18	17	23
Late cases	36	17	21	9
Drug-resistance pattern	INH, RIF (EMB, THA)	INH, SM (RIF, EMB)	INH, RIF, SM (EMB, KM, THA, RBT)	INH, RIF (EMB, THA)
Percent of newly infected patients HIV+	93%	100%	94%	91%
Mortality among HIV+ patients	72%	89%	82%	83%
Median interval from TB diagnosis to death in HIV+ patients	7 weeks	16 weeks	6 weeks	4 weeks

INH = isoniazid; RIF = rifampin; EMB = ethambutol; THA = ethionamide; SM = streptomycin; KM = kanamycin; RBT = rifabutin.
Note: CDC confidentiality forbids naming the hospitals.
Source: Centers for Disease Control.

153. D. E. Snider and W. L. Roper, "The New Tuberculosis," *New England Journal of Medicine* 325 (1992): 703–5; F. Gordin, "Tuberculosis Control—Back to the Future?" *Journal of the American Medical Association* 267 (1992): 2649–50; P. F. Barnes, A. B. Bloch, P. T. Davidson, and D. E. Snider, "Tuberculosis in Patients with Human Immunodeficiency Virus Infection," *New England Journal of Medicine* 324 (1991): 1644–50; and M. A. Fischl, G. L. Daikos, R. B. Uttamchandani, et al., "Clinical Presentation and Outcome of Patients with HIV Infection and Tuberculosis Caused by Multiply-Drug-Resistant Bacilli," *Annals of Internal Medicine* 117 (1992): 184–90.

154. Centers for Disease Control, "Drug-Resistant Tuberculosis Among the Homeless—Boston," *Morbidity and Mortality Weekly Report* 34 (1985): 429–31.

155. Centers for Disease Control, "Multi-Drug-Resistant Tuberculosis—North Carolina," *Morbidity and Mortality Weekly Report* 35 (1987): 785–87.

156. For these and other chilling findings, see a lengthy memo from the CDC to the Department of Health and Human Services, Washington, D.C., dated December 31, 1991; and T. R. Frieden, M. L. Pearson, and J. A. Jereb, "Drug Resistant and Nosocomial Tuberculosis, New York City, 1991," *EPI-AID* (1991): EPI-91-42-2.

157. There were many outbreaks of MDR-TB in the United States and Puerto Rico during the epidemic, which appears to have begun sometime in the mid-1980s and continues at this writing.

MDR-TB OUTBREAKS REPORTED IN THE UNITED STATES AND PUERTO RICO, 1985–92

Location	Drug Resistance	Year/s	Index Case/s	Secondary Case/s	Citation
Texas, California, Pennsylvania	INH, RIF, SM, PZA, EMB	1987	Male, diagnosed with TB in 1971, recalcitrant, in/out of medications. Died in 1987.	9, family and relatives	CDC, "Outbreak of Multidrug-Resistant Tuberculosis—Texas, California, and Pennsylvania," *MMWR* 39 (1990): 369–72.
Mississippi, rural	INH, SM, PAS	1976	High school student	Fellow students and their families	CDC (1977), op. cit.
Boston, homeless shelters	INH, SM	1984, 1985	2 possible, both homeless men	Fellow sheltered homeless	CDC (1985), op. cit.
Miami, outpatient AIDS clinic or HIV ward	INH, RIF, EMB, ETH	1988–1991	1 of the patients	22, HIV patients	M. A. Fischl et al., "An Outbreak of Tuberculosis Caused by Multiple-Drug-

					Resistant Tubercle Bacilli Among Patients with HIV Infection," *Annals of Internal Medicine* 117 (1992): 177–83.
New York State Prison	INH, RIF, PZA, EMB, SM, KM, ETH	1990–1991	Prisoner	7 inmates and 1 prison guard	CDC, "Transmission of Multidrug-Resistant Tuberculosis Among Immunocompromised Persons in a Correctional System—New York, 1991," *MMWR* 41 (1992): 507–9.
New York City Jail, Rikers Island	Various	1988–1992	Prisoners	Spread within jail; diagnosis rate of 500:100,000. Average daily census of jail is 20,000; average stay 65 days.	E. Y. Bellin et al., "Association of Tuberculosis Infection with Increased Time in or Admission to New York City Jail System," *JAMA* 269 (1993): 2228–31.
New York City Jail	Various	1991	Prisoners	720 cases of MDR-TB diagnosed in prisoners	M. McCarthy, "Long-Term TB Detentions in New York," *Lancet* 341 (1993): 751.
Waupun Jail, Wisconsin	NS	1993	Prisoners	22 prisoners	J. Bignall, "TB in Prison," *Lancet* 343 (1993): 49.
Nassau County Jail, NY	NS	1988–1990	Prisoners	45 prisoners	Ibid.
Lincoln Hospital, NYC	INH, RIF, EMB, SM	1991	Noncompliant patient with AIDS	1 patient with AIDS	D. L. Horn et al., "Fatal Hospital-Acquired Multidrug-Resistant Tuberculosis Pericarditis in Two Patients with AIDS," *NEJM* 327 (1992): 1816–17.
7 New York City hospitals	INH, SM, RIF, EMB	1988–1991	Patients	More than 100 patients; 19 health care workers; all but 6 of whom were HIV+.	CDC, "Nosocomial Transmission of Multidrug-Resistant Tuberculosis Among HIV-Infected Persons —Florida and New York, 1988–1991," *MMWR* 40 (1991): 585–90; and, New York City Department of Health reports.
San Juan, Puerto Rico, hospital	12 to INH, RIF, PZA, EMB	1989	Patient(s)	All 17 health care providers on HIV ward infected	S. W. Dooley et al., "Nosocomial Transmission of Tuberculosis in a Hospital Unit for HIV-Infected Patients," *JAMA* 267 (1992): 2632–35.

NYC hospital	NS	1989–1991	Patient(s)	23 patients, 21 of whom were HIV+; 12 health care providers infected, no active cases	M. L. Pearson et al., "Nosocomial Transmission of Multidrug-Resistant *Mycobacterium tuberculosis*," *Annals of Internal Medicine* 117 (1992): 191–96.
NYC hospital	INH, SM, RIF, EMB	1989–1990	Patient(s)	18 patients, all with AIDS	B. R. Edlin et al., "An Outbreak of Multidrug-Resistant Tuberculosis Among Hospitalized Patients with the Acquired Immunodeficiency Syndrome," *NEJM* 326 (1992): 1514–21.
Cook County Hospital, Chicago	NS	1991	Patient(s)	12 health care providers infected, no active cases	L. Cocchiarella et al., presentation to the 1992 International Conference of the American Lung Association, Miami, May 17, 1992.
Miami hospital	INH, RIF	1990–1991	Patient	36 patients, 35 of whom were HIV+	CDC, "Nosocomial Transmission of Multidrug-Resistant Tuberculosis" (1991), op. cit.
Miami hospital	INH, RIF	1987–1990	Patient(s)	29 patients, 13 health care providers (but none developed active TB)	CDC, "Nosocomial Transmission of Multidrug-Resistant Tuberculosis to Health-Care Workers and HIV-Infected Patients in an Urban Hospital—Florida," *MMWR* 39 (1990): 718–22.

INH = isoniazid; RIF = rifampin; EMB = ethambutol; PZA = pyrazinamide; SM = streptomycin; CS = cycloserine; PAS = para-amino-salicylic acid; ETH = ethionamide; KM = kanamycin.

158. R. E. Brown, C. S. Palmer, and K. Simpson, "Estimate of Identifiable Costs of Tuberculosis in the United States in 1991," Battelle Medical Technology Assessment and Policy Research Center, Washington, D.C., 1993.

159. New York State Assembly Committee on Health, "Tuberculosis in New York: The Return of an Epidemic," Report to the Legislature, Albany, 1991.

160. The actual allocation approved by the White House was for $40 million in 1992. CDC director Roper "found" another $14.9 million in funds designated for AIDS education efforts. Of the total $54.9 million, $6.5 million went to the National Institutes of Health for basic TB research and $46.5 million went to CDC tuberculosis control. Of its share, the CDC sent a good portion to New York City.

Already facing a budget crisis, New York City spent more than $40 million on TB control in 1992 alone.

The NIH's TB spending climbed from $3.5 million in 1991 to $46 million in 1994.

161. The cities surveyed were Atlanta, Baltimore, Boston, Chicago, Cleveland, Dallas, Detroit, Honolulu, Jacksonville, Los Angeles, Memphis, Miami, Milwaukee, Nashville, Newark, New Orleans, New York, Philadelphia, Phoenix, San Antonio, San Diego, San Francisco, Seattle, Tampa, and Washington, D.C. See D. R. Leff and A. R. Leff, "Tuberculosis Control Policies in Major Metropolitan Health Departments in the United States: V. Standard of Practice in 1992," *American Reviews of Respiratory Disease* 148 (1993): 1530–36.

162. Tuberculosis Task Force, "Nassau County Policy Plan for Responding to the Metropolitan New York Tuberculosis Epidemic," Report to the Nassau County Health Department, November 1993.

163. M. Goble, M. D. Iseman, L. A. Madsen, et al., "Treatment of 171 Patients with Pulmonary Tuberculosis Resistant to Isoniazid and Rifampin," *New England Journal of Medicine* 328 (1993): 527–32.

164. D. E. Snider, "Shortages of Antituberculosis Drugs, Outbreaks of Multidrug-Resistant Disease and New Drug Development in the U.S.," presentation to the World Health Organization, November 22, 1991.

165. New York City Task Force on Tuberculosis in the Criminal Justice System, "Final Report to Mayor David N. Dinkins and Margaret A. Hamburg, M.D., Commissioner, Department of Health," June 1992.

166. Among the questions TB scientists at the February 10, 1992, meeting identified as most pressing were the following:

Epidemiology
• How long do the bacteria survive in an enclosed airspace, potentially infecting a person?
• How many bacteria does a human need to be exposed to in order to be infected?
• What, precisely, are the relationships between HIV, injecting drug use, poverty, poor housing, and tuberculosis?

Treatment
• How can laboratories shorten the amount of time required to diagnose MDR-TB to some clinically useful period?
• What are the optimal ways to use antitubercular drugs to avoid development of resistance?
• What are the risks for patient reinfection?
• What steps should hospitals and individual health providers take to avoid catching TB from their patients? To prevent patient-to-patient transmission?
• How can surfaces be sterilized for TB?
• Are there any other drugs out there for TB?
• Can anything be done to save HIV-positive people who become infected with MDR-TB?

Pathogenesis/Immunology
• How does the bacteria use complement receptors to gain entry into human cells?
• Why do the bacteria have strong iron-binding capabilities?
• What is the nature of the cell-mediated immune response to the microbe? Do antibodies play any significant role?
• How much of the disease is due to macrophages' release of cytokines and other chemicals following *M. tuberculosis* invasion?
• Why does the human lung self-destruct when infected?
• How do the bacterial colonies form the protective encasements that allow them to hide from the immune system? And what, exactly, causes the liquefaction process that melts those casings and floods the bloodstream with the bacteria?
• Are any of the bovine-based BCG vaccines against TB useful? How useful? For whom?

Resistance
• What is the molecular/genetic basis of drug resistance? Does it always vary from drug to drug, or are there universal resistance mechanisms?
• Is there any way to design a drug that the mycobacteria can't resist?

167. Y. Zhang, B. Heym, B. Allen, et al., "The Catalase-Peroxidase Gene and Isoniazid Resistance of *Mycobacterium tuberculosis*," *Nature* 358 (1992): 591–92; and A. Banerjee, E. Dubnau, A. Quemard, et al., "InhA, a Gene Encoding a Target for Isoniazid and Ethionamide in *Mycobacterium tuberculosis*," *Science* 263 (1993): 227–30.

168. B. J. Culliton, "Drug-Resistant TB May Bring Epidemic," *Nature* 356 (1992): 473.

169. W. R. Jacobs, R. G. Barletta, R. Udani, et al., "Rapid Assessment of Drug Susceptibilities of *Mycobacterium tuberculosis* by Means of Luciferase Reporter Phages," *Science* 260 (1993): 819–22.

170. P. Pancholi, A. Mirza, N. Bhardwaj, and R. M. Steinman, "Sequestration from Immune CD4+ T Cells of Mycobacteria Growing in Human Macrophages," *Science* 260 (1993): 984–86; and S. Arruda, G. Bomfim, R. Knights, et al., "Cloning of a *Mycobacterium tuberculosis* DNA Fragment Associated with Entry and Survival Inside Cells," *Science* 261 (1993): 1454–58.

171. Somewhat after the emergence of MDR-TB in the United States, European communities of poverty witnessed identical chains of social and biological events. In Switzerland, the Netherlands, Italy, Denmark—all over Europe—HIV, poverty, and injecting drug use drove tuberculosis rates upward. According to WHO, Western Europe's increases were as follows:

Country	Time Period	Percent Increase in TB
Switzerland	1986–90	33
Denmark	1984–90	31
Italy	1988–90	28
Norway	1988–91	21
Ireland	1988–90	18
Austria	1988–90	17
Finland	1988–90	17
Netherlands	1987–90	9.5
Sweden	1988–90	4.6
United Kingdom	1987–90	2.0
France, Germany, Belgium	1987–91	stable

Source: World Health Organization, Press Release, June 17, 1992.

See also A. Genewein, A. Telenti, C. Bernasconi, et al., "Molecular Approach to Identifying Route of Transmission of Tuberculosis in the Community," *Lancet* 342 (1993): 841–44; and E. Drucker, "Molecular Epidemiology Meets the Fourth World," *Lancet* 342 (1993): 817–18.

172. K. Brudney and J. Dobkin, "A Tale of Two Cities: Tuberculosis Control in Nicaragua and New York City," unpublished, 1992.

173. D. Wilkinson, "High-Compliance Tuberculosis Treatment Programme in a Rural Community," *Lancet* 173 (1994): 647–48.

174. E. P. Y. Muhondwa, "The Role and Impact of Foreign Aid in Tanzania's Health Development," in M. R. Reich and E. Marvi, eds., *International Cooperation for Health: Problems, Prospects, and Priorities* (Dover, MA: Auburn House, 1989).

175. C. Murray, K. Styblo, and A. Rouillon, "Tuberculosis," in D. T. Jamison and W. H. Mosley, eds., *Disease Control Priorities in Developing Countries* (New York: Oxford University Press, 1991).

176. Sadly, the U.S. Agency for International Development cut its entire overseas tuberculosis budget—all $3 million—the day after Christmas 1993. That represented a third of all funds for the World Health Organization's TB program.

177. Some countries had no tuberculosis control program at all. There, of course, the TB situation could have been worse than in the United States. The World Bank chose to limit comparisons to countries with TB programs and prevalence data.

178. In 1994, the CDC and the New York City Department of Health cautiously announced some success in DOT control of TB, but repeatedly warned physicians and the general public not to misinterpret their findings as indicating that the epidemic was under control. See A. B. Bloch, G. M. Cauthen, I. Onorato, et al., "Nationwide Survey of Drug-Resistant Tuberculosis in the United States," *Journal of the American Medical Association* 271 (1994): 665–71; and S. E. Weis, P. C. Slocum, F. X. Blais, et al., "The Effect of Directly Observed Therapy on the Rates of Drug Resistance and Relapse in Tuberculosis," *New England Journal of Medicine* 330 (1994): 1179–84.

179. National MDR-TB Task Force, "National Action Plan to Combat Multidrug-Resistant Tuberculosis," U.S. Department of Health and Human Services, Washington, D.C., April 1992.

15. All in Good Haste

1. Early accounts of the investigation appear in the CDC's *Morbidity and Mortality Weekly Report*. See Vol. 42: 421–24, 441–43, 477–78, 495–96, 517–19, 570–71, and 612–13. In addition, the CDC published a special *MMWR*: "Hantavirus Infection—Southwestern United States: Interim Recommendations for Risk Reduction," Vol. 42 (No. RR-11), 1993.

2. The list of epidemics then under investigation included:
- Several U.S. outbreaks of lethal food poisoning due to a newly emerging strain of *E. coli* 0157 bacteria, and reports of the same drug-resistant bacteria's spread in South Africa and Swaziland.
- Cryptosporidiosis parasite contamination of the Milwaukee water supply.
- Outbreaks in several hospitals and child care centers of multidrug-resistant pneumonia-producing bacterial disease.
- Reports of vancomycin-resistant stomach infections in New York City hospitals due to enterococcal bacteria.
- A new strain of influenza A spotted in China (A/Beijing/32/92) that some flu experts feared would reach the United States during the coming fall.
- Cholera spreading steadily northward across Latin America since its first arrival in Peru in 1991.

- Yellow fever, for the first time in history, erupting in western Kenya, in an area rife with ethnic warfare.
- A new deadly mutant strain of cholera, causing an epidemic in India and Bangladesh.
- An epidemic of the mosquito-carried disease Rift Valley fever, claiming thousands of victims in southern Egypt.
- Two new outbreaks of the deadly hemorrhagic viral disease Lassa fever in Nigeria.
- The appearance of a new antibiotic-resistant strain of bacteria causing lethal dysentery among children in Burundi.
- Multidrug-resistant strains of tuberculosis surfacing in New York City, Cambodia, and a few other sites in the United States—and rumors of emergences in Switzerland, Paris, Madrid, and London.
- Bolivian hemorrhagic fever, the Machupo viral disease, breaking out again in eastern Bolivia for the first time in thirty years.
- An epidemic of a newly discovered lethal virus in Venezuela, dubbed Venezuelan hemorrhagic fever.

In all of these cases CDC personnel were actively involved in field investigations or laboratory work. In addition, CDC scientists were advising other international and U.S. state agencies on how best to handle several other disease outbreaks, including an out-of-control epidemic of visceral leishmaniasis that had already claimed some 60,000 lives in war-ravaged southern Sudan.

3. For further accounts of these and other events in the outbreak, see L. K. Altman, "Virus That Caused Deaths in New Mexico Is Isolated," *New York Times*, November 21, 1993: A21; L. Garrett, "Medical Gumshoes Confront a Mystery," *Newsday, Discover* section, September 28, 1993: 69–72; D. Grady, "Death at the Corners," *Discover*, December 1993: 82–92; J. Horgan, "Were Four Corners Victims Biowar Casualties?" *Scientific American*, November 1993: 16; J. M. Hughes, C. J. Peters, M. L. Cohen, and B. W. J. Mahy, "Hantavirus Pulmonary Syndrome: An Emerging Infectious Disease," *Science* 262 (1993): 850–51; B. Le Guenno, "Identifying a Hantavirus Associated with Acute Respiratory Illness: A PCR Victory?" *Lancet* 342 (1993): 1438–39; M. D. Lemonick, "Closing In on a Mysterious Killer," *Time*, December 6, 1993: 66–67; R. Levins, P. R. Epstein, M. E. Wilson, et al., "Hantavirus Disease Emerging," *Lancet* 342 (1993): 1292; E. Marshall and R. Stone, "Hantavirus Outbreak Yields to PCR," *Science* 262 (1993): 832–836; and W. F. Rahson, "Indians Doubt Rodents Source of 'Mystery Illness,' " Associated Press, July 1, 1993.

4. The Chinese described an ailment 1,000 years ago that some have interpreted to be a hantaviral disease, and both Russian and Japanese scientists showed during World War II that people suffering hemorrhagic renal failure had an infectious disease. They did so by injecting the filtered urine of disease victims into "volunteers," who subsequently developed the disease.

5. J. W. LeDuc, T. G. Ksiazek, C. A. Rossi, et al., "A Retrospective Analysis of Sera Collected by the Hemorrhagic Fever Commission During the Korean Conflict," *Journal of Infectious Diseases* 162 (1990): 1182–84.

6. H. W. Lee, "Korean Hemorrhagic Fever," in *Ebola Virus Haemorrhagic Fever* (New York: Elsevier, 1978), 331–43.

7. M. Linderholm, B. Settergren, C. Ahlm, et al., "A Swedish Case of Fatal Nephropathia Epidemica," *Scandinavian Journal of Infectious Diseases* 23 (1991): 501–2; B. Settergren, "Nephropathia Epidemica in Scandinavia," *Reviews of Infectious Diseases* 13 (1991): 736–44; and A. Lundkvist, A. Fatouros, and B. Niklasson, "Antigenic Variation of European Haemorrhagic Fever with Renal Syndrome Viral Strains Characterized Using Bank Vole Monoclonal Antibodies," *Journal of General Virology* 72 (1991): 2097–2103.

8. Later one of van der Gröen's graduate students uncovered another hantavirus strain in Ireland, carried by *Rattus rattus*, the world's most common black urban rat. See C. F. Stanford, J. H. Connolly, W. A. Ellis, et al., "Zoonotic Infections in Northern Ireland Farmers," *Epidemiology and Infection* 105 (1990): 565–70.

9. J. E. Childs, G. E. Glass, T. G. Ksiazek, et al., "Human-Rodent Contact and Infection with Lymphocytic Choriomeningitis and Seoul Viruses in an Inner-city Population," *American Journal of Tropical Medicine and Hygiene* 44 (1991): 117–21; and J. E. Childs, G. E. Glass, G. W. Korch, et al., "Evidence of Human Infection with a Rat-Associated Hantavirus in Baltimore, Maryland," *American Journal of Epidemiology* 127 (1988): 875–78.

10. G. E. Glass, A. J. Watson, J. W. LeDuc, et al., "Infection with a Ratborne Hantavirus in U.S. Residents Is Consistently Associated with Hypertensive Renal Disease," *Journal of Infectious Diseases* 167 (1993): 614–20.

11. L. J. Back, R. Yanagihara, C. J. Gibbs, et al., "Leakey Virus: A New Hantavirus Isolated from *Mus musculus* in the United States," *Journal of General Virology* 69 (1988): 3129–32.

12. R. Weiss, "Rat-Borne Virus May Take Secret Toll," *Science* 135 (1993): 292; Glass, Watson,

LeDuc, et al. (1993), op. cit.; and J. W. LeDuc, J. E. Childs, and G. E. Glass, "The Hantaviruses, Etiologic Agents of Hemorrhagic Fever with Renal Syndrome: A Possible Cause of Hypertension and Chronic Renal Disease in the United States," *Annual Review of Public Health* 13 (1992): 79–98.

13. R. R. Arthur, R. S. Lofts, J. Gomez, et al., "Grouping of Hantaviruses by Small S Genome Segment Polymerase Chain Reaction," *American Journal of Tropical Medicine and Hygiene* 47 (1992): 210–24.

14. J. W. Huggins, C. M. Hsiang, T. M. Cosgriff, et al., "Prospective, Double-Blind, Concurrent, Placebo-Controlled Clinical Trials of Intravenous Ribavirin Therapy of Hemorrhagic Fever with Renal Syndrome," *Journal of Infectious Diseases* 164 (1991): 1119–27.

15. Centers for Disease Control and Prevention, "Update: Outbreak of Hantavirus Infection— Southwestern United States, 1993," *Morbidity and Mortality Weekly Report* 42 (1993): 477–78.

16. S. T. Nichol, C. F. Spiropoulou, S. Morzunov, et al., "Genetic Identification of a Hantavirus Associated with an Outbreak of Acute Respiratory Illness," *Science* 262 (1993): 914–18.

17. "Congress Mobilizes Against Hantavirus," *Science* 261 (1993): 415.

18. Centers for Disease Control and Prevention, "Update: Hantavirus Pulmonary Syndrome— United States, 1993," *Morbidity and Mortality Weekly Report* 42 (1993): 816–20.

19. H. Zinsser, "Much About Rats—A Little About Mice," *Rats, Lice and History* (Boston: Little, Brown, 1934), chapter 11.

20. G. Neild, "Mysterious Respiratory Disease in USA" (letter), *Lancet* 342 (1993): 61.

21. J. Pilaski, C. Ellerich, T. Kreutzer, et al., "Haemorrhagic Fever with Renal Syndrome in Germany" (letter), *Lancet* 337 (1991): 111; and P. Kulzer and R. M. Schaefer, "Haemorrhagic Fever with Renal Syndrome, 1993: Endemic or Unrecognised Pandemic?" (letter), *Lancet* 342 (1993): 313.

22. Ibid.; and B. Le Guenno, M. A. Camprasse, J. C. Guilbaut, et al., "Hantavirus Epidemic in Europe, 1993," *Lancet* 343 (1994): 114–15.

23. Le Guenno (1993), op. cit.; J. Clement, P. McKenna, P. Colson, et al., "Hantavirus Epidemic in Europe, 1993," *Lancet* 343 (1994): 114; P. E. Rollin, D. Coudrier, and P. Sureau, "Hantavirus Epidemic in Europe, 1993," *Lancet* 343 (1994): 115–16; and R. A. J. Esselink, M. N. Gerding, P. J. A. M. Brouwers, et al., "Guillain-Barré Syndrome Associated with Hantavirus Infection," *Lancet* 343 (1994): 180–81.

24. Not everyone agreed with this analysis. Some scientists adamantly rejected the hantavirus association with the Four Corners outbreak, and insisted well after all PCR analysis was completed that the ailment was caused by toxic chemicals. See, for example, W. F. Denetclaw, Jr., and T. H. Denetclaw, "Is 'South-West U.S. Mystery Disease' Caused by Hantavirus?" *Lancet* 343 (1994): 53–54.

25. E. Marshall and R. Stone, "Race to Grow Hantavirus Ends in Tie," *Science* 262 (1993): 1509; P. B. Jahrling, presentation to the annual meeting of the American Society of Tropical Medicine and Hygiene, Atlanta, GA, November 1993; and J. S. Duchin, F. T. Koster, C. J. Peters, et al., "Hantavirus Pulmonary Syndrome: A Clinical Description of 17 Patients with a Newly Recognized Disease," *New England Journal of Medicine* 330 (1994): 949–55.

26. For a discussion of Seoul Hantaan, see: C. S. Schmaljohn, S. E. Hasty, and J. M. Dalrymple, "Preparation of Candidate Vaccine-Vectored Vaccines for Haemorrhagic Fever with Renal Syndrome," *Vaccine* 10 (1992): 10–13.

27. Centers for Disease Control and Prevention, "Hantavirus Pulmonary Syndrome—United States, 1993," *Morbidity and Mortality Weekly Report* 43 (1994): 45–48.

28. L. Garrett, "Hantavirus Source Is Still a Mystery," *New York Newsday*, March 25, 1994: A32; E. Lane, "Hantavirus Rodents: A Hard Catch," *New York Newsday*, March 6, 1994: 18; E. Lane, L. Garrett, and A. Smith, "Virus' Deadly Clues," *Newsday*, February 26, 1994: 8; and L. Garrett, E. Lane, J. Mangaliman, et al., "Hantavirus: The Search Is On," *Newsday*, February 25, 1994: 4.

16. Nature and Homo sapiens

1. United Nations Population Fund, "The State of the World Population," United Nations, New York, 1991.

2. For cogent arguments on the relationship between rapid human population growth and environmental destruction and/or human suffering (warfare, economic despair, human rights violations, low quality of life), see P. Kennedy, *Preparing for the Twenty-first Century* (New York: Vintage, 1993); Population Crisis Committee, "Human Suffering Index," Washington, D.C., 1987–93, annually; P. Harrison, *The Third Revolution* (London: I. B. Tauris, 1992); and R. D. Kaplan, "The Coming Anarchy," *Atlantic Monthly*, February 1994: 44–76.

3. E. O. Wilson, *The Diversity of Life* (Cambridge, MA: Harvard University Press, 1992).

4. E. O. Wilson, "Rain Forest Canopy: The High Frontier," *National Geographic*, December 1991: 78–107.

5. Food and Agriculture Organization, "The Forest Resources of the Tropical Zone by Main Ecological Regions," report to the United Nations Conference on Environment and Development by the Forest Resource Assessment 1990 Project, FAO, Rome, 1992.

The FAO results are described in P. Aldous, "Tropical Deforestation: Not Just a Problem in Amazonia," *Science* 259 (1993): 1390. When the loss rate was described in terms of percentages of whole forest areas, FAO found:

Region	% of Forest Lost Annually
South America	0.6
Southeast Asia	1.6
Central America	1.5

6. A. Gentry, presentation to the Neotropical Montane Forests: Biodiversity and Conservatism meeting, New York Botanical Garden, June 21–25, 1993.

7. For a detailed discussion of the conflicting economic and ecological interests at play in the Amazon, see J. de Onís, *The Green Cathedral: Sustainable Development of Amazonia* (New York: Oxford University Press, 1992).

8. D. Skole and C. Tucker, "Tropical Deforestation and Habitat Fragmentation in the Amazon: Satellite Data from 1978 to 1988," *Science* 260 (1993): 1905–10.

9. Centers for Disease Control, "Lyme Disease Surveillance—United States, 1989–1990," *Morbidity and Mortality Weekly Report* 40 (1991): 417–20; and T. F. Tsai, presentation to the Annual Meeting of the American Society of Tropical Medicine and Hygiene, Honolulu, December 10–14, 1989.

10. M. Kirsch, F. L. Ruben, A. C. Steere, et al., "Fatal Adult Respiratory Distress Syndrome in a Patient with Lyme Disease," *Journal of the American Medical Association* 259 (1988): 2737–39; and J. F. Bradley, R. C. Johnson, and J. L. Goodman, "The Persistence of Spirochetal Nucleic Acids in Active Lyme Arthritis," *Annals of Internal Medicine* 120 (1994): 487–89.

11. A. C. Steere, E. Taylor, G. L. McHugh, and E. L. Logigian, "The Overdiagnosis of Lyme Disease," *Journal of the American Medical Association* 269 (1993): 1812–16.

12. A. C. Steere, R. L. Grodzicki, A. N. Kornblatt, et al., "The Spirochetal Etiology of Lyme Disease," *New England Journal of Medicine* 308 (1983): 733–40.

13. A. C. Steere, "Lyme Disease," *New England Journal of Medicine* 321 (1989): 586–96.

14. A. G. Barbour and D. Fish, "The Biology and Social Phenomenon of Lyme Disease," *Science* 260 (1993): 1610–16.

15. H. S. Ginsberg, "Transmission Risk of Lyme Disease and Implications for Tick Management," *American Journal of Epidemiology* 138 (1993): 65–73.

16. J. F. Levine, M. L. Wilson, and A. Spielman, "Mice as Reservoirs of the Lyme Disease Spirochete," *American Journal of Tropical Medicine and Hygiene* 34 (1985): 355–60; J. G. Donahue, J. Piesman, and A. Spielman, "Reservoir Competence of White-Footed Mice for Lyme Disease Spirochetes," *American Journal of Tropical Medicine and Hygiene* 36 (1987): 92–96; J. Piesman, T. N. Mather, G. J. Dammin, et al., "Seasonal Variation of Transmission Risk of Lyme Disease and Human Babeosis," *American Journal of Epidemiology* 126 (1987): 1187–89; S. R. Telford, T. N. Mather, S. I. Moore, et al., "Incompetence of Deer as Reservoirs of the Lyme Disease Spirochete," *American Journal of Tropical Medicine and Hygiene* 39 (1988): 105–9; and T. N. Mather, M. L. Wilson, S. I. Moore, et al., "Comparing the Relative Potential of Rodents as Reservoirs of the Lyme Disease Spirochete (*Borrelia burgdorferia*)," *American Journal of Epidemiology* 130 (1989): 143–50.

17. D. J. White, H. G. Chang, J. L. Benach, et al., "The Geographic Spread and Temporal Increase of the Lyme Disease Epidemic," *Journal of the American Medical Association* 266 (1991): 1230–36.

18. Crucial clues to the history of the emergence of Lyme disease were provided by a study of the 1980–81 appearance of the disease in Ipswich, Massachusetts. See C. C. Lastavica, M. L. Wilson, V. P. Berardi, et al., "Rapid Emergence of a Focal Epidemic of Lyme Disease in Coastal Massachusetts," *New England Journal of Medicine* 320 (1989): 133–37.

19. A. Spielman, "The Emergence of Lyme Disease and Human Babeosis in a Changing Environment," presentation to the Workshop on New Diseases, Woods Hole, MA, November 7–10, 1993.

20. O. L. Phillips and A. H. Gentry, "Increasing Turnover Through Time in Tropical Forests," *Science* 263 (1994): 954–58; and S. L. Pimm and A. M. Sugden, "Tropical Diversity and Global Change," *Science* 263 (1994): 933–34.

21. This has been, and continues to be, a very lively debate that directly affects global treaties still in negotiation. What kind of refrigerants were in American kitchen appliances, the allowable carbon monoxide emissions for Italian cars, incineration policies for Japanese plastics wastes, and forestry plans for Southeast Asia all impinged upon scientific interpretations of available data on ozone depletion and global warming.

As a result, mountains have been written on the topic, and it is well beyond the scope of this book to scrutinize the data at the root of the debate. For a flavor of that debate, see A. Gore, *Earth in the Balance* (New York: Houghton Mifflin, 1992); J. F. Gleason, P. K. Bhartia, J. R. Herman, et al., "Record Low Global Ozone in 1992," *Science* 260 (1993): 523–36; J. Oerlemans, "Quantifying Global Warming from the Retreat of Glaciers," *Science* (1994): 243–45; A. Tabazadeh and R. P. Turco, "Stratospheric Chlorine Injection by Volcanic Eruptions: HCl Scavenging and Implications for Ozone," *Science* 260 (1993): 1082–86; G. Taubes, "The Ozone Backlash," *Science* 260 (1993): 1580–83; M. D. Lemonick, "The Ozone Vanishes," *Time*, February 17, 1992: 60–68; J. B. Kerr and C. T. McElroy, "Evidence for Large Upward Trends of Ultraviolet-B Radiation Limited to Ozone Depletion," *Science* 262 (1993): 1032–34; and E. M. Pokras and A. C. Mix, "Earth's Precession Cycle and Quaternary Climactic Change in Tropical Africa," *Nature* 326 (1987): 486–87.

22. For a detailed discussion of global warming and its expected impact upon potential disease, see T. E. Lovejoy, "Global Change and Epidemiology: Nasty Synergies," Chapter 25 in S. S. Morse, ed., *Emerging Viruses* (Oxford, Eng.: Oxford University Press, 1993).

23. In addition to interviews, details for the section were drawn from the following sources: W. N. Bonner, *The Natural History of Seals* (New York: Facts on File, 1990); R. Dietz, C. T. Ansen, P. Have, and M. P. Heide-Jørgensen, "Clue to Seal Epizootic?" *Nature* 338 (1989): 627; M. Domingo, L. Ferrer, M. Pumarola, et al., "Morbillivirus in Dolphins," *Nature* 348 (1990): 21; M. A. Gracher, V. P. Kumarev, L. V. Mamaev, et al., "Distemper Virus in Baikal Seals," *Nature* 338 (1989): 209; C. B. Goodhart, "Did Virus Transfer from Harp Seals to Common Seals?" *Nature* 336 (1988): 21; J. Harwood, "Lessons from the Seal Epidemic," *New Scientist*, February 18, 1989: 38–42; S. Kennedy, J. A. Smyth, P. F. Cush, et al., "Viral Distemper Found in Porpoises," *Nature* 336 (1988): 21; S. Kennedy, J. A. Smyth, S. J. McCullough, et al., "Confirmation of Cause of Recent Seal Deaths," *Nature* 325 (1988): 404; B. W. J. Mahy, "Seal Plague Virus," Chapter 17 in Morse, ed. (1993), op. cit.; C. Örvell, M. Blixenkrone-Möller, V. Svansson, and P. Have, "Immunological Relationships Between Phocid and Canine Distemper Virus Studied with Monoclonal Antibodies," *Journal of General Virology* 71 (1990): 2085–92; A. D. M. E. Osterhaus, J. Gröen, P. DeVries, et al., "Canine Distemper Virus in Seals," *Nature* 335 (1988): 403–4; S. Pain, "Dolphin Virus Threatens Last Remaining Monk Seals," *New Scientist*, November 3, 1990: 22; I. K. G. Visser, *Morbillivirus Infections in Seals, Dolphins and Porpoises*, Doctoral Thesis, University of Utrecht, Seal Rehabilitation and Research Centre, Zeehondencreche Pieterburen, Netherlands, 1993; and J. Webb, "Dolphin Epidemic Spreads to Greece," *New Scientist*, September 7, 1991: 18.

24. In the Baltic salmon faced extinction in 1993–94 because of just such a mixture of events. The Baltic is heavily polluted, having been a major Soviet dumping ground for more than four decades. Salmon caught in the Baltic or nearby rivers, streams, and lakes are heavily contaminated with chlorinated hydrocarbons.

Since 1993 Baltic salmon mothers have been behaving strangely, swimming improperly when pregnant and laying eggs that hatch fish which can't swim. More than 90 percent of fish hatched since the fall of 1993 have died.

The Baltic salmon species, it turns out, are newly infected with a parasite, *Gyrodactylus*, that may have succeeded in emerging into the Baltic salmon population because of the fish's pollution-induced immunodeficiencies.

25. The estimated death tolls were as follows:

Species	Region	Year(s)	Number Died
Bottlenose dolphin	U.S., east coast	1987–88	2,500
Harbor seal	North Sea and Baltic	1988–89	18,000
Seal	Lake Baikal	1987–88	20,000
Striped & common dolphin	Mediterranean/Ionian	1990–91	6,000
Harbor seal	Long Island, NY	1992	100
Bottlenose dolphin	Gulf of Mexico	1990	300

(from various previously listed sources)

26. For a sampling of Colwell's findings, see R. R. Colwell, J. Kaper, and S. W. Joseph, *"Vibrio cholerae, Vibrio parahaemolyticus* and Other Vibrios: Occurrence and Distribution in Chesapeake Bay,"

Science 198 (1977): 394–96; R. R. Colwell, M. L. Tamplin, P. R. Brayton, et al., "Environmental Aspects of *V. cholerae* in Transmission of Cholera," in R. B. Sack and Y. Zinnaka, eds., *Advances in Research on Cholera and Related Diarrhoeas* (7th ed.; Tokyo: K.T.K. Scientific Publishers, 1990), 327–43; R. R. Colwell, J. A. K. Hasan, A. Hug, et al., "Development and Evaluation of a Rapid, Simple Sensitive Monoclonal Antibody-Based Coagglutination Test for Direct Detection of *V. cholerae* 01," *FEMS Microbiology Letters* 97 (1992): 215–20; and A. Hug, S. Parveen, F. Qadri, and R. R. Colwell, "Comparison of *V. cholerae* Serotype 01 Isolated from Patient and Aquatic Environment," *Journal of Tropical Medicine and Hygiene* 96 (1993): 86–92.

27. S. Pain, "Water Hides a Host of Viruses," *New Scientist*, August 19, 1989: 28.

28. Hundreds of examples could be cited. For just one illustrative case, see Centers for Disease Control, "Multistate Outbreak of Viral Gastroenteritis Related to Consumption of Oysters—Louisiana, Maryland, Mississippi, and North Carolina, 1993," *Morbidity and Mortality Weekly Report* 42 (1993): 945–47.

29. J. L. Melnick and T. G. Metcalf, "Distribution of Viruses in the Water Environment," in B. Fields, M. A. Martin, and D. Kamely, eds., *Genetically Altered Viruses and the Environment* (New York: Cold Spring Harbor Laboratory, 1985), 95–102.

30. World Bank, *World Development Report 1993: Investing in Health* (New York: Oxford University Press, 1993).

31. The horrendous condition of the earth's seas has been described in great detail elsewhere. See, for example, Proceedings, *Dahlem Conference on Ocean Margin Processes in Global Change*, Berlin, March 18–22, 1990 (New York: John Wiley & Sons, 1990); K. Schneider, "Ozone Depletion Harming Sea Life," *New York Times*, November 16, 1991: 19; K. Sherman and L. M. Alexander, *Biomass Yields and Geography of Large Marine Ecosystems* (Boulder, CO: Westview Press, 1989); Groups of Experts on the Scientific Aspects of Marine Pollution (GESAMP), *The State of the Marine Environment*, United Nations Environment Programme Regional Seas Reports and Studies, No. 115, Nairobi, 1990; and J. Pineda, "Predictable Upwelling and the Shoreward Transport of Planktonic Larvae by Internal Tidal Bores," *Science* 253 (1991): 548–49.

32. Center for Marine Conservation, *Global Marine Biological Diversity, a Strategy for Building Conservation into Decision Making*, report to the World Bank, Washington, D.C., 1993.

33. See, for example, C. W. Sullivan, K. R. Arrigo, C. R. McClain, et al., "Distributions of Phytoplankton Blooms in the Southern Ocean," *Science* 262 (1993): 1832–37.

34. A rough estimate of the planet's species distributions, as estimated by Marjorie Readka-Kudla, in a presentation to the annual meeting of the American Association for the Advancement of Science, San Francisco, February 1994, would be as follows:

Species Category	Number of Known Species	% of Total Number of Planetary Species
Terrestrial chordates	2,000	0.1
Insects	750,000	54.0
Fungi	47,000	3.0
Marine chordates	3,000	0.2
Freshwater chordates	1,000	0.08
Viruses and prokaryotes	6,000	4.00
Algae	27,000	2.0
Terrestrial plants	250,000	18.0

Distributed over the planet as follows:

Ecosystem	Km² (in millions)	% of Earth's Surface
Total global surface area	511	100
Global landmass	170.3	33.3
[Rain forests (1990)]	11.9	2.3
Oceans	340.1	66.7
[Coral reefs]	0.6	0.1
[Coastal zones]	40.9	8.0

35. R. I. Glass, M. Claeson, P. A. Blake, et al., "Cholera in Africa: Lessons on Transmission and Control for Latin America," *Lancet* 338 (1991): 791–95.

36. A. K. Siddique, A. H. Baqui, A. Eusof, et al., "Survival of Classic Cholera in Bangladesh," *Lancet* 337 (1991): 1125–27.

37. M. S. Islam, B. S. Drasar, and D. J. Bradley, "Long-Term Persistence of Toxigenic *Vibrio cholerae* 01 in the Mucilaginous Sheath of a Blue-Green Alga, *Anabaena variabilis*," *Journal of Tropical Medicine and Hygiene* 93 (1990): 133–39; and A. Hug, P. A. West, E. B. Small, et al., "Influence of Water Temperature, Salinity and pH on Survival and Growth of Toxigenic *Vibrio cholerae* Copepods in Laboratory Microcosms," *Applied Environmental Microbiology* 48 (1984): 420–24.

Further work demonstrated that the *V. cholerae* could thrive on or inside of a range of freshwater and saltwater algae. See M. J. Islam, "Increased Toxin Production by *Vibrio cholerae* 01 During Survival with a Green Alga, *Rhizoclonium fontanum*, in an Artificial Aquatic Environment," *Microbiology and Immunology* 34 (1990): 557–563; M. S. Islam, B. S. Drasar, and D. J. Bradley, "Attachment of Toxigenic *Vibrio cholerae* 01 to Various Freshwater Plants and Survival with a Filamentous Green Alga, *Rhizoclonium fontanum*," *Journal of Tropical Medicine and Hygiene* 92 (1989): 396–401; and M. S. Islam, B. S. Drasar, and D. J. Bradley, "Survival of Toxigenic *Vibrio cholerae* 01 with a Common Duckweed, *Lemma minor*, in Artificial Aquatic Ecosystems," *Transactions of the Royal Society of Tropical Medicine and Hygiene* 84 (1990): 422–24.

38. For an excellent summary, see R. R. Colwell and W. M. Spira, "The Ecology of *Vibrio cholerae*," Chapter 6 in D. Barua and W. B. Greenough III, eds., *Cholera* (New York: Plenum, 1992).

39. These events are described in the following: P. R. Epstein, T. E. Ford, and R. R. Colwell, "Marine Ecosystems," *Lancet* 342 (1993): 1216–19; A. P. M. Lockwood, "Aliens and Interlopers at Sea," *Lancet* 342 (1993): 942–43; C. Anderson, "Cholera Epidemic Traced to Risk Miscalculation," *Nature* 354 (1991): 255; Colwell and Spira (1992), op. cit.; World Health Organization, "Cholera Alert in Latin America," Press Release, WHO/8/12 February 1991; "Cholera in the Americas," *Bulletin of the Pan American Health Organization* 25 (1991): 267–77; Centers for Disease Control, "Cholera Outbreak—Peru, Ecuador, and Colombia," *Morbidity and Mortality Weekly Report* 40 (1991): 225–27; E. W. Rice and C. H. Johnson, "Cholera in Peru," *Lancet* 338 (1991): 455; V. M. Witt and F. M. Reiff, "Environmental Health Conditions and Cholera Vulnerability in Latin America and the Caribbean," *Journal of Public Health Policy* (Winter 1991): 450–63; M. L. Tamplin and C. C. Parodi, "Environmental Spread of *Vibrio cholerae* in Peru," *Lancet* 338 (1991): 1216–17; J. Sepulveda, H. Gómez-Dantes, and M. Bronfman, "Cholera in the Americas: An Overview," *Infection* 20 (1992): 243–48; and R. V. Tauxe and P. A. Blake, "Cholera Epidemic in Latin America," *Journal of the American Medical Association* 267 (1992): 1388–90.

40. El Tor made its way into Brazil through the Amazon and thence to Rio.

41. Centers for Disease Control, "Cholera—New York, 1991," *Morbidity and Mortality Weekly Report* 40 (1991): 516–18; and Centers for Disease Control, "Cholera Associated with an International Airline Flight," *Morbidity and Mortality Weekly Report* 41 (1992): 134–35.

In September 1991 health authorities in Alabama discovered *V. cholerae* 01 in local seafood. An investigation of the bilge and ballast waters of ships from South America harbored in the Gulf of Mexico revealed that cholera-infested algae were present. See S. A. McCarthy, R. M. McPhearson, A. M. Guarino, and J. L. Gaines, "Toxigenic *Vibrio cholerae* 01 and Cargo Ships Entering Gulf of Mexico," *Lancet* 339 (1992): 624–25.

Polymerase chain reaction analysis of the DNA in various Latin American and Gulf of Mexico isolates of *V. cholerae* didn't match up, however. The cholera found in the Gulf was never matched exactly to cholera strains anywhere else in the world, and its origin remained subject for debate in 1994. See I. K. Wachsmuth, G. M. Evins, P. I. Fields, et al., "The Molecular Epidemiology of Cholera in Latin America," *Journal of Infectious Diseases* 167 (1993): 621–26.

42. The eight drugs were ampicillin, chloramphenicol, colistin, neomycin, kanamycin, gentamicin, tetracycline, and Fansidar. See R. Tabtieng, S. Wattanasri, P. Echeverría, et al., "An Epidemic of *Vibrio cholerae El Tor Inaba* Resistant to Several Antibiotics with a Conjugative Group C Plasmid Coding for Type II Dihydrofolate Reductase in Thailand," *American Journal of Tropical Medicine and Hygiene* 41 (1989): 680–86.

43. A. A. Ries, D. J. Vugia, L. Beingolea, et al., "Cholera in Piura, Peru: A Modern Urban Epidemic," *Journal of Infectious Diseases* 166 (1992): 1429–33; and Editorial, "Of Cabbages and Chlorine: Cholera in Peru," *Lancet* 340 (1992): 20–21.

44. See numerous bulletin reports from officials and scientists working throughout the region, all appearing in *Lancet* 342 (1993): 382–83, 387–90, 430–31, 925–27.

45. P. R. Epstein, T. E. Ford, and R. R. Colwell, "Marine Ecosystems," *Lancet* 342 (1993): 1216–19; and P. R. Epstein, "Cholera and the Environment," *Lancet* 339 (1992): 1167–68.

46. R. Shope, "Global Climate Change and Infectious Diseases," *Environmental Health Perspectives* 96 (1991): 171–74.

47. J. de Zulueta, "Changes in the Geographical Distribution of Malaria Throughout History," *Parasitologia* 29 (1987): 193–205.

48. D. J. Rogers and M. J. Packer, "Vector-Borne Diseases, Models, and Global Change," *Lancet* 342 (1993): 1282–85.

49. The WHO Task Group graded the sensitivities of diseases (or their vectors and transmissibility) in relation to global temperature changes on a scale of 0 to 3, with 0-graded unlikely to be affected at all and 3-graded highly susceptible to temperature variations.

Disease	1990 Prevalence	Grade
Malaria	270 million	3
Lymphatic filariasis	90.2 million	1
Onchocerciasis	17.8 million	1
Schistosomiasis	200 milion	2
Sleeping sickness	25,000/year	1
Leishmaniasis	12 million	1
Dracunculiasis	1 million	0
Dengue	NS	2
Yellow fever	NS	1
Japanese encephalitis	NS	1
St. Louis encephalitis	NS	1
All other insect-borne viral diseases	NS	1

See WHO Task Group, "Potential Health Effects of Climatic Change," report to the World Health Organization, WHO/PEP/90/10, 1990.

50. A. Leaf, "Potential Health Efforts of Global Climatic and Environmental Changes," *New England Journal of Medicine* 321 (1989): 1577–83; A. Jeevan and M. L. Kripke, "Ozone Depletion and the Immune System," *Lancet* 342 (1993): 1159–60; World Health Organization, "Ultraviolet Radiation Can Seriously Damage Your Health," Press Release, WHO/102/17 December 1993; W. Goettsch, J. Garssen, A. Deijins, et al., "UV-B Exposure Impairs Resistance to Infection by *Trichinella spiralis*," *Environmental Health Perspectives* 102 (1994): 298–304; C. Hassett, M. G. Mustafa, W. F. Coulson, and R. M. Elashoff, "Murine Lung Carcinogenesis Following Exposure to Ambient Ozone Concentrations," *Journal of the National Cancer Institute* 75, 4 (1985): 1211–19; L. Calderón-Garcidueñas and G. Roy-Ocotla, "Nasal Cytology in Southwest Metropolitan Mexico City Inhabitants: A Pilot Intervention Study," *Environmental Health Perspectives* 101 (1993): 138–44; and National Research Council, *Biological Markers in Immunotoxicology* (Washington, D.C.: National Academy of Sciences, 1992).

51. M. R. Moser, T. R. Bender, H. S. Margolis, et al., "Aircraft Transmission of Influenza A," *American Journal of Epidemiology* 110 (1979): 1.

52. D. Maki, "Airline Cabin Air Quality," testimony before the Subcommittee on Technology, Environment and Aviation, House Committee on Science, Space and Technology, July 29, 1993; and A. R. Hinman (1993), "Statement," ibid.

53. This was also revealed during the congressional hearings, by Niren Nagda, of ICF Kaiser International, an independent toxicology firm. Office standards for the United States were for twenty cubic feet per minute of fresh air in a standard room. In contrast, sold-out flights had an air intake rate of only nine to fourteen feet per minute.

54. World Health Organization, *Our Planet, Our Health*, WHO Report to the United Nations Earth Summit, Rio de Janeiro, June 1992.

55. In 1990 another new disease—Venezuelan hemorrhagic fever—emerged. Like Machupo and Junín, it was caused by a virus (dubbed Guanarito virus) that was carried by rodents (*Sigmodan hispidus* rats). The rats came in contact with *Homo sapiens* when large numbers of settlers moved into a previously pristine rain forest area. Guanarito killed 30 percent of the people it infected. See R. Salas, N. D. Manzione, R. B. Tesh, et al., "Venezuelan Hemorrhagic Fever," *Lancet* 338 (1991): 1033–36.

Still another hemorrhagic fever virus killed a twenty-five-year-old office worker in São Paulo, Brazil, in 1992. The origin of the Sabía virus, the cause of Brazilian hemorrhagic fever, has yet to be determined. See T. L. M. Coimbra, E. S. Nassar, M. N. Burattini, et al., "New Arenavirus Isolated in Brazil," *Lancet* 343 (1994): 391–92.

56.

COMMERCIAL AIR TRAFFIC
Source: International Air Transportation Association, 1993.

	Year	Millions of Passengers
International	1950	2
	1960	42
	1970	74
	1980	163
	1990	280
Domestic, U.S.A.	1950	17
	1960	38
	1970	153
	1980	273
	1990	424

57. In addition to human beings, hundreds of millions of animals were shipped from continent to continent annually by 1990. House pets, research animals, Thoroughbred horses, breeding livestock, illegally smuggled endangered species, aquarium fish, and a host of other broad categories of animals were routinely shipped overseas aboard airplanes or ocean liners.

58. S. S. Morse, "Emerging Viruses: Defining the Rules for Viral Traffic," *Perspectives in Biology and Medicine* 34 (1991): 387–409; and S. S. Morse, "Origins of Emerging Viruses," Chapter 2 in Morse, ed. (1993), op. cit.

59. B. Fleckenstein and R. C. Desrosiers, "*Herpesvirus saimiri* and *Herpesvirus ateles*," Chapter 6 in B. Roizman, ed., *The Herpesviruses*, Vols. 1 and 2 (New York: Plenum, 1982); J. C. Albrecht and B. Fleckenstein, "Primary Structure of the *Herpesvirus saimiri* Genome," *Journal of Virology* 66 (1992): 5047–58; and B. Biesinger, "Stable Growth Transformation of Human T Lymphocytes by *Herpesvirus saimiri*," *Proceedings of the National Academy of Sciences* 89 (1992): 3116–19.

60. The virus, wrote researchers from the Institut für Klinische und Molekulare Virologie in Erlangen, Germany, "seems to be particularly prone to sequestering cellular genes. The genome appears to function as a spontaneous vector for cellular genes that may have been acquired by a mechanism involving reverse transcription, since most of the viral counterparts have no introns. The uptake of such genes may provide functions necessary for the progression of biological properties and secure a selection advantage in the natural host." See Albrecht and Fleckenstein (1992), op. cit.

61. H. Ludwig, "B Virus," in Roizman, ed. (1982), 1: 417–19; and J. E. Kaplan, "Herpesvirus Simiae (B Virus) Infection in Monkey Handlers," *Journal of Infectious Diseases* 157 (1988): 1090.

62. The Mason-Pfizer virus was found in nineteen French blood donors and in some healthy people in Guinea-Bissau. See V. A. Morozov, F. Saal, A. Gessain, et al., "Antibodies to Gag Gene Coded Polypeptides of Type D Retroviruses in Healthy People from Guinea-Bissau," *Intervirology* 32 (1991): 253–57.

A type D virus similar but not identical to Mason-Pfizer was found in the blood of a Texas AIDS patient. See L. A. Donehower, R. C. Bohannon, R. J. Ford, and R. A. Gibbs, "The Use of Primers from Highly Conserved Pol Regions to Identify Uncharacterized Retroviruses by the Polymerase Chain Reaction," *Journal of Virological Methods* 28 (1990): 33–46.

63. Centers for Disease Control, "Primate Zoonoses Surveillance," Report II, U.S. Department of Health, Education, and Welfare, issued September 1973.

64. Periodically the CDC would manage to spot faults in the monkey importation system before significant numbers of human beings were infected. For example, between June 1990 and May 1993 a West African export company made 249 monkey shipments to the United States: 7 percent of the animals arrived in the United States carrying tuberculosis, some of them suffering active cases of disease. See Centers for Disease Control, "Tuberculosis in Imported Nonhuman Primates—United States, June 1990–May 1993," *Morbidity and Mortality Weekly Report* 42 (1993): 572–75.

65. The data do not include bone marrow transplants. Data derived from F. H. Cate and S. S. Laudicina, "Transplantation White Paper," Annenberg Washington Program, Northwestern University, 1991.

66. C. R. Hitchcock, J. C. Kiser, R. L. Telander, et al., "Baboon Renal Grafts," *Journal of the American Medical Association* 189 (1964): 159; and T. E. Starzl, T. L. Marchioro, G. N. Peters, et al., "Renal Heterotransplantation from Baboon to Man: Experience with Six Cases," *Transplantation* 2 (1964): 752.

67. L. L. Bailey, S. L. Nehlsen-Cannarella, W. Concepción, et al., "Baboon-to-Human Cardiac

Xenotransplantation in a Neonate," *Journal of the American Medical Association* 254 (1985): 3321–29; A. L. Caplan, "Ethical Issues Raised by Research Involving Xenografts," *Journal of the American Medical Association* 254 (1985): 3339–43; taped interview with Bailey conducted by Dr. Norman Swann, Australian Broadcasting Company, 1985; and T. E. Starzl, J. Fung, A. Tzakis, et al., "Baboon-to-Human Liver Transplantation," *Lancet* 341 (1993): 65–71.

68. R. Nowak, "One Baboon Liver, Two Baboon Livers . . . ," *Journal of NIH Research* 5 (1993): 36–37; and R. Nowak, "Hope or Horror? Primate-to-Human Organ Transplants," *Journal of NIH Research* 4 (1992): 37–38.

69. J. D. Meyers, N. Flournoy, and E. D. Thomas, "Risk Factors for Cytomegalovirus Infection After Human Marrow Transplantation," *Journal of Infectious Diseases* 153 (1986): 478–88; and E. Dussaix and C. Wood, "Cytomegalovirus Infection in Pediatric Liver Recipients," *Transplantation* 48 (1989): 272–74.

70. O. Chazouillères, D. Mamish, M. Kim, et al., " 'Occult' Hepatitis B Virus as Source of Infection in Liver Transplant Recipients," *Lancet* 343 (1994): 142–46.

71. A. F. Sheids, R. C. Hackman, K. H. Fife, et al., "Adenovirus Infections in Patients Undergoing Bone-Marrow Transplantations," *New England Journal of Medicine* 312 (1985): 529–33; and B. Koneru, R. Jaffe, C. O. Esquivel, et al., "Adenoviral Infections in Pediatric Liver Transplant Recipients," *Journal of the American Medical Association* 258 (1987): 489–92.

72. S. Euvrard, C. P. Noble, J. Kanitakis, et al., "Brief Report: Successive Occurrence of T-Cell and B-Cell Lymphomas After Renal Transplantation in a Patient with Multiple Cutaneous Squamous-Cell Carcinomas," *New England Journal of Medicine* 327 (1992): 1924–26; and I. J. Spiro, D. W. Yandell, C. Li, et al., "Brief Report: Lymphoma of Donor Origin Occurring in the Porta Hepatitis of a Transplanted Liver," *New England Journal of Medicine* 329 (1993): 27–29.

73. E. K. Kuhn, "Deadly Fly Finds Home in Africa," *New African*, November 1989: 24.

74. W. C. Reeves, *Epidemiology and Control of Mosquito-Borne Arboviruses in California, 1943–1987* (Sacramento, CA: California Mosquito and Vector Control Association, 1990).

75. Centers for Disease Control, "Arboviral Infections of the Central Nervous System—United States, 1985," *Morbidity and Mortality Weekly Report* 35 (1986): 341–49; and Centers for Disease Control, "Arboviral Infections of the Central Nervous System—United States, 1989," *Morbidity and Mortality Weekly Report* 39 (1990): 407–16.

76. Centers for Disease Control, "Arboviral Surveillance—United States, 1990," *Morbidity and Mortality Weekly Report* 39 (1990): 593–98; Centers for Disease Control, "Update: St. Louis Encephalitis—Florida and Texas, 1990," *Morbidity and Mortality Weekly Report* 39 (1990): 756–59; and Centers for Disease Control, "Update: Arboviral Surveillance—Florida, 1990," *Morbidity and Mortality Weekly Report* 39 (1990): 650–51.

77. A. James, "Molecular Biology of the Mosquito Salivary Gland," presentation to the Annual Meeting of the American Society of Tropical Medicine and Hygiene, Seattle, November 16, 1992.

78. A. E. Hussein, R. F. Ramig, F. R. Holbrook, and B. J. Beaty, "Asynchronous Mixed Infection of *Culicoides variipennis* with Bluetongue Virus Serotypes 10 and 17," *Journal of General Virology* 70 (1989): 3355–62.

79. B. J. Beaty, D. W. Trent, and J. T. Roehrig, "Virus Variation and Evolution: Mechanisms and Epidemiological Significance," Chapter 3 in T. Monath, ed., *The Arboviruses: Epidemiology and Ecology* (Boca Raton, FL: CRC Press, 1988).

80. P. K. Anderson and F. J. Morales, "The Emergence of New Plant Diseases: The Case of Insect-Transmitted Plant Viruses," presentation to the Workshop on New Disease, Woods Hole, MA, November 7–10, 1993; and K. Schneider, "Study Finds Risk in Making Plant Viruses Resistant," *New York Times*, March 11, 1994: A16.

81. B. W. Falk and G. Bruening, "Will Transgenic Crops Generate New Viruses and New Diseases?" *Science* 263 (1994): 1395–96.

82. A. E. Greene and R. F. Allison, "Recombination Between Viral RNA and Transgenic Plant Transcripts," *Science* 263 (1994): 1423–25.

83. J. Davies, "Inactivation of Antibiotics and the Dissemination of Resistance Genes," *Science* 264 (1994): 375–82.

84. M. A. McClure, M. S. Johnson, D. F. Feng, and R. F. Doolittle, "Sequence Comparisons of Retroviral Proteins: Relative Rates of Change and General Phylogeny," *Proceedings of the National Academy of Sciences* 85 (1988): 2469–73.

85. H. M. Temin, "Is HIV Unique or Merely Different?" *Journal of Acquired Immune Deficiency Syndromes* 2 (1989): 1–9; and H. M. Temin, "Retrovirus Variation," Chapter 20 in Morse, ed. (1993), op. cit.

86. Temin had no doubt that successful recombination events had occurred in the viral world, and he cited as an example REV-T, a lymphatic cancer virus of birds. REV-T arose in 1946, the result of a recombination event between a retrovirus and a bird oncogene (inherent cancer-causing gene in the bird's DNA).

87. J. H. Strauss and E. G. Strauss, "Evolution of RNA Viruses," *Annual Review of Microbiology* 42 (1988): 657–83.

88. P. J. Cascone, T. F. Haydar, and A. E. Simon, "Sequences and Structures Required for Recombination Between Virus-Associated RNAs," *Science* 260 (1993): 801–5.

89. Hypervariable regions were also found in human DNA. For example, in 1992–94 scientists working separately in laboratories all over the world made the exciting discovery that human DNA contained stretches of long repetitive sequences of what seemed to be garbage. Three nucleotides, such as a CTG, would be repeated over and over, up to fifty or sixty times. For unknown reasons, some people's cells would suddenly expand those repeats, up to 200 or more CTGs, and diseases would occur. Fragile-X Syndrome (Down's Syndrome), Huntington's Disease, Myotonic Dystrophy, and Spinobulbar Muscular Atrophy were all clearly linked to such triple-repeat regions of DNA. There were indications that Alzheimer's Disease was also a triple-repeat disorder.

90. This is the work of Ron Montelaro at Louisiana State University, often in collaboration with Jim Mullins, at Stanford University.

91. M. C. Y. Heny, S. Y. Heng, and S. G. Allen, "Co-Infection and Synergy of Human Immunodeficiency Virus-1 and Herpes Simplex Virus-1," *Lancet* 343 (1994): 255–58.

92. D. Bartels, *New Scientist*, July 30, 1987: 53–54.

93. L. Feigenbaum and G. Khoury, "The Role of Enhancer Elements in Viral Host Range and Pathogenicity," in B. Fields, M. A. Martin, and D. Kamely, eds., *Genetically Altered Viruses and the Environment* (New York: Cold Spring Harbor Laboratory, 1985).

94. B. N. Fields, D. M. Knipe, R. M. Chanock, et al., *Virology* (New York: Raven Press, 1985).

95. For a sense of the microbial ecology issues Fields felt constituted the mysterious "orchestration," see M. L. Nibert, D. B. Furlong, and B. N. Fields, "Mechanisms of Viral Pathogenesis," *Journal of Clinical Investigation* 88 (1991): 727–34; A. Learmouth, *Disease Ecology* (New York: Basil Blackwell, 1988); and D. E. Pomeroy and M. W. Service, "The Ecology of Man," Chapter 8 in *Tropical Ecology* (Essex, Eng.: Longman Scientific and Technical, 1986).

96. J. D. Watson, N. H. Hopkins, J. W. Roberts, et al., *Molecular Biology of the Gene* (4th ed.; Menlo Park, CA: Benjamin/Cummings, 1987).

97. See, for example, J. W. Ajioka and D. L. Hartl, "Population Dynamics of Transposable Elements," Chapter 43 in D. E. Berg and M. M. Howe, *Mobile DNA* (Washington, D.C.: American Society for Microbiology, 1989).

98. S. E. Luria and M. Delbruck, "Mutations of Bacteria from Virus Sensitivity to Virus Resistance," *Genetics* 28 (1943): 491–511; and J. Lederberg and E. M. Lederberg, "Replica Plating and Indirect Selection of Bacterial Mutants," *Journal of Bacteriology* 63 (1952): 399–406.

99. J. Cairns, J. Overbaugh, and S. Miller, "The Origin of Mutants," *Nature* 335 (1988): 142–46.

100. A key counterargument came also from Harvard—the rival Medical School. Researchers showed that *E. coli* which carried an advantageous mutation could take over an apparently stagnant population of bacteria under stress. In their experiment, stressing a colony of bacteria caused the genetically advantaged to replace those that died without changing the overall size of the population. The researchers argued that Cairns and other supporters of the notion that bacteria could selectively mutate under stress were misinterpreting their results; randomness, they argued, was still at play, simply well disguised. See M. M. Zambrano, D. A. Siegele, M. Almirón, et al., "Microbial Competition: *Escherichia coli* Mutants That Take Over Stationary Phase Cultures," *Science* 259 (1993): 1757–58.

101. P. L. Foster, "*Escherichia coli* and *Salmonella typhirium*, Mutagenesis," *Encyclopedia of Microbiology* 3 (1992): 1–8.

102. R. I. Morimoto, "Cells in Stress: Transcriptional Activation of Heat Shock Genes," *Science* 259 (1993): 1409–10.

103. C. T. Walsh, "Vancomycin Resistance: Decoding the Molecular Logic," *Science* 261 (1993): 308–9.

104. S. P. Cohen, W. Yan, and S. B. Levy, "A Multidrug Resistance Regulation Chromosomal Locus Is Widespread Among Enteric Bacteria," *Journal of Infectious Diseases* 168 (1993): 484–88.

105. J. R. Johnson, I. Orskov, F. Orskov, et al., "O, K, and H Antigens Predict Virulence Factors, Carboxylesterase B Pattern, Antimicrobial Resistance, and Host Compromise Among *Escherichia coli* Strains Causing Urosepsis," *Journal of Infectious Diseases* 169 (1994): 119–26.

106. J. Travis, "Possible Evolutionary Role Explored for 'Jumping Genes,' " *Science* 257 (1992): 884–85.

107. A. A. Beaudry and G. F. Joyce, "Directed Evolution of an RNA Enzyme," *Science* 257 (1992): 635–41.

108. R. E. Lenski and J. E. Mittler, "The Directed Mutational Controversy and Neo-Darwinism," *Science* 259 (1993): 188–93.

109. D. A. Watson, "Unusual Mutational Mechanisms and Evolution," Letter, *Science* 260 (1993): 1958.

110. L. D. Hurst, "Unusual Mutational Mechanisms and Evolution," Letter, *Science* 260 (1993): 1959.

111. D. W. Hecht, T. J. Jagielo, and M. H. Malamy, "Conjugal Transfer of Antibiotic Resistance Factors in *Bacteroides fragilis*: The btgA and btgB Genes of Plasmid pBFTM10 Are Required for Its Transfer from *Bacteroides fragilis* and for Its Mobilization by IncP Beta Plasmid R751 in *Escherichia coli*," *Journal of Bacteriology* 173 (1991): 7471–80.

112. D. R. Schaberg, "Evolution of Antimicrobial Resistance and Nosocomial Infection: Lessons from the Vanderbilt Experiment," *American Journal of Medicine* 70 (1981): 445–49; and C. F. Amabile-Cuevas and M. E. Chicurel, "Bacterial Plasmids and Gene Flux," *Cell* 70 (1992): 189–99.

113. S. Schwarz and S. Grolz-Krug, "The Cloramphenicol-Streptomycin-Resistance Plasmid from a Clinical Strain of *Staphylococcus sciuri* and Its Structural Relationship to Other Staphylococcal Resistance Plasmids," *FEMS Microbiology Letters* 66 (1991): 319–22; and T. J. Coffey, C. G. Dowson, M. Daniels, et al., "Horizontal Transfer of Multiple Penicillin-Binding Protein Genes, and Capsular Biosynthetic Genes, in Natural Populations of *Streptococcus pneumoniae*," *Molecular Microbiology* 5 (1991): 2255–60.

114. P. Viljanen and J. Boratynski, "The Susceptibility of Conjugative Resistance Transfer in Gram-Negative Bacteria to Physiochemical and Biochemical Agents," *FEMS Microbiology Reviews* 8 (1991): 43–54.

115. D. S. Thaler, "The Evolution of Genetic Intelligence," *Science* 264 (1994): 224–25; and R. S. Harris, S. Longerich, and S. M. Rosenberg, "Recombination in Adaptive Mutation," *Science* 264 (1994): 258–60.

116. See, for example, R. Levins, T. Awerbach, U. Brinkmann, et al., "The Emergence of New Diseases," *American Scientist* 82 (1994): 52–60; A. Gibbons, "Where Are 'New Diseases' Born?" *Science* 261 (1993): 680–81; K. McAuliffe, "How New Are Today's New Diseases?" *U.S. News & World Report*, November 17, 1986: 75–76; and A. S. Moffat, "Theoretical Ecology: Winning Its Spurs in the Real World," *Science* 263 (1994): 1090–92.

117. N. M. Ampel, "Plagues—What's Past Is Present: Thoughts on the Origin and History of New Infectious Diseases," *Review of Infectious Diseases* 13 (1991): 658–65.

118. P. J. Kanki, K. U. Travers, S. MBoup, et al., "Slower Heterosexual Spread of HIV-2 Than HIV-1," *Lancet* 343 (1994): 943–46; and K. M. DeCock, G. Adjarlolo, E. Ekpini, et al., "Epidemiology and Transmission of HIV-2. Why There Is No HIV-2 Pandemic," *Journal of the American Medical Association* 270 (1993): 2083–86.

119. P. W. Ewald, *Evolution of Infectious Disease* (New York: Oxford University Press, 1993).

120. In Robert Gallo's lab at the National Cancer Institute researchers showed in 1990 that mixing different HIV-1 quasispecies and a mouse retrovirus resulted in an expansion of the range of cell types the viruses were able to infect, suggesting that the various viral strains swapped useful genes. See P. Lusso, M. di Veronese, B. Ensoli, et al., "Expanded HIV-1 Cellular Tropism by Phenotypic Mixing with Murine Endogenous Retroviruses," *Science* 247 (1990): 848–52.

121. R. M. Anderson, R. M. May, M. C. Boily, et al., "The Spread of HIV-1 in Africa: Sexual Contact Patterns and the Predicted Demographic Impact of AIDS," *Nature* 352 (1991): 581–89.

122. P. W. Ewald, "Transmission Modes and the Evolution of Virulence," *Human Nature* 2 (1990): 1–30; and P. W. Ewald, "The Evolution of Virulence," *Scientific American* (April 1993): 86–93.

123. R. B. Johnson, "Human Disease and the Evolution of Pathogen Virulence," *Journal of Theoretical Biology* 122 (1986): 19–24; G. C. Williams and R. M. Neese, "The Dawn of Darwinian Medicine," *The Quarterly Review of Biology* 66 (1991): 1–22; P. W. Ewald, "Pathogen-Induced Cycling of Outbreak Insect Populations," Chapter 11 in *Insect Outbreaks* (New York: Academic Press, 1987); and P. W. Ewald, "Waterborne Transmission and the Evolution of Virulence Among Gastrointestinal Bacteria," *Epidemiology of Infection* 106 (1991): 83–119.

124. E. A. Herre, "Population Structure and the Evolution of Virulence in Nematode Parasites of Fig Wasps," *Science* 259 (1993): 1442–45.

125. J. F. Miller, J. J. Mekalanos, and S. Falkow, "Coordinate Regulation and Sensory Transduction in the Control of Bacterial Virulence," *Science* 243 (1989): 916–22; C. Upton, K. Mossman, and G. McFadden, "Encoding of a Homolog of the IFN-gamma Receptor by Myxoma Virus," *Science* 258 (1992): 1369–72; and L. E. Bermudez, L. S. Young, J. Martinelli, and M. Petrofsky, "Exposure to Ethanol Up-Regulates the Expression of *Mycobacterium avium* Complex Proteins Associated with Bacterial Virulence," *Journal of Infectious Diseases* 168 (1993): 961–68.

126. See, for example, P. L. C. Small, L. Ramakrishnan, and S. Falkow, "Remodeling Schemes of Intracellular Pathogens," *Science* 263 (1994): 637–39; A. McMichael, "Natural Selection at Work on the Surface of Virus-Infected Cells," *Science* 260 (1993): 1771–72; S. Gupta, K. Trenholme, R. M. Anderson, and K. P. Day, "Antigenic Diversity of *Plasmodium falciparum*," *Science* 163 (1994): 961–63; W. W. Stead, "Genetics and Resistance to Tuberculosis," *Annals of Internal Medicine* 116 (1992): 937–41; and M. Barinaga, "Viruses Launch Their Own 'Star Wars,' " *Science* 258 (1992): 1730–31.

127. T. H. Weller, "Science, Society, and Changing Viral-Host Relationships," *Hospital Practice*, March 30, 1988: 113–20.

128. A. Learmouth, *Patterns of Disease and Hunger* (Vancouver, BC: David and Charles, 1978); and P. R. Epstein, "Commentary: Pestilence and Poverty—Historical Transitions and the Great Pandemics," *American Journal of Preventative Medicine* 8 (1992): 263–65.

129. It is this fact—*that vaccines can't prevent infection*—which poses the greatest challenge to scientists who are trying to develop an AIDS vaccine. Since the AIDS virus gets inside cells of the immune system and hides for years on end inside human DNA with close to 100 percent lethal results, *no* level of infection can be acceptable. To date, no one has conceived of a vaccine that will prevent infection with any known microbe.

130. P. E. M. Fine, "Herd Immunity: History, Theory, Practice," *Epidemiologic Reviews* 15 (1993): 265–302.

131. F. L. Black, "Why Did They Die?" *Science* 258 (1992): 1739–40; and F. L. Black, "An Explanation of High Death Rates Among New World Peoples When in Contact with Old World Diseases," *Perspectives in Biology and Medicine* 37 (1994): 292–307.

132. Michel Garenne and Peter Aaby calculated, on the basis of more than 1,000 measles cases, that the first child had a chance of 1.0 of dying of the disease; the second sibling/cousin's chance of dying jumped to 1.9; the third's was 2.3; the fourth had a 3.8 chance of dying. See P. Aaby, J. Bukh, I. M. Lisse, and A. J. Smits, "Measles Mortality, State of Nutrition, and Family Structure: A Community Study from Guinea-Bissau," *Journal of Infectious Diseases* 147 (1983): 693–701; and M. Garenne and P. Aaby, "Pattern of Exposure and Measles Mortality in Senegal," *Journal of Infectious Diseases* 161 (1990): 1088–94.

133. A. Mitchison, "Will We Survive?" *Scientific American*, September 1993: 136–44.

134. C. R. Whitney, "Western Europe's Dreams Turning to Nightmares," *New York Times*, August 8, 1993: A1; and M. Simons, "France Finds Room in Its Heart for the Homeless," *New York Times*, December 9, 1993: A4.

135. M. J. Toole and R. J. Waldman, "An Analysis of Mortality Trends Among Refugee Populations in Somalia, Sudan, and Thailand," *Bulletin of the World Health Organization* 66 (1988): 237–41; M. J. Toole, S. Galson, and W. Brady, "Are War and Public Health Compatible?" *Lancet* 341 (1993): 1193–96; R. Yip and T. W. Sharp, "Active Malnutrition and Childhood Mortality Related to Diarrhea: Lessons from the 1991 Kurdish Refugee Crisis," *Journal of the American Medical Association* 270 (1993): 587–90; Centers for Disease Control, "Status of Public Health—Bosnia and Herzogovinia, August–September 1993," *Morbidity and Mortality Weekly Report* 42 (1993): 973–82; Centers for Disease Control, "Mortality Among Newly Arrived Mozambican Refugees—Zimbabwe and Malawi, 1992," *Morbidity and Mortality Weekly Report* 42 (1993): 468–77; and C. J. Elias, B. H. Alexander, and T. Sokly, "Infectious Disease Control in a Long-Term Refugee Camp: The Role of Epidemiologic Surveillance and Investigation," *American Journal of Public Health* 80 (1990): 824–28.

136. R. W. Kates, "Ending Deaths from Famine," *New England Journal of Medicine* 328 (1993): 1055–57; G. G. Graham, "Starvation in the Modern World," *New England Journal of Medicine* 328 (1993): 1058–60; and G. Heppner, A. J. Magill, R. A. Gasser, et al., "The Threat of Infectious Diseases in Somalia," *New England Journal of Medicine* 328 (1993): 1061–68.

17. Searching for Solutions

1. The five-hour scenario was played out at the American Society of Tropical Medicine and Hygiene, Honolulu, December 11, 1989. See L. Garrett, "Medical War Game," *New York Newsday, Discovery* section, January 23, 1990: 1, 5, 8–9.

2. A few months after Bernard Mandrella's celebrated Lassa case, a Scottish missionary in Zonkwa, Nigeria, contracted the disease. Nobody told him that antiserum was available in nearby Jos. Instead, the Scotsman flew a commercial airline to England in search of a cure, causing panic throughout Europe, and dying on English soil. In February 1976 an American Peace Corps volunteer left Sierra Leone because she was sick. The forty-two-year-old woman traveled to England, spent four hours in London's crowded Heathrow Airport, and went on to Washington, D.C., where she stayed in a popular hotel. Throughout the journey, the Peace Corps volunteer unknowingly respired Lassa virus around 522 strangers. Her diagnosis was only reached afterward, when virus was recovered from her urine. See J. A. Bryan and R. M. Zweighaft, "Surveillance and Transport of Patients with Suspect Viral Hemorrhagic Fevers: The United States Experience," in *Ebola Virus Haemorrhagic Fever* (New York: Elsevier, 1978), 415–25.

By mutual agreement, the U.S. government took responsibility for all the Peace Corps volunteer's contacts during her transatlantic flight and time in Washington, while the British government carried the onus of investigating all her co-passengers aboard the British Airways flight from Sierra Leone and during her layover at Heathrow. At extraordinary expense, the CDC, District of Columbia authorities, and health agencies in twenty-one states tracked down 505 of the Lassa victim's co-passengers, fellow Washington hotel guests, and assorted other individuals with whom she had contact: none tested positive for Lassa infection.

Similarly, the U.K. government issued press releases to airline passengers, tracking down 41 British Airways passengers who had remained in the U.K. and 54 who had traveled on to other countries. Again, none tested Lassa-positive. See Centers for Disease Control, "Possible Lassa Fever—Washington, D.C.," *Morbidity and Mortality Weekly Report* 25 (1976): 64; Centers for Disease Control, "Follow-up on Lassa Fever—Washington, D.C.," *Morbidity and Mortality Weekly Report* 25 (1976): 68; and Centers for Disease Control, "Follow-up on Lassa Fever—Washington, D.C.," *Morbidity and Mortality Weekly Report* 25 (1976): 83.

The cost in work-hours and expenses for these two enormous manhunts has never been published.

3. A good deal has been written about the Reston outbreak. See P. B. Jahrling, T. W. Geisberg, D. W. Dalgard, et al., "Preliminary Report: Isolation of Ebola Virus from Monkeys Imported to USA," *Lancet* 335 (1990): 502–5; M. Sun, "Imported Monkey Puzzle," *Science* 247 (1990): 1538; L. Garrett, "Luck in a Virus Outbreak," *Newsday, Discovery* section, January 23, 1990: 9; J. Palca, "Import Rules Threaten Research on Primates," *Science* 248 (1990): 1071–73; R. Preston, "Crisis in the Hot Zone," *New Yorker*, October 1992: 58–81; and T. J. Moore, "A Virus Emerges," in *Lifespan* (New York: Simon & Schuster, 1993), chapter 6.

4. Similar outbreaks would subsequently be reported in a number of facilities around the world, including Sandoz laboratories in Italy, where the monkey tissue was used for development of polio vaccines. See World Health Organization, "Viral Haemorrhagic Fever in Imported Monkeys," *Weekly Epidemiology Report* 67 (1992): 142–43.

5. W. L. Roper, "Dear Importer," letter to all private importers, March 15, 1990, U.S. Department of Health and Human Services, CDC, Atlanta, GA.

6. Centers for Disease Control, "Update: Filovirus Infection in Animal Handlers," *Morbidity and Mortality Weekly Report* 39 (1990): 221; and Centers for Disease Control, "Update: Ebola-Related Filovirus Infection in Nonhuman Primates and Interim Guidelines for Handling Nonhuman Primates During Transit and Quarantine," *Morbidity and Mortality Weekly Report* 39 (1990): 22–24, 29–30.

The CDC later reported that one of its employees—an animal caretaker—also tested antibody-positive for Reston virus exposure. See Centers for Disease Control. "Update: Evidence of Filovirus Infection in an Animal Caretaker in a Research/Service Facility," *Morbidity and Mortality Weekly Report* 39 (1990): 296–97.

7. A. Sánchez, M. P. Kiley, B. P. Holloway, et al., "The Nucleoprotein Gene of Ebola Virus: Cloning, Sequencing, and *in vitro* Expression," *Virology* 170 (1989): 81–91.

8. D. Axelrod, *Department of Health News*, Albany, NY, March 21, 1990; and *Order for Summary Action, State of New York Department of Health, In the Matter of Filovirus Infections in Monkeys*, Albany, March 21, 1990.

9. Physicians Committee for Responsible Medicine, "Nation-wide Ban Sought for Public Health Reasons," Washington, D.C., March 22, 1990; and American Society for the Prevention of Cruelty to Animals, "ASPCA President Assails CDC's Failure to Impose Immediate Ban on Monkey Imports," Press Release, March 23, 1990.

10. L. Garrett, "Monkey-Import Plans Challenged," *Newsday*, March 24, 1990: 4.

11. Centers for Disease Control, "Update: Filovirus Infection Associated with Contact with Nonhuman Primates or Their Tissues," *Morbidity and Mortality Weekly Report* 39 (1990): 404–5.

12. S. P. Fisher-Hoch, G. F. Pérez-Ornonoz, E. L. Jackson, et al., "Filovirus Clearance in Non-Human Primates," *Lancet* 340 (1992): 451–54.

13. S. P. Fisher-Hoch, T. L. Brammer, S. G. Trappier, et al., "Pathogenic Potential of Filoviruses:

Role of Geographic Origin of Primate Host and Virus Strain," *Journal of Infectious Diseases* 166 (1992): 753–63.

14. Institute of Medicine, *The U.S. Capacity to Address Tropical Infectious Disease Problems* (Washington, D.C.: National Academy Press, 1987); Institute of Medicine, *The Future of Public Health* (Washington, D.C.: National Academy Press, 1988); Institute of Medicine, *Emerging Infections: Microbial Threats to Health in the United States* (Washington, D.C.: National Academy Press, 1992); National Institute of Allergy and Infectious Diseases, "Report of the Task Force on Microbiology and Infectious Diseases," U.S. Department of Health and Human Services, Washington, D.C., 1992; and Centers for Disease Control and Prevention, *Addressing Emerging Infectious Disease Threats: A Prevention Strategy for the United States* (Atlanta, GA: U.S. Department of Health and Human Services, 1994).

15. Division of Communicable Diseases, *Global Surveillance Programme for Recognition and Response to Emerging Diseases, 1994–1995* (Geneva: World Health Organization, 1994); and Conference on Global Monitoring and Response for Emerging Infectious Diseases, Co-sponsored by the Federation of American Scientists and World Health Organization, Geneva, September 11–12, 1993.

16. Classically microbial biology, as a discipline, referred to the study of microbes that grew in unusual settings and affected their physical surroundings. Organisms, for example, that thrived inside highly sulfurous hot springs contributed to the erosion of rocks. What was new was the application of ecological principles to the study of pathogenic microbes.

17. Program for Monitoring Emerging Diseases.

18. Morse has written extensively on both the need for, and likely outlines of, emerging disease surveillance. See, for example, S. S. Morse, "Global Microbial Traffic and the Interchange of Disease," *American Journal of Public Health* 82 (1992): 1326–27; S. S. Morse, ed., *Emerging Viruses* (Oxford, Eng.: Oxford University Press, 1993); and S. S. Morse, "Regulating Viral Traffic," *Issues in Science and Technology* (Fall 1990): 81–84.

19. C. Piller and K. R. Yamamoto, *Gene Wars: Military Control over the New Genetic Technologies* (New York: Beech Tree Books, 1988).

20. R. W. Titball and G. S. Pearson, "BWC Verification Measures: Technologies for the Identification of Biological Warfare Agents," *Politics and Life Sciences*, August 1993: 255–63; B. H. Rosenberg, "North vs. South: Politics and the Biological Weapons Convention," *Politics and Life Sciences*, February 1993: 69–77; and L. A. Cole, "The Worry: Germ Warfare. The Target: Us," *New York Times*, January 25, 1994: A19.

There was ample evidence of past interest in biological weapons, particularly in Lassa. "The U.S. Central Intelligence Agency (CIA) was very interested, apparently as a part of its general intelligence gathering mission," Frame writes. [See J. D. Frame, *"The Story of Lassa Fever Part IV: The Politics of Research,"* *New York State Journal of Medicine*, 93:35–40 (1992). "A CIA representative visited me twice, first to discuss what the outbreaks in Nigeria might mean in terms of infectious disease in West Africa, and later, whether my finding of LV infections in missionaries in Ivory Coast, Mali, and Burkina Faso indicated the spread of the disease up the Niger River." (The Niger River flows through Liberia, Guinea, Mali, Niger, and Nigeria.)

Over the years, Johnson, Casals, McCormick, and Frame were all questioned by CIA representatives regarding Lassa and the other hemorrhagic diseases. Other scientists may have been similarly grilled, though it has not been possible to systematically survey researchers on the matter. CDC director David Sencer told the CIA that no scientist in his agency could be questioned without requests first to his office. Furthermore, he insisted that all CDC personnel had the right to refuse such interrogations. No public records were ever kept of CIA inquiries to CDC personnel, nor is there clear indication of what inspired such interest in the intelligence community. Sencer was adamant in telling CIA investigators that Lassa and other hemorrhagic viruses would make poor biological weapons, as they did not spread through casual contact. Nevertheless, the U.S. military's Medical Research Development Command (USAMRDC) had an active hemorrhagic virus research effort throughout the 1970s and 1980s.

The Soviets were also keenly interested in Lassa and other hemorrhagic viruses. By the mid-1970s, Soviet scientists were openly competing with Frame and Johnson's efforts in Liberia. It is assumed they were debriefed by the KGB.

21. Formally titled "Convention on the Prohibition of the Development, Production, and Stockpiling of Bacteriological (Biological) and Toxin Weapons and Their Destruction," entered into force March 26, 1975.

22. D. A. Henderson, "Surveillance Systems and Intergovernmental Cooperation," Chapter 27 in Morse, ed. (1993), op. cit.

23. "Vigilance Against New Virus Disease Is Urged," *Hospital Practice*, April 1977: 35–36.

24. F. A. Murphy, "New Emerging and Reemerging Infectious Diseases," *Advances in Virus Research* 43 (1994): 1–52.

25. M. L. Zoler, "Emerging Viruses," *Medical World News*, June 26, 1989: 36–42; and Morse (1992), op. cit.

26. Institute of Medicine, *Emerging Infections* (1992), op. cit.

27. Centers for Disease Control, "Update: Surveillance of Outbreaks—United States, 1990," *Morbidity and Mortality Weekly Report* 40 (1991): 173–75.

28. R. L. Berkelman, R. T. Bryan, M. T. Osterholm, et al., "Infectious Disease Surveillance: A Crumbling Foundation," *Science* 254 (1994): 368–70.

29. Such private labs had proven scandalous in many states. In New York City, for example, private labs failed to inform hundreds of women that their Pap smears were precancerous, possibly contributing to many deaths. In several states private labs routinely falsely reported negative results on sexually transmitted disease tests—tests that they never even bothered to perform.

30. Division of Communicable Diseases, *Global Surveillance Programme* (1994), op. cit.

31. L. Garrett, "Crusade on New Disease Sought," *New York Newsday*, April 21, 1994: A21.

32. Global development policies were also held responsible in more direct fashion for disease emergence and spread. For analysis of IMF, World Bank, and other donor policies and their impact on disease, see D. E. Cooper Weil, A. P. Alicbusan, J. F. Wilson, et al., *The Impact of Development Policies on Health* (Geneva: World Health Organization, 1990).

33. UNICEF, *The State of the World's Children—1990* (New York: Oxford University Press, 1990).

34. S. R. Pattyn, *Ebola Virus Haemorrhagic Fever: Proceedings of an International Colloquium on Ebola Virus Infection and Other Haemorrhagic Fevers Held in Antwerp, Belgium, 6–8 December 1977* (Amsterdam: Elsevier/North Holland Biomedical Press, 1978).

35. Ibid.

36. R. C. Baron, J. B. McCormick, and O. A. Zubeir, "Ebola Virus Disease in Southern Sudan: Hospital Dissemination and Intrafamilial Spread," *Bulletin of the World Health Organization* 61 (1983): 997–1003.

37. P. Rollin, L. Wilson, J. Childs, et al., "Lassa Fever Epidemic in Plateau State, Nigeria— 1993," presentation to the annual meeting of the American Society for Tropical Medicine and Hygiene, Atlanta, October 1993; D. E. Carey, et al., "Lassa Fever—Epidemiological Aspects of the 1970 Epidemic, Jos, Nigeria," *Transactions of the Royal Society of Tropical Medicine and Hygiene* 66 (1972): 402–8; and S. P. Fisher-Hoch, O. Tomori, G. I. Pérez-Ornonoz, et al., "Transmission of Lethal Viruses Through Routine Parenteral Drug Administration," personal communication, 1993.

38. "How the Septicemia Trail Led to the IV Bottle Cap," *Hospital Practice*, August 1971: 35– 45, 151–54.

39. M. Burnet and D. O. White, "Hospital Infections and Iatrogenic Disease," in M. Burnet and D. O. White, eds., *Natural History of Infectious Disease* (Cambridge, Eng.: Cambridge University Press, 1972).

40. D. P. Levine and J. D. Sobel, *Infections in Intravenous Drug Abusers* (New York: Oxford University Press, 1991).

41. D. C. DesJarlais and S. R. Friedman, "Research," in J. Stryker and M. D. Smith, eds., *Needle Exchange*, A Kaiser Forum, Henry J. Kaiser Family Foundation, Menlo Park, CA, 1993; P. Lurie, A. L. Reingold, B. Bowser, et al., *The Public Health Impact of Needle Exchange Programs in the United States and Abroad*, Vol. 1, Prepared by the School of Public Health, University of California Berkeley, and the Institute for Health Policy Studies, University of California at San Francisco, for the Centers for Disease Control, 1993; E. J. C. van Ameijden, A. A. R. van den Hoek, and R. A. Coutinho, "Injecting Risk Behavior Among Drug Users in Amsterdam, 1986 to 1992, and Its Relationship to AIDS Prevention Programs," *American Journal of Public Health* 84 (1994): 275–81; and C. F. Turner, H. G. Miller, and L. E. Moses, *AIDS: Sexual Behavior and Intravenous Drug Use* (Washington, D.C.: National Research Council, 1989).

42. The Kaiser Family Foundation convened a special conference on needle exchange issues, bringing together representatives of law enforcement, civil liberties, public health, drug treatment, and political communities. The meeting took place in the Quadras Conference Center, Menlo Park, California, December 10–11, 1992.

43. M. A. Fischl, R. B. Uttamchandani, G. L. Daikos, et al., "An Outbreak of Tuberculosis Caused by Multiple-Drug-Resistant Tubercle Bacilli Among Patients with HIV Infection," *Annals of Internal Medicine* 117 (1992): 177–83.

44. J. W. Kislak, T. C. Kickhoff, and M. Finland, "Hospital-Acquired Infections and Antibiotic Usage in the Boston City Hospital—January 1964," *New England Journal of Medicine* 271 (1969): 834–35; D. R. Schlaberg, "Evolution of Antimicrobial Resistance and Nosocomial Infection," *American Journal of Medicine* 70 (1981): 445; and R. A. Weinstein and S. A. Kabins, "Antimicrobial Resistance," *American Journal of Medicine* 70 (1981): 449.

45. L. L. Livornese, S. Días, C. Samel, et al., "Hospital-Acquired Infection with Vancomycin-Resistant *Enterococcus falcium* Transmitted by Electronic Thermometers," *Annals of Internal Medicine* 117 (1992): 112–16; and R. W. Haley, D. H. Culver, W. M. Morgan, et al., "Increasing Recognition of Infectious Diseases in U.S. Hospitals Through Increased Use of Diagnostic Tests, 1970–76," *American Journal of Epidemiology* 121 (1985): 168–81.

46. "How the Septicemia Trail Led to the IV Bottle Cap" (1971), op. cit.

47. I. Raad, W. Costerton, I. V. Sabharwa, et al., "Ultrastructural Analysis of Indwelling Vascular Catheters: A Quantitative Relationship Between Luminal Colonization and Duration of Placement," *Journal of Infectious Diseases* 168 (1993): 400–7; R. Pallares, M. Pujol, C. Pena, et al., "Cephalosporins as Risk Factors for Nosocomial *Enterococcus falcalis* Bacteremia," *Archives of Internal Medicine* 153 (1993): 1581–86; and W. E. Stam, "Nosocomial Infections: Etiologic Changes, Therapeutic Challenges," *Hospital Practice*, August 1981: 75–88.

48. National Nosocomial Infections Surveillance System, "Nosocomial Infections Rates for Interhospital Comparison: Limitations and Possible Solutions," *Infection Control and Hospital Epidemiology* 112 (1991): 609–21; M. Olson, M. O'Connor, M. D. Schwartz, "Surgical Wound Infections: A 5-Year Prospective Study of 20,193 Wounds at the Minneapolis VA Medical Center," *Annals of Surgery* 199 (1984): 253–59; P. J. E. Cruise, "Wound Infection Surveillance," *Review of Infectious Diseases* 3 (1981): 734–37; and R. W. Haley, D. H. Culver, W. M. Morgan, et al., "Identifying Patients at High Risk of Surgical Wound Infection," *American Journal of Epidemiology* 121 (1985): 206–15.

49. T. M. Stine, A. A. Harris, S. Levin, et al., "A Pseudo-Epidemic Due to Atypical Mycobacteria in a Hospital Water Supply," *Journal of the American Medical Association* 258 (1987): 809–11; and G. A. Harkness, D. W. Bentley, K. J. Roghmann, "Risk Factors for Nosocomial Pneumonia in the Elderly," *American Journal of Medicine* 89 (1990): 457–63.

50. H. C. Neu, "The Crisis in Antibiotic Resistance," *Science* 257 (1992): 1064–73; and B. Jarvis, personal communication, Centers for Disease Control, 1993.

51. Centers for Disease Control, "Summary of the Agency for Toxic Substances and Disease Registry Report to Congress," *Morbidity and Mortality Weekly Report* 39 (1990): 822–24; D. J. Hu, M. A. Kane, and D. L. Heymann, "Transmission of HIV, Hepatitis B Virus and Other Bloodborne Pathogens in Health Care Settings: A Review of Risk Factors and Guidelines for Protection," *Bulletin of the World Health Organization* 69 (1991): 623–30; D. L. Thomas, S. H. Factor, G. D. Kelen, et al., "Viral Hepatitis in Health Care Personnel at the Johns Hopkins Hospital," *Archives of Internal Medicine* 153 (1993): 1705–12; G. P. Kent, J. Brondum, R. A. Keenlyside, et al., "A Large Outbreak of Acupuncture-Associated Hepatitis B," *American Journal of Epidemiology* 127 (1988): 591–98; and F. E. Shaw, C. L. Barrett, R. Hamm, et al., "Lethal Outbreak of Hepatitis B in a Dental Practice," *Journal of the American Medical Association* 255 (1986): 3261–64.

52. Centers for Disease Control, "Nosocomial Enterococci Resistant to Vancomycin—United States, 1988–1993," *Morbidity and Mortality Weekly Report* 42 (1993): 597; and T. R. Frieden, S. S. Munsiff, D. E. Low, et al., "Emergence of Vancomycin-Resistant Enterococci in New York City," *Lancet* 342 (1993): 76–79.

53. J. M. Leclair, J. Freeman, B. F. Sullivan, et al., "Prevention of Nosocomial Respiratory Syncytial Virus Infections Through Compliance with Glove and Gown Isolation Precautions," *New England Journal of Medicine* 317 (1987): 329–34.

54. M. L. Cohen, "Epidemiology of Drug Resistance: Implications for a Post-Antimicrobial Era," *Science* 257 (1992): 1050–55; L. O. Gentry, "Bacterial Resistance," *Orthopedic Clinics of North America* 22 (1991): 379–88; and Neu (1992), op. cit.

55. Neu (1992), op. cit.

56. Ibid.

57. E. E. Mast, H. W. Harmon, S. Gravenstein, et al., "Emergence and Possible Transmission of Amantadine-Resistant Viruses During Nursing Home Outbreaks of Influenza A (H_3N_2)," *American Journal of Epidemiology* 134 (1991): 988–97.

58. A. B. Block, P. T. Davidson, et al., "Tuberculosis in Patients with the Human Immunodeficiency Virus Infection," *New England Journal of Medicine* 324 (1991): 1644–50; M. L. Pearson, J. A. Jereb, T. R. Frieden, et al., "Nosocomial Transmission of Multidrug-Resistant *Mycobacterium tuberculosis*," *Annals of Internal Medicine* 117 (1992): 191–96; B. R. Edlin, J. I. Tokars, M. H. Grieco, et al., "An Outbreak of Multidrug-Resistant Tuberculosis Among Hospitalized Patients with the Acquired Immunodeficiency Syndrome," *New England Journal of Medicine* 326 (1992): 1514–21; and S. Dooley, B. Edlin, M. Pearson, et al., "Multidrug-Resistant Nosocomial Tuberculosis Outbreaks in HIV-Infected Persons," presentation to the VIII International Conference on AIDS, Amsterdam, 1992.

59. G. L. Lattimer and R. A. Ormsbee, *Legionnaires' Disease* (New York: Marcel Dekker, 1981);

A. F. Kaufman, J. E. McDade, C. M. Patton, et al., "Pontiac Fever: Isolation of the Etiologic Agent (*Legionella pneumophila*) and Demonstration of Its Mode of Transmission," *American Journal of Epidemiology* 114 (1981): 337–47; H. M. Foy, P. S. Hayes, M. K. Cooney, et al., "Legionnaires' Disease in a Prepaid Medical-Care Group in Seattle 1963–75," *Lancet*, April 7, 1979: 767–70; L. Saravolatz, L. Arking, B. Wentworth, and E. Quinn, "Prevalence of Antibody to the Legionnaires' Disease Bacterium in Hospital Employees," *Annals of Internal Medicine* 90 (1979): 601–3; C. M. Helms, M. Massanari, R. P. Wenzel, et al., "Legionnaires' Disease Associated with a Hospital Water System," *Journal of the American Medical Association* 259 (1988): 2423–26; K. Nahapetian, O. Challemel, D. Bevrtin, et al., "The Intracellular Multiplication of *Legionella pneumophila* in Protozoa from Hospital Plumbing Systems," *Research in Microbiology* 142 (1991): 677–85; and M. Alary and J. R. Joly, "Factors Contributing to the Contamination of Hospital Water Distribution Systems by Legionellae," *Journal of Infectious Diseases* 165 (1992): 565–69.

60. A. Gibbons, "Where Are Diseases Born?" *Science* 261 (1993): 680–81.

61. N. Chiles, "In Rats' Realm," *New York Newsday*, May 9, 1994: A8.

62. On-line as of mid-1993 were Cameroon, Congo, Ghana (Accra and Navrongo), Kenya, Mozambique, Tanzania, Uganda, Zambia, Zimbabwe, Brazil, Cuba, Canada, Australia, and the United States. By January 1995, Mali, Botswana, Sudan, Malawi, Ethiopia, and Gambia were scheduled to be on-line.

SatelLife is based in Cambridge, Massachusetts.

63. For a synopsis of what might be worth tracking in the environment, see N. Nichols, "Teleconnections and Death," Chapter 16 in M. H. Glantz, R. W. Katz, and N. Nichols, eds., *Teleconnections Linking Worldwide Climate Anomalies* (Cambridge, Eng.: Cambridge University Press, 1991).

64. World Health Organization, "World Immunization Report," released September 28, 1990, Geneva.

65. See E. W. Kitch, "The Vaccine Dilemma," *Issues in Science and Technology* (Winter 1986): 108–22; B. M. Nkowane, S. G. F. Wassilak, W. A. Orenstein, et al., "Vaccine-Associated Paralytic Poliomyelitis," *Journal of the American Medical Association* 257 (1987): 1335–40; J. K. Inglehart, "Compensating Children with Vaccine-Related Injuries," *New England Journal of Medicine* (1987): 1283–88; and R. W. Ellis and R. G. Douglas, "New Vaccine Technologies," *Journal of the American Medical Association* 271 (1994): 929–31.

66. D. A. Henderson, "New Challenges for Tropical Medicine Syndrome," Charles Franklin Lecture at the annual meeting of the American Society of Tropical Medicine and Hygiene, Seattle, November 19, 1992.

67. J. Lederberg, speech before the Irvington Institute for Medical Research, Bankers Trust Company, New York, February 8, 1994.

Acknowledgments

Time spent on preparation of this book constituted time taken from *Newsday/New York Newsday*, and I am extremely grateful for the extraordinary tolerance extended to me by my employers. But of course, I have more than mere tolerance to applaud on *Newsday*'s part: several editors and reporters have been extremely encouraging and directly supportive throughout this effort. I would like to thank Tony Marro, Howard Schneider, and Don Forst for tolerating my absences, and assistant managing editor Les Payne, science editor Liz Bass, deputy science editors Mike Muskal and Reg Gale, and reporter Catherine Woodard for their words of encouragement and insight.

Few American news organizations have demonstrated continued interest in AIDS, developing country issues, public health, or international medical policy issues. *Newsday* has, much to my delight, been a clear exception. And I thank the organization for that as well.

Some of my earliest work in this subject area was carried out while I was a science correspondent for National Public Radio, and I thank science editor Anne Gudenkauf and NPR for their support during that period (1980–88).

Crucial to this project were the Harvard School of Public Health, the Alfred P. Sloan Foundation, and the Kaiser Family Foundation. From September 1992 to June 1994 I had the distinct honor of being a Visiting Fellow under the Harvard Journalism Fellowship for Advanced Studies in Public Health. I received thoughtful assistance from Jay Winsten and his staff in the Center for Health Communication. My stay on the Harvard campus from September 1992 to June 1993 was subsidized by the Alfred P. Sloan Foundation and *Newsday*, and for their generosity I am deeply grateful. At Harvard I received particular encouragement from Bob Meyers, who now heads the Washington Journalism Center.

Because of an occupational injury I am no longer able to use a keyboard. This document had to be written in longhand, and subsequently transcribed by others. I thank Dean Harvey Fineberg of the Harvard School of Public Health for his provision of resources that allowed me to hire graduate student Sue McLaughlin as a transcriber during my tenure at Harvard. And, of course, I thank Sue for her typing, encouragement, and critiques.

The Kaiser Family Foundation generously underwrote some of the transcription costs incurred after my Harvard tenure, allowing me the deeply rewarding opportunity of working with Amy Wollin Benjamin. Having admired her editorial achievements with production of the mammoth *AIDS in the World*, I dared to hope that Benjamin would deign to assist this project. She proved a masterful editor, critic, source of moral rearmament and, of course, transcriber. Quite honestly, this book could not have been produced without her.

I would also like to thank the physical therapists and physicians at the Miller Institute in Manhattan who kept my body fairly functional through this often taxing process, as well as massage therapists Joan Jacob Howe and Jeannette Kossuth.

I have always felt that librarians practiced one of humanity's noblest professions, and researching this book only bolstered that view. I particularly would like to thank the librarians of Harvard's Countway Medical Library and *Newsday/New York Newsday*.

Many people generously assisted along the way, providing reading critiques, research guidance, and crucial insights. I particularly would like to thank Andrea Eagan, Maryse Simonet, Deborah Cotton, Bunmi Makinwa, Jill Hannum, B. D. Colen, Bob Meyers, Jonathan Mann, Andrew Moss, Frank Browning, Bernard Fields, Mark Benjamin, Maya Szalavitz, Stephen Morse, Michael Reich, Barbara Rosencrantz, Penny Duckham, the staffs of the Offices of Public Affairs of the Centers for Disease Control and Prevention, the National Institutes of Health, the World Health Organization, and the

Institut Pasteur, Buki Ponle, Wendy Wertheimer, Günther Haaf, Michael Callen, Uwe Brinkmann, and the members of the Harvard New Diseases Group (Tamara Awerbach, Agnes and Uwe Brinkmann, Richard Cash, Irina Eckardt, Paul Epstein, Timothy Ford, Richard Levins, Najwa Makhoul, Christina de Albuquerque Possas, Charles Puccia, Manuel Sierra, Andrew Spielman, Mary Wilson, and Paul Wise). In the interests of journalistic integrity, no one reviewed sections of the book in which they, their work, or the works of their competitors were discussed.

Special thanks are owed to Jill Hannum, who, in the final crunch, devoted hours to finding ways to reduce the lengthy manuscript.

Anyone who has ever written a book knows what a toll its writing takes on friends and family members. Thanks to all of you, from San Clemente and San Salvador to New York and Boston, for putting up with me throughout this process, especially Bink, Bonnie, Banning, Evelyn, Karen, Bob, Caryl, Jim, Manoli, Lars, Ellen, Angela, Adi, Michael, Lisa, Steve, Larry, Spencer, Frank, and David.

I thank my agent, Charlotte Sheedy, for her tenacity. And Farrar, Straus and Giroux editor John Glusman for working at an unusually frenetic pace in order to hasten this book's publication, and for offering wonderful editorial insight and much appreciated encouragement.

Finally, I deeply regret the untimely deaths of Andrea Eagan, Uwe Brinkmann, and Michael Callen, whose comments during later stages of book production would undoubtedly have immensely improved the manuscript. I only hope I have done justice to their insights.

Index

Aaby, Peter, 590
Aaron Diamond AIDS Research Laboratory, 380
Abrams, Donald, 316
Abscam scandal, 178
acquired immune deficiency syndrome, *see* AIDS
Act Up, 470
acute respiratory distress syndrome (ARDS), 530, 531, 535, 541–45, 548–49
acyclovir, 433–35
adenoviruses, 174, 575
Adolfo Lutz Institute, 63
Aedes: A. aegypti, 66–70, 255–58, 567, 575; *A. albopictus*, 256, 257, 567; *A. pseudoscutellaris*, 204
Afghanistan, 608; heroin in, 492
African-Americans, 262, 507–8, 511, 515–17, 540
African National Congress (ANC), 208
African swine fever virus (ASFV), 323, 383
Agee, Betty, 400, 403
Agency for International Development, U.S., 482, 487
AIDS, 5, 6, 10, 283–389, 390, 409, 413, 459–503, 505–8, 525, 537, 548, 579–81, 587, 601, 604, 606, 611, 617; in Africa, 334–61, 442, 444, 448, 457–59, 482–87, 593; in Asia, 493–98; blood supply and, 308–9, 311–15; drug resistance in, 434–36; in Eastern Europe, 499–503; economic impact of, 477–88, 497–98; first cases of, 283–93; in India, 488–93; origins of epidemic, 361–85; politics of, 298–304, 310–11, 318–19, 328–30; search for method of transmission of, 304–10, search for method of transmission of, 304–10, 319–21, 323–28; simian, *see* simian immunodeficiency virus; social conditions in emergence of, 385–89; social responses to, 472–77; symptoms and mechanisms of, 293–97; tuberculosis and, 515–16, 521; water-borne microbes and, 430; WHO program on, *see* Global Programme on AIDS
air conditioning, 568, 614

Air Force, U.S., 94, 265
air travel, 569–71, 589
Akoi, Juanita, 89
Akosombo Dam (Ghana), 204
Alabama, University of, 375
Albania, 512, 617
Alberta, University of, 586
Albert Einstein School of Medicine, 325, 512, 525
Albuquerque Journal, 533
Alcohol, Drug Abuse, and Mental Health Administration (ADAMHA), 315
alcohol abuse, 276, 327, 328, 503, 505, 506; tuberculosis and, 519, 521–22
Alexander, E. Russell, 170
algal blooms, 561–67
Algeria, 367
Allan, Jon, 575
Alouatta monkeys, 67
Amane Ehumba, 105
amantadine, 433
Amazon rain forest, 551, 552, 556; Indians of, 589
amebiasis, 271, 288
American Academy of Pediatrics, 511
American Association for the Advancement of Science, 443
American College of Obstetrics and Gynecology, 398, 401
American Legion, *see* Legionnaires' Disease
American Psychiatric Association, 274
American Public Health Association, 303
American Red Cross, 315
American Society of Law and Medicine, 476
American Society of Tropical Medicine and Hygiene, 547, 593–96
Amerindians, 41, 213, 241, 246, 386, 589–90; antibiotic resistance among, 421; hantaviruses and, 546; *see also* Navajos
Amherst College, 587
amikacin, 519, 527
Amin, Idi, 206, 209–11, 335, 342, 368–69
aminoglycoside-type antibiotics, 417, 614
Ammann, Arthur, 309

Amnesty International, 495
amodiaquine, 441
amphetamines, 327, 328, 389, 612
amphotericin B, 134–35
ampicillin, 62, 266, 267, 408, 417, 418, 421,
    425, 427, 565, 584
Anderson, Carl, 302
Anderson, Roy, 484, 486
Andrews, Christopher Howard, 158
anemia, 251, 503; malarial, 444, 445, 448;
    sickle-cell, 479
Angola, 206, 208, 368, 608; AIDS in, 367,
    389, 467; malaria in, 441; measles in, 607
Anopheles mosquitoes, 39, 40, 48, 50, 51, 357,
    441, 445, 446, 450, 567–68
ANT70, 377, 379
anthrax, 534, 597
antibiotics, 36–37, 62, 184–85, 211, 214,
    237, 243, 244, 385, 504, 512, 514; inject-
    ing drug users and, 278; resistance to, 63,
    66, 190, 225, 226, 411–32, 437–40, 450,
    519–27, 565, 567, 578, 582, 584–85, 591,
    606, 612, 613, 615; sexually transmitted dis-
    eases and, 264–68, 272; see also specific
    drugs
antibodies, 35–36
Anti Dobola, 102
Anvers, University of, Institute of Tropical
    Medicine, 109–13, 116, 195, 538
Apodemus agrarius mice, 538, 539
arachidonic acid, 584
Aranow, Henry, 395
arboviruses, 605, 616
archeoepidemiology, 373
Argentina: Chagas's disease in, 253; hemor-
    rhagic fever in, 17, 23, 27, 28; yellow fever
    in, 68
Arita, Isao, 7, 45, 46, 604
Armenia, 608, 617
Armour Pharmaceutical Company, 311
Army, U.S., 4, 22, 154, 165, 166, 169–70,
    188, 255, 348, 381, 466, 533, 538–40,
    596, 598, 600; Medical Corps, 47; Medical
    Research and Development Command, 595;
    Medical Research Institute of Infectious Dis-
    eases (USMRIID), 5, 538, 544–45, 547–48
artemether, 453
Artemisia annua, 453
arthritis, 32, 416, 553
Arusha Declaration, 206
ARV (AIDS-Related Virus), 333, 334
asbestos, 393
ascariasis, 253
Ascaris roundworms, 253
Asian Development Bank, 498
Assad, Fakhry, 354, 357–62
Associated Press, 461
Aswan High Dam, 203–5, 251
Ateles monkeys, 67

Atkinson, Bill, 511
ATLV, 229
Auerbach, David, 305–7
Augustin, Father, 100, 106
Aung San Suu Kyi, 495
Australia, 153; AIDS in, 490, 587; antibiotic
    resistance in, 412, 413, 420; influenza in,
    159; malaria in, 450; skin cancer in, 556;
    Toxic Shock Syndrome in, 406
Austria, Black Death in, 238
autoimmune diseases, 405
Avery, Oswald, 34
avian leukosis virus, 227
Axelrod, David, 600
Axelrod, Jeffrey, 182
Azerbaijan, 608, 609
azidothymine (AZT), 436, 479, 482
Aztecs, 41, 245

baboon-to-human transplants, 574–75
Bacchetti, Peter, 300, 302
Bacon, Esther, 86–89
bacteriology, 37, 38
bacteriophages, 224, 227
Bacteroides fragilis, 586
Bahe, Merrill, 528–32, 535
Baltimore, David, 226, 227
Bancha Jarujarett, 496
Bangladesh, 201, 508, 609; AIDS in, 467;
    cholera in, 463, 563, 566; expenditure on
    drugs in, 438; malaria in, 456; smallpox
    eradication in, 42–45, 142
Bantu, 108, 131
Banu, Rahima, 45
barbiturates, 327, 328
Baron, Roy, 215–19
Barr, Y. M., 233
Barré-Sinoussi, Françoise, 321, 325
Bartley, Col. Joseph, 154, 169
Bates, Barbara, 244
Bauer, Gary, 302
Baxter Travenol Laboratories, Inc., 311
B cells, 317, 322, 404, 405
Béata, Sister, 101, 104–6
Beaty, Barry, 577
Beaudry, Amber, 585
Behringwerke AG, 53
Belgium, 112, 290; AIDS in, 291, 320, 344,
    345, 347, 348, 465; hantaviruses in, 538,
    546
Bemba people, 348
Bengal cholera, 565–66
Bennett, William, 470
benzodiazepines, 327
Berg, Paul, 224
Berkelman, Ruth, 532–33, 605, 607
Berlin Wall, fall of, 498–99, 502, 591, 609
Bernard Nocht Institute for Naval and Tropical
    Diseases, 109

Berreth, Don, 184, 185
beta-lactamase, 409, 411–12
Beveridge, W. I. B., 161
Beye, Henry, 17
Bhamarapravati, Natth, 604
Bharat Serums and Vaccine, Ltd., 490
Bhave, Geeta, 492
Biafra, 43, 75
Biggar, Robert, 352, 354–56, 360
bilharzia, 205
biological warfare, 194, 381, 595, 597–98, 602, 603
Bioweapons Convention, 603
Birmingham, University of, 214
birth control pill, 262
birth injury, 479
Bishop, Michael, 227, 228, 230
Black, Francis, 589–90
Black Death, 191, 237–38, 240, 247
blackwater fever, 447–49
Blattner, Bill, 317
bleomycin, 432
blood products and transfusions, 613; AIDS and, 309, 311–15, 332, 333, 343, 349, 360, 362–63, 367, 386, 389, 444, 448, 460, 490, 491; Chagas's disease from, 253
Bloom, Barry, 512, 525
bluetongue virus, 577
B lymphocytes, 293
Boffey, Philip, 169
Bokassa, Emperor, 208
Bolivia, 114, 122, 127, 621–22; Chagas's disease in, 253; cocaine production in, 280; deforestation in, 552; hemorrhagic fever in, see Machupo
Borrelia burgdorferi, 553
Bosnia, 591, 608
Botswana, 208; malaria in, 568
bovine leukemia virus (BLV), 381, 382
Bowen, Otis, 459
Bradford, Marjorie, 398
Brain, Joseph, 569
Brandt, Edward Jr., 298, 302, 314, 315, 349, 350
Brazil, 200; AIDS in, 357, 378, 486; antibiotic resistance in, 419; Chagas's disease in, 253; leishmaniasis in, 254; malaria in, 48, 52, 441; meningitis in, 59–60, 62–66, 617; Ministry of Health, 64–65; schistosomiasis in, 253; yellow fever in, 70
breakbone fever, see dengue fever
breast cancer, 226
Breiman, Rob, 531–32, 534, 537, 540–42, 545
Breman, Joel, 114–15, 119–21, 124–26, 128–30, 135, 138, 139, 146–47, 178, 217, 444, 445
Brès, Paul, 108–10, 126, 142, 144
brevetoxin, 560

Brezhnev, Leonid, 194
Brinkmann, Agnes, 96, 98, 252
Brinkmann, Uwe, 7, 71–73, 94–99, 198, 252, 259, 443, 453, 455, 567–68
bronchitis, 569
Brudney, Karen, 516–20
Brugière, Frédéric, 321, 325
Brun-Vézinet, François, 321, 376
Bryant, Anita, 261–62
bubonic plague, 32, 237, 238, 351, 503, 529, 530, 541
Buckley, Sonja, 80, 83
Buenos Aires, University of, 27
Bumba, Captain General, 129
Burkina Faso, 43, 114, 375; AIDS in, 486
Burkitt, Denis, 233
Burkitt's lymphoma, 352
Burma, 495–97, 617; AIDS in, 498; dengue hemorrhagic fever, 257, 258; heroin in, 491–92; malaria in, 450
Burnet, MacFarlane, 213–14
Burroughs-Wellcome, 433, 438
Burundi, 210, 344, 592; AIDS in, 340, 355, 367, 368, 381, 387, 485, 486, 516; antibiotic resistance in, 422; malaria in, 442, 447; Rwandan refugees in, 593
Bush, George, 349, 469–72, 512, 540
Butler, Jay, 534–37

Cairns, John, 236, 582–86
Califano, Joseph, 162, 181, 186–87
California, University of; AIDS Task Force, 322; Berkeley, 18, 383, 436; Davis, 191, 371, 603; Los Angeles, 229, 283–85, 293, 295, 386, 402, 405; San Diego, 379, 579; San Francisco, 227, 286, 289, 292, 300, 309, 322, 412, 483
California encephalitis, 606
California Institute of Technology, 579
Callen, Michael, 271, 273, 282, 291–93, 329, 361
Calomys field mouse, 25–28
Cambodia, 211, 497, 608; malaria in, 449–52, 454–56, 463
Cambridge University, 34, 385
Cameroon, 367, 375–77
Campbell, Bobbi, 262, 282, 292–93, 329, 330, 361
Campbell, Carlos ("Kent"), 90, 92–94, 97, 444–47, 453, 454
Campylobactr, 424–25
Canada, 153, 506, 596; AIDS in, 306, 325, 380; antibiotic resistance in, 414, 415; International Development Agency of, 484; swine flu in, 170; tampon use in, 392; Toxic Shock Syndrome in, 404, 406; yellow fever in, 66
cancer, 6, 32, 178, 203, 212, 222, 223, 225, 286, 413; AIDS and, 318, 365, 382 (see also Kaposi's sarcoma); chemotherapy drugs, 432,

cancer (*cont.*)
452, 589; chlorine and, 428, 546; hepatitis
B and, 273; oncogenes and, 226–33; skin,
556; viruses and, 317, 573, 581
candida, 38, 134, 278, 284, 286, 293, 295,
347, 364, 612
Cape Verde, 206, 354, 368
Carballo, Manuel, 461, 462, 464, 474–76
carbenicillin, 418, 421, 425
carbovir, 436
carboxymethyl cellulose (CMC), 393, 394, 401
carcinogens, 230
Cardoso, Fernando Henrique, 200
Cargill, Adam, 71, 73, 95, 97, 99
Carrillo, Emilio, 506
Carson, Rachel, 50
Carter, Jimmy, 153, 162, 180, 181, 186, 193,
205, 302, 318
Carter Center for International Peace, 609
Casals, Jordi, 56–57, 76–77, 79–85, 90, 93–
95, 97, 110, 127, 158, 605
CD4 cells, 292, 295, 331, 333, 409, 446, 526
CD8 cells, 295
Ceausescu, Nicolae, 505
cefoxitin, 420
ceftiraxone, 420
Census Bureau, U.S., 485–87, 498
Centers for Disease Control (CDC), 7, 32, 65,
280, 301, 592, 595, 596; and AIDS, 285–
91, 293, 295–99, 302–11, 315, 319, 323–
24, 326–28, 331, 332, 337–38, 341, 344,
345, 348, 350, 351, 353, 355–57, 360,
369–70, 376, 462, 473; and antibiotic resistance, 414, 420, 422, 423, 425; Center for
Infectious Diseases, 532; and chlorine exposure, 429; and cholera, 564; and dengue
hemorrhagic fever, 257; and Ebola virus,
107, 109, 110, 112–16, 118, 122, 126,
132, 141, 142, 146, 149, 153, 219, 220,
598–601; emergency response capabilities
of, 604–7, 609, 610; Epidemic Intelligence
Service, 62, 114, 531; and hantaviruses,
531–37, 539–45, 547–48; and Lassa fever,
72–73, 77, 83, 87–94, 99, 193–95, 214–
15; and Legionnaires' Disease, 172–78,
184–86, 189–91; and Machupo, 18, 621;
and malaria, 444, 445, 448–50; and measles, 511; and sexually transmitted diseases,
265–66, 271–73, 285, 287; and smallpox,
43–44; Special Pathogens and Bacteria
Branch, 60, 62, 119, 215, 533, 540, 598,
600, 601; and Swine Flu, 158–59, 161,
163–67, 169–71, 178–81, 183, 187–89;
and Toxic Shock Syndrome, 394, 396–406,
410; and tuberculosis, 512, 515, 516, 520–
24, 527, 570; Venereal Disease Control Division, 273; and yellow fever, 68–70
Central African Republic, 100, 208, 359, 368,
376; AIDS in, 350, 486; Lassa fever in, 92
Central Asian hemorrhagic fever, 82

Central Intelligence Agency (CIA), 61, 82, 136,
137, 193, 323, 382, 383
cephalosporins, 412, 423
cephalothin, 418
Cercopithecus aethiops, 55
cervical carcinoma, 233
Chabner, Bruce, 308
Chad, 63
Chagas's disease, 205, 253, 613
Chaika, Nikolai, 503
Chan, Roy, 489
chancroid, 267, 268, 270, 336, 347, 478, 507
Chapman, Louisa, 532, 533, 537, 541, 545
Charlemagne, 156
Charles VIII, King of France, 245
Charles River Primates Corporation, 601
Chazov, Yevgeny, 476
Cheek, Jim, 529–32, 534–36
Cheikh Anta Diop University, 358
Chemotherapeutisches Forschunginstitut, 377
chemotherapy drugs, 432, 452, 589
Chen, Irvin, 229, 386
Chernobyl nuclear accident, 9
Cherokee Indians, 536
Chicago, University of, 6, 212
Chikungunya, 211
Children's Defense Fund, 512
Childs, Jamie, 537, 539–41
Chile: skin cancer in, 556; yellow fever in,
66, 68
Chin, Jim, 462, 480–81, 489
China, 32, 208, 247; 512; AIDS in, 466, 467,
476, 496; ancient, 234, 236, 245; antibiotic
resistance in, 420; Black Death in, 237;
dengue fever in, 258; Hantaan disease in,
538, 541; heroin in, 491–92; malaria in, 52,
450, 452–53; smallpox in, 41; tuberculosis
in, 526, 527
chlamydia, 174, 265, 267–71, 283, 347
chloramphenicol, 62, 267, 417, 418, 425, 584
chlorine, 428–30, 562, 564
chloroquine, 47, 49, 52, 100, 102, 432, 440–
46, 448–53, 455, 456, 578
cholera, 208, 210, 241–42, 337, 422, 428,
463, 497, 503, 563–66, 591, 593, 616
cholesterol, 155
choriomeningitis, 174
Christian Broadcasting Network, 469
Christian fundamentalists, 469–70
chronic fatigue syndrome, 383, 532
ciprofloaxin, 420
Citrobactr, 584
civil rights movement, 262, 274
Civil War, 47, 241, 546
Clavel, François, 372
Clethrionomys glareolus voles, 538
clindamycin, 418
Clinton, Bill, 427
Close, William, 106–8, 117, 119
Clostridium, 612

cloxacillin, 418
Clumeck, Nathan, 320, 344–45, 352, 355
cocaine, 280, 297, 327, 363, 389, 507, 509, 552, 612, 621
coccidioidomycosis, 174
Cohen, Mitchell, 414, 425
Cohen, Stanley, 224
cold sores, 433
Cold Spring Harbor Laboratory, 34, 225, 330, 332
Cold War, 45, 161, 193–94, 200, 201, 211, 354, 369, 499, 608
Collas, René, 119
Collins, Joseph, 199
Colombia, 201, 386; cocaine production in, 280; deforestation in, 552; leishmaniasis in, 254; malaria in, 52; yellow fever in, 68
Colorado State University, 577
Columbia University, 76, 84, 92, 415, 431; College of Physicians and Surgeons, 395, 519; Presbyterian Hospital, 78, 80, 83, 94
Columbus, Christopher, 213, 246
Colwell, Rita, 560–66
common cold viruses, 568, 569
Commonwealth Conference, 250
Conant, Marcus, 292–93, 299, 301
condoms, 272–73, 307, 469, 473–74, 479, 494, 496, 499–500, 611
Congo, 371, 375, 376; AIDS in, 474, 486; schistosomiasis in, 252
Congress, U.S., 5, 48–51, 61, 68, 276, 287; and AIDS, 311, 315, 318–19, 345, 471, 472; and hantavirus, 544; and hemophilia, 313; and Legionnaires' Disease, 177; and Swine Flu, 167–69, 171–73, 178, 182, 183, 187
Conn, Del, 139–42
Connecticut, University of, School of Medicine, 177
Conrad, Lyle, 77, 114
Conservative Caucus, 469
contraceptive vaginal sponge, 406
Cooper, Theodore, 167
Coopération Médicale Belge, 106
Corey, Lawrence, 307
Cortez, Hernando, 41, 213
Costa Rica, yellow fever in, 67
Côte d'Ivoire, 252, 375, 609; AIDS in, 486, 516, 587; Lassa fever in, 92; malaria in, 446
Courtois, Dennis, 139
Cox, Nancy, 165
coxiella, 534
crack cocaine, 507, 520, 521, 524–25
Crick, Francis, 34, 223, 225
Crimean–Congo hemorrhagic fever, 82, 211, 534
criminal justice system, 509, 613
Croft, Albert J., 157
Cryptococcus, 291, 293, 344
cryptosporidiosis, 271, 293, 429, 430

c-src, 227
Cuba, 201; AIDS and, 323, 383, 466–67; dengue hemorrhagic fever in, 256, 258; yellow fever in, 67
Culex mosquitoes, 50
Curran, Jim, 273, 285–88, 297–99, 304, 306, 307, 311, 313, 316–17, 324, 326, 327, 331, 332, 337–38, 345, 348, 350, 357
Cutter Biological, 490
Cutter Laboratories, 311
cysticercosis, 38, 252–53
cytomegalovirus (CMV), 267, 271, 284, 287, 293, 295, 317, 322, 324, 344, 364, 367, 478, 613; drugs for, 435; transplants and, 575

dam-related epidemics, 203–5
Dan, Bruce, 396, 402, 403, 405
dapsone, 420
Darrow, Bill, 289, 295–96, 297, 300, 304–7, 326, 347, 364
Darwinian theory, 586
Dauguet, Charles, 325
David, John, 596–97
Davies, Julian, 578
Davis, Dorothy, 77
Davis, Jeffrey, 394
DDT, 28; malaria and, 31, 47–52, 254, 441, 453; yellow fever and, 66, 68
Dean, Andrew, 396
Declaration of Alma-Ata, 211, 619
De Cock, Kevin, 369–70, 587
Decosas, Josef, 484, 487
Defense, U.S. Department of (DOD), 14, 94, 533, 537, 540, 544, 595, 597
Defoe, Daniel, 238–39
deforestation, 551–57
Degefu, Workineh, 355
Delgadillo, René, 111, 112
Delphinus delphis, 558
Delphi Survey technique, 486
delta virus, 612
dementia, AIDS, 344
Democratic Party, 299, 301–3, 318, 489
dengue fever, 254–59, 497, 567
Denmark: antibiotic resistance in, 413, 414; morbilliviruses in, 538; sexually transmitted diseases in, 266; Toxic Shock Syndrome in, 406
Denton, Jeremiah, 330
Denver Veterans Administration Medical Center, 190–91
deoxyfluorothymidine (FLT), 436
deoxyribonucleic acid, see DNA
Des Jarlais, Don, 326–27
Desrosiers (AIDS researcher), 379
Deutsches Rheuma Forschungszentrum, 590
De Vita, Vincent, 230
diarrheal diseases, 251, 420–21, 429, 463, 615, 616; see also specific disorders

didehydrodideoxyguanosine, 436
dideoxycitadine (ddC), 436
dideoxynosine (ddI), 436
Dietrich, Ursula, 377
dioxins, 383, 428
diphtheria, 32, 251–52, 504–5, 617
Directly Observed Therapy (DOT), 527
disinfectants, 428–31
distemper, 558, 559
DNA, 6, 34, 37, 40, 82, 222–28, 230, 231,
    233, 578–86; AIDS and, 321, 373, 376, 436;
    bacterial, 409, 411, 431, 432; of malaria
    parasites, 447; viral, 433–35, 573, 575, 603
Dobkin, Jay, 519, 520
dolphin morbillivirus (DMV), 558–59
Dombe, Sebo, 102
Domestic Policy Council, 303
Dominican Republic, 617; AIDS in, 326, 349
Donatienne, Sister, 117, 125
Doolittle, Russell, 379
Dorado, Einar, 24, 130
Dowdle, Walter, 161, 163–64, 166, 184, 189,
    298
Downs, Wilbur, 76, 81
doxycycline, 420
Dritz, Selma, 288–89, 300, 310
drought, 354–55
Drozdov, S., 382
drunk driving, 503
Dubos, René, 13, 53, 214, 243–45, 390, 475
Duesberg, Peter, 383
Dugas, Gaetan, 306, 307
Duke University, 61, 371
Dulbecco, Renato, 226
Dunlap, Becky, 302
Dwyer, John, 490
dysentery, 254, 421–22

Earth in the Balance (Gore), 550
Earth Summit, 570
Eaves, Julian, 500
Ebola virus, 5, 6, 100–154, 156, 175, 178,
    183, 192, 211, 215–21, 303, 319, 345–46,
    369–70, 384, 463, 477, 501, 538, 572,
    594, 597–601, 603, 606, 607, 610, 612
Ecuador, deforestation in, 552
Edmonda, Sister, 101, 104, 106, 110, 117,
    124
Egypt, 201, 251; ancient, 108, 234, 236, 239,
    245, 390; antibiotic resistance in, 414; in-
    jecting drug users in, 277; schistosomiasis
    in, 204
Ehrlich, Paul and Anne, 555, 557
Eisenhower, Dwight D., 48–49
Ekombe Mongwa, 102, 103
ELISA (enzyme-linked immuno-absorbent as-
    say), 333, 363, 365, 366
Elliott, Luanne, 547
El Niño weather pattern, 560, 564, 616
El Salvador, 511; malaria in, 94

El Tahir, Babiker, 142–45, 148
El Tor cholera strain, 563–66
Elyea, Harry, 86
Emond, Ronald, 134, 135
Emory University, Yerkes Regional Primate
    Center at, 319
encephalitis, 50, 82, 253, 576, 577, 606;
    herpes, 433, 573–74
endocarditis, 278, 408
endotheliomas, 286
"enhancers," 581
Entamoeba histolytica, 270, 271, 293
Enterobactr, 422, 584
Enterococcus, 278, 422–23, 606, 614
enterotoxin B, 407
Environmental Defense Fund, 428
Environmental Protection Agency (EPA), 429,
    430, 564
epitopes, 447
Epstein, Michael, 233
Epstein, Paul, 566, 567
Epstein–Barr virus (EBV), 233, 271, 293, 322,
    324, 364, 478; transplants and, 575
equine infectious anemia virus (EIAV), 331
erythromycin, 415, 417–18
Escherichia coli, 37, 38, 224, 293, 421, 422,
    426–27, 430, 432, 582–86, 591, 606
Esparza, José, 462
Essex, Myron ("Max"), 142, 231–32, 273,
    299, 317, 321, 324–25, 331, 353, 354,
    358, 371–74, 376, 379, 572–73, 587
ethambutol, 522
Ethiopia, 45–46, 208, 354, 355, 375; AIDS
    in, 484; antibiotic resistance in, 412, 420;
    famine relief in, 72; malaria in, 456; sexually
    transmitted diseases in, 268; yellow fever in,
    68
European Economic Community, 465, 466
Evans, John, 212, 213
Evatt, Bruce, 311, 313–14
Ewald, Paul, 587, 588

Factor VIII, 308–9
Faletto, Enzo, 200
famine, 199, 354–55, 368–69, 479, 591, 608
Fansidar, 441, 443, 446, 450, 453
Farley, John, 38
Fauci, Anthony, 317, 525
Federal Bureau of Investigation (FBI), 613
Federal Reserve Board, 508
Federation of American Scientists, 602
feline immunodeficiency virus, 580
feline leukemia virus (FeLV), 231, 232, 317,
    432, 580
Fenner, Frank, 35
Ferlite Company, 598, 599
Field Museum of Natural History, 24
Fields, Bernard, 231, 277, 384, 432, 581–82
Fiji, 269
filoviruses, 600, 601

Fineberg, Harvey, 163, 187
Finland, 505; AIDS in, 465
Finley, Carlos, 67
Finn, Richard, 95
Fischl, Margaret, 307–8
Fish and Wildlife Service, U.S., 563
Fisher-Hoch, Susan, 600, 601, 609–10
Fleming, Alexander, 36
fluorescein, 126
fluorescence-activated cell sorter (FACS), 294
fluoroquinolone antibiotics, 424, 614
Fluss, Sev, 477
foamy viruses, 227
Foege, William, 287, 298, 299, 302, 348, 609
Fogarty International Center, 6
Foley, Tom, 489
Food and Drug Administration (FDA), 159,
    163, 168, 173, 180, 433; AIDS research
    and, 303, 311, 313–15; and antibiotic resis-
    tance, 426, 440; Blood Products Advisory
    Committee, 315; Bureau of Biologics, 189;
    Toxic Shock Syndrome and, 398–400, 402,
    406, 410
Ford, Gerald R., 153, 157, 163, 167–68, 172,
    173, 179, 180, 187
Ford, Timothy, 566
formaldehyde, 569
Forthal, Don, 338–44, 346
foscarnet, 433, 435
Foundation for Economic Trends, 381
Frame, John, 76, 78, 84–86, 92
France, 245, 247; AIDS in, 297, 299, 319–21,
    325, 332–34, 337, 344, 348, 350, 361,
    477; antibiotic resistance in, 412, 420; Black
    Death in, 238; colonialism of, 38, 66; hanta-
    viruses in, 538, 546; leprosy in, 239; morbil-
    liviruses in, 558
Francis, Don, 43–46, 126, 132, 142–46, 219,
    272–73, 280, 288, 298, 299, 311, 313–14,
    317, 319, 324, 326, 331
Francis, Henry, 351
Franco-Italian wars, 246
Frank, André Gunder, 200
Franklin, Rosalind, 34, 223
Frasner, David, 90, 176
Freeman, Harold, 508
Friedman-Kien, Alvin, 286, 306, 308
Fromm, Ernest, 99
fusidic acid, 418

Gabon, 375, 377; Ebola virus in, 345–46
Gajdusek, Carleton, 539
Gallo, Robert, 228, 229, 231, 232, 316–17,
    321–22, 324–26, 330–34, 352, 354, 357,
    365, 372–74, 379
ganciclovir, 433–35
Garbazu, 86, 88, 89
Garenne, Michel, 590
Garrón, Hugo, 14, 16, 20–21
gas chromatography, 174

Gaul, Gilbert, 311
gay men, see homosexuality
Gay Men's Health Crisis (GMHC), 292, 299
Gay-Related Immunodeficiency Disease (GRID),
    292, 293, 295–97, 299–309, 316, 390; see
    also AIDS
Gee, Gayling, 286–88, 290, 299
GenBank, 616, 617; AIDS project, 378–79
genetic engineering, 222–33, 381, 577–78,
    603
GenInfo, 616
genital cancers, 233
Genoveva, Sister, 107, 121, 149, 151–52
gentamicin, 418
Gentry, Al, 556
Germain, Max, 126, 130, 135
German Physicians' Association, 99
Germany, 494; AIDS in, 465, 466, 499–500;
    antibiotic resistance in, 415; Black Death in,
    238; cholera in, 242; Foreign Ministry, 95,
    96; Health Ministry, 99; and Lassa fever,
    71–73, 93–99, 126; Marburg disease in,
    53–57, 59, 60; reunification of, 498–99,
    546; Toxic Shock Syndrome in, 406
Germs That Won't Die (Lappé), 436
Ghana, 208, 376–77; influenza in, 157; oncho-
    cerciasis in, 252; schistosomiasis in, 204
giardia, 271, 288, 429, 430
gibbon ape leukemia virus, 227, 574
Gilada, I. S., 487
Gilligan, John J., 202
Giuliani, Rudolph, 540
Global AIDS Policy Coalition, 485–86
Global Programme on AIDS (GPA), 459, 461–
    68, 472–77, 480–82, 485, 490, 497, 516,
    508
Global Strategy for the Prevention and Control
    of AIDS, 465
global warming, 556–57, 567, 570
Godal, Tore, 452
Goddard Space Flight Center, 552, 556
Godwin, Ronald S., 470
Goedert, Jim, 317
Goff, Paul, 90
Golde, David, 229
Goldfield, Martin, 154–55, 158, 188
Gonda, Matthew, 385
gonorrhea, 32, 211, 264–72, 283, 289, 296,
    347, 420, 478, 490, 503, 506, 507, 606,
    610, 611, 616
Gordin, Fred, 523
Gore, Albert, 9, 550
Gorgas, General William C., 47–48
Gorman, Michael, 300
Gostin, Larry, 476
Gottlieb, Michael, 283–85, 293, 295, 296,
    337, 363
Great Plague, 238–39, 241
Greece; AIDS in, 465; ancient, 235, 236, 245,
    263; malaria in, 48

greenhouse warming effect, 557
Green Monkey Virus, 5
Greenpeace, 428
Green Revolution, 51
Gregg, Michael, 90
Grmek, Mirko, 361
Group of, 77, 200
Groupe de travail français sur le SIDA, 319, 320
Guadeloupe, 420
Guatemala, 615; yellow fever in, 67
Gubler, Duane, 257, 596
Guiana, 365
Guillain-Barré syndrome, 180–83, 188
Guinan, Mary, 285, 287, 297, 306, 326
Guinea, 90, 114, 194, 375
Guinea-Bissau, 206, 368; AIDS in, 389
Gullet, John, 287

Haas, Earle, 391, 392
Habyarimana, Juvénal, 592
Hahn, Beatrice, 374, 375, 379
hairy-cell leukemia, 232
Haiti, AIDS in, 307–10, 314, 319–21, 326, 328, 331, 337, 349, 357, 365, 486, 515
halofantrine, 451, 453
halogen ions, 557
Halstead, Scott, 255, 256
Hamburg, Margaret, 540
Hamburg University Virology Institute, 96
Hannon, Claude, 109
Hansen, Armauer, 239
Hansen's disease, 236, 239, 513
Hantaan virus, 22, 534, 538–39
hantaviruses, 528–49, 606
Harvard University, 9, 46, 91, 142, 146, 160, 252, 317, 324, 353, 371–73, 446, 527, 551, 554, 572, 581, 591, 619; Dana Farber Cancer Institute, 363; Department of Tropical Public Health, 596–97; John F. Kennedy School of Government, 162–63; Medical School, 231, 366, 431, 569, 588, 617; School of Public Health, 48, 50, 51, 163, 232, 443, 451, 485, 566, 583; Working Group on New and Resurgent Diseases, 566, 567
Haseltine, William, 363
Hazelton Research Products, Inc., 598, 600, 601
Health, Education, and Welfare, U.S. Department of, 32
Health and Human Services (HHS), U.S. Department of, 301, 325, 506
Health Research Group, 399
Health Transition, 30–52
Healy, Bernardine, 374
heart disease, 178, 203, 212, 253, 318, 413; Lyme disease and, 553; rheumatic fever and, 416
heavy-metal poisoning, 174

Heckler, Margaret, 318, 332–33
Held, Richard, 613
Helms, Jesse, 472
hemagglutinin, 155, 156, 160
hemophilia, 54–55, 385; AIDS and, 308–9, 311–14, 323, 326, 331, 362–63, 473
Hemophilia Diagnostic and Treatment Center Program Act (1975), 313
Hemophilus: H. ducreyi, 267; H. influenzae, 188, 417, 529, 584
hemorrhagic fevers, 13–29, 55, 56, 82, 546, 605; simian, 598, 601; see also specific diseases
Henderson, D. A., 42, 44–46, 114, 602, 604–5, 609, 617–18
Hensley, George, 307–8
Henson, Jim, 414–15
heparin, 58–59
hepatitis, 6, 232, 233, 270, 271–74, 280, 288, 290, 296, 298, 322, 399, 497, 505, 507, 562, 606, 612–14; in blood supply, 312–15; in drinking water, 429; transplants and, 575
hepatocellular carcinomas, 232
herd immunity, 589
heroin, 274, 276–80, 281–82, 297, 327, 361, 363, 389, 490–94, 496, 500, 507, 509, 521, 524–25, 552, 612, 613
herpes simplex, 233, 266–68, 270, 271, 289, 293, 296, 381, 611; B virus, 573–75, 601; drugs for, 433–35; HIV and, 580–81
Herpesvirus: H. ateles, 573; H. saimiri, 572–73
herpes zoster, 283, 347, 348
Herre, Allen, 587–88
Herzenberg, Len, 294
Heymann, David, 7, 44–46, 147, 149, 175, 176, 178, 444–45, 462
Highton, Barney, 141, 146
Hilliard, Julia, 575
Hinuma, Yorio, 229
Hira, Subhash, 269–70, 283, 347–48, 360, 488, 493, 616
Hirsch, Vanessa, 374
Hispanics, 507, 511
histoplasmosis, 174
Ho, David, 380
Hodgkin's disease, 164, 165
Hoechst AG, 53
Hofnia, 584
Holland, John, 579
Hollwanger, Phebe, 89
Holmberg, Scott, 425
Holmes, King, 611
homelessness, 506–9, 514, 519–21, 525, 591, 615
homicide, 503
homosexuality, 245, 260–63, 266–68, 271–74, 282; AIDS and, 283–93, 295–302, 304–12, 316, 317, 319–34, 337, 347–49, 351, 357, 360, 361, 363, 364, 367, 371,

381, 383, 386, 387, 389, 460, 461, 463–66, 469, 470, 472, 474–76, 500, 587
Hong Kong, 250
Hong Kong flu, 160–62, 165
hookworms, 203
Horn, Joshua, 32
House of Representatives, U.S., 168, 172, 318, 489; Energy and Commerce Subcommittee, 298; Interstate and Foreign Commerce Committee, 177
Howard, Bob, 537
Howard, Greggory, 274, 276, 281–82, 315, 326–28, 361
HTLV, 6, 229–32, 317, 321, 324–26, 330–34, 372–74, 379, 381–82, 385–86, 432, 610, 612
Hubbard, John, 536
Hudson, Edward, 246
Hudson, Robert, 191
Huebner, Robert, 226
Hughes, Jim, 532–35
Human Genome Project, 378
human immunodeficiency virus (HIV), 7, 10, 338, 344, 349, 432, 460, 461, 505, 572, 602, 610–14, 617, 619; in blood supply, 362–63, 444, 448; conspiracy theories about, 380–84; drugs for, 433, 435–36; earliest identified presence of, 363–65, 380; emergence of, 367, 371; genetic variations in, 377–80; Ghanian strain of, 376–77; herpes virus and, 434; iatrogenic spread of, 500–3; incidence rates, 350–52, 356, 360, 479, 481–93, 496–98, 507, 593; Leopoldville strain of, 371, 380; malaria and, 478; measures to prevent spread of, 463–77, 490–91, 494; monkey viruses related to, 358, 371–77, 384–85; mutations of, 578–80, 587; research leading to identification of, 317, 321–22, 324–25, 330–34; social conditions in spread of, 367–71, 380, 385–89; transplants and, 575; tuberculosis and, 515–16, 519–27; see also AIDS
human papillomavirus, 233
human rights, AIDS and, 476–77, 490
Human Rights Watch, 495
Humboldt University, 382
Humphrey, Hubert H., 48
Hungary, antibiotic resistance in, 418
Huntington, Samuel P., 591
Hussein, Saddam, 597
hypertension, 540, 545

iatrogenic diseases, 447–48
Ibadan, University of, 69, 70, 87; Hospital, 72, 73; Medical School, 95
IgG antibodies, 447
immune system, 581–84; anal intercourse and, 272; dengue-2 virus and, 255–56, 258; diphtheria and, 505; herpes viruses and, 573; impact of malnutrition on, 199; of in-

jecting drug users, 277; in Lyme disease, 553; in malaria, 446–47; pollution and, 568, 570; suppression of, 589; in Toxic Shock Syndrome, 404–5, 407–8; tuberculosis and, 513; see also AIDS
Imperial College (London), 484
impetigo, 415
Inaba cholera substrain, 565
India, 201, 251, 603; AIDS in, 357, 467, 468, 488–93, 496–98; ancient, 236, 245; antibiotic resistance in, 420; Black Death in, 237; cholera in, 31, 563, 565–66; dengue hemorrhagic fever in, 257; Health Ministry, 269; hepatitis B in, 272; kala-azar in, 253–54; malaria in, 47, 51, 52, 442, 450; respiratory infections in, 417; smallpox eradication in, 44, 45, 142; tuberculosis in, 527
Indian Health Service (IHS), U.S., 528–30, 532, 534–36
Indonesia, 46, 200, 201, 251; AIDS in, 467, 468; dengue fever in, 258; malaria in, 450, 452, 454; respiratory infections in, 417
Industrial Revolution, 214, 243, 244
infant mortality, 196, 206
influenza, 5, 6, 34, 35, 41, 155–58, 174, 175, 298, 384, 417, 529, 539, 543, 568–70, 574, 578–80, 598, 603, 614; tuberculosis and, 513; see also Swine Flu
injecting drug users, 274–80, 285, 286, 506, 508, 591, 612–13; AIDS in, 297, 307, 308, 310, 312, 314, 315, 325–28, 341, 349, 363–64, 367, 383, 386, 387, 389, 460, 475, 476, 489, 490–94, 496, 500, 502, 507, 587; Staphylococcus infections and, 408, 413; tuberculosis in, 514, 519, 524–25
Inkoo virus, 577
insect vectors, 575–78; see also mosquito-borne agents; specific microbes and diseases
Institute of Medicine (IOM), U.S., 7, 405, 406, 509, 605
Institut Mérieux, 64, 65, 617
International Centre for Diarrhoeal Disease Research, 564
International Commission of Jurists, 209
International Conferences on AIDS, 351, 367, 470–71, 488, 489, 490
International Congress on Tropical Medicine and Malaria, 30–31
International Covenant on Civil and Political Rights, 212
International Covenant on Economic, Social, and Cultural Rights, 212
International Development Advisory Board (IDAB), 48–50
International Development Research Center, 616
International Monetary Fund, 197, 198, 250, 609
International Physicians for the Prevention of Nuclear War, 615–16

International Red Cross/Red Crescent, 607
International Research and Development Corporation, 601
International Union Against Tuberculosis and Lung Diseases, 527
Inuits, 157
Ireland, 92; antibiotic resistance in, 414; morbillivirus in, 558
Isaacson, Margaretha, 58, 117–20, 125, 139–40
Iseman, Michael, 524
isoniazid, 518, 520–22, 524
Israel, 201; sexually transmitted diseases in, 267; Toxic Shock Syndrome in, 406
Italy, 231; antibiotic resistance in, 423; Black Death in, 237; malaria in, 48
ivermectin, 439
Ixodes dammini, 553–55

Jacoby, George, 431
Jaffe, Harold, 273, 287, 289, 297, 298, 304, 306, 308, 326, 331, 364
Jahrling, Peter, 540, 544–45, 547
Japan, 49, 200, 250, 494, 498, 604; AIDS in, 468, 489; expenditure on drugs in, 438; Hantaan disease in, 538; hemophilia in, 313; HTLV in, 231, 321–22, 385, 386; Kawasaki syndrome in, 395–96, 407; malaria in, 450; medieval, 391; National Hospital of, 46; smallpox in, 41; Toxic Shock Syndrome in, 406
Japanese encephalitis, 82, 606
Jarvis, Bill, 422
Jenkins, Carol, 614–15
Jews: ancient, 239; Black Death blamed on, 237; Nazi extermination of, 466
Job Corps, 472
John, Jacob, 489
Johns Hopkins School of Medicine, 49, 539
Johnson, Karl, 13–14, 17–29, 33–35, 38, 56, 62, 70, 81, 82, 87, 110, 112, 113, 119–20, 122, 123, 125–30, 132, 135–41, 146, 147, 154, 178, 192, 193, 197, 346, 538–39, 570, 594, 595, 603, 609, 621–22
Johnson, Lyndon B., 162, 318
Johnson & Johnson, 392
Jordan, 201
Journal of the American Medical Association, 321, 333
Journal of Infectious Diseases, 405
Journal of the National Cancer Institute, 324
Joyce, Gerald, 585
Juliana's disease, 334–37, 340
jumping genes, 225, 226, 585
Junín, 17, 23, 27, 28, 55, 82, 88, 92, 614
Justice Deaprtment, U.S., 182

Kachenko, Sasha, 193
kala-azar, 210, 254
Kalisa Ruti, 345, 351

kanamycin, 432, 616
Kanki, Phyllis, 353, 373, 376, 587
Kansas, University of, 191, 244
Kapita Bila Minlangu, 346, 351, 353–54, 367, 379
Kaposi's sarcoma (KS), 286–88, 290, 292–93, 298, 299, 304, 305, 307, 317, 321, 324, 339, 344, 364, 365–67, 372, 587
Karpas, Abraham, 385
Kaunda, Kenneth, 362
Kawasaki, Tomisaku, 395, 407, 408
Kawasaki syndrome, 395–96, 402–4, 407–8
Kennedy, Edward M., 167, 181
Kennedy, John F., 48, 49, 60, 162
Kenya, 209, 229, 230, 251, 282–83, 344; AIDS in, 339, 340, 352, 354, 359, 366, 368, 486, 493; antibiotic resistance in, 414; expenditure on drugs, 438; malaria in, 441, 442, 445, 446, 448, 568; Marburg virus in, 57; Ministry of Health, 142; Rift Valley fever in, 204; sexually transmitted diseases in, 268, 269; yellow fever in, 575, 606
KGB, 193, 382, 466
Khan, Ali, 543–45
Khmer Rouge, 450, 454, 455, 608
Khorana, Har Gobind, 224
Khouri, Yamil, 365
Kidenya, Jayo, 281, 335–43, 347, 360
kidney disease, 203, 538–40, 545–47
Kiereini, Eunice Muringo, 482–83
Kilama, Wen, 440–42
Kilbourne, Edwin, 159–61, 167
Kimberly-Clark Corporation, 392
King, Martin Luther Jr., 262
King's College (London), 34
Kinshasa, University of, 346
Kirsten sarcoma virus, 227
Kissinger, Henry, 97, 126, 194
Klebsiella, 293, 422, 584
Knowles, John, 506
Koch, Edward, 301–3
Koch, Robert, 244
Koch's Postulate, 403, 408
Kondrusev, Alexander, 501
Koop, C. Everett, 302, 470, 471
Korea, 201, 383; AIDS in, 468; hantaviruses in, 537–39
Korea University Medical School, 538
Korean War, 22, 258, 538, 539
Kotex, 397
Koth, André, 120
Kramer, Larry, 263, 292, 293, 299
Krause, Richard, 5, 345, 414
Krebs, John, 541
Krugman, Paul, 508
Ksiazek, Tom, 527, 540, 541, 543, 545
Kuns, Merl, 18–21, 23–29, 126, 622
Kyoto University, 228, 229

Lagenorhymchus albirostris, 558
Laidlaw, Sir Patrick Playfair, 158
Lallemont, Marc, 474
Lancet, The, 170, 347
Landesman, Sheldon, 308
Landsat satellite imagery, 552
Lane, Cliff, 317
Lang, August, 466
Lange, Michael, 390
Langmuir, Alexander, 160, 186
Laos, 201; dengue hemorrhagic fever in, 257; heroin in, 491–92; malaria in, 450
Lappé, Frances Moore, 199
Lappé, Mark, 436–37
Lassa virus, 71–99, 110, 113, 115, 120, 122–23, 126, 127, 146, 155, 174, 192–97, 211, 215, 218, 220, 252, 257, 298, 384, 463, 572, 597, 602, 606, 609, 610, 612
Leakey virus, 540
Lederberg, Esther, 36–37
Lederberg, Joshua, 5–7, 36–37, 225–27, 431, 579, 602, 607, 619–20
LeDuc, James, 539–41, 545, 596, 598, 606–7
Lee, Ho Wang, 538
Leeuwenhoek, Anton van, 36
Legionella, 189–91, 436, 535
Legionnaires' Disease, 5, 153, 171–78, 184–86, 189–91, 298, 303, 304, 319, 428, 430, 531, 535, 547, 568, 614
Lehmann-Gruber, Fritz, 56, 96
Leifer, Edgar, 80–81
leishmaniasis, 253–54, 259, 356; visceral, 210, 607
lentivirus, 331, 334, 379, 381, 385
Leopold II, King of Belgium, 112
leprosy, 210, 236, 239–40, 420
Lesotho, 208
leukemia, 227, 229, 231, 232, 573; *see also* HTLV
Levins, Dick, 619
Levy, Jay, 321, 330, 333, 334, 367
Levy, Stuart, 424, 427
Lewis, David, 154–55, 158, 159, 163–66, 169, 188, 189
Lewis, Sinclair, 30, 31
Liberia, 192, 194, 608; AIDS in, 375; Lassa fever in, 86–90, 609
Libya, 210, 576
Liebowitch, Jacques, 320, 349
life expectancy, 196, 206
Lisangi Mobago, 123
Lister, Baron Joseph, 90
Little, Arthur D., consulting firm, 70
liver cancer, 273
Lizenge Embale, 102, 103
London Institute of Tropical Medicine and Hygiene, 93, 142, 252, 607
London *Times*, 381
Lootens, Father Germain, 107
Los Alamos National Laboratory, 377–78, 616

Los Angeles County Health Department, 304–5, 400
Louisiana State University, 580
Lovejoy, Frank, 9
Lown, Bernard, 615–16
Luande, Jeff, 366, 467
Lucas, Adetukunbo, 596
Lule, Yusufu, 209, 210
Lumumba, Patrice, 61, 371
lupus, 405
Luwum, Archbishop, 209
Lwangwa, S. K., 480
Lyme disease, 553–55, 606
lymphadenopathy, 319, 325
lymphocytic choriomeningitis (LCM), 82
lymphadenopathy-associated virus (LAV), 331–34, 344, 350, 356, 372
lymphomas, 233, 573

Maalin, Ali Maow, 46
Mabalo Lokela ("Antoine"), 100, 102–4, 128
MacArthur Plan, 200
McClintock, Barbara, 225
McCormick, Joe, 7, 46, 59–66, 122–23, 126, 130–32, 135–36, 138, 143, 145, 146, 148, 192–98, 214–21, 298, 345–51, 357–60, 369–70, 384, 386, 444, 445, 571, 592, 596, 598, 600, 601, 603, 609–10
McDade, Joe, 174, 178, 184–86, 189, 535, 537, 540, 547
McDermott, Jim, 489, 493
McGraw, Pat, 273
MacKenzie, Ron, 14, 16–25, 27–29, 38, 621–22
McKeon, Thomas, 192, 214, 243, 244
McLuhan, Marshall, 9
McNamara, Joseph, 613
McNeill, William H., 6, 41, 212–13, 236, 239
Machupo virus, 13–29, 36, 53, 55, 56, 81, 82, 88, 92, 113, 114, 119, 120, 130, 141, 384, 572, 606, 610, 621
macroecology, 557
macrophages, 580
Madagascar, Rift Valley fever in, 205
Maddox, John, 324
Magellan, Ferdinand, 231
Maguire, Andrew, 167–68
Mahidol University, 565, 604
Mahler, Halfdan, 461, 462, 467, 475
Makuta, 121
malaria, 31, 32, 34, 39–40, 47–53, 66, 78, 92, 94, 100, 195, 205, 210, 222, 251, 254, 269, 355, 356, 357, 360, 432, 459, 463, 595, 612, 613, 616; AIDS and, 478; drug-resistant, 440–56, 497, 578, 582, 608; geographic spread of, 567–68; in injecting drug users, 277–78
Malawi; AIDS in, 482, 485–87, 516; malaria in, 441, 442, 445, 449

Malaysia, 114; AIDS in, 468; malaria in, 449, 450
Mali, 195, 354; Lassa fever in, 92; malaria in, 453; schistosomiasis in, 252
malnutrition, 199, 269, 338, 368, 487, 490, 506; immune system and, 589; tuberculosis and, 513
Malore, Richard, 529
mammary tumor virus, 227
Mandela, Nelson, 515
Mandrella, Bernhard, 71–73, 94–99, 126, 194
Mankerere University, 211
Mann, Dean, 317
Mann, Jonathan, 7, 350–51, 353, 354, 358–60, 389, 459–66, 468, 474–76, 480, 481, 485, 530, 608
Mann, Marie-Paule, 350
Mao Zedong, 32
Marburg virus, 5, 53–60, 77, 79, 80, 112, 115, 117–20, 124, 125, 141, 147, 148, 155, 174, 192, 211, 384, 463, 477, 572, 600, 601, 606
Marcella, Sister, 107, 121–22, 151–52
Mariam, Mengistu Haile, 45
Mariette, Sister, 107, 151–52
marijuana, 276
Marine Fisheries Service, U.S., 563
Marshall, George C., 30–31, 48, 49
Marshall Plan, 200
Martin, Malcolm, 332
Martini, Gustav Adolf, 55, 120
Martinique, 420
Maryland, University of, 383, 560
Masangaya Alola Nzanzu, 104
Mason, James, 302, 348
Masoni africana mosquitoes, 69
Mason-Pfizer virus (M-PMV), 574
Massachusetts Institute of Technology (MIT), 222, 224, 226, 508
Massamba Matondo, 121
mass spectral analysis, 174
Mastomys natalensis rat, 91–92, 192, 194, 196
Masur, Henry, 295, 297, 317, 363–64
Matek, Stanley, 303
Mauritania, 208, 354; Rift Valley fever in, 205
Mayans, 245
Mayinga N'Seka, 106, 117–20, 122, 124, 125
MBoup, Souleymane, 358, 376, 587
Mbuzu Sophie, 100, 102, 103, 106, 128, 129, 135, 140
M cells, 581, 582
measles, 33, 41, 174, 211, 251, 269, 417, 487, 503, 504, 510–12, 558, 559, 590, 591, 603, 607, 614, 617
Mechai Viravaidhya, 496, 497
Médecins du Mond, 607
Médecins Sans Frontières, 43, 607–8
Meedard, Frantz, 527
Meegan, James, 204, 598–99
mefloquine, 441–443, 451, 453, 455

Mekalanos, John, 588
Mendel, Gregor, 34
meningitis, 59–60, 62–66, 553, 617; cryptococcal, 291
mepacrine, 450
Merck & Company, 439
Merson, Michael, 481–82
Mertens, Paul, 88–90
Messier, Jeanne, 544, 547
methadone, 281, 297, 327, 328
methicillin, 278, 412–13, 418, 614
Mexico, 153, 201, 512, 576, 615; antibiotic resistance in, 422; cysticercosis in, 252; expenditure on drugs in, 438; Spanish conquest of, 41; yellow fever in, 67
Meyer, Gene, 465
Mhalu, Fred Solomon, 337
Mhaya people, 335, 340
Micotus: M. californicus, 539; *M pennsylvanicus*, 539
Microbiological Research Establishment (Porton Down), 109, 112, 113, 115–16, 133–35, 600
Middle American Research Unit (MARU), 14, 17, 18, 21, 23, 24, 29
Milk, Harvey, 262
Minnesota, University of, Medical School, 402
Minnesota State Health Department, 425
Missouri Botanical Garden, 555
Mitchell, Sheila, 345
Mitchison, Avrion, 590
Mobutu Sese Seko, 60, 61, 106–8, 112, 117, 129, 137, 208, 345
MO cells, 229
Modcoicar, Prassad, 616
Moi, Daniel Arap, 354
Moke, Germain, 121
Molecular Biology of the Gene, 582
"molecular chaperones," 584
molecular epidemiology, 373
Moloney leukemia virus, 227
Monath, Tom, 5, 69–70, 87–94, 257–58, 575, 605, 609, 612
Mongolia, 237
monkeypox, 146–48
mononucleosis, 271, 293
Montagnier, Luc, 321, 322, 325, 330–32, 334, 348, 352, 354, 372, 373
Moore, Carrie, 86
Moral Majority, 469–70
Morbidity and Mortality Weekly Report, 185, 186, 285, 306
morbilliviruses, 557–60
Morocco, 208
morphine, 612
Morris, Anthony, 189
Morse, Bradford, 355
Morse, Stephen, 5, 572, 602, 610
Moscone, George, 262
mosquito-borne agents, 82, 567–68, 575–76;

AIDS and, 357, 359, 474; dengue fever, 254–58; malaria, 39, 40, 48, 50, 51, 195, 254, 441, 442, 445–46, 449, 450, 453–56; Rift Valley fever, 204–5; yellow fever, 66–70, 195

Moss, Andrew, 289, 300–301, 310, 316, 331–32, 507

Mount Sinai School of Medicine, 159, 579

Mozambique, 195, 206, 208, 354, 368, 371; AIDS in, 366, 389, 484; expenditure on drugs in, 438; gonorrhea in, 616; malaria in, 442; tuberculosis in, 527

Mtu ni Atya Chakula ni Uhai village health campaigns (Tanzania), 206

Muerto Canyon virus, 549, 610

Muhimbili Medical School, 206, 336

Mukhopadyay, A. N., 603

multidrug-resistance (mdr) genes, 452

mumps, 174

Muñoz, Angel, 21–23, 27

Murphy, Frederick, 113, 115, 605

Murphy, John M., 177–78, 184

Murray, Christopher, 526

Mus musculus, 540, 546

Museveni, 592

"mutator alleles," 584

Muyembe Tamfum Lintak, 105, 107

Myanmar, *see* Burma

Mycobacterium, 291; *M. avium*, 293; *M. leprae*, 239, 240, 420, 513; *M. tuberculosis*, 240, 244, 245, 512–15, 518, 521, 526, 569

mycoplasma, 174

Myers, Gerald, 378–80, 387

Myriam, Sister, 101, 104, 105, 106, 117, 124

Nader, Ralph, 168, 181, 399

naficillin, 412

Nagorno-Karabakh, 608

Nakajima, Hiroshi, 481, 485

Namibia, 208, 269, 368; malaria in, 441

Narkevich, Mikhail, 502

National Academy of Sciences, 7, 405, 596

National Aeronautics and Space Administration (NASA), 126, 552, 556, 608

National Association of Public Hospitals, 507

National Cancer Institute, 226, 228–31, 303, 308, 352, 356; AIDS and, 316, 321, 333, 385

National Gay Task Force, 295

National Heart, Lung, and Blood Institute, 311

National Hemophilia Foundation (NHF), 309, 311, 314, 315

National Institute of Allergy and Infectious Diseases (NIAID), 6, 188–89, 317, 332, 345, 351, 434, 523, 525–26, 579

National Institute on Drug Abuse (NIDA), 315, 327, 363

National Institute of Occupational Safety and Health, 401

National Institutes of Health (NIH), 5, 14, 264, 265, 270–71, 290; and AIDS, 297, 300–3, 315–20, 358, 374, 436; and antibiotic resistance, 414; Frederick Laboratory, 230; and hantaviruses, 539, 540; and Machupo, 17, 18, 22, 23, 28–29; and malaria, 449–50; and tuberculosis, 525

National Jewish Center for Immunology and Respiratory Medicine, 408, 524

National Library of Medicine, 616

National Science Foundation, 60

National Swine Flu Immunization Program of 1976, 169, 172–73

Native Americans, *see* Amerindians

Ntural Resources Defense Council, 430

Nature, 324

Navajos, 528–37, 541, 542, 547–49

Navarro, Rose, 21

Navy, U.S., 265; Medical Research Unit, 204

Ne Win, 495

Nebraska, University of, 292

NEC Corporation, 616

Neild, Guy, 546

Neisseria: N. gonorrhoeae, 265, 420, 584; *N. meningitides*, 60

neomycin, 432

Neotoma albigula, 542

Nepal, malaria in, 450, 456

Netherlands: hantaviruses in, 546; morbilliviruses in, 558, 559; plague in, 238

Netter, Tom, 461, 485

Neu, Harold, 415, 431

neuraminidase, 155, 156, 159–60

Neustadt, Richard A., 162–63, 187

nevrapine, 436

New Caledonia, 420

New England Journal of Medicine, 337, 347, 405

New England Medical Center, 408

New England Regional Primate Center, 372–74, 573

New Hampshire, University of, 552

New Jersey Medical School, 325

New Jersey State Health Department, 154–55, 188

New Mexico, University of, 534, 536, 547

New Mexico Department of Health, 529, 531, 534

New Mexico State University, 601

New York Academy of Medicine, 420

New York City Health Department, 328, 414, 423, 507, 518–20, 527

New York City Health and Hospitals Corporation, 506

New York City Police Department, 260–61

New York Native, 323, 383

New York Post, 157, 179

New York State Department of Health, 600

New York Times, The, 160–61, 261, 398

New Zealand, 153; sexually transmitted diseases in, 267; Toxic Shock Syndrome in, 406

Ngaly Bosenge, 351
Ngoi Mushola, 104–5, 121
Ngwété Kikhela, 107, 116, 118, 119, 133
Nicaragua, tuberculosis in, 526
Nichol, Stuart, 541, 548
nickel carbonyl, 177
Nielsen, Godske, 96
Nieuwenhove, Simon van, 130, 138
Nigeria, 194; AIDS in, 473; Lassa fever in, 71–73, 75–77, 79, 82, 85–87, 95–97, 99, 126, 609; sexually transmitted diseases in, 268; yellow fever in, 66, 69, 575, 612
Nimeiri, Jaafar, 215–16
Nixon, Richard M., 153, 194, 318
Njelesani, Evaristo, 348, 355, 360
Noble, Gary, 299, 319
Nolte, Kurt, 547
norfloxacin, 584
North Carolina State Laboratory, 521–22
Norwalk virus, 562
Norway, 561
Ntaryamira, Cyprien, 592
Nussenzweig, Ruth, 595
Nyamuryekunge, Clint, 336–37, 341, 343, 360
Nyerere, Julius, 209, 210, 335
Nzila Nzilambi, 351, 353–54, 370

Obote, Milton, 209, 368
O'Brien, Thomas, 431, 617
OB tampons, 397
ocean pollution, 560–63, 566
Office of Management and Budget (OMB), 167, 182
Ohio University College of Medicine, 273
Okeyo, Mboya, 493
Oleske, James, 325
Olik, Pacifico, 142
Oliveira Bastos, Carlos de, 62
Omombo, 105, 119
Omsk hemorrhagic fever, 82
Onchocerca volvulus, 439
onchocerciasis, 205, 252
oncogenes, 226–33, 581
operons, 584
opium, 491–92
Organization of African Unity, 210
organochlorines, 31
organ transplants, 574–75
Oropouche, 614
Osborn, June, 159, 163, 168, 270–71, 273
Osterholm, Michael, 390, 396, 398–400, 405, 406, 425, 605, 606
otitis medea, 415–16
Over, Mead, 477, 479, 480, 482
Oxfam, 607
Oxford University, 223, 446
ozone layer, depletion of, 556–57, 562

Paintal, A. S., 491
Pakistan, 201, 610; antibiotic resistance in, 419
Palese, Peter, 579
Pan American Health Organization (PAHO), 31, 49, 60, 66–68, 441, 565
Pan American Sanitary Conference, 31, 48
Panama, 114, 386; malaria in, 47–48; yellow fever in, 66–68, 70
Pangu Kaza Asila, 353–54
Panos Institute, 475
Papua New Guinea, 417, 418, 497; Institute of Medical Research, 614–15; malaria in, 450, 452
paramyxovirus simian virus, 5, 574
parasites, 203, 270, 429; drugs for, 439; see also specific diseases
parasitology, 37–38
Parke-Davis, 171, 179, 180
Parmenter, Robert, 536, 542
Parodi, A. S., 27
Pasteur, Louis, 36, 192
Pasteur Institute, 109, 116, 321, 325, 332–34, 344, 350, 357, 372–74, 376, 377, 380, 607
Pattyn, Stefan, 111–13, 119, 126
Paul Ehrlich Institute, 53
Peace Corps, 21, 60, 90, 139, 448, 597
pediculosis, 296
Peloponnesian War, 236
pelvic inflammatory disease (PID), 264–65, 268
penicillin, 4, 36, 37, 62, 73, 76, 267, 433; resistance to, 265–66, 394, 397, 404, 408–12, 414–18, 420, 421, 425, 582–83, 606, 610
Penicillinase-Producing Neisseria gonorrhoeae (PPNG), 265, 268
Pennsylvania Department of Health, 172
pentamidine, 285
Peromyscus, 546; P. leucopus, 554; P. maniculatus, 542, 545, 548
Persian Gulf war, 597–98
pertussis, 174, 617
Peru: AIDS in, 378; cholera in, 428, 563–65; cocaine production in, 280; deforestation in, 552
Peter, Georges, 511
Peters, C. J., 533–35, 537, 540–45
Pfeiffer, Gerd, 466
phagocytic cells, 277
pharmaceutical industry, 438–40, 617
Pharmaceutical Manufacturers Association, 171, 173, 438
pheromones, 432
Philippines, 201, 251, 497; AIDS in, 467, 489, 498; dengue hemorrhagic fever in, 255, 257–59; malaria in, 450; research animals from, 598–601; sexually transmitted diseases in, 265, 266; volcano eruptions in, 557
Phillips, Howard, 469

Phoca sibirica, 557–58
phocine distemper virus (PDV), 558, 572
Phocoena phocoena, 558
phosgene, 530–31
phosphene, 530
phosphorylation, 228
Pinatubo, Mount, 557
Pinching, Anthony, 385
Pine, Seymour, 260
Pinheiro, Francisco, 69
Pinneo, Lily ("Penny"), 73, 75–82, 84–88, 90, 92, 94, 95, 195
Piot, Margarethe, 138–39, 150
Piot, Peter, 7, 110–12, 119–21, 123–24, 127–29, 135–39, 149–51, 217, 290–91, 320, 344–47, 350, 353–54, 365–66, 482, 571
Pittsburgh, University of, 574
plague, 174, 236–38, 459, 472, 531, 543, 598, 615; *see also* bubonic plague; pneumonic plague
Planned Parenthood, 392
plasmids, 225, 226, 431–32, 582, 583, 585, 586, 588, 619
Plasmodium: P. falciparum, 39, 40, 78, 277, 356, 432, 440–43, 446, 449–53, 455, 580; *P. vivax*, 39, 449, 450
Platt, Eileen, 134
Platt, Geoffrey, 115–16, 133–35
Playtex tampons, 392, 397, 400
Pneumococcus, 415–16, 584, 614
Pneumocystis carinii pneumonia (PCP), 284–88, 290, 291, 293, 297–99, 305, 306, 307, 309, 320, 324, 337, 344, 363–66
pneumonia, 169, 186, 188, 254, 296, 347, 414, 416, 529, 569; pediatric, 417, 418; *see also Pneumocystis carinii* pneumonia (PCP)
pneumonic plague, 236–38, 530
Podhista, Chai, 494
Pokrovsky, Vadim, 502
Pokrovsky, Valentin, 501
Pol Pot, 211, 454
Poland, 461, 499; heroin in, 500
polio, 30, 32–35, 66, 82, 167, 182–83, 463, 487, 504, 562, 580, 604, 617; vaccine, AIDS viruses in, 380–81
pollution, 203, 557, 559–63, 566, 568–70, 615
polychlorinated biphenyls (PCBs), 428, 559, 560, 568
polymerase, 579–81
polymerase chain reactions (PCR), 414, 431, 540–42, 544, 545, 548, 601–3, 616
population growth, 551, 571–72
porpoise morbillivirus (PMV), 558–59
Post, Jan, 562–63
poverty, 475, 490; refugee, 593; tuberculosis and, 513; in United States, 506–12
primaquine, 52
Prince Leopold Institute of Tropical Medicine, 109

Principe, 389
Procter & Gamble, 392, 393, 398, 399, 401, 403
Profit, Mrs., 310, 364
proguanil, 449, 450, 453
Project SIDA, 349–51, 353, 355–60, 370, 379, 515
ProMED, 602–4
Prospect Hill virus, 539, 543
prostitution, 245, 263, 265, 266, 269, 283, 503, 591, 593; 611; AIDS and, 288, 297, 305, 310, 340, 347, 352, 366, 368, 370, 372, 373, 387, 467–68, 475, 476, 484, 489–91, 493–96, 500, 502, 507; drug use and, 277
Proteus, 422
Pseudomonas, 191, 278, 422, 612
Ptychodiscus brevis, 560
Public Health Service, U.S., 62, 153, 159, 164, 298, 471, 507, 595
Puerto Rico: AIDS in, 349; yellow fever in, 70
Purtilo, David, 292
Puumala virus, 538, 546
Pygmies, 376
pyrimethamine, 441, 442, 446, 449, 450
pyrogenic exotoxin, 402–5

Q fever, 174
qinghaosu, 453
quinidine, 442, 443
quinine, 47, 48, 52, 278, 442–45, 447–48, 450, 451, 453
Quinn, Sandra Crouse, 383
Quinn, Thomas, 345, 346, 350, 351
quinolone antibiotics, 420, 584

rabies, 82, 448
radon, 569
Raffier, Gilbert, 108–10, 119
Rahman, Sheik Mujibur, 44
rain forests, destruction of, 551–53, 556, 571, 614–15, 619
Rask, Margrethe, 365
Rattus: R. norvegicus, 539, 545; *R. rattus*, 539, 546
Rauscher mouse leukemia virus, 227
Reagan, Ronald, 287, 301, 302, 307, 314, 318, 327, 330, 348, 349, 459, 468–71, 508, 509, 514
recombination, genetic, 577–78, 582–83, 586
Reed, Walter, 67
Reeves, Bill, 18
refugees, 591, 593, 608
Reichman, Lee, 279, 514, 519
Rely tampons, 393–94, 397–403, 410
replicase, 580
Republican Party, 302, 318, 469, 472
respiratory syncytial virus, 417
restriction enzymes, 223
retinitis, CMV, 435

retroviruses, 226–28, 230, 231, 321, 333, 337; African, 348, 357; recombination of, 578; simian, 574; T-lymphotropic, 325
reverse transcriptase, 321, 436
Revlon Cosmetics Corporation, 311
R Factors, 37
rheumatic fever, 415, 416
Rhodesia, 208, 209, 269; Marburg virus in, 58, 59; see also Zimbabwe
ribavirin, 196, 433, 541
ribonucleic acid, see RNA
ribozymes, 585
Richmond, Julius, 211
rickettsia, 174
rifampin, 62, 418, 420, 518, 521, 522, 524
Rift Valley fever, 204–5, 606
rinderpest virus, 558
Risque, Nancy, 302
"Rivet Hypothesis" of diversity, 555
river blindness, 205, 252, 439
RNA, 6, 35, 222, 223, 226, 230, 231, 432, 578–80, 584, 603; AIDS and, 321, 436; of Ebola virus, 346; of hantaviruses, 543; of influenza virus, 155; of Lassa virus, 82; of Marburg virus, 56
Robertson, Pat, 469
Rockefeller Foundation, 33, 56, 69, 70, 82, 127, 506, 602; Laboratories, 27, 595; Virus Program, 33
Rockefeller Institute, 34
Rockefeller University, 5, 6, 36, 227, 418, 572, 579, 584, 586, 602
Rogers, Martha, 324
Roisin, Alain, 308
Roman, Juan, 83, 94
Romana, Sister, 106, 107
Roman Empire, 234–36, 251, 391; smallpox in, 41
Romania, 617; AIDS in, 505, 612; antibiotic resistance in, 419
Roper, William, 601
Rosenberg, Barbara, 602, 603
Rosenberg, David, 548–49
Rouillon, Annik, 527
roundworms, 203, 253
Rous sarcoma, 227
Royal Society of Medicine, 381
Rozeboom, Lloyd, 49–50
Rozenbaum, Willy, 297, 321
rubella, 94
Rubinstein, Arye, 325
Ruppol, Jean-François, 108, 120, 121, 127, 129
Rush, Benjamin, 254
Russell, Paul, 48–52
Russell, Gen. Philip, 164, 166, 595, 597–99
Russia, 502–5; hantaviruses in, 547; snowshoe hare virus in, 577; see also Soviet Union
Rutayuge (Tanzanian hospital administrator), 343–44

Rwanda, 208, 210, 344; AIDS in, 340, 344–45, 350, 352, 359, 367–69, 378, 381, 387, 483; civil war in, 591–93, 608; malaria in, 442, 443, 445, 568

Sabatier, Renée, 362, 475
Sabin, Albert, 167, 168, 171, 182
Sagabiel, Stephanie, 596
SAIDS (simian AIDS), 371–72
Saimiri sciureus, 572
St. Louis encephalitis, 82, 576
St. Mary's Hospital Medical School, 385
St. Petersburg Pasteur Institute, 503
Salk, Jonas, 30, 35, 167, 168
Salmonella, 271, 422, 424, 584, 606; S. newport, 425, 426
Sande, Merle, 436
sandflies, 253–54
San Francisco AIDS Foundation, 299
San Francisco Chronicle, 306
San Francisco Health Department, 273, 288, 300
Santos, Theotonio dos, 200
São Tomé, 389
SatelLife, 615–17
Saudi Arabia, sexually transmitted diseases in, 269
Sauerwald, Egon, 72–73, 95
scabies acariasis roundworms, 38, 296
scarlet fever, 395, 402–3, 415
schistosomiasis, 32, 40, 203–5, 252, 253, 463, 580
Schleissman, Donald, 68
Schlievert, Patrick, 402–5, 407, 408, 414
Schmaljohn, Connie, 540, 544–45, 547, 548
Schmid, George, 396, 400
Schoolnick, Gary, 615
Schweiker, Richard, 302, 318
Scotland, morbilliviruses in, 558
scrapie disease, 323
Scripps Research Institute, 585
Seale, John, 381–82
Segal, Jacob, 382
Selassie, Haile, 72
Senate, U.S., 318, 472; Consumer Protection Subcommittee, 177; Health Subcommittee, 172
Sencer, David, 70, 94, 119, 163, 166–68, 172, 175, 177–78, 181, 183–87, 193, 301, 303, 328
Senegal, 194, 195, 358, 367, 375; antibiotic resistance in, 420; ascariasis in, 253; HIV-2 in, 372, 373; measles in, 590; Rift Valley fever in, 205; yellow fever in, 66
Sennar Dam (Sudan), 204
septicemia, 401, 423
Serbs, 591, 608
Serratia, 191, 614; S. marcescens, 278, 422, 612

Sevilleta Long Term Ecological Research survey, 536
Sewell, C. Mack, 530
sexual conjugation, 225, 582, 586
sexuallay transmitted diseases (STDs), 33, 213, 231, 260–75, 283, 288, 291, 298, 507, 591, 610; drug resistant, 434–35; *see also specific diseases*
Seyfarth-Hermann, Hans, 499–500
Shalala, Donna, 536–37
Shands, Kathryn, 396, 401–3
Sharma, V. P., 454
Shaw, Charlotte, 75–77
Sheffield, University of, Medical School, 170
Shepard, Charles ("Shep"), 174, 175, 178, 184–86, 189
Sherman, Kenneth, 563
Shigella, 427, 429, 574, 584, 588; *S. dysenteriae*, 421–22; *S. flexneri*, 271
Shilts, Randy, 262, 306
shingles, 433
Shope, Richard, 5, 79, 158, 159, 164, 165, 567, 605, 614
Shunkutu, M. R., 616
"sick building syndrome," 568–69
sickle-cell anemia, 479
Siegal, Frederick, 295, 296, 309, 314
Siegert, Rudolf, 55, 120
Sierra Leone, 252, 375; Lassa fever in, 90–93, 122–23, 192, 194–98, 215
Silverman, Mervyn, 329
Silverstein, Arthur, 167, 187–88
simian immunodeficiency virus (SIV), 372–77, 379, 381, 384–85, 574, 575
simian retrovirus (SRV), 574
simian sarcoma-associated viruses (SSAVs), 574
simian T-lymphotropic virus (STLV), 353, 358, 372, 374
Simpson, David, 126, 142–44
Singapore: AIDS in, 468, 489; malaria in, 450
Sisters of Perpetual Indulgence, 262, 282, 293
Sisters of the Holy Heart of Maria, 151
Sisters of the Holy Rosary, 92
skin cancer, 556
Skole, David, 552
Slate, Hal, 329
Slater, A. J., 438
sleeping sickness, 50, 143, 205, 567
"slim disease," 357
smallpox, 10, 32, 40–47, 52, 53, 66, 108, 114, 119, 142, 146–47, 183, 236, 381, 443, 459, 460, 602, 604
Smith, David, 142
Smith, Wilson, 158
Smithsonian Institution, 9, 24; Natural History Museum, 375–76; Tropical Research Institute, 587–88
Snow, John, 242
snowshoe hare virus, 577
Solidarity, 461

Solomon Islands, malaria in, 450
Somalia, 354, 608; smallpox eradication in, 45–46
Sonnabend, Joseph, 291–92, 295
Soper, Fred, 67
Sorbonne, 361
South Africa, 139, 141–42, 208, 209, 269, 368, 371; AIDS in, 484, 485, 492; antibiotic resistance in, 418, 419; ascariasis in, 253; Institute for Medical Research, 58; Lassa fever in, 92; malaria in, 442; Marburg disease in, 58–59, 117–18; organ transplants in, 574; sexually transmitted diseases in, 269; Toxic Shock Syndrome in, 406; tuberculosis in, 244–45, 252, 515, 526
South Wales, University of, 490
Southwest Africa, 208
Southwest Foundation for Biomedical Research, 319, 575
Soviet Union, 49, 147, 200, 208, 209, 369, 498; Africa and, 45; AIDS in, 382, 466, 476, 500–3, 612; biological warfare research in, 595;' Hantaan disease in, 538; hemorrhagic diseases in, 82; and Lassa research, 193–94; morbillivirus in, 557–58; nonmilitary foreign aid policies of, 201; and smallpox eradication campaign, 40, 41
Soyinka, F., 473
Spain: AIDS in, 465; antibiotic resistance in, 418, 419, 424–25; conquest of Mexico by, 41; influenza in, 156; malaria in, 48; morbilliviruses in, 558; syphilis in, 246
spectinomycin, 266, 420
Spielman, Andrew, 7, 49–51, 554–55
spirochetes, 245–47, 264, 553
sporulation, 428
src, 227
Sri Lanka; dengue hemorrhagic fever in, 257; malaria in, 47, 51, 450
Stalin, Joseph, 501
Stall, Mortimer, 191
Stanford University, 224, 555, 615
Staphylococcus, 30, 36, 277, 278, 411–14, 420, 422, 423, 437, 574, 580; methicillin-resistant (MRSA), 278, 612, 614; *S. aureus*, 293, 394–97, 401–10, 412–14, 584
State Department, U.S., 93, 126, 193, 382
Steere, Allen, 553
Steigbeigel, Neil, 326
Stenella coeruleoalba, 558
Stevens, Franklin, 395
Stewart, William H., 33
"sticky sera," 356, 362
Stockman, David, 287
Stone, I. F., 620
Stonewall Riots, 260–61, 328
Straus, Stephen, 434–35
Strauss, Jim and Ellen, 579
Strecker, Robert, 382

Streptococcus, 26, 62, 295, 414–15, 420, 422, 423, 574, 582; *S. pneumoniae*, 415–18

streptomycin, 36, 37, 418, 421, 440, 514, 516, 520–22

Stuart-Harris, Charles, 170

Styblo, Karel, 527

Sudan, 192, 208, 209, 211, 251–52, 354, 369, 608; Ebola virus in, 108–9, 115, 116, 130–34, 136, 142–46, 148, 215–20, 345, 346; Interior Mission (SIM), 76, 85; Lassa fever in, 92; leishmaniasis in, 607; Rift Valley fever in, 205; schistosomiasis in, 204; smallpox eradication in, 44, 142; yellow fever in, 66

Sukato Manzomba, 106, 128, 129

sulfadoxine, 441, 446

sulfamethoxazole, 421, 565

sulfisoxazole, 421

sulfonamides, 267

sulfur drugs, 420

Sumer, ancient, 391

Summit of the Nonaligned Movement, 382

Sunderman, William F. Jr., 177

Sureau, Pierre, 109, 110, 116–25, 127–30, 135, 148, 151–52, 198, 570

Surinam, 231

Surui Indians, 589

SV40, 224

Svahn, Jack, 302

SWAPO (South-West Africa People's Organization), 208

Swaziland: AIDS in, 484; malaria in, 445, 568

Sweden: hantaviruses in, 538; morbilliviruses in, 558; Toxic Shock Syndrome in, 404, 406

Sweet, Bob, 302

sweet wormwood, 452–53

Swine Flu, 146, 153–73, 175, 177–83, 187–89, 192, 303, 304, 319, 572, 587, 617

Switzerland, antibiotic resistance in, 414

Syngyna absorbency assay, 406

syphilis, 32, 236, 245–47, 264, 267–71, 283, 289, 296, 347, 478, 490, 502–3, 506, 507, 611

Syria, 201

Tacaribe virus, 17, 23

Taenia solium, 253

Tahyna virus, 577

Tajikistan, 608

Takatsuki, Kiyoshi, 228–29

Tampax Incorporated, 391, 397, 402

tampons, 391–94, 397–407, 410

Tanzania, 208–10, 592; AIDS in, 281, 334–44, 357, 360, 366–68, 378, 386, 479, 480, 485, 486, 516; GNP per capita in, 478; health-care policy in, 205–6; malaria in, 440–42, 568; Rwandan refugees in, 593; schistosomiasis in, 253; tuberculosis in, 526, 527

tapeworms, 252–53

Tarantola, Daniel, 7, 42–43, 45, 46, 460–62, 464, 476, 485, 608

Tass news agency, 503

Taunay, Augusto, 63

T cells, 228–29, 232, 277, 580; in AIDS, 284, 292, 294, 316, 317, 322, 325, 333, 344, 346, 370; in animal research, 319, 353, 371; LAV and, 331; in malaria, 446–48; in Toxic Shock Syndrome, 404, 409

Teklehaimont, Awash, 455, 456

Temin, Howard, 226, 227, 436, 578, 579

Tempest, Bruce, 529–30, 535

Tester, Patricia, 562, 563

tetanus, 211, 503, 617; neonatal, 479

tetracyclin, 36, 266, 267, 418, 420, 421, 425, 427, 431, 450, 451, 584, 586

Thailand, 251; AIDS in, 378, 467–68, 486, 489, 490, 493–98; cholera in, 565; dengue hemorrhagic fever in, 255, 257, 258; malaria in, 52, 450–52, 455; sexually transmitted diseases in, 267

Thatcher, Margaret, 250

T-helper cells, 292, 295

thiacetazone, 516

Thomas, Stephen, 383

3-thiacytidine (3TC), 436

thrush, 293

thymidine kinase, 434

Tkimalenka, Justhe, 336, 338–42

tobacco, 230, 505

Todaro, George, 226

Today contraceptive vaginal sponge, 406

Todd, James, 394–96, 402, 405, 407

Togo, onchocerciasis in, 252

Tokyo Cancer Institute, 228–29, 231, 232

Tolbert, William R. Jr., 193

Tomasevski, Katarina, 476

Tomasz, Alexander, 418–19, 584

Touré, Sékou, 194

Toxic Shock Syndrome (TSS), 390–411, 414, 425

Toxic Shock-Like Syndrome (TSLS), 414–15

toxoplasmosis, 307

transcriptase, 226

transposons, 225, 432, 582, 583, 585–86, 588, 619

"trench nephritis," 546

Treponema pallidum, 245, 246

trimethoprim, 291, 421, 443, 565

Trinidad, 17, 386

Tropeninstitut (German Tropical Disease Institute), 95, 96, 99

Troup, Jeannette, 75, 76, 84, 95

Trud (newspaper), 501

Trypanosoma, 253

trypanosomiasis, 205, 567

tryptophan, 408, 409

Tsetse flies, 143

T-suppressor cells, 295

tuberculosis, 3, 30, 32, 33, 36, 41, 210, 214,

236, 240–41, 243, 293, 298, 403, 503–4, 512–27, 569, 570, 574, 591, 614, 617; AIDS and, 307, 347, 478; drug-resistant, 437, 440, 497, 508, 512, 517, 519–27, 590, 606, 608, 614; heroin use and, 279
Tucker, Compton, 552
Tufts University, 553; School of Medicine, 424
Tulane University, 323; Medical School, 114
tularemia, 174
Tunisia, 367
Turkana people, 352
Turkey: heroin in, 492; plague in, 238
typhoid fever, 32, 110, 111, 503
typhus, 50, 174, 236, 591

Uganda, 208–11, 283, 338, 344, 592; AIDS in, 334–36, 340–43, 352, 357, 366–69, 378, 386, 485–87, 493, 587; antibiotic resistance in, 414; cancer in, 233; expulsion of Asians from, 206; malaria in, 442; Marburg virus in, 55, 57–59; Rwandan refugees in, 593; sexually transmitted diseases in, 268
Ugawa, 145
ultraviolet radiation, 556, 562, 566–68, 570
UNICEF, 49, 143, 604, 612
United Kingdom, 214, 243, 244; AIDS in, 365, 371, 380, 387, 465, 474, 477, 587; antibiotic resistance in, 412, 414, 415, 418, 420, 423; cholera in, 242; colonialism of, 38, 66, 197, 206, 209, 447; influenza in, 157, 170; Legionnaires' Disease in, 190; leprosy in, 239; plague in, 238, 239, 241; sexually transmitted diseases in, 265, 267
United Nations, 10, 200, 210, 354, 451, 457, 461–64, 467, 476, 505, 512, 562, 570, 571, 591, 592, 602; Children's Fund, 49; Development Program, 486, 498; Economic Council, 465; Food and Agricultural Organization, 484; General Assembly, 212, 389, 465; Office for Emergency Operations in Africa, 355; Population Fund, 551; and Rwanda, 593; Security Council, 608
United States, 147, 208, 369; AIDS in, 283–319, 322–32, 337, 348, 354, 357, 363–65, 367, 380, 468–72, 476, 477, 479, 506–8; air travel in, 571; antibiotic resistance in, 412–16, 418–23, 425–27, 435–36, 437; biological warfare research in, 595; cholera in, 565; cysticercosis in, 253; dengue hemorrhagic fever in, 256, 257; disease eradication in, 32–33; Ebola virus in, 594, 598–601; emergency response capacities of, 595–96; expenditure on drugs in, 438; gay rights movement in, 260–62; GNP per capita in, 478; hantaviruses in, 528–49; health care expenditures in, 205, 206; heroin use in, 613; HTLV in, 385–86; injecting drug users in, 274, 276, 278–80; Legionnaires' Disease in, see Legionnaires' Disease; malaria in, 47–50; morbilliviruses in, 558; organ transplants

in, 574–75; in Persian Gulf War, 597–98; public health in, 505–12; sexually transmitted diseases in, 264–68, 270–73; Soviet Union and, 193–94; subsidies for medical students in, 62; Swine Flu in, see Swine Flu; tampon use in, 391–92; Toxic Shock Syndrome in, 390–410; tuberculosis in, 241, 243, 440, 512, 514–27; water-borne bacteria in, 428–30; yellow fever in, 68, 70; see also specific government agencies
U.S. Agency for International Development, 202
Upper Volta, Lassa fever in, 92
urethritis, nonspecific, 271, 296
Usher, Peter, 354–55

Valium, 274, 276, 327, 328
Valverde Chinel, Luis, 14, 16, 20, 28
vancomycin, 412, 414, 422–23, 584, 606, 614
van der Gröen, Dina, 138–39
van der Gröen, Guido, 111–13, 126–28, 138–39, 147–48, 150–51, 195, 538
Vanuatu, malaria in, 450
Varmus, Harold, 227, 228
venereal diseases, see sexually transmitted diseases
Venezuela: malaria in, 52; yellow fever in, 68
verapamil, 451
Vero cells, 79, 80, 111, 112, 126–27, 547
Vibrio cholerae, 208, 563–66, 619
Vietnam, 201; AIDS in, 496–97; dengue hemorrhagic fever in, 257, 258; malaria in, 450, 451, 453
Vietnam War, 52, 84, 91, 153, 176, 178, 258, 383, 451, 454, 468
"viral trafficking," 572, 610
Virginia State Department of Health, 157
Virgin Islands, yellow fever in, 70
virology, 38
Viseltear, Arthur, 187
visna virus, 227, 331, 379, 381
VLI Corporation, 406
Volberding, Paul, 286, 288–90, 293, 299, 301, 316, 318
v-src, 227

Wain-Hobson, Simon, 377
Walter Reed Army Institute of Research, 453
War on Cancer, 318
warts, venereal, 271, 296
Washington, University of, 307, 371, 611; School of Public Health, 170
wasting syndrome, 344
Watergate scandal, 153
Waterhouse-Friderichsen syndrome, 63
Watson, James, 34, 223, 225, 230
Waxman, Henry, 167–68, 303–4
Webb, Patricia, 22–23, 25–27, 29, 82, 113–15, 118, 119, 125, 146, 147, 192–94, 196, 197, 215, 376, 571, 622

Webster, Robert G., 161
Weill, Jim, 512
Weiss, Ted, 318
Weller, Thomas, 589
West Nile fever, 211
Western Blot test, 363, 365
Western Samoa, influenza in, 157
Wheelis, Mark, 603
White, Dan, 262
White, Harold (Hal), 73, 85, 95
White, Keith, 401
whooping cough, 211, 251, 417, 503, 504
Widdus, Roy, 462
Wieden, Al, 17
Wilson, E. O., 9, 551, 561, 571
Windom, Robert, 302, 459
Wine, Laura, 73, 75, 77
Wirth, Dyann, 451–52
Wisconsin, University of, 159, 226
Wisconsin Division of Health, 394
Wofsy, Constance, 288
Wolfe, Sidney, 169, 399, 400, 406
Wong-Stahl, Flossie, 334
Wood, W. Barry, 395
Woods Hole Oceanographic Institute, 50
World Bank, 9, 51, 197, 198, 200–203, 212,
   215, 248, 250, 251, 437–40, 443, 477,
   479, 482, 485, 486, 491, 498, 551, 562–
   63, 609
World Health Organization (WHO), 200, 211,
   248, 416–17, 421, 439, 512, 540, 601,
   602, 613; and AIDS, 312, 354, 356, 357–
   60, 362, 389, 459–68, 476–77, 481–82,
   484, 486, 489, 490, 492–94, 496, 498,
   515, 593; and antibiotic resistance, 420,
   423, 431; and cholera, 563–65; Constitution
   of, 457; emergency response capacities of,
   595–98, 604, 606–10; and dengue hemorrhagic fever, 257; and Ebola virus, 108–12,
   115–17, 119, 126, 132, 136, 142–48, 215,
   216; Health Legislation Unit, 477; and influenza, 161, 170; and Lassa fever, 87, 95,
   195; and malaria, 31, 49, 50, 447, 449,
   451–53, 455, 456; and Marburg disease, 55,
   57, 58, 80; and meningitis, 64; Regional
   Nursing/Midwifery Task Force, 482; and
   smallpox, 40, 42, 44–47; Task Group on
   global warming, 568, 570; and tuberculosis,
   243, 515, 525; Viral Diseases Branch, 110;
   and yellow fever, 68
World Summit on Children, 512
World Summit of Ministers of Health on Programmes for AIDS Prevention, 459–62, 465,
   467, 472, 475
World War I, 47, 530, 546; influenza pandemic
   of, 154, 157–59, 161, 162, 164, 168, 169,
   175, 179

World War II, 4, 16, 24, 47, 49, 50, 171;
   dengue fever spread by, 257–58; extermina-
   tion of Jews during, 466; recovery of Europe
   and Japan after, 200
Worldwide Primates, 601
Würzburg, University of, 546
Wydess, Robert, 596
Wyngaarden, James, 301–3

Xikrin Indians, 589

Yablokov, A. V., 503
Yale University, 158, 589, 614; Arbovirus Re-
   search Institute, 5, 56, 76, 79, 81–84, 567
Yamarat, Charas, 255, 256
yaws, 245, 247
yeasts, 174, 432, 618
yellow fever, 50, 66–70, 110, 111, 195, 204,
   257, 459, 567, 575, 606, 607, 612, 613
Yemen, 252
Yersinia: Y. enterocolitica, 425; *Y. pestis*, 237,
   238, 240, 615
Yombe Ngongo, 102
Yoshida, Mitsuaki, 229
Young, Frank, 315
Yugoslavia, 591; Marburg disease in, 54, 55,
   57; smallpox in, 43–44

Zah, Peterson, 536
Zaire, 60–61, 192, 208, 211; AIDS in, 291,
   320, 340, 344–51, 353, 355–56, 358, 359,
   365–71, 380, 381, 386–87, 457–59, 479–
   80, 485–87; Ebola virus in, 100–108, 110–
   42, 144, 147–52, 178, 538; GNP per capita
   in, 478; Lassa fever in, 92; malaria in, 442–
   44; Rwandan refugees in, 593; schistosomiasis in, 253; yellow fever in, 68
Zaire, National University of, 105
zalcitabine, 436
Zambia, 208, 354, 616; AIDS in, 306, 347–
   48, 355, 357, 359, 360, 362, 366, 367,
   485–87, 493, 516; malaria in, 441, 442;
   sexually transmitted diseases in, 268, 269–
   70, 488; University of, Medical Library, 615
ZANU (Zimbabwe African National Union), 208
ZAPU (Zimbabwe African People's Union), 208
Zayemba Tshiama, 121
Zidovudine, 436
Ziegler, Jetty, 89
Ziegler, John, 322
Ziegler, Philip, 247
Zimbabwe, 195, 208, 368, 382; AIDS in, 366,
   485, 486; malaria in, 568; schistosomiasis
   in, 253; sexually transmitted diseases in, 268
Zimmermann, Friedrich, 466
zoonosis, 375, 385, 572, 614
Zubeir, Osman, 215, 218, 219